In this technological twenty-first century, mathe... Throughout this book the author discusses how ... arious fields. Here are some examples of the many applications you will encounter.

COMPUTER GRAPHICS

Dynamically displayed data, which involves translations of graphs and figures, play an important role in computer graphics. In early motion pictures, it was common to have the background move to create the appearance that the actors were moving. This same technique is used today on two-dimensional graphics. For example, in video games the background often is translated (shifted in a parallel direction) to give the illusion that the player in the game is moving.

ENVIRONMENTAL ECOLOGY

When quantities such as population and pollution increase rapidly or slowly, exponential and logarithmic equations are used to model their growth. Exponential and logarithmic functions occur in a wide variety of applications, many of which involve our environment. Examples include acid rain, the decline of the bluefin tuna, the life span of robins, air pollution, and earthquakes, to name a few. Modeling these situations helps us predict the future and discover solutions.

COLLEGE ALGEBRA

through Modeling and Visualization

SECOND EDITION

COLLEGE ALGEBRA

through Modeling and Visualization

SECOND EDITION

Gary K. Rockswold
Minnesota State University, Mankato

Addison
Wesley

Boston   San Francisco   New York
London   Toronto   Sydney   Tokyo   Singapore   Madrid
Mexico City   Munich   Paris   Cape Town   Hong Kong   Montreal

Senior Acquisitions Editor: Anne Kelly
Executive Project Manager: Christine O'Brien
Development Editor: Elka Block
Senior Production Supervisor: Karen Wernholm
Production Coordination: Kathy Diamond
Text Design: Barbara T. Atkinson
Marketing Manager: Becky Anderson
Senior Prepress Supervisor: Caroline Fell
Manufacturing Buyer: Evelyn Beaton
Compositor: Nesbitt Graphics, Inc.
Technical Illustration: Techsetters, Inc.
Reflective Illustration: Jim Bryant
Cover Design: Dardani Gasc Design

Photo Credits: Cover: © Corbis; endpapers: courtesy of Siemens AG Medical Solutions, corbisimages.com, courtesy of Nikon, Inc., © Michael S. Yamashita/CORBIS, © Jeffrey L. Rotman/CORBIS; p. 1: © Roger Tully/Stone; p. 13: © David Job/Stone; p. 64: Charles Taylor/© 2000, reprinted with permission of Discover Magazine; p. 157: © Chris Stanford/Allsport; p. 157 (Rick Barry): © 2001 NBA Entertainment, photo by Walter Iooss, Jr.; p. 220: © Donald Johnston/Stone; p. 333: © PhotoDisc; p. 448: © PhotoDisc; p. 546: © PhotoDisc; p. 590: © PhotoDisc

Library of Congress Cataloging-in-Publication Data
Rockswold, Gary K.
   College algebra through modeling and visualization — 2nd ed. / Gary K. Rockswold.
     p. cm.
   Includes bibliographical references and index.
   ISBN 0-321-08137-4, ISBN 0-321-09055-1 (instructor's edition)
     I. Algebra.   I. Title.

QA152.3 .R63 2001
512—dc21   2001033534

3 4 5 6 7 8 9 10—QWT—040302

To my son, Mark

Contents

Chapter 3

Chapter 4

Preface

College Algebra through Modeling and Visualization offers an innovative approach that consistently links mathematical concepts to real-world applications by moving from the concrete to the abstract. It demonstrates the relevance of mathematics and answers the question, "When will I ever need to know this?" This text provides a comprehensive curriculum with the balance and flexibility necessary for today's college algebra courses. The early introduction of functions and graphs allows the instructor to use applications and visualization to present mathematical topics. Real data, graphs, and tables play an important role in the course, giving meaning to the numbers and equations that students encounter. This approach increases students' interest and motivation, and the likelihood for success.

APPROACH

The concept of a function is the unifying theme of the text. Functions and their graphs are frequently used to model data, and students are often asked to interpret their results. Mathematical skill building also plays an important role. Instructors are free to strike their own balance of skills, rule of four, applications, modeling, and technology. With a flexible approach to the rule of four (verbal, graphical, numerical, and symbolic methods), instructors can easily emphasize one rule more than another to meet their students' needs. This approach also extends to modeling, applications, and use of a graphing calculator. The use of technology, which helps students visualize mathematical concepts, is flexible throughout the text, though technology use is not a requirement for students to benefit from this applications-based book. The text contains numerous applications, including models of real-world data with functions and problem-solving strategies. It is not necessary for an instructor to discuss any particular application; rather an instructor has the option to choose from a wide variety of topics.

CHANGES TO THE SECOND EDITION

The second edition contains several important changes, which are a result of the many comments and suggestions received from instructors, students, and reviewers. The second edition is reorganized into eight chapters rather than six. Both the breadth and depth of the discussions of several topics has increased, and hundreds of new examples and exercises that involve mathematical skill building, applications, and modeling have been included. Examples have been added for students to refer to when doing their homework assignments. Many of the real-data applications from the first edition have been updated to make the data current and relevant to students. The book's new organization and seamless presentation facilitate students' understanding of each topic. The extensive exercise sets cover a diverse assortment of topics, and they are carefully graded with several levels of difficulty. The presentation and exercise sets provide instructors with the ability to enhance their courses in several fashions. The chapter summaries have been expanded and presented in an easy-to-read grid.

- **Chapter 1: Introduction to Functions and Graphs**
  This chapter has been streamlined from six sections to four. A basic introduction to

problem solving is presented in Section 1.1 and the distance formula is now given in Section 1.2. Functions and their representation are introduced in Section 1.3, providing the background necessary for future chapters. The material on quadratic functions and transformations of graphs now appears in Chapter 3. Discussion of increasing and decreasing functions is delayed until Chapter 4.

- **Chapter 2: Linear Functions and Equations**
  Section 2.1 is a new section that discusses linear functions and models in detail. Slope is discussed extensively as a rate of change. Equations of lines are now presented before linear equations and inequalities. There is an increased emphasis on problem solving, which now includes several classic word problems. An extensive discussion of piecewise linear functions, including absolute value equations and inequalities, is provided in Section 2.5. Linear regression is introduced in Section 2.6.

- **Chapter 3: Quadratic Functions and Equations**
  This new chapter allows an instructor to introduce quadratic functions and equations before covering the more involved nonlinear functions and equations in Chapter 4. Increased emphasis has been given to problem solving. Quadratic equations and inequalities have been separated into two sections. Transformations of graphs are introduced in Section 3.4.

- **Chapter 4: Nonlinear Functions and Equations**
  In this chapter, the graphs, zeros, and factorization of polynomials are discussed. Section 4.4 has been reorganized so that instructors are able to cover complex numbers and complex solutions to quadratic equations without covering the fundamental theorem of algebra. More opportunities for modeling with nonlinear functions are provided. Section 4.6, Polynomial and Rational Inequalities, and Section 4.7, Power Functions and Radical Equations, can be omitted, but they do provide opportunities for students to study other nonlinear functions, equations, and inequalities.

- **Chapter 5: Exponential and Logarithmic Functions**
  Operations on functions and inverse functions are introduced in Sections 5.1 and 5.2. Section 5.3, Exponential Functions and Models, includes more discussion on the differences between exponential growth and other types of growth. More opportunities for modeling real data with exponential and logarithmic functions are provided throughout the chapter. Logarithmic functions and properties of logarithms are now separated into two sections to provide students with more opportunities to increase skills and work with models. Section 5.7, Constructing Nonlinear Models, is a new section based on nonlinear regression.

- **Chapter 6: Systems of Equations and Inequalities**
  This chapter now provides more examples and exercises for students to practice the skills needed to solve systems of linear and nonlinear equations. The material on linear programming has been increased.

- **Chapter 7: Conic Sections**
  This new chapter has expanded the material on conic sections from one section to three, thereby providing a more comprehensive discussion with additional exercises. More practice solving nonlinear equations has also been included.

- **Chapter 8: Further Topics in Algebra**
  This chapter covers sequences, series, counting, probability, and the binomial theorem. Examples and exercises have been added throughout the chapter. The binomial theorem is presented in its own section. Additional material has been added to Section 8.5, Probability, which includes conditional probability and dependent events.

- **Chapter R: Basic Concepts from Algebra and Geometry**
  This new chapter provides a review of essential material from prerequisite courses, concentrating on skills used in algebra and geometry. Students are referred to this chapter by Algebra and Geometry Review Notes placed in the margins of the text. For example, students who need extra practice factoring trinomials are referred "just in time" to the proper section in Chapter R.

- **Appendix A: A Library of Functions**
  This appendix is new and summarizes many of the basic functions and families of functions.

- **Appendix B: The TI-83/83 Plus Graphing Calculators**
  This appendix has been expanded from the first edition to provide more help with graphing calculators. It contains specific keystrokes for working selected examples from the text. The Graphing Calculator Help notes in the margin of the text refer students to specific pages in Appendix B.

FEATURES

- **Chapter and Section Introductions**
  Many college algebra students have little or no understanding of mathematics. Chapter and section introductions motivate and explain some of the reasons for studying mathematics. (See pages 1, 65, 157, 221, and 268.)

- **Applications and Models**
  Interesting, relevant applications are a major strength of this textbook, and as a result, students become more effective problem solvers and have a better understanding of how mathematics is used in the real world. Because the applications are intuitive and not overly technical, they can be introduced with a minimum amount of class time. Current data are used to create meaningful mathematical models. A unique feature of this text is that the applications and models are woven into both the discussions and the exercises. It is easier for students to learn how to solve applications if they are discussed within the text. (See pages 26, 165–166, 218, 334, and 349–351.) An Index of Applications is included at the end of the book.

- **Sources**
  Since there are numerous applications throughout the text, genuine sources and a comprehensive bibliography are given. These sources reinforce for the student the practical applications of mathematics in real life. (See pages 104–106, and 387–390.)

- **Algebra and Geometry Review Notes**
  Throughout the text, Algebra and Geometry Review Notes are located in the margins, which direct students "just in time" to Chapter R, where important topics in algebra and geometry are reviewed. Instructors can use this chapter for extra review or refer students to it as needed. This feature *frees* instructors from frequently reviewing material from intermediate algebra and geometry. (See pages 77 and 101.)

- **Graphing Calculator Help Notes**
  The Graphing Calculator Help Notes located in the margins direct students "just in time" to Appendix B: The TI-83/83 Plus Graphing Calculators. This appendix shows students the necessary keystrokes to complete specific examples from the text. This feature *frees* instructors from teaching the specifics of the graphing calculator and gives students a convenient reference written specifically for this text. (See pages 7, 79, and 183.)

- **Making Connections**

  This feature occurs throughout the text and helps students see how concepts covered previously are related to new concepts being presented. (See pages 30, 84, and 179.)

- **Putting It All Together**

  This helpful feature at the end of each section summarizes techniques and reinforces the mathematical concepts presented in the section. It is given in an easy-to-follow grid. (See pages 52–53, 319–320, and 383–384.)

- **Checking Basic Concepts**

  This feature includes a small set of exercises provided after every two sections that can be used for review. These exercises require about 15 or 20 minutes to complete and can be used for collaborative learning exercises, if time permits. (See pages 91, 188, and 311.)

- **Class Discussion**

  This feature is included in most sections and poses a question that can be used for either classroom discussion or homework. (See pages 32, 124, and 229.)

- **Extended and Discovery Exercises**

  Extended and Discovery Exercises occur at the end of selected sections and at the end of every chapter. These exercises are usually more complex and challenging, and often require extension of a topic presented or discovery of a new topic. They can be used for either collaborative learning or extra homework assignments. (See pages 172–173, 218–219, and 446–447.)

- **Exercise Sets**

  The exercise sets are the heart of any mathematics text, and this text includes a large variety of instructive exercises. Each set of exercises involves skill building, mathematical concepts, and applications. Graphical interpretation and tables of data are often used to extend students' understanding of mathematical concepts. The exercise sets are graded carefully and categorized according to topic, which makes it easy for an instructor to select appropriate assignments. (See pages 168–173 and 346–352.)

- **Chapter Review Exercises**

  This exercise set contains both skill-building and applied exercises. They stress different techniques for solving problems and provide students with the review necessary to pass a chapter test. (See pages 60–62 and 328–331).

- **Chapter Summaries**

  Chapter summaries are expanded and presented in an easy-to-read grid. They allow students to quickly review key concepts from the chapter. (See pages 213–215 and 324–327.)

SUPPLEMENTS

For the Student

Student's Solutions Manual ISBN 0-321-09162-0. This manual provides complete solutions to all odd-numbered exercises, excluding Checking Basic Concepts and Extended and Discovery Exercises.

Videotape Series ISBN 0-321-09166-3. Keyed to the text, these videotapes feature an engaging team of lecturers who provide comprehensive coverage of each section and every topic (with the exception of Chapter R). In their presentations the lecturers use worked-out examples, visual aids, and manipulatives to emphasize the relevance of the material to students' everyday lives.

Digital Video Tutor ISBN 0-321-09576-6. The videotape series for this text is available on CD-ROM, which makes it convenient for students to watch video segments from a computer at home or on campus. This complete set of videos, now affordable for students and portable, is ideal for distance learning or extra instruction.

InterAct Math® Tutorial Software (CD for Windows) ISBN 0-321-09167-1. Professional software engineers working closely with a team of experienced math educators have designed and developed this software program. It includes exercises that are linked with every objective in the textbook and that require the same computational and problem-solving skills as their companion exercises in the text. Each exercise has an example and an interactive guided solution that are designed to involve students in the solution process and to help them identify precisely where they are having trouble. The software recognizes common student errors and provides students with appropriate customized feedback. With its sophisticated answer-recognition capabilities, *InterAct Math Tutorial Software* recognizes appropriate forms of one answer for any kind of input. It also tracks student activity and scores for each section, which can then be printed out. The software is free to qualifying adopters or can be bundled with books for sale to students.

InterAct MathXL® Software, Text bundled with a 12-month registration coupon ISBN 0-201-71630-5. *InterAct MathXL* provides on-line diagnostic testing and tutorial help, all using *InterAct Math Tutorial Software.* Students can take chapter tests correlated to this textbook, receive individualized study plans based on those test results, work practice problems for areas in which they need improvement, receive tutorial instruction in those topics, and take tests to gauge their progress. With *InterAct MathXL,* instructors can track students' test results, study plans, and practice work. The software is free when the access code is bundled with a new text.

Web Site. The Web site for this text *www.aw.com/rockswold* provides additional resources for both students and instructors. These include downloadable graphing calculator programs for the TI-83/83+, instructions for using the TI-85 and TI-86 model graphing calculators, PowerPoint slides, chapter quizzes, and a tutorial program.

For the Instructor

Instructor's Edition ISBN 0-321-09055-1. This version of the text contains answers to all exercises.

Instructor's Solutions Manual ISBN 0-321-09163-9. This manual contains complete solutions to all text exercises.

Instructor's Testing Manual ISBN 0-321-09165-5. This supplement provides prepared tests for each chapter of the main text.

TestGen-EQ with QuizMaster-EQ® (dual platform for Windows and Macintosh) ISBN 0-321-09169-8. This computerized test generator contains algorithmically defined problems organized specifically for this textbook. Its user-friendly graphical interface enables instructors to select, view, edit, and add test items, then print tests in a variety of fonts and forms. A built-in question editor gives the user the power to create graphs, import graphics, insert mathematical symbols and templates, and insert variable numbers or text. An "Export to HTML" feature lets instructors create practice tests that can be posted to a Web site. Tests created with *TestGen-EQ* can also be displayed on Web pages when the free *TestGen-EQ* plug-in has been installed with Internet

Explorer or Netscape Navigator. Tests created with *TestGen-EQ* can be used with *Quiz-Master-EQ,* which enables students to take exams on a computer network. *QuizMaster-EQ* automatically grades the exams, stores results on disk, and allows the instructor to view or print a variety of reports for individual students, classes, or courses.

MyMathLab. MyMathLab.com is a free one-stop solution for creating and managing an on-line mathematics course for distance learning or to supplement a traditional or self-paced course. It is based on Addison-Wesley mathematics textbooks and integrates interactive multimedia instruction. MyMathLab.com is powered by CourseCompass, a version of Blackboard™ designed specifically for Addison-Wesley, and offers a suite of communication tools that allows users to send e-mail, participate in bulletin board discussions, schedule class chats, provide collaboration tools for students, and more. For more information, visit our web site at www.mymathlab.com or contact your local Addison-Wesley representative for a live demonstration.

Addison-Wesley Math Tutor Center (www.aw.com/tutorcenter). This service provides assistance to students who purchase a textbook published by Addison-Wesley. Help is available via phone, fax, and e-mail from experienced, qualified math instructors.

ACKNOWLEDGMENTS

Many individuals contributed to the development of this textbook. I would like to thank the following reviewers, whose comments and suggestions were invaluable in preparing *College Algebra through Modeling and Visualization, Second Edition.*

Randy Baker	*Fairmont State College*
Rose Cavin	*Chipota Junior College*
Sarah Cook	*Washburn University*
William Fox	*Francis Marion University*
Johanna G. Halsey	*Dutchess Community College*
Libby Higgins	*Greenville Technical College*
Gloria Hitchcock	*Georgia Perimeter College*
Peter Jarvis	*Georgia College and State University*
Tuesday J. Johnson	*New Mexico State University*
Susan Jones	*Nashville State Technical Institute*
Cheryl Kane	*University of Nebraska*
Mike Keller	*St. Johns River Community College*
Dr. Tom Kelly	*Metropolitan State College of Denver*
Mary Kilbride	*Augustana College*
Marlene Kovaly	*Florida Community College, Kent Campus*
Michael Lloyd	*Henderson State University*
Judith Maggiore	*Holyoke Community College*
Mary Martin	*Middle Tennessee State University*
Joe May	*North Hennepin Community College*
Canda Mueller-Engheta	*University of Kansas*
Donna Norman	*Jefferson Community College*
Wing Park	*College of Lake County*
Joan Raines	*Middle Tennessee State University*
Laura Ralston	*Floyd College at North Metro Tech*
Jolene Rhodes	*Valencia Community College*
Judith D. Smalling	*St. Petersburg Junior College*
Christine Wise	*University of Louisiana at Lafayette*

A special thank you is due Elka Block at Twin Prime Editorial for her valuable suggestions, comments, and work with the manuscript, answers, and solutions. I would also like to thank Terry Krieger at Winona State University and Frank Purcell at Twin Prime Editorial for their work with the answers and solutions. Thank you to Steve Ouellette, Namyong Lee, Jodie Fry, Lauri Semarne, Deana Richmond, and Norma James for their excellent work as accuracy checkers.

Without the excellent cooperation from the professional staff at Addison-Wesley, this project would have been impossible. They are, without a doubt, the best. Thanks go to Greg Tobin for his support of this project. Particular recognition is due Christine O'Brien, Bill Poole, and Anne Kelly, who all gave advice, support, assistance, and encouragement. The outstanding contributions of Karen Wernholm, Barbara Atkinson, Becky Anderson, Julia Coen, Jennifer Kerber, Jaime Bailey, Tricia Mescall, and Joe Vetere are much appreciated. The work of Kathy Diamond, Becky Malone, Sandra Scholten, and Naomi Kornhauser was instrumental in the success of this project.

Thanks go to Wendy Rockswold, who proofread the manuscript several times and gave encouragement throughout the project.

A special thank you goes to the many students and instructors who used the first edition. Their suggestions were insightful. Please feel free to contact me at *gary.rockswold@mnsu.edu* or Department of Mathematics, Minnesota State University, Mankato, MN 56001 with your comments. Your opinion is important.

Gary Rockswold

Think you can, think you can't; either way, you'll be right.

—

Henry Ford

—

Introduction to Functions and Graphs

Mathematics is rapidly becoming more and more important to our society. Knowledge of mathematics provides access to today's world because many new technologies have their foundations in mathematics. Today's workplace is increasingly technical, and as a result, college graduates with a solid background in mathematics earn, on average, 38 percent more than their peers without such a background.

Mathematics is vital to understanding and interpreting the data that technology creates. Our society is in transition from an industrial to an informational society. An informational society's primary social and economic activities are the production, storage, and communication of information. Communication skills are essential, and *mathematics is the language of technology.* Mathematics allows society to quantify its experiences.

Decisions by students in high school or college to avoid mathematics classes often affect lifelong vocations, incomes, and lifestyles. Switching majors simply to elude a mathematics prerequisite often prevents students from pursuing their real vocational dreams. Skills in mathematics are becoming as important as skills in language. Remember that neither illiteracy nor innumeracy are desirable for attaining one's full potential.

Source: Department of Education.

1.1 NUMBERS, DATA, AND PROBLEM SOLVING

Sets of Numbers • Scientific Notation • Problem Solving

Introduction

Throughout history, humans have had a desire to communicate information and knowledge. Both languages and number systems have evolved to satisfy this desire. In order to build roads, design CD players, or forecast the weather, numbers and mathematics are necessary. As our society becomes more complex, the problems that it faces also become more complex. Mathematics plays an increasingly important role in solving these problems. Problem-solving strategies are essential for our society to progress. This section discusses sets of numbers and introduces some basic strategies for solving problems.

Sets of Numbers

One important set of numbers is the set of **natural numbers**. This set comprises the *counting numbers* $N = \{1, 2, 3, 4, \ldots\}$. Natural numbers can be used when data are not presented in fractional parts.

The **integers**

$$I = \{\ldots, -3, -2, -1, 0, 1, 2, 3, \ldots\}$$

are a set of numbers that contains the natural numbers. Historically, negative numbers were not readily accepted. They did not appear to have real meaning. However, today when a person opens a personal checking account for the first time, negative numbers quickly take on meaning. There is a significant difference between a positive and a negative balance.

A **rational number** is any number that can be expressed as the ratio of two integers $\frac{p}{q}$, where $q \neq 0$. Rational numbers include the integers. Some examples of rational numbers are

$$\frac{2}{1}, \frac{1}{3}, -\frac{1}{4}, \frac{-50}{2}, \frac{22}{7}, 0, \sqrt{25}, \text{ and } 1.2.$$

Note that 0 and 1.2 are both rational numbers. They can be represented by the fractions $\frac{0}{1}$ and $\frac{12}{10}$. Because two fractions that look different can be equivalent, rational numbers have more than one form. Rational numbers can also be expressed in a decimal form that either *repeats* or *terminates*. For example, $\frac{2}{3} = 0.\overline{6}$, a repeating decimal, and $\frac{1}{4} = 0.25$, a terminating decimal. The overbar indicates that $0.\overline{6} = 0.6666666\ldots$.

■ **CLASS DISCUSSION**

The number 0 was invented well after the natural numbers. Many societies did not have a zero—for example, there is no Roman numeral for 0. Discuss some possible reasons for this. ■

Real numbers can be represented by decimal numbers. Since every fraction has an equivalent decimal form, real numbers include rational numbers. However, some real numbers cannot be expressed by fractions. These numbers are called **irrational numbers**. The numbers $\sqrt{2}$, $\sqrt{15}$, and π are examples of irrational numbers. They can be represented by nonrepeating, nonterminating decimals. Real numbers include both rational and irrational numbers, and can be *approximated* by a terminating decimal. Examples of real numbers include

$$2, -10, -131.3337, \frac{1}{3} = 0.\overline{3}, -\sqrt{5} \approx -2.2361, \text{ and } \sqrt{11} \approx 3.3166.$$

Note: The symbol "$\approx$" means **approximately equal**. This symbol will be used in place of an equals sign whenever two unequal quantities are close in value. For example, $\frac{1}{4} = 0.25$, whereas $\frac{1}{3} \approx 0.3333$.

EXAMPLE 1   *Classifying numbers*

Classify each real number as a natural number, integer, rational number, or irrational number:

$$5, -1.2, \frac{13}{7}, -\sqrt{7}, -12, \sqrt{16}.$$

SOLUTION The numbers 5 and $\sqrt{16} = 4$ are natural numbers. We classify 5, $\sqrt{16}$, and -12 as integers. The rational numbers include 5, $\sqrt{16}$, -12, -1.2, and $\frac{13}{7}$. The only irrational number is $-\sqrt{7}$.

Even though a data set includes only integers, decimals may be used to describe it. One way to analyze data is to find the **average** or **mean**. To calculate the average (or mean) of a set of n numbers, we add the n numbers and then divide the sum by n. The average of a set of integers need not be an integer—sometimes it is a fraction or decimal.

EXAMPLE 2   *Analyzing deaths from AIDS*

Table 1.1 lists the annual deaths caused by AIDS in the United States from 1994 to 1997. Calculate the average number of deaths over this four-year period.

TABLE 1.1

Year	1994	1995	1996	1997
Deaths	48,663	48,371	34,947	14,338

Source: Department of Health and Human Services.

SOLUTION The average number of deaths from AIDS between 1994 and 1997 was equal to

$$\frac{48,663 + 48,371 + 34,947 + 14,338}{4} = \frac{146,319}{4} = 36,579.75.$$

This average of four integers is a rational number *and* a real number, but not an integer.

Real numbers have applications in many areas such as interest rates, grade point averages, and the Consumer Price Index (CPI). The CPI is often referred to as the "cost of living index" and is the numerical scale most commonly used to measure inflation. It tracks the prices of basic consumer goods. Over 40 economists calculate the CPI annually, at a cost of $26 million. The time from 1982 to 1984 is called the *reference period*, where the CPI is defined to be 100.0. If the CPI changes from c_1 to c_2, then the **percent change** is given by $\dfrac{c_2 - c_1}{c_1} \times 100$.

EXAMPLE 3   *Interpreting the Consumer Price Index*

Table 1.2 lists the CPI for selected years from 1950 to 1995. A CPI of 24.1 indicates that a typical sample of consumer goods costing $100 from 1982 to 1984 would have cost $24.10 in 1950.

TABLE 1.2

Year	1950	1955	1960	1965	1970	1975	1980	1985	1990	1995
CPI	24.1	26.8	29.6	31.5	38.8	53.8	82.4	107.6	130.7	153.5

Source: Bureau of the Census.

(a) Discuss the change in the CPI over this 45-year period.
(b) Approximate the percent change in prices from 1970 to 1980.
(c) From 1980 to 1990 the average tuition and fees at public colleges and universities rose from $804 to $1908. Determine if the percent change in tuition and fees was more or less than the percent change in the CPI for the same time period. (Source: The College Board.)

SOLUTION **(a)** The CPI increased during every five-year period between 1950 and 1995. This reflects the fact that prices have consistently risen.
(b) In 1970 the CPI was 38.8 and in 1980 it was 82.4. The percent change was approximately $\dfrac{82.4 - 38.8}{38.8} \times 100 \approx 112\%$.
(c) In 1980 the CPI was 82.4 and in 1990 it was 130.7. The CPI rose by $\dfrac{130.7 - 82.4}{82.4} \times 100 \approx 59\%$, while tuition and fees rose by $\dfrac{1908 - 804}{804} \times 100 \approx 137\%$. The rise in tuition and fees outpaced the CPI.

EXAMPLE 4   *Calculating percent change*

Find the percent change if a quantity changes from price P_1 to price P_2. Round your answer to the nearest tenth of a percent when appropriate.
(a) $P_1 = \$20, P_2 = \$27$ **(b)** $P_1 = \$1.89, P_2 = \$1.56$

SOLUTION **(a)** Percent change $= \dfrac{P_2 - P_1}{P_1} \times 100 = \dfrac{27 - 20}{20} \times 100 = 35\%$.

(b) Percent change $= \dfrac{P_2 - P_1}{P_1} \times 100 = \dfrac{1.56 - 1.89}{1.89} \times 100 \approx -17.5\%$. The negative sign indicates that the price decreased.

Graphing Calculator Help

To display numbers in scientific notation, see Appendix B (page AP-4).

Algebra Review

To review exponents in expressions such as 10^{-6}, see Chapter R (page R-14).

Scientific Notation

Numbers that are large or small in absolute value occur frequently in applications. For simplicity these numbers are often expressed in scientific notation. Table 1.3 lists examples of numbers in **standard form** and in **scientific notation**.

TABLE 1.3

Standard Form	Scientific Notation	Application
93,000,000 mi	9.3×10^7 mi	Distance to sun
12,600	1.26×10^4	Radio stations in 1998
9,000,000,000	9×10^9	Estimated world population in 2050
0.00000538 sec	5.38×10^{-6} sec	Time for light to travel 1 mile
0.000005 cm	5×10^{-6} cm	Size of a typical virus

To write 0.00000538 in scientific notation, start by moving the decimal point to the right of the first nonzero digit 5 to obtain 5.38. Since the decimal point was moved six places to the *right,* the exponent of 10 is -6. Thus, $0.00000538 = 5.38 \times 10^{-6}$. When the decimal point is moved to the *left,* the exponent of 10 is positive, rather than negative. Here is a formal definition of scientific notation.

> ### Scientific notation
>
> A real number r is in **scientific notation** when r is written as $c \times 10^n$, where $1 \le |c| < 10$ and n is an integer.

EXAMPLE 5 *Applying scientific notation*

A compact disc (CD) is an optically readable device that is capable of storing large amounts of data. Some CDs can store 650 million bytes of information. (A **byte** is the amount of memory necessary to store a single character, such as one letter or a blank space. The phrase "math test" requires 9 bytes of storage.) (Source: Grolier Electronic Publishing, Inc.)

(a) Express in scientific notation the number of bytes available on some CDs.
(b) Approximate how many sentences like this one could be stored on a CD. Express your answer in scientific notation.
(c) When playing a CD, a sound channel is sampled (or read) at a rate of 44,100 times per second per channel. If two channels are played for 1 minute of stereo music, find the total number of times the sound channels are sampled. Express your answer in scientific notation.

SOLUTION (a) Move the decimal point in 650,000,000 to the left 8 places to obtain 6.5. Then 650,000,000 can be written as 6.5×10^8.

(b) The sentence
"Approximate how many sentences like this one could be stored on a CD."
is composed of 69 characters including the blank spaces and the period.

$$\frac{650,000,000}{69} \approx 9,400,000 = 9.4 \times 10^6$$

The CD could store the sentence approximately 9.4 million times.

(c) Two channels are being sampled $2 \times 44,100 = 88,200$ times per second. In 60 seconds, they would be sampled $60 \times 88,200 = 5,292,000 = 5.292 \times 10^6$ times.

Calculators often use "E" to express powers of 10. For example, 4.2×10^{-3} might be displayed as 4.2E–3. On some calculators, numbers can be entered in scientific notation with the $\boxed{\text{EE}}$ key, which you can find by pressing $\boxed{\text{2nd}}$ $\boxed{\, ,\,}$.

EXAMPLE 6   *Computing in scientific notation*

Graphing Calculator Help

To enter numbers in scientific notation, see Appendix B (page AP-4).

Use a calculator to approximate each expression. Write your answer in scientific notation.

(a) $(3.1 \times 10^6)(5.6 \times 10^{-4})$ (b) $\left(\dfrac{6 \times 10^3}{4 \times 10^6}\right)(1.2 \times 10^2)$

(c) $\sqrt{4500\pi}\left(\dfrac{103 + 450}{0.233}\right)^3$

SOLUTION

(a) The given expression is entered in two ways in Figure 1.1. Note that in both cases $(3.1 \times 10^6)(5.6 \times 10^{-4}) = 1736 = 1.736 \times 10^3$.

(b) From Figure 1.2, $\left(\dfrac{6 \times 10^3}{4 \times 10^6}\right)(1.2 \times 10^2) = 0.18 = 1.8 \times 10^{-1}$.

(c) Be sure to insert parentheses around 4500π and around the numerator, $103 + 450$, in the ratio. From Figure 1.3 we can see that the result is approximately 1.6×10^{12}.

FIGURE 1.1   **FIGURE 1.2**   **FIGURE 1.3**

EXAMPLE 7   *Computing with a calculator*

Use a calculator to evaluate each expression. Round your answer to the nearest thousandth.

(a) $\sqrt[3]{131}$ (b) $\pi^3 + 1.2^2$ (c) $\dfrac{1 + \sqrt{2}}{3.7 + 9.8}$ (d) $|\sqrt{3} - 6|$

SOLUTION

(a) On some calculators the cube root can be found by using the MATH menu. If your calculator does not have a cube root key, enter 131^(1/3). From the first two lines in Figure 1.4, we see that $\sqrt[3]{131} \approx 5.079$.

(b) Do *not* use 3.14 for the value of π. Instead, use the built-in key to obtain a more accurate value of π. From the bottom two lines in Figure 1.4, $\pi^3 + 1.2^2 \approx 32.446$.

(c) When evaluating this expression be sure to include parentheses around the numerator and around the denominator. Most calculators have a special square root key that can be used to evaluate $\sqrt{2}$. From the first three lines in Figure 1.5 on the next page, $\dfrac{(1 + \sqrt{2})}{(3.7 + 9.8)} \approx 0.179$.

(d) The absolute value can be found on some calculators by using the MATH NUM menus. From the bottom two lines in Figure 1.5, $|\sqrt{3} - 6| \approx 4.268$.

Graphing Calculator Help

To enter expressions such as $\sqrt[3]{131}$, $\sqrt{2}$, π, and $|\sqrt{3} - 6|$, see Appendix B (page AP-5).

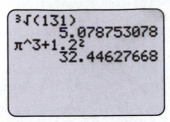

FIGURE 1.4

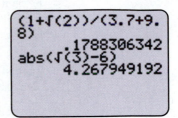

FIGURE 1.5

Problem Solving

Many problem-solving strategies are used in algebra. However, in this subsection we focus on two important strategies that are used frequently: making a sketch and applying one or more formulas. These strategies are illustrated in the next examples.

EXAMPLE 8 *Finding the speed of Earth*

Earth travels in an approximately circular orbit around the sun with an average radius of 93 million miles. If Earth takes 1 year, or about 365 days, to complete one orbit, estimate the orbital speed of Earth in miles per hour.

SOLUTION Speed S equals distance D divided by time T, $S = \dfrac{D}{T}$. We need to find the number of miles Earth travels in 1 year and then divide it by the number of hours in 1 year.

Geometry Review

To find the circumference of a circle, see Chapter R (page R-2).

Distance Traveled A sketch of Earth orbiting the sun is shown in Figure 1.6. In 1 year Earth travels along the circumference of a circle with a radius of 93 million miles. The circumference of a circle is $2\pi r$, where r is the radius, so the distance D is

$$D = 2\pi r = 2\pi(93{,}000{,}000) \approx 584{,}300{,}000 \text{ miles.}$$

Hours in 1 Year The number of hours H in 1 year, or 365 days, equals

$$H = 365 \times 24 = 8760 \text{ hours.}$$

Speed of Earth $\quad S = \dfrac{D}{H} = \dfrac{584{,}300{,}000}{8760} \approx 66{,}700 \text{ miles per hour.}$

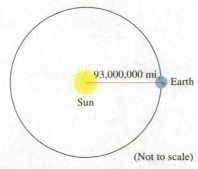

(Not to scale)

FIGURE 1.6

Measuring the thickness of a very thin layer of material can be difficult to do by using a direct measurement. For example, it might be difficult to take a ruler and measure the thickness of a sheet of aluminum foil or to measure the thickness of a coat of paint. However, it can be done indirectly using the following formula.

$$\text{Thickness} = \frac{\text{Volume}}{\text{Area}}$$

That is, the thickness of a thin layer equals the volume of the substance divided by the area that it covers. For example, if a volume of 1 cubic inch of paint is spread over an area of 100 square inches, then the thickness of the paint equals $\frac{1}{100}$ inch. This formula is illustrated in the next example.   (Source: U. Haber-Schaim, *Introductory Physical Science.*)

■ **EXAMPLE 9**   *Calculating the thickness of aluminum foil*

A rectangular sheet of aluminum foil is 15 centimeters by 35 centimeters and weighs 5.4 grams. If 1 cubic centimeter of aluminum weighs 2.7 grams, find the thickness of the aluminum foil.

SOLUTION   Start by making a sketch of a rectangular sheet of aluminum, as shown in Figure 1.7.

Geometry Review
To find the area of a rectangle and the volume of a box, see Chapter R (page R-4).

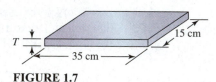

FIGURE 1.7

To complete this problem we need to find the volume V of the aluminum foil and its area A. Then we can determine the thickness T by using the formula $T = \frac{V}{A}$.

Volume   Because the aluminum foil weighs 5.4 grams and each 2.7 grams equals 1 cubic centimeter, the volume of the aluminum foil is

$$\frac{5.4}{2.7} = 2 \text{ cubic centimeters} \qquad \textcolor{blue}{\text{Divide weight by density.}}$$

Area   The aluminum foil is rectangular with an area of $15 \times 35 = 525$ square centimeters.

Thickness   The thickness of 2 cubic centimeters of aluminum foil with an area of 525 square centimeters is

$$T = \frac{V}{A} = \frac{2}{525} \approx 0.0038 \text{ centimeter.}$$

Putting it all Together
1.1

Data and numbers play central roles in a diverse, technological society. Because of the variety of data, it has been necessary to develop different types of numbers. Without numbers, data could be described qualitatively but not quantitatively. For example, we might say that the day seems hot, but would not be able to give an actual number for the temperature. Accurate comparisons with temperatures on other days would be difficult, if not impossible. Problem-solving strategies are used in almost every facet of our lives, providing the procedures needed to systematically complete tasks and perform computations.

The following table summarizes some of the concepts found in this section.

Concept	Examples	Comments
Natural numbers	$1, 2, 3, 4, 5, \ldots$	Sometimes referred to as the *counting numbers*
Integers	$\ldots, -2, -1, 0, 1, 2, \ldots$	Includes the natural numbers, their opposites, and 0.
Rational numbers	$\dfrac{1}{2}, -3, \dfrac{128}{6}, -0.335, 0,$ $0.25 = \dfrac{1}{4}, 0.\overline{33} = \dfrac{1}{3},$ and all fractions	Includes integers; all fractions $\dfrac{p}{q}$, where p and q are integers with $q \neq 0$; all repeating and all terminating decimals
Irrational numbers	$\pi, \sqrt{2}, -\sqrt{5}, \sqrt[3]{7}, \text{ and } \pi^4$	Irrational numbers can be written as nonrepeating, nonterminating decimals. An irrational number cannot be a rational number.
Real numbers	$\pi, \sqrt{7}, -\dfrac{4}{7}, 0, -10,$ $0.\overline{6} = \dfrac{2}{3}, 1000, \sqrt{15}$	Any number that can be expressed in decimal form Includes the rational numbers and irrational numbers
Scientific notation	3.12×10^4 -1.4521×10^{-2} 5×10^9 -3.98715×10^9 1.5987×10^{-6}	Writes a number in the form $c \times 10^n$, where $1 \leq |c| < 10$ and n is an integer Used to represent numbers that are large or small in absolute value
Percent change	If the price of a gallon of gasoline changes from \$1.00 to \$1.40, then the percent change is $\dfrac{1.4 - 1}{1} \times 100 = 40\%.$	If a quantity changes from c_1 to c_2, then the percent change is given by $\dfrac{c_2 - c_1}{c_1} \times 100.$

1.1 EXERCISES

Classifying Numbers

Exercises 1–8: Classify each number as one or more of the following: natural number, integer, rational number, or real number.

1. 400,000 (Minimum cost in dollars to start a McDonald's franchise)

2. $\frac{173}{1000}$ (Fraction of dentists in 1997 who are female)

3. -13 (Percent change in the release of toxic substances into the water between 1988 and 1993)

4. 1,800,000,000 (Sales in dollars for TLC Beatrice International Holding, Inc.)

5. 7.5 (Average number of gallons of water used each minute while taking a shower)

6. 20.7 (Nielsen rating of the TV show *E.R.* during 1997–1998)

7. $90\sqrt{2}$ (Distance in feet from home plate to second base in baseball)

8. -100 (Wind chill when the temperature is $-30°F$ and the wind speed is 40 mph)

Exercises 9 and 10: Classify each number as one or more of the following: natural number, integer, rational number, or irrational number.

9. $\pi, -3, \frac{2}{9}, \sqrt{9}, 1.\overline{3}, -\sqrt{2}$

10. $\frac{3}{1}, -\frac{5}{8}, \sqrt{7}, 0.\overline{45}, 0, 5.6 \times 10^3$

Exercises 11–16: For each measured quantity, state the set of numbers that is most appropriate to describe it. Choose from the natural numbers, integers, or rational numbers. Explain your answer.

11. Shoe sizes

12. Populations of states

13. Gallons of gasoline

14. Speed limits

15. Temperatures given in a winter weather forecast in Montana

16. Numbers of compact disc sales

Real Number Computation

Exercises 17–20: **Percent Change** *(Refer to Example 4.) Find the percent change if a quantity changes from P_1 to P_2. Round your answer to the nearest tenth of a percent when appropriate.*

17. $P_1 = \$8, P_2 = \$13$

18. $P_1 = \$0.90, P_2 = \$13.47$

19. $P_1 = 1.4, P_2 = 0.85$

20. $P_1 = 1256, P_2 = 1195$

Exercises 21–28: Write the number in scientific notation.

21. 187,600 (New lung cancer cases reported in 1999)

22. 109,050 (New AIDS cases reported in New York City during 1998)

23. 0.03983 (Fraction of U.S. deaths attributed to accidents in 1997)

24. 0.001369 (Fraction of U.S. farmland found in New Hampshire in 1998)

25. 2450

26. 105.6

27. 0.56

28. -0.00456

Exercises 29–36: Write the number in standard form.

29. 1×10^{-6} (Approximate wavelength in meters of visible light)

30. 9.11×10^{-31} (Weight in kilograms of an electron)

31. 2×10^8 (Number of years for the sun to complete an orbit in our galaxy)

32. 4.86×10^{12} (Federal debt in dollars in March 1995)

33. 1.567×10^2

34. 0.34×10^4

35. -5.68×10^{-1}

36. 690×10^{-3}

Exercises 37–42: Use a calculator to approximate the expression. Write your result in scientific notation.

37. $(9.87 \times 10^6)(34 \times 10^{11})$

38. $\dfrac{8.947 \times 10^7}{0.00095}$

39. $(56 \times 10^{-7}) + \sqrt{3.4 \times 10^{-2}}$

40. $\sqrt[3]{(2.5 \times 10^{-8}) + 10^{-7}}$

41. $(8.5 \times 10^{-5})(-9.5 \times 10^{7})$

42. $\sqrt{\pi(4.56 \times 10^{4}) + (3.1 \times 10^{-2})}$

Exercises 43–52: Use a calculator to evaluate the expression. Round your result to the nearest thousandth.

43. $\sqrt[3]{192}$

44. $\sqrt{(32 + \pi^3)}$

45. $|\pi - 3.2|$

46. $\dfrac{1.72 - 5.98}{35.6 + 1.02}$

47. $\dfrac{0.3 + 1.5}{5.5 - 1.2}$

48. $3.2(1.1)^2 - 4(1.1) + 2$

49. $\dfrac{1.5}{\sqrt{2 + \pi} - 5}$

50. $4.3 - \dfrac{5}{17}$

51. $15 + \dfrac{4 + \sqrt{3}}{7}$

52. $\dfrac{5 + \sqrt{5}}{2}$

Problem Solving

53. *Consumer Price Index* (Refer to Example 3.) From 1985 to 1998 the average tuition and fees at private colleges and universities rose from $6,121 to $14,508, while the CPI changed from 107.6 to 163.0. Compare the percent change in tuition and fees to the percent change in the CPI. (Sources: Bureau of the Census, The College Board.)

54. *Consumer Price Index* (Refer to Example 3.) Use Table 1.2 to estimate the percent change in prices from 1965 to 1995. In 1965 a gallon of gasoline cost about $0.30. If this price had kept pace with the CPI, what would a gallon of gasoline cost in 1995?

55. *Wages* For some jobs, employees are paid time and a half for every hour worked over 40 hours and double time for all hours on Sunday. Suppose an employee's regular hourly pay is $12. Find the pay for working 47 hours from Monday through Saturday and 8 hours on Sunday.

56. *Storage on a CD* (Refer to Example 5.) Estimate the number of times that the word *mathematics* could be saved on a CD with 650 million bytes of memory. Write your answer in scientific notation.

57. *Orbital Speed* (Refer to Example 8.) The planet Mars travels in a nearly circular orbit around the sun with a radius of 141 million miles. If it requires 1.88 years for Mars to complete one orbit, estimate the orbital speed of Mars in miles per hour.

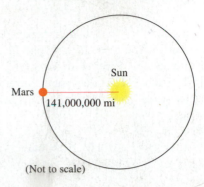

(Not to scale)

58. *Size of the Milky Way* The speed of light is about 186,000 miles per second. The Milky Way galaxy has an approximate diameter of 6×10^{17} miles. Estimate the number of years it takes for light to travel across the Milky Way. (Source: C. Ronan, *The Natural History of the Universe.*)

59. *Federal Debt* The amount of federal debt has changed dramatically during the past 30 years. (Sources: Department of the Treasury, Bureau of the Census.)

(a) In 1970 the population of the United States was 203,000,000 and the federal debt was $370 billion. Find the debt per person.

(b) In 2000 the population of the United States was approximately 281,000,000 and the federal debt was $5.54 trillion. Find the debt per person.

(c) If this trend were to continue, estimate the debt per person in the year 2030. (*Hint:* There may be more than one way to do this estimate.)

60. *Discharge of Water* The Amazon River discharges water into the Atlantic Ocean at an average rate of 4,200,000 cubic feet per second, which is the greatest of any river in the world. Is this more or less than 1 cubic mile of water per day? Explain your calculations. (*Hint:* 1 cubic mile = 5280^3 cubic feet.) (Source: *The Guinness Book of Records 1993.*)

61. *Thickness of an Oil Film* (Refer to Example 9.) A drop of oil measuring 0.12 cubic centimeter is spilled onto a lake. The oil spreads out in a circular shape having a diameter of 23 centimeters. Approximate the thickness of the oil film.

62. *Thickness of Gold Foil*   (Refer to Example 9.) A flat, rectangular sheet of gold foil measures 20 centimeters by 30 centimeters and has a mass of 23.16 grams. If 1 cubic centimeter of gold has a mass of 19.3 grams, find the thickness of the gold foil. (Source: U. Haber-Schaim, *Introductory Physical Science.*)

63. *Analyzing Debt*   A 1-inch high stack of $100 bills contains about 250 bills. In 1995 the federal debt was approximately 5 trillion dollars.
  (a) If the entire federal debt were converted into a stack of $100 bills, how many feet high would it be?
  (b) The distance between Los Angeles and New York is approximately 2500 miles. Could this stack of $100 bills reach between these two cities?

64. *Military Personnel*   The table lists the number of female personnel on active duty in the military for selected years.   (Source: Department of Defense.)
  (a) Discuss the change in the number of active-duty female military personnel between 1970 and 1996.
  (b) Use the concept of an average or mean to estimate the number of female personnel in 1993.

Year	Female Personnel
1970	41,479
1980	171,418
1985	211,606
1990	227,018
1996	197,693

65. *Alcohol Consumption*   The table lists the average annual U.S. consumption in gallons of alcohol per person (14 years old or older) from 1940 to 1994. (Sources: Department of Health and Human Services, Bureau of the Census.)

Year	1940	1950	1960	1970
Alcohol (gallons)	1.56	2.04	2.07	2.52

Year	1980	1990	1994
Alcohol (gallons)	2.76	2.46	2.21

  (a) Discuss the trend in the per capita consumption of alcohol.
  (b) In 1990 there were 199,609,000 people 14 years old or older. Estimate the total gallons of alcohol consumed in the United States in 1990.
  (c) What is the percent change in U.S. consumption of alcohol from 1980 to 1990?

66. *Water in a Lake*   A lake covers 928 acres with an average depth of 17 feet. Estimate the number of cubic feet of water in the lake. If 1 cubic foot equals about 7.48 gallons, approximate the number of gallons of water in the lake. Use scientific notation. (*Hint:* 1 acre equals 43,560 square feet.)

Writing about Mathematics

67. Describe some basic sets of numbers that are used in mathematics.

68. Explain how to calculate percent change.

Extended and Discovery Exercises

1. If you have access to a scale that can weigh in grams, find the thickness of regular and heavy duty aluminum foil. Is heavy duty foil worth the price difference? (*Hint:* Each 2.7 grams of aluminum equals 1 cubic centimeter.)

1.2 VISUALIZATION OF DATA

One-Variable Data • Two-Variable Data • The Distance Formula • Visualizing Data with a Graphing Calculator

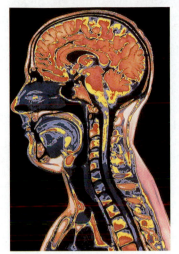

FIGURE 1.8 An MRI Scan

Introduction

Computer technology, the Internet, and electronic communication are creating large amounts of data. The challenge is to convert this data into meaningful information that can be used to solve important problems and create new knowledge. Before conclusions can be drawn about data, it must be analyzed. A powerful tool in this step is visualization. Visual presentations have the capability of communicating vast quantities of information in short periods of time. A typical page of graphics contains one hundred times more information than a page of text. Figure 1.8 shows an MRI scan of a human head and shoulders. Imagine trying to describe this MRI scan with only words. It would be an enormous task, and the final result would not be nearly as understandable as the picture. This section discusses how different types of data can be visualized by using various mathematical techniques.

One-Variable Data

Data often occur in the form of a list. A list of test scores without names is an example. The only variable involved is the score. Data of this type are referred to as **one-variable data**. If the values in a list are unique, they can be represented visually on a number line.

EXAMPLE 1   *Determining the maximum, minimum, and mean of a list of data*

Table 1.4 lists the monthly average temperatures in degrees Fahrenheit at Mould Bay, Canada.   (Source: A. Miller and J. Thompson, *Elements of Meteorology.*)

TABLE 1.4

Temperature (°F)	−27	−31	−26	−9	12	32	39	36	21	1	−17	−24

(a) Plot these temperatures on a number line.
(b) Find the maximum and minimum temperatures.
(c) Determine the mean of these 12 temperatures.

SOLUTION   (a) In Figure 1.9 the numbers −27, −31, −26, −9, 12, 32, 39, 36, 21, 1, −17, and −24 are plotted on a number line.

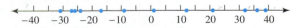

FIGURE 1.9 Monthly Average Temperatures

Graphing Calculator Help

To calculate the maximum, minimum, and mean of a similar data set, see Appendix B (page AP-5).

(b) The maximum temperature of 39°F is plotted farthest to the right in Figure 1.9. Similarly, the minimum temperature of −31°F is plotted farthest to the left.

(c) The sum of the 12 temperatures in Table 1.4 equals 7. The mean or average of these temperatures is $\frac{7}{12} \approx 0.6°\text{F}$.

■ **CLASS DISCUSSION**

In Example 1(c), the mean of the temperatures is approximately 0.6°F. Interpret this temperature. Explain your reasoning. ■

The **median** of a sorted list is equal to the value that is located in the middle of the list. If there is an odd number of data items, the median is the middle data item. If there is an even number of data items, the median is the average of the two middle items. The **range** of a list of data is the difference between the maximum and minimum values. The range is a measure of the *spread* or *dispersion* in the data. Means, medians, and ranges can be found for one-variable data sets.

EXAMPLE 2 *Determining the median and range for a list of data*

The monthly average precipitations in inches for Seattle are 5.7, 4.2, 3.7, 2.4, 1.7, 1.4, 0.8, 1.1, 1.9, 3.5, 5.9, and 5.9. (Source: J. Williams.)
(a) Sort this data in increasing order and display them with a table.
(b) Find the median and the range of this data set.
(c) Interpret the median and the range.

SOLUTION **(a)** The sorted data is displayed in Table 1.5.

TABLE 1.5

Precipitation (inches)	0.8	1.1	1.4	1.7	1.9	2.4	3.5	3.7	4.2	5.7	5.9	5.9

Graphing Calculator Help

To calculate the maximum, minimum, mean, and median of the data in Example 2, see Appendix B (page AP-5).

(b) Since there is an even number of data items, the median is the average of the two middle values, 2.4 and 3.5. This is $\frac{2.4 + 3.5}{2} = 2.95$. The largest value is 5.9 and the smallest value is 0.8. The range is equal to their difference, $5.9 - 0.8 = 5.1$.

(c) In Seattle, half the months have an average precipitation of less than 2.95 inches (the median) and half the months have an average precipitation of greater than 2.95 inches. For any two months, the average precipitations vary by at most 5.1 inches (the range).

Two-Variable Data

It may be possible for a relationship to exist between two lists of data. In Example 2, each monthly average precipitation could be associated with a month. By forming a relationship between the month and the precipitation, more information is communicated. Table 1.6 lists the monthly average precipitation in inches for Portland, Oregon. In this table, 1 corresponds to January, 2 to February, and so on, until 12 represents December. Showing the relationship between a month and its average precipitation is accomplished by combining the two lists, so that corresponding months and precipitations are visually paired.

TABLE 1.6 **Average Precipitation in Portland, Oregon**

Month	1	2	3	4	5	6	7	8	9	10	11	12
Precipitation (inches)	6.2	3.9	3.6	2.3	2.0	1.5	0.5	1.1	1.6	3.1	5.2	6.4

If x is the month and y is the precipitation, then the ordered pair (x, y) represents the average amount of precipitation y during month x. For example, the ordered pair $(5, 2.0)$ indicates that in May the average precipitation is 2.0 inches, whereas the ordered pair $(2, 3.9)$ indicates that the average precipitation in February is 3.9 inches. *Order is important* in an ordered pair.

Since the data in Table 1.6 involve two variables, the month and precipitation, we refer to them as **two-variable data**. It is important to realize that a relation established by two-variable data is between two lists rather than within a single list. January is not related to August and a precipitation of 6.2 inches is not associated with 1.1 inches. Instead January is paired with 6.2 inches, and August is paired with 1.1 inches. We now define the mathematical concept of a relation.

> ## Relation
>
> A **relation** is a set of ordered pairs.

If we denote the ordered pairs in a relation by (x, y), then the set of all x-values is called the **domain** of the relation and the set of all y-values is called the **range**. The relation shown in Table 1.6 has domain

$$D = \{1, 2, 3, 4, 5, 6, 7, 8, 9, 10, 11, 12\},$$

and range

$$R = \{6.2, 3.9, 3.6, 2.3, 2.0, 1.5, 0.5, 1.1, 1.6, 3.1, 5.2, 6.4\}.$$

EXAMPLE 3 *Finding the domain and range of a relation*

A physics class measures the time y that it takes for an object to fall x feet, as shown in Table 1.7.

TABLE 1.7 **Falling Object**

x (feet)	20	20	40	40
y (seconds)	1.2	1.1	1.5	1.6

(a) Express the data as a relation S.
(b) Find the domain and range of S.

SOLUTION **(a)** A relation is a set of ordered pairs, so we can write

$$S = \{(20, 1.2), (20, 1.1), (40, 1.5), (40, 1.6)\}.$$

(b) The domain is the set of x-values of the ordered pairs or $D = \{20, 40\}$. The range is the set of y-values of the ordered pairs or $R = \{1.1, 1.2, 1.5, 1.6\}$.

■ **MAKING CONNECTIONS** ───────

The Different Meanings for Range Notice that the word *range* has different meanings for one-variable and two-variable data. For one-variable data, the range is a *real number* that describes the spread of the data in a single list. For two-variable data, the range is a *set of numbers* associated with a set of ordered pairs. As is often the case in English, we can determine the correct meaning of *range* from the context in which it is used.

To visualize a relation, we often use the **Cartesian coordinate plane** or ***xy*-plane**. The horizontal axis is the ***x*-axis** and the vertical axis is the ***y*-axis**. The axes intersect at the **origin** and determine four regions called **quadrants**, numbered I, II, III, and IV, counterclockwise, as shown in Figure 1.10. We can plot the ordered pair (x, y) using the x- and y-axes. The point $(1, 2)$ is located in quadrant I, $(-2, 3)$ in quadrant II, $(-4, -4)$ in quadrant III, and $(1, -2)$ in quadrant IV. A point lying on a coordinate axis does not belong to any quadrant. The point $(-3, 0)$ is located on the x-axis, whereas the point $(0, -3)$ lies on the y-axis.

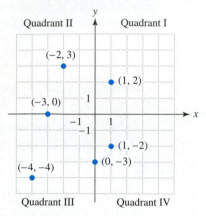

FIGURE 1.10   The xy-plane

The term **scatterplot** is given to a graph in the xy-plane, where distinct data points are plotted. A scatterplot is shown in Figure 1.11.

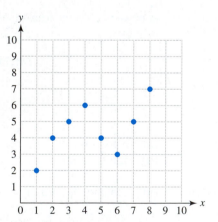

FIGURE 1.11   A Scatterplot

EXAMPLE 4   *Making scatterplots*

Use Table 1.6 to make a scatterplot of the monthly average precipitations in Portland, Oregon.

SOLUTION   Use the x-axis for the months and the y-axis for the precipitation amounts. To make a scatterplot, simply graph the ordered pairs $(1, 6.2)$, $(2, 3.9)$, $(3, 3.6)$, $(4, 2.3)$, $(5, 2.0)$, $(6, 1.5)$, $(7, 0.5)$, $(8, 1.1)$, $(9, 1.6)$, $(10, 3.1)$, $(11, 5.2)$ and $(12, 6.4)$ in the xy-plane as shown in Figure 1.12.

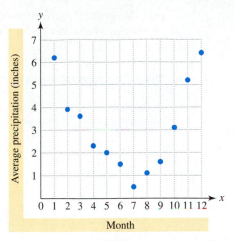

FIGURE 1.12   Monthly Average Precipitation

Sometimes it is helpful to connect the data points in a scatterplot with straight-line segments. This type of graph visually emphasizes changes in the data. It is called a **line graph**.

EXAMPLE 5   *Interpreting a line graph*

The line graph in Figure 1.13 shows the number of college graduates in thousands at 10-year intervals between 1900 and 1990.   (Sources: Department of Education, Center for Educational Statistics.)

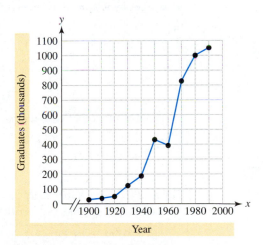

FIGURE 1.13   College Graduates

(a) Did the number of graduates ever decrease during this time period? Explain.
(b) Approximate the number of college graduates in the year 1970.
(c) Determine the 10-year period when there was the greatest increase in the number of college graduates.

SOLUTION   (a) Yes, the number decreased slightly between 1950 and 1960. For this time period, the line segment slopes downward from left to right.
(b) According to the graph, the number of college graduates in 1970 was approximately 830,000.

(c) The greatest increase corresponds to the steepest line segment. This occurs between 1960 and 1970.

The Distance Formula

Geometry Review

To review the Pythagorean theorem, See Chapter R (page R-3).

In the xy-plane, the length of a line segment with endpoints (x_1, y_1) and (x_2, y_2) can be calculated using the Pythagorean theorem. See Figure 1.14.

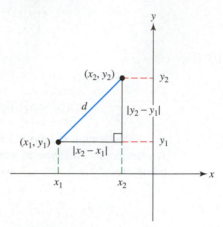

FIGURE 1.14

The lengths of the legs of the right triangle are $|x_2 - x_1|$ and $|y_2 - y_1|$. The distance d is the hypotenuse of the right triangle. Applying the Pythagorean theorem, $d^2 = (x_2 - x_1)^2 + (y_2 - y_1)^2$. Since distance is nonnegative, it follows that $d = \sqrt{(x_2 - x_1)^2 + (y_2 - y_1)^2}$.

Distance formula

The **distance** d between the points (x_1, y_1) and (x_2, y_2) in the xy-plane is

$$d = \sqrt{(x_2 - x_1)^2 + (y_2 - y_1)^2}.$$

Figure 1.15 shows a line segment connecting the points $(-1, 3)$ and $(4, -3)$. Its length is

$$d = \sqrt{(4 - (-1))^2 + (-3 - 3)^2}$$
$$= \sqrt{61}$$
$$\approx 7.81.$$

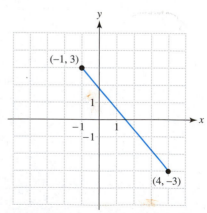

FIGURE 1.15

In the next example the distance between two moving cars is found.

EXAMPLE 6   *Finding the distance between two moving cars*

Suppose that at noon car A is traveling south at 20 miles per hour and is located 80 miles north of car B. Car B is traveling east at 40 miles per hour.
(a) Let (0, 0) be the initial coordinates of car B in the xy-plane, where units are in miles. Plot the location of each car at noon and at 1:30 P.M.
(b) Find the distance between the cars at 1:30 P.M.

SOLUTION   **(a)** If the initial coordinates of car B are (0, 0), then the initial coordinates of car A are (0, 80), because car A is 80 miles north of car B. After 1 hour and 30 minutes or 1.5 hours, car A has traveled $1.5 \times 20 = 30$ miles south, and so it is located 50 miles north of the initial location of car B. Thus its coordinates are (0, 50) at 1:30 P.M. Car B traveled $1.5 \times 40 = 60$ miles east, so its coordinates are (60, 0) at 1:30 P.M. See Figure 1.16 where these points are plotted.
(b) To find the distance between the cars at 1:30 P.M. we must find the distance d between the points (0, 50) and (60, 0).

$$d = \sqrt{(60 - 0)^2 + (0 - 50)^2} \approx 78.1 \text{ miles}.$$

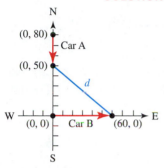

FIGURE 1.16

Visualizing Data with a Graphing Calculator

Graphing calculators provide several features beyond those found on scientific calculators. The bottom rows of keys on graphing calculators are often similar to those found on scientific calculators. However, graphing calculators have additional keys that can be used to create tables, scatterplots, line graphs, and other types of graphs. Another difference between graphing and scientific calculators is that a graphing calculator has a larger screen.

   The **viewing rectangle** or **window** on a graphing calculator is similar to the view finder in a camera. A camera cannot take a picture of an entire scene. It must be centered on a portion of the available scenery. A camera can capture different views of the same scene by zooming in and out. Graphing calculators have similar capabilities. The xy-plane is infinite. The calculator screen can show only a finite, rectangular region of the xy-plane. The viewing rectangle must be specified by setting minimum and maximum values for both the x- and y-axes before a graph can be drawn.

   We will use the following terminology regarding the size of a viewing rectangle. **Xmin** is the minimum x-value and **Xmax** is the maximum x-value along the x-axis. Similarly, **Ymin** is the minimum y-value and **Ymax** is the maximum y-value along the y-axis. Most graphs show an x-scale and y-scale using tick marks on the respective axes. Sometimes the distance between consecutive tick marks is one unit, and other times it might be five units. The distance represented by consecutive tick marks on the x-axis is called **Xscl**, and the distance represented by consecutive tick marks on the y-axis is called **Yscl**. See Figure 1.17 on the next page. This information about the viewing rectangle can be written concisely as [Xmin, Xmax, Xscl] by [Ymin, Ymax, Yscl]. For example, $[-10, 10, 1]$ by $[-10, 10, 1]$ means that Xmin $= -10$, Xmax $= 10$, Xscl $= 1$, Ymin $= -10$, Ymax $= 10$, and Yscl $= 1$. This setting is commonly referred to as the **standard viewing rectangle**.

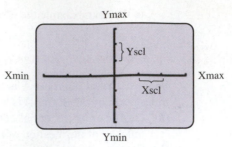

FIGURE 1.17

| EXAMPLE 7 | *Setting the viewing rectangle* |

Show the standard viewing rectangle and the viewing rectangle $[-30, 40, 10]$ by $[-400, 800, 100]$ on your calculator.

SOLUTION   The window settings and viewing rectangles are displayed in Figures 1.18–1.21. Notice that in Figure 1.19, there are 10 tick marks on the positive x-axis, since its length is 10 and the distance between consecutive tick marks is 1.

Graphing Calculator Help

To set a viewing rectangle or window, see Appendix B (page AP-6).

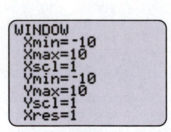

FIGURE 1.18

$[-10, 10, 1]$ by $[-10, 10, 1]$

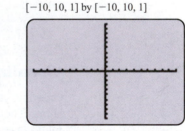

FIGURE 1.19

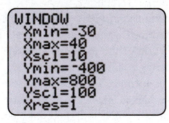

FIGURE 1.20

$[-30, 40, 10]$ by $[-400, 800, 100]$

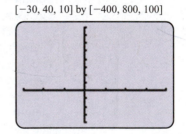

FIGURE 1.21

| EXAMPLE 8 | *Making a scatterplot with a graphing calculator* |

Plot the points $(-5, -5)$, $(-2, 3)$, $(1, -7)$, and $(4, 8)$ in the standard viewing rectangle.

SOLUTION   The standard viewing rectangle is $[-10, 10, 1]$ by $[-10, 10, 1]$. The points $(-5, -5)$, $(-2, 3)$, $(1, -7)$, and $(4, 8)$ are plotted in Figure 1.22.

Graphing Calculator Help

To make the scatterplot in Figure 1.22, see Appendix B (page AP-6).

$[-10, 10, 1]$ by $[-10, 10, 1]$

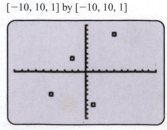

FIGURE 1.22

EXAMPLE 9 *Creating a line graph with a graphing calculator*

Table 1.8 lists the percent share of music sales that compact discs (CDs) held from 1994 to 1998. Make a line graph of these sales in the viewing rectangle [1993, 1999, 1] by [50, 80, 5].

TABLE 1.8

Year	1994	1995	1996	1997	1998
CDs (% share)	58.4	65.0	68.4	70.2	74.8

Source: Recording Industry Association of America.

SOLUTION

Graphing Calculator Help

The steps to make a line graph are similar to the steps to make a scatterplot. See Appendix B (page AP-6).

Enter the data points (1994, 58.4), (1995, 65.0), (1996, 68.4), (1997, 70.2), and (1998, 74.8). A line graph can be created by selecting this option on your graphing calculator. The graph is shown in Figure 1.23.

[1993, 1999, 1] by [50, 80, 5]

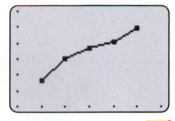

FIGURE 1.23

Putting it all Together

1.2

*T*he following table lists basic concepts in this section.

Concept	Explanation	Examples
Mean or average	To find the mean or average of n numbers, divide their sum by n.	The mean of the four numbers $-3, 5, 6, 9$ is $$\frac{-3 + 5 + 6 + 9}{4} = 4.25.$$
Range	The range of a list of data is the difference between the maximum and the minimum of the numbers or values.	The range of the data $7, 6, -4, 2, 11$ is $$11 - (-4) = 15.$$ (Maximum − Minimum = Range)
Median	The median of a sorted list of numbers equals the value that is located in the middle of the list. Half the data is greater than or equal to the median, and half the data is less than or equal to the median.	The median of $2, 3, 6, 9, 11$ is 6, the middle data item. The median of $2, 3, 6, 9$ is the average of the two middle values: 3 and 6. Therefore the median is $\frac{3 + 6}{2} = 4.5$.
Relation	A relation is a set of ordered pairs (x, y). The set of x-values is called the domain and the set of y-values is called the range.	The relation $S = \{(1, 3), (2, 5), (1, 6)\}$ has domain $D = \{1, 2\}$ and range $R = \{3, 5, 6\}$.

Concept	Explanation	Examples
Distance formula	The distance between (x_1, y_1) and (x_2, y_2) is $$d = \sqrt{(x_2 - x_1)^2 + (y_2 - y_1)^2}.$$	The distance between $(2, -1)$ and $(-1, 3)$ is $$d = \sqrt{(-1 - 2)^2 + (3 - (-1))^2} = 5.$$

The next table summarizes some basic concepts related to one-variable and two-variable data.

Type of Data	Methods of Visualization	Comments
One-variable data	Number line, list, one-column or one-row table	The data items are the same type, and can be described using x-values. Computations of the mean, median, and range are performed on one-variable data.
Two-variable data	Two-column or two-row table, scatterplot, line graph, or other types of graphs in the xy-plane	Two types of data are related, and can be described using ordered pairs (x, y).

1.2 EXERCISES

Data Involving One Variable

Exercises 1–4: Sort each list of numbers from smallest to largest and display them in a table.
  (a) *Determine the maximum and minimum values.*
  (b) *Calculate the mean, median, and range. Round each result to the nearest hundredth when appropriate.*

1. $-10, 25, 15, -30, 55, 61, -30, 45, 5$

2. $-1.25, 4.75, -3.5, 1.5, 2.5, 4.75, 1.5$

3. $\sqrt{15}, 2^{2.3}, \sqrt[3]{69}, \pi^2, 2^\pi, 4.1$

4. $\dfrac{22}{7}, 3.14, \sqrt[3]{28}, \sqrt{9.4}, 4^{0.9}, 3^{1.2}$

Exercises 5–8: ***Geography***   *Each set of numbers contains data about geographic features of the world.*
  (a) *Plot the numbers on a number line.*
  (b) *Calculate the mean, median, and range for the set of numbers. Interpret your results.*
  (c) *Try to identify the geographic feature associated with the largest number in each set.*

5. {840, 87.8, 227, 280, 84.2, 196, 306, 165} (Areas of largest islands in thousands of square miles) (Source: National Geographic Society.)

6. {11.2, 9.91, 31.7, 22.3, 12.3, 26.8, 12.2, 12.0, 11.0, 24.9, 23.0} (Areas of largest freshwater lakes in thousands of square miles) (Source: U.S. National Oceanic and Atmospheric Administration.)

7. {19.3, 18.5, 29.0, 7.31, 16.1, 22.8, 20.3} (Highest elevations of the continents in thousands of feet) (Source: *National Geographic Atlas of the World.*)

8. {4145, 3720, 3360, 3740, 3990, 3650, 3590, 3030} (Lengths of longest rivers in miles) (Source: U.S. National Oceanic and Atmospheric Administration.)

9. ***Basketball Salaries***   In 1999 the average salary of the Portland Trail Blazers was $4,618,669 and the minimum salary was $385,000. Suppose that one player earned $14,795,642, while five other players earned the minimum. (Source: National Basketball Association.)

  (a) Determine the average salary for the six players.
  (b) Find their median salary.
  (c) Discuss why "average salary" can be misleading.

10. *Tennis Earnings*   Martina Navratilova had the following earnings for the first six years of her professional tennis career: $173,668, $128,535, $300,317, $450,757, $747,548, and $749,250. (Source: J. Leder, *Martina Navratilova.*)

(a) Determine her mean and median winnings during this time period.

(b) Find the range of these earnings.

11. *Designing a Data Set*   Find a set of three numbers with a mean of 20 and a median of 18. Is your answer unique?

12. *Designing a Data Set*   Find a set of five numbers with an average of 10, a median of 9, and a range of 15. Is your answer unique?

Distance Formula

Exercises 13–24: Find the distance in the xy-plane between the two points. Round approximate results to the nearest hundredth.

13. $(2, -2), (5, 2)$

14. $(0, -3), (12, -8)$

15. $(7, -4), (9, 1)$

16. $(-1, -6), (-8, -5)$

17. $(-6.5, 2.7), (3.6, -2.9)$

18. $(3.6, 5.7), (-2.1, 8.7)$

19. $(-3, 2), (-3, 10)$

20. $(7, 9), (-1, 9)$

21. $\left(\frac{1}{2}, -\frac{1}{2}\right), \left(\frac{3}{4}, \frac{1}{2}\right)$

22. $\left(-\frac{1}{3}, \frac{2}{3}\right), \left(\frac{1}{3}, -\frac{4}{3}\right)$

23. $(a, b), (a, a)$

24. $(x, y), (1, 2)$

25. *Geometry*   An **isosceles triangle** has at least two sides of equal length. Determine whether the triangle with vertices $(0, 0), (3, 4), (7, 1)$ is isosceles.

26. *Geometry*   An **equilateral triangle** has sides of equal length. Determine whether the triangle with vertices $(-1, -1), (2, 3), (-4, 3)$ is equilateral.

27. *Distance between Cars*   (Refer to Example 6.) At 9:00 A.M. car A is traveling north at 50 miles per hour and is located 50 miles south of car B. Car B is traveling west at 20 miles per hour.

(a) Let $(0, 0)$ be the initial co[...] the *xy*-plane, where units are [...] locations of each car at 9:00 A.M. a[...] A.M.

(b) Find the distance between the cars at 11:[...] A.M.

28. *Distance between Ships*   (Refer to Example 6.) Two ships leave the same harbor at the same time. The first ship heads north at 20 miles per hour and the second ship heads west at 15 miles per hour. Write an expression that gives the distance between the ships after *t* hours.

Data Involving Two Variables

Exercises 29–34: Complete the following.

(a) Find the domain and range of the relation.

(b) Determine the maximum and minimum of the x-values; of the y-values.

(c) Label appropriate scales on the xy-axes.

(d) Plot the relation by hand in the xy-plane.

29. $\{(0, 5), (-3, 4), (-2, -5), (7, -3), (0, 0)\}$

30. $\{(1, 1), (3, 0), (-5, -5), (8, -2), (0, 3)\}$

31. $\{(10, 50), (-35, 45), (20, -55), (75, 25), (-40, 60), (-25, -25)\}$

32. $\{(11, 15), (2, -13), (-5, -14), (-7, 19), (-17, -4), (-1, 1), (5, -11)\}$

33. $\{(0.1, -0.3), (0.5, 0.4), (-0.7, 0.5), (0.8, -0.1), (0.9, 0.9)\}$

34. $\{(-1.2, 1.5), (1.0, 0.5), (-0.3, 1.1), (-0.8, -1.3), (1.4, -1.6)\}$

Exercises 35–42: Show the viewing rectangle on your graphing calculator. Predict the number of tick marks on the positive x-axis and the positive y-axis.

35. Standard viewing rectangle

36. $[-4.7, 4.7, 1]$ by $[-3.1, 3.1, 1]$

37. $[0, 100, 10]$ by $[-50, 50, 10]$

38. $[-30, 30, 5]$ by $[-20, 20, 5]$

39. $[1980, 1995, 1]$ by $[12,000, 16,000, 1000]$

40. $[1800, 2000, 20]$ by $[5, 20, 5]$

41. $[0, 10, 3]$ by $[-4, 5, 2]$

42. $[-0.5, 0.5, 0.3]$ by $[-1.1, 1.1, 0.2]$

Exercises 43–46: Match the settings for a viewing rectangle with the correct figure.

43. $[-9, 9, 1]$ by $[-6, 6, 1]$

44. $[-6, 6, 1]$ by $[-9, 9, 1]$

45. $[-2, 2, 0.5]$ by $[-4.5, 4.5, 0.5]$

46. $[-4, 8, 1]$ by $[-600, 600, 100]$

a.   **b.**

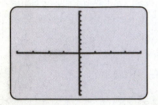   

c.   **d.**

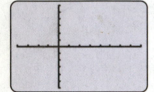   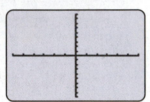

Exercises 47–52: Make a scatterplot of the relation.

47. $\{(1, 3), (-2, 2), (-4, 1), (-2, -4), (0, 2)\}$

48. $\{(6, 8), (-4, -10), (-2, -6), (2, -5)\}$

49. $\{(10, -20), (-40, 50), (30, 60), (-50, -80), (70, 0),$
$(0, -30)\}$

50. $\{(0.01, -0.1), (-0.04, 0.04), (-0.01, 0.09),$
$(0, -1.3)\}$

51. $\{(3.1, 6.2), (-5.1, 10.1), (-0.7, -1.4), (1.8, 3.6),$
$(-4.9, -9.8)\}$

52. $\{(-1.2, 0.6), (1.0, -0.5), (-0.4, 0.2), (-2.8, 1.4),$
$(2.8, -1.4)\}$

Exercises 53–56: Each table contains real data involving two variables.

   (a) *Determine the maximum and minimum values for each variable in the table.*
   (b) *Use your results from part (a) to determine an appropriate viewing rectangle.*
   (c) *Make a scatterplot of the data.*
   (d) *Make a line graph of the data.*

53. Marijuana use by high school seniors

x (year)	1979	1985	1988	1993	1997
y (percent)	37	25	18	15	23

Source: The Substance Abuse and Mental Health Services Administration.

54. Illicit drug use at least once by people ages 18–25

x (year)	1979	1985	1988	1993	1996
y (percent)	46	38	29	24	26

Source: The Substance Abuse and Mental Health Services Administration.

55. Projected population for Asian-Americans in millions

x (year)	1996	1998	2000
y (population)	9.7	10.5	11.2

x (year)	2002	2004
y (population)	12.0	12.8

Source: U.S. Census Bureau.

56. TV cable subscribers

x (year)	1960	1970	1980	1990	1999
y (millions)	0.6	4	18	55	68

Source: National Cable Television Association.

Exercises 57 and 58: **Graduate Degrees** *Create a table that contains the same data as the line graph.*

57. Numbers of doctorate degrees in thousands conferred in the United States from 1950 to 1995

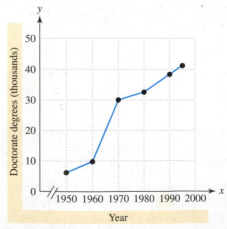

Source: Department of Education.

58. Numbers of master's degrees in thousands conferred in the United States from 1950 to 1995

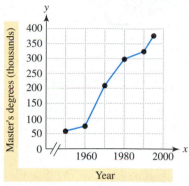

Source: Department of Education.

Writing about Mathematics

59. Describe how one- and two-variable data can be represented. Give one example of each type of data.

60. If a person works at an hourly rate of R dollars for H hours, then the pay P is computed by the equation $P = R \cdot H$. Discuss ways to represent the pay P for various values of R and H using a table. Give examples.

CHECKING BASIC CONCEPTS FOR SECTIONS 1.1 AND 1.2

1. Approximate each expression to the nearest hundredth.
  (a) $\sqrt{4.2(23.1 + 0.5^3)}$
  (b) $\dfrac{23 + 44}{85.1 - 32.9}$

2. Write each number using scientific notation.
  (a) 348,500,000
  (b) -1237.4
  (c) 0.00198

3. Find the distance between the points $(-3, 1)$ and $(3, -5)$.

4. The average depths in feet of the four oceans are 13,215, 12,881, 13,002, and 3953. Calculate the mean, median, and range of these depths. Interpret your results.

5. Make a scatterplot and a line graph with the points $(-5, -4)$, $(-1, 2)$, $(2, -2)$, and $(3, 6)$. State the quadrant in which each point lies.

1.3 FUNCTIONS AND THEIR REPRESENTATIONS

Basic Concepts • Representations of Functions • Formal Definition of a Function • Graphing Calculators and Functions • Identifying Functions

Introduction

Any concept can be made more or less complicated by how it is represented. Many advances throughout history were the result of inventing new representations. For example, the theory of germs provided a new representation of disease that greatly improved medical science.

Similar events have occurred in mathematics. Historically, fractions were developed much earlier than decimals. The introduction of decimals greatly simplified arithmetic involving fractions. For example, sums like $\frac{1}{2} + \frac{1}{4}$ caused difficulty for people because of unlike denominators. However, their decimal equivalents could be easily combined as $0.50 + 0.25 = 0.75$. By developing a new representation, an old problem was made easier.

In this section the concept of a function is introduced. Functions are a special type of relation that can be used to model important phenomena in our world.   (Reference: L. Motz and J. Weaver, *The Story of Mathematics.*)

Basic Concepts

Although thunder is caused by lightning, we sometimes see a flash of lightning before we hear the thunder. This is because light travels at 186,000 *miles* per second, whereas sound travels at about 1050 *feet* per second. Since 1 mile equals 5280 feet, sound takes about 5 seconds to travel 1 mile. It follows that the farther lightning is away, the greater the time lapse between seeing the flash of lightning and hearing the thunder. Table 1.9 lists the *approximate* distance y in miles between a person and a bolt of lightning when there is a time lapse of x seconds between seeing the lightning and hearing the thunder. Note that the value of y can be found by dividing the corresponding value of x by 5.

TABLE 1.9   Distance from a Bolt of Lightning

x (seconds)	5	10	15	20	25
y (miles)	1	2	3	4	5

This table establishes a special type of relation between two sets of numbers, where each valid input x in seconds determines *exactly one* output y in miles. We say that Table 1.9 *represents* a function f, where function f *computes* the distance between an observer and a lightning bolt.

To emphasize that y is a function of x, the notation $y = f(x)$ is used. The expression $f(x)$ does *not* indicate that f and x are multiplied. Rather, the notation $y = f(x)$ is called **function notation**, is read "y equals f of x," and denotes that function f with input x produces output y. That is,

$$f(\text{Input}) = \text{Output},$$

where the input is in seconds and the output is in miles. For example, we write $f(5) = 1$ because if there is a 5-second delay between a lightning bolt and its thunder, then the lightning bolt was about 1 mile away. The expression $f(5)$ represents the output from f when the input is 5. Similarly, $f(10) = 2$ and $f(15) = 3$. Note that a function calculates a set of ordered pairs (x, y), where $y = f(x)$. For example, the ordered pairs (5, 1), (10, 2), and (15, 3) all belong to the relation computed by f. We can think of these ordered pairs as input-output pairs with the form (input, output). A relation that calculates exactly one output for each valid input is a *function*.

The distance y *depends* on the time x, and so y is called the **dependent variable** and x is called the **independent variable**. For example, if we pick the independent variable x to be 10 seconds then the value of y is dependent on x and equal to $\frac{10}{5} = 2$ miles. The y-values depend on the x-values that we choose. The following diagram can help us visualize the computation performed by f.

Input x in seconds $\longrightarrow$ Compute $y = f(x)$ by dividing x by 5. $\longrightarrow$ Output y in miles
(independent variable)   (dependent variable)

The set of valid or meaningful inputs x is called the **domain** of the function and the set of corresponding outputs y is the **range**. For example, suppose a function f

computes the height of a ball thrown into the air after x seconds. Then the domain of f might consist of all times while the ball was in flight, and the range would include all heights attained by the ball.

The following can be computed by functions because they result in *one* output for each valid input.

- Calculating the square of a number x
- Finding the sale price of an item discounted 25% with regular price x
- Naming the biological mother of person x

Representations of Functions

A function can be represented by verbal descriptions, tables, diagrams, symbols, and graphs.

Verbal Representation.   If function f approximates the distance between an observer and a bolt of lightning, then we can verbally describe f with the following sentence: "Divide x seconds by 5 to obtain y miles." We call this a **verbal representation** of f.

Numerical Representation (or Table of Values).   A numerical representation for the function f that calculates the distance between a lightning bolt and an observer is given in Table 1.9. A **numerical representation** is a *table of values* that lists input-output pairs for a function. A different numerical representation for f is shown in Table 1.10.

TABLE 1.10   Distance from a Bolt of Lightning

x (seconds)	1	2	3	4	5	6	7
y (miles)	0.2	0.4	0.6	0.8	1.0	1.2	1.4

One difficulty with a numerical representation is that it is often either inconvenient or impossible to list all possible inputs x. For this reason we sometimes refer to a table of this type as a **partial numerical representation** as opposed to a **complete numerical representation**, which would include all elements from the domain of a function. For example, many valid inputs do not appear in Table 1.10, such as $x = 11$ or $x = 0.75$.

Some tables do not represent a function. For example, Table 1.11 represents a relation, but not a function because input 1 produces two outputs, 3 and 12.

TABLE 1.11   A Relation That Is Not a Function

x	1	2	3	1
y	3	6	9	12

Diagrammatic Representation.   Functions are sometimes represented using **diagrammatic representations** or **diagrams**. Figure 1.24 on the next page is a diagram of a function with domain $D = \{5, 10, 15, 20\}$ and range $R = \{1, 2, 3, 4\}$. An arrow is used to show that input x produces output y. For example, input 5 results in output 1 or $f(5) = 1$. Figure 1.25 shows a relation, but not a function, because input 2 results in two different outputs, 5 and 6.

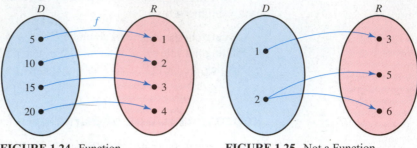

FIGURE 1.24 Function   **FIGURE 1.25** Not a Function

Symbolic Representation.   A formula provides a **symbolic representation** of a function. The computation performed by f is expressed by $f(x) = \dfrac{x}{5}$, where $y = f(x)$. We say that function f is *represented by* or *given by* $f(x) = \dfrac{x}{5}$. It follows that $f(6) = \dfrac{6}{5} = 1.2$.

Similarly, if a function g computes the square of a number x, then g can be represented by $g(x) = x^2$. A formula is an efficient, but less visual, way to define a function.

Graphical Representation.   Leonhard Euler (1707–1783) invented the function notation $f(x)$. He also was the first to allow a function to be represented by a graph, rather than only by a symbolic expression.

A **graphical representation** or **graph** visually pairs an x-input with a y-output. In a graph of a function, the ordered pairs (x, y) are plotted in the xy-plane. The ordered pairs

$$(1, 0.2), (2, 0.4), (3, 0.6), (4, 0.8), (5, 1.0), (6, 1.2), \text{ and } (7, 1.4)$$

from Table 1.10 are plotted in Figure 1.26. This scatterplot suggests a line for the graph of f, as shown in Figure 1.27.

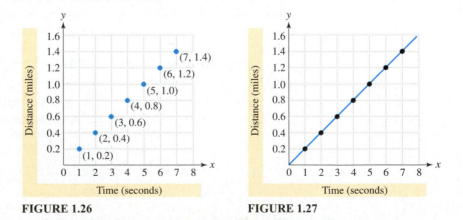

FIGURE 1.26   **FIGURE 1.27**

Note that since $f(5) = 1$, the point $(5, 1)$ lies on the graph of f. Similarly, since $(7, 1.4)$ lies on the graph of f, it follows that $f(7) = 1.4$. In general, the point $(x, f(x))$ lies on the graph of f.

EXAMPLE 1   *Graphing a function by hand*

Graph $f(x) = |x|$ by hand.

SOLUTION   To graph by hand, it is sometimes helpful to make a table of values first. Start by selecting convenient x-values and then substitute them into $f(x) = |x|$, as shown in Table 1.12. For example, when $x = -2$, then $f(-2) = |-2| = 2$ so the point $(-2, 2)$ is located on the graph of $y = f(x)$.

TABLE 1.12

x	-2	-1	0	1	2
$f(x)$	2	1	0	1	2

Next, we plot the points $(-2, 2)$, $(-1, 1)$, $(0, 0)$, $(1, 1)$, and $(2, 2)$ in the xy-plane, as shown in Figure 1.28. The points appear to be V-shaped, and the graph of f shown in Figure 1.29 results if all possible pairs of the form $(x, |x|)$ are plotted.

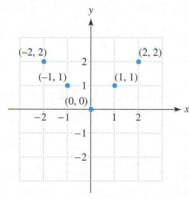

FIGURE 1.28

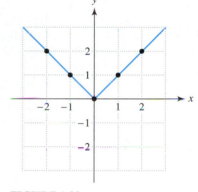

FIGURE 1.29

Although there are many ways to represent a function, there is no *best* representation. It depends on both the individual and the application. Sometimes a formula works best—other times a graph or a table is preferable.

Formal Definition of a Function

Because the idea of a function is a fundamental concept in mathematics, it is important that we define a function precisely. This definition should allow for all representations of a function. The commonality among representations is the concept of an ordered pair.

A relation is a set of ordered pairs. Would it be sufficient to say that a function is a relation? No, because a function must produce only one output for each valid input.

Function

A **function** is a relation in which each element in the domain corresponds to exactly one element in the range.

The set of ordered pairs for a function can be either finite or infinite. The function $f = \{(1, 2), (3, 4), (5, 6)\}$ is a finite set of ordered pairs. In contrast, the function represented by $g(x) = x^2$ with all real numbers as its domain generates an infinite set of ordered pairs.

EXAMPLE 2   *Computing average income as a function of educational attainment*

The function f computes the average 1997 annual household earnings in dollars by educational attainment. This function is defined by $f(N) = 32{,}948$; $f(H) = 46{,}523$; $f(B) = 81{,}026$; $f(M) = 96{,}519$, where N denotes no diploma, H a high school diploma, B a bachelor's degree, and M a master's degree. (Note that these numbers include two-income families.)   (Sources: Bureau of the Census, Department of Commerce.)
(a) Write f as a set of ordered pairs.
(b) Give the domain and range of f.
(c) Discuss the relationship between education and income.

SOLUTION   **(a)** $f = \{(N, 32{,}948), (H, 46{,}523), (B, 81{,}026), (M, 96{,}519)\}$.
(b) The domain and range of f are

$$D = \{N, H, B, M\} \quad \text{and} \quad R = \{32{,}948, 46{,}523, 81{,}026, 96{,}519\}.$$

(c) The greater the educational attainment, the greater the household annual earnings.

■ **MAKING CONNECTIONS**

Relations and Functions   Every function is a relation, whereas every relation is not a function. A function can have only one output for each input. ■

Unless stated otherwise, the domain of a function f is the set of all real numbers for which its symbolic representation is defined. This is called the **implied domain** of a function. The implied domain can be thought of as the set of all valid inputs that make sense in the expression for $f(x)$.

EXAMPLE 3   *Evaluating a function and determining its domain*

Let a function f be represented symbolically by $f(x) = \dfrac{x}{x^2 - 1}$.
(a) Evaluate $f(3)$ and $f(a + 1)$.   **(b)** Find the implied domain of f.

SOLUTION   **(a)** To evaluate $f(3)$, substitute 3 for x in the formula: $f(3) = \dfrac{3}{3^2 - 1} = \dfrac{3}{8}$.

To evaluate $f(a + 1)$, substitute $(a + 1)$ for x in the formula for $f(x)$.

$$f(x) = \frac{(a + 1)}{(a + 1)^2 - 1} \qquad \text{Let } x = (a + 1).$$

$$= \frac{a + 1}{a^2 + 2a + 1 - 1} \qquad \text{Square binomial.}$$

$$= \frac{a + 1}{a^2 + 2a} \qquad \text{Simplify.}$$

Algebra Review
To square a binomial, see
Chapter R (page R-25).

(b) The expression for $f(x)$ is not defined when the denominator $x^2 - 1 = 0$. Therefore the implied domain of f is all real numbers such that $x \neq \pm 1$. The domain can be expressed in set-builder notation as $\{x \mid x \neq \pm 1\}$.

Set-builder notation

The expression $\{x \mid x \neq \pm 1\}$ is written in **set-builder notation**, and represents the set of all real numbers x, such that x does not equal 1 or -1. Another example is $\{y \mid 1 < y < 5\}$, which represents the set of all real numbers y, such that y is greater than 1 *and* less than 5.

EXAMPLE 4   *Evaluating a function symbolically and graphically*

A function f is given by $f(x) = x^2 - 2x$, and its graph is shown in Figure 1.30.
(a) Find the implied domain of f.
(b) Use $f(x)$ to evaluate $f(-1)$.
(c) Use the graph of f to evaluate $f(-1)$.

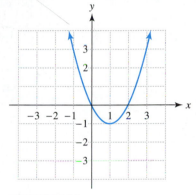

FIGURE 1.30

SOLUTION   **(a)** The implied domain for $f(x) = x^2 - 2x$ includes all real numbers because the formula is defined for all inputs x.

(b) To evaluate $f(-1)$, substitute -1 for x in the given formula.

$$f(-1) = (-1)^2 - 2(-1) = 1 + 2 = 3$$

(c) Refer to Figure 1.31. Begin by finding $x = -1$ on the x-axis. Move upward until the graph of f is reached. Then move across to the y-axis. The y-value corresponding to an x-value of -1 is 3. Thus $f(-1) = 3$.

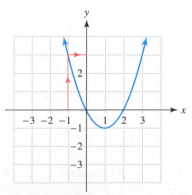

FIGURE 1.31 $f(-1) = 3$

EXAMPLE 5   *Finding the domain and range graphically*

A graph of $f(x) = \sqrt{x - 2}$ is shown in Figure 1.32. Find the domain and range of f.

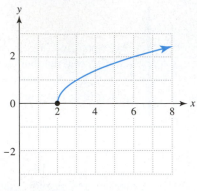

FIGURE 1.32

SOLUTION   The arrow in the graph of f indicates that both the x-values and the y-values increase without reaching a maximum value. In Figure 1.33 the domain and range of f have been labeled by arrows. Note that points appear on the graph for all x greater than or equal to 2. Thus the domain is $D = \{x \mid x \geq 2\}$. The minimum y-value on the graph of f is 0 and it occurs at the point $(2, 0)$. There is no maximum y-value on the graph, so the range is $R = \{y \mid y \geq 0\}$.

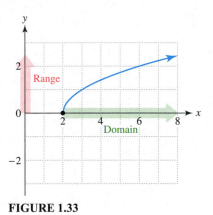

FIGURE 1.33

■ **CLASS DISCUSSION**

Suppose a ball is thrown into the air. Sketch an approximate graph of a function f that outputs the height y (in feet) after x seconds. Use your graph to describe the domain and range of f. ■

Graphing Calculators and Functions

Graphing calculators can create graphical and numerical representations of a function—usually more efficiently and reliably than pencil-and-paper techniques. However, a graphing calculator uses the same basic method that we might use to draw a graph. For example, one way to sketch a graph of $y = x^2$ is to first make a table of values. See Table 1.13.

TABLE 1.13

x	-3	-2	-1	0	1	2	3
y	9	4	1	0	1	4	9

We can plot these points in the xy-plane, as shown in Figure 1.34. Next we might connect the points with a smooth curve, as shown in Figure 1.35.

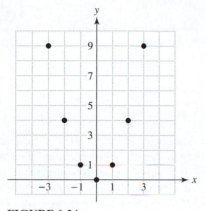

FIGURE 1.34

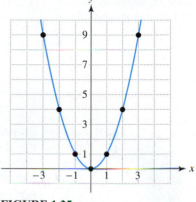

FIGURE 1.35

A graphing calculator typically follows the same process to draw a graph. It plots numerous points and connects them to make a graph.

EXAMPLE 6   *Representing a function*

When the relative humidity is less than 100%, air cools at a rate of 5.4°F for every 1000-foot increase in altitude. Give verbal, symbolic, graphical, and numerical representations of a function f that computes this change in temperature for an increase in altitude of x-thousand feet. Let the domain of f be $0 \le x \le 6$.   (Source: L. Battan, *Weather in Your Life.*)

SOLUTION   *Verbal*   Multiply the input x by -5.4 to obtain the change in temperature.

Symbolic   Let $f(x) = -5.4x$.

Graphical   Since $f(x) = -5.4x$, enter $Y_1 = -5.4X$, as shown in Figure 1.36. Graph Y_1 in a viewing rectangle such as [0, 6, 1] by [−35, 10, 5]. See Figure 1.37.

Graphing Calculator Help

To enter a formula and create a graph, see Appendix B (page AP-8).

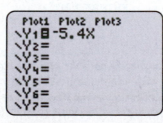

FIGURE 1.36

[0, 6, 1] by [−35, 10, 5]

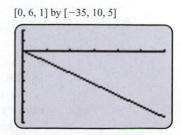

FIGURE 1.37

Numerical   It is impossible to list all inputs x, since $0 \leq x \leq 6$. However, Figures 1.38 and 1.39 show how to create a table for $Y_1 = -5.4X$ with $x = 0, 1, 2, 3, 4, 5, 6$.

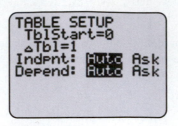

Graphing Calculator Help

To create a table similar to Figure 1.39, see Appendix B (page AP-9).

FIGURE 1.38   **FIGURE 1.39**

EXAMPLE 7   *Evaluating representations of a function*

People who sustain leg injuries often require crutches. It is possible to estimate the proper crutch length without using trial and error. The function given by $f(x) = 0.72x + 2$ can compute the appropriate crutch length in inches for a person with a height of x inches.   (Source: *Journal of the American Physical Therapy Association.*)

(a) Evaluate $f(65)$ symbolically. Interpret your result.
(b) Evaluate $f(65)$ graphically. Use the window $[60, 90, 10]$ by $[40, 80, 10]$.
(c) Make a table of f starting at $x = 60$, incrementing by 1. Evaluate $f(65)$.

SOLUTION   **(a)** *Symbolically*   Since $f(x) = 0.72x + 2$, $f(65) = 0.72(65) + 2 = 48.8$. This means that a person 65 inches tall needs about 49-inch crutches.

(b) *Graphically*   Begin by graphing $Y_1 = 0.72X + 2$, as shown in Figure 1.40. Two ways to evaluate a function with a graphing calculator is to use either the "trace" or "value" utilities. On some calculators the "value" utility is found under the CALCULATE menu. In Figures 1.41 and 1.42 this utility has been used to evaluate the graph at $x = 65$ to obtain 48.8.

Graphing Calculator Help

To make a graph and evaluate a function graphically, see Appendix B (page AP-8).

$[60, 90, 10]$ by $[40, 80, 10]$ $[60, 90, 10]$ by $[40, 80, 10]$

FIGURE 1.40   **FIGURE 1.41**   **FIGURE 1.42**

(c) *Numerically*   To create a table on a graphing calculator, we retain the formula for Y_1 and specify both the *starting value* for x (TblStart) and the *increment* between successive x-values (ΔTbl). Here we start the table at $x = 60$ and increment by 1. (There is no viewing rectangle to set.) Figures 1.43 and 1.44 illustrate these steps. When $x = 65$, $y = 48.8$, so $f(65) = 48.8$.

Graphing Calculator Help

To create the table in Figure 1.44, see Appendix B (page AP-9).

FIGURE 1.43

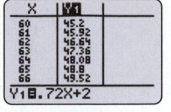

FIGURE 1.44

Identifying Functions

By applying the definition of a function, we can determine if a relation represents a function.

EXAMPLE 8   *Determining if a set of ordered pairs is a function*

Determine if each set of ordered pairs represents a function.
(a) $A = \{(-2, 3), (-1, 2), (0, -3), (-2, 4)\}$
(b) $B = \{(1, 4), (2, 5), (-3, -4), (-1, 7), (0, 9)\}$

SOLUTION   **(a)** Set A does not represent a function because input -2 results in two outputs: 3 and 4.

(b) Set B represents a function because each input (x-value) is paired with exactly one output (y-value.)

Vertical Line Test.   To conclude that a graph represents a function, we must be convinced that it is impossible for two distinct points with the same x-coordinate to lie on the graph. For example, the ordered pairs $(4, 2)$ and $(4, -2)$ are distinct points with the same x-coordinate. These two points could not lie on the graph of the same function because input 4 would result in two outputs, ± 2. A function has exactly one output for each valid input. When the points $(4, 2)$ and $(4, -2)$ are plotted, they lie on the same vertical line, as shown in Figure 1.45. A graph passing through these points intersects the line twice, as illustrated in Figure 1.46.

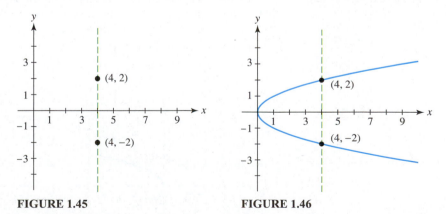

FIGURE 1.45   **FIGURE 1.46**

To determine if a graph represents a function, simply visualize vertical lines in the xy-plane. If each vertical line intersects a graph at no more than one point, then it is a graph of a function. This is called the **vertical line test** for a function.

EXAMPLE 9 *Identifying a function graphically*

Use the vertical line test to determine if the graph of *f* represents a function.

(a)

(b)

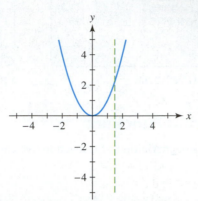

FIGURE 1.47

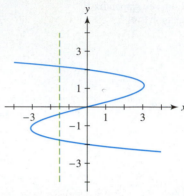

FIGURE 1.48

SOLUTION **(a)** Notice that every vertical line in Figure 1.47 would intersect the graph at most once. Therefore the graph represents a function.

(b) The graph in Figure 1.48 does not represent a function because it is possible for a vertical line to intersect the graph more than once.

EXAMPLE 10 *Identifying a function*

Determine if *y* is a function of *x*.

(a) $x = y^2$ **(b)** $y = x^2 - 2$

SOLUTION **(a)** For *y* to be a function of *x*, each valid *x*-value must result in one *y*-value. If we let $x = 4$, then *y* could be either -2 or 2 since

$$4 = (-2)^2 \quad \text{and} \quad 4 = (2)^2.$$

Therefore, *y* is not a function of *x*. A graph of the equation $x = y^2$ is shown in Figure 1.49. Notice that this graph fails the vertical line test.

(b) In this equation each *x*-value determines exactly one *y*-value, and *y* is a function of *x*. A graph of this equation is shown in Figure 1.50. Notice that this graph passes the vertical line test.

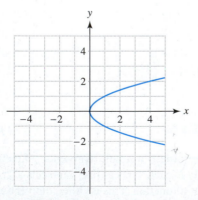

FIGURE 1.49

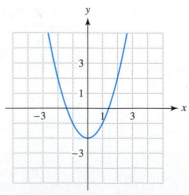

FIGURE 1.50

Putting it all Together

1.3

One of the most important concepts in mathematics is that of a function. A function always produces exactly one output for each element in its domain. The implied domain of a function f consists of the set of all valid x-inputs, and the range is the set of corresponding y-outputs. The following table summarizes some important topics from this section.

Concept	Explanation	Examples
Function	A function is a *relation* in which each valid input results in one output. The *domain* of a function is the set of valid inputs (x-values) and the *range* is the set of resulting outputs (y-values).	$f = \{(1, 3), (2, 6), (3, 9), (4, 12)\}$ The domain is $D = \{1, 2, 3, 4\}$ and the range is $R = \{3, 6, 9, 12\}$.
Implied domain	When a function is represented by a formula, then its implied domain is the set of all valid inputs (x-values) that are defined or make sense in the formula.	$f(x) = \sqrt{x + 4}$ Implied domain of f: $\{x \mid x \geq -4\}$ $g(x) = \dfrac{1}{x - 5}$ Implied domain of g: $\{x \mid x \neq 5\}$
Identifying graphs of functions	The vertical line test can be used to determine if a graph represents a function.	*(graph of an ellipse intersected by a dashed vertical line)* Not a function because a vertical line can intersect the graph more than once
Verbal representation of a function	Words describe precisely what is computed.	A verbal representation of $f(x) = x^2$ is "Square the input x to obtain the output."
Symbolic representation of a function	Mathematical formula	The squaring function is given by $f(x) = x^2$, and the square root function is given by $g(x) = \sqrt{x}$.
Numerical representation of a function	Table of values	A partial numerical representation of $f(x) = 3x$ is shown. | x	0	1	2	3
---	---	---	---	---
$f(x)$	0	3	6	9	

Concept	Explanation	Examples
Graphical representation of a function	Graph of ordered pairs (x, y) that satisfy $y = f(x)$	A graph of $f(x) = 2x$ is shown. Each point on the graph satisfies $y = 2x$.

1.3 EXERCISES

Evaluating and Representing Functions

1. If $f(-2) = 3$, identify a point on the graph of f.

2. If $f(3) = -9.7$, identify a point on the graph of f.

3. If the point $(7, 8)$ lies on the graph of f, then $f(\underline{\quad}) = \underline{\quad}$.

4. If the point $(-3, 2)$ lies on the graph of f, then $f(\underline{\quad}) = \underline{\quad}$.

Exercises 5–14: Graph f by hand by first plotting points to determine the shape of the graph.

5. $f(x) = 2x$ 　　　　**6.** $f(x) = x + 1$

7. $f(x) = 3$ 　　　　　**8.** $f(x) = -2$

9. $f(x) = |x - 1|$ 　　**10.** $f(x) = |0.5x|$

11. $f(x) = 4 - x$ 　　　**12.** $f(x) = 3 + 2x$

13. $f(x) = \dfrac{1}{2}x^2$ 　　　**14.** $f(x) = 2x^2$

Exercises 15–26: Complete the following.
 (a) Evaluate $f(x)$ at the indicated values of x, if possible.
 (b) Find the implied domain of f.

15. $f(x) = x^3$ 　at　 $x = -2, 5$

16. $f(x) = 2x - 1$ 　at　 $x = 8, -1$

17. $f(x) = \sqrt{x}$ 　at　 $x = -1, a + 1$

18. $f(x) = \sqrt{1 - x}$ 　at　 $x = -2, a + 2$

19. $f(x) = \dfrac{1}{x - 1}$ 　at　 $x = -1, a + 1$

20. $f(x) = \dfrac{3x - 5}{x + 5}$ 　at　 $x = -1, a$

21. $f(x) = -7$ 　at　 $x = 6, a - 1$

22. $f(x) = x^2 - x + 1$ 　at　 $x = 1, -2$

23. $f(x) = \dfrac{1}{x^2}$ 　at　 $x = 4, -7$

24. $f(x) = \sqrt{x - 3}$ 　at　 $x = 4, a + 4$

25. $f(x) = \dfrac{1}{x^2 - 9}$ 　at　 $x = 4, a - 5$

26. $f(x) = \dfrac{1}{\sqrt{x - 1}}$ 　at　 $x = 0, a^2 - a + 1$

Exercises 27–34: Use the graph of the function f to estimate its domain and range. Evaluate $f(0)$.

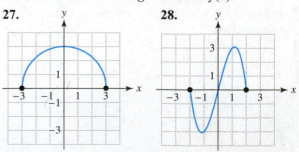

27.　　　　　　　　　　**28.**

29.
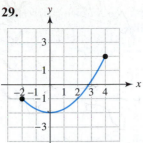

30.
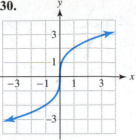

31.
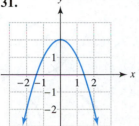

32.
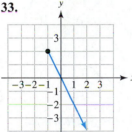

33.
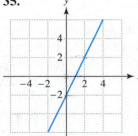

34.
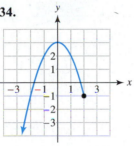

Exercises 35–40: Refer to the graph of f in the figure.

(a) *Evaluate f(0) and f(2).*

(b) *Find all x such that f(x) = 0.*

35.
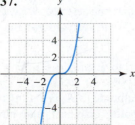

36.
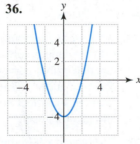

37.
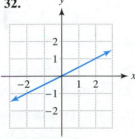

38.
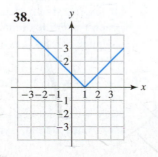

39.
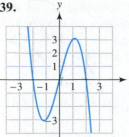

40.
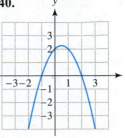

Exercises 41–44: Graph f in the viewing rectangle [−4.7, 4.7, 1] *by* [−3.1, 3.1, 1].

(a) *Use the graph to evaluate f(2).*

(b) *Evaluate f(2) symbolically.*

41. $f(x) = 0.25x^2$

42. $f(x) = 3 - 1.5x^2$

43. $f(x) = \sqrt{x + 2}$

44. $f(x) = |1.6x - 2|$ *absolute value*

Exercises 45–52: Use f(x) to determine verbal, graphical, and numerical representations. For the numerical representation use a table with x = −2, −1, 0, 1, 2.

45. $f(x) = x^2$

46. $f(x) = 2x - 5$

47. $f(x) = |2x + 1|$

48. $f(x) = 8$

49. $f(x) = |x|$

50. $f(x) = 5 - x$

51. $f(x) = \sqrt{x + 1}$

52. $f(x) = x^2 - 1$

Exercises 53 and 54: Express a function f with the specified representation.

53. *Counterfeit Money*   It is estimated that nine out of every one million bills are counterfeit. Give a numerical representation that computes the predicted number of counterfeit bills in a sample of x million bills for x = 0, 1, 2, . . . , 6.   (Source: Department of the Treasury.)

54. *Cost of Driving*   In 1997 the average cost of driving a new car was 41 cents per mile. Give symbolic, graphical, and numerical representations that compute the cost in dollars of driving x miles. For the numerical representation, let x = 1, 2, 3, 4, 5, 6. (Source: Associated Press.)

55. *Radio Stations*   The function f computes the number y in thousands of radio stations on the air during year x.   (Source: M. Street Corporation.)

$$f = \{(1950, 2.8), (1975, 7.7), (1997, 12.8)\}$$

(a) Use a diagram (see Figure 1.24) to represent f.

(b) Evaluate f(1975) and explain what it means.

(c) Identify the domain and range of f.

56. *Unhealthy Air Quality*   The Environmental Protection Agency (EPA) monitors air quality in U.S. cities. The function f, represented by the table, gives the annual number of days with unhealthy air quality in Los Angeles.

x	1991	1992	1993	1994	1995	1996
$f(x)$	184	185	146	136	103	88

(a) Find $f(1992)$. Interpret the result.
(b) Determine when the greatest decrease in the number of unhealthy days occurred.
(c) Discuss the trend of air pollution in Los Angeles.

Identifying Functions

Exercises 57–62: Determine if the graph represents a function. If it represents a function, determine its domain and range.

57.

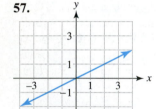

58.

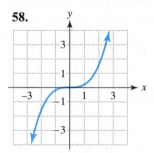

59.

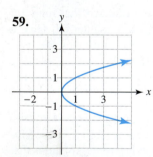

60.

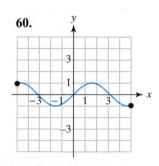

61.

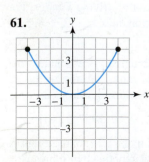

62.

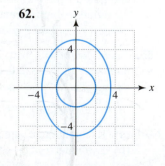

Exercises 63–66: Determine if the following describes a function. Explain your answer.

63. Calculating the cube root of a number

64. Calculating your age

65. Listing the students who passed a given English exam

66. Finding the x-values in the domain of a relation

Exercises 67–72: Determine if S is a function.

67. $S = \{(1, 2), (2, 3), (4, 5), (1, 3)\}$

68. $S = \{(-3, 7), (-1, 7), (3, 9), (6, 7), (10, 0)\}$

69. $S = \{(a, 2), (b, 3), (c, 3), (d, 3), (e, 2)\}$

70. $S = \{(a, 2), (a, 3), (b, 5), (-b, 7)\}$

71. S is given by the table.

x	1	3	1
y	10.5	2	-0.5

72. S is given by the table.

x	1	2	3
y	1	1	1

Exercises 73–76: Determine if y is a function of x.

73. $x = y^4$ **74.** $y^2 = x + 1$

75. $\sqrt{x + 1} = y$

76. $x^2 = y - 7$

Applications

77. *Distance to Lightning*   The speed of sound varies with the temperature of the air. Find a formula for a function f that computes the distance between an observer and a lightning bolt when the speed of sound is 1150 feet per second. Evaluate $f(15)$ and interpret the result.

78. *Distance to Lightning*   Give a reasonable domain for the function f that you found in Exercise 77. Graph f over the domain that you selected. What is the range of your function? (Note that answers may vary.)

79. *Air Temperature*   (Refer to Example 6.) When the relative humidity is 100%, air cools 5.8°F for every 1-mile increase in altitude. (Source: L. Battan.)   Give verbal, symbolic, graphical, and numerical representations of a function f that computes this change in temperature for an increase of x miles. Let the domain of f be $0 \le x \le 3$.

80. *Crutch Length*   (Refer to Example 7.) Determine the crutch length for someone 6 feet 3 inches tall. For each 1-inch increase in a person's height, by how much does the recommended crutch length increase?

Writing about Mathematics

81. Explain how you could use a complete numerical representation of a function to determine its domain and range.

82. Explain in your own words what a function is. How is a function different from a relation?

1.4 TYPES OF FUNCTIONS AND THEIR RATES OF CHANGE

Constant Functions • Linear Functions • Slope as a Rate of Change • Nonlinear Functions • Average Rate of Change

Introduction

A central theme of applied mathematics is modeling real-world phenomena with functions. Because applications involving real data are diverse, mathematicians have created a wide assortment of functions. In fact, professional mathematicians design new functions every day for use in business, education, and government. Mathematics is not static—it is dynamic. It requires both ingenuity and creativity to analyze data and make predictions about the future. This section demonstrates how different types of data require a variety of functions to describe them.

Mathematicians often classify problems into one of two general categories: *linear* or *nonlinear*. Functions that are used to model and solve these problems can also be classified as either linear or nonlinear. Linear functions are usually easier to graph and evaluate than nonlinear functions. A simple example of a linear function is called a *constant* function. For this reason we begin by discussing constant functions.

Constant Functions

The monthly average wind speeds in miles per hour at Hilo, Hawaii, from May through December are listed in Table 1.14.

TABLE 1.14

Month	May	June	July	Aug	Sept	Oct	Nov	Dec
Wind Speed (mph)	7	7	7	7	7	7	7	7

Source: J. Williams, *The Weather Almanac 1995.*

It is apparent that the monthly average wind speed is constant between May and December. This data can be described by a set f of ordered pairs (x, y), where x is the month and y is the wind speed. The months have been assigned the standard numbers.

$$f = \{(5, 7), (6, 7), (7, 7), (8, 7), (9, 7), (10, 7), (11, 7), (12, 7)\}$$

A scatterplot of f is shown in Figure 1.51 on the next page. By the vertical line test f is a function.

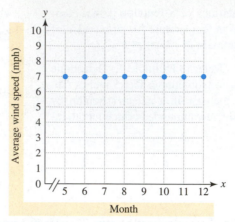

FIGURE 1.51 A Discrete Constant Model

The function f is given by $f(x) = 7$, where $x = 5, 6, 7, \ldots, 12$. The output of f never changes. We say f is a *constant function*.

> ## Constant function
>
> A function f represented by $f(x) = b$, where b is a fixed number, is a **constant function.**

In the previous discussion $f(x) = 7$, so $b = 7$. The range of f is $R = \{7\}$. The range of a constant function contains a single element. The domain of f is $D = \{5, 6, 7, 8, 9, 10, 11, 12\}$. Since f is defined only at individual or discrete values of x, f is called a **discrete function**. The graph of a discrete function suggests a scatterplot.

Sometimes it is more convenient to describe discrete data with a continuous graph. If the domain of $f(x) = 7$ is changed to $D = \{x \mid 5 \le x \le 12\}$, its graph becomes a continuous horizontal line without any breaks. See Figure 1.52. The graph of a **continuous function** can be sketched without picking up the pencil. There are no breaks in the graph of a continuous function. The graph of a continuous constant function is a horizontal line.

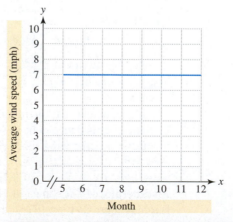

FIGURE 1.52 A Continuous Constant Model

If the domain includes $x = 6.5$, what does $f(6.5) = 7$ represent? The expression $f(6)$ computes the average wind speed in June, and $f(7)$ gives the average wind speed in July. We might interpret $f(6.5)$ to represent the average wind speed from June 15 to July 15. Other interpretations are possible.

As simple as constant functions might appear, they occur frequently in applications. The following are examples that can be modeled by constant functions. In both cases the independent variable (input) is time.

- A thermostat computes a constant function regardless of the weather outside by maintaining a set temperature.
- A cruise control in a car computes a constant function by maintaining a fixed speed, regardless of the type of road or terrain.

Linear Functions

A car is initially located 30 miles north of the Texas border, traveling north on Interstate 35 at 60 miles per hour. The distances between the automobile and the border are listed in Table 1.15 for various times.

TABLE 1.15

Elapsed Time (hours)	0	1	2	3	4	5
Distance (miles)	30	90	150	210	270	330

It can be seen that the distance increases by 60 miles every hour. These data are modeled by a set f of ordered pairs (x, y), where x is the elapsed time and y is the distance from the border.

$$f = \{(0, 30), (1, 90), (2, 150), (3, 210), (4, 270), (5, 330)\}$$

The set f is a function, but not a constant function. A scatterplot of f is shown in Figure 1.53. The scatterplot suggests a line that rises from left to right.

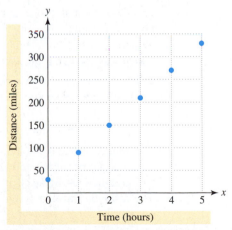

FIGURE 1.53   A Discrete Linear Model

If the car travels for x hours, the distance traveled can be computed by multiplying 60 times x and adding the initial distance of 30 miles. This computation can be expressed as $f(x) = 60x + 30$. The formula is valid for nonnegative values of x. For example, $f(1.5) = 60(1.5) + 30 = 120$ means that the car is 120 miles from the border after 1.5 hours. The graph of $f(x) = 60x + 30$ is shown in Figure 1.54. We call f a *linear function*.

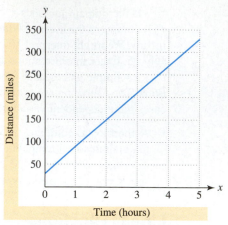

FIGURE 1.54 A Continuous Linear Model

Linear function

A function f represented by $f(x) = ax + b$, where a and b are constants, is a **linear function**.

If $a = 0$ then $f(x) = b$, which defines a constant function. Thus every constant function is also a linear function. In the example of the moving car, $f(x) = 60x + 30$, so $a = 60$ and $b = 30$. The value of a represents the speed of the car, and b corresponds to the initial distance of the car from the border. Other examples of linear functions include the following. The values of the constants a and b are also given.

i.	$f(x) = 1.5x - 6$	$a = 1.5, \quad b = -6$
ii.	$f(x) = 8x$	$a = 8, \quad b = 0$
iii.	$f(x) = 72$	$a = 0, \quad b = 72$
iv.	$f(x) = 1.9 - 3x$	$a = -3, \quad b = 1.9$

A distinguishing feature of a linear function is that each time x increases by one unit, the value of $f(x)$ always changes by an amount equal to a. That is, a linear function f has a **constant rate of change**. (The constant rate of change a is equal to the slope of the graph of f.) The following applications are modeled by linear functions. Try to determine the value of the constant a.

- The wages earned by an individual working x hours at \$6.25 per hour
- The amount of tuition and fees when registering for x credits, if each credit costs \$75 and the fees are fixed at \$56
- The distance traveled by light in x seconds, if the speed of light is 186,000 miles per second

Slope as a Rate of Change

The graph of a (continuous) linear function is a line. Slope is a real number that measures the "tilt" of a line in the xy-plane. If the input x to a linear function increases by 1 unit, then the output y changes by a constant amount that is equal to the slope of its graph. In Figure 1.55 a line passes through the points (x_1, y_1) and (x_2, y_2). The *change in y* is $y_2 - y_1$, and the *change in x* is $x_2 - x_1$. The ratio of the change in y to the change in x is called the *slope*. We sometimes denote the change in y by Δy (delta y) and the change in x by Δx (delta x). That is, $\Delta y = y_2 - y_1$ and $\Delta x = x_2 - x_1$.

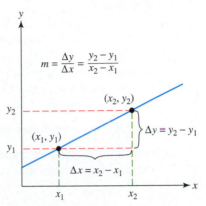

FIGURE 1.55

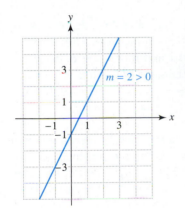

FIGURE 1.56

Slope

The **slope** m of the line passing through the points (x_1, y_1) and (x_2, y_2) is

$$m = \frac{\Delta y}{\Delta x} = \frac{y_2 - y_1}{x_2 - x_1}$$

where $x_1 \neq x_2$.

If the slope of a line is positive, the line *rises* from left to right. If the slope is negative, the line *falls* from left to right. Slope 0 indicates that the line is horizontal. When $x_1 = x_2$, the line is vertical and the slope is undefined. Figures 1.56–1.59 illustrate these situations. Slope 2 indicates that a line rises 2 units for every unit increase in x and slope $-\frac{1}{2}$ indicates that the line *falls* $\frac{1}{2}$ unit for every unit increase in x.

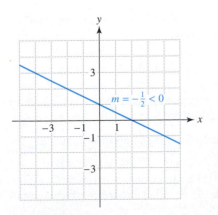

FIGURE 1.57

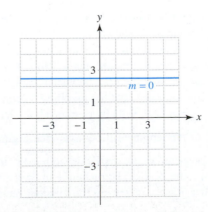

FIGURE 1.58

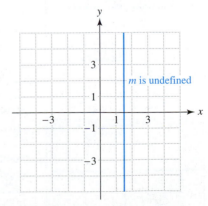

FIGURE 1.59

EXAMPLE 1   *Calculating the slope of a line*

Find the slope of the line passing through the points $(-2, 3)$ and $(1, -2)$. Plot these points together with the line. Interpret the slope.

SOLUTION   The slope is

$$m = \frac{y_2 - y_1}{x_2 - x_1} = \frac{-2 - 3}{1 - (-2)} = -\frac{5}{3}.$$

A graph of the line passing through these two points is shown in Figure 1.60. The change in y is $\Delta y = -5$ and the change in x is $\Delta x = 3$, so $m = \frac{\Delta y}{\Delta x} = -\frac{5}{3}$ indicates that the line falls $\frac{5}{3}$ units for each unit increase in x, or equivalently, the line falls 5 units for each 3-unit increase in x.

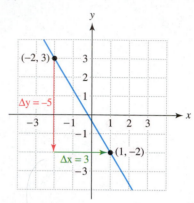

FIGURE 1.60

The graph of $f(x) = ax + b$ is a line. Since $f(0) = b$ and $f(1) = a + b$, the graph of f passes through the points $(0, b)$ and $(1, a + b)$. The slope of this line is

$$m = \frac{y_2 - y_1}{x_2 - x_1} = \frac{a + b - b}{1 - 0} = \frac{a}{1} = a.$$

In applications involving linear functions, slope sometimes is interpreted as a *(constant) rate of change*. In Example 7 of Section 1.3, the recommended crutch length for a person x inches tall was given by $f(x) = 0.72x + 2$. The slope of its graph is 0.72. One interpretation of this slope is that for each 1-inch increase in the height of a person, the crutch length should be increased by 0.72 inch.

Graphing Calculator Help

To set a square viewing rectangle see Appendix B (page AP-9).

$[-10, 10, 1]$ by $[-10, 10, 1]$

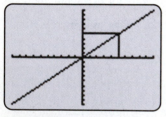

FIGURE 1.61

$[-9, 9, 1]$ by $[-6, 6, 1]$

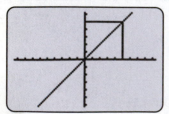

FIGURE 1.62

Squaring a Viewing Rectangle.   The graph of the line $y = x$ in the standard viewing rectangle is shown in Figure 1.61. The visual distance between consecutive tick marks on the two axes are not equal. As a result, the line $y = x$ does not appear to have slope 1. Notice that although the box has 5 units on each side, it appears as a rectangle and not as a square. The visual change in y is less than the visual change in x. The reason for this is that the graphing calculator screen is wider than it is high. If we graph $y = x$ in $[-9, 9, 1]$ by $[-6, 6, 1]$, the length of the y-axis in the viewing rectangle is $\frac{2}{3}$ of the length of the x-axis. See Figure 1.62. There is a uniform spacing between consecutive tick marks on both axes. In this case, the line $y = x$ makes a 45° angle with the x-axis. The rectangle in Figure 1.61 now appears as a *square* in Figure 1.62. This process of setting the viewing rectangle is called *squaring a viewing rectangle*. The ratio of $\frac{2}{3}$ may be different on other calculators. On some calculators a square viewing rectangle can be set automatically by looking under the ZOOM menu.

Nonlinear Functions

Table 1.16 shows the population of the United States at 20-year intervals from 1800 to 1980. Between 1800 and 1820 the population increased by 5 million, between 1820 and 1840 it increased by 7 million, and between 1840 and 1860 the increase was 14 million. If the increases in population for each 20-year period had been equal, or nearly equal, these data could be modeled by a linear function.

TABLE 1.16

Year	Population (millions)	Year	Population (millions)
1800	5	1900	76
1820	10	1920	106
1840	17	1940	132
1860	31	1960	179
1880	50	1980	226

The actual population data are nonlinear. In Figure 1.63 the data together with a nonlinear function f have been plotted. The graph of f is curved rather than straight.

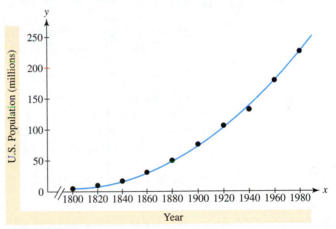

FIGURE 1.63   A Nonlinear Model

If a function is not linear, then it is called a **nonlinear function**. *The graph of a nonlinear function is not a (straight) line.* With a nonlinear function, it is possible for the input x to increase by 1 unit and the output y to change by different amounts. Nonlinear functions *cannot* be written in the form $f(x) = ax + b$. Examples of nonlinear functions include the following.

i. $f(x) = x^2 - 3x + 2$
ii. $f(x) = \sqrt{x}$
iii. $f(x) = x^3$

Real-world phenomena often are modeled using nonlinear functions. The following are some examples of quantities that can be described by nonlinear functions.

- The total number of people who have contracted AIDS beginning in 1980
  (The increase in the number of AIDS cases is not the same each year.)
- The monthly average temperature in Chicago
  (Monthly average temperatures increase and decrease throughout the year.)
- The height of a child between the ages of 2 and 18
  (A child grows faster at certain ages.)

■ CLASS DISCUSSION

The time required to drive a distance of 100 miles depends on the average speed. Let the function f compute this time, given the average speed x as input. For example, $f(50) = 2$, since it would take 2 hours to travel 100 miles at an average speed of 50 miles per hour. Make a table of values for f. Is f linear or nonlinear? ■

There are many examples of nonlinear functions. You may have encountered some of them in previous mathematics courses. In Figures 1.64–1.67 graphs and formulas are given for four common nonlinear functions. Note that each graph is not a line.

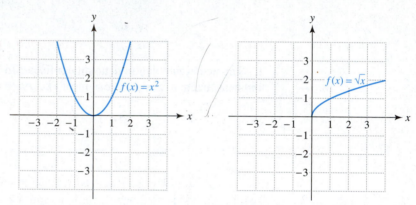

FIGURE 1.64 Square Function   **FIGURE 1.65** Square Root Function

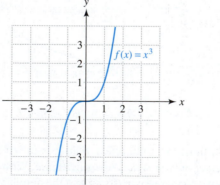   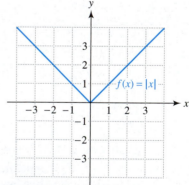

FIGURE 1.66 Cube Function   **FIGURE 1.67** Absolute Value Function

EXAMPLE 2   *Recognizing linear and nonlinear data*

For each table decide whether the data are linear or nonlinear. If the data are linear, state the slope of the line that passes through the data points.

(a)

x	0	5	10	15	20
y	−4	−2	0	2	4

(b)

x	−3	0	3	6	9
y	5	7	10	14	19

non linear

(c)

x	0	1	2	3	4
y	11	11	11	11	11

(d)

x	0	1	3	6	10
y	3	6	9	12	15

SOLUTION **(a)** Between each pair of points, the y-values increase 2 units for each 5-unit increase in x. Therefore the data are linear, but not constant. The slope of the line passing through the data points is $m = \dfrac{2}{5}$.

(b) The y-values do not increase by a constant amount for each 3-unit increase in x. The data are nonlinear.

(c) This data are linear and constant. The slope of the line passing through these points is 0.

(d) Although the y-values increase by 3 units between consecutive data points, the data are nonlinear, because the corresponding increases in the x-values are not constant. For example, the slope of the line passing through the points $(0, 3)$ and $(1, 6)$ is 3, whereas the slope of the line passing through the points $(1, 6)$ and $(3, 9)$ is 1.5.

EXAMPLE 3 *Recognizing linear and nonlinear functions*

Determine if f is a linear or nonlinear function. If f is a linear function state if f is a constant function. Support your results by graphing f in the standard viewing rectangle.

(a) $f(x) = 6 - 4x$ **(b)** $f(x) = 3x^2 - 2$ **(c)** $f(x) = 5$

SOLUTION **(a)** A linear function can be written as $f(x) = ax + b$, where a and b are real numbers. We can write the given formula as $f(x) = -4x + 6$, so f is linear with $a = -4$ and $b = 6$. Figure 1.68 shows that the graph of f is a line.

Graphing Calculator Help

To set a viewing rectangle and graph a function, see Appendix B (pages AP-6 and AP-8).

(b) Function f is nonlinear because its formula contains an x^2-term. Its formula cannot be written in the form $f(x) = ax + b$. The graph of f is not a line, as shown in Figure 1.69.

(c) Function f is both linear and constant. Its graph is a horizontal line, as shown in Figure 1.70.

$[-10, 10, 1]$ by $[-10, 10, 1]$

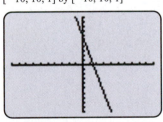

FIGURE 1.68

$[-10, 10, 1]$ by $[-10, 10, 1]$

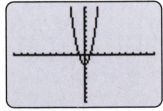

FIGURE 1.69

$[-10, 10, 1]$ by $[-10, 10, 1]$

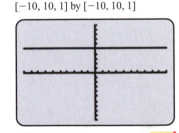

FIGURE 1.70

Average Rate of Change

The graphs of nonlinear functions are not lines, so there is no notion of a single slope. The slope of the graph of a linear function gives its rate of change. With a nonlinear function we speak of an *average* rate of change. Suppose the points (x_1, y_1) and (x_2, y_2) lie on the graph of a nonlinear function f. See Figure 1.71. The slope of the line L passing through these two points represents the *average rate of change of f from x_1 to x_2*. The line L is referred to as a **secant line**. If different values for x_1 and x_2 are selected, then a different secant line and a different average rate of change usually result.

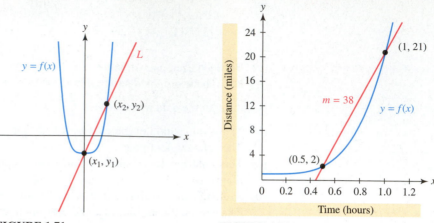

FIGURE 1.71   **FIGURE 1.72**

In applications the average rate of change measures how fast a quantity is changing over an interval of its domain, on average. For example, suppose the graph of the function f in Figure 1.72 represents the distance y in miles that a car has traveled on a straight highway after x hours. The points $(0.5, 2)$ and $(1, 21)$ lie on this graph. Thus after 0.5 hour the car has traveled 2 miles and after 1 hour the car has traveled 21 miles. The slope of the line passing through these two points is

$$m = \frac{y_2 - y_1}{x_2 - x_1} = \frac{21 - 2}{1 - 0.5} = 38.$$

This means that during the half-hour period from 0.5 to 1 hour the average rate of change, or average velocity, was 38 miles per hour. These ideas lead to the following definition.

Average rate of change

Let (x_1, y_1) and (x_2, y_2) be distinct points on the graph of a function f. The **average rate of change of f from x_1 to x_2** is

$$\frac{y_2 - y_1}{x_2 - x_1}.$$

That is, the average rate of change from x_1 to x_2 equals the slope of the line passing through (x_1, y_1) and (x_2, y_2).

If f is a constant function, its average rate of change is zero. For a linear function defined by $f(x) = ax + b$, its average rate of change is equal to a, the slope of its graph. The average rate of change for a nonlinear function varies.

EXAMPLE 4   *Calculating and interpreting average rates of change*

Use Table 1.16 and Figure 1.63 to complete the following.
(a) Calculate the average rates of change in the U.S. population from 1800 to 1840 and from 1900 to 1940. Interpret the results.
(b) Illustrate your results from part (a) graphically.

SOLUTION   (a) In 1800 the population was 5 million and in 1840 it was 17 million. Therefore, the average rate of change in the population from 1800 to 1840 was

$$\frac{17 - 5}{1840 - 1800} = 0.3.$$

In 1900 the population was 76 million and in 1940 it was 132 million. Therefore the average rate of change in the population from 1900 to 1940 was

$$\frac{132 - 76}{1940 - 1900} = 1.4.$$

This means that from 1800 to 1840, the U.S. population increased, on average, by 0.3 million per year and from 1900 to 1940, the U.S. population increased, on average, by 1.4 million per year.

(b) These average rates of change can be illustrated graphically by sketching a line L_1 through the points (1800, 5) and (1840, 17) and another line L_2 through the points (1900, 76) and (1940, 132), as depicted in Figure 1.73. The slope of L_1 is 0.3 and the slope of L_2 is 1.4.

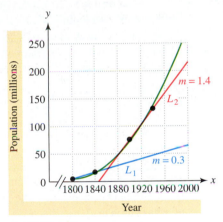

FIGURE 1.73

EXAMPLE 5   *Modeling braking distance for a car*

On wet, level pavement highway engineers sometimes model the braking distance in feet for a car traveling at x miles per hour using $f(x) = \frac{1}{9}x^2$.   (Source: L. Haefner, *Introduction to Transportation Systems.*)

(a) Evaluate $f(30)$ and $f(60)$. Interpret the results.

(b) Calculate the average rate of change of f from 30 to 60. Discuss what the result means.

SOLUTION   (a) $f(30) = \frac{1}{9}(30)^2 = 100$ and $f(60) = \frac{1}{9}(60)^2 = 400$. At 30 miles per hour the braking distance is 100 feet and at 60 miles per hour the braking distance is 400 feet.

(b) The average rate of change of f from 30 to 60 is calculated as follows.

$$\frac{400 - 100}{60 - 30} = 10$$

This means that the braking distance increases, on average, by 10 feet for each 1-mile-per-hour increase in speed between 30 and 60 miles per hour.

*Putting
it all
Together*
1.4

Functions are either linear or nonlinear. A constant function is a special type of linear function. Slope can be used to describe how a linear function and its graph change. Average rate of change can be used to describe how a nonlinear function and its graph change. The following table summarizes these ideas.

Concept	Formula	Examples
Slope of a line passing through (x_1, y_1) and (x_2, y_2)	$$m = \frac{\Delta y}{\Delta x} = \frac{y_2 - y_1}{x_2 - x_1}$$ $\Delta y = y_2 - y_1$ denotes the change in y. $\Delta x = x_2 - x_1$ denotes the change in x.	A line passing through $(-1, 3)$ and $(1, 7)$ has slope $m = \dfrac{7 - 3}{1 - (-1)} = \dfrac{4}{2} = 2$. This slope indicates that the line rises 2 units for each unit increase in x.
Average rate of change of f from x_1 to x_2	If (x_1, y_1) and (x_2, y_2) lie on the graph of f, then the average rate of change from x_1 to x_2 equals $\dfrac{y_2 - y_1}{x_2 - x_1}$.	If $f(x) = 3x^2$, then the average rate of change from $x = 1$ to $x = 3$ is given by $$\frac{27 - 3}{3 - 1} = 12$$ because $f(3) = 27$ and $f(1) = 3$. This means that, on average, $f(x)$ increases by 12 units for each unit increase in x when $1 \le x \le 3$.
Constant function	$f(x) = b$, where b is a fixed number.	$f(x) = 12$, $g(x) = -2.5$, and $h(x) = 0$. Every constant function is a linear function.
Linear function	$f(x) = ax + b$, where a and b are fixed numbers or constants. The graph of f has slope a.	$f(x) = 3x - 1$, $g(x) = -5$, and $h(x) = \dfrac{1}{2} - \dfrac{3}{4}x$. Their graphs have slopes 3, 0, and $-\dfrac{3}{4}$, respectively.
Nonlinear function	Cannot be expressed in the form $f(x) = ax + b$	$f(x) = \sqrt{x + 1}$, $g(x) = 4x^3$, and $h(x) = x^{1.01}$

The following table summarizes important concepts related to constant, linear, and nonlinear functions.

Concept	Constant Function	Linear Function	Nonlinear Function
Slope of graph	Always zero	Always constant	No notion of one slope
Average rate of change	Always zero	Always constant	Can vary
Graph	Horizontal line	Nonvertical line	Not a line

Concept	Constant Function	Linear Function	Nonlinear Function
Examples	$f(x) = 2$	$f(x) = 2x - 2$	$f(x) = x^2 - 3$
Examples	$f(x) = -3$	$f(x) = -x + 3$	$f(x) = x^3 - 4x$

1.4 EXERCISES

Slope

Exercises 1–10: If possible, find the slope of the line passing through each pair of points.

1. $(4, 6), (2, 5)$

2. $(-8, 5), (-3, -7)$

3. $(-0.5, 9.2), (-0.3, 7.6)$

4. $(1.6, 12), (1.6, 5)$

5. $(1997, 5.6), (1994, 7.9)$

6. $(1824, 108), (1900, 380)$

7. $(-5, 6), (-5, 8)$

8. $(17, 7), (19, 7)$

9. $\left(\frac{1}{3}, -\frac{3}{5}\right), \left(-\frac{5}{6}, \frac{7}{10}\right)$

10. $\left(-\frac{13}{15}, -\frac{7}{8}\right), \left(\frac{1}{10}, \frac{3}{16}\right)$

Exercises 11–16: State the slope of the graph of f. Interpret this slope.

11. $f(x) = 2x + 7$

12. $f(x) = 6 - x$

13. $f(x) = 9 - \frac{3}{4}x$

14. $f(x) = \frac{2}{3}x$

15. $f(x) = -5$

16. $f(x) = x + 5$

17. *Age in the U.S.* The median age of the U.S. population for each year x between 1970 and 2010 can be approximated by $f(x) = 0.243x - 450.8$. (Source: Bureau of the Census.)

  (a) Compute the median ages of the U.S. population in 1980 and 2000.

  (b) What is the slope of the graph of f? Interpret the slope.

18. *Commercial Banks* From 1987 to 1997 the numbers of federally insured commercial banks could be modeled by $f(x) = -458x + 923,769$, where x is the year. (Source: Federal Deposit Insurance Corporation.)

  (a) Estimate the number of commercial banks in 1987 and in 1997.

  (b) What is the slope of the graph of f? Interpret the slope.

19. *Women's Olympic Times*   The winning times in seconds for the women's 200-meter dash at the Olympic Games can be approximated by $f(x) = -0.0635x + 147.9$, where x is the year with $1948 \leq x \leq 1996$. (Source: United States Olympic Committee.)
(a) Graph f in [1948, 1996, 4] by [21, 25, 1].
(b) What is the slope of the graph of f? Interpret the slope.
(c) Approximate the average change in the winning times over a 4-year period.

20. *Men's Olympic Times*   The winning times in seconds for the men's 400-meter hurdles at the Olympic Games can be approximated by $f(x) = -0.0975x + 240.8$, where x is the year with $1900 \leq x \leq 1996$. (Source: United States Olympic Committee.)
(a) Graph $f(x)$ in [1900, 1996, 4] by [45, 60, 1].
(b) What is the slope of the graph of f? Interpret the slope.
(c) Approximate the average change in the winning times over a 4-year period.

Linear and Nonlinear Functions

Exercises 21–32: (Refer to Example 3.) Determine if f is a linear or nonlinear function. If f is a linear function, state if f is a constant function. Support your answer by graphing f.

21. $f(x) = -2x + 5$ **22.** $f(x) = -0.01x - 2$

23. $f(x) = 2x^{1.01} - 1$ **24.** $f(x) = 5.123$

25. $f(x) = 1$ **26.** $f(x) = \sqrt{-x}$

27. $f(x) = 0.2x^3 - 2x + 1$

28. $f(x) = 0.5x^2 - 2x - 3$

29. $f(x) = 5x - x^2$ **30.** $f(x) = |2x - 1|$

31. $f(x) = \sqrt[3]{x - 1}$ **32.** $f(x) = \sqrt{x^2 - 4}$

Exercises 33–38: (Refer to Example 2.) Decide whether the data in the table are linear or nonlinear. If the data are linear, state the slope of the line passing through the data points. (To help with your decision, you may want to make a scatterplot.)

33.

x	0	1	2	3	4
y	-1	3	7	11	15

34.

x	-4	-2	0	2	4
y	1.0	-0.5	-2.0	-3.5	-5.0

35.

x	-5	-3	1	3	5
y	-5	-2	1	4	7

36.

x	-4	-2	0	2	4
y	-1	-1	-1	-1	-1

37.

x	-4	0	1	2	5
y	5.0	3.0	2.5	2.0	0.5

38.

x	100	200	250	350	400
y	400	1900	3015	6025	7900

Exercises 39–42: ***Analyzing Real Data***   *For the given data set complete the following.*
(a) *Make a line graph of the data in an appropriate viewing rectangle. Let this graph represent a function f.*
(b) *Decide whether f is linear or nonlinear.*

39. Interest income earned in 1 year on x dollars invested at 7% per year

x (dollars)	500	1000	2000	3500
Interest	$35	$70	$140	$245

40. World energy consumption in quadrillion Btu (*Note*: 1 quadrillion = 1×10^{15})

Year	1990	1991	1992	1993
Consumption	345	345	345	345

Source: Energy Information Administration, *International Energy Annual.*

41. Median incomes of full-time female workers

Year	1970	1980	1990	1998
Income	$5,440	$11,591	$20,591	$26,855

Source: Bureau of the Census, *Current Population Reports.*

42. Median incomes of full-time male workers

Year	1970	1980	1990	1998
Income	$9,184	$19,173	$28,979	$36,252

Source: Bureau of the Census, *Current Population Reports.*

43. *Average Wind Speed*   The table lists the average wind speed in miles per hour at Myrtle Beach, South Carolina. The months are assigned the standard numbers.   (Source: J. Williams.)

Month	1	2	3	4	5	6
Wind (mph)	7	8	8	8	7	7

Month	7	8	9	10	11	12
Wind (mph)	7	7	7	6	6	6

(a) Could these data be modeled exactly by a constant function?

(b) Determine a continuous, constant function f that models these data.

(c) Graph f and the data.

44. *Income Tax Rates*   The table lists the lowest personal income tax rate in percent between 1995 and 2000.   (Source: Internal Revenue Service.)

Year	1995	1996	1997
Rate (%)	15	15	15

Year	1998	1999	2000
Rate (%)	15	15	15

(a) Determine a discrete, constant function f that describes these data. Represent f symbolically.

(b) What is the domain of f?

(c) Graph f.

45. *Circles*   The functions $f, g,$ and h below all compute quantities related to circles, where x is the radius.

(a) Identify each function as constant, linear, or nonlinear.

(b) Find what quantity each function computes.

x	$f(x)$	$g(x)$	$h(x)$
1	2π	π	2
2	4π	4π	4
3	6π	9π	6
4	8π	16π	8
5	10π	25π	10

Geometry Review   To review formulas related to circles, see Chapter R (page R-2).

46. *Income Tax*   The federal income tax rate for single individuals in 1999 was 15% of taxable income between $0 and $25,750, and 28% of taxable income between $25,750 and $62,450.   (Source: Internal Revenue Service.)

(a) Find a linear function that computes the tax on taxable incomes x between $0 and $25,750. Determine the income tax on a taxable income of $9600.

(b) The linear function given by

$$f(x) = \frac{7}{25}x - 3347.5$$

computes the tax on taxable incomes x between $25,750 and $62,450. Determine the income tax on a taxable income of $27,400.

47. Suppose that a car's distance from a service center along a straight highway can be described by a constant function of time. Discuss what can be said about the car's velocity.

48. Suppose that a car's distance in miles from a rest stop along a straight stretch of an interstate highway after x hours can be modeled by $f(x) = ax$. Discuss what can be said about the car's velocity.

Exercises 49–56: Write a symbolic representation for a function f that computes the following.

49. The number of pounds in x ounces

50. The number of dimes in x dollars

51. The distance traveled by a car moving at 50 miles per hour for x hours

52. The monthly electric bill in dollars for using x kilowatt-hours at 6 cents per kilowatt-hour plus a fixed fee of $6.50

53. The cost of downhill skiing x times with a $500 season pass

54. The total number of hours in day x

55. The distance in miles between a runner and home after x hours if the runner is initially 1 mile from home and jogging *away* from home at 6 miles per hour

56. The speed of a car in feet per second after x minutes if its tires are 2 feet in diameter and rotating 14 times per second

Curve Sketching

Exercises 57–62: Critical Thinking   Assume that each function is continuous. Do not use a graphing calculator.

57. Sketch a graph of a linear function f with $f(3) = -1$ and a positive slope.

58. Sketch a graph of a linear function with a slope of $-\frac{2}{3}$, passing through the point $(-2, 3)$.

59. On the same coordinate axes, sketch the graphs of a constant function f and a nonlinear function g that intersect exactly twice.

60. Sketch a graph of a linear function f that intersects a constant function g exactly once.

61. Sketch a graph of a nonlinear function f that has only positive average rates of change.

62. Sketch a graph of a nonlinear function f that has only negative average rates of change.

Average Rates of Change

Exercises 63–66: Use the graph and formula of $f(x)$ to find the average rates of change of f from -3 to -1 and from 1 to 3.

63. $f(x) = 4$

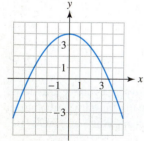

64. $f(x) = 2x - 1$

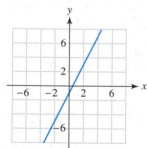

65. $f(x) = -0.3x^2 + 4$

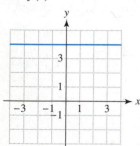

66. $f(x) = 0.3x^2 - 4$

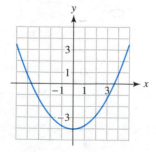

Exercises 67–72: Compute the average rate of change of f from x_1 to x_2. Round your answer to two decimal places when appropriate. Interpret your result graphically.

67. $f(x) = 7x - 2$, $x_1 = 1$, and $x_2 = 4$

68. $f(x) = -8x + 5$, $x_1 = -2$, and $x_2 = 0$

69. $f(x) = x^3 - 2x$, $x_1 = 0.5$, and $x_2 = 4.5$

70. $f(x) = 0.5x^2 - 5$, $x_1 = -1.3$, and $x_2 = 3.7$

71. $f(x) = \sqrt{2x - 1}$, $x_1 = 1$, and $x_2 = 3$

72. $f(x) = \sqrt[3]{x + 1}$, $x_1 = 0.3$, and $x_2 = 1.2$

73. *Remaining Life Expectancy* The accompanying table lists the average *remaining* life expectancy E in years for females at age x. (Source: Department of Health and Human Services.)

x (years)	0	10	20	30	40
E (years)	72.3	69.9	60.1	50.4	40.9

x (years)	50	60	70	80
E (years)	31.6	23.1	15.5	9.2

(a) Make a line graph of the data.
(b) Assume that the graph represents a function f. Compute the average rates of change of f during each 10-year period. Interpret the results.
(c) Determine the life expectancy (not the remaining life expectancy) of a female whose age is 20. What is the life expectancy of a female who is 70 years old? Discuss one reason why these two expectancies are not equal.

74. *Average Life Expectancy* The accompanying table lists the average life expectancy E in years for males born in year x. (Source: Department of Health and Human Services.)

x (years)	1920	1930	1940	1950
E (years)	53.6	58.1	60.8	65.6

x (years)	1960	1970	1980	1990
E (years)	66.6	67.1	70.0	71.8

(a) Make a line graph of the data.
(b) Assume that the graph represents a function f. Compute the average rates of change of f during each 10-year period. Interpret the results.

75. *AIDS Deaths* The table lists reported numbers of AIDS deaths in the United States. (Source: Department of Health and Human Services.)

Year	1981	1985	1990	1995	1997
AIDS Deaths	128	6948	31,269	45,765	14,338

(a) Calculate the average rates of change between consecutive data points in the table.
(b) Interpret these average rates of change.

76. *Temperature Change*   The table gives the outside temperature in degrees Fahrenheit on a summer afternoon.

Time (P.M.)	12:00	1:00	3:15
Temperature (°F)	78	82	88

Time (P.M.)	4:30	6:00
Temperature (°F)	93	91

(a) Calculate the average rate of change between consecutive data points in the table.

(b) Determine the time interval(s) when the thermometer was rising the fastest, on average.

Writing about Mathematics

77. Describe two methods of determining if a data set can be modeled by a linear function. Create or find a data set that consists of four or more ordered pairs. Determine if your data set can be modeled by a linear function.

78. Suppose you are given a graphical representation of a function f. Explain how you would determine whether f was constant, linear, or nonlinear. How would you determine the type if you were given a numerical or symbolic representation? Give examples.

Extended and Discovery Exercises

1. *Geometry*   Suppose that the radius of a circle on a computer monitor is increasing at a constant rate of 1 inch per second.

(a) Does the circumference of the circle increase at a constant rate? If it does, find this rate in inches per second.

(b) Does the area of the circle increase at a constant rate? Explain.

CHECKING BASIC CONCEPTS FOR SECTIONS 1.3 AND 1.4

1. Create symbolic, numerical, and graphical representations of a function f that computes the number of feet in x miles. For the numerical representation let $x = 1, 2, 3, 4, 5$.

2. Sketch the graphs of two relations—one that represents a function and one that does not represent a function. Explain your answer.

3. Find the slope of the line passing through the points $(-2, 4)$ and $(4, -5)$. If the graph of $f(x) = ax + b$ passes through these two points, what is the value of a?

4. Identify each representation of a function f as constant, linear, or nonlinear. Support your answer graphically.

(a) $f(x) = -1.4x + 5.1$
(b) $f(x) = 2x^2 - 5$
(c) $f(x) = 25$

5. The table shows U.S. advertising expenditures y in billions of dollars for year x.

(a) Make a line graph of the data. Let this graph represent a function f.

(b) Compute the average rate of change of f from 1980 to 1990 and from 1990 to 1997. Interpret your results.

x (year)	1980	1990	1997
y ($ billions)	54.8	128.6	187.5

Chapter 1 Summary

CONCEPT	EXPLANATION AND EXAMPLES

Section 1.1

SETS OF NUMBERS

Natural numbers: $N = \{1, 2, 3, 4, \dots\}$

Integers: $I = \{\dots, -3, -2, -1, 0, 1, 2, 3, \dots\}$

Rational numbers: $\frac{p}{q}$, where p and $q \neq 0$ are integers; includes repeating and terminating decimals

Irrational numbers: Includes nonrepeating, nonterminating decimals

Real numbers: Any number that can be expressed in decimal form; includes rational and irrational numbers

SCIENTIFIC NOTATION

A real number r is written as $c \times 10^n$, where $1 \leq |c| < 10$.

Examples: $1234 = 1.234 \times 10^3 \qquad 0.054 = 5.4 \times 10^{-2}$

PERCENT CHANGE

If a quantity changes from c_1 to c_2, then the percent change is

$$\frac{c_2 - c_1}{c_1} \times 100.$$

PROBLEM SOLVING

A problem-solving strategy can help find a solution. Two common examples are making a sketch and applying a formula.

Section 1.2

MEAN (AVERAGE) AND MEDIAN

The mean represents the average of a set of numbers and the median represents the middle of a sorted list.

Example: $4, 6, 9, 13, 15$;   Mean $= \dfrac{4 + 6 + 9 + 13 + 15}{5} = 9.4$; Median $= 9$

RELATION, DOMAIN, AND RANGE

A relation is a set of ordered pairs. The domain is the set of x-values and the range is the set of y-values.

Example: $S = \{(-1, 2), (4, -5), (5, 9)\}$; $D = \{-1, 4, 5\}, R = \{-5, 2, 9\}$

SCATTERPLOT AND LINE GRAPH

A scatterplot consists of a set of ordered pairs plotted in the xy-plane. When the data points are connected with line segments, a line graph results.

Section 1.3

FUNCTION

A function computes exactly one output for each valid input. The set of valid inputs is called the domain and the set of outputs is called the range.

Example: $f(x) = \sqrt{1 - x}$

$$D = \{x \mid x \leq 1\}; R = \{y \mid y \geq 0\}$$

CONCEPT	EXPLANATION AND EXAMPLES
FUNCTION NOTATION	**Example:** $f(x) = x^2 - 4$ $f(3) = 3^2 - 4 = 5$ $f(a + 1) = (a + 1)^2 - 4 = a^2 + 2a - 3$
REPRESENTATIONS OF FUNCTIONS	A function can be represented symbolically (formula), graphically (graph), numerically (table of values), and verbally (words). **Example:** See Example 6, Section 1.3
VERTICAL LINE TEST	If any vertical line intersects a graph at most once, then the graph represents a function.

Section 1.4

CONCEPT	EXPLANATION AND EXAMPLES
SLOPE	The slope m of the line passing through (x_1, y_1) and (x_2, y_2) is $$m = \frac{\Delta y}{\Delta x} = \frac{y_2 - y_1}{x_2 - x_1},$$ where $\Delta y = y_2 - y_1$ is the change in y and $\Delta x = x_2 - x_1$ is the change in x. **Example:** The slope of the line passing through $(1, -1)$ and $(-2, 3)$ is $$m = \frac{3 - (-1)}{-2 - 1} = -\frac{4}{3}.$$
CONSTANT FUNCTION	Given by $f(x) = b$ and its graph is a horizontal line.
LINEAR FUNCTION	Given by $f(x) = ax + b$ and its graph is a nonvertical line. The slope of its graph is equal to a, which is also equal to its constant rate of change. **Example:** The graph of $f(x) = -\frac{2}{3}x + 5$ has slope $-\frac{2}{3}$.
NONLINEAR FUNCTION	The graph of a nonlinear function is not a line and *cannot* be written as $f(x) = ax + b$. **Examples:** $f(x) = x^2 - 4$; $g(x) = \sqrt[3]{x} - 2$
AVERAGE RATE OF CHANGE	If (x_1, y_1) and (x_2, y_2) are points on the graph of f, then the average rate of change from x_1 to x_2 equals the slope of the line passing through these two points. **Example:** $f(x) = x^2$; Because $f(2) = 4$ and $f(3) = 9$, the average rate of change from 2 to 3 is $\frac{9 - 4}{3 - 2} = 5$.

Review Exercises

Exercises 1 and 2: Classify each number listed as one or more of the following: natural number, integer, rational number, or real number.

1. $-2, \frac{1}{2}, 0, 1.23, \sqrt{7}, \sqrt{16}$

2. $55, 1.5, \frac{104}{17}, 2^3, \sqrt{3}, -1000$

Exercises 3–6: Write each number in scientific notation.

3. 1,891,000

4. $-13,850$

5. 0.0000439

6. 0.0001001

Exercises 7 and 8: Write each number in standard form.

7. 1.52×10^4

8. -7.2×10^{-3}

9. Evaluate each expression with a calculator. Round answers to the nearest hundredth.

   (a) $\sqrt[3]{1.2} + \pi^3$

   (b) $\dfrac{3.2 + 5.7}{7.9 - 4.5}$

   (c) $\sqrt{5^2 + 2.1}$

   (d) $1.2(6.3)^2 + \dfrac{3.2}{\pi - 1}$

10. Approximate each expression with a calculator. Write the answer in scientific notation.

    (a) $(9.3 \times 10^3)(2.1 \times 10^{-4})$

    (b) $\dfrac{2.46 \times 10^6}{4.1 \times 10^2}$

11. **Speed of Light** The average distance between the planet Mars and the sun is approximately 228 million kilometers. Estimate the time required for sunlight, traveling at 300,000 kilometers per second, to reach Mars. (Source: C. Ronan, *The Natural History of the Universe.*)

12. **Flow Rate** Niagara Falls has an average of 212,000 cubic feet of water passing over it each second. Is this rate more or less than 10 cubic miles per year? Explain your calculations. (*Hint:* 1 cubic mile = 5280^3 cubic feet.) (Source: National Geographic Society.)

13. **Geometry** Suppose that 0.25 cubic inch of paint is painted onto a circular piece of paper with a diameter of 20 inches. Estimate the thickness of the paint on the paper.

14. **Enclosing a Pool** A rectangular swimming pool that is 25 feet by 50 feet has a 6-foot wide sidewalk around it.

    (a) How much fencing would be needed to enclose both the pool and the sidewalk?

    (b) Find the area of the sidewalk.

Exercises 15 and 16: Sort each list of numbers from smallest to largest and display them in a table.

   (a) *Determine the maximum and minimum values in the list.*

   (b) *Calculate the mean, median, and range.*

15. $-5, 8, 19, 24, -23$

16. $8.9, -1.2, -3.8, 0.8, 1.7, 1.7$

Exercises 17 and 18: Complete the following.

   (a) *Find the domain and range of the relation.*

   (b) *Determine the maximum and minimum of the x-values; of the y-values.*

   (c) *Label appropriate scales on the xy-axes.*

   (d) *Plot the relation by hand in the xy-plane.*

17. $\{(-1, -2), (4, 6), (0, -5), (-5, 3), (1, 0)\}$

18. $\{(10, 20), (-35, -25), (60, 60), (-70, 35), (0, -55)\}$

Exercises 19 and 20: Make a scatterplot of the relation in an appropriate viewing rectangle. Determine if the relation is a function.

19. $\{(10, 13), (-12, 40), (-30, -23), (25, -22), (10, 20)\}$

20. $\{(1.5, 2.5), (0, 2.1), (-2.3, 3.1), (0.5, -0.8), (-1.1, 0)\}$

Exercises 21 and 22: Find the distance between the points.

21. $(-4, 5), (2, -3)$

22. $\left(\dfrac{1}{2}, 0\right), \left(\dfrac{5}{2}, -\dfrac{1}{2}\right)$

23. **Education** The line graph at the top of the next page shows the number of Catholic secondary schools in thousands for various years.

    (a) Discuss any trends from 1960 to 1998.

    (b) Estimate and interpret the slope of each line segment.

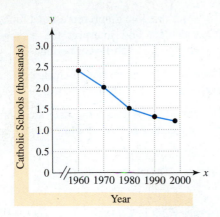

Year

24. Use the graph to determine the domain and range of each function. Evaluate $f(-2)$.

a.

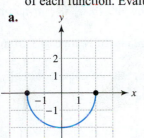

b.

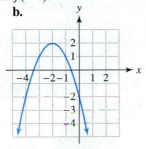

Exercises 25 and 26: Use the verbal representation to express the function f symbolically, graphically, and numerically. Let $y = f(x)$ with $0 \le x \le 100$. For the numerical representation, use $x = 0, 25, 50, 75, 100$.

25. To convert x pounds to y ounces, multiply x by 16.

26. To find the area y of a square, multiply the length x of a side by itself.

27. (a) Use the graph of f to evaluate $f(-3)$ and $f(1)$.
   (b) Find all x such that $f(x) = 0$.

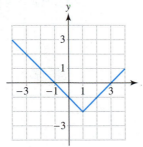

28. *ACT Scores*   The function f defined by

$$f(1987) = 20.6, f(1990) = 20.6, f(1996) = 20.9,$$
$$\text{and } f(1997) = 21.0$$

computes the national average ACT composite score in the year x.   (Source: The American College Testing Program.)

(a) Use ordered pairs to describe f.
(b) Identify the domain and range of f.
(c) Find all x such that $f(x) = 20.6$.

Exercises 29–32: Complete the following for the function f.
   (a) Evaluate $f(x)$ at the indicated values of x.
   (b) Find the implied domain of f.

29. $f(x) = \sqrt[3]{x}$ at $x = -8, 1$

30. $f(x) = 3x + 2$ at $x = -2, 5$

31. $f(x) = \dfrac{1}{x^2 - 4}$ at $x = -3, a + 1$

32. $f(x) = \sqrt{x + 3}$ at $x = 1, a - 3$

33. Determine if y is a function of x in the equation $x = y^2$.

34. Give an example of a relation that involves real data. List some or all of the ordered pairs in the relation.

Exercises 35 and 36: Determine if the graph represents a function.

35.   **36.**

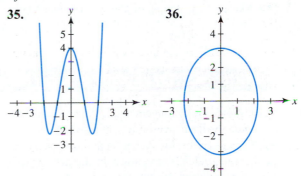

Exercises 37–40: If possible, find the slope of the line passing through each pair of points.

37. $(-1, 7), (3, 4)$ **38.** $(1, -4), (2, 10)$

39. $(8, 4), (-2, 4)$ **40.** $\left(-\dfrac{1}{3}, \dfrac{2}{3}\right), \left(-\dfrac{1}{3}, -\dfrac{5}{6}\right)$

Exercises 41–46: Decide whether the function f is constant, linear, or nonlinear. Support your answer graphically.

41. $f(x) = 8 - 3x$ **42.** $f(x) = 2x^2 - 3x - 8$

43. $f(x) = |x + 2|$ **44.** $f(x) = 2x$

45. $f(x) = 6$ **46.** $f(x) = x^3 - x + 1$

47. Sketch a graph of a 2-hour period showing the distance between two cars meeting on a straight highway, each traveling at 60 miles per hour. Assume that the cars are initially 120 miles apart.

48. Sketch a graph of a function f with $f(-2) = 3$, $f(2) = -3$, and $f(4) = 4$.

49. *Survival Rates*   The survival rates for song sparrows from 100 eggs are shown in the table. The values listed are the numbers of song sparrows that attain a given age. For example, 6 song sparrows reach an age of 2 years from 100 eggs laid in the wild.   (Source: S. Kress, *Bird Life.*)

Age	0	1	2	3	4
Number	100	10	6	3	2

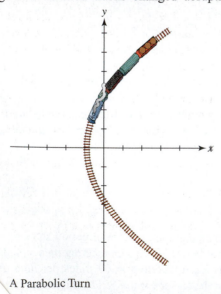

(a) Make a line graph of the data. Interpret the data.

(b) Does this line graph represent a function?

(c) Calculate and interpret the average rate of change for each 1-year period.

50. Compute the average rate of change of $f(x) = x^2 - x + 1$ from $x_1 = 1$ to $x_2 = 3$.

51. Determine if the following data are modeled best by a constant, linear, or nonlinear function.

x	-2	0	2	4
y	50	42	34	26

52. *Average Rate of Change*   Let $f(x) = 0.5x^2 + 50$ compute the outside temperature in degrees Fahrenheit at x P.M. where $1 \le x \le 5$.

(a) Graph f in $[1, 5, 1]$ by $[40, 70, 5]$. Is f linear or nonlinear?

(b) Calculate the average rate of change of f from 1 P.M. to 4 P.M.

(c) Interpret this average rate of change verbally and graphically.

Extended and Discovery Exercises

Curves designed by engineers for highways and railroads frequently have parabolic, rather than circular, shapes. The reason for this is that a parabolic curve becomes sharper gradually, as shown in the accompanying figure. If railroad tracks changed abruptly from straight to circular, the momentum of the locomotive ccould cause a derailment. A second figure illustrates straight track connecting to a circular curve.   (Source: F. Mannering and W. Kilareski, *Principles of Highway Engineering and Traffic Analysis.*)

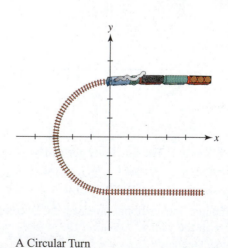

A Parabolic Turn

A Circular Turn

In order to design a curve and estimate its cost, engineers determine the distance around a curve before it is built. In the next figure the distance along a parabolic curve from A to C is approximated by two line segments AB and BC. The distance formula can be used to calculate the length of each segment. The sum of these two lengths gives a crude estimate for the length of the curve, shown below on the left.

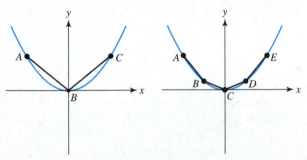

An Estimate of Curve Length   A Better Estimate

A better estimate using four line segments is shown above on the right. As the number of segments increases, so does the accuracy of the approximation.

1. *Curve Length*   Suppose that a curve designed for railroad tracks is represented by the equation $y = 0.2x^2$, where the units are in kilometers. The points $(-3, 1.8)$, $(-1.5, 0.45)$, $(0, 0)$, $(1.5, 0.45)$, and $(3, 1.8)$ lie on the graph of $y = 0.2x^2$. Approximate the length of the curve from $x = 3$ to $x = -3$ by using line segments connecting these points.

Exercises 2–5: Curve Length   Use three line segments connecting the four points to estimate the length of the curve on the graph of f from $x = -1$ to $x = 2$. Graph f and a line graph of the four points in the indicated viewing rectangle.

2. $f(x) = x^2$; $(-1, 1)$, $(0, 0)$, $(1, 1)$, $(2, 4)$; $[-4.5, 4.5, 1]$ by $[-1, 5, 1]$

3. $f(x) = \sqrt[3]{x}$; $(-1, -1)$, $(0, 0)$, $(1, 1)$, $(2, \sqrt[3]{2})$; $[-3, 3, 1]$ by $[-2, 2, 1]$

4. $f(x) = 0.5x^3 + 2$; $(-1, 1.5)$, $(0, 2)$, $(1, 2.5)$, $(2, 6)$; $[-4.5, 4.5, 1]$ by $[0, 6, 1]$

5. $f(x) = 2 - 0.5x^2$; $(-1, 1.5)$, $(0, 2)$, $(1, 1.5)$, $(2, 0)$; $[-3, 3, 1]$ by $[-1, 3, 1]$

6. The distance along the curve of $y = x^2$ from $(0, 0)$ to $(3, 9)$ is approximately 9.747. Use this fact to estimate the distance along the curve of $y = 9 - x^2$ from $(0, 9)$ to $(3, 0)$.

7. Estimate the distance along the curve of $y = \sqrt{x}$, from $(1, 1)$ to $(4, 2)$. (The actual value is approximately 3.168.)

8. *Endangered Species*   The Florida scrub-jay is an endangered species that prefers to live in open landscape with short vegetation. In recent years, NASA has been attempting to create an environment near Kennedy Space Center that supports the habitat for these birds. The following table lists their population for selected years, where $x = 0$ corresponds to 1980, $x = 1$ to 1981, $x = 2$ to 1982, and so on.

x (1980 $\leftrightarrow$ 0)	0	5	9
y (population)	3697	2512	2176

x (1980 $\leftrightarrow$ 0)	11	15	19
y (population)	2100	1689	1127

Source: *Mathematics Explorations II*, NASA–AMATYC–NSF.

(a) Make a scatterplot of the data.

(b) Find a linear function f that models the data approximately.

(c) Graph the data and f in the same viewing rectangle.

(d) Estimate the scrub-jay population in 1987 and in 2003.

9. *Global Warming*   If the global climate were to warm significantly, as a result of the greenhouse effect or other climatic change, the Arctic ice cap would start to melt. It is estimated that this ice cap contains the equivalent of 680,000 cubic miles of water. Over 200 million people currently live on soil that is less than 3 feet above sea level. In the United States, several large cities have low average elevations, such as Boston (14 feet), New Orleans (4 feet), and San Diego (13 feet).   (Sources: Department of the Interior, Geological Survey.)

(a) Devise a plan to determine how much sea level would rise if the Arctic cap melted. (*Hint:* The radius of Earth is 3960 miles and 71% of its surface is covered by oceans.)

(b) Implement your plan to estimate this rise in sea level.

(c) Discuss the implications of your calculation.

(d) Estimate how much sea level would rise if the 6,300,000 cubic miles of water in the Antarctic ice cap melted.

Success is dependent on effort.

— Sophocles —

Chapter 2

Linear Functions and Equations

Today a surgical procedure can be simulated in cyberspace before it is performed on the patient. With the help of computers, engineers are able to model the human body. This new technology gives physicians the ability to tinker with a model of the patient's arteries, experimenting with different placements for arterial bypasses. This virtual reality allows doctors to follow hunches, experiment, and even make mistakes. The software designed to do this task is called ASPIRE—Advanced Surgical Planning Interactive Research Environment. The accompanying figure shows a person's abdominal aorta. Red indicates the greatest blood flow, and blue indicates the least. The areas of blue are of greatest concern to doctors.

Mathematics played an essential role in the development of this software. Modern technology, such as this, would not be possible without mathematics.

Source: William Speed. "Downloading Your Body." *Discover,* September 2000. (Photograph reprinted with permission.)

2.1 LINEAR FUNCTIONS AND MODELS

Exact and Approximate Models • Graphs of Linear Functions • Modeling with Linear Functions

Introduction

Throughout history, people have attempted to explain natural phenomena by creating models. A model is based on observed data. It can be a diagram, an equation, a verbal expression, or some other form of communication. Models are used in diverse areas such as economics, physics, chemistry, astronomy, psychology, religion, and mathematics. Regardless of where it is used, a model is an *abstraction* that has the following two characteristics.

(1) A model is able to explain present phenomena. It should not contradict data and information that is already known to be correct.
(2) A model is able to make predictions about future data or results. It should be able to use current information to forecast future phenomena or create new information.

Mathematical models are used to forecast business trends, design the shapes of airplanes, estimate ecological trends, control highway traffic, describe epidemics, predict weather, and discover new information when human knowledge is inadequate.

Exact and Approximate Models

Not all mathematical models are exact representations of data. Data might appear to be nearly linear, but not exactly linear. In this case a linear function can be used to provide an *approximate model* of the data. In Figure 2.1 the data are modeled exactly by a linear function, whereas in Figure 2.2 the data are modeled approximately. In real applications, an approximate model is much more likely to occur than an exact model.

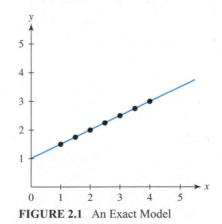

FIGURE 2.1 An Exact Model

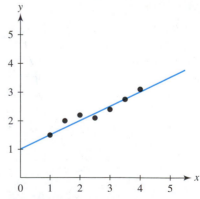

FIGURE 2.2 An Approximate Model

EXAMPLE 1 *Modeling data with a linear function*

Prison populations in millions for the years 1994 to 1997 are listed in Table 2.1 on the next page. These data can be modeled by $f(x) = 0.062x - 122.57$, where x is the year.

TABLE 2.1 **Prison Population**

Year	1994	1995	1996	1997
Inmates (millions)	1.05	1.13	1.18	1.24

Source: Bureau of Justice.

Graphing Calculator Help

To create the graph shown in Figure 2.3, see Appendix B (page AP-10). To create the table in Figure 2.4, see Appendix B (page AP-11).

(a) Graph f and the data in [1992, 1999, 1] by [0, 1.5, 0.5]. Decide if this model is exact or approximate.

(b) Support your conclusion in part (a) by evaluating $f(x)$ at $x = 1994, 1995, 1996,$ and 1997.

(c) Find and interpret the slope of the graph of f.

(d) Use f to estimate the number of inmates in 1998 and compare it to the actual value of 1.30 million.

SOLUTION

(a) Graph $Y_1 = 0.062x - 122.57$ together with the data, as shown in Figure 2.3. This model is approximate because the line does not pass through each point.

[1993, 1998, 1] by [0, 1.5, 0.5]

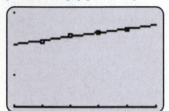

(b) It is often difficult to determine if the graph of a function passes exactly through data points. We can evaluate $f(x)$ at the x-values to confirm our answer in part (a). This can be done with the aid of a calculator, as shown in Figure 2.4.

$$f(1994) = 1.058 \neq 1.05; \quad f(1995) = 1.12 \neq 1.13;$$
$$f(1996) = 1.182 \neq 1.18; \quad f(1997) = 1.244 \neq 1.24$$

Since $f(x)$ does not agree *exactly* with all the values in Table 2.1, f represents an approximate model.

FIGURE 2.3

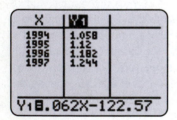

FIGURE 2.4

(c) In Section 1.4 we learned that the slope of the graph of $f(x) = ax + b$ equals a. Since $f(x) = 0.062x - 122.57$, $a = 0.062$. This means that the prison population is increasing, on average, by 0.062 million (62,000) inmates per year.

(d) To estimate the prison population in 1998 evaluate $f(1998)$.

$$f(1998) = 0.062(1998) - 122.57 = 1.306 \text{ million}$$

This differs from the actual value by 0.006 million or 6000 inmates.

Graphs of Linear Functions

Before we can begin to model data with linear functions, it is important to understand their graphs. Consider the graph of the linear function f shown in Figure 2.5. Its graph is a line that intersects each axis once.

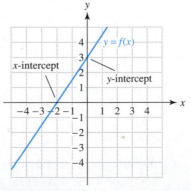

FIGURE 2.5 Graph of a Linear Function

From the graph we can see that when y increases by 3 units, x increases by 2 units. Thus, the change in y is $\Delta y = 3$, the change in x is $\Delta x = 2$, and the slope is $\frac{\Delta y}{\Delta x} = \frac{3}{2}$. The graph of f intersects the y-axis at the point $(0, 3)$. We say that the **y-intercept** on the graph of f is 3. Notice that when we evaluate $f(x) = ax + b$ at $x = 0$ we obtain

$$f(0) = a(0) + b = b.$$

That is, the point $(0, b)$ lies on the graph of f and the value of b represents the y-intercept. Using this information, we can write the symbolic representation of f as

$$f(x) = \frac{3}{2}x + 3.$$

slope y-intercept

Note that a function can have at most one y-intercept because $f(0)$ can have at most one value.

The graph of f in Figure 2.5 intersects the x-axis at the point $(-2, 0)$. We say that the **x-intercept** on the graph of f is -2. Notice that when we substitute $x = -2$ into the formula for f, we obtain

$$f(-2) = \frac{3}{2}(-2) + 3 = 0.$$

An x-intercept corresponds to an x-value that results in an output of 0. We also say that -2 is a *zero* of the function f, since $f(-2) = 0$. A **zero** of a function f corresponds to an x-intercept on the graph of f. If the slope of the graph of a linear function f is not 0, then the graph of f has exactly one x-intercept.

EXAMPLE 2   *Finding a symbolic representation from a graph*

Use the graph of a linear function f in Figure 2.6 to complete the following.

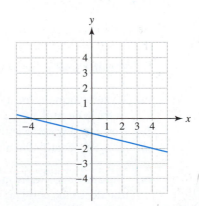

FIGURE 2.6

(a) Find the slope, y-intercept, and x-intercept.
(b) Write a formula for f.
(c) Find any zeros of f.

SOLUTION

(a) The line falls 1 unit each time the *x*-values increase by 4 units. Therefore the slope is $-\frac{1}{4}$. The graph intersects the *y*-axis at the point $(0, -1)$ and intersects the *x*-axis at $(-4, 0)$. Therefore the *y*-intercept is -1, and the *x*-intercept is -4.

(b) Since the slope is $-\frac{1}{4}$ and the *y*-intercept is -1, it follows that

$$f(x) = -\frac{1}{4}x - 1.$$

(c) Zeros of *f* correspond to *x*-intercepts, so the only zero is -4.

Modeling with Linear Functions

Linear functions can be used to model data and physical phenomena that change at a constant rate. For example, the distance traveled by a car can be modeled by a linear function if the car is traveling at a constant speed.

When we are modeling data with a linear function *f*, slope gives us information not only about the "tilt" of the graph of *f*, but also about how fast a quantity is changing or its *rate of change*. If a quantity changes at a constant rate, then we can model the quantity with $f(x) = ax + b$, where the value of *a* corresponds to this constant rate of change and the value of *b* corresponds to the initial amount.

Modeling with a linear function

To model a quantity that is changing at a constant rate with a linear function *f*, the following may be used.

$$f(x) = (\text{constant rate of change})\, x + (\text{initial amount})$$

Note: The constant rate of change corresponds to the slope of the graph of *f* and the initial amount corresponds to the *y*-intercept.

This method is illustrated in the next example.

EXAMPLE 3 *Finding a symbolic representation*

A 100-gallon tank is initially full of water and being drained at a rate of 5 gallons per minute.
(a) Write a formula for a linear function *f* that models the number of gallons of water in the tank after *x* minutes.
(b) How much water is in the tank after 4 minutes?
(c) Graph *f*. Identify the *x*- and *y*-intercepts and interpret each.

SOLUTION

(a) The amount of water in the tank is *decreasing* at a constant rate of 5 gallons per minute, so the constant rate of change is -5. The initial amount of water is equal to 100 gallons.

$$f(x) = (\text{constant rate of change})x + (\text{initial amount})$$
$$= -5x + 100$$

(b) After 4 minutes the tank held $f(4) = -5(4) + 100 = 80$ gallons.

(c) Since $f(x) = -5x + 100$, the graph has y-intercept 100 and slope -5. Its graph is shown in Figure 2.7. The x-intercept is 20, which corresponds to the time in minutes that it took to empty the tank. The y-intercept corresponds to the gallons of water in the tank initially.

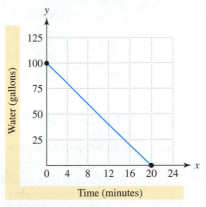

FIGURE 2.7

EXAMPLE 4   *Modeling ice deposits*

A roof has a 0.5-inch layer of ice on it from a previous storm. Another ice storm begins to deposit ice at a rate of 0.25 inch per hour.
(a) Find a formula for a linear function f that models the thickness of the ice on the roof x hours after the ice storm started.
(b) How thick is the ice after 2.5 hours?

SOLUTION   (a) The roof initially has 0.5 inch of ice and ice is accumulating at a constant rate of 0.25 inch per hour.

$$f(x) = \text{(constant rate of change)}x + \text{(initial amount)}$$
$$= 0.25x + 0.5$$

(b) The thickness of the ice after 2.5 hours was

$$f(2.5) = 0.25(2.5) + 0.5 = 1.125 \text{ inches.}$$

If the slope between consecutive data points is always the same, then data can be modeled exactly by a linear function. If the slope between consecutive data points is nearly the same, then the data can be modeled approximately by a linear function. In the next example we model data approximately with a linear function.

EXAMPLE 5   *Modeling fuel consumption*

Table 2.2 shows the distance traveled in miles by a car after burning x gallons of gasoline.

TABLE 2.2

x (gallons)	5	10	15	20
y (miles)	84	169	255	338

(a) Make a scatterplot of the data. Could a linear function be used to model this data?
(b) Find values for a and b so that $f(x) = ax + b$ models the distance traveled on x gallons of gasoline. Graph f and the data in the same viewing rectangle.
(c) Interpret the slope of the graph of f.

SOLUTION (a) A scatterplot of the data is shown in Figure 2.8. The data appear to be nearly linear so a linear function could be used to model the data.

(b) With 0 gallons of gasoline a car could travel 0 miles. So the data point (0, 0) could be included in the table. Thus the y-intercept is 0, and we can let $b = 0$ in the formula $f(x) = ax + b$.

Next we must determine a value for a. The slopes between consecutive points are

$$\frac{84 - 0}{5 - 0} = 16.8, \quad \frac{169 - 84}{10 - 5} = 17.0, \quad \frac{255 - 169}{15 - 10} = 17.2, \quad \text{and} \quad \frac{338 - 255}{20 - 15} = 16.6.$$

Since they are not exactly equal, f cannot model the data exactly. An initial estimate for a might be about 16.9 or 17. If we let $a = 17$ then $f(x) = 17x + 0$ or $f(x) = 17x$. A graph of f and the data are shown in Figure 2.9. (Values for a and b may vary slightly.)

Graphing Calculator Help

To make a scatterplot, see Appendix B (page AP-6). To plot data and graph an equation in the same viewing rectangle, see Appendix B (page AP-10).

[0, 25, 5] by [0, 400, 50]

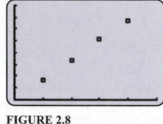

FIGURE 2.8

[0, 25, 5] by [0, 400, 50]

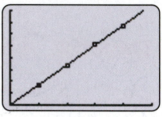

FIGURE 2.9

(c) A slope of 17 indicates that the mileage for this car was 17 miles per gallon.

■ **MAKING CONNECTIONS**

Slope and Approximately Linear Data When modeling data approximately with $f(x) = ax + b$, one way to obtain an initial value for a is to calculate the slope between the first and last data point in the table. The value for a can then be adjusted visually by graphing f and the data in the same viewing rectangle to obtain a slightly better fit. In Example 5 this would have resulted in

$$a = \frac{338 - 84}{20 - 5} \approx 16.93,$$

which compares favorably with our decision to let $a = 17$. ■

Putting it all Together

2.1

Any linear function can be written as $f(x) = ax + b$, and its graph is a line with slope a and y-intercept b. If data increase (or decrease) by nearly the same amount for each unit increase in x, then the data can be modeled by a linear function. One method to determine whether a set of data is linear is to calculate the slope between consecutive data points. If the slopes are nearly equal, it may be possible to model the data with a linear function.

Concept	Description
Exact and approximate models	An exact model describes data precisely. In real life, models are usually approximate. A good model should describe and explain current data. It should also be able to make predictions and forecast phenomena.
Graph of a linear function	The graph of a linear function is a line. Important features include its slope, y-intercept, and x-intercept. If $f(x) = ax + b$, then the slope equals a and the y-intercept equals b. The following graph has slope $\frac{1}{2}$, y-intercept 1, and x-intercept -2. 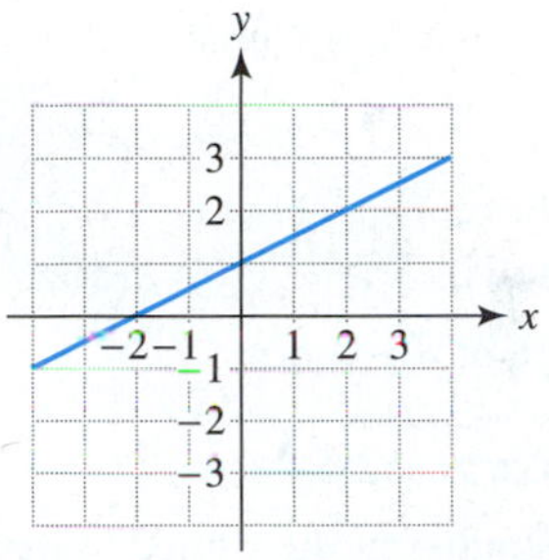
Linear model	If a quantity experiences a constant rate change, then it can be modeled by a linear function in the form $f(x) = ax + b$. The following can be used to determine a linear modeling function f. $$f(x) = (\text{constant rate of change})x + (\text{initial value})$$

2.1 EXERCISES

Exact and Approximate Models

Exercises 1–4: (Refer to Example 1.) For each data table a function f is given. Determine whether f models the data exactly or approximately.

1. $f(x) = 2.1x - 3.7$

x	1	2	3	4
y	−1.6	0.5	2.6	4.7

2. $f(x) = 1 - 0.2x$

x	1.1	2.1	3.1	4.1
y	0.78	0.58	0.38	0.18

3. $f(x) = 3.7 - 1.5x$

x	−6	0	6	12
y	12.7	3.7	−5.4	−14.2

4. $f(x) = 13.3x - 6.1$

x	1	2	5	10
y	7.2	20.6	60.4	127

Exercises 5–6: For each data table a function f is given that models the data approximately, where x = 0 corresponds to 1970, x = 10 to 1980, and so on.

  (a) Give the y-intercept on the graph of f and interpret it.

  (b) Interpret the slope of the graph of f.

5. $f(x) = 5.75x + 25.3$

Average Prices of a New Home

Year (1970 → 0)	0	10	20	28
Price ($ thousands)	27	76	150	182

Source: Bureau of the Census.

6. $f(x) = 0.672x + 52.5$

Number of Families in the United States

Year (1970 → 0)	0	10	20	28
Families (millions)	52	60	66	71

Source: Bureau of the Census.

Graphs of Linear Functions

Exercises 7–12: The graph of a linear function f is shown.

  (a) Identify the slope, y-intercept, and x-intercept.

  (b) Write a formula for f.

  (c) Estimate the zero of f.

7.
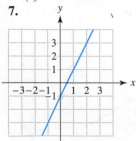

8.
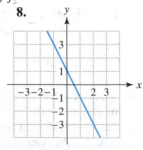

9.
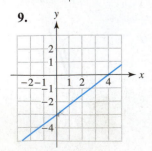

10.
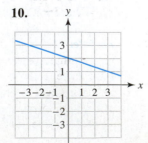

11.
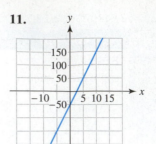

12.
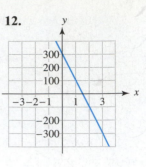

Exercises 13–18: Graph the linear function by hand.

13. $f(x) = 3x + 2$ **14.** $f(x) = -\dfrac{3}{2}x$

15. $f(x) = \dfrac{1}{2}x - 2$ **16.** $f(x) = 3 - x$

17. $g(x) = -2$ **18.** $g(x) = 20 - 10x$

Exercises 19–24: Write a formula for a linear function f whose graph satisfies the conditions.

19. Slope $-\dfrac{3}{4}$, y-intercept $\dfrac{1}{3}$

20. Slope -122, y-intercept 805

21. Slope 15, passing through the origin

22. Slope 1.68, passing through (0, 1.23)

23. Slope 0.5, passing through (1, 4.5)

24. Slope -2, passing through $(-1, 5)$

Modeling with Linear Functions

25. *Filling a Tank* (Refer to Example 3.) A 500-gallon tank initially contains 200 gallons of fuel oil. A pump is filling the tank at a rate of 6 gallons per minute.

  (a) Write a formula for a linear function f that models the number of gallons of fuel oil in the tank after x minutes.

  (b) Graph f. What is an appropriate domain for f?

  (c) Identify the y-intercept and interpret it.

  (d) Does the x-intercept of the graph of f have any physical meaning in this problem? Explain.

26. *HIV Infections* In 1998 there were 47 million people worldwide that had been infected with HIV. At that time the infection rate was 5.8 million people per year. (Source: United Nations AIDS and World Health Organization.)

  (a) Write a formula for a linear function f that models the total number of people that were infected with HIV in year x, where x = 0 corresponds to 1998, x = 1 to 1999, and so on.

(b) Estimate the number of people that may be infected in the year 2002.

27. *Birth Rate*   In 1990 the number of births per 1000 people in the United States was 16.7 and decreasing at 0.326 birth per 1000 people each year. (Source: National Center for Health Statistics.)
  (a) Write a formula for a linear function f that models the birth rate in year x, where $x = 0$ corresponds to 1990, $x = 1$ to 1991, and so on.
  (b) Estimate the birth rate in 1997 and compare it to the actual value of 14.5.

28. *Rainfall*   Suppose that during a storm, rain is falling at a rate of 1 inch per hour. The water coming from a circular roof with a radius of 20 feet is running down a drain spout that can accommodate 400 gallons of water per hour.
  (a) Determine the number of cubic inches of water falling on the roof in one hour.
  (b) One gallon equals about 231 cubic inches. Write a formula for a function g that computes the gallons of water landing on the roof in x hours.
  (c) How many gallons of water land on the roof during a 2.5-hour rain storm?
  (d) Will one down spout be sufficient to handle this type of rainfall? How many down spouts should there be?

Geometry Review   To review formulas related to circles, see Chapter R (page R-2).

Exercises 29–32: Write a formula for a linear function f that models the data exactly.

29.

x	0	1	2	3	4
$f(x)$	−7	−4	−1	2	5

30.

x	−0.4	−0.2	0	0.2	0.4
$f(x)$	−2.54	−2.82	−3.1	−3.38	−3.66

31.

x	−6	−4	1	3
$f(x)$	15	10	−2.5	−7.5

32.

x	15	20	25	30
$f(x)$	30	32	34	36

33. *Passenger Travel*   The table shows the number of miles (in trillions) traveled by passengers of all types for various years, where $x = 0$ corresponds to 1970, $x = 10$ to 1980, and so on.

Year (1970→0)	0	10	20	26
Miles (trillions)	2.2	2.8	3.7	4.4

Source: Department of Transportation.

  (a) Could the data be modeled exactly by a linear function?
  (b) Estimate values for a and b so that $f(x) = ax + b$ models the data. (Answers may vary.)
  (c) Graph f and the data. Interpret the slope.
  (d) Estimate passenger miles in 2003.

34. *School Enrollment*   The table shows the number of students (in millions) attending public school (K–8) for selected years, where $x = 0$ corresponds to 1994, $x = 1$ to 1995, and so on.

Year (1994→0)	0	1	2	3	4
Students (millions)	31.9	32.3	32.8	33.2	33.5

Source: Department of Education.

  (a) Estimate values for a and b so that $f(x) = ax + b$ models the data. (Answers may vary.)
  (b) Graph f and the data. Interpret the slope.
  (c) Predict the public K–8 enrollment in 2002.

35. *Stopping Distance*   The faster a car is traveling, the farther it takes for the car to stop. If a car doubles its speed x, then its braking distance quadruples. Could stopping distance be modeled by $f(x) = ax$? Explain your reasoning.

36. *Computer Memory*   Devices for computers often store information in units called megabytes (MB). For example, discs containing more megabytes can hold more pages of text. Is the amount of text that can be stored on a disc a function of the megabytes available? Is the function a linear function? Explain your reasoning.

Writing about Mathematics

37. How can you recognize a symbolic representation (formula) of a linear function? How can you recognize graphical and numerical representations of a linear function?

38. A student graphs $f(x) = x^2 - x$ in the viewing rectangle [2, 2.1, 0.01] by [1.9, 2.3, 0.1]. Using the graph, the student decides that f is a linear function. How could you convince the student otherwise?

Extended and Discovery Exercises

1. *Temperature and Volume*   The following table shows the relationship between the temperature of a sample of helium and its volume.

Temperature (°C)	0	25	50	75	100
Volume (in.^{3})	30	32.7	35.4	38.1	40.8

   (a) Make a scatterplot of the data.
   (b) Write a formula for a function f that receives the temperature x as input and outputs the volume y of the helium.
   (c) Find the volume of the gas when the temperature is 65°C.

2. *Height and Shoe Size*   In this exercise you will determine if there is a relationship between height and shoe size.
   (a) Have classmates write their sex, shoe size, and height in inches on a slip of paper. When you have enough information, complete the following tables.

Height and Shoe Size for Adult Males

Height (inches)					
Shoe Size					

Height and Shoe Size for Adult Females

Height (inches)					
Shoe Size					

   (b) Make a scatterplot of each table, by letting height correspond to the *x*-axis and shoe size correspond to the *y*-axis. Is there any relationship between height and shoe size? Explain.
   (c) Try to find a linear function that models each data set.

2.2 EQUATIONS OF LINES

Forms for Equations of Lines • Determining Intercepts • Horizontal, Vertical, Parallel, and Perpendicular Lines • Direct Variation

Introduction

Lines are a fundamental geometric concept that have applications in a variety of areas such as computer graphics, business, and science. Any quantity that experiences growth at a constant rate can be modeled by the graph of a linear function, which is a line. In this section we discuss how equations of lines can be determined, and some of their applications.

Forms for Equations of Lines

Suppose that a nonvertical line passes through the point (x_1, y_1) with slope m. If (x, y) is any point on this line with $x \neq x_1$, then the change in y is $\Delta y = y - y_1$, the change in x is $\Delta x = x - x_1$, and the slope equals $m = \dfrac{\Delta y}{\Delta x} = \dfrac{y - y_1}{x - x_1}$, as illustrated in Figure 2.10.

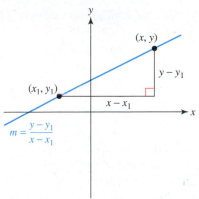

FIGURE 2.10

Using this slope formula, the equation of the line can be found.

$$m = \frac{y - y_1}{x - x_1} \qquad \text{Slope formula}$$

$$y - y_1 = m(x - x_1) \qquad \text{Cross multiply.}$$

$$y = m(x - x_1) + y_1 \qquad \text{Add } y_1 \text{ to both sides.}$$

The equation $y - y_1 = m(x - x_1)$ is traditionally called the *point-slope form* of the equation of a line. Since we think of y as being a function of x, written $y = f(x)$, the equivalent form $y = m(x - x_1) + y_1$ will also be referred to as the point-slope form. The point-slope form is not unique, since any point on the line can be used for (x_1, y_1). However, these point-slope forms are *equivalent*—their graphs are identical.

Point-slope form

The line with slope m passing through the point (x_1, y_1) has an equation

$$y = m(x - x_1) + y_1, \quad \text{or} \quad y - y_1 = m(x - x_1),$$

the **point-slope form** of the equation of a line.

In the next example we find the equation of a line given two points.

EXAMPLE 1 *Determining a point-slope form*

Find an equation of the line passing through the points $(-2, -3)$ and $(1, 3)$. Plot the points and graph the line by hand.

SOLUTION Begin by finding the slope of the line.

$$m = \frac{3 - (-3)}{1 - (-2)} = \frac{6}{3} = 2$$

Substituting $(x_1, y_1) = (1, 3)$ and $m = 2$ into the point-slope form results in

$$y = 2(x - 1) + 3.$$

If we use the point $(-2, -3)$ the point-slope form is

$$y = 2(x + 2) - 3.$$

Note that the point-slope form for a line is *not* unique. A graph of this line passing through the two points is shown in Figure 2.11.

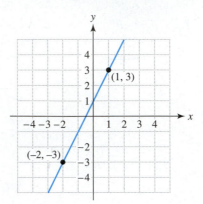

FIGURE 2.11

The two point-slope forms found in Example 1 are equivalent. This can be shown as follows.

$$y = 2(x - 1) + 3 \qquad \text{Point-slope form}$$
$$y = 2x - 2 + 3 \qquad \text{Distributive property}$$
$$y = 2x + 1 \qquad \text{Simplify.}$$

Applying the same steps on the second point slope form we obtain the same result.

$$y = 2(x + 2) - 3 \qquad \text{Point-slope form}$$
$$y = 2x + 4 - 3 \qquad \text{Distributive property}$$
$$y = 2x + 1 \qquad \text{Simplify.}$$

Both point-slope forms simplify to the same equation.

The form $y = mx + b$ is called the *slope-intercept form* and unlike the point-slope form it is *unique*. The real number m represents the slope and the real number b represents the *y*-intercept, as illustrated in Figure 2.12.

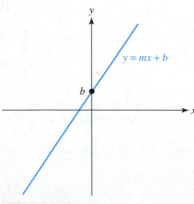

FIGURE 2.12

> ### Slope-intercept form
>
> The line with slope m and y-intercept b has an equation
>
> $$y = mx + b,$$
>
> the **slope-intercept form** of the equation of a line.

Algebra Review

To review the distributive property, see Chapter R (page R-23).

The slope-intercept form can be determined from the point-slope form by applying the **distributive property**. The distributive property states that $a(b + c) = ab + ac$ for all real numbers $a, b,$ and $c.$ When applying this property, it is important to remember that a is multiplied with *both* b and $c.$ Some examples are given.

$$5(x + 6) = 5x + 5(6) = 5x + 30$$
$$-(2x - 3) = -1(2x - 3) = -2x + 3$$
$$2x(7 - 4x) = 14x - 8x^2$$

In the next example we use the point-slope form of a line to model cellular phone growth during the late 1980s.

EXAMPLE 2   *Estimating growth and investment in cellular communication*

Cellular phone use has grown dramatically in the United States. In New York City when there were 25,000 customers, the investment cost per cellular site was $12 million. (A cellular site would include such things as a relay tower to transmit signals between cellular phones.) When the number of customers rose to 100,000, the investment cost rose to $96 million. Although cost usually decreases with additional customers, this was not the case for *early* cellular technology. Instead, cost increased due to the need to purchase expensive real estate and to establish communication among a large number of cellular sites. The relationship between customers and investment costs per site was approximately linear as shown in Figure 2.13.   (Source: M. Paetsch, *Mobile Communications in the U.S. and Europe.*)

(a) Find a point-slope form of the line passing through the points (25,000, 12) and (100,000, 96).

(b) Use this equation to estimate the investment cost per cellular site when there were 70,000 customers in New York City.

(c) Find the slope-intercept form of this line.

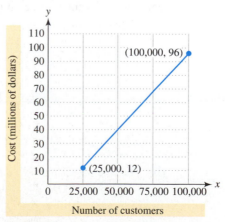

FIGURE 2.13   Cost of a Cellular Site

SOLUTION   (a) The slope m of the line segment passing through the points $(25,000, 12)$ and $(100,000, 96)$ is

$$m = \frac{96 - 12}{100,000 - 25,000} = 0.00112.$$

A point-slope form of the line passing through $(25,000, 12)$ with slope 0.00112 is found as follows.

$$y = m(x - x_1) + y_1 \qquad \text{Point-slope form}$$
$$y = 0.00112(x - 25,000) + 12 \qquad \text{Substitute for } (x_1, y_1) \text{ and } m.$$

In this equation y represents the investment cost for each cellular site in millions of dollars when there were x customers.

(b) For $x = 70,000$ customers, the investment cost was

$$y = 0.00112(70,000 - 25,000) + 12 = 62.4,$$

or $62.4 million per cellular site.

(c) To obtain the slope-intercept form, we apply the distributive property.

$$y = 0.00112(x - 25,000) + 12$$
$$y = 0.00112x - 28 + 12 \qquad \text{Distributive property}$$
$$y = 0.00112x - 16 \qquad \text{Simplify.}$$

EXAMPLE 3   *Finding equations of lines*

Find a point-slope form for the line that satisfies the conditions. Then convert this equation into slope-intercept form.

(a) Slope $-\dfrac{1}{2}$ passing through the point $(-3, -7)$

(b) x-intercept -4, y-intercept 2

SOLUTION   (a) Let $m = -\dfrac{1}{2}$ and $(x_1, y_1) = (-3, -7)$ in the point-slope form.

$$y = m(x - x_1) + y_1 \qquad \text{Point-slope form}$$
$$y = -\frac{1}{2}(x + 3) - 7 \qquad \text{Substitute.}$$

The slope-intercept form can be found by simplifying.

$$y = -\frac{1}{2}(x + 3) - 7 \qquad \text{Point-slope form}$$

$$y = -\frac{1}{2}x - \frac{3}{2} - 7 \qquad \text{Distributive property}$$

$$y = -\frac{1}{2}x - \frac{17}{2} \qquad \text{Slope-intercept form}$$

(b) The line passes through the points $(-4, 0)$ and $(0, 2)$. Its slope is

$$m = \frac{2 - 0}{0 - (-4)} = \frac{1}{2}.$$

Thus a point-slope form is $y = \frac{1}{2}(x + 4) + 0$, where the point $(-4, 0)$ is used for (x_1, y_1). The slope-intercept form is $y = \frac{1}{2}x + 2$.

EXAMPLE 4   *Modeling linear data*

Tuition has risen at private colleges. Table 2.3 lists the average tuition for selected years.

TABLE 2.3   Tuition at Private Colleges

Year	1980	1985	1990	1995
Tuition	$3617	$6121	$9340	$12,432

Source: The College Board.

(a) Make a scatterplot of this data in [1978, 1997, 1] by [0, 13,000, 1000]. What type of model does the scatterplot suggest?

(b) Find a linear function given by $f(x) = m(x - x_1) + y_1$ that models this data. Interpret the slope m.

(c) Use f to estimate tuition in 1987 and in 1998. Compare it to the actual values of $7048 and $14,508, respectively.

Graphing Calculator Help

To make a scatterplot, see Appendix B (page AP-6).

SOLUTION

(a) The scatterplot in Figure 2.14 suggests a linear model.

(b) The data table contains several points that could be used for (x_1, y_1). For example, we could choose the first data point (1980, 3617), then write

$$f(x) = m(x - 1980) + 3617.$$

[1978, 1997, 1] by [0, 13,000, 1000]

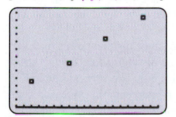

FIGURE 2.14

To estimate a value for m we could choose two points that appear to lie on a line that models the data. For example, if we choose the first point (1980, 3617) and the third data point (1990, 9340), then the slope m is

$$m = \frac{9340 - 3617}{1990 - 1980} = 572.3.$$

This slope indicates that tuition has risen, on average, $572.30 per year.

Figure 2.15 shows the graph of $f(x) = 572.3(x - 1980) + 3617$ and the data. It is important to realize that answers may vary when modeling real data, because if we choose different points, the resulting equation for $f(x)$ would be different.

(c) To estimate the tuition in 1987 and in 1998, evaluate $f(1987)$ and $f(1998)$, respectively.

$$f(1987) = 572.3(1987 - 1980) + 3617 = \$7623.10$$
$$f(1998) = 572.3(1998 - 1980) + 3617 = \$13{,}918.40$$

Both values differ from the actual values by less than $600.

Graphing Calculator Help

To plot data and graph an equation in the same viewing rectangle, see Appendix B (page AP-10).

[1978, 1997, 2] by [0, 13,000, 2000]

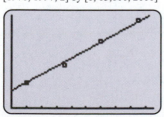

FIGURE 2.15

Determining Intercepts

Sometimes it is necessary to find either the x-intercept or the y-intercept on the graph of $y = f(x)$. The y-intercept can be found by substituting $x = 0$ and solving for y. Similarly, we can find the x-intercept by substituting $y = 0$ and solving for x. This technique is illustrated in the next example.

EXAMPLE 5   *Finding intercepts*

Locate the *x*- and *y*-intercepts on the line whose equation is $4x + 3y = 6$. Use the intercepts to graph the equation.

SOLUTION   To locate the *x*-intercept, let $y = 0$ in the equation.

$$4x + 3(0) = 6 \qquad \text{Let } y = 0.$$
$$x = 1.5 \qquad \text{Divide by 4.}$$

The *x*-intercept is 1.5. Similarly, to find the *y*-intercept, substitute $x = 0$ into the equation.

$$4(0) + 3y = 6 \qquad \text{Let } x = 0.$$
$$y = 2 \qquad \text{Divide by 3.}$$

The *y*-intercept is 2. Therefore the graph of the equation is a line that passes through the points (1.5, 0) and (0, 2), as shown in Figure 2.16.

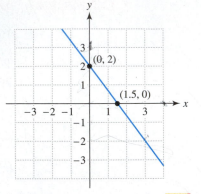

FIGURE 2.16

■ CLASS DISCUSSION
Find the slope-intercept form of the line in Example 5. What are the slope and *y*-intercept? Is your answer consistent with Figure 2.16? ■

Horizontal, Vertical, Parallel, and Perpendicular Lines

The graph of a constant function *f*, represented by $f(x) = b$, is the horizontal line $y = b$. This line has slope 0 and *y*-intercept *b*.

A vertical line cannot be represented by a function, since distinct points on a vertical line have the same *x*-coordinate. In fact, this is the distinguishing feature about points on a vertical line—they all have the same *x*-coordinate. The vertical line in Figure 2.17 is $x = 3$. The equation of a vertical line with *x*-intercept *k* is expressed symbolically as $x = k$, as shown in Figure 2.18. Horizontal lines have a slope equal to 0, and vertical lines have an undefined slope.

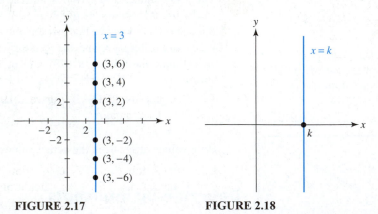

FIGURE 2.17   **FIGURE 2.18**

■ CLASS DISCUSSION
Why do you think that a vertical line sometimes is said to have "infinite slope"? What are some problems with taking this phrase too literally? ■

Equations of horizontal and vertical lines

An equation of the horizontal line with y-intercept b is $y = b$. An equation of the vertical line with x-intercept k is $x = k$.

EXAMPLE 6   *Finding equations of horizontal and vertical lines*

Find equations of vertical and horizontal lines passing through the point $(8, 5)$.

SOLUTION   The x-coordinate of the point $(8, 5)$ is 8. The vertical line $x = 8$ passes through every point in the xy-plane with an x-coordinate of 8, including the point $(8, 5)$. Similarly, the horizontal line $y = 5$ passes through every point with a y-coordinate of 5, including $(8, 5)$.

Slope is an important concept when determining whether two lines are parallel or perpendicular. Two nonvertical parallel lines have equal slopes.

Parallel lines

Two lines with slopes m_1 and m_2, neither of which is vertical, are parallel if and only if $m_1 = m_2$.

Note: The phrase "if and only if" is used when two statements are mathematically equivalent. If two nonvertical lines are parallel, then it is true that $m_1 = m_2$. Conversely, if two nonvertical lines have equal slopes, then they are parallel. Either condition implies the other.

EXAMPLE 7   *Finding parallel lines*

Find the slope-intercept form of a line parallel to $y = -2x + 5$, passing through the point $(-2, 3)$. Support your result graphically.

SOLUTION   Since the line $y = -2x + 5$ has slope -2, any parallel line also has slope -2. The line passing through $(-2, 3)$ with slope -2 is determined as follows.

$$y = -2(x + 2) + 3 \qquad \text{Point-slope form}$$
$$y = -2x - 4 + 3 \qquad \text{Distributive property}$$
$$y = -2x - 1 \qquad \text{Slope-intercept form}$$

Graphical support is shown in Figure 2.19, where the lines $Y_1 = -2X + 5$ and $Y_2 = -2X - 1$ are parallel.

$[-9, 9, 1]$ by $[-6, 6, 1]$

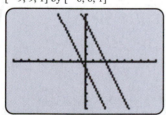

FIGURE 2.19

$[-10, 10, 1]$ by $[-10, 10, 1]$

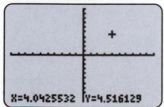

X=4.0425532   Y=4.516129

FIGURE 2.20

Cross hairs, found in telescopes, view finders, and video games, are examples of perpendicular lines. Graphing calculators often use perpendicular line segments for a cursor as shown in Figure 2.20. In video games, provisions sometimes are made for cross hairs to tilt in order to simulate things such as a craft rolling from side to side. In this situation, the cross hairs remain perpendicular, but they are not always in a horizontal-vertical position. Two lines with nonzero slopes are perpendicular if the product of their slopes is equal to -1.

> ### Perpendicular lines
>
> Two lines with nonzero slopes m_1 and m_2 are perpendicular if and only if $m_1 m_2 = -1$.

| **EXAMPLE 8** | *Finding perpendicular lines* |

Graphing Calculator Help

To set a square viewing rectangle, see Appendix B (page AP-9).

Suppose one cross hair is given by the equation $y = -\dfrac{1}{2}x + 9$. If the cross hairs are centered on the point $(8, 5)$, find the equation of the other cross hair. Graph the cross hairs in the viewing rectangle $[0, 15, 1]$ by $[0, 10, 1]$.

SOLUTION   The line $y = -\dfrac{1}{2}x + 9$ has slope $-\dfrac{1}{2}$. The slope of the perpendicular line is 2, since

$$m_1 m_2 = -\frac{1}{2} \cdot 2 = -1.$$

$[0, 15, 1]$ by $[0, 10, 1]$

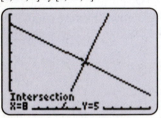

FIGURE 2.21

We find the equation of the line passing through the point $(8, 5)$ with slope 2. A point-slope form is $y = 2(x - 8) + 5$, which simplifies to $y = 2x - 11$.

Graph $Y_1 = -.5X + 9$ and $Y_2 = 2X - 11$, as shown in Figure 2.21. The perpendicular lines intersect at $(8, 5)$ as expected. Why is it important to use a square viewing rectangle?

| **EXAMPLE 9** | *Determining a rectangle* |

In Figure 2.22 a rectangle is outlined by four lines denoted y_1, y_2, y_3, and y_4. Find the equation of each line.

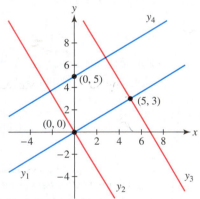

FIGURE 2.22

SOLUTION   *Line y_1:* This line passes through the points $(0, 0)$ and $(5, 3)$, so $m = \dfrac{3}{5}$ and the y-intercept is 0. Its equation is $y_1 = \dfrac{3}{5}x$.

Line y_2: This line passes through the point $(0, 0)$, and is perpendicular to y_1, so its slope is $m = -\dfrac{5}{3}$ and the y-intercept is 0. Its equation is $y_2 = -\dfrac{5}{3}x$.

Line y_3: This line passes through the point $(5, 3)$, and is parallel to y_2, so its slope is $m = -\dfrac{5}{3}$. Using a point-slope form, its equation is $y_3 = -\dfrac{5}{3}(x - 5) + 3$, which is equivalent to $y_3 = -\dfrac{5}{3}x + \dfrac{34}{3}$.

Line y_4: This line passes through the point $(0, 5)$, and is parallel to y_1, so its slope is $m = \dfrac{3}{5}$. Its equation is $y_4 = \dfrac{3}{5}x + 5$.

■ CLASS DISCUSSION

Check the results from Example 9 by graphing the four equations in the same viewing rectangle. How does your graph compare with Figure 2.22? Why is it important to use a square viewing rectangle? ■

Direct Variation

When a change in one quantity causes a proportional change in another quantity, the two quantities are said to *vary directly* or to *be directly proportional*. For example, if we work for $8 per hour, our pay is proportional to the number of hours that we work. Doubling the hours doubles the pay, tripling the hours triples the pay, and so on. This is stated more precisely as follows.

Direct variation

Let x and y denote two quantities. Then y is **directly proportional** to x, or y **varies directly** with x, if there exists a nonzero number k such that

$$y = kx.$$

The number k is called the **constant of proportionality** or the **constant of variation**.

If a person earns $57.75 working for 7 hours, the constant of proportionality k is the hourly pay rate. If y represents the pay in dollars and x the hours worked, then k is found by substituting values for x and y into the equation $y = kx$ and solving for k. That is,

$$57.75 = k(7) \qquad \text{or} \qquad k = \frac{57.75}{7} = 8.25,$$

so the hourly pay rate is $8.25.

Given a set of data points, one method to determine if y is directly proportional to x is to graph the ordered pairs (x, y). If the points lie on a line that passes through the origin, then y varies directly with x and the constant of proportionality k is equal to the slope of the line. A second method is to compute the ratio $\dfrac{y}{x}$ for each ordered pair. The equation $y = kx$ implies that $k = \dfrac{y}{x}$, so each ratio $\dfrac{y}{x}$ will equal k.

EXAMPLE 10   *Modeling storage requirements for recording music*

Recording music requires an enormous amount of storage. A compact disc (CD) can hold approximately 650 million bytes. One million bytes is commonly referred to as a **megabyte** (MB). (See Example 5, Section 1.1 for an explanation of a byte.) Table 2.4 on the next page lists the megabytes x needed to record y seconds of music.

TABLE 2.4

x (MB)	0.129	0.231	0.415	0.491
y (sec)	6.010	10.74	19.27	22.83

x (MB)	0.667	1.030	1.160	1.260
y (sec)	31.00	49.00	55.25	60.18

Source: Gateway 2000 System CD.

(a) Compute the ratio $\frac{y}{x}$ for each musical segment in the table. Interpret these ratios.

(b) Approximate a constant of proportionality k satisfying the equation $y = kx$. Graph the data and the equation together.

(c) Estimate the maximum number of seconds of music that can be placed on a 1.44-megabyte floppy disc.

SOLUTION   **(a)** The ratios are shown in Table 2.5. For example, $\frac{6.010}{0.129} \approx 46.6$. These ratios represent the number of seconds that can be recorded on 1 megabyte. They are approximately equal, which indicates direct variation.

TABLE 2.5

x (MB)	0.129	0.231	0.415	0.491
y (sec)	6.010	10.74	19.27	22.83
y/x	46.6	46.5	46.4	46.5

x (MB)	0.667	1.030	1.160	1.260
y (sec)	31.00	49.00	55.25	60.18
y/x	46.5	47.6	47.6	47.8

Graphing Calculator Help

To plot data and graph an equation in the same viewing rectangle, see Appendix B (page AP-10).

$[-0.1, 1.5, 0.25]$ by $[0, 70, 10]$

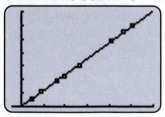

FIGURE 2.23

(b) From Table 2.5 it appears that approximately 47 seconds of music can be recorded on 1 megabyte. Therefore let the constant of proportionality be $k = 47$. The data points and the equation $y = 47x$ are graphed in Figure 2.23.

(c) Since $y = 47x$, a 1.44-megabyte floppy disc could store at most $47 \cdot 1.44 = 67.68 \approx 68$ seconds of music.

■ **MAKING CONNECTIONS**

Direct Variation and Linear Functions   If a set of data points (x, y) can be modeled by a linear function f that passes through the origin, then y varies directly with x. In this case the y-intercept is 0 and f can be written as $f(x) = ax$, where a is the constant of variation.

Putting it all Together 2.2

The following table summarizes some important topics.

Concept	Comments	Example
Point-slope form $$y = m(x - x_1) + y_1$$ or $$y - y_1 = m(x - x_1)$$	Is used to find the equation of a line, given two points or one point and the slope	Given two points $(5, 1)$ and $(4, 3)$, first compute $m = \dfrac{3 - 1}{4 - 5} = -2$. An equation of this line is $y = -2(x - 5) + 1$.
Slope-intercept form $$y = mx + b$$	Is a unique equation for a line, determined by the slope m and the y-intercept b	An equation of the line with slope $m = 5$ and y-intercept $b = -4$ is $y = 5x - 4$.
Direct variation	y is directly proportional to x or varies directly with x if $y = kx$ for some constant k. Constant k is the constant of proportionality or the constant of variation.	The sales tax on a purchase is directly proportional to the amount of the purchase. If the sales tax rate is 7%, sales tax y on a purchase of x dollars is given by $y = 0.07x$. The sales tax on \$125 is given by $y = 0.07(125) = \$8.75$.

The following table summarizes the important concepts concerning special types of lines.

Concept	Equation(s)	Example
Horizontal line	$y = b$, where b is a constant	A horizontal line with y-intercept 7 has the equation $y = 7$.
Vertical line	$x = k$, where k is a constant	A vertical line with x-intercept -8 has the equation $x = -8$.
Parallel lines	$y = m_1 x + b_1$ and $y = m_2 x + b_2$, where $m_1 = m_2$	The lines $y = -3x - 1$ and $y = -3x + 5$ are parallel because they both have slope -3.
Perpendicular lines	$y = m_1 x + b_1$ and $y = m_2 x + b_2$, where $m_1 m_2 = -1$	The lines $y = 2x - 5$ and $y = -\dfrac{1}{2}x + 2$ are perpendicular because $m_1 m_2 = 2\left(-\dfrac{1}{2}\right) = -1$.

2.2 EXERCISES

Equations of Lines

Exercises 1–6: Match the equation to its graph (a–f).

1. $y = m(x - x_1) + y_1, m > 0$

2. $y = m(x - x_1) + y_1, m < 0$

3. $y = mx, m > 0$

4. $y = mx + b, m < 0$ and $b > 0$

5. $x = k, k > 0$

6. $y = b, b < 0$

a.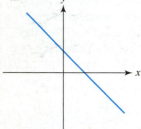

b.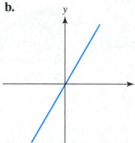

c.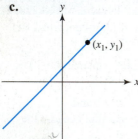

d.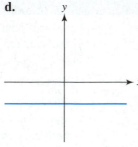

e.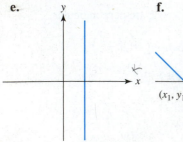

f.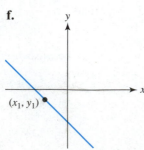

Exercises 7–10: Find the slope-intercept form for the line in the figure.

7.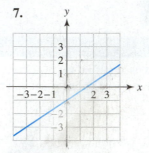

8.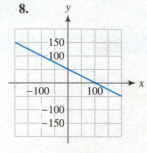

9.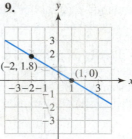

10.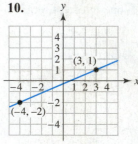

Exercises 11–16: Find a point-slope form for the equation of the line satisfying the conditions.

11. Slope -2.4, passing through $(4, 5)$

12. Slope 1.7, passing through $(-8, 10)$

13. Passing through $(1, -2)$ and $(-9, 3)$

14. Passing through $(-6, 10)$ and $(5, -12)$

15. Passing through $(1980, 5)$ and $(1990, 25)$

16. Passing through $(1990, -3)$ and $(1996, 19)$

Exercises 17–34: Find the slope-intercept form for the line satisfying the conditions.

17. y-intercept 5, slope -7.8

18. y-intercept -155, slope 5.6

19. y-intercept 45, x-intercept 90

20. x-intercept -6, y-intercept -8

21. Parallel to $y = 4x + 16$, passing through $(-4, -7)$

22. Parallel to $y = -\dfrac{3}{4}(x - 100) - 99$, passing through $(1.5, \sqrt{3})$

23. Perpendicular to $y = -\frac{2}{3}(x - 1980) + 5$, passing through $(1980, 10)$

24. Perpendicular to $y = 6x - 1000$, passing through $(15, -7)$

25. Slope -3, passing through $(0, 5)$

26. Slope $\frac{1}{3}$, passing through $\left(\frac{1}{2}, -2\right)$

27. Passing through $(0, -6)$ and $(4, 0)$

28. Passing through $\left(\frac{3}{4}, -\frac{1}{4}\right)$ and $\left(\frac{5}{4}, \frac{7}{4}\right)$

29. Parallel to $y = \frac{2}{3}x + 3$, passing through $(0, -2.1)$

30. Parallel to $y = -4x - \frac{1}{4}$, passing through $(2, -5)$

31. Perpendicular to $2x + y = 0$, passing through $(-2, 5)$

32. Perpendicular to $y = -\frac{6}{7}x + \frac{3}{7}$, passing through $(3, 8)$

33. Perpendicular to $x + y = 4$, passing through $(15, -5)$

34. Parallel to $2x - 3y = -6$, passing through $(4, -9)$

Exercises 35–42: Find an equation of the line satisfying the following conditions.

35. Vertical, passing through $(-5, 6)$

36. Vertical, passing through $(1.95, 10.7)$

37. Horizontal, passing through $(-5, 6)$

38. Horizontal, passing through $(1.95, 10.7)$

39. Perpendicular to $y = 15$, passing through $(4, -9)$

40. Perpendicular to $x = 15$, passing through $(1.6, -9.5)$

41. Parallel to $x = 4.5$, passing through $(19, 5.5)$

42. Parallel to $y = -2.5$, passing through $(1985, 67)$

Determining Intercepts

Exercises 43–52: Determine symbolically the x- and y-intercepts on the graph of the equation. Support your results graphically, except for Exercise 52.

43. $y = 8x - 5$

44. $y = -1.5x + 15$

45. $y = 3(x - 2) - 5$

46. $y = -2(x + 1) + 7$

47. $y - 3x = 7$

48. $4x - 3y = 6$

49. $0.2x + 0.4y = 0.8$

50. $\frac{2}{3}y - x = 1$

51. $\frac{x}{5} + \frac{y}{7} = 1$

52. $\frac{x}{a} + \frac{y}{b} = 1$

Applications

53. *Distance* A person is riding a bicycle along a straight highway. The accompanying graph shows the rider's distance y in miles from an interstate highway after x hours.

(a) How fast is the biker traveling?

(b) Find the slope-intercept form of the line.

(c) How far was the biker from the interstate highway initially?

(d) How far was the biker from the interstate highway after 1 hour and 15 minutes?

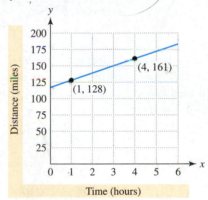

54. *Water in a Tank* The graph shows the amount of water y in a 100-gallon tank after x minutes have elapsed.

(a) Is water entering or leaving the tank? How much water is in the tank after 3 minutes?

(b) Find both the x- and y-intercepts. Interpret their meanings.

(c) Find the slope-intercept form of the equation of the line. Interpret the slope.

(d) Estimate the x-coordinate of the point $(x, 50)$ that lies on the line.

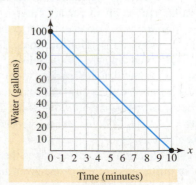

55. *Tuition and Fees*   The graph models average tuition and fees in dollars at public 4-year colleges from 1981 to 1995.   (Source: The College Board.)

(a) The graph passes through (1984, 1225) and (1987, 1621). Interpret these two points.

(b) Find a point-slope form for this line. Interpret the slope.

(c) Estimate tuition and fees in 2000.

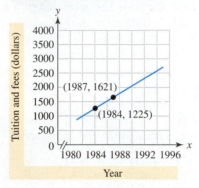

56. *Tuition and Fees*   (Refer to Example 4 and Exercise 55.) The following table lists average tuition and fees in dollars at public 4-year colleges for selected years.

Year	1980	1985	1990
Tuitions and Fees	$804	$1318	$1908

Year	1995	1998
Tuitions and Fees	$2811	$3243

Source: The College Board.

(a) Make a scatterplot of the data in the viewing rectangle [1978, 2000, 2] by [0, 3500, 500].

(b) Find a linear function given by $f(x) = m(x - x_1) + y_1$ that models the data. Interpret the slope m.

(c) Use f to estimate tuition in 1987 and 2002.

57. *Antarctic Ozone Layer*   Stratospheric ozone occurs in the atmosphere between altitudes of 12 and 18 miles and is an important filter of ultraviolet light from the sun. Ozone in the stratosphere is frequently measured in Dobson units. The Dobson scale is linear where 300 Dobson units is a midrange value that corresponds to an ozone layer 3 millimeters thick. In 1991 the reported minimum in the Antarctic *ozone hole* was about 110 Dobson units.   (Source: R. Huffman, *Atmospheric Ultraviolet Remote Sensing.*)

(a) The equation $y = kx$ describes the thickness y in millimeters of the ozone layer that is x Dobson units. Find k.

(b) How thick was the ozone layer over the Antarctic in 1991?

(c) Was the ozone hole actually a *hole* in the ozone layer?

58. *Farm Pollution*   In 1988 the number of farm pollution incidents reported in England and Wales was 4000. This number had increased roughly at a rate of 280 per year since 1979.   (Source: C. Mason, *Biology of Freshwater Pollution.*)

(a) Find an equation of a line $y = m(x - x_1) + y_1$ that models these data, where y represents the number of pollution incidents during the year x.

(b) Estimate the number of incidents in 1975.

59. *Cost of Driving*   The cost of driving a car includes both fixed costs and mileage costs. Assume that it costs $350 per month for insurance and car payments, and it costs $0.29 per mile for gasoline, oil, and routine maintenance.

(a) Determine a linear function that computes the annual cost of driving this car x miles.

(b) What does the y-intercept on the graph of f represent?

60. *Average Wages*   The average hourly wage (adjusted to 1982 dollars) was $8.03 in 1970 and $7.75 in 1998.   (Source: Department of Commerce.)

(a) Find an equation of a line that passes through (1970, 8.03) and (1998, 7.75).

(b) Interpret the slope.

(c) Approximate the hourly wage in 1990. Compare it to the actual value of $7.52.

Exercises 61–64: Determine values for m, x_1, and y_1 so that the data are modeled exactly by $f(x) = m(x - x_1) + y_1$. Let (x_1, y_1) correspond to the first point in the data table.

61.

x	1	2	3	4
y	3	5	7	9

62.

x	3	4	5	6
y	4	2	0	-2

63.

x	-6	-3	0	3
y	1	-8	-17	-26

64.

x	-5	5	15	25
y	-15	25	65	105

Exercises 65–68: ***Modeling Real Data***   *Each table contains real data that can be modeled approximately by $f(x) = m(x - x_1) + y_1$.*

(a) *Make a scatterplot of the data. (Do not try to plot the undetermined point in the table.)*

(b) *Fit f to the data by approximating values for the constants m, x_1, and y_1. Graph f together with the data on the same coordinate axes.*

(c) *Interpret the slope m.*

(d) *Use f to approximate the undetermined value in the table.*

65. Municipal solid waste in millions of tons

Year	1960	1965	1970	1975
Waste	87.8	103.4	121.9	128.1

Year	1980	1985	1990	1995
Waste	151.5	164.4	195.7	?

Source: Environmental Protection Agency.

66. Population in millions of the western region of the United States

Year	1950	1960	1970
Population	20.2	28.1	34.8

Year	1980	1990	2000
Population	43.2	52.8	?

Source: Bureau of the Census.

67. Projected Asian-American population in millions

Year	1996	1997	1998	1999
Population	9.7	10.1	10.5	10.9

Year	2000	2001	2002	2005
Population	11.2	11.6	12.0	?

Source: Bureau of the Census.

68. State and federal prison inmates in thousands

Year	1992	1993	1994	1995
Inmates	884	970	1054	1126

Year	1996	1997	1998	1999
Inmates	1183	1242	1302	?

Source: Bureau of Justice *Statistics Bulletin.*

Perspectives and Viewing Rectangles

69. Graph $y = \dfrac{1}{1024}x + 1$ in $[0, 3, 1]$ by $[-2, 2, 1]$.

(a) Is the graph a horizontal line?

(b) Conjecture why the calculator screen appears as it does.

70. Graph $y = 1000x + 1000$ in the standard viewing rectangle.

(a) Is the graph a vertical line?

(b) Explain why the calculator screen appears as it does.

71. *Square Viewing Rectangle* Graph the lines $y = 2x$ and $y = -\dfrac{1}{2}x$ in the standard viewing rectangle.

(a) Do the lines appear to be perpendicular?

(b) Graph the lines in the following viewing rectangles.

   i. $[-15, 15, 1]$ by $[-10, 10, 1]$

   ii. $[-10, 10, 1]$ by $[-3, 3, 1]$

   iii. $[-3, 3, 1]$ by $[-2, 2, 1]$

Do the lines appear to be perpendicular in any of these viewing rectangles?

(c) Determine the viewing rectangles where perpendicular lines will appear perpendicular. (Answers may vary depending on the model of graphing calculator used.)

72. Continuing with Exercise 71, make a conjecture about which viewing rectangles result in the graph of a circle with radius 5 and center at the origin appearing circular.

   i. $[-9, 9, 1]$ by $[-6, 6, 1]$

   ii. $[-5, 5, 1]$ by $[-10, 10, 1]$

   iii. $[-5, 5, 1]$ by $[-5, 5, 1]$

   iv. $[-18, 18, 1]$ by $[-12, 12, 1]$

Test your conjecture by graphing this circle in each viewing rectangle. (*Hint:* Graph the equations

$$y_1 = \sqrt{25 - x^2} \text{ and } y_2 = -\sqrt{25 - x^2}$$

to create the circle.)

Graphing a Rectangle

Exercises 73–76: (Refer to Example 9.) A rectangle is determined by the stated conditions. Find the slope-intercept form of the four lines that outline the rectangle.

73. Vertices $(0, 0)$, $(2, 2)$, and $(1, 3)$

74. Vertices $(1, 1)$, $(5, 1)$, and $(5, 5)$

75. Vertices $(4, 0)$, $(0, 4)$, $(0, -4)$, and $(-4, 0)$

76. Vertices $(1, 1)$, $(2, 3)$, and the point $(3.5, 1)$ lies on a side of the rectangle

Direct Variation

Exercises 77–80: Find the constant of proportionality k and the undetermined value in the table if y is directly proportional to x. Support your answer by graphing the equation y = kx and the data points.

77.

x	3	5	6	8
y	7.5	12.5	15	?

78.

x	1.2	4.3	5.7	?
y	3.96	14.19	18.81	23.43

79. Sales tax *y* on a purchase of *x* dollars

x	$25	$55	?
y	$1.50	$3.30	$5.10

80. Cost *y* of buying *x* compact discs having the same price

x	3	4	5
y	$41.97	$55.96	?

81. *Cost of Tuition*   The cost of tuition is directly proportional to the number of credits taken. If 11 credits cost $720.50, find the cost of taking 16 credits. What is the constant of proportionality?

82. *Strength of a Beam*   The maximum load that a horizontal beam can carry is directly proportional to its width. If a beam 1.5 inches wide can support a load of 250 pounds, find the load that a beam of the same type can support if its width is 3.5 inches.

83. *Hooke's Law*   The distance that a spring stretches is directly proportional to the size of the weight hung on the spring. Suppose a 15-pound weight stretches a spring 8 inches, as shown in the figure.

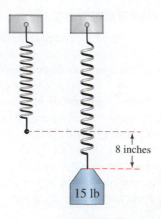

(a) The equation $y = kx$ models this situation. Find k.

(b) How far will a 25-pound weight stretch this spring?

84. *Distance Traveled*   Suppose a car travels at 60 miles per hour for *x* hours. Then the distance *y* that the car travels is directly proportional to *x*. Find the constant of proportionality. How far does the car travel in 3 hours?

85. *Stopping Distance*   The approximate stopping distances *y* in feet for a car traveling at *x* miles per hour are listed in the table. Determine if stopping distance is directly proportional to speed. Interpret your results.   (Source: L. Haefner, *Introduction to Transportation Systems.*)

x (mph)	0	20	40	60	80
y (ft)	0	118	324	620	1004

86. *Electrical Resistance*   The electrical resistance of a wire varies directly with its length. If a 255-foot wire has a resistance of 1.2 ohms, find the resistance of 135 feet of the same type of wire. Interpret the constant of proportionality in this situation.

Writing about Mathematics

87. Compare the slope-intercept form with the point-slope form. Give examples of each.

88. Give an example of two quantities in real life that vary directly. Explain your answer. Use an equation to describe the relationship between the two quantities.

Extended and Discovery Exercises

*Exercises 1–2: **Estimating Populations** Biologists sometimes use direct variation to estimate the number of fish in small lakes. This is done by tagging a small number of fish and then releasing them. Biologists assume that over a period of time, the tagged fish distribute themselves evenly throughout the lake. Later a second sample is collected. The total number of fish in the sample is counted along with the number of tagged fish. To determine the total population of fish in the lake, biologists assume that the proportion of tagged fish in the second sample is equal to the proportion of tagged fish in the entire lake. This technique can also be used to count other types of animals such as birds, when they are not migrating.*

1. Eighty-five fish are tagged and released into a pond. Later a sample of 94 fish from the pond contains 13 tagged fish. Estimate the number of fish in the pond.

2. Sixty-three blackbirds are tagged and released. Later it is estimated that out of a sample of 32 blackbirds, only 8 are tagged. Estimate the population of blackbirds in the area.

CHECKING BASIC CONCEPTS FOR SECTIONS 2.1 AND 2.2

1. Graph $f(x) = 4 - 2x$ by hand. Identify the slope, the x-intercept, and the y-intercept.

2. The death rate from heart disease for ages 15 through 24 is 2.7 per 100,000 people.
   (a) Write a function f that models the number of deaths in a population of x million people who are 15–24 years old.
   (b) There are about 39 million people in the United States who are 15–24 years old. Use f to estimate the number of deaths from heart disease in this age group.

3. A driver of a car is initially 50 miles south of home, driving 60 miles per hour south. Write a function f that models the distance between the driver and home.

4. Find an equation of the line passing through the points $(-3, 4)$ and $(5, -2)$. Give equations of lines that are parallel and perpendicular to this line.

5. Find equations of horizontal and vertical lines that pass through the point $(-4, 7)$.

2.3 LINEAR EQUATIONS

Equations • The Intersection-of-Graphs Method • Solving Linear Equations Symbolically • Problem-Solving Strategies

Introduction

A primary objective of both theoretical and applied mathematics is solving equations. Billions of dollars are spent each year on computers and personnel to solve equations that hold the answers for creating better products. Equations occur in a wide variety of forms. Some equations can be solved easily, whereas others require enormous amounts of resources. Without graphical, numerical, and symbolic techniques for solving equations, our society would not have televisions, CD players, satellites, fiber optics, CAT scans, computers, or accurate weather forecasts.

Chapter 1 introduced the concept of a function and its representations. In this section we see that applications involving functions lead to equations. For example,

$f(x) = 572.3(x - 1980) + 3617$ models the average tuition at private colleges from 1980 to 2000. (See Example 4, Section 2.2.) One way to determine the year x when the average tuition was \$14,000 would be to solve the equation

$$572.3(x - 1980) + 3617 = 14,000.$$

Equations

An **equation** is a statement that two mathematical expressions are equal. Equations always contain an equals sign. Some examples of equations include

$$x + 15 = 9x - 1, \qquad x^2 - 2x + 1 = 2x, \qquad z + 5 = 0,$$
$$xy + x^2 = y^3 + x, \qquad \text{and} \qquad 1 + 2 = 3.$$

The first three equations have one variable, the fourth equation has two variables, and the fifth equation contains only constants. For now, our discussion concentrates on equations with one variable.

To **solve** an equation means to find all values for the variable that make the equation a true statement. Such values are called **solutions**. The set of all solutions is the **solution set**. The solutions to the equation $x^2 - 1 = 0$ are 1 or -1, written as $x = \pm 1$. Either value for x **satisfies** the equation. The solution set is $\{-1, 1\}$. Two equations are **equivalent** if they have the same solution set. For example, the equations $x + 2 = 5$ and $x = 3$ are equivalent.

If an equation has no solutions, then its solution set is empty and the equation is called a **contradiction**. The equation $x + 2 = x$ has no solutions and is a contradiction. However, if every (meaningful) value for the variable is a solution, then the equation is an **identity**. The equation $x + x = 2x$ is an identity because every value for x makes the equation true. Any equation that is satisfied by some, but not all, values of the variable is a **conditional equation**. The equation $x^2 - 1 = 0$ is a conditional equation. Only the values -1 and 1 for x make this equation a true statement.

As with functions, equations can be either *linear* or *nonlinear*. A linear equation is one of the simplest types of equations.

Linear equation in one variable

A **linear equation** in one variable is an equation that can be written in the form

$$ax + b = 0,$$

where a and b are real numbers with $a \neq 0$.

If an equation is not linear, then we say that it is **nonlinear**. The following are examples of linear equations. In each case, rules of algebra could be used to write the equation in the form $ax + b = 0$.

$$x - 12 = 0, \quad 2x - 4 = -x, \quad 2(1 - 4x) = 16, \quad x - 5 + 3(x - 1) = 0$$

The Intersection-of-Graphs Method

The equation $f(x) = g(x)$ results whenever the formulas for two functions f and g are set equal to each other. A solution to this equation corresponds to the x-coordinate of a point where the graphs of f and g intersect. This technique is called the **intersection-of-graphs method**. If the graphs of f and g are lines with different slopes, then their graphs intersect once. For example, if $f(x) = 2x + 1$ and

$g(x) = -x + 4$, then the equation $f(x) = g(x)$ becomes $2x + 1 = -x + 4$. To apply the intersection-of-graphs method, we graph $y_1 = 2x + 1$ and $y_2 = -x + 4$, as shown in Figure 2.24.

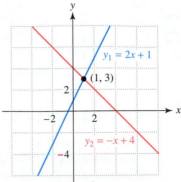

FIGURE 2.24

Their graphs intersect at the point $(1, 3)$. Since the variable in the equation $2x + 1 = -x + 4$ is x, the solution is $x = 1$, the x-coordinate of the point of intersection. When $x = 1$, the functions f and g both assume the value 3, that is, $f(1) = 2(1) + 1 = 3$ and $g(1) = -1 + 4 = 3$. The value 3 is the y-coordinate of the point of intersection $(1, 3)$.

The intersection-of-graphs method is summarized in the following.

Intersection-of-graphs method

The intersection-of-graphs method can be used to solve an equation graphically with a calculator. To implement this procedure follow these steps.

STEP 1: Set Y_1 equal to the left side of the equation, and set Y_2 equal to the right side of the equation.

STEP 2: Graph Y_1 and Y_2 in the same viewing rectangle.

STEP 3: Locate any points of intersection. The x-values of these points correspond to solutions to the equation.

EXAMPLE 1   *Applying the intersection-of-graphs method*

The percent share of music sales (in dollars) that compact discs (CDs) held from 1987 to 1998 could be modeled by $f(x) = 5.91x + 13.7$. During the same time period the percent share of music sales that cassette tapes held could be modeled by $g(x) = -4.71x + 64.7$. In these formulas $x = 0$ corresponds to 1987, $x = 1$ to 1988, and so on. Use the intersection-of-graphs method to estimate the year when sales of CDs equaled sales of cassettes.   (Source: Recording Industry Association of America.)

SOLUTION   We must solve the linear equation $f(x) = g(x)$, or equivalently,

$$5.91x + 13.7 = -4.71x + 64.7.$$

Graph $Y_1 = 5.91X + 13.7$ and $Y_2 = -4.71X + 64.7$, as shown in Figure 2.25 on the next page. In Figure 2.26 their graphs intersect near the point $(4.8, 42.1)$. Since $x = 0$ corresponds to 1987 and $1987 + 4.8 \approx 1992$, it follows that in 1992 sales of CDs and cassette tapes were approximately equal. Both shared about 42.1% of the sales in 1992.

Graphing Calculator Help

To find the point of interesection in Figure 2.26, see Appendix B (page AP-11).

[0, 12, 2] by [0, 100, 10]

[0, 12, 2] by [0, 100, 10]

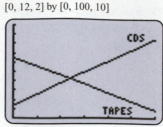

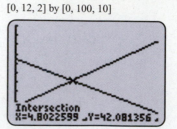

FIGURE 2.25

FIGURE 2.26

Solving Linear Equations Symbolically

To solve an equation symbolically, we transform the given equation into equivalent equations that are usually simpler. A symbolic solution provides an exact answer. This is illustrated in the next two examples.

EXAMPLE 2 *Solving a linear equation symbolically*

Solve the equation $3(x - 4) = 2x - 1$. Check your answer.

SOLUTION

$$3(x - 4) = 2x - 1 \qquad \text{Given equation}$$
$$3x - 12 = 2x - 1 \qquad \text{Distributive property}$$
$$3x - 2x = 12 - 1 \qquad \text{Subtract } 2x \text{ and add 12.}$$
$$x = 11 \qquad \text{Simplify.}$$

We can check our answer as follows.

$$3(x - 4) = 2x - 1 \qquad \text{Given equation}$$
$$3(11 - 4) \stackrel{?}{=} 2 \cdot 11 - 1 \qquad \text{Let } x = 11.$$
$$21 = 21 \qquad \text{The answer checks.}$$

EXAMPLE 3 *Solving a linear equation symbolically*

Solve $3(2x - 5) = 10 - (x + 5)$. Check your answer.

SOLUTION

$$3(2x - 5) = 10 - (x + 5) \qquad \text{Given equation}$$
$$6x - 15 = 10 - x - 5 \qquad \text{Distributive property}$$
$$6x - 15 = 5 - x \qquad \text{Simplify.}$$
$$7x - 15 = 5 \qquad \text{Add } x \text{ to both sides.}$$
$$7x = 20 \qquad \text{Add 15 to both sides.}$$
$$x = \frac{20}{7} \qquad \text{Divide both sides by 7.}$$

To check this answer let $x = \frac{20}{7}$ in the given equation and simplify.

$$3(2x - 5) = 10 - (x + 5) \qquad \text{Given equation}$$
$$3\left(2 \cdot \frac{20}{7} - 5\right) \stackrel{?}{=} 10 - \left(\frac{20}{7} + 5\right) \qquad \text{Let } x = \frac{20}{7}.$$
$$\frac{15}{7} = \frac{15}{7} \qquad \text{The answer checks.}$$

In Example 1 we solved an equation graphically and in Examples 2 and 3 we solved equations symbolically. In the next example we solve the equation presented in the introduction to this section symbolically and graphically.

EXAMPLE 4   *Solving an application symbolically and graphically*

The linear function given by $f(x) = 572.3(x - 1980) + 3617$ models average tuition at private colleges from 1980 to 2000. Solve the equation $f(x) = 14,000$ symbolically and graphically to determine the year when tuition reached $14,000.

SOLUTION   *Symbolic Solution*

$$f(x) = 14,000 \qquad \text{Given equation}$$

$$572.3(x - 1980) + 3617 = 14,000 \qquad \text{Substitute for } f(x).$$

$$572.3(x - 1980) = 14,000 - 3617 \qquad \text{Subtract 3617.}$$

$$x - 1980 = \frac{14,000 - 3617}{572.3} \qquad \text{Divide by 572.3.}$$

$$x = 1980 + \frac{14,000 - 3617}{572.3} \qquad \text{Add 1980.}$$

$$x \approx 1998.1426 \qquad \text{Approximate.}$$

Tuition at private colleges reached $14,000 in 1998.

Graphing Calculator Help

To find a point of intersection, see Appendix B (page AP-11).

[1980, 2000, 5] by [0, 18,000, 1000]

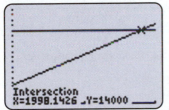

FIGURE 2.27

Graphical Solution
Graph $Y_1 = 572.3(X - 1980) + 3617$ and $Y_2 = 14,000$, as shown in Figure 2.27. Their graphs intersect near the point (1998.1426, 14,000). Notice that the graphical solution agrees with the symbolic solution.

Hand-held personal computers (PCs) are becoming increasingly popular. In the next example we estimate the year when sales of hand-held PCs could reach 12 million.

EXAMPLE 5   *Modeling hand-held PCs*

In 1998 worldwide sales of hand-held PCs reached 4 million, and in 2000 this number was 7.7 million. Use a linear function f to estimate the year when this number could reach 12 million.   (Source: Dataquest.)

SOLUTION   The graph of f must pass through the points (1998, 4.0) and (2000, 7.7), and its slope is

$$m = \frac{7.7 - 4.0}{2000 - 1998} = 1.85.$$

Thus, $f(x) = 1.85(x - 1998) + 4$ models the data. To determine when sales could reach 12 million solve the equation $f(x) = 12$.

$$1.85(x - 1998) + 4 = 12 \qquad f(x) = 12$$

$$1.85(x - 1998) = 8 \qquad \text{Subtract 4.}$$

$$x - 1998 = \frac{8}{1.85} \qquad \text{Divide by 1.85.}$$

$$x = 1998 + \frac{8}{1.85} \qquad \text{Add 1998.}$$

$$x \approx 2002.3 \qquad \text{Approximate.}$$

This linear model predicts that 12 million hand-held PCs will be sold during 2002.

Equations can be solved numerically as well as symbolically and graphically. When we solve an equation numerically, we use a table of values to estimate the solution. In the next example we approximate the year when there were 10,000 radio stations on the air. Graphical, symbolic, and numerical solutions are presented.

EXAMPLE 6   *Solving a linear equation graphically, symbolically, and numerically*

Figure 2.28 shows the graph of a linear function f that models the number of radio stations on the air from 1950 to 1995.   (Source: National Association of Broadcasters.)

(a) Visually apply the intersection-of-graphs method to estimate when 10,000 stations were on the air.

(b) A symbolic representation of f is $f(x) = 214.2(x - 1950) + 2322$. Determine an equation whose solution is the year when there were 10,000 radio stations on the air. Solve this equation symbolically.

(c) Solve the equation from part (b) numerically.

(d) Compare the graphical, symbolic, and numerical solutions.

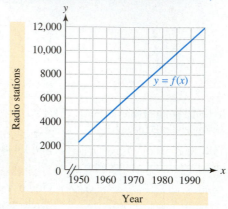

FIGURE 2.28 Radio Stations on the Air

SOLUTION   (a) *Graphical Solution*   Visually estimate the x-value where the graphs of $y = 10,000$ and $y = f(x)$ intersect as shown in Figure 2.29. This x-value is approximately 1986.

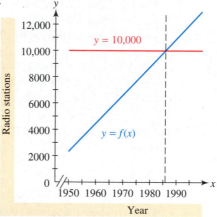

FIGURE 2.29

(b) *Symbolic Solution*   Solve the linear equation $f(x) = 10,000$.

$214.2(x - 1950) + 2322 = 10{,}000$	Given equation
$214.2(x - 1950) = 7678$	Subtract 2322 from both sides.
$x - 1950 = \dfrac{7678}{214.2}$	Divide both sides by 214.2.
$x = \dfrac{7678}{214.2} + 1950$	Add 1950 to both sides.
$x \approx 1985.845$	Approximate.

According to this model, there were approximately 10,000 radio stations on the air in 1986.

(c) *Numerical Solution*   Table $Y_1 = 214.2(X - 1950) + 2322$. Approximate the value of x where Y_1 is equal to 10,000. Figure 2.30 shows that the number of radio stations was approximately 10,000 in 1986. If desired, we could increment by 0.1 to obtain a more accurate solution. See Figure 2.31.

Graphing Calculator Help

To make a table of values, see Appendix B (page AP-11).

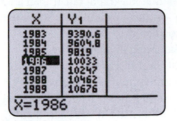

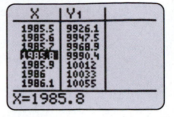

FIGURE 2.30   **FIGURE 2.31**

(d) The graphical, symbolic, and numerical solutions all give similar results.

■ **MAKING CONNECTIONS** ──────

Graphical, Numerical, and Symbolic Solutions   It is important to realize that regardless of the method used to solve an equation, the results should be approximately equal. Symbolic solutions give exact answers, whereas graphical and numerical solutions often give approximate answers. For example, if the true solution is $\frac{2}{3}$, then graphical and numerical methods may give a decimal approximation, such as 0.6667.

In Example 6, the number of radio stations on the air is modeled by a continuous linear function f. In part (c), it is assumed that since $Y_1 < 10,000$ when $x = 1985$ and $Y_1 > 10,000$ when $x = 1986$, there is an x-value between 1985 and 1986 where $Y_1 = 10,000$. This is true because the graph of Y_1 is *continuous* with no breaks. This basic concept is referred to as the *intermediate value property*.

Intermediate value property

Let (x_1, y_1) and (x_2, y_2) with $y_1 \neq y_2$ be two points on the graph of a continuous function f. Then, on the interval $x_1 \leq x \leq x_2$, f takes on every value between y_1 and y_2 at least once.

The intermediate value property is illustrated in Figure 2.32. The points $(2, -2)$ and $(7, 6)$ lie on the graph of a function f. The value 3 lies between the y-values of -2 and 6. Because f is continuous, $f(x)$ must equal 3 for some x on the interval $2 \leq x \leq 7$.

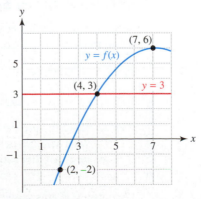

FIGURE 2.32

This *x*-value is 4, since the point (4, 3) lies on the graph of *f*. Loosely speaking, the intermediate value property is saying that we cannot draw a continuous curve that connects the points (2, −2) and (7, 6) without crossing the line *y* = 3 at least once. The only way not to cross this line would be to pick up the pencil. However, this creates a discontinuous graph, rather than a continuous one.

There are many examples of the intermediate value property. Physical motion is usually considered to be continuous. Suppose at one time a car is traveling at 20 miles per hour and at another time it is traveling at 40 miles per hour. It is logical to assume that the car traveled 30 miles per hour at least once between these times. In fact, by the intermediate value property, the car must have assumed all speeds between 20 and 40 miles per hour at least once. Similarly, if a jet airliner takes off and flies at an altitude of 30,000 feet, then by the intermediate value property, we may conclude that the airliner assumed all altitudes between ground level and 30,000 feet at least once.

Percentages.   Applications involving percentages often result in linear equations, because percentages can be computed by linear functions. Taking *P* percent of *x* is performed by $f(x) = Px$, where *P* is in decimal form. For example, to calculate 35% of *x*, let $f(x) = 0.35x$. Then 35% of $150 can be computed by $f(150) = 0.35(150) = 52.5$ or $52.50.

EXAMPLE 7   *Solving an application involving percentages*

A survey found that 76% of bicycle riders do not wear helmets. (Source: Opinion Research Corporation for Glaxo Wellcome, Inc.)

(a) Find a symbolic representation for a function that computes the number of people who do not wear helmets among *x* bicycle riders.

(b) There are approximately 38.7 million riders of all ages who do not wear helmets. Write a linear equation whose solution gives the total number of bicycle riders. Solve this equation.

SOLUTION   **(a)** A linear function *f* that computes 76% of *x* is given by $f(x) = 0.76x$.

(b) We must find the *x*-value for which $f(x) = 38.7$ million. The linear equation is $0.76x = 38.7$. Solving gives $x = \dfrac{38.7}{0.76} \approx 50.9$ million bike riders.

Problem-Solving Strategies

To become more proficient at solving problems, we need to establish a procedure to guide our thinking. The following steps may be helpful in solving application problems.

Solving application problems

STEP 1: Read the problem and make sure you understand it. Assign a variable to what you are being asked to find. If necessary, write other quantities in terms of this variable.

> **STEP 2:**   Try to write an equation that relates the quantities described in the problem. You may need to sketch a diagram and refer to known formulas.
>
> **STEP 3:**   Solve the equation and determine the solution.
>
> **STEP 4:**   Look back and check your solution. Does it seem reasonable?

These steps are applied to the next examples.

EXAMPLE 8   *Working together*

A large pump can empty a tank of gasoline in 5 hours, and a smaller pump can empty the same tank in 9 hours. If both pumps are used to empty the tank, how long will it take?

SOLUTION   **STEP 1:**   We are asked to find the time it takes for both pumps to empty the tank. Let this time be x.

$$x: \text{Time to empty the tank}$$

STEP 2:   In 1 hour the large pump will empty $\frac{1}{5}$ of the tank and the smaller pump will empty $\frac{1}{9}$ of the tank. The fraction of the tank that they will empty together in 1 hour is $\frac{1}{5} + \frac{1}{9}$. In 2 hours the large pump will empty $\frac{2}{5}$ of the tank and the smaller pump will empty $\frac{2}{9}$ of the tank. The fraction of the tank that they will empty together in 2 hours is $\frac{2}{5} + \frac{2}{9}$. Similarly, in x hours the fraction of the tank that the two pumps can empty is $\frac{x}{5} + \frac{x}{9}$. Since the tank is empty when this fraction reaches 1, we must solve the following equation.

$$\frac{x}{5} + \frac{x}{9} = 1$$

STEP 3:   Find a common denominator and then add the terms on the left side.

$$\frac{9x}{45} + \frac{5x}{45} = 1 \qquad \text{A common denominator is 45.}$$

$$\frac{14x}{45} = 1 \qquad \text{Add fractions.}$$

$$14x = 45 \qquad \text{Multiply by 45.}$$

$$x = \frac{45}{14} \approx 3.21 \qquad \text{Divide by 14 and approximate.}$$

Working together, the two pumps can empty the pool in about 3.21 hours.

STEP 4:   This sounds reasonable. Working together the two pumps should be able to empty the tank faster than the large pump working alone, but not twice as fast.

EXAMPLE 9   *Solving an application involving motion*

In 1 hour an athlete travels 10.1 miles by running at 8 miles per hour and then at 11 miles per hour. How long did the athlete run at each speed?

SOLUTION   **STEP 1:**   We are asked to find the time spent running at each speed. If we let x represent the time in hours running at 8 miles per hour, then $1 - x$ represents the time spent running at 11 miles per hour because the total running time was 1 hour.

$$x: \text{Time spent running at 8 miles per hour}$$
$$1 - x: \text{Time spent running at 11 miles per hour}$$

STEP 2:   Distance d equals rate r times time t: that is, $d = rt$. In this example we have two rates (or speeds) and two times. The total distance must sum to 10.1 miles.

$$d = r_1t_1 + r_2t_2$$
$$10.1 = 8x + 11(1 - x)$$

STEP 3:   We can solve this equation symbolically.

$10.1 = 8x + 11 - 11x$	Distributive Property.
$10.1 = 11 - 3x$	Combine like terms.
$3x = 0.9$	Add $3x$; subtract 10.1.
$x = 0.3$	Divide by 3.

The athlete runs 0.3 hour (18 minutes) at 8 miles per hour and 0.7 hour (42 minutes) at 11 miles per hour.

STEP 4:   We can check this solution as follows.

$$8(0.3) + 11(0.7) = 10.1 \qquad \text{It checks.}$$

This sounds reasonable. The runner's average speed was 10.1 miles per hour so the runner must have run longer at 11 miles per hour than at 8 miles per hour.

Similar triangles are often used in applications involving geometry. Similar triangles are used to solve the next application.

EXAMPLE 10   *Solving an application involving similar triangles*

A person 6 feet tall stands 17 feet from the base of a streetlight, as illustrated in Figure 2.33. If the person's shadow is 8 feet, estimate the height of the street light.

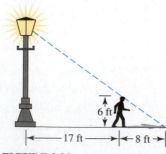

6 ft

├── 17 ft ──┤├─ 8 ft ─┤

FIGURE 2.33

SOLUTION   **STEP 1:**   We are being asked to find the height of the streetlight in Figure 2.33. Let x represent this height.

$$x: \text{Height of the streetlight}$$

STEP 2:   In Figure 2.34, triangle ACD is similar to triangle BCE. Thus, corresponding sides are proportional.

$$\frac{AD}{DC} = \frac{BE}{EC}$$

$$\frac{x}{17 + 8} = \frac{6}{8}$$

Geometry Review
To review similar triangles, see Chapter R (page R-6).

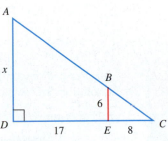

FIGURE 2.34

STEP 3:   We can solve this equation symbolically.

$$\frac{x}{25} = \frac{3}{4} \qquad \textcolor{blue}{\text{Simplify.}}$$

$$x = \frac{25 \cdot 3}{4} \qquad \textcolor{blue}{\text{Multiply by 25.}}$$

$$x = 18.75 \qquad \textcolor{blue}{\text{Simplify.}}$$

The street light is 18.75 feet tall.

STEP 4:   This sounds reasonable. The streetlight should be taller than the person.

EXAMPLE 11   *Mixing Acid in Chemistry*

Pure water is being added to a 30% solution of 153 milliliters of hydrochloric acid. How much water should be added to reduce it to a 13% mixture?

SOLUTION   **STEP 1:**   We are asked to find the amount of water that should be added to 153 milliliters of 30% acid to make it a 13% solution. Let this amount of water be equal to x.

$$x: \text{Amount of pure water to be added}$$

$$x + 153: \text{Final volume of 13\% solution}$$

STEP 2:   Since only water is added, the total amount of acid in the solution after adding the water must equal the amount of acid before the water is added. The volume of pure acid after the water is added equals 13% of $x + 153$ milliliters, and the volume of pure acid before the water is added equals 30% of 153 milliliters. We must solve the following equation.

$$0.13(x + 153) = 0.30(153)$$

STEP 3:   Begin by dividing both sides by 0.13.

$$x + 153 = \frac{0.30(153)}{0.13} \qquad \text{Divide by 0.13 and rewrite.}$$

$$x = \frac{0.30(153)}{0.13} - 153 \qquad \text{Subtract 153.}$$

$$x \approx 200.08 \qquad \text{Approximate.}$$

We should add about 200 milliliters of pure water.

STEP 4:   This sounds reasonable. If we added 153 milliliters of water we would have diluted the acid to half its concentration, which would be 15%. It follows that we should add more than 153 milliliters, but not much more.

Putting it all Together 2.3

A linear equation can be solved graphically, symbolically, and numerically. Symbolic solutions are exact, whereas graphical and numerical solutions are often approximate. For example, if the exact solution is $\frac{1}{3}$, a graphical or numerical solution might give 0.333333. The following table summarizes some important concepts in this section.

Concept	Description
Linear equation	A linear equation can be written in the form $ax + b = 0$, $a \neq 0$, and has one solution. To solve an equation symbolically we simplify the equation until we have $x = k$ for some k. *Example:* $2x - 3 = 5$ $\qquad\qquad 2x = 8$ Add 3 to both sides. $\qquad\qquad\; x = 4$ Divide both sides by 2.
Intersection-of-graphs method	This is a graphical method for solving equations. Set Y_1 equal to the left side and Y_2 equal to the right side. Find the point of intersection. The solution is the x-coordinate of this point. *Example:* $3x = x - 2$ Graph $Y_1 = 3X$ and $Y_2 = X - 2$. The point of intersection is $(-1, -3)$, so the solution is $x = -1$. $[-9, 9, 1]$ by $[-6, 6, 1]$ Intersection X=-1 Y=-3
Problem solving	Becoming a proficient problem solver takes practice. A general step-by-step procedure for solving applications is found in the subsection headed Problem-Solving Strategies.

2.3 EXERCISES

Identifying Linear and Nonlinear Equations

Exercises 1–6: Determine whether the equation is linear or nonlinear by trying to write it in the form $ax + b = 0$.

1. $3x - 1.5 = 7$

2. $100 - 23x = 20x$

3. $1.2x^2 - 3x = x^2 + 1$

4. $x^2 - 3x = x^3$

5. $7x - 55 = 3(x - 8) + x$

6. $2x^3 - 8x = 2(x^3 - 8) + 2x$

Solving Linear Equations Graphically

Exercises 7 and 8: A linear equation can be solved using the intersection-of-graphs method. Find the solution by interpreting the graph.

7.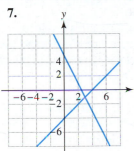

8.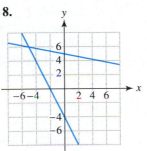

Exercises 9–18: Solve the linear equation with the intersection-of-graphs method. Approximate the solution to the nearest thousandth whenever appropriate.

9. $5x - 1.5 = 5$

10. $8 - 2x = 1.6$

11. $3x - 1.7 = 1 - x$

12. $\sqrt{2}x = 4x - 6$

13. $3.1(x - 5) = \frac{1}{5}x - 5$

14. $65 = 8(x - 6) - 5.5$

15. $\frac{6 - x}{7} = \frac{2x - 3}{3}$

16. $\pi(x - \sqrt{2}) = 1.07x - 6.1$

17. $\sqrt{5}(3 - 2.1x) = 18$

18. $3(x - 4.7) + 5(x + 1.1^2) = 15$

Solving Linear Equations Symbolically

Exercises 19–32: Complete the following.

(a) *Solve the equation symbolically.*
(b) *Check your answer whenever possible.*
(c) *Classify each equation as being a contradiction, an identity, or a conditional equation.*

19. $3x - 7 = 8$

20. $12 = 13 - 17x$

21. $5x - 1 = 5x + 4$

22. $7 - 9x = 1.5x$

23. $3(x - 1) = 5$

24. $22 = -2(2x + 1.4)$

25. $0.5(x - 2) + 5 = 0.5x + 4$

26. $\frac{1}{2}x - 2(x - 1) = 0$

27. $\frac{x + 1}{2} = x - 1$

28. $\frac{2x + 1}{3} = \frac{2x - 1}{3}$

29. $\frac{1 - 2x}{4} = \frac{3x - 1.5}{-6}$

30. $3(x - 2) + x = 2(x + 3)$

31. $0.5(x - 1980) + 5 = 10$

32. $5(x - 1995) - 15 = 65$

Exercises 33–38: If possible, solve for x.

33. $x^2 + 3x - 1 = x^2 - 2$

34. $9x - 3x^2 = 3(4 + x - x^2)$

35. $\frac{1}{2}(x - 3) = \sqrt{2}(4 + x)$

36. $\frac{2}{3}(4 - 3x) + \frac{1}{6}x = \frac{1}{3}x + 10$

37. $\pi(2x - 2.1) = 16 + 2\pi x$

38. $2x^2 + 4x - 6 = x(4 + 2x) - 6$

Solving Linear Equations Numerically

Exercises 39 and 40: Let $Y_1 = f(x)$ and $Y_2 = g(x)$, where f and g are linear functions. Use the table to solve the linear equation $f(x) = g(x)$.

39.

X	Y₁	Y₂
-3	3	7
-2	4	4
-1	5	1
0	6	-2
1	7	-5
2	8	-8
3	9	-11

X = -3

40.

X	Y₁	Y₂
0	-1	1
.1	-.8	.7
.2	-.6	.4
.3	-.4	.1
.4	-.2	-.2
.5	0	-.5
.6	.2	-.8

X = 0

Exercises 41–46: Use tables to solve the equation numerically to the nearest tenth.

41. $2x - 7.2 = 10$

42. $5.8x - 8.7 = 0$

43. $5x - 16 = 5 - x$

44. $\frac{3 - x}{2} + 2x = 96$

45. $6x + \frac{3}{4} = \frac{x - 5}{9}$

46. $8 - 2.1x = 3(x - \sqrt{2}) + \pi$

Solving Linear Equations by More than One Method

Exercises 47–50: Solve the equation (a) symbolically, (b) graphically, and (c) numerically.

47. $6x - 8 = -7x + 18$

48. $5 - 8x = 3(x - 7) + 37$

49. $5.5x - 16 = 2.3x + 8$ **50.** $\dfrac{x + 1}{3} = 3x$

Applications

51. *Income* The per capita (per person) income from 1970 to 2000 can be modeled by

$$f(x) = 604.8(x - 1970) + 2280,$$

where x is the year. Determine the year when the per capita income was $16,190. (Source: Bureau of the Census.)

52. *Median Age* The median age in the United States from 1970 to 2000 can be modeled by

$$f(x) = 0.264(x - 1970) + 27.7,$$

where x is the year. (Source: Bureau of the Census.)
(a) Graph f in [1970, 2000, 10] by [20, 40, 5]. Describe what has happened to the median age during this time period.
(b) Determine graphically when the median age reached 35 years.
(c) Solve part (b) numerically.
(d) Solve part (b) symbolically.

53. *Classroom Ventilation* Ventilation is an effective method for removing indoor air pollutants. According to the American Society of Heating, Refrigerating and Air-Conditioning Engineers (ASHRAE), a classroom should have a ventilation rate of 900 cubic feet per hour for each person in the classroom. (Source: ASHRAE.)
(a) Find a function V that computes the ventilation rate in cubic feet per hour that is necessary for a classroom containing x people.
(b) Determine the hourly ventilation rate necessary for a classroom containing 50 people.
(c) If a classroom contains 10,000 cubic feet with 50 people, how many times in 1 hour should all of the air in the classroom be replaced?
(d) In areas like bars and lounges that allow smoking, the ventilation rate should be increased to 3000 cubic feet per hour for each person. Compared to classrooms, by what factor should the ventilation rate be increased in smoking areas?

54. *Celestial Orbits* To escape the gravity of any celestial body, such as Earth or the sun, a spacecraft must attain a certain velocity, called the *escape velocity*, denoted by v_e. In order to go into a circular orbit, a slower velocity v_c is necessary. The relation between v_e and v_c is described by $v_e = \sqrt{2}v_c$. (Source: H. Karttunen, *Fundamental Astronomy*.)

(a) The velocity for a circular orbit around Earth is 17,700 miles per hour. Approximate the escape velocity for Earth.
(b) A spacecraft from Earth requires a velocity of 94,000 miles per hour in order to escape the solar system. What velocity is needed to travel in a circular orbit around the sun?

55. *Population Density* (Refer to Example 5.) In 1980 the population density of the United States was 64 people per square mile and in 1990 it was 70 people per square mile. Use a linear function to estimate when the U.S. population density reached 72.4 people per square mile.

56. *Shadow Length* (Refer to Example 10.) A person 66 inches tall is standing 15 feet from a streetlight. If the person casts a shadow 84 inches long, how tall is the streetlight?

57. *Height of a Tree* In the accompanying figure, a person 5 feet tall casts a shadow 4 feet long. A nearby tree casts a shadow that is 33 feet long. Find the height of the tree by solving a linear equation.

5 ft

4 ft

33 ft

58. Two-Cycle Engines   Two-cycle engines, used in snowmobiles, chain saws, and outboard motors, require a mixture of gasoline and oil. For certain engines the amount of oil in pints that should be added to x gallons of gasoline is computed by $f(x) = 0.16x$. (Source: Johnson Outboard Motor Company.)
(a) Why is it reasonable to expect that f is linear?
(b) Evaluate $f(3)$ and interpret the answer.
(c) How much gasoline should be mixed with 1 quart of oil?

59. Grades   In order to receive an A in a college course is it necessary to obtain an average of 90% correct on three 1-hour exams of 100 points each and on one final exam of 200 points. If a student scores 82, 88, and 91 on the 1-hour exams, what is the minimum score on the final exam that the person can receive and still earn an A?

60. Conical Water Tank   A water tank in the shape of an inverted cone has a height of 11 feet and a radius of 3.5 feet, as illustrated in the accompanying figure.

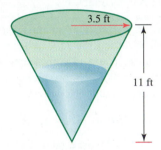

If the volume of the cone is

$$V = \frac{1}{3}\pi r^2 h,$$

find the volume of the water in the tank when the water is 7 feet deep. (*Hint*: Consider using similar triangles.)

Geometry Review To review similar triangles and to review formulas related to cones, see Chapter R (pages R-6 and R-5).

61. Deaths in the United States   There were 2.31 million deaths in the United States during the year 1997. This number represented 8.6 deaths per 1000 people. Approximate the population of the United States in 1997. (Source: Bureau of the Census.)

62. Skin Cancer   In 2000 the population of the United States was approximately 281 million and there were 54,000 cases of skin cancer reported. (Sources: American Cancer Society, Bureau of the Census.)
(a) Determine the rate of skin cancer per 1000 people.
(b) Estimate the population in a region where 15,200 cases of skin cancer were reported.

63. French Fry Production   The United States, the top producer of frozen french fries, increased its exports in 1995 by 36% over the 1994 level. In 1995, 48.5% of the 360,184 tons of french fries exported were sold to Japan. (Source: Department of Agriculture.)
(a) How many tons were exported to all countries during 1994?
(b) Determine the total amount of french fries sold to Japan in 1995.
(c) Suppose that an order of french fries contains 4 ounces. Use scientific notation to express the total number of orders of fries exported to Japan in 1995.
(d) In 1995 the population of Japan was approximately 125 million. How many orders of fries is this for each person?

64. Corporate Downsizing   In 1991 and 1992, millions of workers lost their jobs due to corporate downsizing. In a 1994 survey of these workers, 32%, or 1.76 million workers, found full-time work for equal or higher pay, 24% were jobless, while 28% were working full time for less pay. The remaining were either self-employed or working part time. How many of these workers were jobless? (Source: Department of Labor.)

65. Sale Price   A store is discounting all regularly priced merchandise by 25%. Find a function f that computes the sale price of an item having a regular price of x. If an item normally costs $56.24, what is its sale price?

66. To continue Exercise 65, use f to find the regular price of an item that costs $19.62 on sale.

67. Motion   (Refer to Example 9.) A car travels 372 miles in 6 hours, traveling at 55 miles per hour and at 70 miles per hour. How long did the car travel at each speed?

68. *Perimeter*   Find the length of the longest side of the rectangle if its perimeter is 25 feet.

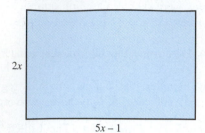

2x

5x − 1

Geometry Review   To review formulas related to rectangles, see Chapter R (page R-1).

69. *Working Together*   (Refer to Example 8.) Suppose that a lawn can be raked by one gardener in 3 hours and it can be raked by a second gardener in 5 hours.
 (a) Mentally estimate how long it will take them to rake the lawn working together.
 (b) Solve part (a) symbolically.

70. *Chemistry*   (Refer to Example 11.) Determine how much pure water should be mixed with 5 liters of a 40% solution of sulfuric acid to make a 15% solution of sulfuric acid.

71. *Mixing Candy*   Candy sells for $2.50 per pound and $4.00 per pound. A store clerk is trying to make a 5-pound mixture of candy that is worth $17.60. How much of each type of candy should be added to the mixture?

72. *Running*   At 2:00 P.M. a runner heads north on a highway jogging at 10 miles per hour. At 2:30 P.M. a driver heads north on the same highway to pick up the runner. If the car travels at 55 miles per hour, how long will it take the driver to catch the runner?

73. *Window Dimensions*   A rectangular window has a length that is 18 inches more than its width. If its perimeter is 180 inches, find its dimensions.

74. *Modeling Data*   The following data can be modeled by a linear function. Estimate the value of x when $y = 2.99$.

x	0	1	2	3
y	−5.38	−1.66	2.06	5.78

75. *Geometry*   A 174-foot long fence is being placed around the perimeter of a rectangular swimming pool with a 3-foot wide sidewalk around it. The ac-

tual swimming pool without the sidewalk is twice as long as it is wide. Find the dimensions of the pool without the sidewalk.

76. *Housing Costs*   The following table shows the average costs of a new house in the South for three selected years.

Average Sale Prices of New One-Family Homes

Year	1996	1997	1998
Price ($ thousands)	144	151	160

Source: Bureau of the Census.

 (a) Find the percent increase in sale price from 1996 to 1997 and from 1997 to 1998.
 (b) If a house was worth $200 thousand in 1998, estimate its worth in 1996.

77. *Investments*   A total of $5000 is invested in two accounts. One pays 5% annual interest, and the second pays 7% annual interest. If the interest at the end of the first year is $325, how much was invested in each account?

78. *Business*   A company manufactures compact discs with recorded music. The master disc costs $2000 to produce and all copies cost $0.45 each. If a company spends $2990 producing compact discs, how many copies did the company manufacture?

79. *Temperature Scales*   The Celsius and Fahrenheit scales are related by the equation $C = \dfrac{5}{9}(F - 32)$. These scales have the same temperature reading at a unique value, where $F = C$. Find this temperature.

80. *Indoor Air Pollution*   Formaldehyde is an indoor air pollutant found in plywood, foam insulation, and carpeting. When concentrations in the air reach 33 micrograms per cubic foot ($\mu g/ft^3$) eye irritation can occur. One square foot of new plywood can emit 140 μg per hour.   (Source: A. Hines, *Indoor Air Quality & Control.*)
 (a) A room has 100 square feet of new plywood flooring. Find a linear function f that computes the amount of formaldehyde in micrograms emitted in x hours.
 (b) The room contains 800 cubic feet of air and has no ventilation. Determine how long it takes for concentrations to reach 33 $\mu g/ft^3$.

Intermediate Value Property

81. The points $(1, -4)$ and $(5, 6)$ lie on the graph of $f(x) = 2.5x - 6.5$. Find a point $(x, 3.5)$ that lies on the graph of f. Sketch a graph to illustrate how the intermediate value property applies in this situation.

82. Sketch a graph of a function f that passes through the points $(-2, 3)$ and $(1, -2)$ but never takes on a value of 0. What must be true about the graph of f?

Exercises 83–86: Use the intermediate value property to show that $f(x) = 0$ for some x on the given interval for x.

83. $f(x) = x^2 - 5, 2 \le x \le 3$ (*Hint:* Evaluate $f(2)$ and $f(3)$ and then apply the intermediate value property.)

84. $f(x) = x^3 - x - 1, 1 \le x \le 2$

85. $f(x) = 2x^3 - 1, 0 \le x \le 1$

86. $f(x) = 4x^2 - x - 1, -1 \le x \le 0$

Writing about Mathematics

87. Describe a basic graphical method used to solve a linear equation. Give examples.

88. Describe verbally how to solve the linear equation $ax + b = 0$. What assumptions have you made about the value of a?

Extended and Discovery Exercises

1. *Geometry* Suppose that two rectangles are similar and the sides of the first rectangle are twice as long as the corresponding sides of the second rectangle.
  (a) Is the perimeter of the first rectangle twice the perimeter of the second rectangle? Explain.
  (b) Is the area of the first rectangle twice the area of the second rectangle? Explain.

2. *Geometry* Repeat the previous exercise for an equilateral triangle. Try to make a generalization. (*Hint:* The area of an equilateral triangle is $A = \frac{\sqrt{3}}{4}x^2$, where x is the length of a side.) What will happen to the circumference of a circle if the radius is doubled? What will happen to its area?

2.4 LINEAR INEQUALITIES

Inequalities • Techniques for Solving Inequalities • Compound Inequalities • Interval Notation

Introduction

If a person who weighs 143 pounds needs to purchase a life preserver for whitewater rafting, it is doubtful that there is one designed exactly for this weight. Life preservers are manufactured to support a range of body weights. A vest that is approved for weights between 120 and 160 pounds would be appropriate. Every airplane has a maximum weight allowance. It is important that this weight limit is accurately determined. However, most people feel more comfortable at takeoff if that maximum has not been reached. This is because any weight that is less than the maximum is also safe and allows for a greater margin of error. Both of these situations involve the concept of inequality.

In mathematics much effort is expended toward solving equations and determining equality. One reason for this is that equality is frequently a boundary between *greater than* and *less than*. The solution to an inequality often can be found by first locating where two expressions are equal. Since equality and inequality are closely related, many of the techniques used to solve equations also can be applied to inequalities.

Inequalities

Inequalities result whenever the equals sign in an equation is replaced with any one of the symbols $<$, $\leq$, $>$, or $\geq$. Some examples of inequalities include

$$x + 15 < 9x - 1, \qquad x^2 - 2x + 1 \geq 2x, \qquad z + 5 > 0,$$
$$xy + x^2 \leq y^3 + x, \qquad \text{and} \qquad 2 + 3 > 1.$$

The first three inequalities have one variable, the fourth inequality contains two variables, and the fifth inequality has only constants. As with linear equations, our discussion focuses on inequalities with one variable.

To **solve** an inequality means to find all values for the variable that make the inequality a true statement. Such values are **solutions** and the set of all solutions is the **solution set** of the inequality. Two inequalities are **equivalent** if they have the same solution set. It is common for an inequality to have infinitely many solutions. For instance, the inequality $x - 1 > 0$ has infinitely many solutions because any real number x satisfying $x > 1$ is a solution. The solution set is $\{x \mid x > 1\}$.

Like functions and equations, inequalities in one variable can be classified as linear or nonlinear.

Linear inequality in one variable

A **linear inequality** in one variable is an inequality that can be written in the form

$$ax + b > 0,$$

where $a \neq 0$. (The symbol $>$ may be replaced by $\geq$, $<$, or $\leq$.)

Examples of linear inequalities include

$$3x - 4 < 0, \qquad 7x + 5 \geq x, \qquad x + 6 > 23, \qquad \text{and} \qquad 7x + 2 \leq -3x + 6.$$

Using techniques from algebra, each of these inequalities can be transformed into one of the forms $ax + b > 0$, $ax + b \geq 0$, $ax + b < 0$, or $ax + b \leq 0$. For example, by subtracting x from both sides of $7x + 5 \geq x$, we obtain the equivalent inequality $6x + 5 \geq 0$.

Techniques for Solving Inequalities

Linear inequalities may be solved graphically, symbolically, and numerically. We begin our discussion with graphical solutions.

Graphical Solutions.   The intersection-of-graphs method can be extended to solve inequalities. Figure 2.35 shows the velocity of two cars in miles per hour after x minutes. V_A denotes the velocity of car A, and V_B denotes the velocity of car B.

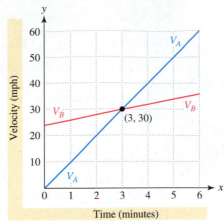

FIGURE 2.35 Velocities of Two Cars

At 3 minutes $V_A = V_B$, and both cars are traveling at 30 miles per hour. To the left of $x = 3$, the graph of V_A is below the graph of V_B so car A is traveling slower than car B. Thus

$$V_A < V_B \quad \text{when} \quad 0 \le x < 3.$$

To the right of $x = 3$ the graph of V_A is above the graph of V_B, so car A is traveling faster than car B. Thus

$$V_A > V_B \quad \text{when} \quad 3 < x \le 6.$$

This technique is used in the next example.

EXAMPLE 1   *Solving a linear inequality graphically*

Graph $y_1 = \dfrac{1}{2}x + 2$ and $y_2 = 2x - 1$ by hand. Use the graph to solve the inequality $\dfrac{1}{2}x + 2 > 2x - 1$.

SOLUTION   The graph of $y_1 = \dfrac{1}{2}x + 2$ is a line with slope $\dfrac{1}{2}$ and y-intercept 2; the graph of $y_2 = 2x - 1$ is a line with slope 2 and y-intercept -1. Their graphs are shown in Figure 2.36.

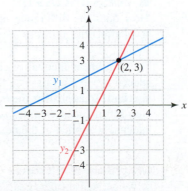

FIGURE 2.36

Note that the graphs intersect at the point (2, 3). The graph of $y_1 = \frac{1}{2}x + 2$ is above the graph of $y_2 = 2x - 1$ to the left of the point of intersection or when $x < 2$. Thus the solution set to the inequality $\frac{1}{2}x + 2 > 2x - 1$ is $\{x \mid x < 2\}$.

In the next example, we use graphical techniques to solve an application from meteorology.

EXAMPLE 2   *Using the intersection-of-graphs method*

When the air temperature reaches the dew point, fog may form. This phenomenon also causes clouds to form at higher altitudes. Both the air temperature and the dew point decrease at a constant rate as the altitude above ground level increases. If the ground-level temperature and dew point are T_0 and D_0, the air temperature can be approximated by $T(x) = T_0 - 29x$ and the dew point by $D(x) = D_0 - 5.8x$ at an altitude of x miles.

(a) If $T_0 = 75°F$ and $D_0 = 55°F$, determine the altitudes where clouds will not form. See Figure 2.37.

(b) The slopes of the graphs for the functions T and D are called *lapse rates*. Interpret their meanings. Explain how these two slopes ensure a strong likelihood of clouds forming. (Source: A. Miller and R. Anthes, *Meteorology.*)

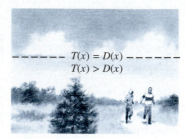

FIGURE 2.37

SOLUTION   **(a)** Since $T_0 = 75$ and $D_0 = 55$, let $T(x) = 75 - 29x$ and $D(x) = 55 - 5.8x$. Graph $Y_1 = 75 - 29X$ and $Y_2 = 55 - 5.8X$, as shown in Figure 2.38. The graphs intersect near (0.86, 50). This means that the air temperature and dew point are both 50°F at about 0.86 mile above ground level. Clouds will not form below this altitude or when the graph of Y_1 is above the graph of Y_2. The solution set is $\{x \mid 0 \leq x < 0.86\}$, where the endpoint 0.86 has been approximated.

(b) The slope of the graph of T is -29. This means that for each 1-mile increase in altitude, the air temperature *decreases* by 29°F. Similarly, the slope of the graph of D is -5.8. The dew point *decreases* by 5.8°F for every 1-mile increase in altitude. As the altitude increases, the air temperature decreases at a faster rate than the dew point. As a result, the air temperature typically cools to the dew point at higher altitudes. At or above this altitude clouds may form.

[0, 2, 1] by [0, 80, 10]

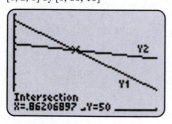

FIGURE 2.38

Graphing Calculator Help

To find a point of intersection, see Appendix B (page AP-11).

■ **CLASS DISCUSSION**

How does the difference between the air temperature and the dew point at ground level affect the altitudes at which clouds may form? Explain. ■

If a linear inequality can be written in the form $y_1 > 0$, where $>$ may be replaced by $\geq$, $\leq$, or $<$, then we can solve this inequality by using the **x-intercept method**. To apply this method for $y_1 > 0$, graph y_1 and find the x-intercept. The solution set includes x-values where the graph of y_1 is above the x-axis. The x-intercept method is applied in the next example.

EXAMPLE 3   *Applying the x-intercept method*

Solve the inequality $1 - x > \frac{1}{2}x - 2$.

SOLUTION   Start by subtracting $\frac{1}{2}x - 2$ from both sides to obtain $1 - x - \left(\frac{1}{2}x - 2\right) > 0$. Then graph $Y_1 = 1 - X - (0.5X - 2)$, as shown in Figure 2.39, where the x-intercept is 2. The graph of Y_1 is above the x-axis when $x < 2$. Therefore the solution set to $Y_1 > 0$ is $\{x \mid x < 2\}$.

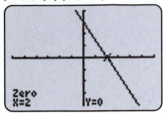

$[-6, 6, 1]$ by $[-4, 4, 1]$

FIGURE 2.39

Graphing Calculator Help

To locate a zero or x-intercept on the graph of a function, see Appendix B (page AP-12).

Symbolic Solutions.   To solve linear inequalities symbolically, the following properties of inequalities may be used.

Properties of inequalities

Let a, b, and c be real numbers.

1. $a < b$ and $a + c < b + c$ are equivalent.
   (The same number may be added to both sides of an inequality.)
2. If $c > 0$, then $a < b$ and $ac < bc$ are equivalent.
   (Both sides of an inequality may be multiplied by the same positive number.)
3. If $c < 0$, then $a < b$ and $ac > bc$ are equivalent.
   (Both sides of an inequality may be multiplied by the same negative number provided the inequality symbol is reversed.)

Note: Replacing $<$ with $\leq$ and $>$ with $\geq$ results in similar properties. The word "multiplied" may be replaced with "divided" in Properties 2 and 3.

The following examples illustrate each property.

Property 1: To solve $x - 5 < 6$, add 5 to both sides to obtain $x < 11$.

Property 2: To solve $5x < 10$, divide both sides by 5 to obtain $x < 2$.

Property 3: To solve $-5x < 10$, divide both sides by -5 to obtain $x > -2$. (Whenever you multiply or divide an inequality by a negative number, reverse the inequality sign.)

EXAMPLE 4   *Solving a linear inequality symbolically*

Use the properties of inequalities to solve each inequality.

(a) $2x - 3 < \dfrac{x + 2}{-3}$ **(b)** $-3(4x - 4) \geq 4 - (x - 1)$

SOLUTION   **(a)** Use Property 3 by multiplying each side by -3 to clear fractions. Remember to reverse the inequality sign when multiplying by a negative number.

$$2x - 3 < \frac{x + 2}{-3} \qquad \text{Given inequality}$$

$$-6x + 9 > x + 2 \qquad \text{Property 3: Multiply by } -3 \text{ and reverse the inequality sign.}$$

$$9 > 7x + 2 \qquad \text{Property 1: Add } 6x.$$

$$7 > 7x \qquad \text{Property 1: Add } -2 \text{ (or subtract 2.)}$$

$$1 > x \qquad \text{Property 2: Divide by 7.}$$

The solution set is $\{x \mid x < 1\}$. We can also express the solution set by $x < 1$.

(b) Begin by applying the distributive property.

$$-3(4x - 4) \geq 4 - (x - 1) \qquad \text{Given inequality}$$

$$-12x + 12 \geq 4 - x + 1 \qquad \text{Distributive property}$$

$$-12x + 12 \geq -x + 5 \qquad \text{Simplify.}$$

$$-12x + x \geq 5 - 12 \qquad \text{Property 1: Add } x \text{ and } -12.$$

$$-11x \geq -7 \qquad \text{Simplify.}$$

$$x \leq \frac{7}{11} \qquad \text{Property 3: Divide by } -11 \text{ and reverse inequality sign.}$$

The solution set is $\left\{ x \mid x \leq \dfrac{7}{11} \right\}$.

Numerical Solutions.   Inequalities are sometimes solved using a table of values. The following example helps to explain the mathematical concept behind this method.

Suppose that it costs $5x + 200$ dollars for a company to produce x pair of headphones and that the company receives $15x$ dollars for selling x pair of headphones. Then the profit P from selling x pair of headphones is $P = 15x - (5x + 200) = 10x - 200$. A value of $x = 20$ results in $P = 0$, and so $x = 20$ is called the **boundary number** because it represents the boundary between making money and losing money. To make money the profit must be positive, and the inequality

$$10x - 200 > 0$$

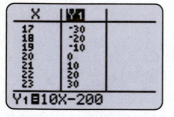

FIGURE 2.40

must be satisfied. A table of values for $Y_1 = 10X - 200$ in Figure 2.40 shows the boundary number $x = 20$ along with several **test values**. On one hand, the test values of $x = 17, 18,$ and 19 result in a negative profit and the company is losing money. On the other hand, the test values of $x = 21, 22,$ and 23 result in a positive profit. Therefore the solution set to $10x - 200 > 0$ is $\{x \mid x > 20\}$. These concepts are applied in the next example.

EXAMPLE 5   *Solving a linear inequality with test values*

Solve $3(6 - x) + 5 - 2x < 0$ numerically.

SOLUTION   Begin by making a table of $Y_1 = 3(6 - X) + 5 - 2X$, as shown in Figure 2.41. We can see that the boundary number for this inequality lies between $x = 4$ and $x = 5$. In Figure 2.42 the increment is changed from 1 to 0.1 and the boundary number for the inequality is $x = 4.6$. The test values of $x = 4.7, 4.8,$ and 4.9 indicate that when $x > 4.6$, the inequality $Y_1 < 0$ is true. The solution set is $\{x \mid x > 4.6\}$.

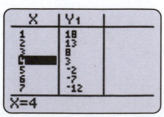

FIGURE 2.41

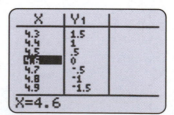

FIGURE 2.42

Compound Inequalities

Sometimes a variable must satisfy two inequalities. For example, on a freeway there may be a minimum speed limit of 40 miles per hour and a maximum speed limit of 70 miles per hour. If the variable x represents the speed of a vehicle, then x must satisfy the compound inequality

$$x \geq 40 \quad \text{and} \quad x \leq 70.$$

A compound inequality occurs when two inequalities are connected by the word *and* or *or*. When the word *and* connects two inequalities, it can sometimes be written as a **three-part inequality**. For example, the previous compound inequality may be written as the three-part inequality

$$40 \leq x \leq 70.$$

Compound inequalities involving the word *or* are discussed in the next section.

EXAMPLE 6 *Solving a three-part inequality symbolically*

Solve the inequality. Graph the solution set on the number line.

(a) $-4 \leq 5x + 1 < 21$ (b) $\dfrac{1}{2} < \dfrac{1 - 2x}{4} < 2$

SOLUTION (a) Use properties of inequalities to simplify the three-part inequality.

$-4 \leq 5x + 1 < 21$	Given inequality
$-5 \leq 5x < 20$	Add -1 to each part.
$-1 \leq x < 4$	Divide each part by 5.

The solution set is $\{x \mid -1 \leq x < 4\}$. A number line graph of the solution set is shown in Figure 2.43. Note that a bracket at $x = -1$ indicates that the endpoint is included, whereas a parenthesis at $x = 4$ indicates that the endpoint is not included.

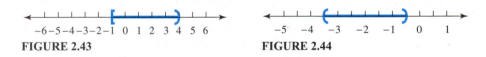

FIGURE 2.43   **FIGURE 2.44**

(b) Begin by multiplying each part by 4 to clear fractions.

$\dfrac{1}{2} < \dfrac{1 - 2x}{4} < 2$	Given inequality
$2 < 1 - 2x < 8$	Multiply by 4.
$1 < -2x < 7$	Add -1 to each part.
$-\dfrac{1}{2} > x > -\dfrac{7}{2}$	Divide by -2; reverse inequalities.
$-\dfrac{7}{2} < x < -\dfrac{1}{2}$	Rewrite the inequality.

The solution set is $\left\{ x \mid -\dfrac{7}{2} < x < -\dfrac{1}{2} \right\}$. See Figure 2.44.

Three-part inequalities occur in applications and can also be solved graphically. This is demonstrated in the next example.

EXAMPLE 7   *Solving a compound inequality graphically and symbolically*

As the altitude above ground level increases, the air temperature becomes cooler. The function given by $T(x) = T_0 - 29x$ models the Fahrenheit temperature x miles high, where T_0 is the ground temperature. (Source: A. Miller.)

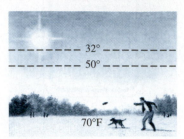

(a) Suppose $T_0 = 70°F$. Use the intersection-of-graphs method to determine the altitudes where the air temperature is from 32°F to 50°F. See Figure 2.45.

FIGURE 2.45

(b) Solve part (a) symbolically.

SOLUTION   (a) *Graphical Solution*   Let $T(x) = 70 - 29x$. The solution set consists of x-values where the compound inequality $32 \leq T(x) \leq 50$ is true. Graph $Y_1 = 32$, $Y_2 = 70 - 29X$, and $Y_3 = 50$. Their graphs intersect near the points $(0.69, 50)$ and $(1.31, 32)$, as shown in Figures 2.46 and 2.47. The air temperature is from 32°F to 50°F whenever the graph of $Y_2 = 70 - 29X$ is between the graphs of $Y_1 = 32$ and $Y_3 = 50$. Thus the solution set is $\{x \mid 0.69 \leq x \leq 1.31\}$, where the endpoints have been approximated. This means that the air temperature is between 32°F and 50°F from about 0.69 mile to 1.31 miles above ground level.

Graphing Calculator Help

To find a point of intersection, see Appendix B (page AP-11).

[0, 3, 1] by [0, 80, 10]

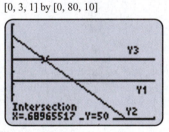

[0, 3, 1] by [0, 80, 10]

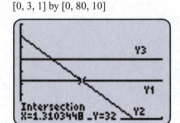

FIGURE 2.46   **FIGURE 2.47**

(b) *Symbolic Solution*   The air temperature is from 32°F to 50°F for altitudes x satisfying the following.

$$32 \leq 70 - 29x \leq 50 \qquad \text{Compound inequality}$$

$$-38 \leq -29x \leq -20 \qquad \text{Subtract 70 from each expression.}$$

$$\frac{-38}{-29} \geq x \geq \frac{-20}{-29} \qquad \text{Divide by } -29. \text{ Reverse each inequality.}$$

$$\frac{38}{29} \geq x \geq \frac{20}{29} \qquad \text{Simplify.}$$

$$\frac{20}{29} \leq x \leq \frac{38}{29} \qquad \text{Rewrite the inequality.}$$

The solution set is $\left\{ x \mid \dfrac{20}{29} \leq x \leq \dfrac{38}{29} \right\}$. Both the graphical and symbolic methods produce similar results since $\dfrac{20}{29} \approx 0.69$ and $\dfrac{38}{29} \approx 1.31$. Notice that the exact values were found using a symbolic approach rather than a graphical approach.

Interval Notation

The solution set in Example 7 consists of all real numbers x satisfying $\frac{20}{29} \le x \le \frac{38}{29}$. This solution set can be graphed using a number line, as shown in Figure 2.48.

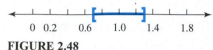

FIGURE 2.48

A convenient notation for number line graphs is called **interval notation**. Instead of drawing the entire number line, as in Figure 2.48, the solution set can be expressed as $\left[\frac{20}{29}, \frac{38}{29}\right]$. Because the solution set includes the endpoints $\frac{20}{29}$ and $\frac{38}{29}$, the interval is a **closed interval** and brackets are used. A solution set that includes all real numbers satisfying $-2 < x < 3$ would be expressed as the **open interval** $(-2, 3)$. Parentheses indicate that the endpoints are not included in the solution set. An example of a **half-open interval** is $[0, 4)$, which represents the inequality $0 \le x < 4$.

Table 2.6 provides some examples of interval notation. The symbol ∞ refers to infinity. It does not represent a real number. The notation $(1, \infty)$ means $\{x \mid x > 1\}$ or simply $x > 1$. Since this interval has no maximum x-value, ∞ is used in the position of the right endpoint. A similar interpretation holds for the symbol $-\infty$, which represents negative infinity.

TABLE 2.6   Interval Notation

Inequality	Interval Notation	Graph
$-2 < x < 2$	$(-2, 2)$ open interval	
$-1 < x \le 3$	$(-1, 3]$ half-open interval	
$-3 \le x \le 2$	$[-3, 2]$ closed interval	
$x > -3$	$(-3, \infty)$ infinite interval	
$x \le 1$	$(-\infty, 1]$ infinite interval	
$-\infty < x < \infty$ (entire number line)	$(-\infty, \infty)$ infinite interval	

■ **MAKING CONNECTIONS**

Points and Intervals   The expression $(2, 5)$ has two possible meanings. The first is that $(2, 5)$ is an ordered pair, which can be plotted as a point on the xy-plane. The second is that $(2, 5)$ represents the open interval $2 < x < 5$. To alleviate confusion, phrases like "the point $(2, 5)$" or "the interval $(2, 5)$" may be used.

EXAMPLE 8   *Solving inequalities symbolically*

Solve the linear inequalities symbolically. Express the solution set using interval notation.

(a) $-\dfrac{x}{2}+1 \le 3$ (b) $-8 < \dfrac{3x-1}{2} \le 5$ (c) $5(x-6) < 2x - 2(1-x)$

SOLUTION   (a) Simplify the inequality as follows.

$$-\frac{x}{2} + 1 \le 3 \qquad \text{Given inequality}$$

$$-\frac{x}{2} \le 2 \qquad \text{Add } -1 \text{ or subtract } 1.$$

$$x \ge -4 \qquad \text{Multiply by } -2. \text{ Reverse the inequality.}$$

In interval notation the solution set is $[-4, \infty)$.

(b) The two parts of this compound inequality can be solved simultaneously.

$$-8 < \frac{3x-1}{2} \le 5 \qquad \text{Given inequality}$$

$$-16 < 3x - 1 \le 10 \qquad \text{Multiply by } 2.$$

$$-15 < 3x \le 11 \qquad \text{Add } 1.$$

$$-5 < x \le \frac{11}{3} \qquad \text{Divide by } 3.$$

The solution set is $\left(-5, \dfrac{11}{3}\right]$.

(c) Start by applying the distributive property on both sides of the inequality.

$$5(x-6) < 2x - 2(1-x) \qquad \text{Given inequality}$$

$$5x - 30 < 2x - 2 + 2x \qquad \text{Distributive property}$$

$$5x - 30 < 4x - 2 \qquad \text{Simplify.}$$

$$x - 30 < -2 \qquad \text{Subtract } 4x.$$

$$x < 28 \qquad \text{Add } 30.$$

The solution set is $(-\infty, 28)$.

Putting it all Together

2.4

Applications involving linear functions result in both linear equations and inequalities. Like a linear equation, the solution set for a linear inequality can be found graphically, numerically, and symbolically. One common strategy for solving a linear inequality is first to locate the x-value that results in equality. This boundary number represents the boundary between *greater than* and *less than*. Using this value, the solution set for the linear inequality can be found.

The following table lists methods to solve a linear inequality in the form $h(x) > 0$ or $f(x) > g(x)$. Inequalities involving $<$, $\le$, and $\ge$ are solved in a similar manner.

Concept	Explanation	Example
Symbolic method	Use properties of inequalities to simplify $f(x) > g(x)$ to either $x > k$ or $x < k$ for some real number k.	$\frac{1}{2}x + 1 > 3 - \frac{3}{2}x$ $2x + 1 > 3$ Add $\frac{3}{2}x$. $2x > 2$ Subtract 1. $x > 1$ Divide by 2.
Intersection-of-graphs method	To solve $f(x) > g(x)$, graph $y_1 = f(x)$ and $y_2 = g(x)$. Find the point of intersection. The solution set includes x-values where the graph of y_1 is above the graph of y_2.	$\frac{1}{2}x + 1 > 3 - \frac{3}{2}x$ Graph $y_1 = \frac{1}{2}x + 1$ and $y_2 = 3 - \frac{3}{2}x$. The solution set for $y_1 > y_2$ is $x > 1$.
The x-intercept method	Write the inequality as $h(x) > 0$. Graph $y_1 = h(x)$. Solutions occur where the graph is above the x-axis.	$\frac{1}{2}x + 1 > 3 - \frac{3}{2}x$ Graph $y_1 = \frac{1}{2}x + 1 - \left(3 - \frac{3}{2}x\right)$. The solution set for $y_1 > 0$ is $x > 1$.
Numerical method	Write the inequality as $h(x) > 0$. Create a table for $y_1 = h(x)$ and find the boundary number $x = k$ such that $h(k) = 0$. Use the test values in the table to determine if the solution set is $x > k$ or $x < k$.	$\frac{1}{2}x + 1 > 3 - \frac{3}{2}x$ Table $y_1 = \frac{1}{2}x + 1 - \left(3 - \frac{3}{2}x\right)$. The solution set for $y_1 > 0$ is $x > 1$.

The following table summarizes compound inequalities and interval notation.

Concept	Explanation	Example
Compound inequality	Two inequalities connected by the words *and* or *or*	$x \leq 4$ or $x \geq 10$ $x \geq -3$ and $x < 4$ The inequality $x > 5$ and $x \leq 20$ can be written as the three-part inequality $5 < x \leq 20$.
Interval notation	An efficient notation to write solutions to inequalities	$x \leq 6$ is equivalent to $(-\infty, 6]$. $x > 3$ is equivalent to $(3, \infty)$. $2 < x \leq 5$ is equivalent to $(2, 5]$.

2.4 EXERCISES

Solving Linear Inequalities Graphically

1. *Interest* The function f computes the annual interest y on a loan of x dollars with an interest rate of 10%. The graphs of f and the horizontal line $y = 100$ are shown in the figure. Determine the loan amounts that result in the following. Express your answers verbally.

   (a) An annual interest equal to $100
   (b) An annual interest of more than $100
   (c) An annual interest of less than $100
   (d) An annual interest of $100 or more

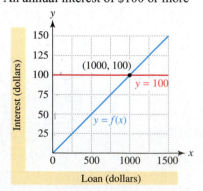

2. *U.S. Population* The function f models the population of the United States from 1970 to 1995. The graphs of f and the horizontal line $y = 226$ are shown in the figure. Use the graphs to determine when each of the following were satisfied. Express your answers verbally.

   (a) A population equal to 226 million
   (b) A population of 226 million or less
   (c) A population of 226 million or more

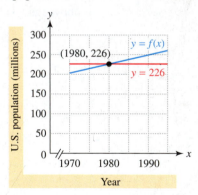

3. Use the figure for Exercise 1 to determine the x-values that satisfy the equation or inequality. Express your answers symbolically.

   (a) $f(x) = 100$
   (b) $f(x) > 100$
   (c) $f(x) < 100$
   (d) $f(x) \geq 100$

4. Use the figure for Exercise 2 to determine the x-values that satisfy the equation or inequality. Express your answers symbolically.

   (a) $f(x) = 226$
   (b) $f(x) \leq 226$
   (c) $f(x) \geq 226$

5. ***Distance between Cars*** Cars A and B are both traveling in the same direction. Their distances in miles north of St. Louis after x hours are computed by the functions f_A and f_B, respectively. The graphs of f_A and f_B are shown in the figure for $0 \le x \le 10$.

   (a) Which car is traveling faster? Explain.
   (b) How many hours elapse before the two cars are the same distance from St. Louis? How far are they from St. Louis when this occurs?
   (c) During what time interval is car B farther from St. Louis than car A?

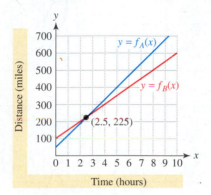

6. The graphs of two linear functions f and g are shown in the figure.

   (a) Solve the equation $g(x) = f(x)$.
   (b) Solve the inequality $g(x) > f(x)$.

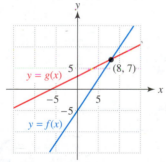

7. ***Distance*** The linear function f computes the distance y in miles between a car and the city of Omaha after x hours, where $0 \le x \le 6$. The graphs of f and the horizontal lines $y = 100$ and $y = 200$ are shown in the accompanying figure. Use the graphs to answer the following.

   (a) Is the car moving toward or away from Omaha? Explain.
   (b) Determine the times when the car is 100 miles or 200 miles from Omaha.
   (c) When is the car from 100 to 200 miles from Omaha?

   (d) When is the car's distance from Omaha greater than 100 miles?

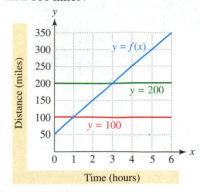

8. Use the figure to solve each equation or inequality.

   (a) $f(x) = g(x)$
   (b) $g(x) = h(x)$
   (c) $f(x) < g(x) < h(x)$
   (d) $g(x) > h(x)$

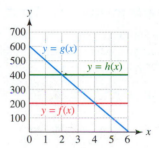

Exercises 9–14: Solve the linear inequality graphically.

9. $5x - 4 > 10$
10. $-3x + 6 \le 9$
11. $-2(x - 1990) + 55 \ge 60$
12. $\sqrt{2}x > 10.5 - 13.7x$
13. $\sqrt{5}(x - 1.2) - \sqrt{3}x < 5(x + 1.1)$
14. $1.238x + 0.998 \le 1.23(3.987 - 2.1x)$

Exercises 15–20: Solve the compound linear inequality graphically.

15. $3 \le 5x - 17 < 15$
16. $-4 < \dfrac{55 - 3.1x}{4} < 17$
17. $1.5 \le 9.1 - 0.5x \le 6.8$
18. $0.2x < \dfrac{2x - 5}{3} < 8$
19. $x - 4 < 2x - 5 < 6$
20. $-3 \le 1 - x \le 2x$

Interval Notation

Exercises 21–32: Express each of the following in interval notation.

21. $x \geq 5$

22. $x < 100$

23. $4 \leq x < 19$

24. $-4 < x < -1$

25. $x \leq -37$

26. $\{x \mid x \leq -3\}$

27. $\{x \mid -1 \leq x\}$

28. $17 > x \geq -3$

29.

30.

31.

32.

Solving Linear Inequalities Symbolically

Exercises 33–56: Solve each inequality symbolically. Express the solution set in interval notation.

33. $2x + 6 \geq 10$

34. $-4x - 3 < 5$

35. $-2(x - 10) + 1 > 0$

36. $3(x + 5) \leq 0$

37. $\dfrac{x + 2}{3} \geq 5$

38. $\dfrac{2 - x}{6} < 0$

39. $-3(x - 4) \geq 2(1 - 2x)$

40. $-\dfrac{1}{4}(2x - 6) + \dfrac{1}{2}x \geq 5$

41. $\dfrac{1 - x}{4} < \dfrac{2x - 2}{3}$

42. $\dfrac{3x}{4} < x - \dfrac{x + 2}{2}$

43. $2x - 3 > \dfrac{1}{2}(x + 1)$

44. $5 - (2 - 3x) \leq -5x$

45. $-1 \leq 2x \leq 4$

46. $5 < 4x - 1 \leq 11$

47. $3 \leq 4 - x \leq 20$

48. $-5 < 1 - 2x < 40$

49. $0 < \dfrac{7x - 5}{3} \leq 4$

50. $-7 \leq \dfrac{1 - 4x}{7} < 12$

51. $5 > 2(x + 4) - 5 > -5$

52. $\dfrac{8}{3} \geq \dfrac{4}{3} - (x + 3) \geq \dfrac{2}{3}$

53. $3 \leq \dfrac{1}{2}x + \dfrac{3}{4} \leq 6$

54. $-4 \leq 5 - \dfrac{4}{5}x < 6$

55. $3x - 1 < 2(x - 3) + 1$

56. $5x - 2(x + 3) \geq 4 - 3x$

Solving Linear Inequalities Numerically

Exercises 57–60: Assume Y_1 and Y_2 represent linear functions with the set of real numbers for their domains. Use the table to solve the inequalities.

57. $Y_1 > 0$, $Y_1 \leq 0$

58. $Y_1 < 0$, $Y_1 \geq 0$

59. $Y_1 \geq Y_2$, $Y_1 < Y_2$

60. $Y_1 > Y_2$, $Y_1 \leq Y_2$

Exercises 61–64: Solve each inequality numerically. Approximate endpoints to the nearest tenth.

61. $-4x - 6 > 0$

62. $(\sqrt{11} - \pi)x - 5.5 \leq 0$

63. $\dfrac{3x - 1}{5} < 15$

64. $1.5(x - 0.7) + 1.5x < 1$

You Decide the Method

Exercises 65–68: Solve the inequality. Write the solution set in interval notation. Approximate the endpoints of intervals to the nearest thousandth when appropriate.

65. $2x - 8 > 5$

66. $\pi x - 5.12 \leq \sqrt{2}x - 5.7(x - 1.1)$

67. $5.1x - \pi \geq \sqrt{3} - 1.7x$

68. $5 < 4x - 2.5$

Applications

69. *Temperature and Altitude*   (Refer to Example 7.) Suppose the temperature x miles above ground level is given by $T(x) = 85 - 29x$.

(a) Use the intersection-of-graphs method to estimate the altitudes where the temperature is below freezing. Assume that the domain of T is $0 \le x \le 6$.

(b) What does the x-intercept on the graph of T represent?

(c) Solve part (a) symbolically.

70. *Clouds and Temperature* (Refer to Example 2.) Suppose the ground-level temperature is 65°F and the dew point is 50°F.

(a) Use the intersection-of-graphs method to estimate the altitudes where clouds will not form.

(b) Solve part (a) symbolically.

71. *Prices of Homes* The median prices of a single-family home from 1980 to 1990 can be approximated by $P(x) = 3421x + 61{,}000$, where $x = 0$ corresponds to 1980 and $x = 10$ to 1990. (Source: Department of Commerce.)

(a) Interpret the slope of the graph of P.

(b) Estimate the years when the median price range was from \$71,000 to \$88,000.

72. *Motorcycles* The number of Harley-Davidson motorcycles manufactured between 1985 and 1995 can be approximated by $N(x) = 6409(x - 1985) + 30{,}300$, where x is the year. (Source: Harley-Davidson.)

(a) Has the demand for Harley-Davidson motorcycles increased or decreased over this time period? Explain your reasoning.

(b) Estimate the years when production was between 56,000 and 75,000.

73. *Indoor Air Pollution* Kitchen gas ranges are a source of indoor pollutants such as carbon monoxide and nitrogen dioxide. One of the most effective ways to remove contaminants from the air while cooking is to operate a range hood that is vented. If a range hood is capable of removing x liters of air per second from a kitchen, then the percentage of the contaminants that are also removed may be calculated by $f(x) = 1.06x + 7.18$ where $10 \le x \le 75$. A graph of f is shown in the accompanying figure. (Source: R. L. Rezvan, "Effectiveness of Local Ventilation in Removing Simulated Pollutants from Point Sources.")

(a) How does increasing the amount of ventilation x affect the percentage of pollutants removed from the kitchen? Interpret the slope of the graph of f.

(b) Estimate the x-values where 50% to 70% of the pollutants are removed.

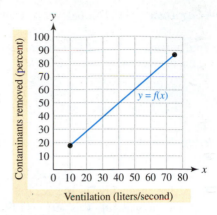

74. *Oil Consumption* From 1973 to 2005, oil consumption in million metric tons of oil equivalent (Mtoe) by the Middle East can be modeled by $f(x) = 10.1x - 19{,}904$, where x is the year. Similarly, oil consumption by Eastern Europe can be modeled by $g(x) = 2.89x - 5622$. (Source: International Energy Agency, *Global Energy: The Changing Outlook.*)

(a) Determine the year when the oil consumption by the Middle East and Eastern Europe were approximately equal.

(b) Determine the years when oil consumption by the Middle East exceeds that of Eastern Europe.

75. *Error Tolerances* Suppose that an aluminum can is manufactured so that its radius r can vary from 1.99 inches to 2.01 inches. What range of values is possible for the circumference C of the can? Express your answer using a three-part inequality.

76. *Error Tolerances* Suppose that a square picture frame has sides that vary between 9.9 inches and 10.1 inches. What range of values is possible for the perimeter P of the picture frame? Express your answer using a three-part inequality.

77. *Modeling Data* The following data are exactly linear.

(a) Find a linear function f that models the data.

(b) Solve the inequality $f(x) > 2.25$.

x	0	2	4	6
y	-1.5	4.5	10.5	16.5

78. *Modeling Data*   The data below are exactly linear.
   (a) Find a linear function f that models the data.
   (b) Solve the inequality $2 \leq f(x) \leq 8$.

x	1	2	3	4	5
y	0.4	3.5	6.6	9.7	12.8

79. *Modeling Salaries*   The following data shows a person's salary for 4 years.
   (a) Find a linear function f that models the data.
   (b) Solve the inequality $f(x) \leq 52{,}700$. Assume that the domain of f is $D = \{1997, 1998, 1999, \ldots, 2008\}$. Interpret your answer.

Year	1997	1998	1999
Salary ($)	47,975	48,650	49,325

Year	2000	2001
Salary ($)	50,000	50,675

80. *Modeling Real Data*   The following table lists numbers of visitors to national parks.

Year	1979	1985	1990	1999
Visitors (millions)	60	68	79	88

Source: National Park Service

   (a) Find a linear function f that models the data.
   (b) Solve the inequality $65 \leq f(x) \leq 75$. Interpret your answer.

Writing about Mathematics

81. Suppose the solution to the equation $ax + b = 0$ with $a > 0$ is $x = k$. Discuss how the value of k may be used to help solve the inequalities $ax + b > 0$ and $ax + b < 0$. Illustrate this graphically. How would the solution sets change if $a < 0$?

82. Describe how to numerically solve the linear inequality $ax + b \leq 0$. Give an example.

Extended and Discovery Exercises

1. *Arithmetic Mean*   The arithmetic mean of two numbers a and b is given by $\dfrac{a + b}{2}$. Use properties of inequalities to show that if $a < b$, then $a < \dfrac{a + b}{2} < b$.

2. *Geometric Mean*   The geometric mean of two numbers a and b is given by $\sqrt{ab}$. Use properties of inequalities to show that if $0 < a < b$, then $a < \sqrt{ab} < b$.

CHECKING BASIC CONCEPTS FOR SECTIONS 2.3 AND 2.4

1. Use each technique to solve the linear equation $4(x - 2) = 2(5 - x) - 3$. Compare your results.
   (a) Graphically   **(b)** Numerically
   (c) Symbolically

2. Solve the inequality $2(x - 4) > 1 - x$.

3. Solve the compound inequality $-2 \leq 1 - 2x \leq 3$. Use interval notation to express the solution set.

2.5   PIECEWISE-DEFINED FUNCTIONS

Evaluating and Graphing Piecewise-Defined Functions • The Greatest Integer Function • The Absolute Value Function • Equations and Inequalities Involving Absolute Values

Introduction

When a function f models real data, it is possible that no single formula can conveniently represent f. An example is first-class postage. In 2000 a first-class letter weighing up to one ounce cost $0.33 to mail, whereas a letter weighing 1.1 ounces cost $0.55 to mail. It is common to define a postage function in pieces that depend on the weight of the letter. Functions defined in pieces are called **piecewise-defined functions**. They typically use different formulas on various intervals of their domains. Our discussion in this section centers on piecewise-defined functions where each piece is linear.

Evaluating and Graphing Piecewise-Defined Functions

In 2000 postage rates in dollars for first-class letters weighing up to 5 ounces could be computed by the piecewise-defined function f.

$$f(x) = \begin{cases} 0.33 & \text{if } 0 < x \le 1 \\ 0.55 & \text{if } 1 < x \le 2 \\ 0.77 & \text{if } 2 < x \le 3 \\ 0.99 & \text{if } 3 < x \le 4 \\ 1.21 & \text{if } 4 < x \le 5 \end{cases}$$

The function f is not a constant function because the cost varies according to the weight x of a letter. However, each piece of f is a constant. Therefore, f is a **piecewise-constant function**. Since each constant piece is linear, f also can be called a **piecewise-linear function**.

The graph of f is composed of horizontal line segments. When $0 < x \le 1$, $f(x) = 0.33$, as shown in Figure 2.49. A small circle occurs at the point $(0, 0.33)$ because $x = 0$ is not in the domain of f. The other pieces of f are graphed similarly. Since there are breaks in the graph, f is not a continuous function. It is **discontinuous** at $x = 1, 2, 3,$ and 4. This type of function is called a **step function**.

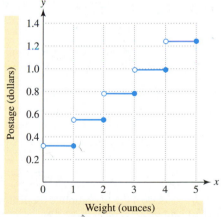

FIGURE 2.49 First-Class Postage in 2000

EXAMPLE 1 *Evaluating a graphical representation*

Figure 2.50 on the next page depicts a graph of a piecewise-linear function f. It models the amount of water in thousands of gallons in a swimming pool after x hours have elapsed.
(a) Use the graph to evaluate $f(0)$, $f(25)$, and $f(40)$. Interpret the results.
(b) Discuss how the amount of water in the pool changed. Is f a continuous function on the interval $[0, 50]$?
(c) Interpret the slope of each line segment in the graph.

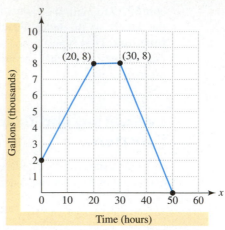

FIGURE 2.50   Water in a Swimming Pool

SOLUTION   **(a)** From Figure 2.50, when $x = 0$, $y = 2$, so $f(0) = 2$. Initially, there are 2 thousand gallons of water in the pool. The points $(25, 8)$ and $(40, 4)$ lie on the graph of f. Therefore, $f(25) = 8$ and $f(40) = 4$. After 25 hours there were 8 thousand gallons in the pool, and after 40 hours there were 4 thousand gallons.

(b) During the first 20 hours, water in the pool increases from 2 thousand to 8 thousand gallons. For the next 10 hours, the water level is constant. Between 30 and 50 hours, the pool is drained. Since there are no breaks in the graph, f is a continuous function on the interval $[0, 50]$.

(c) Slope indicates the rate at which water is entering or leaving the pool. The first line segment connects the points $(0, 2)$ and $(20, 8)$. Its slope is $m = \dfrac{8 - 2}{20 - 0} = 0.3$.

Since the units are thousands of gallons and hours, water is entering the pool during this time at a rate of 300 gallons per hour. Between 20 and 30 hours the slope of the line segment is 0. Water is neither entering nor leaving the pool. The slope of the line segment connecting the points $(30, 8)$ and $(50, 0)$ is $m = \dfrac{0 - 8}{50 - 30} = -0.4$.

On this interval, the amount of water is *decreasing* at a rate of 400 gallons per hour.

■ CLASS DISCUSSION

Suppose that a function g models the amount of water in a swimming pool. Discuss whether g must be continuous. If g were discontinuous at some point in time, what must happen to the amount of water in the pool? ■

EXAMPLE 2   *Evaluating and graphing a piecewise-defined function*

Use $f(x)$ to complete the following.

$$f(x) = \begin{cases} x - 1 & \text{if } -4 \le x < 2 \\ -2x & \text{if } \ \ 2 \le x \le 4 \end{cases}$$

(a) What is the domain of f?   **(b)** Evaluate f at $x = -3, 2, 4$, and 5.
(c) Sketch a graph of f.   **(d)** Is f a continuous function?
(e) Find all x-values where $f(x) = -4$.

SOLUTION   **(a)** Function f is defined for x-values satisfying either $-4 \le x < 2$ or $2 \le x \le 4$. Thus the domain of f is $D = \{x \mid -4 \le x \le 4\}$.

(b) For x-values satisfying $-4 \le x < 2$, $f(x) = x - 1$ and so $f(-3) = -3 - 1 = -4$. Similarly, if $2 \le x \le 4$, then $f(x) = -2x$. Thus $f(2) = -2 \cdot 2 = -4$ and

$f(4) = -2 \cdot 4 = -8$. The expression $f(5)$ is undefined because 5 is not in the domain of f.

(c) Let $y_1 = x - 1$. The graph of y_1 is a line with slope 1 and y-intercept -1. However, we must restrict this line to the region where $-4 \le x < 2$. When $x = -4$, $y = -5$, so place a dot at $(-4, -5)$. Similarly, when $x = 2$, $y_1 = 1$, so place an open circle at $(2, 1)$. Connect the points with a line segment. See Figure 2.51.

Similarly, graph the line $y_2 = -2x$ for x-values satisfying $2 \le x \le 4$ by placing dots at $(2, -4)$ and $(4, -8)$, and then connecting them with a line segment. See Figure 2.52 for the graph of f.

(d) The function f is not continuous because there is a break in its graph at $x = 2$.

(e) To determine all x-values that satisfy the equation $f(x) = -4$, visualize the line $y = -4$ on the graph of f, as shown in Figure 2.53. We can see that this line intersects the graph of f when $x = -3$ or when $x = 2$.

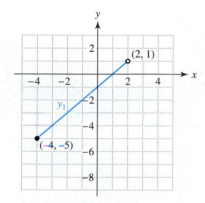

FIGURE 2.51

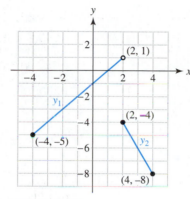

FIGURE 2.52

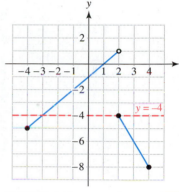

FIGURE 2.53

EXAMPLE 3   *Analyzing pollution and brown trout*

Due to acid rain, the percentage of lakes in Scandinavia that lost their population of brown trout increased dramatically between 1940 and 1975. Based on a sample of 2850 lakes, this percentage can be approximated by the piecewise-linear function f. (Source: C. Mason, *Biology of Freshwater Pollution.*)

$$f(x) = \begin{cases} \dfrac{11}{20}(x - 1940) + 7 & \text{if } 1940 \le x < 1960 \\[2mm] \dfrac{32}{15}(x - 1960) + 18 & \text{if } 1960 \le x \le 1975 \end{cases}$$

(a) Determine the percentage of lakes that lost brown trout by 1947 and by 1972.
(b) Sketch a graph of f.
(c) Is f a continuous function?

SOLUTION   **(a)** Since $1940 \le 1947 < 1960$, $f(1947)$ is calculated using the first formula.

$$f(1947) = \frac{11}{20}(1947 - 1940) + 7 = 10.85 \text{ (percent)}.$$

Similarly, $f(1972)$ is found using the second formula.

$$f(1972) = \frac{32}{15}(1972 - 1960) + 18 = 43.6 \text{ (percent)}.$$

Therefore by 1947 approximately 11% of the lakes had lost their population of brown trout. By 1972 this percentage had increased to about 44%.

(b) Because f is a piecewise-linear function, the graph of f consists of two line segments. The first segment is determined by $y = \frac{11}{20}(x - 1940) + 7$. If $x = 1940$ then $y = 7$, and if $x = 1960$ then $y = 18$. Place a dot at $(1940, 7)$ and a small circle at $(1960, 18)$, since the year 1960 is not in the interval $1940 \leq x < 1960$. These points are endpoints of the line segment.

The second line segment is determined by $y = \frac{32}{15}(x - 1960) + 18$. If $x = 1960$ then $y = 18$, and if $x = 1975$ then $y = 50$. This segment includes $(1960, 18)$ and $(1975, 50)$, so place a dot at both endpoints. Notice that this fills in the small circle at $(1960, 18)$ from the first line segment. The graph of f is shown in Figure 2.54.

(c) The graph has no breaks, so f is continuous.

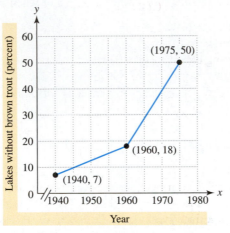

FIGURE 2.54   Lakes Losing Brown Trout

The Greatest Integer Function

A common piecewise-defined function is the greatest integer function, denoted $f(x) = [\![x]\!]$. The **greatest integer function** is defined as follows.

$$[\![x]\!] \text{ is the greatest integer less than or equal to } x.$$

Some examples that evaluate $[\![x]\!]$ include

$$[\![6.7]\!] = 6, \quad [\![3]\!] = 3, \quad [\![-2.3]\!] = -3, \quad [\![-10]\!] = -10, \quad \text{and} \quad [\![-\pi]\!] = -4.$$

The graph of $y = [\![x]\!]$ is shown in Figure 2.55 with its symbolic representation. The greatest integer function is both a piecewise-constant function and a step function.

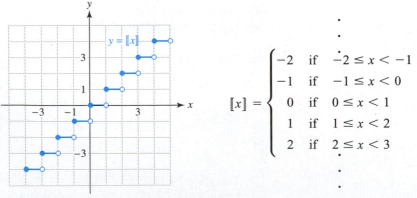

$$[\![x]\!] = \begin{cases} \vdots \\ -2 & \text{if} & -2 \leq x < -1 \\ -1 & \text{if} & -1 \leq x < 0 \\ 0 & \text{if} & 0 \leq x < 1 \\ 1 & \text{if} & 1 \leq x < 2 \\ 2 & \text{if} & 2 \leq x < 3 \\ \vdots \end{cases}$$

FIGURE 2.55   The Greatest Integer Function

Graphing Calculator Help

To access the greatest integer function, see Appendix B (page AP-12).

[0, 10, 1] by [31, 35, 1]

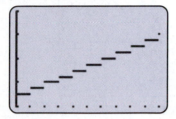

FIGURE 2.56

Graphing Calculator Help

To set a calculator in dot mode, see Appendix B (page AP-13).

[0, 10, 1] by [31, 35, 1]

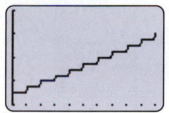

FIGURE 2.57

In some applications, fractional parts are either not allowed or ignored. Framing lumber for houses is measured in 2-foot multiples, and mileage charges for rental cars may be calculated to the mile.

Suppose a car rental company charges $31.50 per day plus $0.25 for each mile driven, where fractions of a mile are ignored. The function given by $f(x) = 0.25[\![x]\!] + 31.50$ calculates the cost of driving x miles in one day. For example, the cost of driving 100.4 miles is

$$f(100.4) = 0.25[\![100.4]\!] + 31.50 = 0.25(100) + 31.50 = \$56.50.$$

On some calculators and computers, the greatest integer function is denoted by int(X). A graph $Y_1 = 0.25*int(X) + 31.5$ is shown in Figure 2.56.

■ **MAKING CONNECTIONS**

Connected and Dot Modes   It is common for graphing calculators to connect points in order to make a graph look continuous. However, if a graph has breaks in it, a graphing calculator may connect points where there should be breaks. In dot mode, points are plotted but not connected. Figure 2.57 is the same graph shown in Figure 2.56, except that it is plotted in connected mode. Notice that in this instance connected mode generates an inaccurate graph.

The Absolute Value Function

The graph of $y = |x|$ is shown in Figure 2.58. It is V-shaped and cannot be represented by a single linear function. However, it can be represented by the lines $y = x$ (when $x \geq 0$) and $y = -x$ (when $x < 0$.) This suggests that the absolute value function can be defined symbolically using a piecewise-linear function.

$$|x| = \begin{cases} -x & \text{if } x < 0 \\ x & \text{if } x \geq 0 \end{cases}$$

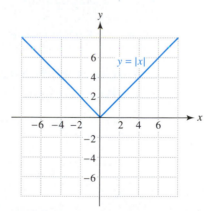

FIGURE 2.58   The Absolute Value Function

■ **EXAMPLE 4**   *Analyzing the graph of $y = |ax + b|$*

For each linear function f, graph $y = f(x)$ and $y = |f(x)|$ separately. Discuss how the absolute value affects the graph of f.
(a) $f(x) = x + 2$ **(b)** $f(x) = -2x + 4$

SOLUTION   **(a)** On some graphing calculators, the absolute value function is denoted abs(X). The graphs of $Y_1 = X + 2$ and $Y_2 = $ abs(X + 2) are shown in Figures 2.59 and 2.60 respectively, on the next page. The graph of Y_1 is a line with x-intercept -2. The graph of Y_2 is V-shaped. The graphs are identical to the right of the x-intercept. To the left of the x-intercept, the graph of $Y_1 = f(x)$ passes below the x-axis, and the graph of $Y_2 = |f(x)|$ is the reflection of $Y_1 = f(x)$ across the x-axis. The graph of $Y_2 = |f(x)|$ does not dip below the x-axis, because an absolute value is never negative.

Graphing Calculator Help

To access the absolute value function, see Appendix B (page AP-13).

$[-9, 9, 1]$ by $[-6, 6, 1]$

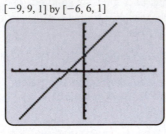

FIGURE 2.59

$[-9, 9, 1]$ by $[-6, 6, 1]$

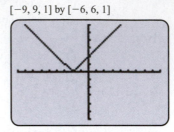

FIGURE 2.60

(b) The graphs of $Y_1 = -2X + 4$ and $Y_2 = \text{abs}(-2X + 4)$ are shown in Figures 2.61 and 2.62, respectively. Again, the graph of Y_2 is V-shaped. The graph of $Y_2 = |f(x)|$ is the reflection of f across the x-axis whenever the graph of $Y_1 = f(x)$ is below the x-axis.

$[-9, 9, 1]$ by $[-6, 6, 1]$

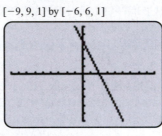

FIGURE 2.61

$[-9, 9, 1]$ by $[-6, 6, 1]$

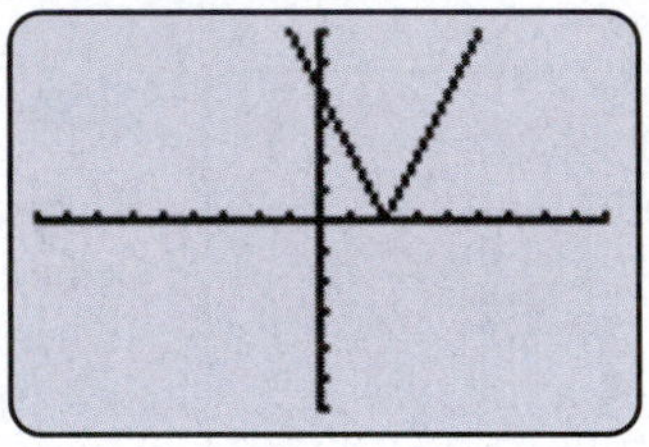

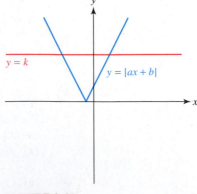

FIGURE 2.62

■ **CLASS DISCUSSION**

Graph $f(x) = \sqrt{x^2}$ in $[-15, 15, 1]$ by $[-10, 10, 1]$. Find a second symbolic representation for f. Explain your reasoning. ■

Equations and Inequalities Involving Absolute Values

The equation $|x| = 5$ has two solutions, $x = \pm 5$. This is shown visually in Figure 2.63, where the graph of $y = |x|$ intersects the horizontal line $y = 5$ at the points $(\pm 5, 5)$.

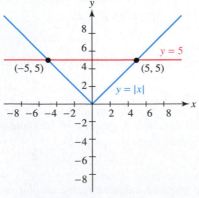

FIGURE 2.63

FIGURE 2.64

In general, the graph of $y = |ax + b|$ with $a \neq 0$ is V-shaped. It intersects the horizontal line $y = k$ twice whenever $k > 0$, as illustrated in Figure 2.64. Thus there are two solutions to the equation $|ax + b| = k$.

Absolute value equations

Let k be a positive number.

$|ax + b| = k$ is equivalent to $ax + b = \pm k$.

EXAMPLE 5   *Solving an equation involving absolute value*

Solve the equation $|2x + 5| = 2$ graphically, numerically, and symbolically.

SOLUTION   ***Graphical Solution***   Graph $Y_1 = \text{abs}(2X + 5)$ and $Y_2 = 2$. The V-shaped graph of Y_1 intersects the horizontal line at the points $(-3.5, 2)$ and $(-1.5, 2)$, as shown in Figures 2.65 and 2.66. The solutions are -3.5 and -1.5.

Numerical Solution   Table $Y_1 = \text{abs}(2X + 5)$ and $Y_2 = 2$, as in Figure 2.67. The solutions to $Y_1 = Y_2$ are -3.5 and -1.5.

$[-9, 9, 1]$ by $[-6, 6, 1]$ $[-9, 9, 1]$ by $[-6, 6, 1]$

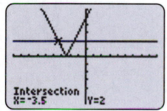   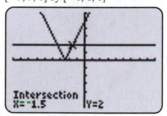

FIGURE 2.65   **FIGURE 2.66**   **FIGURE 2.67**

Graphing Calculator Help

To find a point of intersection, see Appendix B (page AP-11).

Symbolic Solution   The equation $|2x + 5| = 2$, is satisfied when $2x + 5 = \pm 2$.

$$2x + 5 = 2 \qquad \text{or} \qquad 2x + 5 = -2$$
$$2x = -3 \qquad \text{or} \qquad 2x = -7$$
$$x = -\frac{3}{2} \qquad \text{or} \qquad x = -\frac{7}{2}$$

Sometimes more than one absolute value sign can occur in an equation. For example, an equation might be in the form

$$|ax + b| = |cx + d|.$$

In this case there are two possibilities: either

$$ax + b = cx + d \qquad \text{or} \qquad ax + b = -(cx + d).$$

This symbolic technique is demonstrated in the next example.

EXAMPLE 6   *Solving an equation involving two absolute values*

Solve the equation $|x - 2| = |1 - 2x|$.

SOLUTION   We must solve both of the following equations.

$$x - 2 = 1 - 2x \qquad \text{or} \qquad x - 2 = -(1 - 2x)$$

$$3x = 3 \qquad\qquad \text{or} \qquad x - 2 = -1 + 2x$$

$$x = 1 \qquad\qquad \text{or} \qquad -1 = x$$

There are two solutions: $x = -1$ or $x = 1$.

In Figure 2.68 the solutions to $|ax + b| = k$ are labeled s_1 and s_2. The V-shaped graph of $y = |ax + b|$ is below the horizontal line $y = k$ between s_1 and s_2 or when $s_1 < x < s_2$. The solution set for the inequality $|ax + b| < k$ is green on the x-axis. In Figure 2.69 the V-shaped graph is above the horizontal line $y = k$ left of s_1 or right of s_2. That is, when $x < s_1$ or $x > s_2$. The solution set for the inequality $|ax + b| > k$ is green on the x-axis. Notice that in both figures, equality (determined by s_1 and s_2) is the boundary between *greater than* and *less than*. For this reason, s_1 and s_2 are called *boundary numbers*.

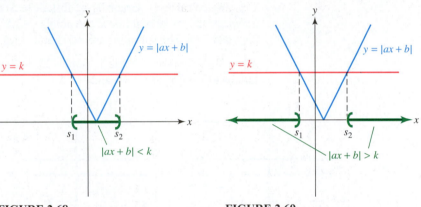

FIGURE 2.68   **FIGURE 2.69**

For example, the graphs of $y = |x + 1|$ and $y = 2$ are shown in Figure 2.70. These graphs intersect at the points $(-3, 2)$ and $(1, 2)$. It follows that the two solutions to

$$|x + 1| = 2$$

are $s_1 = -3$ and $s_2 = 1$. The solutions to $|x + 1| < 2$ lie between $s_1 = -3$ and $s_2 = 1$, which can be written as $-3 < x < 1$. Furthermore, the solutions to $|x + 1| > 2$ lie "outside" $s_1 = -3$ and $s_2 = 1$. This can be written as $x < -3$ or $x > 1$.

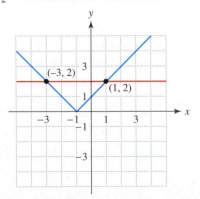

FIGURE 2.70

These results are generalized on the next page.

Absolute value inequalities

Let the solutions to $|ax + b| = k$ be s_1 and s_2, where $s_1 < s_2$ and $k > 0$.

1. $|ax + b| < k$ is equivalent to $s_1 < x < s_2$.

2. $|ax + b| > k$ is equivalent to $x < s_1$ or $x > s_2$.

Similar statements can be made for inequalities involving $\leq$ or $\geq$.

Note: The **union symbol** $\cup$ may be used to write $x < s_1$ or $x > s_2$ in interval notation. For example, $x < -2$ or $x > 5$ is written as $(-\infty, -2) \cup (5, \infty)$ in interval notation. This indicates that the solution set includes all real numbers in either $(-\infty, -2)$ or $(5, \infty)$.

EXAMPLE 7 *Analyzing the temperature range in Santa Fe*

The inequality $|T - 49| \leq 20$ describes the range of monthly average temperatures T in degrees Fahrenheit for Santa Fe, New Mexico. (Source: A. Miller and J. Thompson, *Elements of Meteorology.*)

(a) Solve this inequality graphically and symbolically.

(b) The high and low monthly average temperatures satisfy the absolute value equation $|T - 49| = 20$. Use this fact to interpret the results from part (a).

SOLUTION

Graphing Calculator Help

To find a point of intersection, see Appendix B (page AP-11).

(a) *Graphical Solution*   Graph $Y_1 = \text{abs}(X - 49)$ and $Y_2 = 20$, as in Figure 2.71. The V-shaped graph of Y_1 intersects the horizontal line at the points $(29, 20)$ and $(69, 20)$. See Figures 2.72 and 2.73. The graph of Y_1 is below the graph of Y_2 between these two points. Thus, the solution set consists of all temperatures T satisfying $29 \leq T \leq 69$.

[20, 80, 5] by [0, 35, 5]

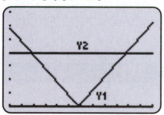

FIGURE 2.71

[20, 80, 5] by [0, 35, 5]

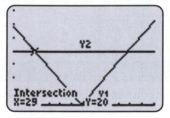

FIGURE 2.72

[20, 80, 5] by [0, 35, 5]

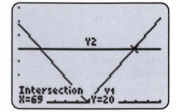

FIGURE 2.73

Symbolic Solution   First solve the related equation $|T - 49| = 20$.

$$T - 49 = -20 \quad \text{or} \quad T - 49 = 20$$
$$T = 29 \quad \text{or} \quad T = 69$$

Thus by our previous discussion, the inequality $|T - 49| \leq 20$ is equivalent to $29 \leq T \leq 69$.

(b) The solutions to $|T - 49| = 20$ are 29 and 69. Therefore, the monthly average temperatures in Sante Fe vary between a low of 29°F (January) and a high of 69°F (July). The monthly averages are always within 20 degrees of 49°F.

EXAMPLE 8 *Solving inequalities involving absolute values symbolically*

Solve the inequality symbolically. Write the solution set in interval notation.

(a) $|2x - 5| \leq 6$ **(b)** $|5 - x| > 3$

SOLUTION   (a) Begin by solving $2x - 5 = \pm 6$.

$$2x - 5 = 6 \quad \text{or} \quad 2x - 5 = -6$$
$$2x = 11 \quad \text{or} \quad 2x = -1$$
$$x = \frac{11}{2} \quad \text{or} \quad x = -\frac{1}{2}$$

The solutions to $|2x - 5| = 6$ are $-\frac{1}{2}$ and $\frac{11}{2}$. The solution set for the inequality $|2x - 5| \leq 6$ includes all real numbers x satisfying $-\frac{1}{2} \leq x \leq \frac{11}{2}$. In interval notation this is written as $\left[-\frac{1}{2}, \frac{11}{2}\right]$.

(b) Begin by solving $5 - x = \pm 3$.

$$5 - x = 3 \quad \text{or} \quad 5 - x = -3$$
$$-x = -2 \quad \text{or} \quad -x = -8$$
$$x = 2 \quad \text{or} \quad x = 8$$

The solutions to $|5 - x| = 3$ are 2 and 8. The solution set for $|5 - x| > 3$ includes all real numbers x left of 2 and right of 8. Thus $|5 - x| > 3$ is equivalent to $x < 2$ or $x > 8$. In interval notation this is written as $(-\infty, 2) \cup (8, \infty)$.

Putting it all Together 2.5

*P*iecewise-defined functions are used when it is not convenient to define a function using one formula. The greatest integer function and the absolute value function are two examples of piecewise-defined functions.

The following table summarizes some important concepts from this section.

Concept	Explanation	Examples
Piecewise-defined function	A function is piecewise-defined if it has different formulas on different intervals of its domain.	$f(x) = \begin{cases} 2x - 3 & \text{if } -3 \leq x \leq 1 \\ x + 5 & \text{if } 1 \leq x \leq 5 \end{cases}$ When $x = 2$ then $f(x) = x + 5$, so $f(2) = 2 + 5 = 7$.
Absolute value function	$f(x) = |x|$	$f(4) = |4| = 4$ $f(-2) = |-2| = 2$ The graph of $y = |x|$ is V-shaped and never goes below the x-axis.

Concept	Explanation	Examples
Absolute value equations	$\lvert ax + b \rvert = k$ with $k > 0$ is equivalent to $ax + b = \pm k$.	Solve $\lvert 3x - 5 \rvert = 4$ $3x - 5 = 4$ or $3x - 5 = -4$ $3x = 9$ or $3x = 1$ $x = 3$ or $x = \dfrac{1}{3}$
	If $k < 0$ then the absolute value equation has no solutions.	$\lvert 4x + 9 \rvert = -2$ has no solutions.
Absolute value inequalities	Let the solutions to $\lvert ax + b \rvert = k$, $k > 0$, be s_1 and s_2 with $s_1 < s_2$. The solution set to $\lvert ax + b \rvert < k$ is $s_1 < x < s_2$ and the solution set to $\lvert ax + b \rvert > k$ is $x < s_1$ or $x > s_2$.	To solve $\lvert x - 5 \rvert < 4$ and $\lvert x - 5 \rvert > 4$, first solve $x - 5 = 4$ and $x - 5 = -4$. The solutions are $x = 1$ and 9. Thus the solution set for $\lvert x - 5 \rvert < 4$ is $1 < x < 9$, and the solution set for $\lvert x - 5 \rvert > 4$ is $x < 1$ or $x > 9$.

2.5   EXERCISES

Evaluating and Graphing Piecewise-Defined Functions

1. **Speed Limits**   The graph of $y = f(x)$ gives the speed limit y along a rural highway after traveling x miles.
   (a) What are the maximum and minimum speed limits along this stretch of highway?
   (b) Estimate the miles of highway with a speed limit of 55 miles per hour.
   (c) Evaluate $f(4)$, $f(12)$, and $f(18)$.
   (d) At what x-values is the graph discontinuous? Interpret each discontinuity.

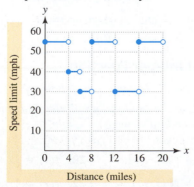

2. **ATM**   The graph of $y = f(x)$ depicts the amount of money y in dollars in an automatic teller machine (ATM) after x minutes.
   (a) Determine the initial and final amounts of money in the ATM.
   (b) Evaluate $f(10)$ and $f(50)$. Is f continuous?
   (c) How many withdrawals occurred during this time period?
   (d) When did the largest withdrawal occur? How much was it?
   (e) How much was deposited into the machine?

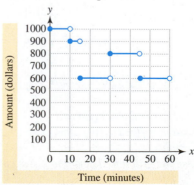

3. *Swimming Pool Levels* The graph of $y = f(x)$ represents the amount of water in thousands of gallons remaining in a swimming pool after x days.

(a) Estimate the initial and final amounts of water contained in the pool.

(b) When did the amount of water in the pool remain constant?

(c) Approximate $f(2)$ and $f(4)$.

(d) At what rate was water being drained from the pool when $1 \le x \le 3$?

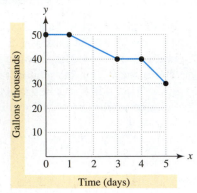

4. *Gasoline Usage* The graph shows the gallons of gasoline y in the gas tank of a car after x hours.

(a) Estimate how much gasoline was in the gas tank when $x = 3$.

(b) Interpret the graph.

(c) During what time interval did the car burn gasoline at the fastest rate?

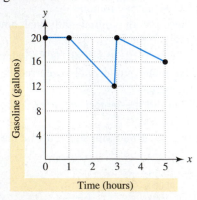

Exercises 5–10: (Refer to Example 2.) Complete the following for $f(x)$.

(a) Determine the domain of f.

(b) Evaluate $f(-2)$, $f(0)$, and $f(3)$.

(c) Graph f.

(d) Use the graph to determine if f is continuous.

5. $f(x) = \begin{cases} 2 & \text{if } -5 \le x \le -1 \\ x + 3 & \text{if } -1 < x \le 5 \end{cases}$

6. $f(x) = \begin{cases} 2x + 1 & \text{if } -3 \le x < 0 \\ x - 1 & \text{if } 0 \le x \le 3 \end{cases}$

7. $f(x) = \begin{cases} 3x & \text{if } -1 \le x < 1 \\ x + 1 & \text{if } 1 \le x \le 2 \end{cases}$

8. $f(x) = \begin{cases} -2 & \text{if } -6 \le x < -2 \\ 0 & \text{if } -2 \le x < 0 \\ 3x & \text{if } 0 \le x \le 4 \end{cases}$

9. $f(x) = \begin{cases} x & \text{if } -3 \le x \le -1 \\ 1 & \text{if } -1 < x < 1 \\ 2 - x & \text{if } 1 \le x \le 3 \end{cases}$

10. $f(x) = \begin{cases} 3 & \text{if } -4 \le x \le -1 \\ x - 2 & \text{if } -1 < x \le 2 \\ 0.5x & \text{if } 2 < x \le 4 \end{cases}$

Exercises 11–14: Use $f(x)$ to complete the following.

$$f(x) = \begin{cases} 3x - 1 & \text{if } -5 \le x < 1 \\ 4 & \text{if } 1 \le x \le 3 \\ 6 - x & \text{if } 3 < x \le 5 \end{cases}$$

11. Evaluate f at $x = -3, 1, 2$, and 5.

12. On what interval is f constant?

13. Sketch a graph of f. Is f continuous?

14. Find the x-value(s) where $f(x) = 2$.

Exercises 15–18: Use $g(x)$ to complete the following.

$$g(x) = \begin{cases} -2x - 6 & \text{if } -8 \le x \le -2 \\ x & \text{if } -2 < x < 2 \\ 0.5x + 1 & \text{if } 2 \le x \le 8 \end{cases}$$

15. Evaluate g at $x = -8, -2, 2$, and 8.

16. For what x-values does $g(x)$ have a positive slope?

17. Sketch a graph of g. Is g continuous?

18. Find the x-value(s) where $g(x) = 0$.

Greatest Integer Function

Exercises 19–22: Complete the following.

(a) Use dot mode to graph the function f in the standard viewing rectangle.

(b) Evaluate $f(-3.1)$ and $f(1.7)$.

19. $f(x) = [\![2x - 1]\!]$ **20.** $f(x) = [\![x + 1]\!]$

21. $f(x) = 2[\![x]\!] + 1$ **22.** $f(x) = [\![-x]\!]$

23. *Lumber Costs*   Lumber that is used to frame walls of houses is frequently sold in multiples of 2 feet. If the length of a board is not exactly a multiple of 2 feet, there is often no charge for the additional length. For example, if a board measures at least 8 feet but less than 10 feet, then the consumer is charged for only 8 feet.

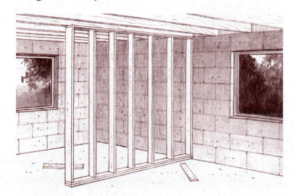

   (a) Suppose that the cost of lumber is $0.80 every 2 feet. Find a symbolic representation of a function f that computes the cost of a board x feet long for $6 \le x \le 18$.
   (b) Graph f.
   (c) Determine the costs of boards with lengths of 8.5 feet and 15.2 feet.

24. *Cellular Phone Bills*   Suppose that the charges for a cellular phone call are $0.50 for the first minute and $0.25 for each additional minute. Assume that a fraction of a minute is rounded up.
   (a) Determine the cost of a phone call lasting 3.5 minutes.
   (b) Find a symbolic representation for a function f that computes the cost of a telephone call x minutes long, where $0 < x \le 5$. (*Hint:* Express f as a piecewise-constant function.)
   (c) The greatest integer function is not convenient to use in part (b). Explain why.

Absolute Value Equations and Inequalities

Exercises 25–30: (Refer to Example 4.)
   (a) Graph $y = f(x)$.
   (b) Use the graph of $y = f(x)$ to sketch a graph of the equation $y = |f(x)|$.
   (c) Determine the x-intercept for the graph of $y = |f(x)|$.

25. $y = 2x$ **26.** $y = 0.5x$

27. $y = 3x - 3$ **28.** $y = 2x - 4$

29. $y = 6 - 2x$ **30.** $y = 2 - 4x$

Exercises 31 and 32: Let $f(x) = |ax + b|$ where $a \ne 0$ and $g(x) = k$, where $k > 0$ is a constant. The points where the graphs of f and g intersect are shown in the viewing rectangles. Solve each equation and inequality.

31. (a) $f(x) = g(x)$ **(b)** $f(x) < g(x)$
     (c) $f(x) > g(x)$

$[-10, 10, 1]$ by $[-10, 10, 1]$ $[-10, 10, 1]$ by $[-10, 10, 1]$

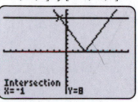

32. (a) $f(x) = g(x)$ **(b)** $f(x) \le g(x)$
     (c) $f(x) \ge g(x)$

$[-6, 12, 1]$ by $[-6, 6, 1]$ $[-6, 12, 1]$ by $[-6, 6, 1]$

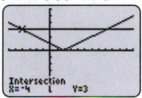

Exercises 33–36: Solve the equation
   (a) graphically,
   (b) numerically, and
   (c) symbolically.

Then solve the related inequality.

33. $|2x - 5| = 10,$ $|2x - 5| < 10$

34. $|3x - 4| = 8,$ $|3x - 4| \le 8$

35. $|5 - 3x| = 2,$ $|5 - 3x| > 2$

36. $|4x - 7| = 5,$ $|4x - 7| \ge 5$

Exercises 37–42: Solve the equation symbolically. Then solve the related inequality.

37. $|2.1x - 0.7| = 2.4,$ $|2.1x - 0.7| \ge 2.4$

38. $\left|\dfrac{1}{2}x - \dfrac{3}{4}\right| = \dfrac{7}{4},$ $\left|\dfrac{1}{2}x - \dfrac{3}{4}\right| \le \dfrac{7}{4}$

39. $|3x| + 5 = 6,$ $|3x| + 5 > 6$

40. $|x| - 10 = 25,$ $|x| - 10 < 25$

41. $\left|\dfrac{2}{3}x - \dfrac{1}{2}\right| = -\dfrac{1}{4},$ $\left|\dfrac{2}{3}x - \dfrac{1}{2}\right| \le -\dfrac{1}{4}$

42. $|5x - 0.3| = -4,$ $|5x - 0.3| > -4$

Exercises 43–50: Solve the absolute value equation.

43. $|-2x| = 4$ **44.** $|3x| = -6$

45. $|4x - 5| + 3 = 2$

46. $|4.5 - 2x| + 1.1 = 9.7$

47. $|2x - 9| = |8 - 3x|$

48. $|x - 3| = |8 - x|$

49. $\left|\frac{3}{4}x - \frac{1}{4}\right| = \left|\frac{3}{4} - \frac{1}{4}x\right|$

50. $|20x - 40| = |80 - 20x|$

You Decide the Method

Exercises 51–62: Solve the inequality. Write the solution in interval notation.

51. $|3x - 1| < 8$

52. $|15 - x| < 7$

53. $|7 - 4x| \leq 11$

54. $|-3x + 1| \leq 5$

55. $|0.5x - 0.75| < 2$

56. $|2.1x - 5| \leq 8$

57. $|2x - 3| > 1$

58. $|5x - 7| > 2$

59. $|-3x + 8| \geq 3$

60. $|-7x - 3| \geq 5$

61. $|0.25x - 1| > 3$

62. $|-0.5x + 5| \geq 4$

Sketching Graphs

63. Sketch a graph that depicts the amount of water in a 100-gallon tank. The tank is initially empty and then filled at a rate of 5 gallons per minute. Immediately after it is full, a pump is used to empty the tank at 2 gallons per minute.

64. Sketch a graph showing the distance that a person is from home after x hours if that individual drives on a straight road at 40 mph to a park 20 miles away, remains at the park for 2 hours, and then returns home at a speed of 20 mph.

65. Sketch a graph showing the amount of money y that a person makes after working x hours in one day, where $0 \leq x \leq 12$. Assume that the worker earns \$8 per hour and receives time-and-a-half pay for all hours over eight.

66. Suppose that two cars, both traveling at a constant speed of 60 miles per hour, approach each other on a straight highway. If they are initially 4 miles apart, sketch a graph of the distance between the cars after x minutes, where $0 \leq x \leq 4$.

Applications

Exercises 67–72: Average Temperatures (Refer to Example 7.) The inequality describes the range of monthly average temperatures T in degrees Fahrenheit at a certain location.

   (a) Solve the inequality.

   (b) If the high and low monthly average temperatures satisfy equality, interpret the inequality.

67. $|T - 43| \leq 24$, Marquette, Michigan

68. $|T - 62| \leq 19$, Memphis, Tennessee

69. $|T - 50| \leq 22$, Boston, Massachusetts

70. $|T - 10| \leq 36$, Chesterfield, Canada

71. $|T - 61.5| \leq 12.5$, Buenos Aires, Argentina

72. $|T - 43.5| \leq 8.5$, Punta Arenas, Chile

73. *Shoe Size* Professional basketball player Shaquille O'Neal is 7 feet 1 inch tall and weighs 325 pounds. The table lists his shoe sizes at certain ages. (Source: *USA Today.*)

Age	20	21	22	23
Shoe Size	19	20	21	22

  (a) Write a formula that gives his shoe size at age $x = 20, 21, 22,$ and 23.

  (b) Suppose that after age 23 his shoe size did not change. Sketch a graph of a continuous, piecewise-linear function f that models his shoe size from ages 20 to 26.

74. *ACT Scores* The table lists the average composite scores on the American College Test (ACT) for selected years. (Source: The American College Testing Program.)

Year	1975	1980	1985	1990	1995	1999
Score	18.6	18.5	18.6	20.6	20.8	21.0

   (a) Make a line graph of the data in [1970, 2000, 5] by [15, 25, 1].

   (b) If this graph represents a piecewise-linear function f, find a symbolic representation for the piece of f on the interval [1990, 1995].

   (c) Evaluate $f(1993)$. (The actual average composite score in 1993 was 20.7.)

75. *Violent Crimes*   The table lists victims of violent crime per 1000 people in 1997 by age group using interval notation.   (Source: Department of Justice.)

Age	[12, 16)	[16, 19)	[19, 24)	[24, 35)
Crime Rate	88	96	68	47

Age	[35, 50)	[50, 65)	[65, 90)
Crime Rate	32	15	4

   (a) Sketch the graph of a piecewise-constant function that models the data, where x represents age.

   (b) Discuss the impact that age has on the likelihood of being a victim of a violent crime.

76. *New Homes*   The table lists the numbers in millions of houses built for various time intervals from 1940 to 1990 using interval notation.   (Source: Bureau of the Census, *American Housing Survey*.)

Year	[1940, 1950)	[1950, 1960)	[1960, 1970)
Houses	8.5	13.6	16.1

Year	[1970, 1980)	[1980, 1990)
Houses	11.6	8.1

   (a) Sketch the graph of a piecewise-constant function that models this data, where x represents the year.

   (b) Discuss the trends in housing starts between 1940 and 1990.

77. *Perspectives and Viewing Rectangles*   The table lists the average math scores on the Scholastic Aptitude Test (SAT) for selected years.   (Source: The College Board.)

Year	1970	1975	1980	1985	1990	1995
Score	488	472	466	475	476	482

   (a) Make a line graph of the data in [1968, 1997, 5] by [460, 490, 5].

   (b) If this graphical representation of a piecewise-linear function appeared in a newspaper, how might people interpret the college entrance math scores?

   (c) The math scores can vary between 200 and 800. Remake the line graph in [1968, 1997, 5] by [200, 800, 100]. Discuss how interpretation of data can be influenced by changing the scale on the y-axis.

78. *Student Loans*   The table lists the amounts in millions of dollars of government-guaranteed student loans from 1990 to 1994.   (Source: *USA Today*.)

Year	1990	1991	1992	1993	1994
Loans ($ millions)	12.3	13.5	14.7	16.5	18.3

   (a) Make a line graph of the data. Can a linear function model the data exactly?

   (b) Find values for the constants a, b, c, and d so that $f(x)$ models the data.

$$f(x) = \begin{cases} a(x - 1990) + b & \text{if } 1990 \le x < 1992 \\ c(x - 1992) + d & \text{if } 1992 \le x \le 1994 \end{cases}$$

   (c) Evaluate $f(1991)$ and $f(1993)$. Do the results agree with the actual values listed in the table?

79. *Cesarean Births*   The percentages of babies delivered by Cesarean birth are listed in the table.   (Source: S. Teutsch.)

Year	1970	1975	1980	1985	1990
Cesarean Births (%)	5	11	17	23	23

   (a) Make a line graph of the data. Interpret the graph.

   (b) Find values for the constants a, b, and c so that $f(x)$ models the data.

$$f(x) = \begin{cases} a(x - 1970) + b & \text{if } 1970 \le x < 1985 \\ c & \text{if } 1985 \le x \le 1990 \end{cases}$$

   (c) Evaluate $f(1978)$ and $f(1988)$. Interpret the results.

80. *Flow Rates*   A water tank has an inlet pipe with a flow rate of 5 gallons per minute and an outlet pipe with a flow rate of 3 gallons per minute. A pipe can be either closed or completely open. The graph shows the number of gallons of water in the tank after x minutes have elapsed. Use the concept of slope to interpret each piece of this graph.

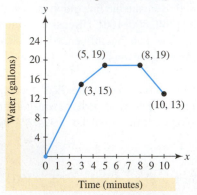

Writing about Mathematics

81. Find a piecewise-constant function that describes movie prices for children, adults, and senior citizens. Represent your function graphically and symbolically. Explain your results.

82. Explain why some functions are defined piecewise. Give an example.

Extended and Discovery Exercises

1. *Measurement*   The circumference of a circular garden with a radius r is to have a 150-foot fence around it. The length of the fence can vary from 150 feet by at most ± 2 feet. What range of values are possible for r?

2.6 LINEAR APPROXIMATION

Midpoint Formula • Interpolation and Extrapolation • Linear Regression

Introduction

In previous sections, several instances were presented where real data was approximated using linear functions and graphs. The approximation of data with lines and linear functions is called **linear approximation**. It is an essential technique used in both elementary and advanced mathematics. In this section we discuss some of the many types of linear approximation.

Midpoint Formula

A common way to make estimations is to average data values. For example, in 1980 the average cost of tuition and fees at public colleges and universities was $800, whereas in 1982 it was $1000. One might estimate the cost of tuition and fees in 1981 to be $900. This averaging is referred to as finding the midpoint. If a line segment is drawn between two data points, then its *midpoint* is the unique point on the line segment that is equidistant from the endpoints.

On a real number line, the midpoint M of two data points x_1 and x_2 is calculated by averaging their coordinates, as shown in Figure 2.74. For example, the midpoint of -3 and 5 is $M = \dfrac{-3 + 5}{2} = 1$.

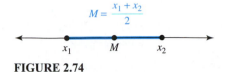

FIGURE 2.74

The midpoint formula in the xy-plane is similar to the formula for the real number line, except that both coordinates are averaged. Figure 2.75 shows midpoint M located on the line segment connecting the two data points (x_1, y_1) and (x_2, y_2). The x-coordinate of the midpoint M is located halfway between x_1 and x_2 and is $\dfrac{x_1 + x_2}{2}$. Similarly, the y-coordinate of M is the average of y_1 and y_2. For example, if we let (x_1, y_1) be $(-3, 1)$ and (x_2, y_2) be $(-1, 3)$ in Figure 2.75, then the midpoint is computed by

$$\left(\frac{-3 + -1}{2}, \frac{1 + 3}{2}\right) = (-2, 2).$$

The midpoint formula is summarized as follows.

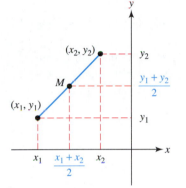

FIGURE 2.75

> ## Midpoint formula in the *xy*-plane
>
> The **midpoint** of the line segment with endpoints (x_1, y_1) and (x_2, y_2) in the xy-plane is $\left(\dfrac{x_1 + x_2}{2}, \dfrac{y_1 + y_2}{2}\right)$.

The population of the United States from 1800 to 1990 is modeled by the nonlinear function f in Figure 2.76. The line segment connecting the points $(1800, 5)$ and $(1990, 249)$ does not accurately describe the population over the time period from 1800 to 1990. However, over a shorter time interval a linear approximation may be more accurate.

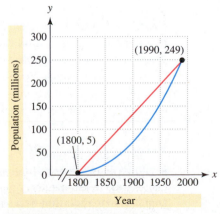

FIGURE 2.76   U.S. Population

▌**EXAMPLE 1**   *Using the midpoint formula*

In 1970 the population of the United States was 203 million and in 1990 it was 249 million. (Source: Bureau of the Census.)
(a) Use the midpoint formula to estimate the population in 1980.
(b) Describe this approximation graphically.

SOLUTION   **(a)** The U.S. populations in 1970 and 1990 can be represented by the data points (1970, 203) and (1990, 249). The midpoint M of the line segment connecting these points is

$$M = \left(\frac{1970 + 1990}{2}, \frac{203 + 249}{2}\right) = (1980, 226).$$

The midpoint formula estimates a population of 226 million in 1980. (The actual population was 226.5 million.)
(b) Figure 2.77 shows the U.S. population and a line segment with midpoint M connecting the data points (1970, 203) and (1990, 249). Notice that the line segment appears to be an accurate approximation over a relatively short period of 20 years.

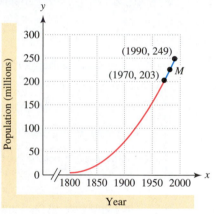

FIGURE 2.77 A Linear Approximation

Interpolation and Extrapolation

Total energy production in the Asia-Pacific region of the world is predicted to increase substantially between 1990 and 2005. Table 2.7 lists this energy production in million metric tons of oil equivalent (Mtoe) for selected years. (Source: R. Andre-Pascal, *Global Energy: The Changing Outlook.*)

TABLE 2.7

Year	1990	1995	2000	2005
Energy (Mtoe)	415	530	645	760

If the data in Table 2.7 is used to estimate energy production during 1998, this would be an example of **interpolation** since 1998 lies between 1990 and 2005. If this data is used to estimate energy production in either 1980 or 2010, this would be an example of **extrapolation** because these years do not lie between 1990 and 2005. Interpolation usually provides more reliable results than extrapolation. Extrapolation should be used cautiously.

EXAMPLE 2 *Estimating past and future energy production*

(a) Use Table 2.7 to determine an equation of a line that models the data. Interpret the slope.

(b) Use this line to estimate energy production in 1998 and 1970. Discuss the results.

SOLUTION

(a) The data increases by exactly 115 Mtoe every 5 years, or by $\frac{115}{5} = 23$ Mtoe per year. Thus the line with slope 23 passing through (1990, 415) models these data exactly. Its equation is $y = 23(x - 1990) + 415$. The slope indicates that energy production in the Asia-Pacific region is expected to increase at a rate of 23 Mtoe per year.

(b) To estimate energy production in 1998 and 1970, substitute 1998 and 1970 for x.

$$y = 23(1998 - 1990) + 415 = 599 \text{ Mtoe}$$

$$y = 23(1970 - 1990) + 415 = -45 \text{ Mtoe}$$

The interpolated value for 1998 is reasonable, since it lies between the 1995 and 2000 values of 530 and 645 Mtoe, respectively. The extrapolated value for 1970 is not reasonable because it is negative. It might be acceptable to use extrapolation for estimating energy production in 1988 or 2007, since these years lie relatively close to the domain of the data. However, extrapolating 20 years back provides an incorrect result.

■ CLASS DISCUSSION

In general, which do you think would be a more accurate method to make estimates: extrapolation or interpolation? Why? ■

Linear Regression

Throughout Chapters 1 and 2, linear functions and lines have been used to model data involving the variables x and y. Unknown values for y were predicted at given values of x. Problems where one variable is used to predict the behavior of a second variable are called **regression** problems. If a line is used to approximate the data, then this is referred to as **linear regression**.

We have already solved numerous problems using linear regression by selecting a line that *visually* fits the data in a scatterplot. However, this technique has some disadvantages. First, it does not produce a unique line. Different people may arrive at different lines to fit the same data. Second, the line is not determined automatically by a calculator or computer. A person must view the data and adjust the line until it "fits." By contrast, a statistical method used to determine a unique regression line is based on **least squares**.

Most graphing calculators have the capability to calculate the least-squares regression line automatically after the data points have been entered. When determining the least-squares line, calculators often compute a real number r, called the **correlation coefficient**, where $-1 \leq r \leq 1$. When r is positive and near 1, low x-values correspond to low y-values and high x-values correspond to high y-values. For example, there is a positive correlation between years of education x and income y. More years of education correlate with higher income. When r is near -1, the reverse is true. Low x-values correspond to high y-values and high x-values correspond to low y-values. An example of this would be the relation between latitude and average yearly temperature. As latitude increases (moving toward either pole) the average yearly temperature decreases. Therefore, there will be a negative correlation between latitude and average annual temperature. If $r \approx 0$, then there is little or no correlation between the data points. In this case, a least-squares regression line does not provide a suitable model. A summary of these concepts regarding correlation is shown in Table 2.8.

TABLE 2.8   Correlation Coefficient r, $(-1 \leq r \leq 1)$

Value of r	Comments	Sample Scatterplot
$r = 1$	There is an exact linear fit. The line passes through all data points and has a positive slope.	
$r = -1$	There is an exact linear fit. The line passes through all data points and has a negative slope.	
$0 < r < 1$	There is a positive correlation. As the x-values increase so do the y-values. The fit is not exact.	
$-1 < r < 0$	There is a negative correlation. As the x-values increase the y-values decrease. The fit is not exact.	
$r = 0$	There is no correlation. The data has no tendency toward being linear. A regression line predicts poorly.	

■ **MAKING CONNECTIONS**

Correlation and Causation   When geese begin to fly north, summer is coming and the weather becomes warmer. Geese flying north correlate with warmer weather. However, geese flying north clearly do not *cause* warmer weather. It is important to remember that correlation does not always indicate the cause.

In the next example we use a graphing calculator to find the line of least-squares fit that passes through three data points.

EXAMPLE 3   *Determining a line of least-squares fit*

Find the line of least-squares fit passing through the data points (1, 1), (2, 3), and (3, 4). What is the correlation coefficient? Plot the data and graph the line.

SOLUTION   Begin by entering the three data points into the calculator. Refer to Figures 2.78–2.81. Select the LinReg(ax + b) option from the STAT CALC menu. From the home screen we can see that the equation of the line of least-squares is $y = 1.5x - \dfrac{1}{3}$. The correlation coefficient is $r \approx 0.98$. Since $r \neq 1$, the line does not provide an exact model of the data.

Graphing Calculator Help

To find a line of least squares fit, see Appendix B (page AP-13).

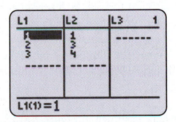

FIGURE 2.78

FIGURE 2.79

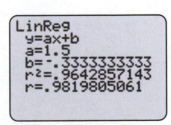

FIGURE 2.80

[0, 5, 1] by [0, 5, 1]

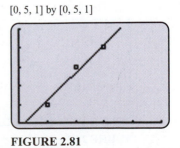

FIGURE 2.81

In the next example we use regression to predict the increase in the number of airline passengers.

EXAMPLE 4 *Predicting airline passenger growth*

Table 2.9 lists the estimated numbers in millions of airline passengers at some of the fastest growing airports in 1992 and 2005. (Source: Federal Aviation Administration.)

TABLE 2.9 Airline Passengers (millions)

Airport	1992	2005
Harrisburg International	0.7	1.4
Dayton International	1.1	2.4
Austin Robert Mueller	2.2	4.7
Milwaukee General Mitchell	2.2	4.4
Sacramento Metropolitan	2.6	5.0
Fort Lauderdale–Hollywood	4.1	8.1
Washington Dulles	5.3	10.9
Greater Cincinnati	5.8	12.3

(a) Make a scatterplot of the data using the 1992 data for *x*-values and the corresponding 2005 data for *y*-values. Predict whether the correlation coefficient will be positive or negative.
(b) Use a calculator to find the least-squares regression line.
(c) Raleigh-Durham International had 4.9 million passengers in 1992. Assuming Raleigh-Durham International follows the same trend in growth as the other airports, estimate the number of passengers it may have in 2005. Compare this result with the Federal Aviation Administration (FAA) estimation of 10.3 million passengers.

SOLUTION **(a)** Plot the data as shown in Figure 2.82. Since increasing *x*-values correspond to increasing *y*-values, the correlation coefficient will be positive.

[0, 6, 1] by [0, 14, 1]

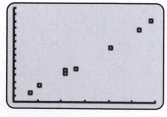

FIGURE 2.82

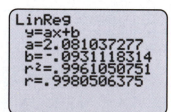

FIGURE 2.83

[0, 6, 1] by [0, 14, 1]

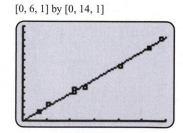

FIGURE 2.84

(b) The equation of the regression line is shown in Figure 2.83. The line $y = 2.081x - 0.0931$ and the data are graphed together in Figure 2.84. Note that $r \approx 0.998$, which is positive as predicted.
(c) When $x = 4.9$, $y = 2.081(4.9) - 0.0931 \approx 10.1$ million passengers. This is quite close to the FAA prediction of 10.3 million.

Putting it all Together
2.6

Linear approximation, a fundamental tool in mathematics, usually becomes more accurate as the interval of approximation becomes smaller. One important type of linear approximation is a regression line. A regression line can be found by at least three techniques. One is using trial and error to visually fit a line to a set of data. A second is to choose two data points and find the equation of the line passing through these points. The third is to find the least-squares regression line using technology.

The following table outlines some important ideas presented in this section.

Concept	Comments	Example
Midpoint formula $\left(\dfrac{x_1 + x_2}{2}, \dfrac{y_1 + y_2}{2}\right)$	Computes a linear approximation	Given $(2, -3)$ and $(4, 5)$, then $M = \left(\dfrac{2 + 4}{2}, \dfrac{-3 + 5}{2}\right) = (3, 1)$.
Interpolation	Estimates values that are between two or more known data values	If the data points $(0, 1)$ and $(2, 3)$ are used to estimate the value of y when $x = 1$, then this would involve interpolation.
Extrapolation	Estimates values that are not between two known data values	If the data points $(0, 1)$ and $(2, 3)$ are used to estimate the value of y when $x = 4$, then this would involve extrapolation.
Correlation coefficient	The value of r gives a measure of how good the line fits the data.	The values of r satisfy $-1 \le r \le 1$, where a line fits the data better if r is near -1 or 1. A value near 0 indicates a poor fit.
Least-squares regression line	A method for finding a unique regression line with a calculator	The line of least-squares fit for the data points $(1, 3)$, $(2, 5)$, and $(3, 6)$ is $$y = \frac{3}{2}x + \frac{5}{3}$$ and $r \approx 0.98$. Try verifying this with a calculator.

2.6 EXERCISES

Midpoint Formula

Exercises 1–6: Use the midpoint formula to complete the following.

1. **Charge Card Balances**   The balances that were outstanding on revolving credit totaled $500 billion in 1996 and $560 billion in 1998. Estimate this debt in 1997. (Actual amount was $531 billion.) (Sources: BankCard Holders of America, Federal Reserve Board.)

2. **State and Federal Inmates**   In 1996 there were 1,183,368 inmates in state and federal prisons and in 1998 there were 1,302,019. Estimate the number of inmates in 1997. (Actual number was 1,242,153.) (Source: Department of Justice.)

3. **State and Federal Inmates**   In 1980 there were 329,821 inmates in state and federal prisons and in 1983 there were 436,885. Estimate the number of inmates in 1986. (Actual number was 544,972.) (Source: Department of Justice.)

4. **Population Estimates**   The population of the United States was 151 million in 1950 and 179 million in 1960. Estimate the population in 1970. (Actual population was 203 million.)   (Source: Bureau of the Census.)

5. **Olympic Times**   In the Olympic Games, the 200-meter dash is run in approximately 20 seconds. Estimate the time to run the 100-meter dash.

6. Between any two real numbers a and b there is always another real number. How could such a number be found?

Exercises 7–14: Find the midpoint of the line segment connecting the points.

7. $(1, 2), (5, -3)$

8. $(-6, 7), (9, -4)$

9. $(1.5, 2.9), (-5.7, -3.6)$

10. $(9.4, -4.5), (-7.7, 9.5)$

11. $(\sqrt{2}, \sqrt{5}), (\sqrt{2}, -\sqrt{5})$

12. $(\sqrt{7}, 3\sqrt{3}), (-\sqrt{7}, -\sqrt{3})$

13. $(a, b), (-a, 3b)$

14. $(-a, b), (3a, b)$

Interpolation and Extrapolation

Exercises 15–18: The table lists data that are exactly linear.

(a) *Find the slope-intercept form of the line that passes through these data points.*

(b) *Predict y when $x = -2.7$ and 6.3. Decide whether these calculations involve interpolation or extrapolation.*

15.

x	-3	-2	-1	0	1
y	-7.7	-6.2	-4.7	-3.2	-1.7

16.

x	-2	-1	0	1	2
y	10.2	8.5	6.8	5.1	3.4

17.

x	5	23	32	55	61
y	94.7	56.9	38	-10.3	-22.9

18.

x	-11	-8	-7	-3	2
y	-16.1	-10.4	-8.5	-0.9	8.6

19. **Air Safety Inspectors**   The number of air safety inspectors is shown in the accompanying table for selected years.

Year	1998	1999	2000
Inspectors	3305	3185	3089

Source: Federal Aviation Administration.

(a) Find a linear function f that models these data. Is f exact or approximate?

(b) Use f to estimate the number of inspectors in 1996. Compare your answer to the actual value of 2776. Did your estimate involve interpolation or extrapolation?

20. **Deaths on School Grounds**   Nationwide deaths on school grounds for school years ending in year x are shown in the table.

x (year)	1998	1999	2000
y (deaths)	43	26	9

Source: FBI.

(a) Find a linear function f that models these data. Is f exact or approximate?

(b) Use f to estimate the number of deaths on school grounds in 1997. Compare your answer to the actual value of 26. Did your estimate involve interpolation or extrapolation?

Exercises 21–24: Complete the following.

(a) *Conjecture whether the correlation coefficient r for the data will be positive, negative, or zero.*

(b) *Use a calculator to find the equation of least-squares regression line with the form $y = ax + b$ and the value of r.*

(c) *Use the regression line to predict y when $x = 2.4$.*

21.

x	−3	−2	−1	0
y	−11.1	−8.4	−5.7	−2.6

x	1	2	3
y	1.1	3.9	7.3

22.

x	−4	−2	0	2	4
y	1.2	2.8	5.3	6.7	9.1

23.

x	1	3	5	7	10
y	5.8	−2.4	−10.7	−17.8	−29.3

24.

x	−4	−3	−1	3	5
y	37.2	33.7	27.5	16.4	9.8

Linear Approximation Applications

25. *Jail Population*   The table lists the average daily jail population in thousands for three selected years.   (Source: Department of Justice.)

Year	1983	1988	1993
Population (thousands)	228	336	466

(a) Make a line graph of this data in [1982, 1994, 1] by [200, 500, 100].

(b) Make a conjecture whether the midpoint formula gives an estimate for 1988 that is high or low.

(c) Compute the midpoint estimate. Was your conjecture correct?

26. *Coverting Units*   The following table lists the number of tablespoons y in x quarts. Find a linear function f that model these data exactly. Use f to determine the number of tablespoons in 20 quarts.

x	3	5	7	9
y	192	320	448	576

27. *Federal Debt*   The table lists the total federal debt in billions of dollars for three years.

Year	1985	1987	1989
Debt ($ billions)	1828	2354	2881

Source: Office of Management and Budget.

(a) Use the midpoint method to predict whether the data are exactly linear.

(b) Find a linear function f whose graph passes through the two points (1985, 1828) and (1989, 2881).

(c) Evaluate $f(1987)$. How does this value compare to the midpoint estimate? Explain.

28. (Continuation of Exercise 27)

(a) Use f to estimate the federal debt in 1984, 1988, and 1996.

(b) The actual debts in 1984, 1988, and 1996 were 1572, 2602, and 5020 billion dollars, respectively. Compare the accuracies of extrapolation and interpolation.

29. *Distant Galaxies*   In the late 1920s the famous observational astronomer Edwin P. Hubble (1889–1953) determined both the distance to several galaxies and the velocity at which they were receding from Earth. Four galaxies with their distances in light-years and velocities in miles per second are listed in the table.   (Sources: A. Acker and C. Jaschek, *Astronomical Methods and Calculations*; A. Sharov and I. Novikov, *Edwin Hubble, The Discoverer of the Big Bang Universe.*)

Galaxy	Distance	Velocity
Virgo	50	990
Ursa Minor	650	9300
Corona Borealis	950	15,000
Bootes	1700	25,000

(a) Let x represent distance and y velocity. Plot the four data points in [−100, 1800, 100] by [−1000, 28,000, 1000].

(b) Find the least-squares regression line that models these data.

(c) If the galaxy Hydra is receding at a speed of 37,000 miles per second, estimate its distance.

30. *Cellular Phones*   One of the early problems with cellular phones was the delay involved with placing a call when the system was busy. One study analyzed this delay. The table shows that as the number of calls increased by P percent, the average delay time D to put through a call also increased. (Source: A. Mehrotra, *Cellular Radio: Analog and Digital Systems.*)

P (%)	0	20	40	60	80	100
D (minutes)	1	1.6	2.4	3.2	3.8	4.4

(a) Let P correspond to x-values and D to y-values. Find the least-squares regression line that models these data. Plot the data and the regression line.

(b) Estimate the delay for a 50% increase in the number of calls.

31. *Women in Politics* The table lists percentages of women in state legislatures for past years. (Source: National Women's Political Caucus.)

Year	1975	1977	1979	1981
Percent	8.0	9.1	10.3	12.1

Year	1983	1985	1987	1989
Percent	13.3	14.8	15.7	17.0

Year	1991	1993	1995
Percent	18.3	20.5	20.7

(a) Make a scatterplot of the data.

(b) Find the least-squares regression line that models these data.

(c) Interpret the slope.

(d) Assuming that trends continued, use this line to estimate the percentage of women in state legislatures in 1998.

32. *Average Household Size* The following table lists the average number y of people per household for various years x.

x	1940	1950	1960	1970
y	3.67	3.37	3.33	3.14

x	1980	1990	1998
y	2.76	2.63	2.62

Source: Bureau of the Census.

(a) Make a scatterplot of the data.

(b) Decide whether the correlation coefficient is positive, negative, or zero.

(c) Find a least-squares regression line that models these data. What is the correlation coefficient?

(d) Estimate the number of people per household in 1975 and compare it to the actual value of 2.94. Did this estimate involve interpolation or extrapolation?

Writing about Mathematics

33. Explain the difference between interpolation and extrapolation. Sketch a graph of $y = x^2$ and the line passing through the points $(0, 0)$ and $(1, 1)$. Use the graph to explain why linear interpolation tends to be more accurate than linear extrapolation.

34. What type of data can the midpoint formula be used to approximate? Explain your answer with the aid of a graph.

Extended and Discovery Exercises

Exercises 1 and 2: Graph the function f in the standard viewing rectangle.

(a) Choose any curved portion of the graph of f and repeatedly zoom in. Describe how the graph appears. Repeat this on different portions of the graph.

(b) Under what circumstances could a linear approximation be used to accurately model a nonlinear graph?

1. $f(x) = 4x - x^3$ **2.** $f(x) = x^4 - 5x^2$

CHECKING BASIC CONCEPTS FOR SECTIONS 2.5 AND 2.6

1. Sketch a graph of

$$f(x) = \begin{cases} 2 - x & \text{if } -5 \le x < 0 \\ 2x - 1 & \text{if } \ \ 0 \le x \le 5. \end{cases}$$

(a) What is the domain of f?
(b) Evaluate $f(-2)$, $f(0)$, and $f(3)$.
(c) Graph f.   (d) Is f continuous?

2. (a) Solve the equation $|2x - 1| = 5$.
   (b) Use part (a) to solve the absolute value inequalities $|2x - 1| \le 5$ and $|2x - 1| > 5$.

3. Find the midpoint of the line segment connecting the points $(-2, 3)$ and $(4, 2)$.

4. According to the 1996 federal *Dietary Guidelines for Americans*, the recommended minimum weight for a person 58 inches tall is 91 pounds, and for a person 64 inches tall it is 111 pounds.

(a) Find an equation of the line that passes through the points $(58, 91)$ and $(64, 111)$.
(b) Use this line to estimate the minimum weight for someone 61 inches tall. Then use the midpoint formula to find this minimum weight. Are the results the same? Why?

5. The following table lists per capita income.

Year	1970	1980	1990	1998
Income (per capita)	$4072	$10,029	$19,142	$26,412

Source: Bureau of Economic Analysis.

(a) Find the least-squares regression line for the data.
(b) Estimate the per capita income in 1995. Did this calculation involve interpolation or extrapolation?

Chapter 2 Summary

CONCEPT	EXPLANATION AND EXAMPLES

Section 2.1

LINEAR FUNCTION

A linear function can be written as $f(x) = ax + b$. Its graph is a line.

Example: $f(x) = -2x + 5$;
slope $= -2$, y-intercept $= 5$

LINEAR MODEL

If a quantity increases or decreases by a constant amount for each unit increase in x, then it can be modeled by a linear function given by

$$f(x) = (\text{constant rate of change})x + (\text{initial amount}).$$

Section 2.2

SLOPE

Given points (x_1, y_1) and (x_2, y_2) then $m = \dfrac{y_2 - y_1}{x_2 - x_1}$.

The change in y is denoted $\Delta y = y_2 - y_1$, the change in x is denoted $\Delta x = x_2 - x_1$, and the slope is also given by $m = \dfrac{\Delta y}{\Delta x}$.

Example: If $(3, 0)$ and $(2, -4)$ lie on a line, then the slope of the line is

$$m = \frac{-4 - 0}{2 - 3} = 4.$$

CONCEPT	EXPLANATION AND EXAMPLES

Section 2.2 (continued)

POINT-SLOPE FORM

If a line passes through (x_1, y_1) with slope m, then
$$y = m(x - x_1) + y_1.$$

Example: $y = -\dfrac{3}{4}(x + 4) + 5$ indicates slope $-\dfrac{3}{4}$ and the line passes through $(-4, 5)$.

SLOPE-INTERCEPT FORM

If a line has slope m and y-intercept b, then
$$y = mx + b.$$

Example: $y = 3x - 4$ indicates slope 3 and y-intercept -4.

HORIZONTAL AND VERTICAL LINES

A horizontal line passing through the point (h, k) is given by $y = k$, and vertical line passing through (h, k) is $x = h$.

Examples: The vertical line $x = 4$ passes through $(4, -2)$.
The horizontal line $y = -3$ passes through $(6, -3)$.

PARALLEL AND PERPENDICULAR LINES

Parallel lines have equal slopes satisfying $m_1 = m_2$, and perpendicular lines have slopes satisfying $m_1 \cdot m_2 = -1$.

Examples: The lines $y_1 = 3x - 1$ and $y_2 = 3x + 4$ are parallel.

The lines $y_1 = 3x - 1$ and $y_2 = -\dfrac{1}{3}x + 4$ are perpendicular.

Section 2.3

LINEAR EQUATION

Can be written as $ax + b = 0$ with $a \neq 0$ and has one solution.

Example: The solution to $2x - 4 = 0$ is $x = 2$.

INTERSECTION-OF-GRAPHS METHOD

Set Y_1 equal to the left side of the equation, and set Y_2 equal to the right side. The x-coordinate of a point of intersection is a solution.

Example: The graphs of $y_1 = 2x$ and $y_2 = x + 1$ intersect at $(1, 2)$, so the solution to $2x = x + 1$ is $x = 1$.

Section 2.4

LINEAR INEQUALITY

Can be written as $ax + b > 0$ with $a \neq 0$, where $>$ can be replaced by $<$, $\leq$, or $\geq$.

Example: $3x - 1 \leq 2$ is linear since it can be written as $3x - 3 \leq 0$. The solution set is $\{x \mid x \leq 1\}$.

COMPOUND INEQUALITY

Example: $x \geq -2$ and $x \leq 4$, or equivalently, $-2 \leq x \leq 4$. This is called a three-part inequality.

CONCEPT	EXPLANATION AND EXAMPLES

Section 2.4 (continued)

INTERVAL NOTATION

A concise way to express intervals on the number line

Example: $x < 4$ is expressed as $(-\infty, 4)$.
$-3 \le x < 1$ is expressed as $[-3, 1)$.
$x \le 2$ or $x \ge 5$ is expressed as $(-\infty, 2] \cup [5, \infty)$.

Section 2.5

PIECEWISE-DEFINED FUNCTIONS

A function defined by more than one formula on its domain

Example: Step function, greatest integer function, absolute value function, and

$$f(x) = \begin{cases} 2 - x & \text{if } -4 \le x < 1 \\ 3x & \text{if } 1 \le x \le 5. \end{cases}$$

ABSOLUTE VALUE EQUATIONS

$|ax + b| = k$ is equivalent to $ax + b = \pm k$.

Example: $|2x - 3| = 4$ is equivalent to $2x - 3 = 4$ or $2x - 3 = -4$.
The solutions are $x = 3.5$ and $x = -0.5$.

ABSOLUTE VALUE INEQUALITIES

Example: The solutions to $|2x + 1| = 5$ are $x = -3$ and $x = 2$.
The solutions to $|2x + 1| < 5$ are $-3 < x < 2$.
The solutions to $|2x + 1| > 5$ are $x < -3$ or $x > 2$.

Section 2.6

MIDPOINT METHOD FUNCTIONS

The midpoint of the line segment connecting (x_1, y_1) and (x_2, y_2) is

$$\left(\frac{x_1 + x_2}{2}, \frac{y_1 + y_2}{2} \right).$$

Example: The midpoint of the line segment connecting $(1, 2)$ and $(-3, 6)$ is
$$\left(\frac{1 + -3}{2}, \frac{2 + 6}{2} \right) = (-1, 4).$$

LINEAR CURVE FITTING

One method to determine a line that models linear data is to use the method of least squares. This method determines a unique line and can be found with a calculator. The correlation coefficient r, $(-1 \le r \le 1)$, measures how well a line fits the data.

Example: The line of least squares modeling the data $(1, 1)$, $(3, 4)$, and $(4, 6)$ is given by $y \approx 1.643x - 0.714$, with $r \approx 0.997$.

Review Exercises

Exercises 1–2: The graph of a linear function f is shown.

(a) *Identify the slope, y-intercept, and x-intercept.*
(b) *Write a symbolic representation for f.*
(c) *Find any zeros of f.*

1.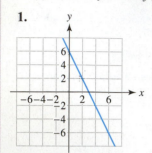

2.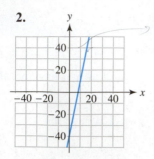

Exercises 3 and 4: Find f(x) = ax + b so that f models the data exactly.

3.

x	1	2	3	4
$f(x)$	2.5	0	−2.5	−5

4.

x	−3	6	15	24	33
$f(x)$	−1.65	−1.2	−0.75	−0.3	0.15

Exercises 5–10: Solve the linear equation symbolically or graphically.

5. $5x - 22 = 10$

6. $5(4 - 2x) = 16$

7. $-2(3x - 7) + x = 2x - 1$

8. $\pi x + 1 = 6$

9. $5x - \dfrac{1}{2}(4 - 3x) = \dfrac{3}{2} - (2x + 3)$

10. $\dfrac{x - 4}{2} = x + \dfrac{1 - 2x}{3}$

Exercises 11 and 12: Use a table to solve each linear equation numerically to the nearest tenth.

11. $3.1x - 0.2 = 2(x - 1.7)$

12. $\sqrt{7} - 3x = 2.1(1 + x)$

13. *Median Income* The median U.S. family income between 1980 and 1993 can be modeled by $f(x) = 1321.7(x - 1980) + 21{,}153$, where x is the year. (Source: Department of Commerce.)
   (a) Solve the equation $f(x) = 25{,}000$ graphically. Interpret the solution.
   (b) Solve part (a) symbolically.

14. *Course Grades* In order to receive a B grade in a college course, it is necessary to have an overall average of 80% correct on two 1-hour exams of 75 points each and one final exam of 150 points. If a person scores 55 and 72 on the 1-hour exams, what is the minimum score on the final exam that the person can receive and still earn a B?

15. The graphs of two linear functions f and g are shown in the figure. Solve each equation or inequality.
   (a) $f(x) = g(x)$ (b) $f(x) < g(x)$
   (c) $f(x) > g(x)$

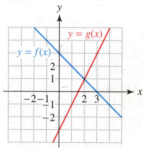

16. The graphs of three linear functions f, g, and h with domains $D = \{x \mid 0 \le x \le 7\}$ are shown in the figure. Solve each equation or inequality.
   (a) $f(x) = g(x)$ (b) $g(x) = h(x)$
   (c) $f(x) < g(x) < h(x)$ (d) $g(x) > h(x)$

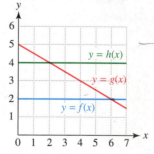

Exercises 17–22: Solve the linear inequality symbolically or graphically. Write the solution set in interval notation.

17. $3x - 4 \le 2 + x$

18. $-2x + 6 \le -3x$

19. $\dfrac{2x - 5}{2} < \dfrac{5x + 1}{5}$

20. $-5(1 - x) > 3(x - 3) + \dfrac{1}{2}x$

21. $-2 \le 5 - 2x < 7$

22. $-1 < \dfrac{3x - 5}{-3} < 3$

23. *Medicare Costs* Future estimates for Medicare costs in billions of dollars can be modeled by $f(x) = 18x - 35{,}750$, where $1995 \le x \le 2007$. Determine when Medicare costs are expected to be from 268 to 358 billion dollars. (Source: Office of Management and Budget.)

24. *Temperature Scales* The table shows equivalent temperatures in degrees Celsius and degrees Fahrenheit.

°F	−40	32	59	95	212
°C	−40	0	15	35	100

(a) Plot the data by having the x-axis correspond to Fahrenheit temperature and the y-axis to Celsius temperature. What type of relation exists between the data?

(b) Find a function C that receives the Fahrenheit temperature x as input and outputs the corresponding Celsius temperature. Interpret the slope.

(c) If the temperature is 83°F, what is this temperature in degrees Celsius?

Exercises 25–30: Find the slope-intercept form of the equation of a line satisfying the following conditions.

25. Slope 7, passing through $(-3, 9)$

26. Passing through $(2, -4)$ and $(7, -3)$

27. Passing through $(1, -1)$, parallel to $y = -3x + 1$

28. Passing through $(-2, 1)$, perpendicular to $y = 2(x + 5) - 22$

29. Parallel to the line segment connecting $(0, 3.1)$ and $(5.7, 0)$, passing through $(1, -7)$

30. Perpendicular to $y = -\frac{5}{7}x$, passing through $\left(\frac{6}{7}, 0\right)$

Exercises 31–36: Find an equation of the line satisfying the following conditions.

31. Parallel to the y-axis, passing through $(6, -7)$

32. Parallel to the x-axis, passing through $(-3, 4)$

33. Horizontal, passing through $(1, 3)$

34. Vertical, passing through $(1.5, 1.9)$

35. Horizontal with y-intercept -8

36. Vertical with x-intercept 2.7

37. *Distance from Home* The graph depicts the distance y that a person driving a car on a straight road

is from home after x hours. Interpret the graph. What speeds did the car travel?

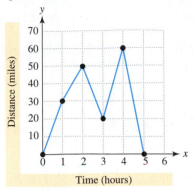

38. *Distance between Bicyclists* The graph shows the distance after x hours between two bicyclists traveling in opposite directions along a straight road.

(a) At what time did the bicycle riders meet?

(b) When were they 20 miles apart?

(c) Find the times when they were less than 20 miles apart.

(d) Estimate the sum of the speeds of the two bicyclists.

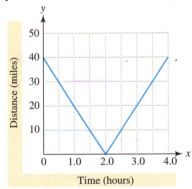

39. Use $f(x)$ to complete the following.

$$f(x) = \begin{cases} 8 + 2x & \text{if } -3 \le x \le -1 \\ 5 - x & \text{if } -1 < x \le 2 \\ x + 1 & \text{if } 2 < x \le 5 \end{cases}$$

(a) Evaluate f at $x = -2, -1, 2,$ and 3.

(b) Sketch a graph of f. Is f continuous?

(c) Determine the x-value(s) where $f(x) = 3$.

40. *Graphical Model* A 500-gallon water tank is initially full, then emptied at a constant rate of 50 gallons per minute. Then the tank is filled by a pump that outputs 25 gallons of water per minute. Sketch a graph that depicts the amount of water in the tank after x minutes.

41. Graph $f(x) = [\![x]\!]$ without a calculator on the interval $-4 \le x \le 4$.

42. If $f(x) = [\![2x - 1]\!]$, evaluate $f(-3.1)$ and $f(2.5)$.

Exercises 43–46: Solve the equation. Use the solutions to help solve the related inequality.

43. $|x| = 3$, $|x| > 3$

44. $|-3x + 1| = 2$, $|-3x + 1| < 2$

45. $|3x - 7| = 10$, $|3x - 7| > 10$

46. $|4 - x| = 6$, $|4 - x| \le 6$

47. Solve $|x + 1| = |3x + 2|$.

48. Solve $\left|\dfrac{1}{2}x - 3\right| + 5 > 17$.

49. **ACT Scores** The table lists the average composite ACT scores for selected years. (Source: The American College Testing Program.)

Year	1989	1990	1991	1992	1993	1994
Score	20.6	20.6	20.6	20.6	20.7	20.8

   (a) Make a line graph of the data.

   (b) Let this graph represent a piecewise-linear function f. Find a symbolic representation for f.

50. **Midpoint** Find the midpoint of the line segment connecting the points $(-1, 3)$ and $(13, 23)$.

51. **Population Estimates** In 1996 the population of a city was 143,247 and in 2000 it was 167,933. Estimate the population in 1998 using the midpoint formula.

52. **Distance** A driver of a car is initially 455 miles from home, traveling toward home on a straight freeway at 70 miles per hour.

   (a) Write a formula for a linear function f that models the distance between the driver and home after x hours.

   (b) Graph f. What is an appropriate domain for f?

   (c) Identify the x- and y-intercepts. Interpret each.

53. **Working Together** Suppose that one worker can shovel snow from a storefront sidewalk in 50 minutes and another worker can shovel it in 30 minutes. How long will it take if they work together?

54. **Antifreeze** Initially, a tank contains 20 gallons of a 30% antifreeze solution. How many gallons of an 80% antifreeze solution should be added to the tank in order to have the concentration of the antifreeze in the tank increase to 50%?

55. **Running** An athlete travels 13.5 miles in 1 hour 48 minutes, jogging at 7 miles per hour and then at 8 miles per hour. How long did the runner jog at each speed?

56. **Geometry** A rectangle is twice as long as it is wide and has a perimeter of 78 inches. Find the width and length of this rectangle.

57. **Math SAT Scores** The table lists average math SAT scores for selected years.

Year	1994	1995	1996	1998	2000
Score	504	506	508	512	514

Source: The College Board.

   (a) Find a linear function f that models these data. (Answers may vary.)

   (b) Use f to approximate the average math SAT score in 1997. Did this estimate involve interpolation or extrapolation?

58. The table lists data that is exactly linear.

   (a) Determine the slope-intercept form of the line that passes through these data points.

   (b) Predict y when $x = -1.5$ and 3.5. State whether these calculations involve interpolation or extrapolation.

x	-3	-2	-1	0	1	2	3
y	6.6	5.4	4.2	3	1.8	0.6	-0.6

59. **Flow Rates** (Refer to Exercise 80 in Section 2.5.) Suppose the tank is modified so that it has a second inlet pipe, which flows at a rate of 2 gallons per minute. Interpret the graph by determining when each inlet and outlet pipe is open or closed.

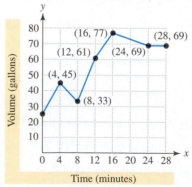

60. **Air Temperature** For altitudes up to 4 kilometers, moist air will cool at a rate of about 6°C per kilometer. If the ground temperature is 25°C, at what altitudes would the air temperature be from 5°C to 15°C? (Source: A. Miller and R. Anthes, *Meteorology.*)

61. *Water Pollution*   At one time, the Thames River in England supported an abundant community of fish. By 1915, however, pollution destroyed all the fish in a 40-mile stretch near its mouth for a 45-year period. Since then, improved sewage treatment facilities and other ecological steps have resulted in a dramatic increase in the number of different fish present. The number of species present from 1967 to 1978 can be modeled by $f(x) = 6.15x - 12,059$, where x is the year.
  (a) Find the year when the number of species first exceeded 70.
  (b) The domain of f is $1967 \le x \le 1978$. Discuss whether f would continue to give good approximations beyond 1978. What would likely happen to the number of species of fish present after time passes and pollution disappears? (Source: C. Mason, *Biology of Freshwater Pollution.*)

62. *Ozone Layer*   Ozone is the only gas capable of screening out lethal ultraviolet light. Without ozone, ultraviolet light would kill terrestrial life. Although the ozone layer can vary in thickness, it is typically only 3 millimeters thick. Worldwide, it has been estimated that for each 0.03 millimeter decrease in the thickness of the ozone layer, there would be an additional 100,000 cases of blindness annually, due to cataracts. (Source: R. K. Turner, *Environmental Economics.*)
  (a) Find a linear function that computes the estimated number of additional cases of blindness

each year from cataracts for a decrease of x millimeters in the ozone layer.
  (b) Approximate the additional annual cases of blindness that may occur due to a decrease in the ozone layer of 0.05 millimeter.

63. *Least-Squares Fit*   The accompanying table lists the actual annual cost y to drive a midsize car 15,000 miles per year for selected years x.

x (year)	1960	1970	1980	1990	2000
y (cost)	$1394	$1763	$3176	$5136	$6880

Source: Runzheimer International.

  (a) Predict whether the correlation coefficient is positive, negative, or zero.
  (b) Find a least-squares regression line that models these data. What is the correlation coefficient?
  (c) Estimate the annual cost of driving a midsize car in 1995.

Extended and Discovery Exercises

1. *Indoor Air Pollution*   Today, indoor air pollution has become more hazardous as people spend 80 to 90% of their time in tightly sealed, energy-efficient buildings that often lack proper ventilation. Many contaminants such as tobacco smoke, formaldehyde, radon, lead, and carbon monoxide often are allowed to increase to unsafe levels. Mathematics plays a central role in risk assessment of pollutants. If x people are exposed to a contaminant, then the resulting number of cancer cases can sometimes be estimated by $f(x) = Cx$. In this formula, C is a constant that represents the excess cancer risk per year. For example, if $C = 0.01$ then an exposed individual has a 1% greater chance of developing cancer during a period of 1 year. (Source: A. Hines, *Indoor Air Quality & Control.*)

  (a) Environmental tobacco smoke is a health hazard. The EPA has estimated that the annual excess cancer risk is $C = 0.00002$. The population of the United States was approximately 265 million in 1996. Assuming that everyone is exposed to tobacco smoke to some degree, estimate the annual number of cancer cases that are caused by environmental tobacco smoke.
  (b) Formaldehyde is a volatile organic compound that has come to be recognized as a highly toxic indoor air pollutant. Its source is found in building materials such as fiberboard, plywood, foam insulation, and carpeting. Some studies have estimated that $C = 0.0007$ for formaldehyde. In a population of 100,000

people exposed to formaldehyde, how many are likely to develop cancer during a year?

(c) Uranium occurs in small traces in most soils throughout the world. Radon gas occurs naturally in homes and is produced when uranium decays radioactively into lead. Radon enters buildings through basements, water supplies, and natural gas lines. Exposure to radon gas is a known factor in lung cancer risk. According to the EPA, radon causes between 5000 and 20,000 lung cancer cases each year in the United States. Approximate a range for the constant C, if the population of the United States was 250 million when the EPA study was performed.

(d) Discuss why it is reasonable that f is a linear function.

2. *Archeology*   It is possible for archeologists to estimate the height of an adult based only on the length of the humerus, a bone located between the elbow and the shoulder. The approximate relationship between the height y of an individual and the length x of the humerus is shown in the table for both males and females. All measurements are recorded in inches. Tables like this are the result of measuring bones from many skeletons. Individuals may vary from these values.

x	8	9	10	11
y (females)	50.4	53.5	56.6	59.7
y (males)	53.0	56.0	59.0	62.0

x	12	13	14
y (females)	62.8	65.9	69.0
y (males)	65.0	68.0	71.0

(a) Find the estimated height of a female with a 12-inch humerus.

(b) Plot the ordered pairs (x, y) for both sexes. What type of relation exists between the data?

(c) For each 1-inch increase in the length of the humerus, what are the corresponding increases in the heights of females and of males?

(d) Determine linear functions f and g that model these data for females and males, respectively.

(e) Suppose a humerus is estimated to be between 9.7 and 10.1 inches long. If the sex is not known, use f and g to approximate the range for the height of an individual of each sex.

3. Continuing with Exercise 2, have members of the class measure their heights and the lengths of their humeri (plural of humerus) in inches.

(a) Make a table of the results.

(b) Find regression lines that fit the data points for males and females.

(c) Compare your results with the table in the previous exercise.

4. *Comparing Ages*   The age of Earth is approximately 4.45 billion years old. The earliest evidence of dinosaurs dates back 200 million years, whereas the earliest evidence for *Homo sapiens* dates back 300,000 years. If the age of Earth was condensed into one year, determine the approximate times when dinosaurs and *Homo sapiens* first appeared.

200 million   300,000
years ago   years ago

5. *A Puzzle*   Three people leave for a city 15 miles away. The first person walks 4 miles per hour and the other two people ride in a car that travels 28 miles per hour. After some time, the second person gets out of the car and walks 4 miles per hour to the city while the driver goes back and picks up the first person. The driver takes the first person to the city. If all three people arrive in the city at the same time, how far did each person walk?

> The work of mathematicians from even a millennium ago is still routinely used today, and the work of mathematicians from today will be part of the math of the future.
>
> —
>
> *Terry Tao*
>
> —

Chapter

3

Quadratic Functions and Equations

The last basketball player in the NBA to shoot foul shots underhand was Rick Barry, who retired in 1980. On average, he was able to make about 9 out of 10 shots. Since then, every NBA player has used the overhand style of shooting foul shots—even though, this style has often resulted in lower free-throw percentages.

According to Peter Brancazio, a physics professor emeritus from Brooklyn College and author of *Sports Science,* there are good reasons for shooting underhand. An underhand shot obtains a higher arc, and as the ball approaches the hoop, it has a better chance of going through the hoop than does a ball with a flatter arc. If a basketball is tossed at an angle of 32 degrees or less, it will likely hit the back of the

Rick Barry

rim and bounce out. The optimal angle is greater than 45 degrees and depends on the height at which the ball is released. Lower release points require steeper arcs and increase the chances of the ball passing through the hoop. (See the Extended and Discovery Exercise at the end of this chapter to model the arc of a basketball.)

Mathematics plays an important role in analyzing applied problems such as shooting foul shots. Whether NBA players choose to agree with Professor Brancazio is another question, but mathematics tells us that steeper arcs are necessary for accurate foul shooting.

Source: Curtis Rist. "The Physics of Foul Shots." *Discover,* October 2000. (Photograph reprinted with permission.)

3.1 QUADRATIC FUNCTIONS AND MODELS

Basic Concepts • Vertex Form • Models and Applications • Quadratic Regression

Introduction

Sometimes when data lie on a line or nearly lie on a line, they can be modeled with a linear function and are called linear data. Data that are not linear are called **nonlinear data** and require a nonlinear function to model them. There are many types of nonlinear data, and mathematicians can choose from a wide variety of nonlinear functions to create models. One of the simplest types of nonlinear functions is a quadratic function. Figures 3.1–3.3 illustrate three sets of nonlinear data that can be modeled by a quadratic function.

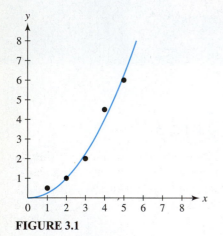

FIGURE 3.1

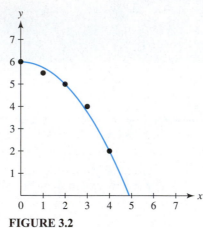

FIGURE 3.2

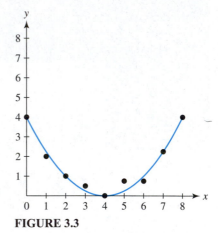

FIGURE 3.3

Basic Concepts

Recall that a linear function can be written as $f(x) = ax + b$ (or $f(x) = mx + b$). The symbolic representation of a quadratic function is different from that of a linear function because it contains an x^2-term. Examples of quadratic functions include

$$f(x) = 2x^2 - 4x - 1, \qquad g(x) = 4 - x^2, \qquad \text{and} \qquad h(x) = \frac{1}{3}x^2 + \frac{2}{3}x + 1.$$

The following defines a quadratic function.

> ### Quadratic function
>
> Let a, b, and c be real numbers with $a \neq 0$. A function represented by $f(x) = ax^2 + bx + c$ is a **quadratic function**.

We refer to the form $f(x) = ax^2 + bx + c$ as **standard form** for a quadratic function. The implied domain of a quadratic function includes all real numbers. That is, there are no real number inputs for which a quadratic function is not defined. The leading coefficient of a quadratic function is a. For the previous three functions, the leading coefficients are $a = 2$ for $f(x) = 2x^2 - 4x - 1$, $a = -1$ for $g(x) = 4 - x^2$, and $a = \frac{1}{3}$ for $h(x) = \frac{1}{3}x^2 + \frac{2}{3}x + 1$.

The graph of a quadratic function is a **parabola**—a U-shaped graph that opens either upward or downward. Graphs of the three quadratic functions f, g, and h are shown in Figures 3.4–3.6, respectively. A parabola opens upward if a is positive and opens downward if a is negative. For example, since $a = -1 < 0$ for $g(x) = 4 - x^2$, the graph of g opens downward. The highest point on a parabola that opens downward or the lowest point on a parabola that opens upward is called the **vertex**. The vertical line passing through the vertex is called the **axis of symmetry**. The vertex and axis of symmetry are shown in each figure.

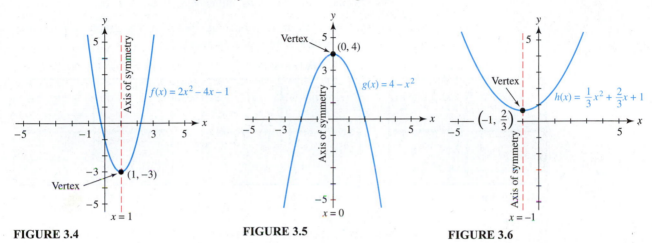

FIGURE 3.4   **FIGURE 3.5**   **FIGURE 3.6**

The leading coefficient a of a quadratic function not only determines whether its graph opens upward or downward, but it also controls the width of the parabola. Larger values of $|a|$ result in a narrower parabola, and smaller values of $|a|$ result in a wider parabola. This concept is illustrated in Figures 3.7 and 3.8.

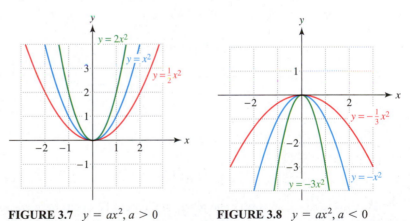

FIGURE 3.7 $y = ax^2, a > 0$ **FIGURE 3.8** $y = ax^2, a < 0$

■ **CLASS DISCUSSION**

What does the graph of a quadratic function resemble if its leading coefficient is nearly 0? ■

EXAMPLE 1   *Identifying quadratic functions*

Identify the function as linear, quadratic, or neither. If it is quadratic, identify the leading coefficient and evaluate $f(2)$. Support your result graphically.

(a) $f(x) = 3 - 2^2 x$ **(b)** $g(x) = 5 + x - 3x^2$ **(c)** $h(x) = \dfrac{3}{x^2 + 1}$

SOLUTION   **(a)** Because $f(x) = 3 - 2^2x$ can be written as $f(x) = -4x + 3$, f is linear. In Figure 3.9 the graph of f is a line.

(b) Function g can be written as $g(x) = -3x^2 + x + 5$, so g is a quadratic function with $a = -3$, $b = 1$, and $c = 5$. The leading coefficient is $a = -3$ and

$$f(2) = 5 + 2 - 3(2)^2 = -5.$$

Figure 3.10 shows that the graph of g is a parabola opening downward.

(c) $h(x) = \dfrac{3}{x^2 + 1}$ cannot be written as either $h(x) = ax + b$ or $h(x) = ax^2 + bx + c$, so h is neither a linear nor a quadratic function. The graph of h is shown in Figure 3.11 and is neither a line nor a parabola.

[-9, 9, 1] by [-6, 6, 1]   [-9, 9, 1] by [-6, 6, 1]   [-6, 6, 1] by [-4, 4, 1]

   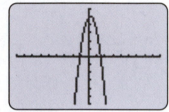   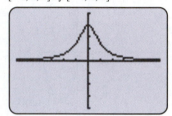

FIGURE 3.9   **FIGURE 3.10**   **FIGURE 3.11**

EXAMPLE 2   *Analyzing graphs of quadratic functions*

Use the graph of each quadratic function to determine the sign of the leading coefficient, its vertex, and the equation of the axis of symmetry.

(a)   **(b)**

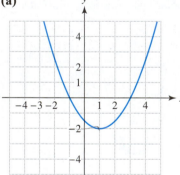   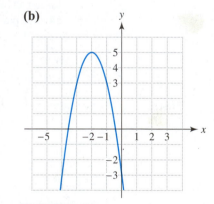

FIGURE 3.12   **FIGURE 3.13**

SOLUTION   **(a)** The graph in Figure 3.12 opens upward, so the leading coefficient a is positive. The vertex is the lowest point on the graph and is located at $(1, -2)$. The axis of symmetry is a vertical line passing through the vertex with equation $x = 1$.

(b) The graph in Figure 3.13 opens downward, so the leading coefficient a is negative. The vertex is the highest point on the graph and is located at $(-2, 5)$. The axis of symmetry is a vertical line through the vertex with equation $x = -2$.

Vertex Form

A well-conditioned athlete's heart rate can reach 200 beats per minute (bpm) during strenuous physical activity. Upon stopping an activity, a typical heart rate decreases rapidly at first and then gradually levels off, as shown in Table 3.1.

TABLE 3.1   Athlete's Heart Rate

Time (min)	0	2	4	6	8
Heart Rate (bpm)	200	150	110	90	80

Adapted from: V. Thomas, *Science and Sport*.

The data are not linear because for each 2-minute interval the heart rate does not decrease by a fixed amount. A scatterplot of the data is shown in Figure 3.14. In Figure 3.15 the data are modeled with a nonlinear function. Note that the graph of this function resembles the left half of a parabola that opens upward.

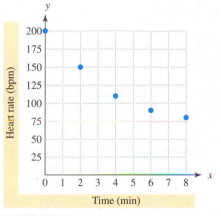

FIGURE 3.14

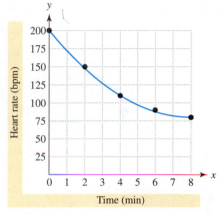

FIGURE 3.15

When a quadratic function is expressed as $f(x) = ax^2 + bx + c$, the coordinates of the vertex are not apparent. However, if a quadratic function is written as $f(x) = a(x - h)^2 + k$, then the vertex is located at (h, k), as illustrated in Figure 3.16. For example, the graph of $f(x) = -2(x - 1)^2 + 2$ is a parabola opening downward with vertex $(1, 2)$. See Figure 3.17.

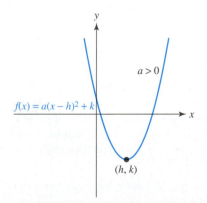

FIGURE 3.16   A Parabola with Vertex (h, k)

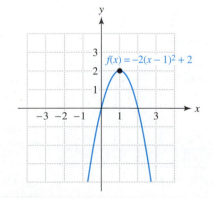

FIGURE 3.17   A Parabola with Vertex $(1, 2)$

Vertex form

The parabolic graph of $f(x) = a(x - h)^2 + k$ with $a \neq 0$ has vertex (h, k). Its graph opens upward when $a > 0$ and opens downward when $a < 0$.

| **EXAMPLE 3** | *Converting to standard form* |

Write $f(x) = 2(x - 1)^2 + 4$ in standard form.

SOLUTION   Begin by expanding the binomial $(x - 1)^2$.

Algebra Review
To review squaring a binomial, see Chapter R (page R-25).

$$2(x - 1)^2 + 4 = 2(x^2 - 2x + 1) + 4 \quad \text{Expand binomial.}$$
$$= 2x^2 - 4x + 2 + 4 \quad \text{Distributive property}$$
$$= 2x^2 - 4x + 6 \quad \text{Add terms.}$$

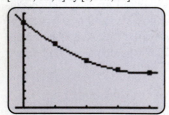

Thus $f(x) = 2(x - 1)^2 + 4$ is equivalent to $f(x) = 2x^2 - 4x + 6$.

| **EXAMPLE 4** | *Modeling an athlete's heart rate* |

Find a quadratic function f expressed in vertex form that models the data in Table 3.1. Support your result by graphing f and the data together.

SOLUTION   To model the data we use the left half of a parabola. Since the minimum heart rate of 80 beats per minute occurs when $x = 8$, let $(8, 80)$ be the vertex and write

$$f(x) = a(x - 8)^2 + 80.$$

Next, we must determine a value for the leading coefficient a. One possibility is to let the graph of f pass through the first data point $(0, 200)$, or equivalently, let $f(0) = 200$.

[−0.5, 8.5, 2] by [0, 220, 20]

FIGURE 3.18

$$f(0) = 200 \quad \text{Have the graph pass through } (0, 200).$$
$$a(0 - 8)^2 + 80 = 200 \quad \text{Let } x = 0 \text{ in } f(x).$$
$$a(0 - 8)^2 = 120 \quad \text{Subtract 80.}$$
$$a = \frac{120}{64} \quad \text{Divide by } (0 - 8)^2 = 64.$$
$$a = 1.875 \quad \text{Write as a decimal.}$$

Thus $f(x) = 1.875(x - 8)^2 + 80$ can be used to model the athlete's heart rate. A graph of f and the data are shown in Figure 3.18, which is similar to Figure 3.15. Figure 3.19 shows a table of $Y_1 = 1.875(X - 8)^2 + 80$. Although the table in Figure 3.19 does not match Table 3.1 exactly, it gives reasonable approximations. (Note that formulas for $f(x)$ may vary. For example, if we had selected the point $(2, 150)$ rather than $(0, 200)$, then $a \approx 1.94$. You may wish to verify this result.)

X	Y1
0	200
2	147.5
4	110
6	87.5
8	80

Y1◾1.875(X−8)^2...

FIGURE 3.19

Given the standard form for a quadratic function f, the following formula can be used to find the vertex.

Graphing Calculator Help
To plot data and graph an equation in the same viewing rectangle, see Appendix B (page AP-10). To create a table similar to Figure 3.19, see Appendix B (page AP-11).

Vertex formula

The *vertex* of the graph of $f(x) = ax^2 + bx + c$ with $a \neq 0$ is the point

$$\left(-\frac{b}{2a}, f\left(-\frac{b}{2a} \right) \right).$$

EXAMPLE 5 *Using the vertex formula*

Find the vertex of the graph of $f(x) = 1.5x^2 - 6x + 4$ symbolically. Support your answer graphically and numerically.

SOLUTION *Symbolic* If $f(x) = 1.5x^2 - 6x + 4$, then $a = 1.5$, $b = -6$, and $c = 4$. The x-coordinate of the vertex is

$$x = -\frac{b}{2a} = -\frac{(-6)}{2(1.5)} = 2.$$

The y-coordinate of the vertex can be found by evaluating $f(2)$.

$$y = f(2) = 1.5(2)^2 - 6(2) + 4 = -2$$

Thus the vertex is $(2, -2)$.

Graphical This result can be supported by graphing $Y_1 = 1.5X^2 - 6X + 4$ as shown in Figure 3.20. The vertex is the lowest point on the graph.

Numerical Numerical support is shown in Figure 3.21. The minimum y-value of -2 occurs when $x = 2$. Notice that if we move an equal distance left or right of the axis of symmetry, the y-values are equal. For example, when $x = 1$ or $x = 3$, $Y_1 = -0.5$. Similarly when $x = 0$ or 4, $Y_1 = 4$.

Graphing Calculator Help

To find a minimum point on a graph, see Appendix B (page AP-14).

$[-4.7, 4.7, 1]$ by $[-3.1, 3.1, 1]$

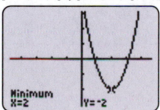

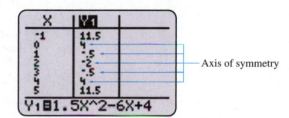

FIGURE 3.20   **FIGURE 3.21**

■ **MAKING CONNECTIONS**

Graphing and the Axis of Symmetry If the graph of a quadratic function f is folded along the axis of symmetry, the two halves of the parabola match exactly. If the axis of symmetry is $x = a$, then $f(a + k) = f(a - k)$ for every value of k. This is illustrated in Figure 3.22. For example, if the axis of symmetry is $x = 1$, then $f(0) = f(2)$ and $f(-1) = f(3)$.

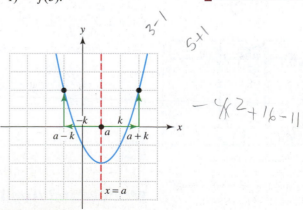

FIGURE 3.22

EXAMPLE 6   *Converting to vertex form*

Write $f(x) = 3x^2 + 12x + 7$ in vertex form. Support your answer graphically.

SOLUTION   Begin by finding the vertex on the graph of f.

$$x = -\frac{b}{2a} \qquad \text{Vertex formula}$$

$$= -\frac{12}{2(3)} \qquad \text{Let } a = 3 \text{ and } b = 12.$$

$$= -2 \qquad \text{Simplify.}$$

Since $f(-2) = 3(-2)^2 + 12(-2) + 7 = -5$, the vertex is $(-2, -5)$. The leading coefficient is $a = 3$, so $f(x)$ can be written in vertex form as $f(x) = 3(x + 2)^2 - 5$. In Figures 3.23 and 3.24, the graphs of $Y_1 = 3X^2 + 12X + 7$ and $Y_2 = 3(X + 2)^2 - 5$ appear identical.

$[-10, 10, 1]$ by $[-10, 10, 1]$ $\qquad$ $[-10, 10, 1]$ by $[-10, 10, 1]$

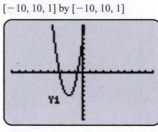

FIGURE 3.23   **FIGURE 3.24**

EXAMPLE 7   *Modeling quadratic data*

Use $f(x) = a(x - h)^2 + k$ to model the data in Table 3.2 exactly.

TABLE 3.2

x	2	3	4	5	6
y	10	1	-2	1	10

SOLUTION   Notice that the smallest y-value in Table 3.2 is $y = -2$ when $x = 4$. If x decreases or increases by 1 unit to $x = 3$ or $x = 5$, the y-value increases to 1. If x decreases or increases by 2 units to $x = 2$ or $x = 6$, the y-value increases to 10. This symmetry about $x = 4$ indicates that the axis of symmetry is $x = 4$ and the vertex is $(4, -2)$. It follows that $f(x) = a(x - 4)^2 - 2$. To determine the value of a, we can use any point in the table because f models the data exactly. For example, if we make the graph of f pass through $(5, 1)$, then $f(5) = 1$.

$$f(5) = 1 \qquad \text{Have the graph pass through } (5, 1).$$

$$a(5 - 4)^2 - 2 = 1 \qquad \text{Let } x = 5 \text{ in } f(x) = a(x - 4)^2 - 2.$$

$$a(5 - 4)^2 = 3 \qquad \text{Add 2.}$$

$$a = 3 \qquad \text{Divide by } (5 - 4)^2 = 1.$$

Graphing Calculator Help

To create a table similar to Figure 3.25, see Appendix B (page AP-11).

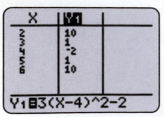

FIGURE 3.25

Thus $f(x) = 3(x - 4)^2 - 2$ models the data, which is supported in Figure 3.25 where $Y_1 = 3(X - 4)^2 - 2$.

Models and Applications

Quadratic functions occur in a number of real-life applications. One example is a falling object. If air resistance is ignored, then the quadratic function given by

$$s(t) = -16t^2 + v_0 t + h_0$$

models the height in feet of a falling object after t seconds, when its initial velocity is v_0 feet per second and its initial height is h_0 feet.

EXAMPLE 8 *Modeling the flight of a baseball*

A baseball is hit upward with an initial velocity of $v_0 = 80$ feet per second (or about 55 miles per hour) and leaves the bat with an initial height of $h_0 = 3$ feet, as shown in Figure 3.26.

(a) Write a formula $s(t)$ that models the height of the baseball after t seconds.
(b) How high was the baseball after 2 seconds?
(c) Find the maximum height of the baseball. Support your answer graphically.

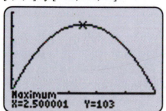

FIGURE 3.26

SOLUTION **(a)** Because $v_0 = 80$ and the initial height is $h_0 = 3$,

$$s(t) = -16t^2 + v_0 t + h_0$$
$$= -16t^2 + 80t + 3.$$

Graphing Calculator Help

To find a maximum point on a graph, see Appendix B (page AP-14).

[0, 5, 1] by [−20, 120, 20]

FIGURE 3.27

(b) Since $s(2) = -16(2)^2 + 80(2) + 3 = 99$, the baseball is 99 feet in the air after 2 seconds.

(c) Because $a < -16$, the graph of s is a parabola opening downward. The vertex is the highest point on the graph, with an x-coordinate of

$$x = -\frac{b}{2a} = -\frac{80}{2(-16)} = 2.5.$$

The corresponding y-coordinate of the vertex is

$$f(2.5) = -16(2.5)^2 + 80(2.5) + 3 = 103 \text{ feet.}$$

Thus the vertex is (2.5, 103), and the maximum height of the baseball was 103 feet. Graphical support is shown in Figure 3.27, where $Y_1 = -16X^2 + 80X + 3$ and the vertex is located at (2.5, 103).

The number of women in the work force increased dramatically during the past century. This growth can be modeled with a quadratic function, as demonstrated in the next example.

EXAMPLE 9 *Modeling the number of women in the work force*

The number of women that were gainfully employed in the work force for selected years is shown in Table 3.3.

TABLE 3.3

Year	1900	1910	1920	1930	1940	1950	1960	1970	1980	1990	1998
Work Force (millions)	5.3	7.4	8.6	10.8	12.8	18.4	23.2	31.5	45.5	56.6	63.7

Source: Department of Labor.

(a) Make a scatterplot of the data. Are the data linear or nonlinear?
(b) Find a quadratic function f so that $f(x) = a(x - h)^2 + k$ models the data. Support your result graphically.
(c) Predict the number of women in the labor force in 2010.

SOLUTION

(a) A scatterplot of the data is shown in Figure 3.28. The data are nonlinear.

(b) Since the smallest number of women in the work force occurred in 1900, we let the vertex be (1900, 5.3). It follows that

$$f(x) = a(x - 1900)^2 + 5.3.$$

[1890, 2010, 20] by [0, 70, 10]

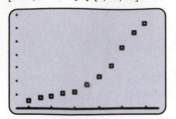

FIGURE 3.28

Next, we must determine a value for the leading coefficient a. One possibility is to let the graph of f pass through the last data point (1998, 63.7), or equivalently, let $f(1998) = 63.7$. (Note that if a different data point is used, the formula for $f(x)$ will usually change, giving slightly different results. When modeling, it is important to realize that more than one modeling function may be possible.)

[1890, 2010, 20] by [0, 70, 10]

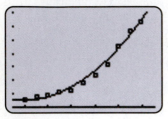

FIGURE 3.29

$$a(1998 - 1900)^2 + 5.3 = 63.7 \qquad \text{Let } f(1998) = 63.7.$$
$$a(98)^2 = 58.4 \qquad \text{Subtract 5.3 and simplify.}$$
$$a = \frac{58.4}{98^2} \qquad \text{Divide by } 98^2.$$
$$a \approx 0.006081 \qquad \text{Approximate } a.$$

So $f(x) = 0.006081(x - 1900)^2 + 5.3$. A graph of f and the data is shown in Figure 3.29.

(c) Because $f(2010) = 0.006081(2010 - 1900)^2 + 5.3 \approx 78.9$, f predicts that there may be 78.9 million women in the work force in 2010.

Graphing Calculator Help

To create a scatterplot, see Appendix B (page AP-6). To plot data and graph an equation in the same viewing rectangle, see Appendix B (page AP-10).

■ **MAKING CONNECTIONS**

Standard Form, Vertex Form, and Modeling When modeling quadratic data by hand, it is often easier to use vertex form: $f(x) = a(x - h)^2 + k$, rather than standard form: $f(x) = ax^2 + bx + c$. Because (h, k) corresponds to the vertex of a parabola, it may be appropriate to let (h, k) correspond to either the highest data point or the lowest data point in the scatterplot. Then a value for a can be found by substituting a data point into the formula for $f(x)$. In the next subsection, least-squares regression provides a quadratic modeling function in standard form.

Quadratic Regression

In Chapter 2 we discussed how a regression line could be found by the method of least squares. This method can also be applied to quadratic data and is illustrated in the next example.

EXAMPLE 10 *Finding a quadratic regression model*

In one study the efficiency of photosynthesis in an Antarctic species of grass was investigated. Table 3.4 lists results for various temperatures. The temperature x is in degrees Celsius and the efficiency y is given as a percent. The purpose of the research was to determine the temperature at which photosynthesis is most efficient. (Source: D. Brown and P. Rothery, *Models in Biology: Mathematics, Statistics and Computing.*)

TABLE 3.4

x (°C)	−1.5	0	2.5	5	7	10	12	15	17	20	22	25	27	30
y (%)	33	46	55	80	87	93	95	91	89	77	72	54	46	34

(a) Plot the data. Discuss reasons why a quadratic function might model the data.

(b) Find a least-squares function f given by $f(x) = ax^2 + bx + c$ that models the data.

(c) Graph f and the data in $[-5, 35, 5]$ by $[20, 110, 10]$. Discuss the fit.

(d) Determine the temperature at which f predicts that photosynthesis is most efficient. Compare the results with the findings of the experiment.

SOLUTION

(a) A plot of the data is shown in Figure 3.30. The y-values first increase and then decrease as the temperature x increases. The data suggest a parabolic shape opening downward.

$[-5, 35, 5]$ by $[20, 110, 10]$

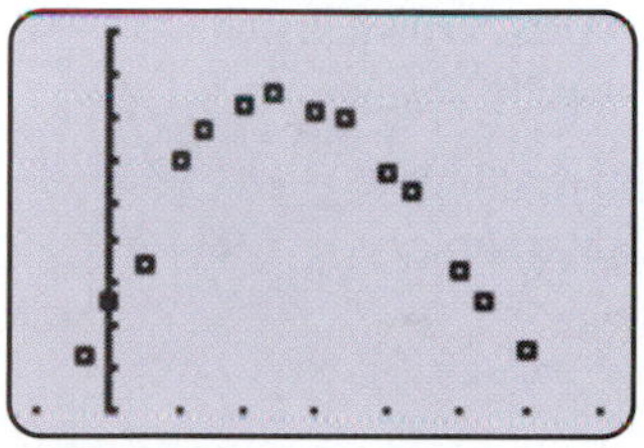

FIGURE 3.30

Graphing Calculator Help

To find an equation of least-squares fit, see Appendix B (page AP-15).

(b) Enter the data into your calculator, and then select quadratic regression from the menu, as shown in Figures 3.31 and 3.32. In Figure 3.33, the modeling function f is given (approximately) by $f(x) = -0.249x^2 + 6.77x + 46.37$.

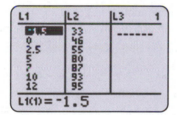

FIGURE 3.31

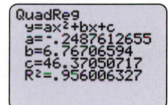

FIGURE 3.32

FIGURE 3.33

(c) The graph of $Y_1 = -.249X^2 + 6.77X + 46.37$ with the data is shown in Figure 3.34. Although the model is not exact, the parabola describes the general trend in the data. Notice that the parabola is opening downward since $a = -0.249 < 0$.

$[-5, 35, 5]$ by $[20, 110, 10]$

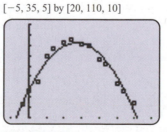

FIGURE 3.34

$[-5, 35, 5]$ by $[20, 110, 10]$

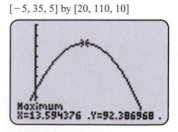

FIGURE 3.35

(d) The vertex is the highest point on the graph. Its coordinates are approximately (13.6, 92.4), as shown in Figure 3.35. This quadratic model predicts that the highest efficiency is about 92.4%, and it occurs near 13.6°C. Although there are percentages in the table higher than 92.4%, 13.6°C is a reasonable estimate for the optimum temperature. The function f is attempting to model the general trend in the data and predict future results.

Putting it all Together 3.1

The following table summarizes some important topics from this section.

Concept	Symbolic Representation	Comments and Examples
Quadratic function	$f(x) = ax^2 + bx + c$, where a, b, and c are constants with $a \neq 0$, (standard form).	It models data that are not linear. Its graph is a parabola (U-shape) that opens either upward ($a > 0$) or downward ($a < 0$).
Parabola	The graph of $y = ax^2 + bx + c$ is a parabola.	 Vertex: (h, k) Axis of symmetry: $x = h$ Maximum (or minimum) y-value on the graph: k
Vertex formula	The vertex for $f(x) = ax^2 + bx + c$ is the point $\left(-\dfrac{b}{2a}, f\left(-\dfrac{b}{2a}\right)\right).$	The x-value of the vertex on the graph of $f(x) = x^2 - 2x + 3$ is $$x = -\frac{-2}{2(1)} = 1.$$ y-value of vertex: $f(1) = 2$. Vertex: $(1, 2)$. Axis of symmetry: $x = 1$.
Vertex form	The vertex form for a quadratic function is $f(x) = a(x - h)^2 + k$, with vertex (h, k).	Let $f(x) = 2(x + 3)^2 - 5$. Parabola opens upward: $a = 2 > 0$ Vertex: $(-3, -5)$ Axis of symmetry: $x = -3$ Minimum y-value on graph: -5

3.1 EXERCISES

Basics of Quadratic Functions

Exercises 1–6: Sketch a graph of f.

1. $f(x) = x^2$

2. $f(x) = -2x^2$

3. $f(x) = -\dfrac{1}{2}x^2$

4. $f(x) = 4 - x^2$

5. $f(x) = x^2 - 3$

6. $f(x) = (x + 1)^2 - 2$

Exercises 7–14: *Identify f as being linear, quadratic, or neither. If f is quadratic, identify the leading coefficient and evaluate f(−2). Support your results graphically.*

7. $f(x) = 1 - 2x + 3x^2$

8. $f(x) = -5x + 11$

9. $f(x) = \dfrac{1}{x^2 - 1}$

10. $f(x) = (x^2 + 1)^2$

11. $f(x) = \dfrac{1}{2} - \dfrac{3}{10}x$ **12.** $f(x) = \dfrac{1}{5}x^2$

13. $f(x) = -3x^2 + 9$

14. $f(x) = 5x^2 - \sqrt{2}x$

Exercises 15–18: *The formulas for f(x) and g(x) are identical except for their leading coefficients a. Compare the graphs of f and g. You may want to support your answers by graphing f and g together.*

15. $f(x) = x^2,$ $g(x) = 2x^2$

16. $f(x) = \dfrac{1}{2}x^2,$ $g(x) = -\dfrac{1}{2}x^2$

17. $f(x) = 2x^2 + 1,$ $g(x) = -\dfrac{1}{3}x^2 + 1$

18. $f(x) = x^2 + x,$ $g(x) = \dfrac{1}{4}x^2 + x$

Exercises 19–22: *Use the graph of the quadratic function to determine the sign of the leading coefficient, its vertex, and the equation of the axis of symmetry.*

19.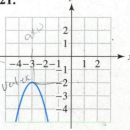

20.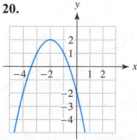

21.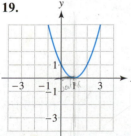

22.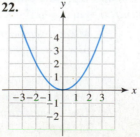

Vertex Form

Exercises 23–28: *Identify the vertex and leading coefficient. Then write f(x) in standard form.*

23. $f(x) = -3(x - 1)^2 + 2$

24. $f(x) = 5(x + 2)^2 - 5$

25. $f(x) = 5 - 2(x - 4)^2$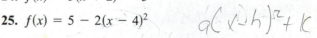

26. $f(x) = \dfrac{1}{2}(x + 3)^2 - 5$

27. $f(x) = \dfrac{3}{4}(x + 5)^2 - \dfrac{7}{4}$

28. $f(x) = -5(x - 4)^2$

Algebra Review To review squaring binomials, see Chapter R (page R-25).

Exercises 29–36: *Use the vertex formula to determine the vertex of the graph of f. Support your results graphically.*

29. $f(x) = 6 - x^2$ **30.** $f(x) = 2x^2 - 2x + 1$

31. $f(x) = x^2 - 6x$ **32.** $f(x) = -2x^2 + 4x + 5$

33. $f(x) = x^2 - 3.8x - 2$

34. $f(x) = -4x^2 + 16x$

35. $f(x) = 1.5 - 3x - 6x^2$

36. $f(x) = 0.25x^2 - 1.5x + 1$

Exercises 37–42: *Identify the leading coefficient and find the vertex. Then write f(x) in vertex form.*

37. $f(x) = x^2 + 4x - 5$

38. $f(x) = x^2 + 10x + 7$

39. $f(x) = 2x^2 - 8x - 1$

40. $f(x) = -\dfrac{1}{2}x^2 - x$

41. $f(x) = 2 - 9x - 3x^2$

42. $f(x) = 6 + 5x - 10x^2$

Exercises 43–44: *Use the graph of the quadratic function f to write it in vertex form.*

43.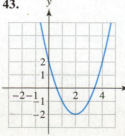

44.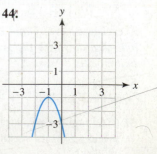

Exercises 45–48: Find $f(x) = a(x - h)^2 + k$ so that f models the data exactly.

45.

x	-1	0	1	2	3
y	5	-1	-3	-1	5

46.

x	-4	-3	-2	-1	0
y	1	-0.5	-1	-0.5	1

47.

x	-2	-1	0	1	2
y	2	4	2	-4	-14

48.

x	-4	-3	-2	-1	0
y	8	9	8	5	0

Models and Applications

49. *Hitting a Baseball* A baseball is hit so that its height in feet after t seconds is $s(t) = -16t^2 + 44t + 4$.
(a) How high was the baseball after 1 second?
(b) Find the maximum height of the baseball. Support your answer graphically.

50. *Throwing a Stone* A stone is thrown *downward* with a velocity of 66 feet per second (45 miles per hour) from a bridge that is 120 feet above a river, as illustrated in the accompanying figure.
(a) Write a formula $s(t)$ that models the height of the stone after t seconds.
(b) Does the stone hit the water within the first 10 seconds? Explain.

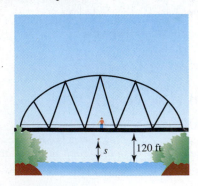

Exercises 51–54: Quadratic Models Match the physical situation with the graph of the quadratic function that models it best.

51. The height y of a stone thrown from ground level after x seconds

52. The number of people attending a popular movie x weeks after its opening

53. The temperature after x hours in a house where the furnace quits and a repairperson fixes it

54. The cumulative number of reported AIDS cases from 1980 to year x

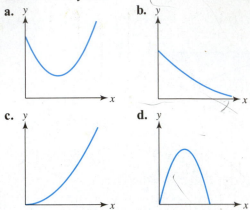

Exercises 55–58: Maximizing Altitude If air resistance is ignored, the height h of a projectile above the ground after x seconds is given by $h(x) = -\frac{1}{2}gx^2 + v_0x + h_0$. In this formula g is the acceleration due to gravity, v_0 is the projectile's initial velocity, and h_0 is its initial height. The leading coefficient is negative since gravity pulls downward. If v_0 is positive the projectile is propelled upward, and if v_0 is negative it is propelled downward. This formula is valid not only for Earth, but also for other celestial bodies. A ball is thrown straight up at 88 feet per second from a height of 25 feet.
(a) For the given value of g, graphically estimate both the maximum height and the time when this occurs.
(b) Solve part (a) symbolically.

55. $g = 32$ (Earth)

56. $g = 5.1$ (Moon)

57. $g = 13$ (Mars)

58. $g = 88$ (Jupiter)

Exercises 59–62: Modeling Real Data Each table contains real data that can be modeled approximately by $f(x) = a(x - h)^2 + k$.
(a) Make a scatterplot of the data. (Do not try to plot the undetermined point(s) listed in the table.)
(b) Find values for a, h, and k. Graph f together with the data in the same viewing rectangle.
(c) Use f to approximate the undetermined value(s) in the table.

59. Poverty income thresholds in dollars for one person

Year	1960	1965	1970	1975
Income	1490	1582	1954	2724

Year	1980	1985	1990	1995
Income	4190	5469	6652	?

Source: Bureau of the Census.

60. Poverty income thresholds in dollars for two people

Year	1960	1965	1970	1975
Income	1924	2048	2525	3506

Year	1980	1985	1990	1995
Income	5363	6998	8509	?

Source: Bureau of the Census.

61. Total cumulative AIDS cases in the United States

Year	1982	1984	1986	1988
Cases	1586	10,927	41,910	106,304

Year	1990	1992	1994	1998
Cases	196,576	329,205	441,528	?

Source: Department of Health and Human Services.

62. U.S. population in millions

Year	1800	1820	1840	1860
Population	5	10	17	31

Year	1870	1880	1900	1920
Population	?	50	76	106

Year	1940	1960	1980	2000
Population	132	179	226	?

Source: Bureau of the Census.

63. *Maximizing Revenue* Suppose the revenue R in thousands of dollars that a company receives from producing x-thousand compact disc players is given by $R(x) = x(40 - 2x)$.
  (a) Evaluate $R(2)$ and interpret the results.
  (b) How many CD players should the company produce to maximize its revenue?
  (c) What is the maximum revenue?

64. *Minimizing Cost* Suppose the cost C in dollars of producing x items is given by
$$C(x) = \frac{1}{2}x^2 - 50x + 2000.$$
  (a) Evaluate $C(20)$ and interpret the results.

  (b) How many of the items should the company produce to minimize its cost?
  (c) What is the minimum cost?

65. *Maximizing Area* A home owner has 80 feet of fence to enclose a rectangular garden. What dimensions for the garden give a maximum area?

Geometry Review To review formulas related to rectangles, see Chapter R (page R-1).

66. *Suspension Bridge* The cables that support a suspension bridge, such as the Golden Gate Bridge, can be modeled by parabolas. Suppose that a 300-foot long suspension bridge has towers at its ends that are 120 feet tall, as illustrated in the accompanying figure. If the cable comes within 20 feet of the road in the center of the bridge, find a quadratic function that models the height of the cable above the road a distance of x feet from the center of the bridge.

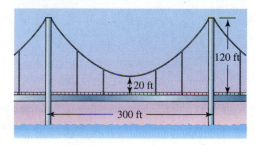

Quadratic Regression

Exercises 67–70: Quadratic Models Use least-squares regression to find a quadratic function f that models the data given in the table. Then estimate $f(3.5)$ to the nearest hundredth.

67.

x	0	2	4	6
$f(x)$	−1	16	57	124

68.

x	10	20	30	40
$f(x)$	4.2	24.3	84.1	184

69.

x	1	2	3	4
$f(x)$	4.14	10.12	18.92	30.56

70.

x	−2	−1	0	1	2
$f(x)$	−2.88	1.95	1.5	−4.25	−15.28

71. *Household Size*   The following table lists the average number of people aged 18 or older in a U.S. household for selected years.

Year	1940	1950	1960	1970
Household Size	2.53	2.31	2.12	2.05

Year	1980	1990	1998
Household Size	1.97	1.93	1.92

Source: Bureau of the Census.

(a) Find a quadratic function that models the data. Support your result graphically.
(b) Estimate the average number of people aged 18 or older in a household during 1975. Compare it to the actual value of 2.01.

72. *Consumer Price Index*   The following table lists the consumer price index (CPI) for selected years.

Year	1950	1960	1970	1980	1990	1998
CPI	24.1	29.6	38.8	82.4	130.7	163.0

Source: Bureau of the Census.

(a) Find a quadratic function that models the data. Support your result graphically.
(b) Estimate the CPI in 1995 and compare it to the actual value of 152.4.

Writing about Mathematics

73. How do the values of a, h, and k affect the graph of $f(x) = a(x - h)^2 + k$?

74. Explain why the vertex is important when you are trying to find either the maximum y-value or the minimum y-value on the graph of a quadratic function.

Extended and Discovery Exercises

Exercises 1–4: Linear and Quadratic Data   *Suppose we are given a table that lists ordered pairs (x, y), where the x-values increase by a fixed amount between consecutive data points. If the data are linear then the differences between consecutive y-values are always the same. For example, the following table contains linear data generated by $y = 3x + 1$. Notice that the increases in consecutive y-values are $7 - 4 = 3$, $10 - 7 = 3, 13 - 10 = 3$, and $16 - 13 = 3$. The increases always equal 3 for each unit increase in x. Thus when $x = 6, y = 16 + 3 = 19$, and this agrees with the formula $y = 3x + 1$.*

x	1	2	3	4	5
y	4	7	10	13	16

Quadratic data can be determined in a similar manner. The following table contains quadratic data generated by $y = x^2 + 2$. The increases in consecutive y-values are $6 - 3 = 3, 11 - 6 = 5, 18 - 11 = 7$, and $27 - 18 = 9$. The sequence 3, 5, 7, 9 is not constant, but determined by adding 2 to the previous increase. According to this pattern, the next increase should be $9 + 2 = 11$ so when $x = 6, y = 27 + 11 = 38$. This agrees with the formula $y = x^2 + 2$ when $x = 6$.

x	1	2	3	4	5
y	3	6	11	18	27

Use this method to determine if the data in the table are linear, quadratic, or neither.
If the data are either linear or quadratic, find the next y-value in the table.

1.

x	1	2	3	4	5	6
y	-6	-9	-16	-27	-42	?

2.

x	5	10	15	20	25	30
y	9	8	7	6	5	?

3.

x	-2	0	2	4	6	8
y	-15	-7	1	18	43	?

4.

x	-4	-2	0	2	4	6
y	-9	-7	-1	9	23	?

3.2   QUADRATIC EQUATIONS AND PROBLEM SOLVING

Basic Concepts • Solving Quadratic Equations • Problem Solving

Introduction

In Chapter 2, applications involving linear functions led to linear equations. In a similar manner, applications involving nonlinear functions often result in nonlinear equations. Linear equations can always be solved symbolically. This is not the case with some types of nonlinear equations. Their solutions must be approximated using numerical and graphical techniques.

Quadratic equations are one of the simplest types of nonlinear equations. They can be solved symbolically, as well as numerically and graphically.

Basic Concepts

The scatterplot in Figure 3.36 on the following page shows cumulative numbers of AIDS deaths in the United States from 1984 to 1994. In Figure 3.37, the data are modeled by a quadratic function D, given by

$$D(x) = 2375x^2 + 5134x + 5020,$$

where $x = 0$ corresponds to 1984, $x = 1$ to 1985, and so on until $x = 10$ corresponds to 1994. (Note that function D was found using quadratic regression.)

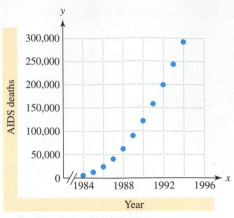

FIGURE 3.36 AIDS Deaths   **FIGURE 3.37** Modeling AIDS Deaths

Once the function D has been found, it is natural to use D to obtain information about AIDS. For example, to estimate the year when the number of AIDS deaths might reach 500,000, we could solve the following equation.

$$D(x) = 500,000$$

$2375x^2 + 5134x + 5020 = 500,000$ Substitute for $D(x)$.

$2375x^2 + 5134x - 494,980 = 0$ Subtract 500,000 from both sides.

The equation $2375x^2 + 5134x - 494,980 = 0$ is a *quadratic equation*. We solve this equation in Example 6.

Quadratic equation

A **quadratic equation** in one variable is an equation that can be written in the form

$$ax^2 + bx + c = 0,$$

where a, b, and c are real numbers with $a \neq 0$.

Solving Quadratic Equations

Quadratic equations can be solved graphically, numerically, and symbolically. Four basic symbolic strategies include factoring, the square root property, completing the square, and the quadratic formula. We begin by discussing factoring.

Factoring.   Factoring is a common technique used to solve equations. It is based on the **zero-product property**, which states that if $ab = 0$, then either $a = 0$ or $b = 0$. It is important to remember that this property only works for 0. For example, if $ab = 1$, then this equation does *not* imply that either $a = 1$ or $b = 1$, since $a = \frac{1}{2}$ and $b = 2$ also satisfies $ab = 1$.

EXAMPLE 1   *Solving a quadratic equation with factoring*

Solve the quadratic equation. Check your results.
(a) $2x^2 + 2x - 11 = 1$ **(b)** $12x^2 = x + 1$

SOLUTION   **(a)** Start by writing the equation in the form $ax^2 + bx + c = 0$.

$$2x^2 + 2x - 11 = 1 \qquad \text{Given equation}$$
$$2x^2 + 2x - 12 = 0 \qquad \text{Subtract 1 from both sides.}$$
$$x^2 + x - 6 = 0 \qquad \text{Divide both sides by 2. (Optional step)}$$
$$(x + 3)(x - 2) = 0 \qquad \text{Factor.}$$
$$x + 3 = 0 \quad \text{or} \quad x - 2 = 0 \qquad \text{Zero-product property}$$
$$x = -3 \quad \text{or} \quad x = 2 \qquad \text{Solve.}$$

Algebra Review
To review factoring, see
Chapter R (page R-27).

These solutions can be checked by substituting them into the given equation.

$$2(-3)^2 + 2(-3) - 11 \overset{?}{=} 1 \qquad\qquad 2(2)^2 + 2(2) - 11 \overset{?}{=} 1$$
$$1 = 1 \qquad\qquad\qquad\qquad\qquad 1 = 1$$

(b) Write the equation in the form $ax^2 + bx + c = 0$.

$$12x^2 = x + 1 \qquad \text{Given equation}$$
$$12x^2 - x - 1 = 0 \qquad \text{Subtract } x \text{ and 1.}$$
$$(3x - 1)(4x + 1) = 0 \qquad \text{Factor.}$$
$$3x - 1 = 0 \quad \text{or} \quad 4x + 1 = 0 \qquad \text{Zero-product property}$$
$$x = \frac{1}{3} \quad \text{or} \quad x = -\frac{1}{4} \qquad \text{Solve.}$$

To check these solutions, substitute them into the given equation.

$$12\left(\frac{1}{3}\right)^2 = \frac{1}{3} + 1 \qquad\qquad 12\left(-\frac{1}{4}\right)^2 = -\frac{1}{4} + 1$$
$$\frac{4}{3} = \frac{4}{3} \qquad\qquad\qquad\qquad \frac{3}{4} = \frac{3}{4}$$

EXAMPLE 2   *Solving a quadratic equation symbolically, graphically, and numerically*

On the shoreline surrounding certain types of recreational lakes in Minnesota, the Department of Natural Resources (DNR) restricts the sizes of buildings. Storage buildings built within 100 feet of the shoreline are sometimes limited to a maximum area of 120 square feet. Suppose a person is building a rectangular shed that has an area of 120 square feet and is 7 feet longer than it is wide. See Figure 3.38. Determine its dimensions symbolically, graphically, and numerically.

FIGURE 3.38

SOLUTION *Symbolic Solution* Let x be the width of the shed. Then $x + 7$ represents the length. Since area is equal to width times length, we solve the following equation.

$$x(x + 7) = 120 \qquad \text{Area is equal to 120 square feet.}$$
$$x^2 + 7x = 120 \qquad \text{Distributive property}$$
$$x^2 + 7x - 120 = 0 \qquad \text{Subtract 120 from both sides.}$$
$$(x + 15)(x - 8) = 0 \qquad \text{Factor.}$$
$$x + 15 = 0 \quad \text{or} \quad x - 8 = 0 \qquad \text{Zero-product property}$$
$$x = -15 \quad \text{or} \quad x = 8 \qquad \text{Solve.}$$

Since dimensions cannot be negative, the solution that has meaning is $x = 8$. The length is 7 feet longer than the width, so the dimensions of the shed are 8 feet by 15 feet.

Graphical Solution The graphs of $Y_1 = X(X + 7)$ and $Y_2 = 120$ intersect at the points $(-15, 120)$ and $(8, 120)$, as shown in Figures 3.39 and 3.40. The solutions to the equation are -15 and 8.

Numerical Solution Table $Y_1 = X(X + 7)$ and $Y_2 = 120$. The solution of 8 is shown in Figure 3.41, and the solution of -15 may be found by scrolling through the table. The equation $Y_1 = Y_2$ is satisfied when $x = -15$ or 8. The symbolic, graphical, and numerical solutions agree.

Graphing Calculator Help
To find a point of intersection, see Appendix B (page AP-11).

$[-20, 20, 5]$ by $[-20, 150, 10]$

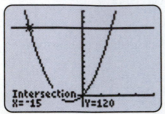

FIGURE 3.39

$[-20, 20, 5]$ by $[-20, 150, 10]$

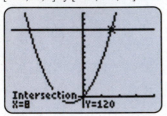

FIGURE 3.40

X	Y1	Y2
5	60	120
6	78	120
7	98	120
8	120	120
9	144	120
10	170	120
11	198	120

X=8

FIGURE 3.41

The Square Root Property. Some types of quadratic equations can be written as $x^2 = k$, where k is a positive number. The solutions to this equation are $\pm\sqrt{k}$. For example, $x^2 = 16$ has two solutions, ± 4. We refer to this as the **square root property**.

EXAMPLE 3 *Using the square root property*

If a metal ball is dropped 100 feet from a water tower, its distance above the ground after x seconds is given by $h(x) = 100 - 16x^2$. See Figure 3.42. Determine how long it takes the ball to hit the ground.

SOLUTION The ball strikes the ground when the equation $100 - 16x^2 = 0$ is satisfied.

$$100 - 16x^2 = 0$$
$$100 = 16x^2 \qquad \text{Add } 16x^2 \text{ to both sides.}$$
$$\frac{100}{16} = x^2 \qquad \text{Divide both sides by 16.}$$
$$\pm\sqrt{\frac{100}{16}} = x \qquad \text{Square root property}$$
$$\pm\frac{10}{4} = x \qquad \text{Simplify.}$$
$$x = \pm 2.5 \qquad \text{Rewrite.}$$

FIGURE 3.42

In this example only positive values for time are valid, so after 2.5 seconds the ball strikes the ground.

Completing the Square. Another technique that may be used to solve a quadratic equation is **completing the square**. If a quadratic equation can be written in the form $x^2 + kx = d$, where k and d are constants, then the equation can be solved using

$$x^2 + kx + \left(\frac{k}{2}\right)^2 = \left(x + \frac{k}{2}\right)^2.$$

For example, $k = 6$ in $x^2 + 6x = 7$, so add $\left(\frac{k}{2}\right)^2 = \left(\frac{6}{2}\right)^2 = 9$ to both sides.

$x^2 + 6x = 7$	Given equation
$x^2 + 6x + 9 = 7 + 9$	Add 9 to both sides.
$(x + 3)^2 = 16$	Factor the perfect square.
$x + 3 = \pm 4$	Square root property
$x = -3 \pm 4$	Subtract 3 from both sides.
$x = 1$ or $x = -7$	Simplify.

Algebra Review

To review factoring perfect square trinomials, see Chapter R (page R-31).

Completing the square is useful when solving quadratic equations that do not factor easily.

EXAMPLE 4   *Completing the square*

Solve the equation
(a) $x^2 - 8x + 9 = 0$ **(b)** $2x^2 - 8x = 7$

SOLUTION   **(a)** Start by writing the equation in the form $x^2 + kx = d$.

$x^2 - 8x + 9 = 0$	Given equation
$x^2 - 8x = -9$	Subtract 9 from both sides.
$x^2 - 8x + 16 = -9 + 16$	Add $\left(\frac{k}{2}\right)^2 = \left(\frac{-8}{2}\right)^2 = 16.$
$(x - 4)^2 = 7$	Factor the perfect square.
$x - 4 = \pm\sqrt{7}$	Square root property
$x = 4 \pm \sqrt{7}$	Add 4 to both sides.

(b) Divide both sides by 2 to obtain a 1 for the leading coefficient.

$2x^2 - 8x = 7$	Given equation
$x^2 - 4x = \dfrac{7}{2}$	Divide by 2.
$x^2 - 4x + 4 = \dfrac{7}{2} + 4$	Add $\left(\dfrac{-4}{2}\right)^2 = 4$ to both sides.
$(x - 2)^2 = \dfrac{15}{2}$	Factor the perfect square.
$x - 2 = \pm\sqrt{\dfrac{15}{2}}$	Square root property
$x = 2 \pm \sqrt{\dfrac{15}{2}}$	Add 2 to both sides.
$x \approx 4.74$ or $x \approx -0.74$	Approximate if desired.

```
2+√(15/2)
         4.738612788
2-√(15/2)
         -.7386127875
```

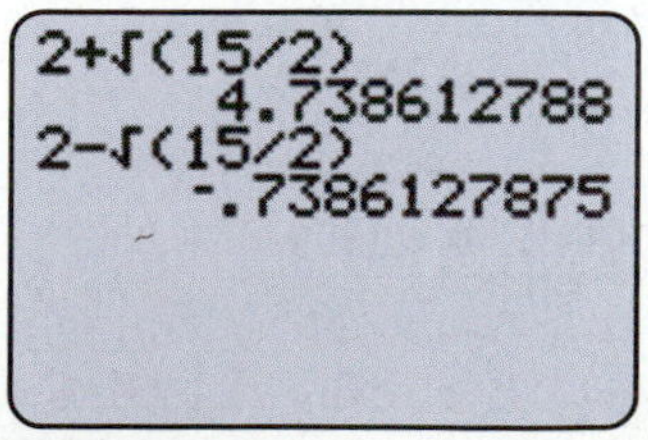

FIGURE 3.43

Figure 3.43 illustrates how to approximate the solutions.

The Quadratic Formula. The quadratic formula can be used to find the solutions to any quadratic equation. Its derivation is based on completing the square and left as an exercise.

> ### Quadratic formula
>
> The solutions of the quadratic equation $ax^2 + bx + c = 0$, where $a \neq 0$, are
> $$x = \frac{-b \pm \sqrt{b^2 - 4ac}}{2a}.$$

EXAMPLE 5 *Using the quadratic formula*

Solve the equation $3x^2 - 6x + 2 = 0$.

SOLUTION Let $a = 3$, $b = -6$, and $c = 2$.

$$x = \frac{-b \pm \sqrt{b^2 - 4ac}}{2a} \qquad \text{Quadratic formula}$$

$$x = \frac{-(-6) \pm \sqrt{(-6)^2 - 4(3)(2)}}{2(3)} \qquad \text{Substitute for } a, b, \text{ and } c.$$

$$x = \frac{6 \pm \sqrt{12}}{6} \qquad \text{Simplify.}$$

$$x = 1 \pm \frac{1}{6}\sqrt{12} \qquad \text{Divide.}$$

■ **CLASS DISCUSSION**

Use the results of Example 5 to evaluate each expression mentally.

$$3\left(1 + \frac{1}{6}\sqrt{12}\right)^2 - 6\left(1 + \frac{1}{6}\sqrt{12}\right) + 2$$

$$3\left(1 - \frac{1}{6}\sqrt{12}\right)^2 - 6\left(1 - \frac{1}{6}\sqrt{12}\right) + 2 \quad ■$$

EXAMPLE 6 *Estimating AIDS deaths*

Solve the quadratic equation $2375x^2 + 5134x - 494{,}980 = 0$ (as discussed earlier) to estimate when the number of AIDS deaths might reach 500,000. Use symbolic and graphical methods.

SOLUTION *Symbolic Solution* Apply the quadratic formula with $a = 2375$, $b = 5134$, and $c = -494{,}980$.

$$x = \frac{-b \pm \sqrt{b^2 - 4ac}}{2a}$$

$$x = \frac{-5134 \pm \sqrt{5134^2 - 4(2375)(-494{,}980)}}{2(2375)}$$

$$x = \frac{-5134 \pm \sqrt{4{,}728{,}667{,}956}}{4750}$$

$$x \approx 13.4 \qquad \text{or} \qquad x \approx -15.6$$

Since $x = 0$ corresponds to 1984, $x = -15.6$ represents $1984 - 15.6 = 1968.4$, and $x = 13.4$ represents $1984 + 13.4 = 1997.4$. AIDS was unknown in 1968, so the model estimates that about 500,000 AIDS deaths were reported by 1997.

Graphical Solution   Graph of $Y_1 = 2375X^2 + 5134X - 494980$. Two x-intercepts, or zeros, are located at $x \approx -15.6$ and $x \approx 13.4$, as shown in Figures 3.44 and 3.45. These graphs support our symbolic solutions.

Graphing Calculator Help

To find a zero, or x-intercept, see Appendix B (page AP-12).

$[-30, 30, 10]$ by $[-6 \times 10^5, 6 \times 10^5, 10^5]$

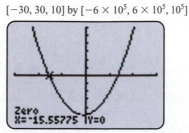

FIGURE 3.44

$[-30, 30, 10]$ by $[-6 \times 10^5, 6 \times 10^5, 10^5]$

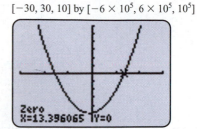

FIGURE 3.45

■ CLASS DISCUSSION

The number of reported AIDS deaths through 1997 was actually 390,242. Discuss a reason why the number of deaths was not 500,000, as was predicted by function D in Example 6. What are some problems with making predictions with modeling functions? ■

■ MAKING CONNECTIONS

Quadratic equations and graphical solutions   When solving a quadratic equation graphically, such as $4x^2 - 2x = 3$, we can graph $Y_1 = 4X^2 - 2X$ and $Y_2 = 3$. The x-coordinate of each point of intersection represents a solution. This *intersection-of-graphs method* was used in Example 2. A second approach is to write the equation as $4x^2 - 2x - 3 = 0$ and graph $Y_1 = 4X^2 - 2X - 3$. Each x-intercept, or zero, represents a solution. This *x-intercept method* was used in Example 6. You can use either graphical method to solve a quadratic equation.

The Discriminant.   One important difference between linear and nonlinear equations is that nonlinear equations may have more than one solution. If the quadratic equation $ax^2 + bx + c = 0$ is solved graphically, the parabola $y = ax^2 + bx + c$ can intersect the x-axis 0, 1, or 2 times as illustrated in Figures 3.46–3.48, respectively. Each x-intercept is a real solution to the quadratic equation $ax^2 + bx + c = 0$.

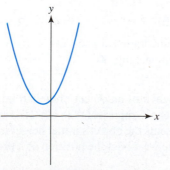

FIGURE 3.46 No Real Solutions

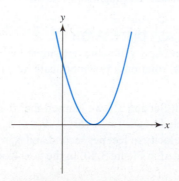

FIGURE 3.47 One Real Solution

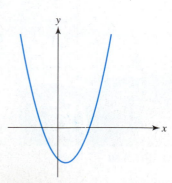

FIGURE 3.48 Two Real Solutions

The quantity $b^2 - 4ac$ in the quadratic formula is called the **discriminant**. It provides information about the number of solutions to a quadratic equation.

Quadratic equations and the discriminant

To determine the number of real solutions to $ax^2 + bx + c = 0$, with $a \neq 0$, evaluate the discriminant $b^2 - 4ac$.

1. If $b^2 - 4ac > 0$, there are two real solutions. *positive*
2. If $b^2 - 4ac = 0$, there is one real solution.
3. If $b^2 - 4ac < 0$, there are no real solutions.

Note: When $b^2 - 4ac < 0$, the solutions to a quadratic equation may be expressed as complex numbers. This is discussed in Section 4.4.

In Example 5 the discriminant is

$$b^2 - 4ac = (-6)^2 - 4(3)(2) = 12 > 0.$$

Since the discriminant is positive, there are two real solutions.

EXAMPLE 7   *Using the discriminant*

Use the discriminant to determine the number of solutions to the quadratic equation $9x^2 - 12.6x + 4.41 = 0$. Then solve the equation using the quadratic formula. Support your result graphically.

SOLUTION   *Symbolic Solution*   Let $a = 9$, $b = -12.6$, and $c = 4.41$. The discriminant is given by

$$b^2 - 4ac = (-12.6)^2 - 4(9)(4.41) = 0.$$

[0, 1.5, 0.1] by [−0.5, 1, 0.1]

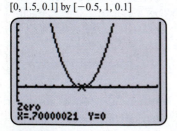

Zero
X=.70000021 Y=0

FIGURE 3.49

Since the discriminant is 0, there is one solution.

$$x = \frac{-b \pm \sqrt{b^2 - 4ac}}{2a} \qquad \text{Quadratic formula}$$

$$x = \frac{-(-12.6) \pm \sqrt{0}}{18} \qquad \text{Substitute.}$$

$$x = 0.7 \qquad \text{Simplify.}$$

The only solution is 0.7.

Graphical Solution   A graph of $y = 9x^2 - 12.6x + 4.41$ is shown in Figure 3.49. Notice that the graph suggests that there is one x-intercept corresponding to 0.7.

■ **CLASS DISCUSSION**
Explain how you could determine that the quadratic equation $ax^2 + bx + c = 0$ has no real solutions (a) symbolically and (b) graphically. ■

FIGURE 3.50

In Section 3.1 we learned that the height s of an object propelled into the air is modeled by $s(t) = -16t^2 + v_0 t + h_0$, where v_0 represents the object's (upward) initial velocity in feet per second and h_0 represents the object's initial height in feet, as illustrated in Figure 3.50. In the next example we model the position of a projectile.

EXAMPLE 8   *Modeling projectile motion*

Table 3.5 shows the height of a projectile fired into the air.

TABLE 3.5   Height of a Projectile

t (seconds)	0	2	4	6	8
$s(t)$ (feet)	10	346	554	634	586

(a) Use $s(t) = -16t^2 + v_0 t + h_0$ to model the data.
(b) After how many seconds did the projectile strike the ground?

SOLUTION   (a) If $t = 0$, then $s(0) = 10$ so the initial height is $h_0 = 10$, and

$$s(t) = -16t^2 + v_0 t + 10.$$

The value of v_0 can be determined by noting that when $t = 2$, $s(2) = 346$. (Other values for t and $s(t)$ could be used.) Substituting gives the following result.

$$-16(2)^2 + v_0(2) + 10 = 346 \qquad s(2) = 346$$
$$2v_0 = 400 \qquad \text{Add 54 to both sides.}$$
$$v_0 = 200 \qquad \text{Divide both sides by 2.}$$

Thus $s(t) = -16t^2 + 200t + 10$ models the height of the projectile. A table of values in Figure 3.51 supports this result.

(b) The projectile strikes the ground when $s(t) = 0$, or when

$$-16t^2 + 200t + 10 = 0,$$

which we solve graphically. (You can also solve this equation with the quadratic formula.) In Figure 3.52, the graph of $Y_1 = -16X^2 + 200X + 10$ is shown with a positive x-intercept at $x \approx 12.55$. Thus the projectile strikes the ground after about 12.55 seconds.

Graphing Calculator Help

To create a table similar to Figure 3.51, see Appendix B (page AP-11).

```
   X  |  Y1
  0   |  10
  2   | 346
  4   | 554
  6   | 634
  8   | 586

Y1 ⊟ -16X²+200X+10
```

FIGURE 3.51

Graphing Calculator Help

To locate a zero, or x-intercept, see Appendix B (page AP-12).

$[-4, 20, 2]$ by $[-500, 1000, 100]$

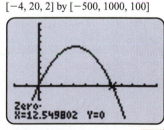

Zero
X=12.549802 Y=0

FIGURE 3.52

Problem Solving

Many types of applications involve quadratic equations. To solve these problems we use the steps for "Solving Application Problems" from Section 2.3. In the next example the dimensions of a box are found by solving a quadratic equation.

EXAMPLE 9   *Solving a construction problem*

A box is being constructed by cutting 2-inch squares from the corners of a rectangular piece of cardboard that is 6 inches longer than it is wide, as illustrated in Figure 3.53 on the next page. If the box is to have a volume of 224 cubic inches, find the dimensions of the cardboard.

Geometry Review

To review formulas related to boxes, see Chapter R (page R-4).

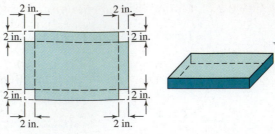

FIGURE 3.53

SOLUTION   **STEP 1:**   We are asked to find the dimensions of a rectangular piece of cardboard that is 6 inches longer than it is wide. Let x be its width and $x + 6$ be its length.

$$x: \text{ Width of the cardboard in inches}$$

$$x + 6: \text{ Length of the cardboard in inches}$$

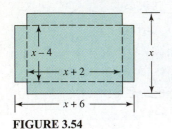

FIGURE 3.54

STEP 2:   First make a drawing of the box with the appropriate labeling, as shown in Figure 3.54. The width of the bottom of the box is $x - 4$ inches, because two square corners, with sides of 2 inches have been removed. Similarly, the length of the bottom of the box is $x + 2$ inches. Because the height times the width times the length must equal the volume, or 224 cubic inches, it follows that

$$2(x - 4)(x + 2) = 224$$

or

$$(x - 4)(x + 2) = 112.$$

STEP 3:   Write the quadratic equation in the form $ax^2 + bx + c = 0$ and factor.

$x^2 - 2x - 8 = 112$	Equation to be solved
$x^2 - 2x - 120 = 0$	Subtract 112.
$(x - 12)(x + 10) = 0$	Factor.
$x = 12 \quad \text{or} \quad x = -10$	Zero-product property

Since the dimensions cannot be negative, the width of the cardboard is 12 inches and the length is 6 inches more, or 18 inches.

STEP 4:   After the 2-inch square corners are cut out, the dimensions of the bottom of the box are $12 - 4 = 8$ inches by $18 - 4 = 14$ inches. The volume of the box is then $2 \cdot 8 \cdot 14 = 224$ cubic inches, which checks.

When compact discs (CDs) are manufactured, a discount is sometimes given to a customer who makes a large order. Discounts affect the revenue that a company receives, and we discuss this situation in the next example.

EXAMPLE 10   *Determining revenue*

A company charges $5 to burn (make) one CD, but it reduces this cost by $0.05 per CD for each additional CD ordered, up to a maximum of 60 CDs. For example, the price for one CD is $5, the price for two CDs is 2($4.95) = $9.90, the price for 3 CDs is 3($4.90) = $14.70, and so on. If the total price is $126, how many CDs were ordered?

SOLUTION   **STEP 1:**   We are asked to find the number of CDs that result in an order of $126. Let this number be x.

$$x: \text{ Number of CDs ordered}$$

STEP 2:   Revenue equals the number of CDs sold times the price of each CD. If x CDs are sold, then the price in dollars of each CD is $5 - 0.05(x - 1)$. (Note that when $x = 1$ the price is $5 - 0.05(1 - 1) = \$5$.) The revenue R is given by

$$R(x) = x(5 - 0.05(x - 1))$$

and we must solve the equation

$$x(5 - 0.05(x - 1)) = 126.$$

STEP 3:   Although this equation could be solved by using the quadratic formula, it is easier to solve it graphically by letting $Y_1 = X(5 - 0.05(X - 1))$ and $Y_2 = 126$. In Figures 3.55 and 3.56, their graphs intersect at $(45, 126)$ and $(56, 126)$. Either 45 or 56 compact discs were ordered.

Graphing Calculator Help

To find a point of intersection, see Appendix B (page AP-11).

[0, 60, 10] by [0, 150, 50]

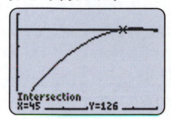

[0, 60, 10] by [0, 150, 50]

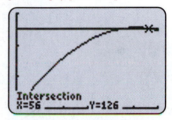

FIGURE 3.55   **FIGURE 3.56**

STEP 4:   If 45 compact discs are ordered, then the cost of each disc would be $5 - 0.05(44) = \$2.80$ and the total revenue would be $45 \cdot 2.80 = \$126$. The solution for 56 compact discs can be checked in a similar manner.

Putting it all Together

3.2

*T*he following table summarizes important topics related to quadratic equations.

Concept	Explanation	Examples
Quadratic equation	$ax^2 + bx + c = 0$, where a, b, and c are constants with $a \neq 0$.	A quadratic equation can have 0, 1, or 2 real solutions.
		$x^2 = -5$ (No real solutions)
		$(x - 2)^2 = 0$ (One real solution)
		$x^2 - 4 = 0$ (Two real solutions)
Factoring	A symbolic technique for solving equations; based on the zero-product property: if $ab = 0$, then either $a = 0$ or $b = 0$	$x^2 - 3x = -2$
		$x^2 - 3x + 2 = 0$
		$(x - 1)(x - 2) = 0$
		$x - 1 = 0$ or $x - 2 = 0$
		$x = 1$ or $x = 2$

Concept	Explanation	Examples
Square root property	The solutions to $x^2 = k$ are $x = \pm\sqrt{k}$.	$x^2 = 9$ is equivalent to $x = \pm 3$. $x^2 = 11$ is equivalent to $x = \pm\sqrt{11}$.
Completing the square	To solve $x^2 + kx = d$ symbolically, add $\left(\dfrac{k}{2}\right)^2$ to both sides to obtain a perfect trinomial square. Use the square root property to solve.	$$x^2 - 6x = 1$$ $$x^2 - 6x + 9 = 1 + 9 \qquad \left(\frac{-6}{2}\right)^2 = 9$$ $$(x-3)^2 = 10$$ $$x - 3 = \pm\sqrt{10}$$ $$x = 3 \pm\sqrt{10}$$
Quadratic formula	The solutions to $ax^2 + bx + c = 0$ are $$x = \frac{-b \pm \sqrt{b^2 - 4ac}}{2a}.$$	To solve $2x^2 - x - 4 = 0$, let $a = 2$, $b = -1$, and $c = -4$. $$x = \frac{-(-1) \pm \sqrt{(-1)^2 - 4(2)(-4)}}{2(2)}$$ $$= \frac{1 \pm \sqrt{33}}{4} \approx 1.69 \text{ or } -1.19$$
Graphical solution	To solve $ax^2 + bx + c = 0$, let Y_1 equal the left side of the equation and graph Y_1. The solutions correspond to the x-intercepts.	To solve $0.5x^2 + 1.5x - 0.75 = 0$, graph $Y_1 = 0.5X\text{^}2 + 1.5X - 0.75$. The solutions are $x \approx 0.44$ or $x \approx -3.44$. $[-6, 6, 1]$ by $[-4, 4, 1]$
Numerical solution	To solve $ax^2 + bx + c = 0$, let Y_1 equal the left side of the equation and create a table for Y_1. The zeros of Y_1 are the solutions.	To solve $2x^2 + x - 1 = 0$, make a table for $Y_1 = 2X\text{^}2 + X - 1$. The solutions are $x = -1$ or $x = 0.5$.

3.2 EXERCISES

Quadratic Equations

Exercises 1–4: The graph of $f(x) = ax^2 + bx + c$ is shown in the figure.

  (a) State whether $a > 0$ or $a < 0$.
  (b) Solve the equation $ax^2 + bx + c = 0$.
  (c) Determine if the discriminant is positive, negative, or zero.

1. 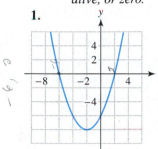   **2.**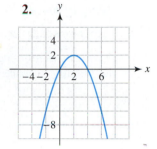

3. 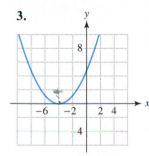   **4.**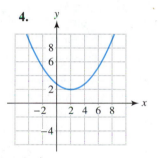

Exercises 5–10: Solve each quadratic equation

  (a) graphically,
  (b) numerically, and
  (c) symbolically.

Express graphical and numerical solutions to the nearest tenth when appropriate.

5. $x^2 - x - 6 = 0$ **6.** $2x^2 + 5x - 3 = 0$

7. $2x^2 = 6$ **8.** $x^2 - 225 = 0$

9. $4x^2 - 12x = -9$ **10.** $-4x(x - 1) = 1$

Algebra Review To review factoring, see Chapter R (page R-27).

Exercises 11–16: Solve the quadratic equation graphically.

11. $20x^2 + 11x = 3$ **12.** $x^2 - 3.1x = 0.32$

13. $2.5x^2 = 4.75x - 2.1$ **14.** $-2x^2 + 4x = 1.595$

15. $x(x + 24) = 6912$

16. $x(31.4 - 2x) = -348$

Exercises 17–32: Complete the following.

  (a) Write each equation as $ax^2 + bx + c = 0$.
  (b) Calculate the discriminant $b^2 - 4ac$ and determine the number of real solutions.
  (c) Solve each equation.

17. $3x^2 = 12$ **18.** $8x^2 - 2 = 14$

19. $x^2 - 2x = -1$ **20.** $6x^2 = 4x + 2$

21. $4x - 2 = x^2$ **22.** $2x^2 - 3x - 3 = 0$

23. $x^2 + 1 = x$ **24.** $2x^2 + x = 2$

25. $2x^2 + 3x = 12 - 2x$ **26.** $3x^2 - 2 = 5x$

27. $\frac{1}{4}x^2 + 3x = x - 4$ **28.** $9x(x - 4) = -36$

29. $x\left(\frac{1}{2}x + 1\right) = -\frac{13}{2}$ **30.** $4x = 6 + x^2$

31. $3x^2 = 1 - x$ **32.** $x(5x - 3) = 1$

Exercises 33–40: Solve the equation by completing the square.

33. $x^2 + 4x = 6$ **34.** $x^2 - 10x = 1$

35. $x^2 + 5x - 4 = 0$ **36.** $2x^2 + 4x - 5 = 0$

37. $3x^2 - 6x - 2 = 0$ **38.** $2x^2 - 3x + 1 = 0$

39. $x^2 - 8x = 10$ **40.** $5x^2 - 10x = 2$

Algebra Review To review factoring perfect squares, see Chapter R (page R-31).

Applications Involving Quadratic Equations

41. *Safe Runway Speed* The road (or taxiway) used by aircraft to exit from the runway should not contain sharp curves. The safe radius for any curve depends on the taxiing speed of an airplane. The table lists the recommended minimum radius R of these exit curves, where the taxiing speed of the airplane is x miles per hour. (Source: Federal Aviation Administration.)

x (mph)	10	20	30	40	50	60
R (ft)	50	200	450	800	1250	1800

  (a) If the taxiing speed x of the plane doubles, what happens to the minimum radius R of the curve?
  (b) The Federal Aviation Administration (FAA) used $R(x) = ax^2$ to compute the values in the table. Determine the constant a.
  (c) If $R = 500$, find x. Interpret your results.

42. *Falling Object* The table lists the velocity and distance traveled by a falling object.

Elapsed Time (sec)	0	1	2	3	4	5
Velocity (ft/sec)	0	32	64	96	128	160
Distance (ft)	0	16	64	144	256	400

(a) Make a scatterplot of the ordered pairs (time, velocity) and (time, distance) in the same viewing rectangle $[-1, 6, 1]$ by $[-10, 450, 50]$.

(b) Find a function v that models the velocity.

(c) The distance is modeled by $d(x) = ax^2$. Find the constant a.

(d) Use the table to mentally estimate the elapsed time when the distance traveled by the falling object is 200 feet. Determine this time symbolically using d. Find the velocity at this time.

43. *Airline Passengers* The number of worldwide airline passengers y between 1950 and 1990 in millions is approximated in the table. The variable x represents the year, where $x = 0$ corresponds to 1950 and $x = 40$ to 1990. (Source: International Civil Aviation Organization.)

x	0	10	20	30	40
y	31	106	386	734	1273

The data may be modeled by $f(x) = 0.68x^2 + 3.8x + 24$.

(a) Graph f and the data in $[-5, 45, 5]$ by $[0, 1500, 100]$.

(b) Determine graphically the year when there were 1 billion passengers.

(c) Solve part (b) symbolically.

44. *Projectile Motion* (Refer to Example 8.) The accompanying table shows the height of a projectile that is fired into the air.

t (seconds)	0	1	2	3	4
s (feet)	32	176	288	368	416

(a) Find $s(t) = -16t^2 + v_0 t + h_0$ so that s models the data.

(b) After how many seconds did the projectile strike the ground?

45. *Biology* Some types of worms have a remarkable capacity to live without moisture. The table shows the number of worms y surviving after x days in one study. (Source: D. Brown and P. Rothery, *Models in Biology: Mathematics, Statistics and Computing.*)

x (days)	0	20	40	80	120	160
y (worms)	50	48	45	36	20	3

(a) Find a quadratic function in the form $f(x) = a(x - h)^2 + k$ that models this data.

(b) Solve the quadratic equation $f(x) = 0$. Do both solutions have real meaning? Explain.

46. *Stopping Distance* Braking distance for cars may be approximated by $D(x) = \dfrac{x^2}{30k}$. The input x is the car's velocity in miles per hour and the output $D(x)$ is the braking distance in feet. The positive constant k is a measure of the traction of the tires. Small values of k indicate a slippery road. (Source: L. Haefner, *Introduction to Transportation Systems.*)

(a) Let $k = 0.3$. Evaluate $D(60)$ and interpret the result.

(b) If $k = 0.25$, find the velocity x that corresponds to a braking distance of 300 feet.

47. *Screen Dimensions* The width of a rectangular computer screen is 2.5 inches more than its height. If the area of the screen is 93.5 square inches, determine its dimensions.

48. *Maximizing Area* A rectangular pen for a pet is under construction using 100 feet of fence.

(a) Determine the dimensions that result in an area of 576 square feet.

(b) Find the dimensions that give maximum area.

49. *Construction* (Refer to Example 9.) A box is being constructed by cutting 4-inch squares from the corners of a rectangular sheet of metal that is 10 inches longer than it is wide. If the box is to have a volume of 476 cubic inches, find the dimensions of the metal sheet.

50. *Construction* A cylindrical aluminum can is being constructed to have a height h of 4 inches, as illustrated in the accompanying figure. If the can is to have a volume of 28 cubic inches, approximate its radius r. (*Hint*: $V = \pi r^2 h$.)

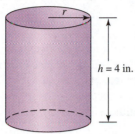

Geometry Review To review formulas related to cylinders, see Chapter R (page R-4).

51. *Picture Frame* A frame for a picture is 2 inches wide. The picture inside the frame is 4 inches longer than it is wide. See the accompanying figure. If the area of the picture is 320 square inches, find the outside dimensions of the picture frame.

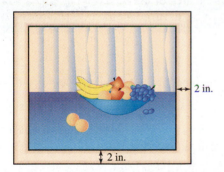

52. *Ticket Prices* One airline ticket costs $250. For each additional airline ticket sold to a group, the price of all the tickets is reduced by $2. For example, 2 tickets cost $2 \cdot 248 = \$496$ and 3 tickets cost $3 \cdot 246 = \$738$.

(a) Write a quadratic function that gives the total cost of buying x tickets.

(b) What is the cost of 5 tickets?

(c) How many tickets were sold if the total cost was $5200?

(d) What number of tickets sold results in the greatest cost?

53. *Window Dimensions* A window is comprised of a square with sides of length x and a semicircle with diameter x, as shown in the accompanying figure. If the total area of the window is 463 square inches, estimate the value of x.

Geometry Review To review formulas related to circles, see Chapter R (page R-2).

54. *Quadratic Formula* Prove the quadratic formula by completing the following.

(a) Write $ax^2 + bx + c = 0$ as $x^2 + \dfrac{b}{a}x = -\dfrac{c}{a}$.

(b) Complete the square to obtain
$$\left(x + \frac{b}{2a}\right)^2 = \frac{b^2 - 4ac}{4a^2}.$$

(c) Use the square root property and solve for x.

Writing about Mathematics

55. Discuss three symbolic methods for solving a quadratic equation. Make up a quadratic equation and use each method to find the solution set.

56. Explain how to solve a quadratic equation graphically.

CHECKING BASIC CONCEPTS FOR SECTIONS 3.1 AND 3.2

1. Graph $f(x) = (x - 1)^2 - 4$. Identify the vertex, axis of symmetry, and x-intercepts.

2. A graph of $y = ax^2 + bx + c$ is shown.
   (a) Which of the following is true: $a > 0$, $a < 0$, or $a = 0$?
   (b) Find the vertex and axis of symmetry.
   (c) Solve $ax^2 + bx + c = 0$.

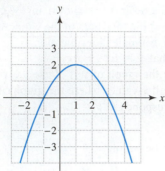

3. Use $f(x) = a(x - h)^2 + k$ to model the data in the table exactly.

x	−3	−2	−1	0	1
$f(x)$	11	5	3	5	11

4. Find the vertex on the graph of $y = 3x^2 - 9x - 2$.

5. Solve the quadratic equations.
   (a) $16x^2 = 81$
   (b) $2x^2 + 3x = 2$
   (c) $x^2 = x - 3$

6. A rectangle is 4 inches longer than it is wide and has an area of 165 square inches. Find its dimensions.

3.3 QUADRATIC INEQUALITIES

Graphical Solutions • Symbolic and Numerical Solutions

Introduction

Highway engineers often use quadratic functions to model safe stopping distances for cars. For example, $f(x) = \frac{1}{12}x^2 + \frac{11}{5}x$ is sometimes used to model the stopping distance for a car traveling at x miles per hour on dry, level pavement. If a driver can see 200 feet ahead on a highway with a sharp curve, then safe driving speeds x satisfy the quadratic inequality

$$\frac{1}{12}x^2 + \frac{11}{5}x \le 200$$

or equivalently,

$$\frac{1}{12}x^2 + \frac{11}{5}x - 200 \le 0.$$

A quadratic equation can be written as $ax^2 + bx + c = 0$. If the equals sign is replaced by $>$, $\ge$, $<$, or $\le$ a **quadratic inequality** results. Since equality is (usually) the boundary between *greater than* and *less than*, a first step in solving a quadratic inequality is to determine the x-values where equality occurs. These x-values are the *boundary numbers*. We begin by discussing graphical solutions of quadratic inequalities.

Graphical Solutions

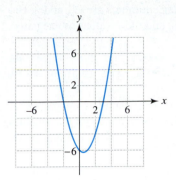

FIGURE 3.57

The graph of a quadratic function is a parabola that opens either upward or downward. For example, Figure 3.57 shows the parabola $y = x^2 - x - 6$.

Since $a = 1 > 0$, this parabola opens upward. It has x-intercepts -2 and 3, which satisfy the equation $x^2 - x - 6 = 0$. The parabola lies below the x-axis between the intercepts (or boundary numbers), so the solution set to the inequality $x^2 - x - 6 < 0$ is $\{x \mid -2 < x < 3\}$. Similarly, the solutions to $x^2 - x - 6 > 0$ include x-values either left of $x = -2$ or right of $x = 3$, where the parabola is above the x-axis. This solution set is $\{x \mid x < -2 \text{ or } x > 3\}$.

■ MAKING CONNECTIONS

Graphs and Inequalities   Suppose that $y = ax^2 + bx + c$. Then the solution set to $y > 0$ includes all x-values where the graph of y is *above* the x-axis. Similarly, the solution set to $y < 0$ includes all x-values where the graph of y is *below* the x-axis. This is because the x-axis corresponds to $y = 0$.   ■

These concepts are applied in the next example.

EXAMPLE 1   *Solving quadratic inequalities graphically*

Use the graphs of $y = ax^2 + bx + c$ in Figures 3.58–3.61 to solve the inequalities:
i. $ax^2 + bx + c \le 0$ and **ii.** $ax^2 + bx + c > 0$.

(a)

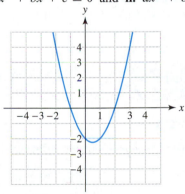

FIGURE 3.58

(b)

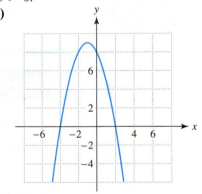

FIGURE 3.59

(c)

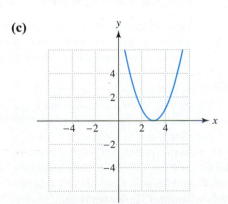

FIGURE 3.60

(d)

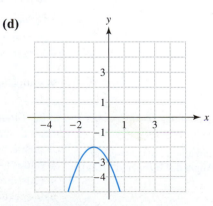

FIGURE 3.61

SOLUTION **(a)** **i.** $ax^2 + bx + c \le 0$ The x-intercepts or boundary numbers are -1 and 2. Between these x-values, the graph is below the x-axis, and the y-values are negative. The solution set is $\{x \mid -1 \le x \le 2\}$.

ii. $ax^2 + bx + c > 0$ The graph is above the x-axis, and the y-values are positive, either left of $x = -1$ or right of $x = 2$. The solution set is $\{x \mid x < -1 \text{ or } x > 2\}$.

(b) **i.** $ax^2 + bx + c \le 0$ The parabola opens downward with x-intercepts or boundary numbers of -4 and 2. The graph is below the x-axis, and the y-values are negative, either left of $x = -4$ or right of $x = 2$. The solution set is $\{x \mid x \le -4 \text{ or } x \ge 2\}$.

ii. $ax^2 + bx + c > 0$ Between $x = -4$ and $x = 2$, the graph is above the x-axis and the y-values are positive. The solution set is $\{x \mid -4 < x < 2\}$.

(c) **i.** $ax^2 + bx + c \le 0$ The graph never dips below the x-axis so the y-values are never negative. There is one x-intercept at $x = 3$, where $y = 0$. The solution set is $\{x \mid x = 3\}$.

ii. $ax^2 + bx + c > 0$ The graph is always above the x-axis except at $x = 3$. Thus every real number except $x = 3$ is a solution. The solution set is $\{x \mid x \ne 3\}$.

(d) **i.** $ax^2 + bx + c \le 0$ The graph always lies below the x-axis so the y-values are negative and always satisfy the given inequality. The solution set includes all real numbers or $\{x \mid -\infty < x < \infty\}$.

ii. $ax^2 + bx + c > 0$ The graph is never above the x-axis so the solution set is empty.

■ **CLASS DISCUSSION**

Sketch a graph of $y = ax^2 + bx + c$ if the quadratic inequality $ax^2 + bx + c < 0$ satisfies the following conditions.

(a) $a > 0$, solution set: $\{x \mid -1 < x < 3\}$

(b) $a < 0$, solution set: $\{x \mid x \ne 1\}$

(c) $a < 0$, solution set: $\{x \mid x < -2 \text{ or } x > 2\}$ ■

EXAMPLE 2   *Determining safe speeds*

In the introduction to this section the quadratic inequality

$$\frac{1}{12}x^2 + \frac{11}{5}x \le 200$$

was introduced. Recall that $f(x) = \frac{1}{12}x^2 + \frac{11}{5}x$ computes the stopping distance in feet for a car traveling x miles per hour on dry, level pavement. Solve this inequality to determine safe speeds on a curve where a driver can see the road ahead for at most 200 feet. What might be a safe speed limit to drive on this curve?

SOLUTION   We can solve this inequality by graphing $Y_1 = X^2/12 + 11X/5$ and $Y_2 = 200$, as shown in Figure 3.62. Since we are interested in *positive* speeds, we need to locate only the point of intersection where x is positive. This occurs when $x \approx 37.5$. For positive x-values to the left of $x \approx 37.5$, $Y_1 < 200$. Thus safe speeds are less than 37.5 miles per hour. A reasonable speed limit might be 35 miles per hour.

Graphing Calculator Help

To find a point of intersection, see Appendix B (page AP-11).

$[-100, 100, 50]$ by $[-300, 300, 100]$

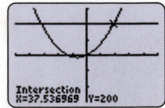

FIGURE 3.62

In the next example we model a person's heart rate.

EXAMPLE 3 *Modeling heart rate*

Suppose that a person's heart rate, x minutes after vigorous exercise has stopped, can be modeled by $f(x) = \frac{4}{5}(x - 10)^2 + 80$. The output is in beats per minute and $0 \le x \le 10$.

(a) Evaluate $f(0)$ and $f(2)$. Interpret the result.

(b) Estimate the times when the person's heart rate was between 100 and 120 beats per minute, inclusively.

SOLUTION **(a)** $f(0) = \frac{4}{5}(0 - 10)^2 + 80 = 160$ and $f(2) = \frac{4}{5}(2 - 10)^2 + 80 = 131.2$. Initially when the person stops exercising the heart rate is 160 beats per minute, and after 2 minutes the heart rate has dropped to about 131 beats per minute.

(b) Graph $Y_1 = 100$, $Y_2 = (.8)(X - 10)\wedge 2 + 80$, and $Y_3 = 120$. Then find their points of intersection, as shown in Figures 3.63 and 3.64. The person's heart rate is between 100 and 120, when the graph of Y_2 is between the graphs of Y_1 and Y_3. This occurs between 2.9 minutes (approximately) and 5 minutes after the person stops exercising. Numerical support is shown in Figure 3.65, where $Y_1 \approx 120$ when $x = 3$, and $Y_1 = 100$ when $x = 5$.

[0, 10, 1] by [60, 180, 20]

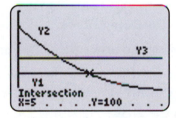

FIGURE 3.63

[0, 10, 1] by [60, 180, 20]

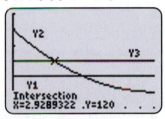

FIGURE 3.64

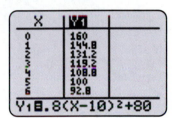

FIGURE 3.65

Symbolic and Numerical Solutions

The reasoning that we used to solve quadratic inequalities graphically can be applied to symbolic solutions. For symbolic solutions we also rely on numerical help. The first step in solving a quadratic inequality is to find the boundary numbers. For example, to solve

$$2x^2 - 5x - 12 < 0,$$

replace $<$ with $=$ and solve the equation.

$$2x^2 - 5x - 12 = 0 \qquad \text{Quadratic equation}$$
$$(2x + 3)(x - 4) = 0 \qquad \text{Factor.}$$
$$2x + 3 = 0 \quad \text{or} \quad x - 4 = 0 \qquad \text{Zero-product property}$$
$$x = -\frac{3}{2} \quad \text{or} \quad x = 4 \qquad \text{Solve.}$$

The boundary numbers $-\frac{3}{2}$ and 4 separate the number line into three disjoint intervals:

$-\infty < x < -\frac{3}{2}$, $-\frac{3}{2} < x < 4$, and $4 < x < \infty$, as illustrated in Figure 3.66 on the next page.

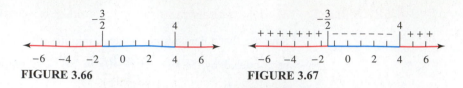

FIGURE 3.66   **FIGURE 3.67**

The expression $2x^2 - 5x - 12$ is either always positive or always negative on each interval. To determine where $2x^2 - 5x - 12 < 0$ we can use the **test values** shown in Table 3.6. For example, since the test value -2 lies in the interval $-\infty < x < -\frac{3}{2}$, and $2x^2 - 5x - 12$ evaluated at $x = -2$ equals $6 > 0$, it follows that the expression $2x^2 - 5x - 12$ is always positive whenever $-\infty < x < -\frac{3}{2}$. This interval has $+$ signs on the number line in Figure 3.67.

TABLE 3.6   A Table of Test Values

x	-2	$-\dfrac{3}{2}$	0	4	6
$2x^2 - 5x - 12$	6	0	-12	0	30

Graphing Calculator Help

To create a table similar to Figure 3.68, see Appendix B (page AP-11).

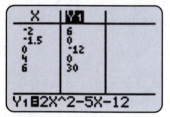

FIGURE 3.68

From Table 3.6 we can see that the expression $2x^2 - 5x - 12$ is negative when $x = 0$, so it is always negative between the boundary numbers of $-\frac{3}{2}$ and 4. Negative signs are shown on the number line in Figure 3.67 in this interval. Finally, when $x = 6$, $2x^2 - 5x - 12 = 30 > 0$ and $+$ signs are placed along the x-axis in Figure 3.67 when $x > 4$. Therefore, the solution set for $2x^2 - 5x - 12 < 0$ is $\left\{ x \mid -\frac{3}{2} < x < 4 \right\}$.

Table 3.6 can also be generated with a graphing calculator, as illustrated in Figure 3.68. Note that different test values could be used. However, it is important to choose one test value less than $-\frac{3}{2}$, one test value between $-\frac{3}{2}$ and 4, and one test value greater than 4. Graphical support is shown in Figure 3.69, where the graph of $y = 2x^2 - 5x - 12$ is below the x-axis between the x-intercepts of $-\frac{3}{2}$ and 4.

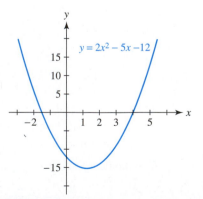

FIGURE 3.69

This symbolic and numerical procedure is summarized verbally in the following.

Solving quadratic inequalities symbolically

To solve a quadratic inequality, such as $ax^2 + bx + c > 0$, use the following steps. (Note that $>$ may be replaced by $<$, $\leq$, or $\geq$.)

STEP 1: Replace the inequality sign with an equals sign and solve the resulting equation, $ax^2 + bx + c = 0$. The solutions to this equation are called *boundary numbers*.

STEP 2: Use the boundary numbers to separate the number line into disjoint intervals.

STEP 3: Make a table of test values. Be sure to select one test value from each interval.

STEP 4: If the expression $ax^2 + bx + c$ is positive when evaluated at a test value, then it is positive on the entire interval containing the test value. Similarly, if the expression $ax^2 + bx + c$ is negative when evaluated at a test value, then it is negative on the entire interval containing the test value. Choose the interval(s) that satisfy your inequality. This is your solution set.

EXAMPLE 4   *Solving a quadratic inequality symbolically*

Solve $x^2 \geq 2 - x$ symbolically.

SOLUTION   **STEP 1:** Rewrite the inequality as $x^2 + x - 2 \geq 0$ and then solve $x^2 + x - 2 = 0$.

$$x^2 + x - 2 = 0$$
$$(x + 2)(x - 1) = 0 \qquad \text{Factor.}$$
$$x = -2 \quad \text{or} \quad x = 1 \qquad \text{Zero-product property}$$

STEP 2: These boundary numbers separate the number line into the following disjoint intervals.

$$-\infty < x < -2, \quad -2 < x < 1, \quad \text{and} \quad 1 < x < \infty.$$

STEP 3: We choose the test values $x = -3$, $x = 0$, and $x = 2$. From Table 3.7, the expression $x^2 + x - 2$ is positive when $x = -3$ and $x = 2$.

TABLE 3.7   A Table of Test Values

x	-3	-2	0	1	2
$x^2 + x - 2$	4	0	-2	0	4

STEP 4: The expression $x^2 + x - 2$ is positive when $x < -2$ or $x > 1$. Thus the solution set is $\{x \mid x \leq -2 \text{ or } x \geq 1\}$. The boundary numbers, -2 and 1, are included because the inequality involves $\geq$ rather than $>$.

Putting it all Together

3.3

*I*n this section we solved different types of quadratic inequalities, both graphically and symbolically with numerical help. The following table summarizes important concepts related to solving quadratic inequalities.

Concept	Description
Quadratic inequality	Can be written as $ax^2 + bx + c < 0$, where $<$ may be replaced by $>$, $\leq$, or $\geq$. **Example:** $x^2 - x < 2$ is a quadratic inequality because it can be written as $x^2 - x - 2 < 0$.
Graphical solution	Write the inequality in the form $ax^2 + bx + c < 0$, where $<$ may be replaced by $>$, $\leq$, or $\geq$. Graph $y = ax^2 + bx + c$, and locate the x-intercepts, or boundary numbers. Use these numbers to determine x-values where the graph is below (above) the x-axis. In the accompanying figure, $y < 0$ when $-1 < x < 2$, and $y > 0$ when either $x < -1$ or $x > 2$.
Symbolic solution	Write the inequality as $ax^2 + bx + c < 0$, where $<$ may be replaced by $>$, $\leq$, or $\geq$. Solve the equation $ax^2 + bx + c = 0$. To determine where $y = ax^2 + bx + c$ is positive or negative, use a table of test values. **Example:** $x^2 - x - 2 < 0$ Solving $x^2 - x - 2 = 0$ results in $x = -1$ or $x = 2$. From the table the solution set is $\{x \mid -1 < x < 2\}$.

x	-2	-1	0	2	3
$x^2 + x - 2$	4	0	-2	0	4

3.3 EXERCISES

Quadratic Inequalities

Exercises 1–6: The graph of $f(x) = ax^2 + bx + c$ is shown in the figure. Solve each inequality.

1. (a) $f(x) < 0$
  (b) $f(x) \geq 0$

2. (a) $f(x) > 0$
  (b) $f(x) < 0$

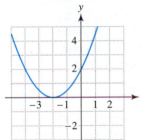 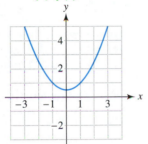

3. (a) $f(x) \leq 0$
  (b) $f(x) > 0$

4. (a) $f(x) \geq 0$
  (b) $f(x) \leq 0$

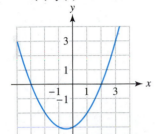

5. (a) $f(x) > 0$
  (b) $f(x) < 0$

6. (a) $f(x) \geq 0$
  (b) $f(x) < 0$

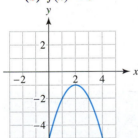 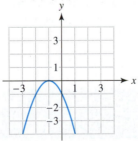

Exercises 7–8: The accompanying table contains test values for a quadratic function $f(x) = ax^2 + bx + c$. Use the table to determine the solution to each inequality.

  (a) $f(x) > 0$ (b) $f(x) \leq 0$

7.

x	-2	-1	0	1	2
$f(x)$	3	0	-1	0	3

8.

x	-6	-4	1	5	8
$f(x)$	-22	0	20	0	-36

Exercises 9–34: Solve the inequality.

9. $x^2 - 3x - 4 < 0$

10. $2x^2 + 5x + 2 \leq 0$

11. $x^2 + x > 6$

12. $-3x \geq 9 - 12x^2$

13. $x^2 \leq 4$

14. $2x^2 > 16$

15. $x(x - 4) \geq -4$

16. $x^2 - 3x - 10 < 0$

17. $-x^2 + x + 6 \leq 0$

18. $-x^2 - 2x + 8 > 0$

19. $6x^2 - x < 1$

20. $5x^2 \leq 10 - 5x$

21. $(x + 4)(x - 10) \leq 0$

22. $(x - 3.1)(x + 2.7) > 0$

23. $x^2 + 4x + 3 < 0$

24. $2x^2 + x + 4 < 0$

25. $x^2 + 2x \geq 35$

26. $9x^2 + 4 > 12x$

27. $x^2 \geq x$

28. $x^2 \geq -3$

29. $x(x - 1) \geq 6$

30. $x^2 - 9 < 0$

31. $x^2 - 5 \leq 0$

32. $0.5x^2 - 3.2x > -0.9$

33. $7x^2 + 515.2 \geq 179.8x$

34. $-10 < 3x - x^2$

Algebra Review To review factoring, see Chapter R (page R-27).

35. *AIDS Deaths* The formula $f(x) = 2375x^2 + 5134x + 5020$ may be used to estimate the number of AIDS deaths from 1984 to 1994, where $x = 0$ corresponds to 1984, $x = 1$ to 1985, and so on. Estimate the years when the number of AIDS deaths was from 90,000 to 200,000.

36. *Safe Driving Speeds* The stopping distance d in feet for a car traveling at x miles per hour is given by $d(x) = \dfrac{1}{12}x^2 + \dfrac{11}{9}x$. Determine the driving speeds that correspond to stopping distances between 300 and 500 feet, inclusively. Round speeds to the nearest mile per hour.

37. *Cellular Phones* Our society is in transition from an industrial to an informational society. Cellular communication has played an increasingly large role in this transition. The number of cellular subscribers in thousands from 1985 to 1991 can be modeled by $f(x) = 163x^2 - 146x + 205$, where x is the year and $x = 0$ corresponds to 1985. (Source: M. Paetsch, *Mobile Communication in the U.S. and Europe.*)

(a) Write a quadratic inequality whose solution set represents the years when there were over 2 million subscribers.

(b) Solve this inequality.

38. *Air Density* As the altitude increases, air becomes thinner or less dense. An approximation of the density of air at an altitude of x meters above sea level is given by

$$d(x) = (3.32 \times 10^{-9})x^2 - (1.14 \times 10^{-4})x + 1.22.$$

The output is the density of air in kilograms per cubic meter. The domain of d is $0 \leq x \leq 10,000$. (Source: A. Miller and J. Thompson, Elements of Meteorology.)

(a) Denver is sometimes referred to as the mile-high city. Compare the density of air at sea level and in Denver. (*Hint:* 1 ft $\approx$ 0.305 m)

(b) Determine the altitudes where the density ranges from 0.5 to 1 kilogram per cubic meter.

39. *Modeling Water Flow* A cylindrical container measuring 16 centimeters high and 12.7 centimeters in diameter with a 0.5-centimeter hole at the bottom was completely filled with water. As water leaked out, the height of the water level inside the container was measured every 15 seconds. The results of the experiment appear in the accompanying table.

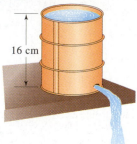

16 cm

Time (sec)	0	15	30	45	60
Height (cm)	16	13.8	11.6	9.8	8.1

Time (sec)	75	90	105	120	135
Height (cm)	6.6	5.3	4.1	3.1	2.3

Time (sec)	150	165	180
Height (cm)	1.4	0.8	0.5

(a) Explain why the data cannot be modeled by a linear function.

(b) Use the table to estimate when the height of the water was from 5 to 10 centimeters.

(c) The data can be modeled by

$$f(x) = 0.0003636x^2 - 0.1511x + 15.92,$$

where x represents the time. Graph f and the data in the same viewing rectangle.

(d) Solve part (b) using $f(x)$.

40. *Heart Rate* The accompanying table shows a person's heart rate after exercising.

Time (min)	0	2	4
Heart Rate (bpm)	154	106	90

(a) Find values for a, h, and k so that $f(x) = a(x - h)^2 + k$ models the data, where x represents time. (*Hint:* Let (4, 90) be the vertex.)

(b) Evaluate $f(1)$ and interpret the result.

(c) Estimate the times when the heart rate was from 115 to 125 beats per minute.

Writing about Mathematics

41. Explain how a table of values can be used to help solve a quadratic inequality once the boundary numbers are known.

42. Explain how to determine the solution set for the inequality $ax^2 + bx + c < 0$, where $a > 0$. How would the solution set change if $a < 0$?

3.4 TRANSFORMATIONS OF GRAPHS

Vertical and Horizontal Translations • Stretching and Shrinking • Reflection of Graphs • Dynamically Displayed Data: An Application

Introduction

Graphs are often used to model different types of phenomena. For example, when a cold front moves across the United States, we might use a circular arc on a weather map to describe its shape. (See Exercise 7 in the Extended and Discovery Exercises at the end of this chapter.) If the front does not change its shape significantly, we could model the movement of the front on a television weather map by translating the circular arc. Before we can portray a cold front on a weather map, we need to discuss how to transform graphs of functions. (Sources: S. Hoggar, *Mathematics for Computer Graphics;* A. Watt, *3D Computer Graphics.*)

Vertical and Horizontal Translations

Graphs of two functions defined by $f(x) = x^2$, and $g(x) = \sqrt{x}$ will be used to demonstrate translations in the xy-plane. Symbolic, numerical, and graphical representations of these functions are shown in Figures 3.70 and 3.71, respectively. Points listed in the table are plotted on the graph.

x	-2	-1	0	1	2
y	4	1	0	1	4

x	0	1	4
y	0	1	2

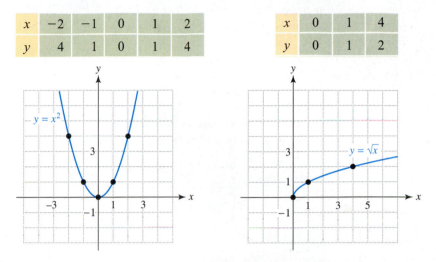

FIGURE 3.70   **FIGURE 3.71**

Vertical Shifts.   If 2 is added to the formula of each function, the original graphs are shifted upward 2 units. The graphs of $y = x^2 + 2$ and $y = \sqrt{x} + 2$ are shown in Figures 3.72 and 3.73. A table of points is included, together with the graph of the original function. Notice how the y-values in both the graphical and numerical representations increase by 2 units.

x	-2	-1	0	1	2
y	6	3	2	3	6

x	0	1	4
y	2	3	4

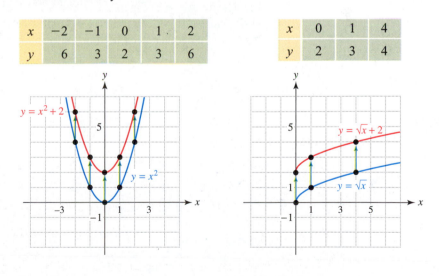

FIGURE 3.72   **FIGURE 3.73**

If 2 is subtracted from each of the original formulas, the graphs would translate downward 2 units. Verify this by graphing $y = x^2 - 2$ and $y = \sqrt{x} - 2$. Translations of this type are called **vertical shifts**. They do not alter the shape of the graph—only its position. A vertical shift is an example of an **isometry**, because distance between any two points on the graph does not change. The original and shifted graphs are congruent.

Horizontal Shifts. If the variable x is replaced by $(x - 2)$ in each of the formulas for f and g, a different type of shift results. Figures 3.74 and 3.75 show the graphs and tables of $y = (x - 2)^2$ and $y = \sqrt{(x - 2)}$ together with the graphs of the original functions.

x	0	1	2	3	4
y	4	1	0	1	4

x	2	3	6
y	0	1	2

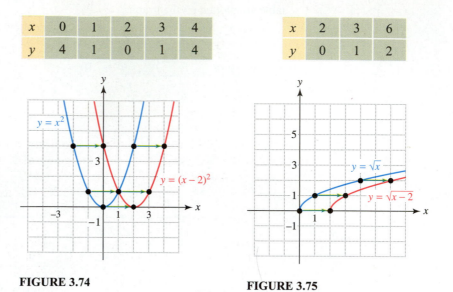

FIGURE 3.74　　　　　**FIGURE 3.75**

Each new graph suggests a shift of the original graph to the *right* by 2 units. Notice that a table for a graph shifted right 2 units can be obtained from the original table by *adding* 2 to each x-value.

If the variable x is replaced by $(x + 3)$ in each equation, the original graphs are translated to the *left* 3 units. The graphs of $y = (x + 3)^2$ and $y = \sqrt{(x + 3)}$ and their tables are shown in Figures 3.76 and 3.77. This type of translation is a **horizontal shift**. Horizontal shifts are also examples of isometries. The table for a graph shifted left 3 units is obtained from the original table by *subtracting* 3 from each x-value.

x	−5	−4	−3	−2	−1
y	4	1	0	1	4

x	−3	−2	1
y	0	1	2

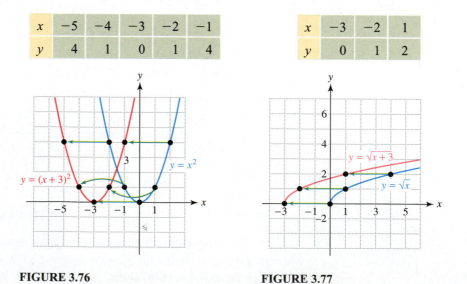

FIGURE 3.76　　　　　**FIGURE 3.77**

These ideas are summarized in the following box.

Vertical and horizontal shifts

Let f be a function, and let k be a positive number.

To graph	Shift the graph of $y = f(x)$ by k units
$y = f(x) + k$	upward
$y = f(x) - k$	downward
$y = f(x - k)$	right
$y = f(x + k)$	left

Shifts can be combined to translate a graph of $y = f(x)$ both vertically and horizontally. For example, to shift the graph of $y = f(x)$ to the right 2 units and downward 4 units, we graph $y = f(x - 2) - 4$. If $y = |x|$, then $y = f(x - 2) - 4 = |x - 2| - 4$, as shown in Figures 3.78–3.80. This technique is illustrated in the next example.

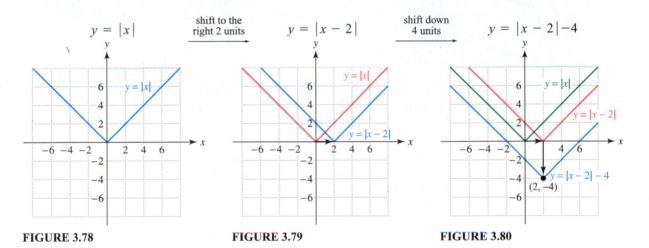

FIGURE 3.78   **FIGURE 3.79**   **FIGURE 3.80**

EXAMPLE 1 *Combining vertical and horizontal shifts*

Find an equation that shifts the graph of $f(x) = \frac{1}{2}x^2 - 4x + 6$ to the left 8 units and upward 4 units. Support this result by graphing.

SOLUTION To shift the graph of f to the left 8 units, replace x with $(x + 8)$ in the formula for $f(x)$. This results in

$$y = f(x + 8) = \frac{1}{2}(x + 8)^2 - 4(x + 8) + 6.$$

To shift the graph of this new equation upward 4 units, add 4 to the formula to obtain

$$y = f(x + 8) + 4 = \frac{1}{2}(x + 8)^2 - 4(x + 8) + (6 + 4).$$

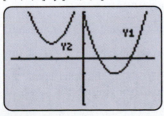

$[-9, 9, 2]$ by $[-6, 6, 2]$

FIGURE 3.81

The graphs of $Y_1 = .5X^2 - 4X + 6$ and $Y_2 = .5(X + 8)^2 - 4(X + 8) + 10$ are shown in Figure 3.81. The vertex of Y_2 is 8 units to the left and 4 units upward compared to the vertex of Y_1.

Stretching and Shrinking

The graph of $f(x) = \sqrt{x}$ in Figure 3.82 can be stretched or shrunk vertically. On one hand, the graph of $y = 2f(x)$, or $y = 2\sqrt{x}$, in Figure 3.83 represents a vertical stretching of the graph of $y = \sqrt{x}$. On the other hand, the graph of $y = \frac{1}{2}f(x)$, or $y = \frac{1}{2}\sqrt{x}$, in Figure 3.84 represents a vertical shrinking of the graph of $y = \sqrt{x}$.

x	0	1	4
$f(x)$	0	1	2

x	0	1	4
$2f(x)$	0	2	4

x	0	1	4
$\frac{1}{2}f(x)$	0	$\frac{1}{2}$	1

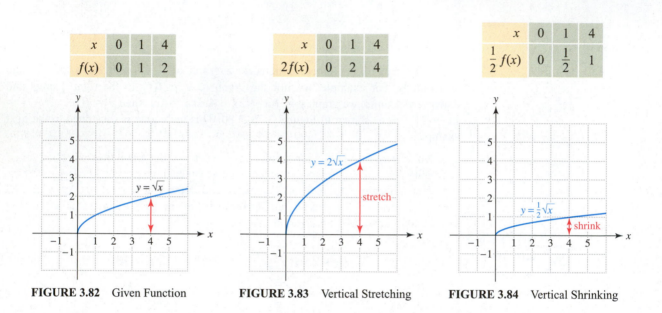

FIGURE 3.82   Given Function   **FIGURE 3.83**   Vertical Stretching   **FIGURE 3.84**   Vertical Shrinking

These results can be generalized for any function f.

> ### Vertical stretching and shrinking
>
> If the point (x, y) lies on the graph of $y = f(x)$, then the point (x, cy) lies on the graph of $y = cf(x)$. If $c > 1$, the graph of $y = cf(x)$ is a vertical stretching of the graph of $y = f(x)$, whereas if $0 < c < 1$, the graph of $y = cf(x)$ is a vertical shrinking of the graph of $y = f(x)$.

For example, if the point $(4, 2)$ is on the graph of $y = f(x)$, then the point $(4, 4)$ is on the graph $y = 2f(x)$, and the point $(4, 1)$ is on the graph of $y = \frac{1}{2}f(x)$.

The line graph in Figure 3.85 can be stretched or shrunk horizontally. On one hand, if the line graph represents the graph of a function f, then the graph of $y = f\left(\frac{1}{2}x\right)$ in Figure 3.86 is a horizontal stretching of the graph of $y = f(x)$. On the other hand, the graph of $y = f(2x)$ in Figure 3.87 represents a horizontal shrinking of the graph of $y = f(x)$.

x	−2	−1	1	2
$f(x)$	3	−3	3	−3

x	−4	−2	2	4
$f\left(\frac{1}{2}x\right)$	3	−3	3	−3

x	−1	$-\frac{1}{2}$	$\frac{1}{2}$	1
$f(2x)$	3	−3	3	−3

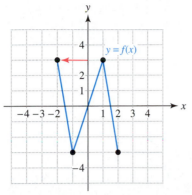

FIGURE 3.85 Given Function

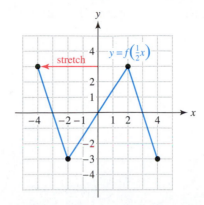

FIGURE 3.86 Horizontal Stretching

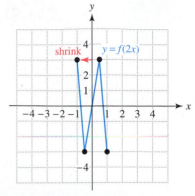

FIGURE 3.87 Horizontal Shrinking

If one were to imagine the graph of $y = f(x)$ as a flexible wire, then a horizontal stretching would happen if the wire were pulled on each end, and a horizontal shrinking would happen if the wire were compressed. Note that horizontal stretching and shrinking does not change the height (y-values) of the graph. Horizontal stretching and shrinking can be generalized for any function f.

> ### Horizontal stretching and shrinking
>
> If the point (x, y) lies on the graph of $y = f(x)$, then the point $\left(\frac{x}{c}, y\right)$ lies on the graph of $y = f(cx)$. If $c > 1$, the graph of $y = f(cx)$ is a horizontal shrinking of the graph of $y = f(x)$, whereas if $0 < c < 1$, the graph of $y = f(cx)$ is a horizontal stretching of the graph of $y = f(x)$.

For example, if the point $(8, 3)$ is on the graph of $y = f(x)$, then the point $(4, 3)$ is on the graph of $y = f(2x)$, and the point $(16, 3)$ is on the graph of $y = f\left(\frac{1}{2}x\right)$.

The next example illustrates the previous discussion.

EXAMPLE 2 *Stretching and shrinking of a graph*

Use the graph of $y = f(x)$ in Figure 3.88 to sketch a graph of each equation.

(a) $y = 3f(x)$ **(b)** $y = f\left(\frac{1}{2}x\right)$

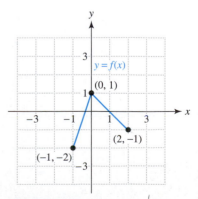

FIGURE 3.88

SOLUTION   **(a)** The graph of $y = 3f(x)$ is a vertical stretching of the graph of $y = f(x)$ shown in Figure 3.88 (which is repeated here) and can be obtained by multiplying each y-coordinate on the graph of $y = f(x)$ by 3.

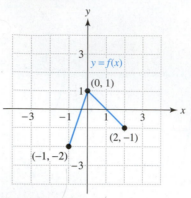

FIGURE 3.88

When the y-coordinates of the points $(-1, -2)$, $(0, 1)$, and $(2, -1)$ are multiplied by 3, the following results.

$$(-1, 3 \cdot -2) = (-1, -6)$$
$$(0, 3 \cdot 1) = (0, 3)$$
$$(2, 3 \cdot -1) = (2, -3)$$

Next we sketch a line graph with the points $(-1, -6)$, $(0, 3)$, and $(2, -3)$, as shown in Figure 3.89.

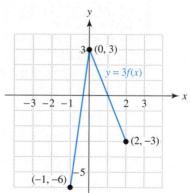

FIGURE 3.89   Vertical Stretching

(b) The graph of $y = f\left(\frac{1}{2}x\right)$ is a horizontal stretching of the graph of $y = f(x)$ and can be obtained by dividing each x-coordinate on the graph of $y = f(x)$ by $\frac{1}{2} = 0.5$. For the points $(-1, -2)$, $(0, 1)$, and $(2, -1)$, we obtain the following.

$$\left(\frac{-1}{0.5}, -2\right) = (-2, -2)$$
$$\left(\frac{0}{0.5}, 1\right) = (0, 1)$$
$$\left(\frac{2}{0.5}, -1\right) = (4, -1)$$

Next we sketch a line graph with the points $(-2, -2)$, $(0, 1)$, and $(4, -1)$, as shown in Figure 3.90.

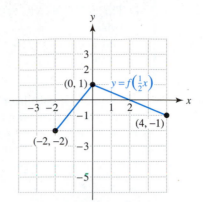

FIGURE 3.90 Horizontal Stretching

Reflection of Graphs

Another type of translation is called a **reflection**. For example, the reflection of the graph of $f(x) = x^2 - x + 2$ across the x-axis is shown in Figure 3.91 as a red curve.

If (x, y) is a point on the graph of f, then $(x, -y)$ lies on the graph of its reflection across the x-axis, as shown in Figure 3.91. Thus a reflection of $y = f(x)$ is given by $-y = f(x)$, or equivalently, $y = -f(x)$. For example, a reflection of the graph of $f(x) = x^2 - x + 2$ is obtained by graphing $y = -f(x) = -(x^2 - x + 2)$.

If a point (x, y) lies on the graph of a function f, then the point $(-x, y)$ lies on the graph of its reflection across the y-axis, as shown in Figure 3.92. Thus a reflection of $y = f(x)$ is given by $y = f(-x)$. For example, a reflection of the graph of $f(x) = (x - 2)^2$ across the y-axis is shown in Figure 3.93 as a red curve. Its reflection is given by

$$f(-x) = (-x - 2)^2.$$

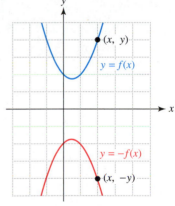

FIGURE 3.91

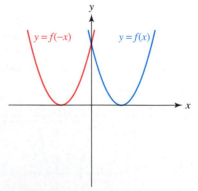

FIGURE 3.92   **FIGURE 3.93**

These results are summarized on the next page.

> ### Reflections of graphs across the *x*- and *y*-axes
>
> **1.** The graph of $y = -f(x)$ is a reflection of the graph of $y = f(x)$ across the *x*-axis.
> **2.** The graph of $y = f(-x)$ is a reflection of the graph of $y = f(x)$ across the *y*-axis.

If a graphing calculator is capable of using function notation, equations for reflections of a function f can be entered easily. For example, if $f(x) = (x - 4)^2$, then let $Y_1 = (X - 4)^2$, $Y_2 = -Y_1$, and $Y_3 = Y_1(-X)$. The graph of Y_2 is the reflection of f across the *x*-axis, and Y_3 is the reflection of Y_1 across the *y*-axis. See Figures 3.94 and 3.95. However, it is not necessary to have this feature to plot reflections.

Graphing Calculator Help

To access the variable Y_1, as shown in Figure 3.94, see Appendix B (page AP-15).

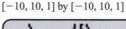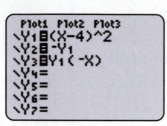

$[-10, 10, 1]$ by $[-10, 10, 1]$

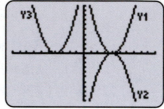

FIGURE 3.94   **FIGURE 3.95**

EXAMPLE 3   *Reflecting graphs of functions*

For each representation of a function f, graph its reflection across the *x*-axis and across the *y*-axis.
(a) $f(x) = x^2 + 2x - 3$
(b) The graph of f is a line graph determined by Table 3.8.

TABLE 3.8

x	-2	-1	0	3
$f(x)$	1	-3	-1	2

SOLUTION   **(a)** The graph of $f(x) = x^2 + 2x - 3$ is a parabola with vertex $(-1, -4)$ and *x*-intercepts -3 and 1, as shown in Figure 3.96. To obtain its reflection across the *x*-axis, graph $y = -f(x)$, or $y = -(x^2 + 2x - 3)$, as shown in Figure 3.97. The vertex is now $(-1, 4)$, and the *x*-intercepts have not changed. To obtain the reflection of f across the *y*-axis, let $y = f(-x)$, or $y = (-x)^2 + 2(-x) - 3$ and graph, as shown in Figure 3.98. Note that the vertex is now $(1, -4)$, and the *x*-intercepts have changed to -1 and 3.

$[-9, 9, 1]$ by $[-6, 6, 1]$

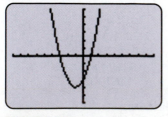

FIGURE 3.96 $y = f(x)$

$[-9, 9, 1]$ by $[-6, 6, 1]$

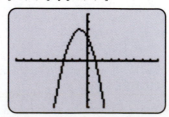

FIGURE 3.97 $y = -f(x)$

$[-9, 9, 1]$ by $[-6, 6, 1]$

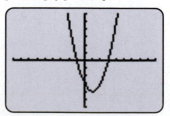

FIGURE 3.98 $y = f(-x)$

(b) The graph of $y = f(x)$ is a line graph, which is shown in Figure 3.99. To graph the reflection of f across the x-axis, start by making a table of values for $y = -f(x)$. See Table 3.9, where each point (x, y) in Table 3.8 has been changed to $(x, -y)$. Plot the points $(-2, -1)$, $(-1, 3)$, $(0, 1)$, and $(3, -2)$. Note that each point in Figure 3.99 has been reflected across the x-axis in Figure 3.100.

TABLE 3.9

x	-2	-1	0	3
$f(x)$	-1	3	1	-2

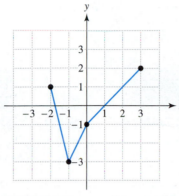

FIGURE 3.99 $y = f(x)$

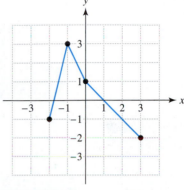

FIGURE 3.100 $y = -f(x)$

To graph the reflection of f across the y-axis, make a table of values for $y = f(-x)$. To do this, change each point (x, y) in Table 3.8 to $(-x, y)$, as shown in Table 3.10. Plot the points $(2, 1)$, $(1, -3)$, $(0, -1)$, and $(-3, 2)$. Note that each point in Figure 3.99 has been reflected across the y-axis in Figure 3.101.

TABLE 3.10

x	2	1	0	-3
$f(x)$	1	-3	-1	2

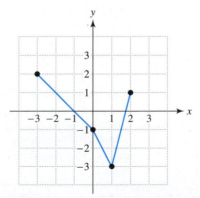

FIGURE 3.101 $y = f(-x)$

Dynamically Displayed Data: An Application

With the introduction of computer graphics, dynamic displays of data are routine. Translations of graphs and figures play an important role in this new technology. In early motion pictures, it was common to have the background move to create the appearance that the actors and actresses were moving. This same technique is used today in two-dimensional graphics.

In video games, the background often is translated to give the illusion that the player in the game is moving. A simple scene of a mountain and an airplane is shown in Figure 3.102. In order to make it appear as though the airplane is flying to the right, the image of the mountain could be translated horizontally to the left, as shown in Figure 3.103.   (Reference: C. Pokorny and C. Gerald, *Computer Graphics.*)

FIGURE 3.102   FIGURE 3.103

EXAMPLE 4   *Using translations to model movement*

Suppose the mountain in Figure 3.102 can be described by the quadratic function represented by $f(x) = -0.4x^2 + 4$ and that the airplane is located at the point $(1, 5)$.
(a) Graph f in $[-4, 4, 1]$ by $[0, 6, 1]$, where the units are in kilometers. Plot a point to mark the position of the airplane.
(b) Assume that the airplane is moving horizontally to the right at 0.4 kilometer per second. To give a video player the illusion that the airplane is moving, graph the image of the mountain and the position of the airplane after 5 seconds and after 10 seconds.

SOLUTION   (a) The graph of $y = f(x) = -0.4x^2 + 4$ and the position of the airplane at $(1, 5)$ are shown in Figure 3.104. The "mountain" has been shaded to emphasize its position.
(b) Five seconds later, the airplane has moved $5(0.4) = 2$ kilometers right. In 10 seconds it has moved $10(0.4) = 4$ kilometers right. To graph these new positions, translate the graph of the mountain 2 and 4 kilometers to the left. Replace x with $(x + 2)$ and graph

$$y = f(x + 2) = -0.4(x + 2)^2 + 4$$

together with the point $(1, 5)$. In a similar manner graph

$$y = f(x + 4) = -0.4(x + 4)^2 + 4.$$

Graphing Calculator Help

To shade below a parabola, see Appendix B (page AP-16).

The results are shown in Figures 3.105 and 3.106. The position of the airplane at $(1, 5)$ has not changed. However, it has the appearance that it has moved to the right.

$[-4, 4, 1]$ by $[0, 6, 1]$ $[-4, 4, 1]$ by $[0, 6, 1]$ $[-4, 4, 1]$ by $[0, 6, 1]$

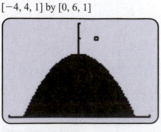   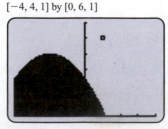   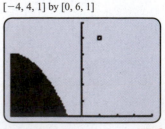

FIGURE 3.104   FIGURE 3.105   FIGURE 3.106

■ **CLASS DISCUSSION**
Discuss how one might create the illusion of the airplane moving to the left and *gaining altitude* as it passes over the mountain. ■

*Putting
it all
Together*
3.4

In this section we discussed several transformations of graphs. The following table summarizes how these transformations affect the graph of $y = f(x)$.

Equation	Effect on Graph of $y = f(x)$
Let $k > 0$. $y = f(x) + k$ $y = f(x) - k$ $y = f(x + k)$ $y = f(x - k)$	Graph is shifted upward k units. Graph is shifted downward k units. Graph is shifted to the left k units. Graph is shifted to the right k units. **Examples:** **(a)** $y = f(x) - 1$ **(b)** $y = f(x + 1)$
Let $c > 0$. $y = cf(x)$	If (x, y) lies on the graph of $y = f(x)$, then (x, cy) lies on the graph of $y = cf(x)$. The graph is vertically stretched if $c > 1$ and vertically shrunk if $0 < c < 1$. **Examples:** **(a)** $y = 2f(x)$ **(b)** $y = \frac{1}{2}f(x)$

Equation	Effect on Graph of $y = f(x)$
Let $c > 0$. $y = f(cx)$	If (x, y) lies on the graph of $y = f(x)$, then $\left(\dfrac{x}{c}, y\right)$ lies on the graph of $y = f(cx)$. The graph is horizontally shrunk if $c > 1$ and horizontally stretched if $0 < c < 1$. **Examples:** **(a)** $y = f(2x)$ **(b)** $y = f\left(\dfrac{1}{2}x\right)$
$y = -f(x)$ $y = f(-x)$	Graph is reflected across the x-axis. Graphs is reflected across the y-axis. **Examples:** **(a)** $y = -f(x)$ **(b)** $y = f(-x)$

3.4 EXERCISES

Vertical and Horizontal Translations

Exercises 1–10: Write the equation of the graph. (Note: the given graph represents a translation of the graph of one of the following equations: $y = x^2$, $y = \sqrt{x}$, *or* $y = |x|$.)

1.
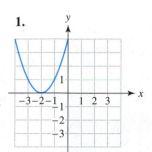

2.
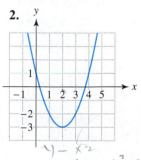

$y - x_2$
$y - (x-2)^2 - 3$

3.
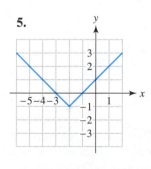

4.
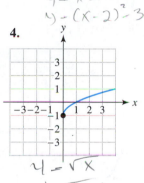

$y = \sqrt{x}$
$y = \sqrt{x-1}$

5.
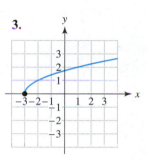

6.
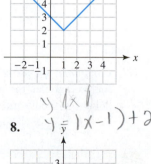

$y |x|$
$y = |x-1| + 2$

7.
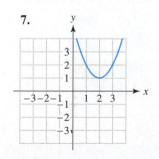

8.
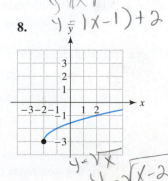

$y - \sqrt{x}$
$y - \sqrt{(x-2)} - 3$

9.
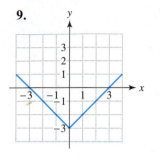

10.
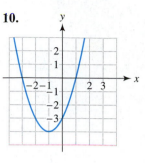

Transforming Graphical Representations

Exercises 11–26: Use transformations of graphs to sketch a graph of f.

11. $f(x) = x^2 - 3$ **12.** $f(x) = -x^2$

13. $f(x) = (x + 4)^2$ **14.** $f(x) = (x - 5)^2 + 3$

15. $f(x) = -\sqrt{x}$

16. $f(x) = 2(x - 1)^2 + 1$

17. $f(x) = -x^2 + 4$

18. $f(x) = \sqrt{-x}$

19. $f(x) = \sqrt{x + 1}$ **20.** $f(x) = \sqrt{x} + 1$

21. $f(x) = |x| - 4$ **22.** $f(x) = |x - 4|$

23. $f(x) = \sqrt{x - 3} + 2$ **24.** $f(x) = |x + 2| - 3$

25. $f(x) = \sqrt{2x}$ **26.** $f(x) = \dfrac{1}{2}(x + 2)^2$

Exercises 27–32: Use the accompanying graph of $y = f(x)$ *to sketch a graph of each equation.*

27. **(a)** $y = f(x) + 2$ **28.** **(a)** $y = f(x + 1)$
    (b) $y = f(x - 2) - 1$ **(b)** $y = -f(x)$
    (c) $y = -f(x)$ **(c)** $y = 2 f(x)$

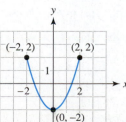

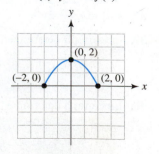

29. (a) $y = f(x + 3) - 2$
(b) $y = f(-x)$
(c) $y = \frac{1}{2}f(x)$

30. (a) $y = f(x - 1) - 2$
(b) $y = -f(x) + 1$
(c) $y = f\left(\frac{1}{2}x\right)$

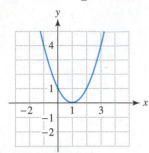
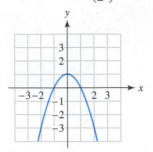

31. (a) $y = f(x) - 2$
(b) $y = f(x - 1) + 2$
(c) $y = 2f(x)$

32. (a) $y = f(x) + 1$
(b) $y = -f(x) - 1$
(c) $y = 2f\left(\frac{1}{2}x\right)$

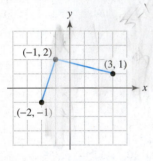
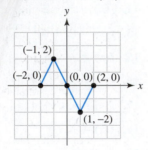

Exercises 33–36: (Refer to Example 1.) Find an equation that shifts the graph of f by the desired amounts. Graph f and the shifted graph in the same viewing rectangle.

33. $f(x) = x^2$;   right 2 units, downward 3 units

34. $f(x) = 3x - 4$;   left 3 units, upward 1 unit

35. $f(x) = x^2 - 4x + 1$;   left 6 units, upward 4 units

36. $f(x) = 5 - 3x - \frac{1}{2}x^2$;   right 5 units, downward 8 units

Exercises 37–40: (Refer to Example 3.) For the given representation of a function f, graph its reflection across the x-axis and across the y-axis.

37. $f(x) = x^2 - 2x - 3$ **38.** $f(x) = 4 - 7x - 2x^2$

39. Line graph determined by the table.

x	-3	-1	1	2
f(x)	2	3	-1	-2

40. Line graph determined by the table.

x	-4	-2	0	1
f(x)	-1	-4	2	2

Transforming Numerical Representations

Exercises 41–46: Two functions f and g are related by the given equation. Use the numerical representation of f to make a numerical representation of g.

41. $g(x) = f(x) + 7$

x	1	2	3	4	5
f(x)	5	1	6	2	7

42. $g(x) = f(x) - 10$

x	0	5	10	15	20
f(x)	-5	11	21	32	47

43. $g(x) = f(x - 2)$

x	-4	-2	0	2	4
f(x)	5	2	-3	-5	-9

44. $g(x) = f(x + 50)$

x	-100	-50	0	50	100
f(x)		25	80	120	150	100

45. $g(x) = f(x + 1) - 2$

x	1	2	3	4	5
f(x)	2	4	3	7	8

46. $g(x) = f(x - 3) + 5$

x	-3	0	3	6	9
f(x)	3	8	15	27	31

Applications

47. *Consumer Spending*   The numerical representation for f computes the average consumer spending in dollars on media and entertainment for the year x.

x	1990	1991	1992	1993
f(x)	$360	$380	$410	$430

x	1994	1995	1996
f(x)	$460	$490	$520

Source: Veronis, Suhler & Associates.

(a) Give a numerical representation for a function g that computes the average consumer spending in dollars on media and entertainment during the year x, where $x = 0$ corresponds to 1990 and $x = 6$ to 1996.

(b) What are the domains of f and g?

(c) Write an equation that shows the relationship between $f(x)$ and $g(x)$.

48. *Public College Tuition*   The function f, represented by $f(x) = 122.8x + 786.2$, models the average tuition at public four-year colleges from 1981 to 1994.   (Source: The College Board.) The value $x = 1$ corresponds to 1981, $x = 2$ to 1982, and so on. Use transformations to determine a function g that computes the average tuition at four-year colleges during the year x, where $x = 1981, 1982, \ldots, 1994$.

49. *Daylight Hours*   The accompanying figure shows a partial graph of the number of daylight hours at latitude 60°N. The variable x is measured in months where $x = 0$ corresponds to December 21, the day with the least amount of daylight. The domain is $-6 \le x \le 6$. For example, $x = 2$ represents February 21 and $x = -3$ represents September 21. (Source: J. Williams, *The Weather Almanac 1995.*)

(a) Estimate the number of daylight hours on February 21.

(b) Make a conjecture about the number of daylight hours on October 21 when $x = -2$.

(c) Sketch the left side of the graph for $-6 \le x \le 0$.

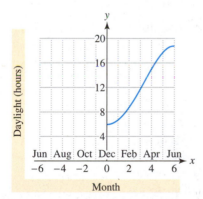

50. *Daylight Hours*   (Continuation of Exercise 49.) Sketch a graph that shows the daylight hours at a latitude of 60°S, which is in the Southern Hemisphere rather than in the Northern Hemisphere.

Using Transformations to Represent Data Dynamically

51. *Computer Graphics*   (Refer to Example 4.) Suppose that the airplane in Figure 3.102 is flying at 0.2 kilometer per second to the left, rather than to

the right. If the position of the airplane is fixed at $(-1, 5)$, graph the image of the mountain and the position of the airplane after 15 seconds.

52. *Computer Graphics*   (Refer to Example 4.) Suppose that the airplane in Figure 3.102 is traveling to the right at 0.1 kilometer per second and gaining altitude at 0.05 kilometer per second. If the airplane's position is fixed at $(-1, 5)$, graph the image of the mountain and the position of the airplane after 20 seconds.

53. *Modeling a Weather Front*   Suppose a cold front is passing through the United States at noon with a shape described by the function $y = \dfrac{1}{20}x^2$. Each unit represents 100 miles. Des Moines, Iowa, is located at $(0, 0)$, and the positive y-axis points north. See the accompanying figure.

(a) If the cold front is moving south at 40 miles per hour and retains its present shape, graph its location at 4 P.M.

(b) Suppose that by midnight the vertex of the front has moved 250 miles south and 210 miles east of Des Moines, maintaining the same shape. Columbus, Ohio, is located approximately 550 miles east and 80 miles south of Des Moines. Plot the locations of Des Moines and Columbus together with the new position of the cold front. Determine whether the cold front has reached Columbus by midnight.

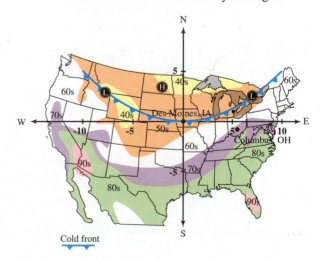

Cold front

54. *Modeling Motion in Computer Graphics*   The first figure contains a picture that is composed of lines and curves. In this exercise we will model only the semicircle that outlines the top of the silo. In order to make it appear that the person is walking to the right, the background must be translated horizontally to the left as shown in the second figure.

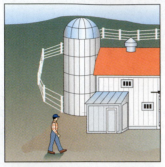

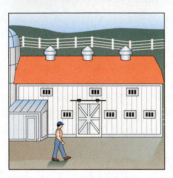

The semicircle at the top of the silo in the first figure is described by $f(x) = \sqrt{9 - x^2} + 12$.

(a) Graph f in the (square) viewing rectangle $[-12, 12, 1]$ by $[0, 16, 1]$.

(b) To give the illusion that the person is walking to the right at 2 units per second, graph the top of the silo after 1 second, and after 4 seconds.

Writing about Mathematics

55. Explain how to graph the reflection of $y = f(x)$ across the x-axis. Give an example.

56. Let k be a positive number. Explain how to shift the graph of $y = f(x)$ upward, downward, left, or right k units. Give examples.

Extended and Discovery Exercises

Exercises 1–4: Commutative Property and Transformations   In the following exercises you will determine if transformations of a graph are commutative. That is, does the order in which transformations of graphs are applied affect the final graph? To answer this question, start by determining if the two sequences of transformations are equivalent for all possible graphs. You may want to try some examples to help decide.

1. Reflect across the x-axis, shift upward 1 unit.
   Shift upward 1 unit, reflect across the x-axis.

2. Shift upward 2 units, shift left 3 units.
   Shift left 3 units, shift upward 2 units.

3. Reflect across the x-axis, reflect across the y-axis.
   Reflect across the y-axis, reflect across the x-axis.

4. Stretch in the vertical direction, shift downward 2 units.
   Shift downward 2 units, stretch in the vertical direction.

CHECKING BASIC CONCEPTS FOR SECTIONS 3.3 AND 3.4

1. A graph of $y = f(x)$ is shown in the accompanying figures. Solve the inequalities $f(x) \leq 0$ and $f(x) > 0$ for each figure.

(a) **(b)**

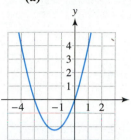 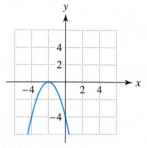

2. Solve each quadratic inequality.
(a) $2x^2 + 7x - 4 \leq 0$ **(b)** $2x^2 + 7x - 4 > 0$

3. Graph $f(x) = x^2$. Then predict how the graph of each equation will appear compared to the graph of

f. Test your prediction by graphing f and the equation in the same viewing rectangle.
(a) $y = (x + 4)^2$ **(b)** $y = x^2 - 3$
(c) $y = (x - 5)^2 + 3$

4. A graph of $f(x)$ is shown in the figure. Sketch a graph of each figure.

(a) $y = -2f(x)$ **(b)** $y = f\left(-\frac{1}{2}x\right)$
(c) $y = f(x - 1) + 1$

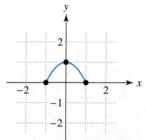

Chapter 3 Summary

CONCEPT	EXPLANATION AND EXAMPLES

Section 3.1

QUADRATIC FUNCTION

Standard form: $f(x) = ax^2 + bx + c, a \neq 0$

Examples: $f(x) = x^2$; $f(x) = -3x^2 + x + 5$

PARABOLA

The graph of a quadratic function is a parabola.

Vertex form: $f(x) = a(x - h)^2 + k$

Leading coefficient: a

Vertex: (h, k)

Axis of symmetry: $x = h$

Example:

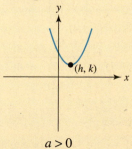

$a > 0$ $a < 0$

CONCEPT	EXPLANATION AND EXAMPLES

Section 3.1 (continued)

VERTEX FORMULA

x-coordinate: $x = -\dfrac{b}{2a}$

y-coordinate: $f\left(-\dfrac{b}{2a}\right)$

Example: $f(x) = 2x^2 + 4x - 4$

$$x = -\frac{4}{2(2)} = -1, \quad y = 2(-1)^2 + 4(-1) - 4 = -6$$

Vertex: $(-1, -6)$

Section 3.2

QUADRATIC EQUATION

$ax^2 + bx + c = 0, a \neq 0$
A quadratic equation can have 0, 1, or 2 real solutions.

Examples: $x^2 + 1 = 0, x^2 + 2x + 1 = 0, x(x - 1) = 20$

FACTORING

Write an equation in the form $ab = 0$.
Example:
$$x^2 - 3x = -2$$
$$x^2 - 3x + 2 = 0$$
$$(x - 1)(x - 2) = 0$$
$$x = 1 \quad \text{or} \quad x = 2$$

SQUARE ROOT PROPERTY

If $x^2 = k$ and $k > 0$, then $x = \pm\sqrt{k}$.

Example: $x^2 = 16$ implies $x = \pm 4$.

COMPLETING THE SQUARE

If $x^2 + kx = d$, then add $\left(\dfrac{k}{2}\right)^2$ to both sides.

Example:
$$x^2 - 4x = 2 \qquad k = -4$$
$$x^2 - 4x + 4 = 2 + 4 \qquad \left(\frac{-4}{2}\right)^2 = 4$$
$$(x - 2)^2 = 6 \qquad \text{Factor.}$$
$$x - 2 = \pm\sqrt{6} \qquad \text{Square root property}$$
$$x = 2 \pm \sqrt{6} \qquad \text{Add 2.}$$

QUADRATIC FORMULA

$$x = \frac{-b \pm \sqrt{b^2 - 4ac}}{2a} \qquad \text{Always works}$$

Example: $2x^2 - 5x - 3 = 0$

$$x = \frac{-(-5) \pm \sqrt{(-5)^2 - 4(2)(-3)}}{2(2)} = \frac{5 \pm 7}{4} = 3, -\frac{1}{2}$$

Section 3.3

QUADRATIC INEQUALITY

$ax^2 + bx + c < 0$, where $<$ may be replaced by $\leq$, $>$, or $\geq$.

Example: $3x^2 - x + 1 \leq 0$

CONCEPT	EXPLANATION AND EXAMPLES

Section 3.3 (continued)

GRAPHICAL SOLUTION

Graph $y = ax^2 + bx + c$ and locate the x-intercepts; then determine x-values where the inequality is satisfied.

Example: $-x^2 - x + 2 > 0$.

The x-intercepts are -2 and 1.

Solution set is $\{x \mid -2 < x < 1\}$.

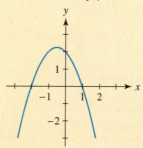

SYMBOLIC SOLUTION

Solve $ax^2 + bx + c = 0$ and use a table of values to determine the x-values where the inequality is satisfied.

Example: $x^2 - 4 \geq 0$.

$x^2 - 4 = 0$ implies $x = \pm 2$.

Solution set is $\{x \mid x \leq -2 \text{ or } x \geq 2\}$.

x	-3	-2	0	2	3
y	5	0	-4	0	5

Section 3.4

VERTICAL SHIFTS
$k > 0$

$y = f(x) + k$ shifts the graph of $y = f(x)$ upward k units.
$y = f(x) - k$ shifts the graph of $y = f(x)$ downward k units.

HORIZONTAL SHIFTS
$k > 0$

$y = f(x - k)$ shifts the graph of $y = f(x)$ to the right k units.
$y = f(x + k)$ shifts the graph of $y = f(x)$ to the left k units.

VERTICAL STRETCHING AND SHRINKING

$y = cf(x)$ vertically stretches the graph of $y = f(x)$ when $c > 1$ and shrinks the graph when $0 < c < 1$.

HORIZONTAL STRETCHING AND SHRINKING

$y = f(cx)$ horizontally shrinks the graph of $y = f(x)$ when $c > 1$ and stretches the graph when $0 < c < 1$.

REFLECTIONS

$y = -f(x)$ is a reflection of $y = f(x)$ across the x-axis.

$y = f(-x)$ is a reflection of $y = f(x)$ across the y-axis.

Review Exercises

Exercises 1 and 2: Use the graph of the quadratic function to determine the sign of the leading coefficient, its vertex, and the equation of the axis of symmetry.

1.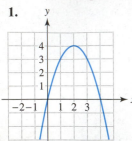

2.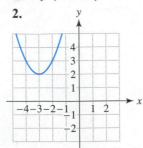

Exercises 3 and 4: Write f(x) in standard form, f(x) = ax² + bx + c, and identify the leading coefficient.

3. $f(x) = -2(x - 5)^2 + 1$

4. $f(x) = \frac{1}{3}(x + 1)^2 - 2$

Exercises 5 and 6: Write f(x) in vertex form, f(x) = a(x − h)² + k, and identify the vertex.

5. $f(x) = x^2 + 6x - 1$

6. $f(x) = 2x^2 + 4x - 5$

Exercises 7 and 8: Use the vertex formula to determine the vertex on the graph of f.

7. $f(x) = -3x^2 + 2x - 4$

8. $f(x) = x^2 + 8x - 5$

Exercises 9 and 10: The graph of f(x) = ax² + bx + c is given.

   (a) State whether $a > 0$ or $a < 0$.
   (b) Estimate the real solutions to $ax^2 + bx + c = 0$.
   (c) Determine if the discriminant is *positive*, *negative*, or *zero*.

9.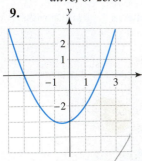

10.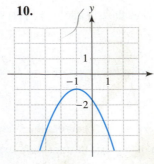

11. Solve $x^2 - x - 20 = 0$.

12. Solve $3x^2 + 4x = 1$ symbolically.

13. Solve the quadratic inequality $x^2 - 3x + 2 > 0$.

14. Solve $2x^2 + 1.3x \le 0.4$ graphically.

15. Use the graph of $y = f(x)$ to solve the inequality.
   (a) $f(x) > 0$
   (b) $f(x) \le 0$

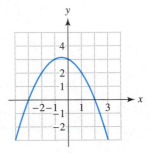

16. The accompanying table contains test values for a quadratic function $f(x)$. Use the table to determine the solution to each inequality.
   (a) $f(x) < 0$
   (b) $f(x) \ge 0$

x	-6	-5	0	3	4
$f(x)$	9	0	-15	0	9

17. If $f(x) = 2x^2 - 3x + 1$, graph $y = -f(x)$ and $y = f(-x)$.

18. Use the given graph of $y = f(x)$ to sketch a graph of each expression.
   (a) $y = f(x + 1) - 2$
   (b) $y = -2f(x)$
   (c) $y = f(2x)$

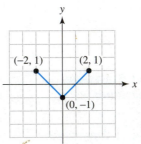

Exercises 19–22: Use transformations of graphs to sketch a graph of f.

19. $f(x) = x^2 - 4$

20. $f(x) = 3|x - 2| + 1$

21. $f(x) = (x + 4)^2 - 2$

22. $f(x) = \sqrt{x + 4}$

Exercises 23 and 24: Two functions f and g are related by the given equation. Use the numerical representation of f to construct a numerical representation of g.

23. $g(x) = f(x) + 2$

x	1	2	3	4
$f(x)$	−3	3	4	7

24. $g(x) = f(x - 5)$

x	0	5	10	15
$f(x)$	−5	11	21	32

25. Sketch a graph of a quadratic function with axis of symmetry $x = 2$ and that passes through the points $(2, 3)$ and $(4, -1)$.

26. Find a function given by $f(x) = a(x - h)^2 + k$ that models the data exactly.

x	2	3	4	5
y	4	7	16	31

27. *Projectile* A slingshot is used to propel a stone upward so that its height h in feet after t seconds is given by $h(t) = -16t^2 + 88t + 5$.
 (a) Evaluate $h(0)$ and interpret the result.
 (b) How high was the stone after 2 seconds?
 (c) Find the maximum height of the stone.
 (d) At what time(s) was the stone 117 feet high?

28. *Modeling Water Flow* Water is leaking out of a hole in the bottom of a 100-gallon tank. The table lists the volume V of the water after t minutes. Decide whether $f(t) = 0.4t^2 - 9.9t + 100$ or $g(t) = 100 - 9.5t$ models the data better.

t (min)	0	1	2	3	4
V (gal)	100	90.5	81.9	74.1	67.0

29. *World Population* The function given by $f(x) = 0.000478x^2 - 1.813x + 1720.1$ models the world population in billions between 1950 and 1995, where x is the year.
 (a) Evaluate $f(1985)$ and interpret the result.
 (b) Use f to estimate world population in the year 2000.
 (c) When does this model predict that world population may reach 7 billion?

30. *Construction* A box is being constructed by cutting 3-inch squares from the corners of a rectangular sheet of metal that is 4 inches longer than it is wide. If the box is to have a volume of 135 cubic inches, find the dimensions of the metal sheet.

31. *Room Prices* Room prices are regularly $100, but for each additional room rented by a group, the price is reduced by $3 for each room. For example, 1 room costs $100, 2 rooms cost $2 \times \$97 = \$194$, 3 rooms cost $3 \times \$94 = \$282$, and so on.
 (a) Write a quadratic function C that gives the total cost of renting x rooms.
 (b) What is the total cost of renting 6 rooms?
 (c) How many rooms are rented if the total cost is $730?
 (d) What number of rooms rented results in the greatest total cost?

32. *Irrigation and Yield* The table shows how irrigation of rice crops affects yield, where x represents the percent of total area that is irrigated and y is the rice yield in tons per hectare. (1 hectare $\approx$ 2.47 acres.) (Source: D. Grigg, *The World Food Problem.*)

x	0	20	40	60	80	100
y	1.6	1.8	2.2	3.0	4.5	6.1

 (a) Use least-squares regression to find a quadratic function that models the data.
 (b) Solve the equation $f(x) = 3.7$. Interpret the results.

33. *Credit Card Debt* The table lists the outstanding balances on Visa and MasterCard credit cards in billions of dollars. (Sources: Bankcard Holders of America, Federal Reserve Board.)

Year	1980	1984	1988	1992	1996
Debt	82	108	172	254	444

 (a) Explain why a linear function does not model the data.
 (b) The data can be modeled by

$$f(x) = 1.3(x - 1980)^2 + 82,$$

 where x is the year. Solve the equation $f(x) = 212$. Interpret the results.

34. *Internet* Downloading files from the Internet can be both slow and unreliable when standard phone lines are being used. Phone modems can transmit 56,000 bits per second, whereas cable TV modems are capable of transmitting 2–10 million bits per second. Demand for this new technology increased from 1996 to 2000. The table shows this increase in millions of accounts. (Source: *New York Times Service, Pioneer Press Research.*)

Year	1996	1997	1998	1999	2000
Accounts (millions)	0.5	1	2	4	7

(a) Let $f(x) = a(x - 1996)^2 + 0.5$. Approximate a so that f models the data.

(b) Solve the equation $f(x) = 10$ and interpret the result.

Extended and Discovery Exercises

1. *Shooting a Foul Shot* (Refer to the introduction of this chapter.) To make a foul shot in basketball, the ball must follow a parabolic arc. This arc depends on both the angle and velocity with which the basketball is released. If a person shoots the basketball overhand from a position 8 feet above the floor, then the path can sometimes be modeled by the parabola

$$y = \frac{-16x^2}{0.434v^2} + 1.15x + 8,$$

where v is the velocity of the ball in feet per second, as illustrated in the accompanying figure. (Source: C. Rist, "The physics of foul shots.")

(a) If the basketball hoop is 10 feet high and located 15 feet away, what initial velocity v should the basketball have?

(b) Check your answer from part (a) graphically. Plot the point $(0, 8)$ where the ball is released and the point $(15, 10)$ where the basketball hoop is. Does your graph pass through both points?

(c) What is the maximum height of the basketball?

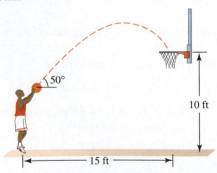

2. (Continuation of Exercise 1) If a person releases a basketball underhand from a position 3 feet above

the floor, it often has a steeper arc than if it is released overhand and sometimes may be modeled by

$$y = \frac{-16x^2}{0.117v^2} + 2.75x + 3.$$

See the accompanying figure. Complete parts (a), (b), and (c) for Exercise 1. Then compare the paths for an overhand shot and an underhand shot.

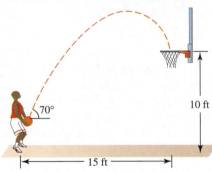

Exercises 3–6: Reflecting Functions Computer graphics frequently use reflections. Reflections can speed up the generation of a picture or create a figure that appears perfectly symmetrical. (Source: S. Hoggar, *Mathematics for Computer Graphics.*)

(a) For the given $f(x)$, constant k, and viewing rectangle, graph
$$x = k, \quad y = f(x), \quad \text{and} \quad y = f(2k - x).$$

(b) Generalize how the graph of $y = f(2k - x)$ compares to the graph of $y = f(x)$.

3. $f(x) = \sqrt{x}$, $k = 2$, $[-1, 8, 1]$ by $[-4, 4, 1]$

4. $f(x) = x^2$, $k = -3$, $[-12, 6, 1]$ by $[-6, 6, 1]$

5. $f(x) = x^4 - 2x^2 + 1$, $k = -6$, $[-15, 3, 1]$ by $[-3, 9, 1]$

6. $f(x) = 4x - x^3$, $k = 5$, $[-6, 18, 1]$ by $[-8, 8, 1]$

7. *Modeling a Cold Front*   A weather map of the United States on April 22, 1996, is shown in the figure. There was a cold front roughly in the shape of a circular arc passing north of Dallas and west of Detroit. The center of the arc was located near Pierre, South Dakota, with a radius of about 750 miles. If Pierre has the coordinates $(0, 0)$ and the positive y-axis points north, then the equation of the front can be modeled by

$$f(x) = -\sqrt{750^2 - x^2},$$

where $0 \le x \le 750$.   (Source: AccuWeather, Inc.)

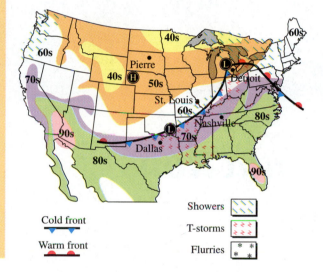

Cold front

Warm front

Showers

T-storms

Flurries

(a) St. Louis is located at $(535, -400)$ and Nashville is at $(730, -570)$. Plot these points and graph f in [0, 1200, 100] by [−800, 0, 100]. Did the cold front reach these cities?

(b) During the next 12 hours, the center of the front moved approximately 110 miles south and 160 miles east. Assuming the cold front did not change shape, use transformations of graphs to determine an equation that models its new location.

(c) Use graphing to determine visually if the cold front reached both cities.

*The purpose of
computing is insight,
not numbers.*

—

Richard Hamming

—

4 Nonlinear Functions and Equations

The location of a parked car can be described by a constant function, while the position of a car traveling at a constant velocity can be modeled by a linear function. However, the location of a vehicle moving with a changing velocity must be modeled by a nonlinear function. In real-world applications, quantities frequently do not change at a constant rate. As a result, nonlinear functions are necessary to describe and predict events.

Both this chapter and the next present phenomena that are modeled by nonlinear functions. Some examples are the amount of air in a person's lungs, weather, concentrations of a drug in the bloodstream, growth of *E. coli* bacteria, use of marijuana among high school seniors, numbers of endangered species, and growth in computer technology.

Until the development of calculators and computers, many nonlinear problems were either too difficult or impossible to solve by hand. While technology and visualization provide new ways for us to investigate mathematical problems, they are not replacements for mathematical understanding. The human mind is capable of mathematical insight and decision making, but does not usually perform long arithmetic calculations proficiently. On the other hand, computers and calculators are incapable of mathematical insight, but are excellent at performing arithmetic and other routine computation. In this way, technology complements the human mind.

Reference: R. Wolff and L. Yaeger, *Visualization of Natural Phenomena.*

4.1 NONLINEAR FUNCTIONS AND THEIR GRAPHS

Polynomial Functions • Increasing and Decreasing Functions • Extrema of Nonlinear Functions • Symmetry

Introduction

Monthly average high temperatures at Daytona Beach are shown in Table 4.1.

TABLE 4.1   Monthly Average High Temperatures at Daytona Beach

Month	1	2	3	4	5	6	7	8	9	10	11	12
Temperature (°F)	69	70	75	80	85	88	90	89	87	81	76	70

Source: J. Williams, *The USA Weather Almanac, 1995.*

Figure 4.1 shows a scatterplot of the data. A linear function would not model this data because the data do not lie on a line. One possibility is to model the data with a quadratic function, as shown in Figure 4.2. However, a better fit can be obtained with the nonlinear function f whose graph is shown in Figure 4.3 and is given by

$$f(x) = 0.0145x^4 - 0.426x^3 + 3.53x^2 - 6.23x + 72.0,$$

where $x = 1$ corresponds to January, $x = 2$ to February, and so on. (Least-squares regression was used to determine f.) Function f is a *polynomial function* with degree 4.

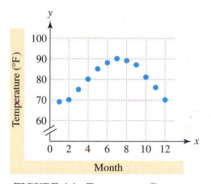

FIGURE 4.1   Temperature Data

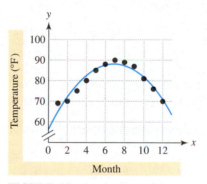

FIGURE 4.2   Quadratic Model

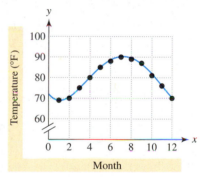

FIGURE 4.3   New Nonlinear Model

Polynomial Functions

Polynomial functions are frequently used to approximate data. The implied domain of a polynomial function is all real numbers, and its graph is continuous.

Polynomial function

A **polynomial function f of degree n in the variable x** can be represented by

$$f(x) = a_n x^n + \cdots + a_2 x^2 + a_1 x + a_0$$

where each coefficient a_k is a real number, $a_n \neq 0$, and n is a nonnegative integer. The **leading coefficient** is a_n and the **degree** is n.

Symbolic representations, degrees, and leading coefficients of some polynomial functions are given on the following page.

Algebra Review
To review polynomials, see Chapter R (page R-21).

Symbolic Representation	Degree	Leading Coefficient
i. $f(x) = 10$	0	$a_0 = 10$
ii. $f(x) = 2x - 3.7$	1	$a_1 = 2$
iii. $f(x) = 1 - 1.4x + 3x^2$	2	$a_2 = 3$
iv. $f(x) = -\frac{1}{2}x^6 + 4x^4 + x$	6	$a_6 = -\frac{1}{2}$

A polynomial function of degree 2 or higher is a nonlinear function. Functions (i) and (ii) are linear, whereas functions (iii) and (iv) are nonlinear. As a result, polynomial functions are used to model both linear and nonlinear data.

Functions that contain radicals, ratios, or absolute values of variables are not polynomials. For example, $f(x) = 2\sqrt{x}$, $g(x) = \dfrac{1}{x-1}$, and $h(x) = |2x + 5|$ are *not* polynomials.

EXAMPLE 1   *Working with polynomials*

For each polynomial $f(x)$, state its degree and leading coefficient. Then graph f.

(a) $f(x) = 1 + x - 3x^2$ **(b)** $f(x) = \frac{1}{2}x^3 - 2x - 1$ **(c)** $f(x) = x^4 - 3x^2 + 1$

SOLUTION   **(a)** Since we can write $f(x) = -3x^2 + x + 1$, its degree and leading coefficient are 2 and -3, respectively. The graph of $Y_1 = 1 + X - 3X^\wedge2$ is a parabola opening downward and is shown in Figure 4.4.

(b) The degree and leading coefficient are 3 and $\dfrac{1}{2}$, respectively. The graph of $Y_1 = .5X^\wedge3 - 2X - 1$ is shown in Figure 4.5. Note that its graph is neither a line nor a parabola.

(c) The degree and leading coefficient are 4 and 1, respectively. The nonlinear graph of $Y_1 = X^\wedge4 - 3X^\wedge2 + 1$ is shown in Figure 4.6.

$[-6, 6, 1]$ by $[-4, 4, 1]$

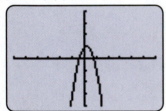

FIGURE 4.4

$[-6, 6, 1]$ by $[-4, 4, 1]$

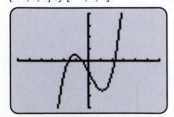

FIGURE 4.5

$[-6, 6, 1]$ by $[-4, 4, 1]$

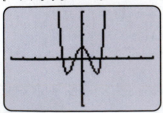

FIGURE 4.6

Increasing and Decreasing Functions

The concepts of increasing and decreasing relate to whether the graph of a function rises or falls. Intuitively, if we could walk *from left to right* along the graph of an increasing function, it would be uphill. For a decreasing function, we would walk downhill. We sometimes speak of a function f increasing or decreasing over an interval of its domain. For example, in Figure 4.7 the function is decreasing (the graph falls) when $-2 \leq x \leq 0$ and increasing (the graph rises) when $0 \leq x \leq 2$.

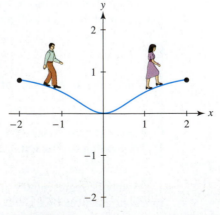

FIGURE 4.7

Increasing and decreasing functions are defined as follows.

> ## Increasing and decreasing functions
>
> Suppose that a function f is defined over an interval I on the number line. If x_1 and x_2 are in I,
>
> **(a)** f **increases** on I if, whenever $x_1 < x_2$, $f(x_1) < f(x_2)$;
> **(b)** f **decreases** on I if, whenever $x_1 < x_2$, $f(x_1) > f(x_2)$.

Figures 4.8 and 4.9 illustrate these concepts.

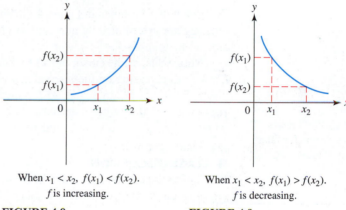

When $x_1 < x_2$, $f(x_1) < f(x_2)$.
f is increasing.

FIGURE 4.8

When $x_1 < x_2$, $f(x_1) > f(x_2)$.
f is decreasing.

FIGURE 4.9

EXAMPLE 2   *Determining where a function is increasing or decreasing*

A graph of a function f shown in Figure 4.10 gives the tides at Clearwater Beach, Florida, x hours after midnight on August 28, 1998.   (Source: D. Pentcheff, *WWW Tide and Current Predictor.*) Use the graph to determine when water levels were increasing and when they were decreasing for $0 \leq x \leq 27$.

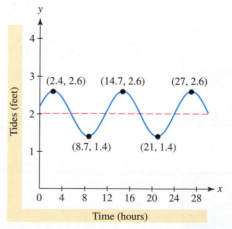

FIGURE 4.10   Tides at Clearwater Beach

SOLUTION   Water levels are rising when the graph of f increases from left to right. This occurs during the following time intervals:

$$0 \leq x \leq 2.4, \quad 8.7 \leq x \leq 14.7, \quad \text{and} \quad 21 \leq x \leq 27.$$

For example, the tides rose from 1.4 feet to 2.6 feet between 8.7 and 14.7 hours after midnight. Similarly, the tides were falling during the following time intervals:

$$2.4 \leq x \leq 8.7 \quad \text{and} \quad 14.7 \leq x \leq 21.$$

Note: When determining where a function is increasing and where it is decreasing, it is important to give *x*-intervals and not *y*-intervals.

EXAMPLE 3 *Determining where a function is increasing or decreasing*

Graph $f(x) = 4x - \frac{1}{3}x^3$ in $[-9.4, 9.4, 1]$ by $[-6.2, 6.2, 1]$. Use the trace feature to approximate the *x*-intervals where *f* is increasing or decreasing.

SOLUTION The graph of *f* is shown in Figure 4.11. Since *x*-intervals where *f* is increasing or decreasing are determined by moving from left to right, start with the cursor at the left side of the graph.

Notice that, as the cursor moves to the right, the *y*-values decrease, until $x = -2$. Thus *f* is decreasing on the interval $x \leq -2$. See Figure 4.12. Then the *y*-values increase from $x = -2$ to $x = 2$, so *f* is increasing on the interval $-2 \leq x \leq 2$. To the right of $x = 2$ the *y*-values decrease. See Figure 4.13. It follows that *f* is decreasing when either $x \leq -2$ or $x \geq 2$, and *f* is increasing when $-2 \leq x \leq 2$.

Graphing Calculator Help

The window used in Figures 4.11–4.13 is a decimal window. See Appendix B (page AP-16).

$[-9.4, 9.4, 1]$ by $[-6.2, 6.2, 1]$

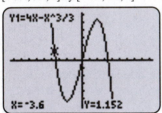

FIGURE 4.11

$[-9.4, 9.4, 1]$ by $[-6.2, 6.2, 1]$

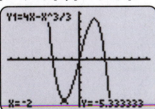

FIGURE 4.12

$[-9.4, 9.4, 1]$ by $[-6.2, 6.2, 1]$

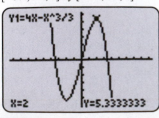

FIGURE 4.13

■ **CLASS DISCUSSION**

How is the slope of the graph of a linear function related to whether the function is increasing, decreasing, or neither? Explain. ■

Extrema of Nonlinear Functions

In Figure 4.3 the minimum monthly average temperature of 69°F occurs in January ($x = 1$) and the maximum average monthly temperature of 90°F occurs in July ($x = 7$). Minimum and maximum *y*-values on the graph of a function often represent important data. Graphs of polynomial functions often have "hills" or "valleys." For example, consider the two polynomial functions *f* and *g* given by

$$f(x) = \frac{1}{2}x^2 - 2x - 4 \quad \text{and} \quad g(x) = -\frac{1}{4}x^4 + \frac{2}{3}x^3 + \frac{5}{2}x^2 - 6x.$$

Their graphs are shown in Figures 4.14 and 4.15. (Some values have been rounded.)

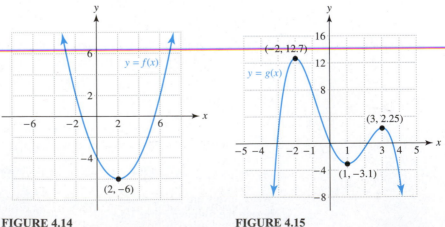

FIGURE 4.14   **FIGURE 4.15**

Figure 4.14 shows the graph of a parabola. If one traces along this graph from left to right, the y-values decrease until the vertex $(2, -6)$ is reached. To the right of the vertex, the y-values increase. The minimum y-value on the graph of f is -6. It is called the *absolute minimum* of f. The function f has no *absolute maximum* because there is no largest y-value on a parabola opening upward.

In Figure 4.15 the highest "hill" on the graph of g is located at $(-2, 12.7)$. Therefore the absolute maximum of g is 12.7. There is a smaller peak located at the point $(3, 2.25)$. In a small interval near $x = 3$, the y-value of 2.25 is locally the largest. We say that g has a *local maximum* of 2.25. Similarly, a "valley" occurs on the graph of g. The lowest point is $(1, -3.1)$. The value -3.1 is not the smallest y-value on the entire graph of g. Therefore it is not an absolute minimum. Rather, -3.1 is a *local minimum*.

Maximum and minimum values that are either absolute or local are called **extrema** (plural of extremum). A function may have several local extrema, but at most one absolute maximum and one absolute minimum. However, it is possible for a function to assume an absolute extremum at two values of x. In Figure 4.16 the absolute maximum is 11. It occurs at $x = \pm 2$. Note that 11 is also a local maximum, because near $x = -2$ and $x = 2$ it is the largest y-value.

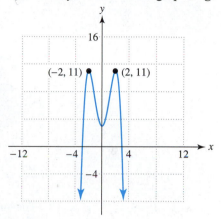

FIGURE 4.16

This discussion is summarized in the following.

Absolute and local extrema

Let c be in the domain of f.

$f(c)$ is an **absolute maximum** if $f(c) \geq f(x)$ *for all x in the domain of f.*

$f(c)$ is an **absolute minimum** if $f(c) \leq f(x)$ *for all x in the domain of f.*

$f(c)$ is a **local maximum** if $f(c) \geq f(x)$ when x is *near c.*

$f(c)$ is a **local minimum** if $f(c) \leq f(x)$ when x is *near c.*

Note: The expression "near c" means that there is an open interval in the domain of f containing c, where $f(c)$ satisfies the inequality.

EXAMPLE 4 *Identifying and interpreting extrema*

Figure 4.17 on the next page shows the graph of a function f that models the volume of air in a person's lungs measured in liters after x seconds.   (Adapted from: V. Thomas, *Science and Sport.*)

(a) Determine the absolute maximum and the absolute minimum of f. Interpret the results.

(b) Identify two local maxima (plural of maximum) and two local minima (plural of minimum) of f. Interpret the results.

SOLUTION

(a) The absolute maximum is 4 liters and occurs at C. The absolute minimum is 1 liter and occurs at D. At C a deep breath has been taken and the lungs are more inflated. After C, the person exhales beyond normal breathing until the lungs contain only 1 liter of air at D.

(b) One local maximum is 3 liters. It occurs at A and E and represents the amount of air in a person's lungs after inhaling normally. One local minimum is 2 liters. It occurs at B and F and represents the amount of air after exhaling normally. Another local maximum is 4 liters, which was also the absolute maximum. Similarly, 1 liter is a local minimum and also the absolute maximum.

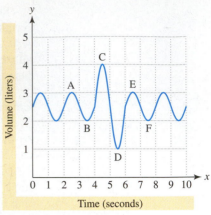

FIGURE 4.17 Volume of Air in a Person's Lungs

EXAMPLE 5 *Modeling ocean temperatures*

The monthly average ocean temperature in degrees Fahrenheit at Bermuda can be modeled by $f(x) = 0.0215x^4 - 0.648x^3 + 6.03x^2 - 17.1x + 76.4$, where $x = 1$ corresponds to January and $x = 12$ to December. The domain of f is $D = \{x \mid 1 \le x \le 12\}$. (Source: J. Williams, *The Weather Almanac 1995*.)

(a) Graph f in [1, 12, 1] by [50, 90, 10].

(b) Estimate the absolute extrema. Interpret the results.

SOLUTION

(a) The graph of $Y_1 = .0215X^4 - .648X^3 + 6.03X^2 - 17.1X + 76.4$ is shown in Figure 4.18.

(b) Many graphing calculators have the capability to find maximum and minimum y-values. The points associated with absolute extrema are shown in Figures 4.18 and 4.19. An absolute minimum of about 61.5 corresponds to the point (2.01, 61.5). This means that the average ocean temperature is coldest during the month of February ($x = 2$) when it reaches a minimum of about 61.5°F.

An absolute maximum of approximately 82 corresponds to the point (7.61, 82.0). The warmest average ocean temperature occurs during August ($x \approx 8$) when it reaches a maximum of 82°F. We might also say that this maximum occurs in late July, since $x \approx 7.61$.

Graphing Calculator Help

To find a minimum or maximum point on a graph, see Appendix B (page AP-14).

[1, 12, 1] by [50, 90, 10]

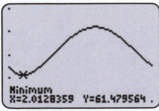

FIGURE 4.18

[1, 12, 1] by [50, 90, 10]

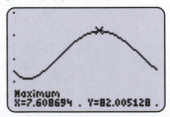

FIGURE 4.19

■ **MAKING CONNECTIONS**

Local and Absolute Extrema   Extrema can be either local or absolute. Also, it is possible for an extremum to be both local and absolute. When using a graph to identify either local or absolute extrema, it is important to remember that an extremum is a *y-value*—not a point (x, y).

Symmetry

Symmetry is used frequently in art, mathematics, and computer graphics. Many objects are symmetric along a vertical line so that the left and right sides are mirror images. If an automobile is viewed from the front, the left side is typically a mirror image of the right side. Similarly, animals and people usually have an approximate left-right symmetry. Graphs of functions may also exhibit this type of symmetry, as shown in Figures 4.20–4.22.

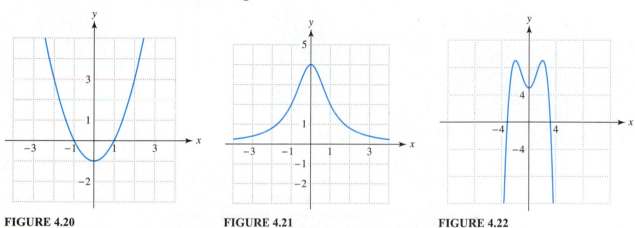

FIGURE 4.20   **FIGURE 4.21**   **FIGURE 4.22**

If each graph were folded along the y-axis, the left and right halves would match. These graphs are **symmetric with respect to the y-axis**. A function whose graph satisfies this characteristic is called an *even function*.

Figure 4.23 shows a graph of an even function f. Since the graph is symmetric with respect to the y-axis, the points (x, y) and $(-x, y)$ both lie on the graph of f. Thus $f(x) = y$ and $f(-x) = y$, and so $f(x) = f(-x)$ for an even function. This means that if we change the sign of the input, the output does not change. For example, if $g(x) = x^2$, then $g(2) = g(-2) = 4$. Since this is true for every input, g is an even function.

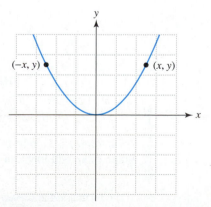

FIGURE 4.23   An Even Function

Even function

A function f is an **even function** if $f(-x) = f(x)$ for every x in its domain. The graph of an even function is symmetric with respect to the y-axis.

A second type of symmetry is shown in Figures 4.24–4.26. If we could spin or rotate the graph about the origin, the original graph would reappear after half a turn. These graphs are **symmetric with respect to the origin** and represent *odd functions*.

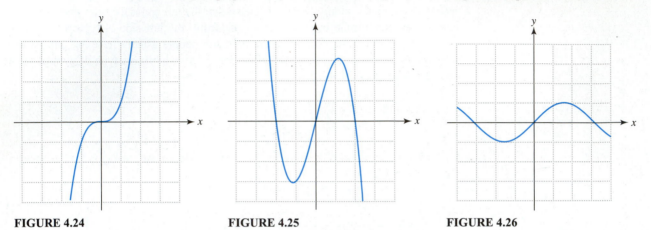

FIGURE 4.24   **FIGURE 4.25**   **FIGURE 4.26**

In Figure 4.27 the point (x, y) lies on the graph of an odd function f. If this point spins half a turn or 180° around the origin, its new location is $(-x, -y)$. Thus $f(x) = y$ and $f(-x) = -y$. It follows that $f(-x) = -y = -f(x)$ for any odd function f. Changing the sign of the input only changes the sign of the output. For example, if $g(x) = x^3$ then $g(3) = 27$ and $g(-3) = -27$. Since this is true for every input, g is an odd function.

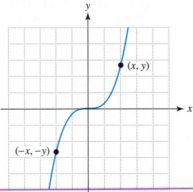

FIGURE 4.27   An Odd Function

Odd function

A function f is an **odd function** if $f(-x) = -f(x)$ for every x in its domain. The graph of an odd function is symmetric with respect to the origin.

The terms *odd* and *even* have special meaning when they are applied to a polynomial function f. If $f(x)$ contains terms that have only odd powers of x, then f is an odd function. Similarly, if $f(x)$ contains terms that have only even powers of x (and possibly a constant term), then f is an even function. For example, $f(x) = x^6 - 4x^4 - 2x^2 + 5$ is an even function, whereas $g(x) = x^5 + 4x^3$ is an odd function. This can be shown symbolically as follows.

$$f(-x) = (-x)^6 - 4(-x)^4 - 2(-x)^2 + 5 \qquad \text{Substitute } -x \text{ for } x.$$
$$= x^6 - 4x^4 - 2x^2 + 5 \qquad \text{Simplify.}$$
$$= f(x) \qquad f \text{ is an even function.}$$
$$g(-x) = (-x)^5 + 4(-x)^3 \qquad \text{Substitute } -x \text{ for } x.$$
$$= -x^5 - 4x^3 \qquad \text{Simplify.}$$
$$= -g(x) \qquad g \text{ is an odd function.}$$

■ CLASS DISCUSSION

If 0 is in the domain of an odd function f, what point must lie on its graph? Explain your reasoning. ■

It is important to remember that the graphs of many functions exhibit no symmetry with respect to either the y-axis or the origin. As a result, these functions are neither odd nor even.

EXAMPLE 6   *Identifying odd and even functions*

For each representation of a function f, identify whether f is odd, even, or neither.

(a)   TABLE 4.2

x	-3	-2	-1	0	1	2	3
$f(x)$	10.5	2	-0.5	-2	-0.5	2	10.5

(b)

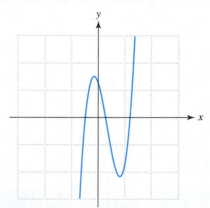

FIGURE 4.28

(c) $f(x) = x^3 - 5x$ **(d)** f is the cube root function.

SOLUTION   **(a)** The function defined by Table 4.2 has domain $D = \{-3, -2, -1, 0, 1, 2, 3\}$. Notice that $f(-3) = 10.5 = f(3)$ and $f(-2) = 2 = f(2)$. The function f satisfies the statement $f(-x) = f(x)$ for every x in D. Thus f is an even function.

(b) If we fold the graph in Figure 4.28 on the y-axis, the two halves do not match, so f is not an even function. Similarly, f is not an odd function since spinning its graph half a turn about the origin does not result in the same graph. The function f is neither odd nor even.

(c) Since f is a polynomial containing only odd powers of x, it is an odd function. This also can be shown symbolically as follows.

$$
\begin{aligned}
f(-x) &= (-x)^3 - 5(-x) && \text{Substitute } -x \text{ for } x. \\
&= -x^3 + 5x && \text{Simplify.} \\
&= -(x^3 - 5x) && \text{Distributive property} \\
&= -f(x) && f \text{ is an odd function.}
\end{aligned}
$$

This result is supported graphically in Figure 4.29. Spinning the graph $180°$ about the origin results in the same graph.

$[-9, 9, 1]$ by $[-6, 6, 1]$ $[-6, 6, 1]$ by $[-4, 4, 1]$

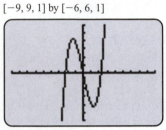   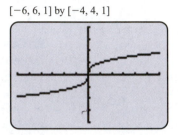

FIGURE 4.29   **FIGURE 4.30**

Graphing Calculator Help

On some calculators, the cube root can be found under the MATH menu. See Appendix B (page AP-5).

(d) Since $f(x) = \sqrt[3]{x}$, graph $Y_1 = \sqrt[3]{(X)}$ or equivalently $Y_1 = X^{\wedge}(1/3)$, as shown in Figure 4.30. Spinning the graph half a turn about the origin results in the same graph so f is an odd function.

■ **CLASS DISCUSSION**
Discuss the possibility of the graph of a function being symmetric with respect to the x-axis. ■

Putting it all Together

4.1

Nonlinear functions have graphs that can increase or decrease over their domains. Extrema and symmetry are used to describe nonlinear graphs. Polynomial functions of degree 2 or higher are examples of nonlinear functions. The following table summarizes some important concepts related to the graphs of nonlinear functions.

Concept	Explanation	Graphical Example
Increasing	f increases on an interval, if whenever $x_1 < x_2$, then $f(x_1) < f(x_2)$.	f is increasing for $x \le -1$ or $0 \le x \le 1$.

Concept	Explanation	Graphical Example
Decreasing	f decreases on an interval, if whenever $x_1 < x_2$, then $f(x_1) > f(x_2)$.	f is decreasing for $-2 \le x \le 2$.
Even function	$f(-x) = f(x)$ Graph is symmetric with respect to the y-axis.	
Odd function	$f(-x) = -f(x)$ Graph is symmetric with respect to the origin.	
Absolute maximum (minimum)	The maximum (minimum) y-value on the graph $y = f(x)$	
Local maximum (minimum)	A maximum (minimum) y-value on the graph $y = f(x)$ in an open interval of the domain of f	

4.1 EXERCISES

Polynomials

Exercises 1–10: Determine if f is a polynomial function. If it is, state its degree and leading coefficient.

1. $f(x) = 2x^3 - x + 5$ **2.** $f(x) = -x^4 + 1$

3. $f(x) = \sqrt{x}$

4. $f(x) = 2x^3 - \sqrt[3]{x}$

5. $f(x) = 1 - 4x - 5x^4$

6. $f(x) = \dfrac{1}{1-x}$

7. $f(x) = \dfrac{1}{x^2 + 3x - 1}$ **8.** $f(x) = 5 - 4x$

9. $f(x) = 22$ **10.** $f(x) = |2x|$

Increasing and Decreasing Functions

Exercises 11–16: Use the graph of f to determine intervals where f is increasing and where f is decreasing.

11.
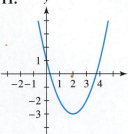

12.
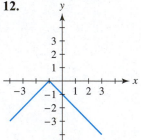

13.
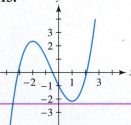

14.
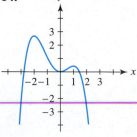

15.
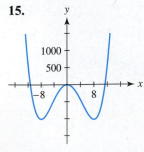

16.
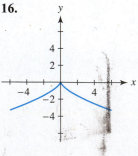

Exercises 17–24: Identify intervals where f is increasing and where f is decreasing. Approximate endpoints to the nearest hundredth when appropriate. (Hint: You may want to graph f.)

17. $f(x) = 2x - 1$

18. $f(x) = x^2 - 3x - 1$

19. $f(x) = 4 - x^2$ **20.** $f(x) = x^4 - 4x^2$

21. $f(x) = x^3 - 2x + 1$ **22.** $f(x) = |x - 1|$

23. $f(x) = 3(x + 1)^2 - 5$

24. $f(x) = \sqrt[3]{x^2}$

Absolute and Local Extrema

25. *Wiretaps*   The line graph shows the approximate number of state and federal wiretaps approved from 1968 to 1992. If the line graph represents a function f, find the local extrema. (*Hint:* Local extrema do not occur at endpoints.)   (Source: Administrative Office of the U.S. Courts.)

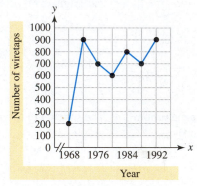

26. *Tides*   The following figure shows tide levels x hours after midnight.
  (a) Estimate the x-values when the tides were rising on the interval $0 \le x \le 26.8$.
  (b) Identify all local extrema and interpret each.

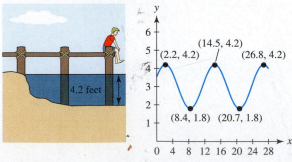

Exercises 27–36: Use the graph of f to estimate the

  (a) local extrema, and

  (b) absolute extrema.

(*Hint*: Local extrema cannot occur at endpoints, but absolute extrema can.)

27.

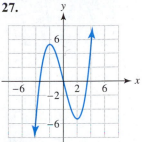

28.

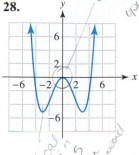

29.

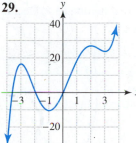

30.

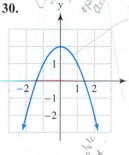

31.

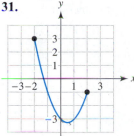

32.

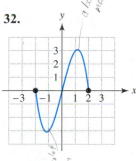

33.

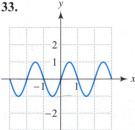

34.

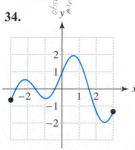

35.

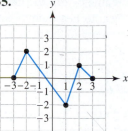

36.

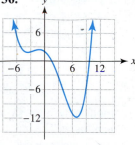

Exercises 37–42: Determine graphically any

  (a) local extrema, and

  (b) absolute extrema.

37. $f(x) = -x^2 + 3x + 3$

38. $f(x) = \frac{1}{9}x^3 - 3x$

39. $f(x) = 0.5x^4 - 5x^2 + 4.5$

40. $f(x) = 0.01x^5 + 0.02x^4 - 0.35x^3 - 0.36x^2 + 1.8x$

41. $f(x) = \dfrac{8}{1 + x^2}$

42. $f(x) = x^{1/3}$

43. *Energy*   The U.S. consumption of energy by residential and commercial users from 1950 to 1980 can be modeled by

$$f(x) = -0.00113x^3 + 0.0408x^2 - 0.0432x + 7.66,$$

where $x = 0$ corresponds to 1950 and $x = 30$ to 1980. Consumption is measured in quadrillion Btu. (Source: Department of Energy.)

  (a) Evaluate $f(5)$ and interpret the result.

  (b) Graph f in [0, 30, 5] by [6, 16, 1]. Describe the energy usage during this time period.

  (c) Approximate the local maximum and interpret the result.

44. *Natural Gas*   The U. S. consumption of natural gas from 1965 to 1980 can be modeled by

$$f(x) = 0.0001234x^4 - 0.005689x^3 + 0.08792x^2 - 0.5145x + 1.514,$$

where $x = 5$ corresponds to 1965 and $x = 20$ to 1980. Consumption is measured in trillion cubic feet.   (Source: Department of Energy.)

  (a) Evaluate $f(10)$ and interpret the result.

  (b) Graph f in [5, 20, 5] by [0.4, 0.8, 0.1]. Describe the energy usage during this time period.

  (c) Determine the local extrema and interpret the results.

45. *Heating Costs* In colder climates of the United States the cost for natural gas to heat homes can vary dramatically from one month to the next. The polynomial function given by

$$f(x) = -0.1213x^4 + 3.462x^3 - 29.22x^2 + 64.68x + 97.69$$

models the monthly cost in dollars of heating a typical home. The input x represents the month, where $x = 1$ corresponds to January and $x = 12$ to December. (Source: Minnegasco, A NORAM Energy Company.)

(a) Where might the absolute extrema occur for $1 \le x \le 12$?

(b) Graph f in [1, 12, 1] by [0, 150, 10]. Identify the absolute extrema in this graph and interpret the results.

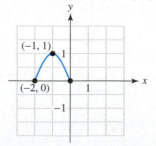

46. *Daylight Hours* The accompanying graph shows the daylight hours at 60°N latitude, where $x = 1$ corresponds to January 1, $x = 2$ to February 1, and so on.

(a) Estimate when daylight hours are increasing for $1 \le x \le 12$.

(b) Identify and interpret the local maximum.

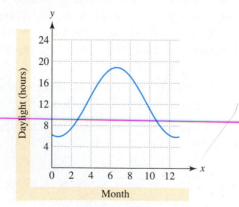

Symmetry

Exercises 47–48: Use the graph to determine whether the function f is odd or even. Then use the point shown to find a second point on the graph.

47. **48.**

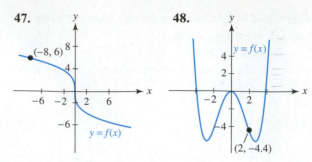

49. The table is a complete representation of f. Decide if f is even, odd, or neither.

x	-100	-10	-1	0	1	10	100
$f(x)$	56	-23	5	0	-5	23	-56

50. The table is a complete representation of f. Decide if f is even, odd, or neither.

x	-5	-3	-1	1	2	3
$f(x)$	-4	-2	1	1	-2	-4

51. Complete the table if f is an even function.

x	-3	-2	-1	0	1	2	3
$f(x)$	21	$?$	-25	$?$	$?$	-12	$?$

52. Complete the table if f is an odd function.

x	-5	-3	-2	0	2	3	5
$f(x)$	13	$?$	-5	$?$	$?$	-1	$?$

53. Complete the table so the function f is neither odd nor even. Answers may vary.

x	-2	-1	0	1	2
$f(x)$	5	$?$	$?$	3	$?$

54. Complete the table if the function f is odd.

x	-2	0	2
$f(x)$	5	$?$	$?$

55. A partial graph of an odd function with domain $D = \{x \mid -2 \le x \le 2\}$ is shown in the figure. Make a sketch of the complete graph.

56. A partial graph of an even function f with domain $D = \{x \mid -10 \le x \le 10\}$ is shown in the figure. Make a sketch of the complete graph.

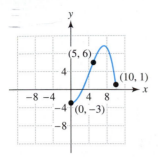

57. *Average Temperature* The accompanying graph approximates the monthly average temperatures in degrees Fahrenheit in Austin, Texas. In this graph x represents the month, where $x = 0$ corresponds to July.

(a) Is this a graph of an odd or even function?

(b) June corresponds to $x = -1$ and August to $x = 1$. The average temperature in June is 83°F. What is the average temperature in August?

(c) March corresponds to $x = -4$ and November to $x = 4$. According to the graph, how do their average temperatures compare?

(d) Interpret what this type of symmetry implies about average temperatures in Austin.

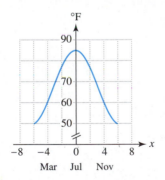

58. *Average Temperature* The accompanying graph shows the deviation of the monthly average temperatures from the yearly average temperature in Anchorage, Alaska. In this graph x represents the month, where $x = 0$ corresponds to October. For example, the average temperature in October is equal to the average yearly temperature, so its deviation is 0°F. The deviation in July ($x = -3$) is 22°F above the yearly average.

(a) October and what other month have a temperature deviation of 0°F from the yearly average?

(b) Is this a graph of an odd or even function?

(c) September corresponds to $x = -1$ and November to $x = 1$. The average temperature in No-

vember is 11°F below the average yearly temperature. What does this indicate about the average temperature in September?

(d) August corresponds to $x = -2$ and December to $x = 2$. How do their temperatures compare to the average yearly temperature?

(e) Interpret what this type of symmetry implies about the average monthly temperatures in Anchorage.

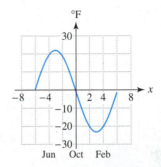

59. *Height of a Projectile* If a projectile is shot into the air, it attains a maximum height and then falls back to the ground. Suppose that $x = 0$ corresponds to the time when the projectile's height is maximum. If air resistance is ignored, its height h above the ground at any time x may be modeled by $h(x) = -16x^2 + h_{max}$, where h_{max} is the projectile's maximum height above the ground. Height is measured in feet and time in seconds. Let $h_{max} = 400$ feet.

(a) Evaluate $h(-2)$ and $h(2)$. Interpret these results.

(b) Evaluate $h(-5)$ and $h(5)$. Interpret these results.

(c) Graph h in $[-5, 5, 1]$ by $[0, 500, 100]$. Is h an even or odd function?

(d) How do $h(x)$ and $h(-x)$ compare when $-5 \le x \le 5$? What does this indicate about the flight of a projectile?

60. *Velocity of a Projectile* (Refer to Exercise 59.) The velocity in feet per second of a projectile shot into the air is given by $v(t) = -32t$, where $-5 \le t \le 5$ and $t = 0$ corresponds to the time when the maximum height is attained. A positive velocity indicates that the projectile is traveling up, whereas a negative velocity indicates that it is falling.

(a) Evaluate $v(-2)$ and $v(2)$. Interpret each result.

(b) Graph v in $[-5, 5, 1]$ by $[-200, 200, 100]$. Is v an even or odd function?

(c) How do $v(t)$ and $v(-t)$ compare when $-5 \le t \le 5$? What does this mean about the velocity of a projectile?

Exercises 61–70: Determine if f is even, odd, or neither. Support your answer graphically.

61. $f(x) = 3x^4 - 1.5x^2 - 9$

62. $f(x) = 4x^3 - 18x$ **63.** $f(x) = 3x^5 + 6x$

64. $f(x) = 3x^6 + x^4 - 4$

65. $f(x) = 7x^2 - 4x + 1$

66. $f(x) = 3x^3 - 2x + 1$

67. $f(x) = \sqrt{1 - x^2}$ **68.** $f(x) = \dfrac{1}{x^3}$

69. $f(x) = \sqrt[3]{x^2}$ **70.** $f(x) = -7x$

Concepts

71. Sketch a graph of a linear function that is an odd function.

72. Sketch a graph of a linear function that is an even function.

73. Does there exist a continuous odd function that is always increasing and whose graph passes through the points $(-3, -4)$ and $(2, 5)$? Explain.

74. Is there an even function whose domain is all real numbers and is always decreasing? Explain.

Translations of Graphs

Exercises 75–78: Use the graph of $f(x) = 4x - \dfrac{x^3}{3}$ and translations of graphs to sketch the graph of the equation.

75. $y = f(x + 1)$ **76.** $y = f(x) - 2$

77. $y = 2f(x)$ **78.** $y = f\left(\dfrac{1}{2}x\right)$

Writing about Mathematics

79. Explain the difference between a local maximum and an absolute maximum. Are extrema x-values, y-values, or points (x, y)?

80. Describe ways to determine if a polynomial function is odd, even, or neither. Give examples.

Extended and Discovery Exercise

Find the dimensions of a rectangle of maximum area that can be inscribed in a semicircle with radius 3. Assume that the rectangle is positioned as shown in the accompanying figure.

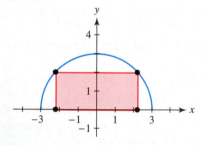

Geometry Review
To review formulas for circles, see Chapter R (page R-2).

4.2 POLYNOMIAL FUNCTIONS AND MODELS

Graphs of Polynomial Functions • Polynomial Regression • Piecewise-Defined Polynomial Functions

Introduction

The study of higher degree polynomial equations dates back to Old Babylonian civilization in about 1800–1600 B.C. Gottfried Leibniz (1646–1716) was the first mathematician to generalize polynomial functions of degree n. Many eighteenth-century mathematicians devoted their lives to studying polynomial equations. Today, polynomial functions are used to model a wide variety of real data. (References: *Historical Topics for the Mathematics Classroom, Thirty-first Yearbook,* NCTM; L. Motz and J. H. Weaver, *The Story of Mathematics.*)

For example, the consumption of natural gas by the United States has varied throughout past decades. Table 4.3 lists energy consumption (in quadrillion Btu) for selected years, which increased, decreased, and then increased again. A scatterplot of the data is shown in Figure 4.31, and one possibility for a polynomial modeling function f is shown in Figure 4.32. Notice that f is neither linear nor quadratic. What degree of polynomial might we use to model this data? We learn the answer to this question in this section.

TABLE 4.3 Natural Gas Consumption

Year	1960	1970	1980	1990	1997
Consumption	12.4	21.8	20.4	19.3	22.6

Source: Department of Energy.

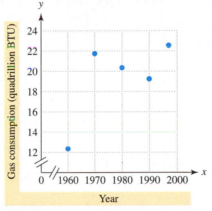

FIGURE 4.31 A Scatterplot

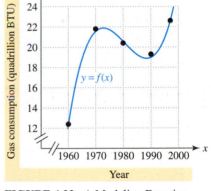

FIGURE 4.32 A Modeling Function

Graphs of Polynomial Functions

In Section 4.1 polynomial functions were defined. Polynomial functions have been used in several applications. Their graphs are continuous; they have no breaks or sharp corners. The domain of a polynomial function is all real numbers. A polynomial function of degree n can be expressed as

$$f(x) = a_n x^n + \cdots + a_2 x^2 + a_1 x + a_0,$$

where each coefficient a_k is a real number, $a_n \neq 0$, and n is a nonnegative integer. The *leading coefficient* is a_n.

The expression $a_n x^n + \cdots + a_2 x^2 + a_1 x + a_0$ is a **polynomial of degree n** and the equation $a_n x^n + \cdots + a_2 x^2 + a_1 x + a_0 = 0$ is a **polynomial equation of degree n**. Thus, f is a polynomial function, $f(x)$ is a polynomial, and $f(x) = 0$ is a polynomial equation. For example, the function f, given by $f(x) = x^3 - 3x^2 + x - 5$, is a polynomial function, $x^3 - 3x^2 + x - 5$ is a polynomial, and $x^3 - 3x^2 + x - 5 = 0$ is a polynomial equation.

A **turning point** occurs whenever the graph of a polynomial function changes from increasing to decreasing or from decreasing to increasing. Turning points are associated with "hills" or "valleys" on a graph. The y-value at a turning point is either a

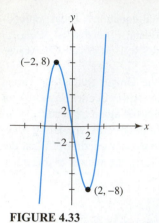

FIGURE 4.33

local maximum or local minimum of the function. In Figure 4.33 the graph has two turning points, $(-2, 8)$ and $(2, -8)$. A local maximum is 8 and a local minimum is -8.

■ MAKING CONNECTIONS

Turning Points and Local Extrema A turning point (x, y) is a point on the graph of a function that is located where the graph changes from increasing to decreasing or from decreasing to increasing. A local extrema is a y-value, not a point, and often corresponds to the y-value of a turning point.

We discuss the graphs of polynomial functions, starting with degree 0 and continuing to degree 5.

Constant Polynomial Functions. If $f(x) = a$ and $a \neq 0$, then f is both a constant function and a polynomial function of degree 0. (If $a = 0$ then f has an undefined degree.) Its graph is a horizontal line that does not coincide with the x-axis. Graphs of $f(x) = 4$ and $g(x) = -3$ are shown in Figures 4.34 and 4.35. A graph of a polynomial function of degree 0 has no x-intercepts or turning points.

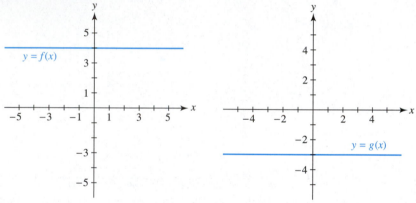

FIGURE 4.34 Constant, $a > 0$ **FIGURE 4.35** Constant, $a < 0$

Linear Polynomial Functions. If $f(x) = ax + b$ and $a \neq 0$, then f is both a linear function and a polynomial function of degree 1. Its graph is a line that is neither horizontal nor vertical. The graphs of $f(x) = 2x - 3$ and $g(x) = -1.6x + 5$ are shown in Figures 4.36 and 4.37. A polynomial function of degree 1 has one x-intercept and no turning points.

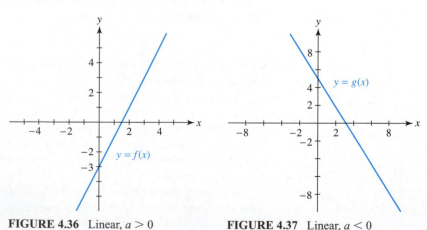

FIGURE 4.36 Linear, $a > 0$ **FIGURE 4.37** Linear, $a < 0$

The graph of $f(x) = ax + b$ with $a > 0$ is a line sloping upward from left to right. As one traces from left to right, the y-values become larger without a maximum. We say that the **end behavior** of the graph tends to $-\infty$ on the left and ∞ on the right. (Strictly speaking, the graph of a polynomial has infinite length and does not have an end.)

If $a < 0$, then the end behavior is reversed. The line slopes downward from left to right. The y-values on the left side of the graph become large positive values without a maximum and the y-values on the right side become negative without a minimum. The end behavior tends to ∞ on the left and $-\infty$ on the right.

Quadratic Polynomial Functions.   If $f(x) = ax^2 + bx + c$ and $a \neq 0$, then f is both a quadratic function and a polynomial function of degree 2. Its graph is a parabola that either opens upward ($a > 0$) or downward ($a < 0$). The graphs of $f(x) = 0.5x^2 + 2$, $g(x) = x^2 + 4x + 4$, and $h(x) = -x^2 + 3x + 4$ are shown in Figures 4.38–4.40, respectively. Quadratic functions can have zero, one, or two x-intercepts. A parabola has exactly one turning point, which is also the vertex.

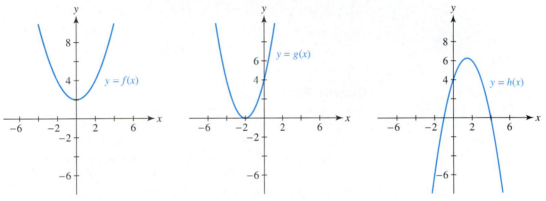

FIGURE 4.38 Quadratic, $a > 0$ **FIGURE 4.39** Quadratic, $a > 0$ **FIGURE 4.40** Quadratic, $a < 0$

If $a > 0$, then both sides of the graph go up. The end behavior tends to ∞ on both sides. If $a < 0$, as in Figure 4.40, then the end behavior is reversed and tends to $-\infty$ on both sides.

Cubic Polynomial Functions.   If $f(x) = ax^3 + bx^2 + cx + d$ and $a \neq 0$, then f is both a **cubic function** and a polynomial function of degree 3. The graph of a cubic function can have zero or two turning points. The graph of $f(x) = -x^3 + 4x + 1$ in Figure 4.41 has two turning points, whereas the graph of $g(x) = \frac{1}{4}x^3$ in Figure 4.42 has no turning points.

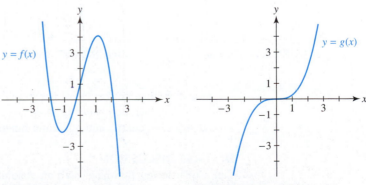

FIGURE 4.41 Cubic, $a < 0$ **FIGURE 4.42** Cubic, $a > 0$

If $a > 0$, the graph of a cubic function falls to the left and rises to the right, as in Figure 4.42. If $a < 0$, its graph rises to the left and falls to the right, as in Figure 4.41. The end behavior of a cubic function is similar to that of a linear function, and tends to ∞ on one side and $-\infty$ on the other. Therefore its graph must cross the x-axis at least once. A cubic function can have up to three x-intercepts. The graph of $h(x) = 0.1x^3 - 0.1x^2 - 2.1x + 4.5$ in Figure 4.43 has two x-intercepts.

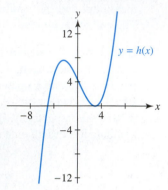

FIGURE 4.43 Cubic, $a > 0$

Quartic Polynomial Functions. If $f(x) = ax^4 + bx^3 + cx^2 + dx + e$ and $a \neq 0$, then f is both a **quartic function** and a polynomial function of degree 4. The graph of a quartic function can have up to four x-intercepts and three turning points, and the graph of $f(x) = 0.1x^4 - 0.3x^3 - 1.2x^2 + 1.8x + 2$ in Figure 4.44 is an example. The graph of $g(x) = -x^4 + 2x^3$ in Figure 4.45 has one turning point and two x-intercepts, and the graph of $h(x) = -x^4 + x^3 + 3x^2 - x - 2$ in Figure 4.46 has three turning points and three x-intercepts.

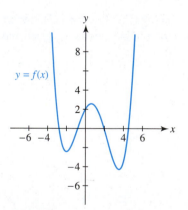

FIGURE 4.44 Quartic, $a > 0$

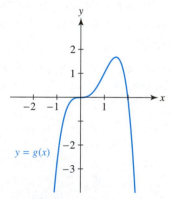

FIGURE 4.45 Quartic, $a < 0$

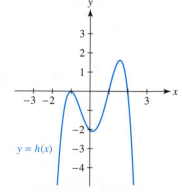

FIGURE 4.46 Quartic, $a < 0$

If $a > 0$, then both sides of the graph of a quartic function go up, as in Figure 4.44. If $a < 0$, then both sides of its graph go down, as in Figures 4.45 and 4.46. The end behaviors of quartic and quadratic functions are similar.

■ **CLASS DISCUSSION**

Can a quartic function have both an absolute maximum and an absolute minimum? Explain. ■

Quintic Polynomial Functions.   If $f(x) = ax^5 + bx^4 + cx^3 + dx^2 + ex + k$ and $a \neq 0$, then f is both a **quintic function** and a polynomial function of degree 5. The graph of a quintic function may have up to five x-intercepts and four turning points. An example is shown in Figure 4.47, given by

$$f(x) = 0.01x^5 + 0.03x^4 - 0.63x^3 - 0.67x^2 + 8.46x - 7.2.$$

Other quintic functions are shown in Figures 4.48 and 4.49. They are defined by

$$g(x) = \frac{1}{5}x^5 - 3 \qquad \text{and}$$

$$h(x) = -0.02x^5 - 0.14x^4 + 0.04x^3 + 0.92x^2 - 1.3x + 0.5.$$

The function g has one x-intercept and no turning points. The graph of h appears to have two x-intercepts and two turning points. Notice that the end behavior of a quintic function is similar to linear and cubic functions.

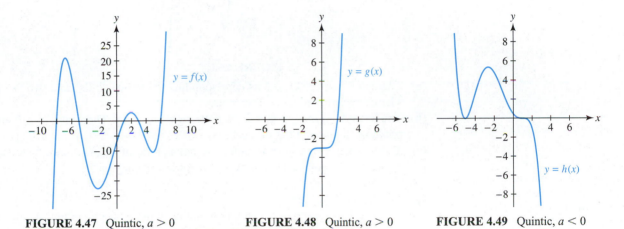

FIGURE 4.47   Quintic, $a > 0$ FIGURE 4.48   Quintic, $a > 0$ FIGURE 4.49   Quintic, $a < 0$

The end behavior of polynomial functions is summarized in the following.

End behavior of polynomial functions

Let f be a polynomial function with leading coefficient a and degree n.

1. $n \geq 2$ is even.
   $a > 0$ implies the graph of f rises both to the left and to the right.
   $a < 0$ implies the graph of f falls both to the left and to the right.
2. $n \geq 1$ is odd.
   $a > 0$ implies the graph of f falls to the left and rises to the right.
   $a < 0$ implies the graph of f rises to the left and falls to the right.

The maximum number of x-intercepts and turning points on the graph of a polynomial function are summarized in the following.

Degree, x-intercepts, and turning points

The graph of a polynomial function of degree $n \geq 1$ has at most n x-intercepts and at most $n - 1$ turning points.

EXAMPLE 1   *Analyzing the graph of a polynomial function*

Figure 4.50 shows the graph of a polynomial function f.
(a) How many turning points and x-intercepts are there?
(b) Is the leading coefficient a positive or negative? Is the degree odd or even?
(c) Determine the minimum possible degree of f.

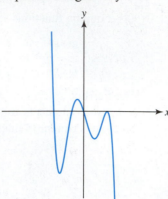

FIGURE 4.50

SOLUTION   **(a)** There are four turning points corresponding to the two "hills" and two "valleys." There appear to be four x-intercepts.
(b) The left side of the graph rises and the right side falls. Therefore, $a < 0$ and the polynomial function has odd degree.

Graphing Calculator Help

To find a minimum or a maximum point on a graph, see Appendix B (page AP-14).

(c) The graph has four turning points. A polynomial of degree n can have at most $n - 1$ turning points. Therefore, f must be at least degree 5.

■ **CLASS DISCUSSION**

Can you sketch the graph of a quadratic function with no turning points, a cubic function with one turning point, or a quartic function with two turning points? Explain. ■

EXAMPLE 2   *Analyzing the graph of a polynomial function*

Graph $f(x) = x^3 - 2x^2 - 5x + 6$, and then complete the following.
(a) Identify the x-intercepts.
(b) Approximate the coordinates of any turning points to the nearest hundredth.
(c) Use the turning points to approximate any local extrema.

SOLUTION   **(a)** A graph of f, shown in Figure 4.51, appears to intersect the x-axis at the points $(-2, 0)$, $(1, 0)$, and $(3, 0)$. Therefore the x-intercepts are -2, 1, and 3.
(b) There are two turning points. From Figures 4.52 and 4.53 their coordinates are approximately $(-0.79, 8.21)$ and $(2.12, -4.06)$.
(c) There is a local maximum of about 8.21 and a local minimum of about -4.06.

$[-10, 10, 1]$ by $[-10, 10, 1]$

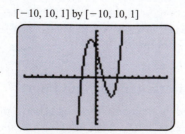

FIGURE 4.51

$[-10, 10, 1]$ by $[-10, 10, 1]$

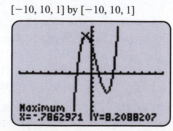

FIGURE 4.52

$[-10, 10, 1]$ by $[-10, 10, 1]$

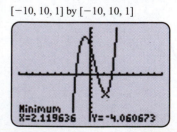

FIGURE 4.53

Polynomial Regression

We now have the mathematical understanding to model the data presented in the introduction to this section. The polynomial modeling function f shown in Figure 4.32 on page 237 falls to the left and rises to the right, so it has odd degree and the leading coefficient is positive. Since the graph of f has two turning points, it must be at least degree 3. A cubic polynomial $f(x)$ is a possible choice, where

$$f(x) = ax^3 + bx^2 + cx + d.$$

Trial and error would be a difficult method to find values for a, b, c, and d. Instead, we can use least-squares regression, which was also discussed in Sections 2.6 and 3.1, for linear and quadratic functions. The next example illustrates cubic regression.

EXAMPLE 3   *Determining a cubic modeling function*

The data in Table 4.3 list natural gas consumption in the United States.
(a) Find a polynomial function of degree 3 that models the data.
(b) Graph f and the data together.
(c) Estimate the natural gas consumption in 1982.

SOLUTION
(a) Enter the five data points (1960, 12.4), (1970, 21.8), (1980, 20.4), (1990, 19.3), and (1997, 22.6) into your calculator. Then select cubic regression, as shown in Figure 4.54. The equation for $f(x)$ is shown in Figure 4.55.
(b) A graph of f and a scatterplot of the data are shown in Figure 4.56.
(c) To estimate natural gas consumption in 1975, evaluate $f(1982)$ to obtain about 20.1 quadrillion Btu.

[1955, 2005, 5] by [10, 25, 5]

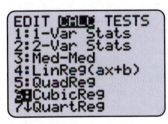

FIGURE 4.54

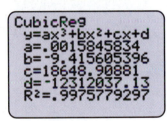

FIGURE 4.55

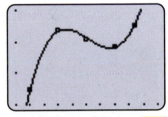

FIGURE 4.56

Graphing Calculator Help

To find an equation of least-squares fit, see Appendix B (page AP-15). To copy a regression equation into Y_1, see (page AP-17.)

Piecewise-Defined Polynomial Functions

In Section 2.5 piecewise-defined functions were discussed. If each piece is a polynomial, then the function is a **piecewise-defined polynomial function** or **piecewise-polynomial function**. An example is given by $f(x)$.

$$f(x) = \begin{cases} x^3 & \text{if } x < 1 \\ x^2 - 1 & \text{if } x \geq 1 \end{cases}$$

The graph of f can be determined by graphing $y = x^3$ when $x < 1$ and graphing $y = x^2 - 1$ when $x \geq 1$. At $x = 1$ there is a break in the graph, where the graph of f is discontinuous. See Figure 4.57 on the next page.

Algebra Review
To review piecewise-linear
functions, see Section 2.5.

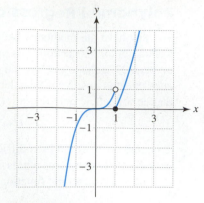

FIGURE 4.57

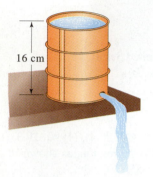

16 cm

FIGURE 4.58

A cylindrical container has a height of 16 centimeters. Water entered the container at a constant rate until it was completely filled. Then, water was allowed to leak out through a small hole in the bottom. See Figure 4.58. The height of the water in the container was recorded every half minute in Table 4.4 over a 5-minute period.

TABLE 4.4

Time (min)	0	0.5	1.0	1.5	2.0	2.5	3.0	3.5	4.0	4.5	5.0
Height (cm)	0	4	8	12	16	11.6	8.1	5.3	3.1	1.4	0.5

During the first 2 minutes, the water level rose at a constant rate of 4 centimeters every half minute or 8 centimeters per minute. After 2 minutes, water was drained at a nonconstant rate. A scatterplot of the data in Table 4.4 is shown in Figure 4.59.

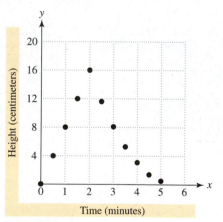

FIGURE 4.59

Water entering the container can be modeled with a linear formula, and the water leaking out can be modeled with a nonlinear formula. The function f models the water level in this example.

$$f(x) = \begin{cases} 8x & \text{if } 0 \le x \le 2 \\ 1.32x^2 - 14.4x + 39.39 & \text{if } 2 < x \le 5 \end{cases}$$

The graph of f together with a scatterplot of the data are shown in Figure 4.60. By defining f in pieces, f is capable of modeling water flowing into and out of the container. (Quadratic regression was used to find the formula for $f(x)$ when $2 < x \le 5$.)

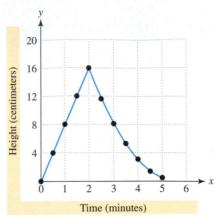

FIGURE 4.60

EXAMPLE 4   *Analyzing a piecewise-defined nonlinear function*

Use the previous function f to complete the following.
(a) Approximate the water level in the container after 1.25 and after 3.2 minutes.
(b) Estimate the time when the water level was 5 centimeters.

SOLUTION   **(a)** When $x = 1.25$, water is flowing into the tank and $f(x) = 8x$.

$$f(1.25) = 8(1.25) = 10 \text{ cm}$$

When $x = 3.2$ minutes, water is flowing out of the tank and $f(x) = 1.32x^2 - 14.4x + 39.39$.

$$f(3.2) = 1.32(3.2)^2 - 14.4(3.2) + 39.39 = 6.8268 \approx 7 \text{ cm}$$

[2, 5, 1] by [0, 20, 5]

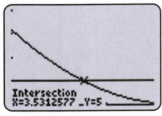

FIGURE 4.61

(b) The water level in Figure 4.60 equals 5 centimeters twice—once when water is entering the container and once when it is leaking out. To find these times, solve the equations $8x = 5$ and $1.32x^2 - 14.4x + 39.39 = 5$.

$$8x = 5 \qquad \text{or} \qquad x = \frac{5}{8} = 0.625$$

Thus, after 0.625 minute or 37.5 seconds, the water level was 5 centimeters. The equation $1.32x^2 - 14.4x + 39.39 = 5$ is solved graphically in Figure 4.61. The solution satisfying $2 < x \le 5$ is $x \approx 3.53$ minutes.

Putting it all Together

4.2

*H*igher degree polynomials generally have more complicated graphs. Each additional degree allows the graph to have possibly one more turning point and one more x-intercept. The graph of a polynomial function is continuous and smooth; it has no breaks or sharp corners. Its domain includes all real numbers. The end behavior of a polynomial always tends to either ∞ or $-\infty$. End behavior describes what happens to the y-values as $|x|$ becomes large.

The following summary of polynomial functions shows important concepts regarding graphs of polynomial functions.

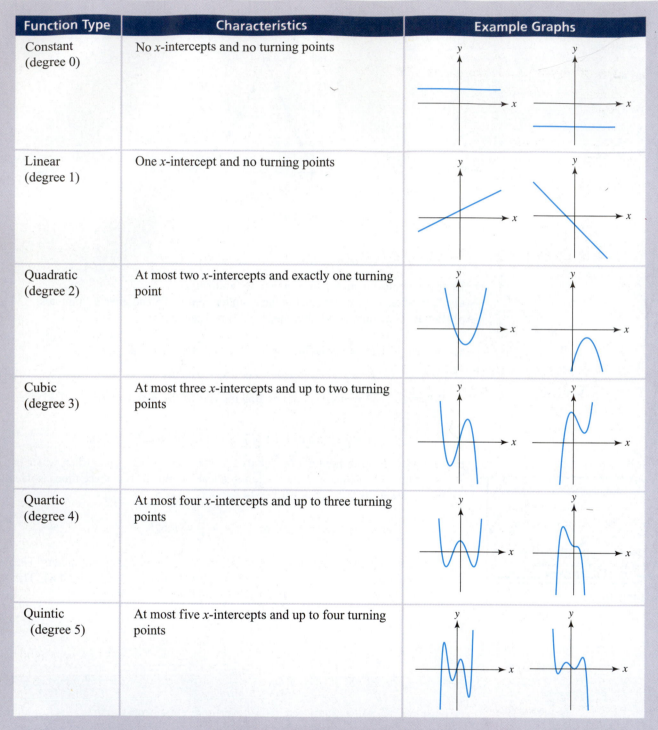

Function Type	Characteristics	Example Graphs
Constant (degree 0)	No x-intercepts and no turning points	
Linear (degree 1)	One x-intercept and no turning points	
Quadratic (degree 2)	At most two x-intercepts and exactly one turning point	
Cubic (degree 3)	At most three x-intercepts and up to two turning points	
Quartic (degree 4)	At most four x-intercepts and up to three turning points	
Quintic (degree 5)	At most five x-intercepts and up to four turning points	

Piecewise-defined functions occur when a function is defined by using two or more formulas for different intervals of the domain. If each formula is a polynomial, it is called a piecewise-polynomial function. Piecewise-polynomial functions are valuable when describing data that changes its character over time.

4.2 EXERCISES

Graphs of Polynomial Functions

1. A runner is working out on a straight track. The graph shows the runner's distance y in hundreds of feet from the starting line after x minutes.
 (a) Estimate the turning points.
 (b) Interpret each turning point.

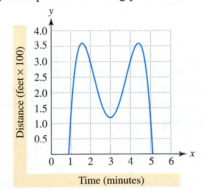

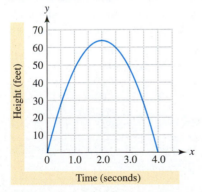

2. A stone is thrown into the air. Its height y in feet after x seconds is shown in the graph.
 (a) Estimate the turning point.
 (b) Interpret this point.

Exercises 3–8: Use the graph of the polynomial function f to complete the following.

 (a) Estimate the x-intercept(s).
 (b) State whether the leading coefficient is positive or negative.
 (c) Determine the minimum degree of f.

3.

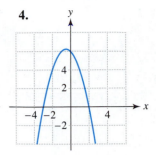

4.

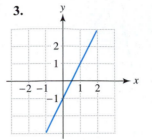

5.

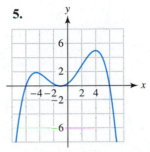

6.

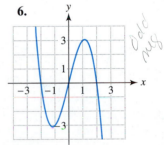

7.

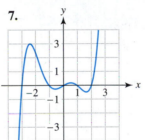

8.

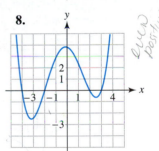

Exercises 9–14: Complete the following without a calculator.

 (a) Match the equation with its graph (a–f).
 (b) Identify the turning points.
 (c) Estimate the x-intercepts.
 (d) Estimate any local maxima or local minima.
 (e) Estimate any absolute maxima or absolute minima.

9. $f(x) = 1 - 2x + x^2$

10. $f(x) = 3x - x^3$

11. $f(x) = x^3 + 3x^2 - 9x$

12. $f(x) = x^4 - 8x^2$

13. $f(x) = 8x^2 - x^4$

14. $f(x) = x^5 + \dfrac{5}{2}x^4 - \dfrac{5}{3}x^3 - 5x^2$

a.
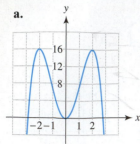

b.
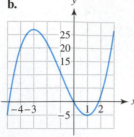

c.
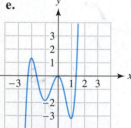

d.
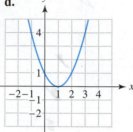

e.
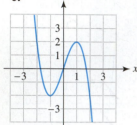

f.
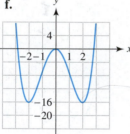

Exercises 15–22: For each polynomial f(x) complete the following.

(a) Graph $f(x)$ in the standard viewing rectangle.
(b) Approximate the coordinates of each turning point.
(c) Use the turning points to estimate any local extrema.

15. $f(x) = \dfrac{1}{9}x^3 - 3x$

16. $f(x) = x^2 - 4x - 3$

17. $f(x) = 0.025x^4 - 0.45x^2 - 5$

18. $f(x) = -\dfrac{1}{8}x^4 + \dfrac{1}{3}x^3 + \dfrac{5}{4}x^2 - 3x + 3$

19. $f(x) = 1 - 2x + 3x^2$ **20.** $f(x) = 4x - \dfrac{1}{3}x^3$

21. $f(x) = \dfrac{1}{3}x^3 + \dfrac{1}{2}x^2 - 2x$

22. $f(x) = \dfrac{1}{4}x^4 + \dfrac{2}{3}x^3 - \dfrac{1}{2}x^2 - 2x + 1$

Exercises 23–28: For each polynomial function f complete the following.

(a) State the degree and leading coefficient.
(b) Predict the end behavior of the graph of f.
(c) Support your answer in part (b) by graphing f.

23. $f(x) = x^2 - x^3 - 4$

24. $f(x) = x^4 - 4x^3 + 3x^2 - 3$

25. $f(x) = 0.1x^5 - 2x^2 - 3x + 4$

26. $f(x) = 3x^3 - 2 - x^4$

27. $f(x) = 4 + 2x - \dfrac{1}{2}x^2$

28. $f(x) = -0.2x^5 + 4x^2 - 3$

Exercises 29 and 30: Graph the functions f, g, and h in the same viewing rectangle. What happens to their graphs as the size of the viewing rectangle increases? Give an explanation.

29. $f(x) = 2x^4$, $g(x) = 2x^4 - 5x^2 + 1$, and $h(x) = 2x^4 + 3x^2 - x - 2$

  i. $[-4, 4, 1]$ by $[-4, 4, 1]$

  ii. $[-10, 10, 1]$ by $[-100, 100, 10]$

  iii. $[-100, 100, 10]$ by $[-10^6, 10^6, 10^5]$

30. $f(x) = -x^3$, $g(x) = -x^3 + x^2 + 2$, and $h(x) = -x^3 - 2x^2 + x - 1$

  i. $[-4, 4, 1]$ by $[-4, 4, 1]$

  ii. $[-10, 10, 1]$ by $[-100, 100, 10]$

  iii. $[-100, 100, 10]$ by $[-10^5, 10^5, 10^4]$

Exercises 31–34: The data table has been generated by a linear, quadratic, or cubic function f. All zeros of f are real numbers located in the interval $[-3, 3]$.

(a) Make a line graph of the data.
(b) Conjecture the degree of f.

31.

x	-3	-2	-1	0	1	2	3
$f(x)$	11	9	7	5	3	1	-1

32.

x	-3	-2	-1	0	1	2	3
$f(x)$	3	-8	-7	0	7	8	-3

33.

x	-3	-2	-1	0	1	2	3
$f(x)$	14	7	2	-1	-2	-1	2

34.

x	-3	-2	-1	0	1	2	3
$f(x)$	-13	-6	-1	2	3	2	-1

Sketching Graphs of Polynomials

Exercises 35–42: Sketch a graph of a polynomial that satisfies the conditions.

35. Degree 3 with three real zeros and a positive leading coefficient

36. Degree 4 with four real zeros and a negative leading coefficient

37. Linear with a negative leading coefficient

38. Cubic with one real zero and a positive leading coefficient

39. Degree 3 with turning points $(-1, 2)$ and $\left(1, \dfrac{2}{3}\right)$

40. Quartic with turning points $(-1, -1)$, $(0, 0)$, and $(1, -1)$

41. Degree 2 with turning point $(-1, 2)$, passing through $(-3, 4)$ and $(1, 4)$

42. Degree 5 and an odd function with five x-intercepts and a negative leading coefficient

Applications

43. *Highway Design* In order to allow enough distance for cars to pass on two-lane highways, engineers calculate minimum sight distances between curves and hills. See the accompanying figure. The table shows the minimum sight distance y in feet for a car traveling at x miles per hour. (Source: L. Haefner, *Introduction to Transportation Systems.*)

x (mph)	20	30	40	50
y (ft)	810	1090	1480	1840

x (mph)	60	65	70
y (ft)	2140	2310	2490

(a) Make a scatterplot of the data. As x increases, describe how y changes.

(b) What minimum degree polynomial might model this data? Find such a polynomial, $D(x)$.

(c) If a car is traveling at 43 miles per hour, estimate the minimum sight distance.

44. *Sunspots* The table shows the number of sunspots during the month of April from 1950 to 1969. (Source: R. Wolff and L. Yaeger, *Visualization of Natural Phenomena.*)

Year	1950	1951	1952	1953	1954
Sunspots	106	109	23	13	1

Year	1955	1956	1957	1958	1959
Sunspots	29	137	165	175	172

Year	1960	1961	1962	1963	1964
Sunspots	120	51	44	43	10

Year	1965	1966	1967	1968	1969
Sunspots	24	45	87	127	120

(a) Make a line graph of the data.

(b) If you were asked to find a polynomial whose graph models the data, what degree polynomial might you try first? Explain your reasoning.

45. *Modeling Temperature* In the accompanying figure the monthly average temperature in degrees Fahrenheit from January to December in Minneapolis is modeled by a polynomial function f, where $x = 1$ corresponds to January and $x = 12$ to December. (Source: A. Miller and J. Thompson, *Elements of Meteorology.*)

(a) Estimate the turning points.

(b) Interpret each turning point.

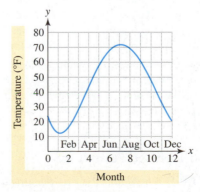

46. *Endangered Species* The total numbers y of endangered and threatened species in the United States are given in the table for various years x. (Source: Fish and Wildlife Services.)

x (yr)	1980	1985	1990	1995
y (total)	786	941	1181	1599

(a) Find a polynomial function f that models the data.

(b) Use f to predict the number of endangered and threatened species in 2000. Compare your answer to the actual value of 1741.

47. *Marijuana Use* The table lists the percentage y of high school seniors that had used marijuana within the previous month in the United States for various years x. In this table $x = 0$ corresponds to 1975 and $x = 20$ to 1995. (Source: Health and Human Services Department.)

x (yr)	0	3	5	10	15	20
y (%)	27	37	33	25	14	21

(a) Use least-squares regression to find a polynomial function f that models the data.

(b) Use f to estimate marijuana use in 1997. Compare your answer to the actual value of 23%.

48. *Aging of America* The table lists the numbers in thousands of Americans expected to be over 100 years old for selected years. (Source: Bureau of the Census.)

Year	Number (thousands)
1994	50
1996	56
1998	65
2000	75
2002	94
2004	110

(a) Use least-squares regression to find a function f that models the data. Let $x = 0$ correspond to 1994.

(b) Estimate the number of Americans over 100 in 2008.

Piecewise-Defined Functions

Exercises 49–52: Evaluate $f(x)$ at the given values of x.

49. $x = -2$ and 1

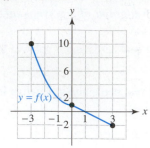

50. $x = -1, 0,$ and 3

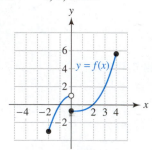

51. $x = -2, 1,$ and 2

$$f(x) = \begin{cases} x^2 + 2x + 6 & \text{if } -5 \le x < 0 \\ x + 6 & \text{if } 0 \le x < 2 \\ x^3 + 1 & \text{if } 2 \le x \le 5 \end{cases}$$

52. $x = 1975, 1980,$ and 1998

$$f(x) = \begin{cases} 0.2(x - 1970)^3 + 60 & \text{if } 1970 \le x < 1980 \\ 190 - (x - 1980)^2 & \text{if } 1980 \le x < 1990 \\ 2(x - 1990) + 100 & \text{if } 1990 \le x \le 2000 \end{cases}$$

Exercises 53–56: Complete the following.

(a) Sketch a graph of f.

(b) Determine if f is continuous.

53. $f(x) = \begin{cases} 4 - x^2 & \text{if } -3 \le x \le 0 \\ x^2 - 4 & \text{if } 0 < x \le 3 \end{cases}$

54. $f(x) = \begin{cases} x^2 & \text{if } -2 \le x < 0 \\ x + 1 & \text{if } 0 \le x \le 2 \end{cases}$

55. $f(x) = \begin{cases} 2x & \text{if } -5 \le x < -1 \\ -2 & \text{if } -1 \le x < 0 \\ x^2 - 2 & \text{if } 0 \le x \le 2 \end{cases}$

56. $f(x) = \begin{cases} 0.5x^2 & \text{if } -4 \le x \le -2 \\ x & \text{if } -2 < x < 2 \\ x^2 - 4 & \text{if } 2 \le x \le 4 \end{cases}$

57. *Modeling* An object is lifted rapidly into the air at a constant speed and then dropped. Its height after x seconds is listed in the table.

Time (sec)	0	1	2	3
Height (ft)	0	36	72	108

Time (sec)	4	5	6	7
Height (ft)	144	128	80	0

(a) Make a line graph of the data. At what time does it appear that the object was dropped?

(b) Identify the time interval when the height could be modeled by a linear function. When could it be modeled by a nonlinear function?

(c) Determine values for the constants m, a, and b so that f models the data.

$$f(x) = \begin{cases} mx & \text{if } 0 \le x \le 4 \\ a(x-4)^2 + b & \text{if } 4 < x \le 7 \end{cases}$$

(d) Solve the equation $f(x) = 100$ and interpret your answer.

58. *Modeling* (Refer to Example 4.) A tank of water is filled with a hose and then drained. The accompanying table shows the number of gallons y in the tank after t minutes.

t (min)	0	1	2	3	4	5	6	7
y (gal)	0	9	18	27	36	16	4	0

The following piecewise-polynomial function f models the data in the accompanying table.

$$f(t) = \begin{cases} 9t & \text{if } 0 \le t \le 4 \\ 4t^2 - 56t + 196 & \text{if } 4 < t \le 7 \end{cases}$$

Solve the equation $f(t) = 12$ and interpret the results.

Writing about Mathematics

Exercises 59–62: Discuss possible local or absolute extrema on the graph of f. Assume that $a > 0$.

59. $f(x) = ax + b$

60. $f(x) = ax^2 + bx + c$

61. $f(x) = ax^3 + bx^2 + cx + d$

62. $f(x) = a|x|$

CHECKING BASIC CONCEPTS FOR SECTIONS 4.1 AND 4.2

1. Graph $f(x) = \frac{1}{4}x^4 - \frac{2}{3}x^3 - \frac{5}{2}x^2 + 6x$.

(a) Identify any local extrema.

(b) Find any absolute extrema.

(c) Approximate the zeros of f.

2. Make a table of $f(x) = \sqrt{25 - x^2}$ starting at $x = -5$, incrementing by 1.

(a) Is f an odd or even function? Graph f in $[-9, 9, 1]$ by $[-6, 6, 1]$ to support your answer.

(b) Find the domain and range of f.

3. Sketch graphs of cubic polynomials with negative leading coefficients or 1, 2, or 3 x-intercepts.

4. Plot the data in the accompanying table.

x	−3.2	−2	0	2	3.2
y	−11	15	−10	15	−11

(a) What minimum degree polynomial function f would be needed to model this data? Explain.

(b) Should function f be odd, even, or neither? Explain.

(c) Should the leading coefficient of f be positive or negative? Explain.

5. Use least-squares regression to find a polynomial that models the data in Exercise 4.

4.3 REAL ZEROS OF POLYNOMIAL FUNCTIONS

Division of Polynomials • Factoring Polynomials • Graphs and Multiple Zeros • Rational Zeros • Polynomial Equations

Introduction

In Section 3.2 the quadratic formula was used to solve $ax^2 + bx + c = 0$. Are there similar formulas for higher degree polynomial equations? One of the most spectacular mathematical achievements during the sixteenth century was the discovery of formulas for solving cubic and quartic equations. This was accomplished by the Italian mathematicians Tartaglia, Cardano, Fior, del Ferro, and Ferrari between 1515 and 1545. These formulas are quite complicated and typically used only in computer software packages. Between 1750 and 1780 both Euler and Lagrange failed at finding symbolic solutions to the quintic equation $ax^5 + bx^4 + cx^3 + dx^2 + ex + k = 0$. Later, in about 1805, the Italian physician Ruffini proved that formulas for quintic or higher degree equations were impossible. His results make it necessary for us to rely on numerical and graphical methods.   (Source: H. Eves, *An Introduction to the History of Mathematics.*)

Division of Polynomials

Some species of birds, such as robins, have two nesting periods each summer. Because the survival rate for young birds is low, bird populations can fluctuate greatly during the summer months.   (Source: S. Kress, *Bird Life.*)

The graph of $f(x) = x^3 - 61x^2 + 839x + 4221$ shown in Figure 4.62 models a population of birds in a small county, where $x = 1$ corresponds to June 1, $x = 2$ to June 2, $x = 3$ to June 3, and so on.

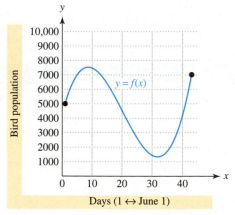

FIGURE 4.62   A Summer Bird Population

If we want to determine the dates when the population was 5000, we can solve the equation

$$x^3 - 61x^2 + 839x + 4221 = 5000,$$

which can also be written as

$$x^3 - 61x^2 + 839x - 779 = 0.$$

From the graph of f it appears that there were 5000 birds around June 1 ($x = 1$), June 20 ($x = 20$), and July 10 ($x = 40$). An exact symbolic solution can be obtained by factoring the polynomial $g(x) = x^3 - 61x^2 + 839x - 779$.

Division of polynomials is an important tool for factoring polynomials. We begin by reviewing how to divide a monomial into a polynomial.

EXAMPLE 1   *Dividing by a monomial*

Divide $6x^3 - 3x^2 + 2$ by $2x^2$.

SOLUTION Write the problem as $\dfrac{6x^3 - 3x^2 + 2}{2x^2}$. Then divide $2x^2$ into *every* term in the numerator.

$$\frac{6x^3 - 3x^2 + 2}{2x^2} = \frac{6x^3}{2x^2} - \frac{3x^2}{2x^2} + \frac{2}{2x^2}$$

$$= 3x - \frac{3}{2} + \frac{1}{x^2}$$

Before dividing a polynomial by the expression $x - k$ for some constant k, we briefly review division of natural numbers.

$$
\begin{array}{r}
\text{quotient} \to 58 \\
\text{divisor} \to 3\overline{)175} \leftarrow \text{dividend} \\
\underline{15} \\
25 \\
\underline{24} \\
1 \leftarrow \text{remainder}
\end{array}
$$

This result is checked as follows: $3 \cdot 58 + 1 = 175$. The quotient and remainder also can be expressed as $58\dfrac{1}{3}$. Since 3 does not divide into 175 evenly, 3 is not a factor of 175. When the remainder is 0, the divisor is a factor of the dividend. Division of polynomials is similar to division of natural numbers.

EXAMPLE 2 *Dividing polynomials*

Divide $2x^3 - 3x^2 - 11x + 7$ by $x - 3$. Check the result.

SOLUTION Begin by dividing x into $2x^3$.

$$
\begin{array}{r}
2x^2 \\
x - 3\overline{)2x^3 - 3x^2 - 11x + 7} \\
\underline{2x^3 - 6x^2} \\
3x^2 - 11x
\end{array}
\qquad
\begin{array}{l}
\dfrac{2x^3}{x} = 2x^2 \\[4pt]
2x^2(x - 3) = 2x^3 - 6x^2 \\
\text{Subtract. Bring down } -11x.
\end{array}
$$

In the next step, divide x into $3x^2$.

$$
\begin{array}{r}
2x^2 + 3x \\
x - 3\overline{)2x^3 - 3x^2 - 11x + 7} \\
\underline{2x^3 - 6x^2} \\
3x^2 - 11x \\
\underline{3x^2 - 9x} \\
-2x + 7
\end{array}
\qquad
\begin{array}{l}
\dfrac{3x^2}{x} = 3x \\[4pt]
3x(x - 3) = 3x^2 - 9x \\
\text{Subtract. Bring down } 7.
\end{array}
$$

Now divide x into $-2x$.

$$
\begin{array}{r}
2x^2 + 3x - 2 \\
x - 3\overline{)2x^3 - 3x^2 - 11x + 7} \\
\underline{2x^3 - 6x^2} \\
3x^2 - 11x \\
\underline{3x^2 - 9x} \\
-2x + 7 \\
\underline{-2x + 6} \\
1
\end{array}
\qquad
\begin{array}{l}
\dfrac{-2x}{x} = -2 \\[10pt] \\[10pt] \\
-2(x - 3) = -2x + 6 \\
\text{Subtract. Remainder is } 1.
\end{array}
$$

The quotient is $2x^2 + 3x - 2$ and the remainder is 1. This can be written as $2x^2 + 3x - 2 + \dfrac{1}{x - 3}$ in the same manner as $175 \div 3$ was expressed as $58\frac{1}{3}$. Polynomial division is checked by multiplying the divisor with the quotient and then adding the remainder.

Algebra Review
To review multiplication of polynomials, see Chapter R (page R-23).

$$(x - 3)(2x^2 + 3x - 2) + 1 = x(2x^2 + 3x - 2) - 3(2x^2 + 3x - 2) + 1$$
$$= 2x^3 + 3x^2 - 2x - 6x^2 - 9x + 6 + 1$$
$$= 2x^3 - 3x^2 - 11x + 7$$

This process illustrates an algorithm for division of polynomials by $x - k$.

> ### Division algorithm for polynomials
>
> For any polynomial $f(x)$ with degree $n \geq 1$ and any number k, there exists a unique polynomial $q(x)$ and a number r such that
>
> $$f(x) = (x - k)\,q(x) + r.$$
>
> The degree of $q(x)$ is one less than the degree of $f(x)$ and r is called the *remainder*.

Note: By the division algorithm it follows that

$$\text{dividend} = (\text{divisor}) \cdot (\text{quotient}) + (\text{remainder}).$$

Synthetic Division.   There is a shortcut called **synthetic division** that can be used to divide $x - k$ into a polynomial. For example, to divide $x - 2$ into $f(x) = 3x^4 - 7x^3 - 4x + 5$, we perform the following steps. The equivalent steps involving long division are shown to the right.

$$
\begin{array}{r|rrrrr}
2 & 3 & -7 & 0 & -4 & 5 \\
  &   & 6 & -2 & -4 & -16 \\
\hline
  & 3 & -1 & -2 & -8 & -11
\end{array}
$$

$$
\begin{array}{r}
3x^3 - x^2 - 2x - 8 \\
x - 2\overline{\smash{)}3x^4 - 7x^3 + 0x^2 - 4x + 5} \\
\underline{3x^4 - 6x^3} \\
-1x^3 + 0x^2 \\
\underline{-1x^3 + 2x^2} \\
-2x^2 - 4x \\
\underline{-2x^2 + 4x} \\
-8x + 5 \\
\underline{-8x + 16} \\
-11
\end{array}
$$

Notice how the highlighted numbers in the expression for long division correspond to the third row in synthetic division. The remainder is -11, which is the last number in the third row. The quotient, $3x^3 - x^2 - 2x - 8$, is one degree less than $f(x)$. Its coefficients are 3, -1, -2, and -8 and are found in the third row. The steps to divide a polynomial $f(x)$ by $x - k$ using synthetic division are summarized.

1. Write k to the left and the coefficients of $f(x)$ to the right in the top row. If any power of x does not appear in $f(x)$, include a 0 for that term. In this example an x^2-term did not appear, so a 0 is included in the first row.

2. Copy the leading coefficient of $f(x)$ into the third row and multiply it by k. Write the result below the next coefficient of $f(x)$ in the second row. Add the numbers

in the second column and place the result in the third row. Repeat the process. In this example, the leading coefficient is 3 and $k = 2$. Since $3 \cdot 2 = 6$, 6 is placed below -7. Then add to obtain $-7 + 6 = -1$. Multiply -1 by 2 and repeat.

3. The last number in the third row is the remainder. If the remainder is 0, then $x - k$ is a factor of $f(x)$. The other numbers in the third row are the coefficients of the quotient in descending powers.

EXAMPLE 3   *Performing synthetic division*

Use synthetic division to divide $2x^3 + 4x^2 - x + 5$ by $x + 2$.

SOLUTION   Let $k = -2$ and perform the following.

$$
\begin{array}{r|rrrr}
-2 & 2 & 4 & -1 & 5 \\
   &   & -4 & 0 & 2 \\
\hline
   & 2 & 0 & -1 & 7
\end{array}
$$

The remainder is 7 and the quotient is $2x^2 - 1$. This result is expressed by the equation

$$\frac{2x^3 + 4x^2 - x + 5}{x + 2} = 2x^2 - 1 + \frac{7}{x + 2}.$$

If we let $x = k$ in the division algorithm, $f(x) = (x - k)q(x) + r$, then

$$f(k) = (k - k)q(k) + r = r.$$

Thus $f(k)$ is equal to the remainder obtained in synthetic division. In Example 3 when $f(x) = 2x^3 + 4x^2 - x + 5$ was divided by $x + 2$, the remainder was 7. It follows that $f(-2) = 7$. This result is summarized by the following theorem.

Remainder theorem

If a polynomial $f(x)$ is divided by $x - k$, the remainder is $f(k)$.

Factoring Polynomials

The polynomial $f(x) = x^2 - 3x + 2$ can be factored as $f(x) = (x - 1)(x - 2)$. Notice that $f(1) = 0$ and $(x - 1)$ is a factor of $f(x)$. Similarly, $f(2) = 0$ and $(x - 2)$ is a factor. This concept is known as the factor theorem and is a consequence of the remainder theorem.

Factor theorem

A polynomial $f(x)$ has a factor $x - k$ if and only if $f(k) = 0$.

EXAMPLE 4   *Applying the factor theorem*

Use the graph of $f(x) = x^3 - 2x^2 - 5x + 6$ in Figure 4.63 on the next page and the factor theorem to list the factors of $f(x)$.

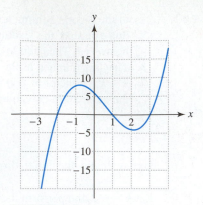

FIGURE 4.63

SOLUTION   Figure 4.63 shows that the zeros (or x-intercepts) of f are -2, 1, and 3. Since $f(-2) = 0$, the factor theorem states that $(x + 2)$ is a factor of $f(x) = x^3 - 2x^2 - 5x + 6$. Similarly, $f(1) = 0$ implies that $(x - 1)$ is a factor, and $f(3) = 0$ implies that $(x - 3)$ is a factor. The factors of $x^3 - 2x^2 - 5x + 6$ are $(x + 2)$, $(x - 1)$, and $(x - 3)$. ▨ ▮

   The polynomial $f(x) = x^2 + 4x + 4$ can be written as $f(x) = (x + 2)^2$. Since the factor $(x + 2)$ occurs twice in $f(x)$, the zero -2 is called a **zero of multiplicity two**. The polynomial $g(x) = (x + 1)^3(x - 2)$ has zeros -1 and 2 with *multiplicities* 3 and 1, respectively. A graph of g is shown in Figure 4.64, where the x-intercepts coincide with the zeros of g. *Counting multiplicities,* a polynomial of degree n has at most n real zeros. For $g(x)$, the sum of the multiplicities is $3 + 1 = 4$, which equals its degree.

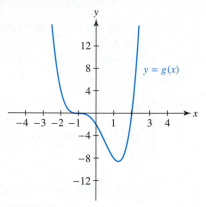

FIGURE 4.64

   These concepts together with the factor theorem can be used to find the *complete factored form* of a polynomial.

Complete factored form

Suppose a polynomial

$$f(x) = a_n x^n + \cdots + a_2 x^2 + a_1 x + a_0$$

has n real zeros $c_1, c_2, c_3, \ldots, c_n$, where distinct zeros are listed as many times as their multiplicities. Then $f(x)$ can be written in **complete factored form** as

$$f(x) = a_n(x - c_1)(x - c_2)(x - c_3) \cdots (x - c_n).$$

EXAMPLE 5 *Finding a complete factorization*

Write the complete factorization for each polynomial with the given zeros.

(a) $f(x) = 13x^2 - \dfrac{91}{2}x + 39$;   zeros: $\dfrac{3}{2}$ and 2

(b) $f(x) = 7x^3 - 21x^2 - 7x + 21$;   zeros: $-1, 1,$ and 3

(c) $f(x) = x^4 - 4x^3 + 16x - 16$;   zeros: $-2,$ and 2 with multiplicity 3

SOLUTION (a) The leading coefficient is 13 and its zeros are $\dfrac{3}{2}$ and 2. By the factor theorem, $\left(x - \dfrac{3}{2}\right)$ and $(x - 2)$ are factors. The complete factorization is

$$f(x) = 13\left(x - \frac{3}{2}\right)(x - 2).$$

(b) The leading coefficient is 7 and its zeros are $-1, 1,$ and 3. The complete factorization is

$$f(x) = 7(x + 1)(x - 1)(x - 3).$$

(c) The leading coefficient is 1 and its zeros are $-2, 2, 2, 2,$ where we have listed the zero 2 with multiplicity 3, three times. The complete factorization is as follows.

$$f(x) = 1(x + 2)(x - 2)(x - 2)(x - 2)$$
$$= (x + 2)(x - 2)^3$$

EXAMPLE 6 *Factoring a polynomial graphically*

Use graphing to factor $f(x) = 2x^3 - 4x^2 - 10x + 12$.

SOLUTION A graph of $Y_1 = 2X\text{\textasciicircum}3 - 4X\text{\textasciicircum}2 - 10X + 12$ is shown in Figure 4.65. Its zeros are $-2, 1,$ and 3. Since the leading coefficient is 2, the complete factorization is

$$f(x) = 2(x + 2)(x - 1)(x - 3).$$

$[-5, 5, 1]$ by $[-25, 25, 5]$

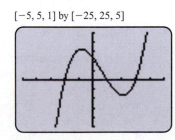

FIGURE 4.65

EXAMPLE 7 *Factoring a polynomial symbolically*

The polynomial $f(x) = 2x^3 - 2x^2 - 34x - 30$ has a zero of -1. Express $f(x)$ in complete factored form.

SOLUTION   If -1 is a zero, then by the factor theorem $(x + 1)$ is a factor of $2x^3 - 2x^2 - 34x - 30$. To factor $f(x)$, divide $x + 1$ into $2x^3 - 2x^2 - 34x - 30$ using synthetic division.

$$
\begin{array}{r|rrrr}
-1 & 2 & -2 & -34 & -30 \\
   &   & -2 & 4   & 30 \\
\hline
   & 2 & -4 & -30 & 0
\end{array}
$$

The remainder is 0, so $x + 1$ divides evenly into the dividend. By the division algorithm,

$$2x^3 - 2x^2 - 34x - 30 = (x + 1)(2x^2 - 4x - 30).$$

The quotient $2x^2 - 4x - 30$ can be factored further.

$$
\begin{aligned}
2x^2 - 4x - 30 &= 2(x^2 - 2x - 15) \\
&= 2(x + 3)(x - 5)
\end{aligned}
$$

Thus the complete factored form is $f(x) = 2(x + 1)(x + 3)(x - 5)$.

In the next example we factor the polynomial $g(x) = x^3 - 61x^2 + 839x - 779$, which was presented earlier, and then use the complete factorization to determine the days when the bird population was 5000.

EXAMPLE 8   *Factoring a polynomial*

Factor $g(x) = x^3 - 61x^2 + 839x - 779$. Use the zeros of $g(x)$ to determine when the bird population was 5000. Support your findings graphically.

SOLUTION   From Figure 4.62 on page 252, it appears that the bird population was 5000 when $x = 1$. If we substitute $x = 1$ into this polynomial the result is 0.

$$g(1) = 1^3 - 61(1)^2 + 839(1) - 779 = 0$$

By the factor theorem, $(x - 1)$ is one factor of $g(x)$. Begin by applying synthetic division with $k = 1$. This is equivalent to dividing $x^3 - 61x^2 + 839x - 779$ by $x - 1$.

$$
\begin{array}{r|rrrr}
1 & 1 & -61 & 839 & -779 \\
  &   & 1   & -60 & 779 \\
\hline
  & 1 & -60 & 779 & 0
\end{array}
$$

By the division algorithm, we can write

$$x^3 - 61x^2 + 839x - 779 = (x - 1)(x^2 - 60x + 779).$$

Since it is not obvious how to factor $x^2 - 60x + 779$, we can use the quadratic formula to find its zeros and then write its factors.

$$
\begin{aligned}
x &= \frac{-b \pm \sqrt{b^2 - 4ac}}{2a} \\
&= \frac{-(-60) \pm \sqrt{(-60)^2 - 4(1)(779)}}{2(1)} \\
&= \frac{60 \pm 22}{2} \\
&= 41 \text{ or } 19
\end{aligned}
$$

The zeros of $g(x) = x^3 - 61x^2 + 839x - 779$ are 1, 19, and 41, and its leading coefficient is 1. The complete factorization is $g(x) = (x - 1)(x - 19)(x - 41)$. The bird population equals 5000 on June 1 ($x = 1$), June 19 ($x = 19$), and July 11 ($x = 41$). The graph of $g(x)$ is shown in Figures 4.66–4.68. Notice that the zeros, or x-intercepts, are 1, 19, and 41.

Graphing Calculator Help

To find a zero of a function, see Appendix B (page AP-12).

$[0, 45, 5]$ by $[-5000, 5000, 1000]$

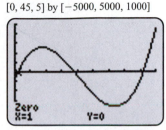

FIGURE 4.66

$[0, 45, 5]$ by $[-5000, 5000, 1000]$

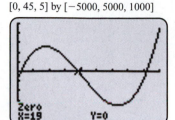

FIGURE 4.67

$[0, 45, 5]$ by $[-5000, 5000, 1000]$

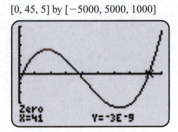

FIGURE 4.68

Graphs and Multiple Zeros

The polynomial $f(x) = 0.02(x + 3)^3(x - 3)^2$ has zeros -3 and 3 with multiplicities 3 and 2, respectively. At the zero of even multiplicity the graph does not cross the x-axis, whereas the graph does cross the x-axis at the zero of odd multiplicity. See Figure 4.69.

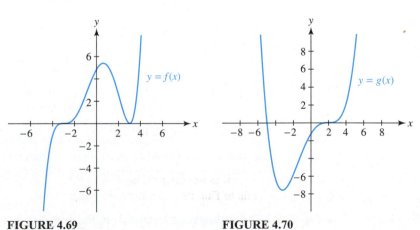

FIGURE 4.69   **FIGURE 4.70**

The graph of $g(x) = 0.03(x + 5)(x - 2)^3$ is shown in Figure 4.70. The zeros of g are -5 and 2 with multiplicities of 1 and 3, respectively. Since both zeros have odd multiplicity, the graph crosses the x-axis at -5 and 2. Notice that the zero 2 has a higher multiplicity and the graph levels off more near $x = 2$ than it does near $x = -5$. The higher the multiplicity of a zero, the more the graph of a polynomial levels off near the zero.

■ **MAKING CONNECTIONS**

Zeros and multiplicity   If a zero of a polynomial $f(x)$ has odd multiplicity, then its graph crosses the x-axis at the zero. If a zero has even multiplicity, then its graph intersects, but does not cross, the x-axis at the zero. The higher the multiplicity of a zero, the more the graph levels off at the zero. These concepts are illustrated in the following figures.

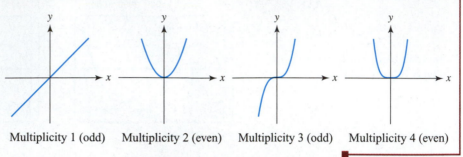

Multiplicity 1 (odd)   Multiplicity 2 (even)   Multiplicity 3 (odd)   Multiplicity 4 (even)

EXAMPLE 9   *Finding the multiplicity of a zero graphically*

Figure 4.71 shows the graph of a sixth-degree polynomial $f(x)$ with leading coefficient 1. All zeros are integers. Write $f(x)$ in complete factored form.

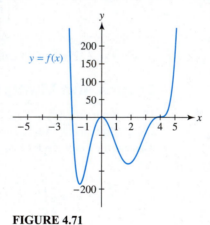

FIGURE 4.71

SOLUTION   The x-intercepts or zeros of f are -2, 0, and 4. Since the graph crosses the x-axis at -2 and 4, these zeros have odd multiplicity. The graph of f levels off more at $x = 4$ than at $x = -2$, so 4 has a higher multiplicity than -2. At $x = 0$ the graph of f does not cross the x-axis. Thus, 0 has even multiplicity. If -2 has multiplicity 1, 0 has multiplicity 2, and 4 has multiplicity 3, then the sum of the multiplicities is $1 + 2 + 3 = 6$, which equals the degree of $f(x)$. List the zeros as

$$-2, \quad 0, \quad 0, \quad 4, \quad 4, \quad \text{and} \quad 4.$$

The leading coefficient is 1, so the complete factorization of $f(x)$ is

$$f(x) = 1(x + 2)(x - 0)(x - 0)(x - 4)(x - 4)(x - 4) \qquad \text{or}$$
$$f(x) = x^2(x + 2)(x - 4)^3.$$

Multiple zeros sometimes have physical significance in an application. The next example shows how a multiple zero represents the boundary between an object's floating or sinking.

EXAMPLE 10   *Interpreting a multiple zero*

The polynomial $f(x) = \frac{\pi}{3}x^3 - 5\pi x^2 + \frac{500\pi d}{3}$ can be used to find the depth that a ball, 10 centimeters in diameter, sinks in water. The constant d is the density of the ball, where the density of water is 1. The smallest *positive* zero of $f(x)$ equals the depth that the sphere sinks. Approximate this depth for each material and interpret the results.

(a) A wood ball with $d = 0.8$

(b) A solid aluminum sphere with $d = 2.7$

(c) A water balloon with $d = 1$

SOLUTION

(a) Let $d = 0.8$ and graph $Y_1 = (\pi/3)X^3 - 5\pi X^2 + 500\pi/3(0.8)$. In Figure 4.72 the smallest positive zero is near 7.13. This means that the 10-centimeter wood ball sinks about 7.13 centimeters into the water.

(b) Let $d = 2.7$ and graph $Y_1 = (\pi/3)X^3 - 5\pi X^2 + 500\pi/3(2.7)$. In Figure 4.73 there is no smallest positive zero. The aluminum sphere is more dense than water and sinks.

(c) Let $d = 1$ and graph $Y_1 = (\pi/3)X^3 - 5\pi X^2 + 500\pi/3$. The graph in Figure 4.74 has one positive zero of 10 with multiplicity 2. The water balloon has the same density as water and "floats" even with the surface. The value of $d = 1$ represents the boundary between sinking and floating. If the ball floats, $f(x)$ has two positive zeros; if it sinks, $f(x)$ has no positive zeros. With the water balloon there is one positive zero with multiplicity 2 that represents a transition point between floating and sinking.

[−20, 20, 5] by [−300, 500, 100]

FIGURE 4.72

[−20, 20, 5] by [−500, 2000, 500]

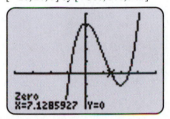

FIGURE 4.73

[−20, 20, 5] by [−300, 600, 100]

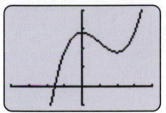

FIGURE 4.74

■ CLASS DISCUSSION

Make a conjecture about the depth that a ball with a 10-centimeter diameter will sink in water if $d = 0.5$. Test your conjecture graphically. ■

Rational Zeros

If a polynomial has a zero that is rational, it can be found using the rational zero test.

> ### Rational zero test
>
> Let $f(x) = a_n x^n + \cdots + a_2 x^2 + a_1 x + a_0$, where $a_n \neq 0$, represent a polynomial function f with integer coefficients. If $\frac{p}{q}$ is a rational number written in lowest terms and if $\frac{p}{q}$ is a zero of f, then p is a factor of the constant term a_0 and q is a factor of the leading coefficient a_n.

The following example illustrates how to find rational zeros by using this test.

EXAMPLE 11   *Finding rational zeros of a polynomial*

Find all rational zeros of $f(x) = 6x^3 - 5x^2 - 7x + 4$. Write $f(x)$ in complete factored form.

SOLUTION If $\frac{p}{q}$ is a rational zero in lowest terms, then p is a factor of the constant term 4 and q is a factor of the leading coefficient 6. The possible values for p and q are as follows.

$$p: \quad \pm 1, \quad \pm 2, \quad \pm 4$$
$$q: \quad \pm 1, \quad \pm 2, \quad \pm 3, \quad \pm 6$$

As a result, any rational zero of $f(x)$ in the form $\frac{p}{q}$ must occur in the list

$$\pm\frac{1}{6}, \quad \pm\frac{1}{3}, \quad \pm\frac{1}{2}, \quad \pm\frac{2}{3}, \quad \pm\frac{1}{1}, \quad \pm\frac{4}{3}, \quad \pm\frac{2}{1}, \quad \text{or} \quad \pm\frac{4}{1}.$$

Evaluate $f(x)$ at each value in the list. See Table 4.5.

TABLE 4.5

x	$f(x)$	x	$f(x)$	x	$f(x)$	x	$f(x)$
$\frac{1}{6}$	$\frac{49}{18}$	$\frac{1}{2}$	0	1	-2	2	18
$-\frac{1}{6}$	5	$-\frac{1}{2}$	$\frac{11}{2}$	-1	0	-2	-50
$\frac{1}{3}$	$\frac{4}{3}$	$\frac{2}{3}$	$-\frac{10}{9}$	$\frac{4}{3}$	0	4	280
$-\frac{1}{3}$	$\frac{50}{9}$	$-\frac{2}{3}$	$\frac{14}{3}$	$-\frac{4}{3}$	$\frac{88}{9}$	-4	-432

From Table 4.5 there are three rational zeros of -1, $\frac{1}{2}$, and $\frac{4}{3}$. Since a third degree polynomial has at most three zeros, the complete factored form of $f(x)$ is

$$f(x) = 6(x + 1)\left(x - \frac{1}{2}\right)\left(x - \frac{4}{3}\right).$$

Although $f(x)$ in Example 11 had only rational zeros, it is important to realize that many polynomials have irrational zeros. Irrational zeros cannot be found using the rational zero test.

Polynomial Equations

In Section 3.2, factoring was used to solve quadratic equations. Factoring also can be used to solve polynomial equations with degrees greater than 2. This is illustrated in the next example.

EXAMPLE 12 *Solving a cubic equation*

Solve $x^3 + 3x^2 - 4x = 0$ symbolically. Support your answer graphically and numerically.

SOLUTION *Symbolic Solution*

Algebra Review
To factor polynomials, see
Chapter R (page R-27).

$x^3 + 3x^2 - 4x = 0$	Given equation
$x(x^2 + 3x - 4) = 0$	Factor out x.
$x(x + 4)(x - 1) = 0$	Factor the quadratic expression.
$x = 0, x + 4 = 0, \text{ or } x - 1 = 0$	Zero-product property
$x = 0, -4, \text{ or } 1$	Solve.

Graphical Solution Graph $Y_1 = X\char94 3 + 3X\char94 2 - 4X$ as in Figure 4.75. The x-intercepts are $-4, 0,$ and $1,$ which correspond to the solutions.

Numerical Solution Table $Y_1 = X\char94 3 + 3X\char94 2 - 4X$ as in Figure 4.76. The zeros of Y_1 occur at $x = -4, 0,$ and 1.

$[-5, 5, 1]$ by $[-15, 15, 5]$

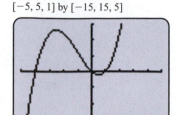

FIGURE 4.75

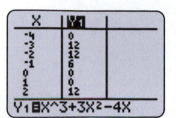

FIGURE 4.76

■ **MAKING CONNECTIONS**

Functions and Equations Each time a new type of function is defined, we can define a new type of equation. For example, cubic functions given by $f(x) = ax^3 + bx^2 + cx + d$ can be used to define cubic equations of the form $ax^3 + bx^2 + cx + d = 0.$ This concept will be used to define other types of equations.

EXAMPLE 13 *Solving a polynomial equation*

Find all real solutions for the equation $4x^4 - 5x^2 - 9 = 0$ symbolically.

SOLUTION The expression $4x^4 - 5x^2 - 9$ can be factored in a manner similar to the way quadratic expressions are factored.

$$4x^4 - 5x^2 - 9 = 0 \qquad \text{Given equation}$$
$$(4x^2 - 9)(x^2 + 1) = 0 \qquad \text{Factor.}$$
$$4x^2 - 9 = 0 \quad \text{or} \quad x^2 + 1 = 0 \qquad \text{Zero-product property}$$
$$4x^2 = 9 \quad \text{or} \quad x^2 = -1 \qquad \text{Add 9 or subtract 1.}$$
$$x^2 = \frac{9}{4} \quad \text{or} \quad x^2 = -1 \qquad \text{Divide by 4.}$$
$$x = \pm\frac{3}{2} \quad \text{or} \quad x^2 = -1 \qquad \text{Square root property}$$

Since $x^2 \geq 0$ for all $x,$ the equation $x^2 = -1$ has no real solutions. Thus the only solutions are $x = \pm\frac{3}{2}.$

Some types of polynomial equations cannot be solved easily with symbolic methods. The next example illustrates how we can obtain an approximate solution graphically.

EXAMPLE 14 *Finding a solution graphically*

Solve the equation $\frac{1}{2}x^3 - 2x - 4 = 0$ graphically. Round any solutions to the nearest hundredth.

SOLUTION   A graph of $Y_1 = .5X^3 - 2X - 4$ is shown in Figure 4.77. Since there is only one x-intercept, the equation has one solution of $x \approx 2.65$.

$[-9, 9, 1]$ by $[-6, 6, 1]$

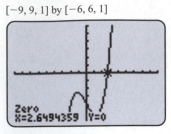

Graphing Calculator Help

To find a zero of a function, see Appendix B (page AP-12).

FIGURE 4.77

Putting it all Together

4.3

*T*he following table lists some important concepts related to a polynomial $f(x)$.

Concept	Explanation	Example
Factor theorem	$(x - k)$ is a factor of $f(x)$ if and only if $f(k) = 0$.	$f(x) = x^2 + 3x - 4$ and $f(1) = 0$ implies that $(x - 1)$ is a factor of $f(x)$. That is, $f(x) = (x - 1)(x + 4)$.
Complete factored form	$f(x) = a_n(x - c_1)(x - c_2)(x - c_3) \cdots (x - c_n)$, where the c_k are zeros of f, listed as many times as their multiplicities.	$f(x) = 3(x - 5)(x + 3)(x + 3)$ $\qquad = 3(x - 5)(x + 3)^2$ $c_1 = 5, \quad c_2 = -3, \quad c_3 = -3$
Division algorithm	$f(x) = (x - k)q(x) + r$, where r is the remainder.	$f(x) = (x - 1)(x^2 - x - 2) + 3$ $q(x) = x^2 - x - 2$ and $r = 3$
Zero with odd multiplicity	The graph of $y = f(x)$ crosses the x-axis at a zero of odd multiplicity.	$f(x) = (x + 1)^3(x - 3)$ Both zeros of -1 and 3 have odd multiplicity.
Zero with even multiplicity	The graph of $y = f(x)$ intersects but does not cross the x-axis at a zero of even multiplicity.	$f(x) = (x + 1)^2(x - 3)^4$ Both zeros of -1 and 3 have even multiplicity.
Factoring a polynomial graphically (Only real zeros)	Graph $y = f(x)$ and locate all the zeros, or x-intercepts. If leading coefficient is a, and the zeros are c_1, c_2, and c_3, then $f(x) = a(x - c_1)(x - c_2)(x - c_3)$.	$f(x) = 2x^3 + 4x^2 - 2x - 4$ has zeros -2, -1, and 1 with leading coefficient 2. Thus $$f(x) = 2(x + 2)(x + 1)(x - 1).$$

4.3 EXERCISES

Division of Polynomials

Exercises 1–6: Divide the expression.

1. $\dfrac{5x^4 - 15}{10x}$

2. $\dfrac{x^2 - 5x}{5x}$

3. $\dfrac{3x^4 - 2x^2 - 1}{3x^3}$

4. $\dfrac{5x^3 - 10x^2 + 5x}{15x^2}$

5. $\dfrac{x^3 - 4}{4x^3}$

6. $\dfrac{2x^4 - 3x^2 + 4x - 7}{-4x}$

Exercises 7–12: Divide the first polynomial by the second. State the quotient and remainder.

7. $x^3 - 2x^2 - 5x + 6$ $x - 3$

8. $3x^3 - 10x^2 - 27x + 10$ $x + 2$

9. $2x^4 - 7x^3 - 5x^2 - 19x + 17$ $x + 1$

10. $x^4 - x^3 - 4x + 1$ $x - 2$

11. $3x^3 - 7x + 10$ $x - 1$

12. $x^4 - 16x^2 + 1$ $x + 4$

Exercises 13 and 14: Use the division algorithm to complete the following.

13. $\dfrac{x^3 - 8x^2 + 15x - 6}{x - 2} = x^2 - 6x + 3$ implies
$(x - 2)(x^2 - 6x + 3) = $ _____ .

14. $\dfrac{x^4 - 15}{x + 2} = x^3 - 2x^2 + 4x - 8 + \dfrac{1}{x + 2}$
implies $x^4 - 15 = (x + 2) \times$ _____ $+$ _____ .

Exercises 15–20: Divide the expression.

15. $\dfrac{x^4 - 3x^3 - x + 3}{x - 3}$

16. $\dfrac{x^3 - 2x^2 - x + 3}{x + 1}$

17. $\dfrac{4x^3 - x^2 - 5x + 6}{x - 1}$

18. $\dfrac{x^4 + 3x^3 - 4x + 1}{x + 2}$

19. $\dfrac{x^3 + 1}{x + 1}$

20. $\dfrac{x^5 + 3x^4 - x - 3}{x + 3}$

Factor Theorem

Exercises 21–24: Use the factor theorem to decide if $x - c$ is a factor of $f(x)$.

21. $f(x) = x^3 - 6x^2 + 11x - 6$, $c = 2$

22. $f(x) = x^3 + x^2 - 14x - 24$, $c = -3$

23. $f(x) = x^4 - 2x^3 - 13x^2 - 10x$, $c = 3$

24. $f(x) = 2x^4 - 11x^3 + 9x^2 + 14x$, $c = \dfrac{1}{2}$

Factoring Polynomials

Exercises 25–30: Use the given zeros to write the complete factored form of $f(x)$.

25. $f(x) = 2x^2 - 25x + 77$; zeros: $\dfrac{11}{2}$ and 7

26. $f(x) = 6x^2 + 21x - 90$; zeros: -6 and $\dfrac{5}{2}$

27. $f(x) = x^3 - 2x^2 - 5x + 6$; zeros: -2, 1, and 3

28. $f(x) = x^3 + 6x^2 + 11x + 6$;
zeros: -3, -2, and -1

29. $f(x) = -2x^3 + 3x^2 + 59x - 30$; zeros: -5, $\dfrac{1}{2}$, and 6

30. $f(x) = 3x^4 - 8x^3 - 67x^2 + 112x + 240$;
zeros: -4, $-\dfrac{4}{3}$, 3, and 5

Exercises 31–34: The graph of a polynomial $f(x)$ with leading coefficient ± 1 and integer zeros is shown in the figure. Write its complete factored form.

31.
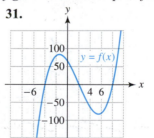

32.
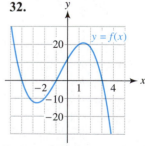

33.
34.
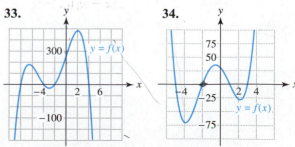

Exercises 35–40: (Refer to Example 6.) Use graphing to factor $f(x)$.

35. $f(x) = 10x^2 + 17x - 6$

36. $f(x) = 2x^3 + 7x^2 + 2x - 3$

37. $f(x) = -3x^3 - 3x^2 + 18x$

38. $f(x) = \frac{1}{2}x^3 + \frac{5}{2}x^2 + x - 4$

39. $f(x) = x^4 + \frac{5}{2}x^3 - 3x^2 - \frac{9}{2}x$

40. $f(x) = 10x^4 + 7x^3 - 27x^2 + 2x + 8$

Exercises 41–46: (Refer to Example 7.) Write the complete factored form of the polynomial $f(x)$, given that c is a zero.

41. $f(x) = x^3 - 9x^2 + 23x - 15,$ $\qquad c = 1$

42. $f(x) = 2x^3 + x^2 - 11x - 10,$ $\qquad c = -2$

43. $f(x) = -4x^3 - x^2 + 51x - 36,$ $\qquad c = -4$

44. $f(x) = 3x^3 - 11x^2 - 35x + 75,$ $\qquad c = 5$

45. $f(x) = 2x^4 - x^3 - 13x^2 - 6x,$ $\qquad c = -2$

46. $f(x) = 35x^4 + 48x^3 - 41x^2 + 6x,$ $\qquad c = \frac{3}{7}$

Graphs and Multiple Zeros

Exercises 47–50: The graph of a polynomial $f(x)$ is shown in the figure. Estimate the zeros and state whether their multiplicities are odd or even. State the minimum degree of $f(x)$.

47.

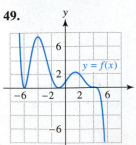

48.

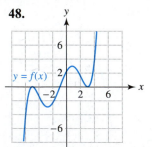

49.

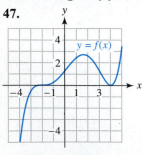

50.

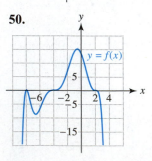

Exercises 51–54: Write a polynomial $f(x)$ in complete factored form that satisfies the conditions. Let the leading coefficient be 1. Support your answer graphically.

51. Degree 3; zeros: -1 with multiplicity 2, and 6 with multiplicity 1

52. Degree 4; zeros: 5 and 7, both with multiplicity 2

53. Degree 4; zeros: 2 with multiplicity 3, and 6 with multiplicity 1

54. Degree 5; zeros: -2 with multiplicity 2, and 4 with multiplicity 3

Exercises 55–58: The graph of either a cubic or quartic polynomial $f(x)$ with a leading coefficient of ± 1 and integer zeros is shown. Write the complete factored form of $f(x)$.

55.

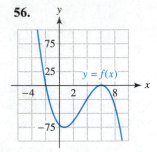

56.

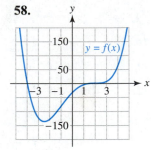

57.

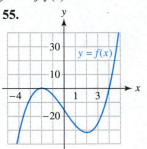

58.

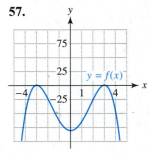

Rational Zeros

Exercises 59–66: (Refer to Example 11.)

(a) Use the rational zero test to find any rational zeros of the polynomial $f(x)$.

(b) Write the complete factored form of $f(x)$.

59. $f(x) = 2x^3 + 3x^2 - 8x + 3$

60. $f(x) = x^3 - 7x + 6$

61. $f(x) = 2x^4 + x^3 - 8x^2 - x + 6$

62. $f(x) = 2x^4 + x^3 - 19x^2 - 9x + 9$

63. $f(x) = 3x^3 - 16x^2 + 17x - 4$

64. $f(x) = x^3 + 2x^2 - 3x - 6$

65. $f(x) = x^3 - x^2 - 7x + 7$

66. $f(x) = 2x^3 - 5x^2 - 4x + 10$

Polynomial Equations

Exercises 67–72: Find the zeros of f(x)

   (a) *symbolically,*
   (b) *graphically, and*
   (c) *numerically.*

67. $f(x) = x^3 + x^2 - 6x$ **68.** $f(x) = 2x^2 - 8x + 6$

69. $f(x) = x^4 - 1$ **70.** $f(x) = x^4 - 5x^2 + 4$

71. $f(x) = -x^3 + 4x$ **72.** $f(x) = 6 - 4x - 2x^2$

Exercises 73–84: Solve each polynomial equation symbolically. Support your results graphically or numerically.

73. $x^3 - 25x = 0$ **74.** $x^4 - x^3 - 6x^2 = 0$

75. $x^4 - x^2 = 2x^2 + 4$ **76.** $x^4 + 5 = 6x^2$

77. $x^3 - 3x^2 - 18x = 0$ **78.** $x^4 - x^2 = 0$

79. $2x^3 = 4x^2 - 2x$ **80.** $x^3 = x$

81. $12x^3 = 17x^2 + 5x$ **82.** $3x^3 + 3x = 10x^2$

83. $9x^4 + 4 = 13x^2$ **84.** $4x^4 + 7x^2 - 2 = 0$

Exercises 85–90: (Refer to Example 14.) Solve the equations graphically. Round your answers to the nearest hundredth.

85. $x^3 - 1.1x^2 - 5.9x + 0.7 = 0$

86. $x^3 + x^2 - 18x + 13 = 0$

87. $-0.7x^3 - 2x^2 + 4x + 2.5 = 0$

88. $3x^3 - 46x^2 + 180x - 99 = 0$

89. $2x^4 - 1.5x^3 + 13 = 24x^2 + 10x$

90. $-x^4 + 2x^3 + 20x^2 = 22x + 41$

Applications

91. *Water Pollution*   In one study, freshwater mussels were used to monitor copper discharge into a river from an electroplating works. Copper in high doses can be lethal to aquatic life. The table lists copper concentrations in mussels after 45 days at various distances downstream from the plant. The concentration C is measured in micrograms of copper per gram of mussel x kilometers downstream. (Sources: R. Foster and J. Bates, "Use of mussels to monitor point source industrial discharges"; C. Mason, *Biology of Freshwater Pollution.*)

x	5	21	37	53	59
C	20	13	9	6	5

   (a) Describe the relationship between x and C.
   (b) This data is modeled by
      $C(x) = -0.000068x^3 + 0.0099x^2 - 0.653x + 23$.
      Graph C and the data.
   (c) Concentrations above 10 are lethal to mussels. Locate this region in the river.

92. *Dog Years*   There is an old saying that every year of a dog's life is equal to 7 years for a human. A more accurate approximation is given by the graph of f. Given a dog's age x, where $x \geq 1$, $f(x)$ models the equivalent age in human years. According to the Bureau of the Census, middle age for people begins at age 45. (Source: J. Brearley and A. Nicholas, *This Is the Bichon Frise.*)

   (a) Use the graph of f to estimate the equivalent age for dogs.
   (b) Use $f(x) = -0.001183x^4 + 0.05495x^3 - 0.8523x^2 + 9.054x + 6.748$ to solve part (a) graphically or numerically.

Dog's age (years)

93. *Floating Ball*   (Refer to Example 10.) If a ball has a 20-centimeter diameter, then

$$f(x) = \frac{\pi}{3}x^3 - 10\pi x^2 + \frac{4000\pi d}{3}$$

can be used to determine the depth that it sinks in water. Find the depth that this size ball sinks when $d = 0.6$.

94. *Floating Ball*   (Refer to Example 10.) Determine the depth that a pine ball with a 10-centimeter diameter sinks in water, if $d = 0.55$.

95. *Modeling Temperature*   Complete the following.
   (a) Approximate the complete factored form of $f(x) = -0.184x^3 + 1.45x^2 + 10.7x - 27.9$.
   (b) The cubic polynomial $f(x)$ models monthly average temperature at Trout Lake, Canada, in degrees Fahrenheit, where $x = 1$ corresponds to January and $x = 12$ represents December. Interpret the zeros of f.

96. *Average High Temperatures*   The monthly average high temperatures in degrees Fahrenheit at Daytona Beach can be modeled by $f(x) = 0.0151x^4 - 0.438x^3 + 3.60x^2 - 6.49x + 72.5$, where $x = 1$ corresponds to January and $x = 12$ represents December.

 (a) Find the average high temperature during March and July.

 (b) Graph f in $[0.5, 12.5, 1]$ by $[60, 100, 10]$. Interpret the graph.

 (c) Estimate graphically and numerically when the average high temperature is $80°F$.

97. *Bird Populations*   (Refer to Example 8.) A bird population can be modeled by the polynomial $f(x) = x^3 - 66x^2 + 1052x + 652$, where $x = 1$ corresponds to June 1, $x = 2$ to June 2, and so on. Find the days when f estimates that there were 2500 birds.

98. *Geometry*   A rectangular box has sides with lengths x, $x + 1$, and $x + 2$. If the volume of the box is 504 cubic inches, find the dimensions of the box.

Writing about Mathematics

99. Suppose that $f(x)$ is a quintic polynomial with distinct real zeros. Assuming you have access to technology, explain how to factor $f(x)$ approximately. Have you used the factor theorem? Explain.

100. Explain how to determine graphically whether a zero of a polynomial is a multiple zero. Sketch examples.

4.4 THE FUNDAMENTAL THEOREM OF ALGEBRA

Complex Numbers • Quadratic Equations with Complex Solutions • Fundamental Theorem of Algebra • Polynomial Equations with Complex Solutions

Introduction

Mathematics can be both abstract and applied. Abstract mathematics is focused on axioms, theorems, and proofs. It can be derived independently of empirical evidence. Theorems that were proved centuries ago are still valid today. In this sense, abstract mathematics transcends time. Yet, even though mathematics can be developed in an abstract setting—separate from science and all measured data—it also has countless applications.

There is a common misconception that theoretical mathematics is unimportant, yet many of the ideas that eventually had great practical importance were first born in the abstract. For example, in 1854 George Boole published *Laws of Thought,* which outlined the basis for Boolean algebra. This was 85 years before the invention of the first digital computer. However, Boolean algebra became the basis by which modern computer hardware operates.

In this section we discuss some important abstract topics in algebra that have had an impact on society. We are privileged to read in a few hours what took people centuries to discover. To ignore either the abstract beauty or the profound applicability of mathematics is like seeing a rose, but never smelling one.

Complex Numbers

Throughout history, people have invented new numbers to solve equations and describe data. Often these new numbers were met with resistance and were regarded as being imaginary or unreal. The number 0 was not invented at the same time as the natural numbers. There was no Roman numeral for 0. No doubt there were skeptics who wondered why a number was needed to represent nothing. Negative numbers also met strong resistance. After all, how could one possibly have -6 apples? The same was true for complex numbers. However, complex numbers are no more imaginary than any other number created by mathematics. Much like Boolean algebra, the purpose of complex numbers was at first theoretical. However, today complex numbers are used in the design of electrical circuits, ships, and airplanes. Basic quantum mechanics and certain features of the theory of relativity would not have been developed without complex numbers. Even the *fractal image* shown in Figure 4.78 would not have been discovered without complex numbers.

FIGURE 4.78   The Cube Roots of Unity
Source: D. Kincaid and W. Cheney, *Numerical Analysis,*
© 1991 by Brooks/Cole Publishing Co. Used with permission.

If we graph $y = x^2 + 1$, there are no x-intercepts. See Figure 4.79. Therefore the equation $x^2 + 1 = 0$ has no real solutions.

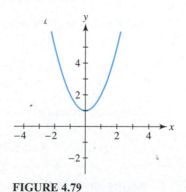

FIGURE 4.79

If we attempt to solve $x^2 + 1 = 0$ symbolically, it results in $x^2 = -1$. Since $x^2 \geq 0$ for any real number x, there are no real number solutions. However, we can invent a solution.

$$x^2 = -1$$
$$x = \pm\sqrt{-1}$$

We now define a new number called the **imaginary unit**, denoted i.

> ### Properties of the imaginary unit i
> $$i = \sqrt{-1}, \qquad i^2 = -1$$

By inventing the number i, the solutions to the equation $x^2 + 1 = 0$ are i and $-i$. Using the real numbers and the imaginary unit i, complex numbers can be defined. A **complex number** can be written in **standard form** as $a + bi$, where a and b are real numbers. The **real part** is a and the **imaginary part** is b. Every real number a is also a complex number because it can be written as $a + 0i$. A complex number $a + bi$ with $b \neq 0$ is an **imaginary number**. Table 4.6 lists several complex numbers with their real and imaginary parts.

TABLE 4.6

$a + bi$	$-3 + 2i$	5	$-3i$	$-1 + 7i$	$-5 - 2i$	$4 + 6i$
a	-3	5	0	-1	-5	4
b	2	0	-3	7	-2	6

Using the imaginary unit i, square roots of negative numbers can be written as complex numbers. For example, $\sqrt{-3} = i\sqrt{3}$, and $\sqrt{-16} = i\sqrt{16} = 4i$. This can be summarized as follows.

> ### The expression $\sqrt{-a}$
> If $a > 0$, then $\sqrt{-a} = i\sqrt{a}$.

Arithmetic operations are also defined for complex numbers.

Addition and Subtraction. To add the complex numbers $(-2 + 3i)$ and $(4 - 6i)$, simply combine the real and imaginary parts.

$$(-2 + 3i) + (4 - 6i) = -2 + 4 + 3i - 6i$$
$$= 2 - 3i$$

This same process works for subtraction.

$$(5 - 7i) - (8 + 3i) = 5 - 8 - 7i - 3i$$
$$= -3 - 10i$$

Multiplication. Two complex numbers can be multiplied. The property $i^2 = -1$ is applied when appropriate.

Algebra Review

Before multiplying complex numbers, you may want to review multiplication of binomials; see Chapter R (page R-23).

$$(-5 + i)(7 - 9i) = -5(7) + -5(-9i) + (i)(7) + (i)(-9i)$$
$$= -35 + 45i + 7i - 9i^2$$
$$= -35 + 52i - 9(-1)$$
$$= -26 + 52i$$

Note: When performing arithmetic with complex numbers, express the result in standard form $a + bi$.

Division. The **conjugate** of $a + bi$ is $a - bi$. To find the conjugate, change the sign of the imaginary part b. Table 4.7 lists examples of complex numbers and their conjugates.

TABLE 4.7

$a + bi$	$2 + 5i$	$6 - 3i$	$-2 + 7i$	$-1 - i$	5	$-4i$
$a - bi$	$2 - 5i$	$6 + 3i$	$-2 - 7i$	$-1 + i$	5	$4i$

To simplify the quotient $\dfrac{3 + 2i}{5 - i}$, first multiply both the numerator and the denominator by the conjugate of the denominator.

$$\frac{3 + 2i}{5 - i} = \frac{(3 + 2i)(5 + i)}{(5 - i)(5 + i)} \qquad \text{Multiply by conjugate.}$$

$$= \frac{3(5) + (3)(i) + (2i)(5) + (2i)(i)}{(5)(5) + (5)(i) + (-i)(5) + (-i)(i)} \qquad \text{Expand.}$$

$$= \frac{15 + 3i + 10i + 2i^2}{25 + 5i - 5i - i^2} \qquad \text{Simplify.}$$

$$= \frac{15 + 13i + 2(-1)}{25 - (-1)} \qquad i^2 = -1$$

$$= \frac{13 + 13i}{26} \qquad \text{Simplify.}$$

$$= \frac{1}{2} + \frac{1}{2}i \qquad \frac{a + bi}{c} = \frac{a}{c} + \frac{b}{c}i$$

The last step expresses the quotient as a complex number in standard form.

Evaluating Complex Arithmetic with Technology. Some graphing calculators perform complex arithmetic. The evaluation of the previous examples are shown in Figures 4.80 and 4.81.

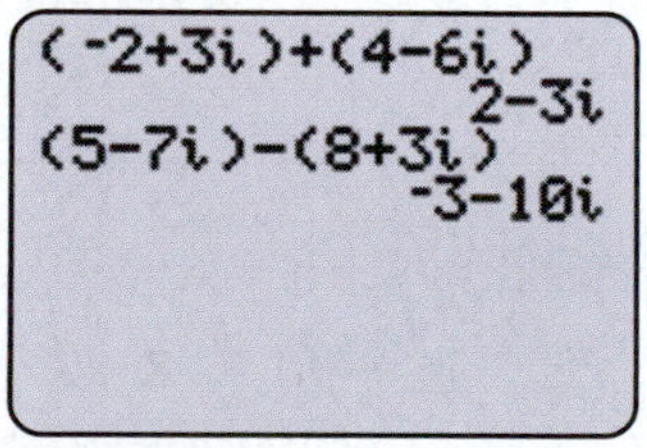

```
( -2+3i )+(4-6i )
              2-3i
(5-7i )-(8+3i )
             -3-10i
```

```
( -5+i )(7-9i )
           -26+52i
(3+2i )/(5-i )
             .5+.5i
Ans▸Frac
          1/2+1/2i
```

FIGURE 4.80 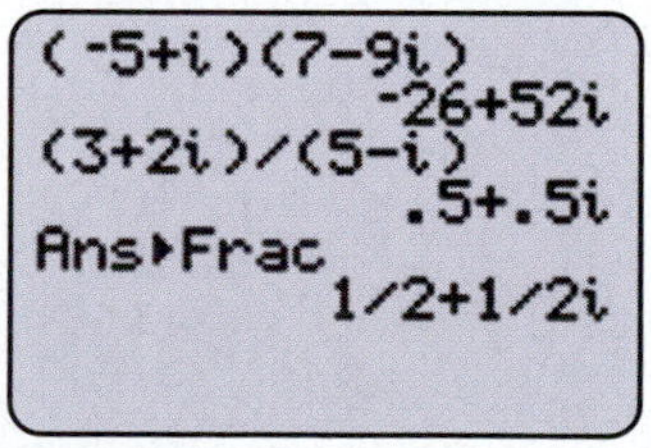   **FIGURE 4.81**

EXAMPLE 1   *Performing complex arithmetic*

Write each expression in standard form. Support your results using a calculator.
(a) $(-3 + 4i) + (5 - i)$
(b) $(-7i) - (6 - 5i)$
(c) $(-3 + 2i)^2$
(d) $\dfrac{17}{4 + i}$

SOLUTION
(a) $(-3 + 4i) + (5 - i) = -3 + 5 + 4i - i = 2 + 3i$

(b) $(-7i) - (6 - 5i) = -6 - 7i + 5i = -6 - 2i$

(c) $(-3 + 2i)^2 = (-3 + 2i)(-3 + 2i)$
$$= 9 - 6i - 6i + 4i^2$$
$$= 9 - 12i + 4(-1)$$
$$= 5 - 12i$$

(d) $\dfrac{17}{4 + i} = \dfrac{17}{4 + i} \cdot \dfrac{4 - i}{4 - i}$
$$= \dfrac{68 - 17i}{16 - i^2}$$
$$= \dfrac{68 - 17i}{17}$$
$$= 4 - i$$

Standard forms can be found using a calculator. See Figures 4.82 and 4.83.

Graphing Calculator Help

To perform arithmetic on complex numbers, see Appendix B (page AP-17).

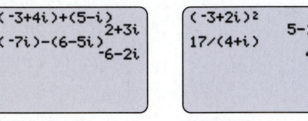

FIGURE 4.82　　　　**FIGURE 4.83**

■ **MAKING CONNECTIONS**

Complex, Real, and Imaginary Numbers　The following diagram illustrates the relationship among complex, real, and imaginary numbers, where a and b are real numbers. Note that complex numbers comprise two disjoint sets of numbers: the real numbers and the imaginary numbers.

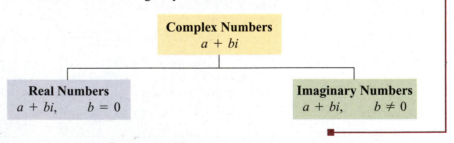

Quadratic Equations with Complex Solutions

We can use the quadratic formula to solve the quadratic equation $ax^2 + bx + c = 0$. If the discriminant, $b^2 - 4ac$, is negative, then there are no real solutions, and the graph of $y = ax^2 + bx + c$ does not intersect the x-axis. However, there are solutions that can be expressed as imaginary numbers. This is illustrated in the next example.

EXAMPLE 2   *Solving quadratic equations with imaginary solutions*

Solve the quadratic equation

(a) $\frac{1}{2}x^2 + 17 = 5x$ **(b)** $x^2 + 3x + 5 = 0$ **(c)** $2x^2 + 3 = 0$

SOLUTION   **(a)** Rewrite the equation as $\frac{1}{2}x^2 - 5x + 17 = 0$, and let $a = \frac{1}{2}$, $b = -5$, and $c = 17$ in the quadratic formula.

$$x = \frac{-b \pm \sqrt{b^2 - 4ac}}{2a}$$

$$= \frac{5 \pm \sqrt{(-5)^2 - 4(0.5)(17)}}{2(0.5)}$$

$$= 5 \pm \sqrt{-9}$$

$$= 5 \pm 3i$$

Algebra Review

To review the quadratic formula, see Section 3.2.

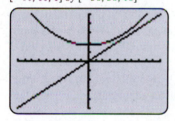

FIGURE 4.84
[−10, 10, 1] by [−50, 50, 10]

Notice that if we graph $Y_1 = .5X\hat{}2 + 17$ and $Y_2 = 5X$, as shown in Figure 4.84, their graphs do not intersect. This indicates that there are no real solutions. However, there are two complex solutions of $5 + 3i$ and $5 - 3i$.

(b) Let $a = 1$, $b = 3$, and $c = 5$ and apply the quadratic formula.

$$x = \frac{-b \pm \sqrt{b^2 - 4ac}}{2a}$$

$$= \frac{-3 \pm \sqrt{3^2 - 4(1)(5)}}{2(1)}$$

$$= \frac{-3 \pm \sqrt{-11}}{2}$$

$$= \frac{-3 \pm i\sqrt{11}}{2}$$

$$= -\frac{3}{2} \pm \frac{i\sqrt{11}}{2}$$

(c) Rather than use the quadratic formula for this equation, we apply the square root property because the equation contains no x-term.

$2x^2 + 3 = 0$	Given equation
$2x^2 = -3$	Subtract 3.
$x^2 = -\dfrac{3}{2}$	Divide by 2.
$x = \pm\sqrt{-\dfrac{3}{2}}$	Square root property
$x = \pm i\sqrt{\dfrac{3}{2}}$	$\sqrt{-a} = i\sqrt{a}$

■ **CLASS DISCUSSION**

What is the result if each expression is evaluated? (See Example 2(b).)

$$\left(-\frac{3}{2} + \frac{i\sqrt{11}}{2}\right)^2 + 3\left(-\frac{3}{2} + \frac{i\sqrt{11}}{2}\right) + 5$$

$$\left(-\frac{3}{2} - \frac{i\sqrt{11}}{2}\right)^2 + 3\left(-\frac{3}{2} - \frac{i\sqrt{11}}{2}\right) + 5 \quad ■$$

Fundamental Theorem of Algebra

One of the most brilliant mathematicians of all time, Carl Friedrich Gauss, at age 20 proved the fundamental theorem of algebra as part of his doctoral thesis. Although his theorem and proof were completed in 1797, they are still valid today.

> ### Fundamental theorem of algebra
>
> A polynomial $f(x)$ of degree $n \geq 1$ has at least one complex zero.

The fundamental theorem of algebra guarantees that every polynomial has a complete factorization, if we are allowed to use complex numbers. (A complex number can be written as $a + bi$. If $b = 0$, then the complex number $a + bi$ is also a real number.)

If $f(x)$ is a polynomial of degree 1 or higher, then by the fundamental theorem of algebra there is a zero c_1 such that $f(c_1) = 0$. By the factor theorem $(x - c_1)$ is a factor of $f(x)$ and

$$f(x) = (x - c_1)q_1(x)$$

for some polynomial $q_1(x)$. If $q_1(x)$ has positive degree, then by the fundamental theorem of algebra there exists a zero c_2 of $q_1(x)$. By the factor theorem $q_1(x)$ can be written as

$$q_1(x) = (x - c_2)q_2(x).$$

Then,

$$f(x) = (x - c_1)q_1(x)$$
$$= (x - c_1)(x - c_2)q_2(x).$$

If $f(x)$ has degree n, this process can be continued until $f(x)$ is written in the complete factored form

$$f(x) = a_n(x - c_1)(x - c_2) \cdots (x - c_n),$$

where a_n is the leading coefficient and the c_k are complex zeros of $f(x)$. If each c_k is distinct, then $f(x)$ has n zeros. However, in general the c_k may not be distinct since multiple zeros are possible.

> ### Number of zeros theorem
>
> A polynomial of degree n has at most n distinct zeros.

EXAMPLE 3   *Classifying zeros*

All zeros for the given polynomials are distinct. Use Figures 4.85–4.87 to determine graphically the number of real zeros and the number of imaginary zeros.
(a) $f(x) = 3x^3 - 3x^2 - 3x - 5$
(b) $g(x) = 2x^2 + x + 1$
(c) $h(x) = -x^4 + 4x^2 + 4$

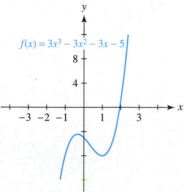

FIGURE 4.85

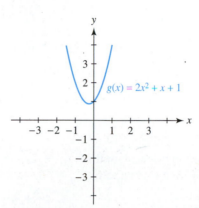

FIGURE 4.86

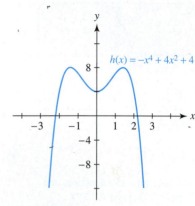

FIGURE 4.87

SOLUTION   (a) The graph of $f(x)$ in Figure 4.85 crosses the x-axis once so there is one real zero. Since f is degree 3 and all zeros are distinct, there are two imaginary zeros.
(b) The graph of $g(x)$ in Figure 4.86 never crosses the x-axis. Since g is degree 2, there are no real zeros and two imaginary zeros.
(c) The graph of $h(x)$ is shown in Figure 4.87. Since h is degree 4, there are two real zeros and the remaining two zeros are imaginary.

EXAMPLE 4   *Constructing a polynomial with prescribed zeros*

Represent a polynomial $f(x)$ of degree 4 with a leading coefficient of 2 and zeros of $-3, 5, i,$ and $-i$ in
(a) complete factored form, and
(b) expanded form.

SOLUTION   (a) Let $a_n = 2, c_1 = -3, c_2 = 5, c_3 = i,$ and $c_4 = -i.$ Then,

$$f(x) = 2(x + 3)(x - 5)(x - i)(x + i).$$

(b) To expand this expression for $f(x)$, perform the following.

$$
\begin{aligned}
2(x + 3)(x - 5)(x - i)(x + i) &= 2(x + 3)(x - 5)(x^2 + 1) \\
&= 2(x + 3)(x^3 - 5x^2 + x - 5) \\
&= 2(x^4 - 5x^3 + x^2 - 5x + 3x^3 \\
&\quad - 15x^2 + 3x - 15) \\
&= 2(x^4 - 2x^3 - 14x^2 - 2x - 15) \\
&= 2x^4 - 4x^3 - 28x^2 - 4x - 30
\end{aligned}
$$

Algebra Review
To review multiplication of polynomials, see Chapter R (page R-23).

Thus $f(x) = 2x^4 - 4x^3 - 28x^2 - 4x - 30.$

EXAMPLE 5   *Factoring a cubic polynomial with imaginary zeros*

Determine the complete factored form for $f(x) = x^3 + 2x^2 + 4x + 8$.

SOLUTION   One zero of the polynomial $f(x)$ is -2. The other two zeros are imaginary. See Figure 4.88, where $Y_1 = X\char94 3 + 2X\char94 2 + 4X + 8$. By the factor theorem $x + 2$ is a factor of $f(x)$. Divide $x + 2$ into $x^3 + 2x^2 + 4x + 8$. (Synthetic division could also be used.)

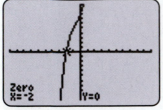

$[-10, 10, 1]$ by $[-10, 10, 1]$

FIGURE 4.88

$$
\begin{array}{r}
x^2 + 0x\ + 4 \\
x + 2\overline{)x^3 + 2x^2 + 4x + 8} \\
\underline{x^3 + 2x^2} \\
0x^2 + 4x \\
\underline{0x^2 + 0x} \\
4x + 8 \\
\underline{4x + 8} \\
0
\end{array}
$$

Graphing Calculator Help

To find a zero of a function, see Appendix B (page AP-12).

By the division algorithm, $x^3 + 2x^2 + 4x + 8 = (x + 2)(x^2 + 4)$. To factor $x^2 + 4$, first find its zeros.

$$x^2 + 4 = 0$$
$$x^2 = -4$$
$$x = \pm\sqrt{-4}$$
$$x = \pm 2i$$

The zeros of $f(x)$ are -2, $2i$, and $-2i$. Its complete factored form is

$$f(x) = (x + 2)(x - 2i)(x + 2i).$$

Notice that in Example 5 both $2i$ and $-2i$ were zeros of $f(x)$. The numbers $2i$ and $-2i$ are conjugates. This result can be generalized for any polynomial with real coefficients.

Conjugate zeros theorem

If a polynomial $f(x)$ has only real coefficients and if $a + bi$ is a zero of $f(x)$, then the conjugate $a - bi$ is also a zero of $f(x)$.

EXAMPLE 6   *Constructing a polynomial with prescribed zeros*

Find a cubic polynomial $f(x)$ with real coefficients, a leading coefficient of 2, and zeros of 3 and $5i$. Express f in
(a) complete factored form, and
(b) expanded form.

SOLUTION   **(a)** Since $f(x)$ has real coefficients, it must also have a third zero of $-5i$, the conjugate of $5i$. Let $c_1 = 3$, $c_2 = 5i$, $c_3 = -5i$, and $a_n = 2$. The complete factored form is

$$f(x) = 2(x - 3)(x - 5i)(x + 5i).$$

(b) To expand $f(x)$ perform the following steps.

$$2(x - 3)(x - 5i)(x + 5i) = 2(x - 3)(x^2 + 25)$$
$$= 2(x^3 - 3x^2 + 25x - 75)$$
$$= 2x^3 - 6x^2 + 50x - 150$$

Polynomial Equations with Complex Solutions

Every polynomial equation of degree n can be written in the form

$$a_n x^n + \cdots + a_2 x^2 + a_1 x + a_0 = 0.$$

If we let $f(x) = a_n x^n + \cdots + a_2 x^2 + a_1 x + a_0$ and write $f(x)$ in complete factored form as

$$f(x) = a_n(x - c_1)(x - c_2) \cdots (x - c_n),$$

then the solutions to the polynomial equation are $c_1, c_2, \ldots, c_n$. Solving a polynomial equation with this technique is illustrated in the next two examples.

EXAMPLE 7   *Solving a polynomial equation*

Solve $x^3 = 3x^2 - 7x + 21$.

SOLUTION   Write the equation as $f(x) = 0$, where $f(x) = x^3 - 3x^2 + 7x - 21$. Then graph $f(x)$. Figure 4.89 shows that 3 is a zero of $f(x)$. By the factor theorem, $x - 3$ is a factor of $f(x)$. Using synthetic division, we divide $x - 3$ into $f(x)$.

$$
\begin{array}{r|rrrr}
3 & 1 & -3 & 7 & -21 \\
  &   & 3 & 0 & 21 \\
\hline
  & 1 & 0 & 7 & 0
\end{array}
$$

[−5, 5, 1] by [−30, 30, 10]

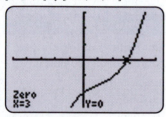

FIGURE 4.89

Thus, $x^3 - 3x^2 + 7x - 21 = (x - 3)(x^2 + 7)$.

$x^3 - 3x^2 + 7x - 21 = 0$	$f(x) = 0$
$(x - 3)(x^2 + 7) = 0$	Factor.
$x - 3 = 0$ or $x^2 + 7 = 0$	Zero-product property
$x = 3$ or $x^2 = -7$	Solve.
$x = 3$ or $x = \pm i\sqrt{7}$	Property of i

The solutions are 3 and $\pm i\sqrt{7}$.

EXAMPLE 8   *Solving a polynomial equation*

Solve $x^4 + x^2 = x^3$.

SOLUTION   Write the equation as $f(x) = 0$, where $f(x) = x^4 - x^3 + x^2$.

$x^4 - x^3 + x^2 = 0$	$f(x) = 0$
$x^2(x^2 - x + 1) = 0$	Factor out x^2.
$x^2 = 0$ or $x^2 - x + 1 = 0$	Zero-product property

The only solution to $x^2 = 0$ is 0. To solve $x^2 - x + 1 = 0$, use the quadratic formula.

$$x = \frac{-b \pm \sqrt{b^2 - 4ac}}{2a}$$

$$= \frac{1 \pm \sqrt{(-1)^2 - 4(1)(1)}}{2(1)}$$

$$= \frac{1 \pm \sqrt{-3}}{2}$$

$$= \frac{1 \pm i\sqrt{3}}{2}$$

$$= \frac{1}{2} \pm \frac{\sqrt{3}}{2}i$$

The solutions are 0 and $\frac{1}{2} \pm \frac{\sqrt{3}}{2}i$.

$[-3, 3, 1]$ by $[-2, 2, 1]$

FIGURE 4.90

The graphs of $Y_1 = X^4 + X^3$ and $Y_2 = X^3$ appear in Figure 4.90. Notice that they appear to intersect only at the origin. This indicates that the only real solution is $x = 0$.

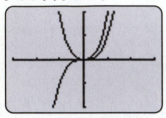

Putting it all Together
4.4

Some of the important topics in this section are summarized in the following table.

Concept	Explanation	Comments and Examples
Imaginary unit	$i = \sqrt{-1}, i^2 = 1$	The imaginary unit i allows us to define a new set of numbers called the complex numbers.
The expression $\sqrt{-a}$ with $a > 0$	$\sqrt{-a} = i\sqrt{a}$	$\sqrt{-4} = 2i$ $\sqrt{-5} = i\sqrt{5}$
Complex number	$a + bi$, where a and b are real numbers	Every real number is a complex number. $5 - 4i, 5, 2 + i$, and $-9i$ are examples of complex numbers.
Standard form of a complex number	$a + bi$, where a and b are real numbers	Converting to standard form: $\frac{3 \pm 4i}{2} = \frac{3}{2} + 2i$ or $\frac{3}{2} - 2i$
Conjugates	The conjugate of $a + bi$ is $a - bi$.	**Number** **Conjugate** $5 - 6i$ $5 + 6i$ $12i$ $-12i$

Concept	Explanation	Comments and Examples
Arithmetic operations on complex numbers	Complex numbers may be added, subtracted, multiplied, or divided.	$(2 + 3i) + (-3 - i) = -1 + 2i$ $(5 + i) - (3 - i) = 2 + 2i$ $(1 + i)(5 - i) = 5 - i + 5i - i^2$ $= 6 + 4i$ $\dfrac{3 + i}{1 - i} = \dfrac{(3 + i)(1 + i)}{(1 - i)(1 + i)} = 1 + 2i$
Number of zeros theorem	A polynomial of degree n has at most n distinct zeros.	The cubic polynomial, $$ax^3 + bx^2 + cx + d,$$ has *at most* 3 distinct zeros.
Fundamental theorem of algebra	A polynomial of degree $n \geq 1$ has at least one complex zero.	This theorem guarantees that we can always factor a polynomial $f(x)$ into complete factored form: $$f(x) = a_n(x - c_1) \cdots (x - c_n),$$ where the c_k are complex numbers.
Conjugate zeros theorem	If a polynomial has real coefficients and $a + bi$ is a zero, then $a - bi$ is also a zero.	Since $\dfrac{1}{2} + \dfrac{1}{2}i$ is a zero of $2x^2 - 2x + 1$, it follows that $\dfrac{1}{2} - \dfrac{1}{2}i$ is also a zero.

4.4 EXERCISES

Complex Numbers

Exercises 1–8: Simplify the expression using the imaginary unit i.

1. $\sqrt{-4}$
2. $\sqrt{-16}$
3. $\sqrt{-100}$
4. $\sqrt{-49}$
5. $\sqrt{-23}$
6. $\sqrt{-11}$
7. $\sqrt{-12}$
8. $\sqrt{-32}$

Exercises 9–28: Write the expression in standard form.

9. $3i + 5i$
10. $-7i + 5i$
11. $(3 + i) + (-5 - 2i)$
12. $(-4 + 2i) + (7 + 35i)$
13. $(12 - 7i) - (-1 + 9i)$
14. $(2i) - (-5 + 23i)$
15. $(3) - (4 - 6i)$
16. $(7 + i) - (-8 + 5i)$
17. $(2)(2 + 4i)$

18. $(-5)(-7 + 3i)$
19. $(1 + i)(2 - 3i)$
20. $(-2 + i)(1 - 2i)$
21. $(-3 + 2i)(-2 + i)$
22. $(2 - 3i)(1 + 4i)$
23. $\dfrac{1}{1 + i}$
24. $\dfrac{1 - i}{2 + 3i}$
25. $\dfrac{4 + i}{5 - i}$
26. $\dfrac{10}{1 - 4i}$
27. $\dfrac{2i}{10 - 5i}$
28. $\dfrac{3 - 2i}{1 + 2i}$

Exercises 29–34: Evaluate the expression with a calculator.

29. $(23 - 5.6i) + (-41.5 + 93i)$
30. $(-8.05 - 4.67i) + (3.5 + 5.37i)$
31. $(17.1 - 6i) - (8.4 + 0.7i)$

32. $\left(\dfrac{3}{4} - \dfrac{1}{10}i\right) - \left(-\dfrac{1}{8} + \dfrac{4}{25}i\right)$

33. $(-12.6 - 5.7i)(5.1 - 9.3i)$

34. $(7.8 + 23i)(-1.04 + 2.09i)$

Exercises 35 and 36: Perform the complex division using a calculator. Express your answer as a + bi, where a and b are rounded to the nearest thousandth.

35. $\dfrac{17 - 135i}{18 + 142i}$ **36.** $\dfrac{141 + 52i}{102 - 31i}$

Exercises 37–42: **Electricity** *Complex numbers are used in the study of electrical circuits, such as the current found in a household outlet. Impedance Z (or the opposition to the flow of electricity), voltage V, and current I can all be represented by complex numbers. They are related by the equation $Z = \dfrac{V}{I}$. Use this equation to find the value of the missing variable.*

37. $V = 50 + 98i$ $I = 8 + 5i$

38. $V = 30 + 60i$ $I = 8 + 6i$

39. $I = 1 + 2i$ $Z = 3 - 4i$

40. $I = \dfrac{1}{2} + \dfrac{1}{4}i$ $Z = 8 - 9i$

41. $Z = 22 - 5i$ $V = 27 + 17i$

42. $Z = 10 + 5i$ $V = 10 + 8i$

Quadratic Equations with Complex Solutions

Exercises 43–54: Solve the quadratic equation. Write complex solutions in standard form.

43. $x^2 + 5 = 0$ **44.** $4x^2 - 3 = 0$

45. $5x^2 + 1 = 3x^2$ **46.** $x(3x + 1) = -1$

47. $3x = 5x^2 + 1$ **48.** $4x^2 = x - 1$

49. $x^2 - 4x + 5 = 0$ **50.** $2x^2 + x + 1 = 0$

51. $x^2 = 3x - 5$ **52.** $3x - x^2 = 5$

53. $6 = x^2 + 2x + 10$ **54.** $x(x - 4) = -8$

Zeros of Quadratic Polynomials

Exercises 55–60: The graph and the formula for a quadratic polynomial f(x) are given.

(a) *Use the graph to predict the number of real zeros and the number of imaginary zeros.*

(b) *Find these zeros using the quadratic formula.*

55. $f(x) = 2x^2 - x - 3$

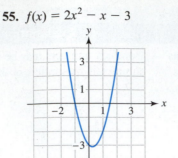

56. $f(x) = -x^2 + 4.6x - 5.29$

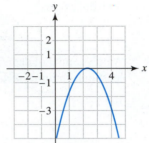

57. $f(x) = x^2 + x + 2$

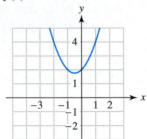

58. $f(x) = -2x^2 + 2x - 3$

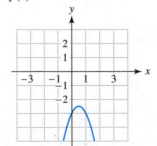

59. $f(x) = 5x^2 + 4x + 1$

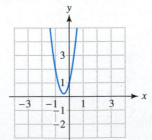

60. $f(x) = 9x^2 - 12x + 4$

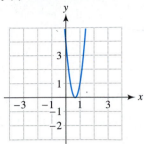

Zeros of Polynomials

Exercises 61–66: The graph of a polynomial f(x) is given. Determine the number of real zeros and the number of imaginary zeros. Assume that all zeros of f(x) are distinct.

61. $f(x) = x^3 - 5x^2 + 5x - 15$

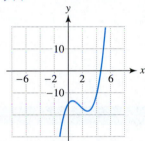

62. $f(x) = -x^3 + 5x^2 - x - 5$

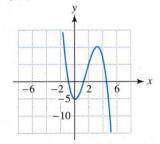

63. $f(x) = x^4 + 2x^3 - 3x^2 + 2x - 3$

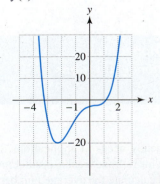

64. $f(x) = x^4 - 5x^2 + 10$

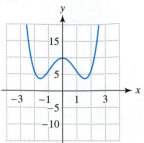

65. $f(x) = x^5 - 6x^4 + 6x^3 - 8x^2 + 9x + 70$

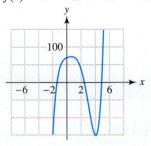

66. $f(x) = 4x^5 + 8x^4 - 31x^3 - 67x^2 - 33x - 100$

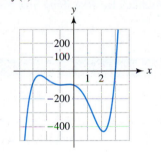

Exercises 67–72: Let a_n denote the leading coefficient of a polynomial f(x).

    (a) Find the complete factored form of f(x) that satisfies the given conditions.

    (b) Express f(x) in expanded form.

67. Degree 2; $a_n = 1$;   zeros $6i$ and $-6i$

68. Degree 3; $a_n = 5$;   zeros 2, i, and $-i$

69. Degree 3; $a_n = -1$;   zeros -1, $2i$, and $-2i$

70. Degree 4; $a_n = 3$;   zeros -2, 4, i, and $-i$

71. Degree 4; $a_n = 10$;   zeros 1, -1, $3i$, and $-3i$

72. Degree 2; $a_n = -5$;   zeros $1 + i$ and $1 - i$

Exercises 73–80: Express f(x) in complete factored form.

73. $f(x) = x^2 + 25$ **74.** $f(x) = x^2 + 11$

75. $f(x) = 3x^3 + 3x$ **76.** $f(x) = 2x^3 + 10x$

77. $f(x) = x^4 + 5x^2 + 4$ **78.** $f(x) = x^4 + 4x^2$

79. $f(x) = x^4 + 2x^3 + x^2 + 8x - 12$

80. $f(x) = x^3 + 2x^2 + 16x + 32$

Exercises 81–92: Solve the polynomial equation.

81. $x^3 + x = 0$ **82.** $2x^3 - x + 1 = 0$

83. $x^3 = 2x^2 - 7x + 14$ **84.** $x^2 + x + 2 = x^3$

85. $x^4 + 5x^2 = 0$

86. $x^4 - 2x^3 + x^2 - 2x = 0$

87. $x^4 = x^3 - 4x^2$

88. $x^5 + 9x^3 = x^4 + 9x^2$

89. $x^4 + x^3 = 16 - 8x - 6x^2$

90. $x^4 + 2x^2 = x^3$

91. $3x^3 + 4x^2 + 6 = x$

92. $2x^3 + 5x^2 + x + 12 = 0$

Writing about Mathematics

93. Could a cubic function have only imaginary zeros? Explain.

94. Give an example of a polynomial function that has only imaginary zeros and a polynomial function that has only real zeros. Explain how to determine graphically if a function has only imaginary zeros.

Extended and Discovery Exercise

The properties of the imaginary unit are $i = \sqrt{-1}$ and $i^2 = -1$.

(a) Begin simplifying the expressions $i, i^2, i^3, i^4, i^5, \ldots$, until a pattern is discovered. For example, $i^3 = i \cdot i^2 = i \cdot (-1) = -i$.

(b) Summarize your findings by describing how to simplify i^n for any natural number n.

CHECKING BASIC CONCEPTS FOR SECTIONS 4.3 AND 4.4

1. Solve $x^3 - 2x^2 - 15x = 0$.

2. Determine graphically the zeros of $f(x) = x^4 - x^3 - 18x^2 + 16x + 32$. Write $f(x)$ in complete factored form.

3. Divide $x^3 - x^2 + 4x - 4$ by $x - 1$.

4. Find a quadratic polynomial $f(x)$ with zeros $\pm 4i$ and leading coefficient 3. Write $f(x)$ in complete factored form and expanded form.

5. Sketch a graph of a quartic function (degree 4) with a negative leading coefficient, two real zeros, and two imaginary zeros.

6. Write $f(x) = x^3 - x^2 + 4x - 4$ in complete factored form.

4.5 RATIONAL FUNCTIONS AND MODELS

Rational Functions • Vertical Asymptotes • Horizontal Asymptotes • Identifying Asymptotes • Rational Equations • Variation

Introduction

Rational functions are nonlinear functions that frequently occur in applications. For example, rational functions are used to design curves for railroad tracks, predict mortality rates, determine stopping distances on hills, and calculate the average number of people waiting in a line.

Rational Functions

A rational number can be expressed as a ratio $\frac{p}{q}$, where p and q are integers and $q \neq 0$. A rational function is defined similarly by using the concept of a polynomial.

Rational function

A function f represented by $f(x) = \frac{p(x)}{q(x)}$, where $p(x)$ and $q(x)$ are polynomials and $q(x) \neq 0$, is a **rational function**.

The domain of a rational function includes all real numbers *except* the zeros of the denominator $q(x)$. The graph of a rational function is continuous except at x-values where $q(x) = 0$.

EXAMPLE 1 *Identifying rational functions*

Determine if the function is rational and state its domain.

(a) $f(x) = \dfrac{2x - 1}{x^2 + 1}$ **(b)** $g(x) = \dfrac{1}{\sqrt{x}}$ **(c)** $h(x) = \dfrac{x^3 - 2x^2 + 1}{x^2 - 3x + 2}$

SOLUTION **(a)** Both the numerator, $2x - 1$, and the denominator, $x^2 + 1$, are polynomials so f is a rational function. The domain of f includes all real numbers because $x^2 + 1 \neq 0$ for any real number x.

(b) Since $\sqrt{x}$ is not a polynomial, g is not a rational function. The domain of g is $\{x \mid x > 0\}$.

(c) Both the numerator and the denominator are polynomials, so h is a rational function. Because

$$x^2 - 3x + 2 = (x - 1)(x - 2) = 0$$

when $x = 1$ or $x = 2$, the domain of h is $\{x \mid x \neq 1 \text{ or } x \neq 2\}$.

■ **CLASS DISCUSSION**
Is an integer a rational number? Is a polynomial function a rational function? ■

Vertical Asymptotes

If cars arrive randomly at the exit of a parking ramp, then the average length of the line depends on two factors: the average traffic volume exiting the ramp and the average rate that a parking attendant can wait on cars. For instance, if the average traffic volume is three cars per minute and the parking attendant is serving four cars per minute, then at times a line may form if cars arrive in a random manner. The **traffic intensity** x is the ratio of the average traffic volume to the average working rate of the attendant. In this example, $x = \dfrac{3}{4}$. (Source: F. Mannering and W. Kilareski, *Principles of Highway Engineering and Traffic Control.*)

EXAMPLE 2 *Estimating the length of parking ramp lines*

If the traffic intensity is x, then the average number of cars waiting to exit a parking ramp can be computed by $f(x) = \dfrac{x^2}{2 - 2x}$, where $0 \leq x < 1$.

(a) Evaluate $f(0.5)$ and $f(0.9)$. Interpret the results.
(b) Graph f in $[0, 1, 0.1]$ by $[0, 10, 1]$.
(c) Explain what happens to the length of the line as the traffic intensity approaches 1.

SOLUTION
(a) $f(0.5) = \dfrac{0.5^2}{2 - 2(0.5)} = 0.25$ and $f(0.9) = \dfrac{0.9^2}{2 - 2(0.9)} = 4.05$. This means that if the traffic intensity is 0.5, there is little waiting in line. As the traffic intensity increases to 0.9, the average line has more than four cars.

(b) Graph $Y_1 = X^2/(2 - 2X)$ as in Figure 4.91.

(c) As the traffic intensity x approaches 1 from the left, the graph of f increases rapidly without bound. Numerical support is given in Figure 4.92. With a traffic intensity slightly less than 1, the attendant has difficulty keeping up. If cars occasionally arrive in groups, long lines will form. At $x = 1$ the denominator, $2 - 2x$, equals 0 and $f(x)$ is undefined. See Figure 4.93.

$[0, 1, 0.1]$ by $[0, 10, 1]$

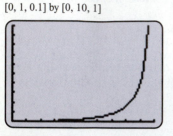

FIGURE 4.91

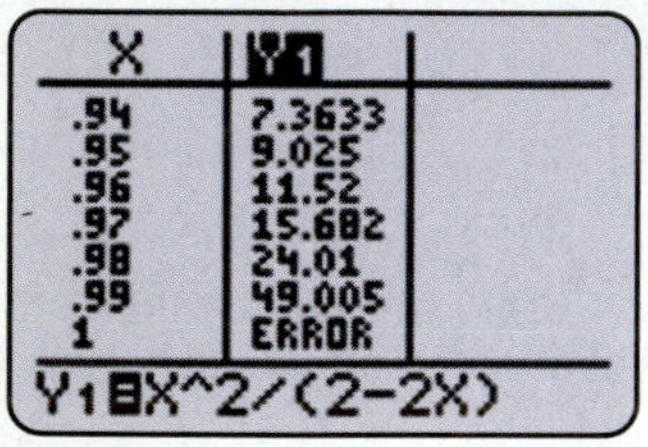

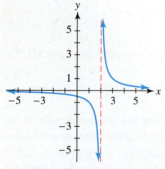

FIGURE 4.92

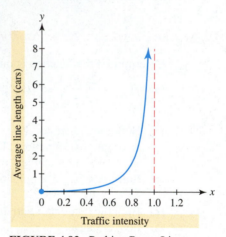

FIGURE 4.93 Parking Ramp Lines

In Figure 4.93, the vertical line $x = 1$ is a *vertical asymptote* of the graph of f. A graph of a different rational function f is shown in Figure 4.94. The $f(x)$-values *decrease without bound* as x approaches 2 from the left. This is denoted by $f(x) \to -\infty$ as $x \to 2^-$. Similarly, the $f(x)$-values *increase without bound* as x approaches 2 from the right. This is expressed as $f(x) \to \infty$ as $x \to 2^+$. The line $x = 2$ is a vertical asymptote of the graph of f.

FIGURE 4.94

Vertical asymptote

The line $x = k$ is a **vertical asymptote** of the graph of f if $f(x) \to \infty$ or $f(x) \to -\infty$, as x approaches k from either the left or the right.

If $x = k$ is a vertical asymptote of the graph of f, then k is not in the domain of f. Furthermore, the graph of a rational function f does *not* cross a vertical asymptote.

Horizontal Asymptotes

Probability sometimes can be described with rational functions. Suppose that a container holds ten identical balls numbered 1 through 10, where one ball has the winning number. Then the probability or likelihood of drawing the winning ball is 1 in 10.

This probability can be expressed as the ratio $\frac{1}{10}$. Probabilities vary between 0 and 1, where 1 represents an event that is certain to occur.

If there are x balls, the probability of drawing the winning ball is given by $P(x) = \frac{1}{x}$, where x is a natural number. A graph of P is shown in Figure 4.95. As the number of balls increases, the probability of drawing the winning ball decreases toward 0, without becoming 0. Even if there are one million balls in the container, there still is a slight chance of drawing the winning ball. As a result, the graph of P comes closer and closer to the x-axis ($y = 0$) without ever actually touching it. The x-axis is a *horizontal asymptote* of the graph of P.

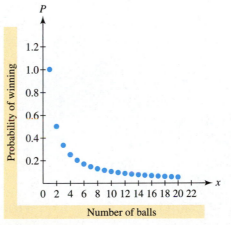

FIGURE 4.95

Horizontal asymptote

The line $y = b$ is a **horizontal asymptote** of the graph of f, if $f(x) \to b$ as x approaches either ∞ or $-\infty$.

The graph of f in Figure 4.96 is an example of a von Bertalanffy growth curve. It models the length in millimeters of a small fish after x weeks. After several weeks the length of the fish begins to level off near 25 millimeters. Thus $y = 25$ is a horizontal asymptote of the graph of f. This is denoted by $f(x) \to 25$ as $x \to \infty$. (Source: D. Brown and P. Rothery, *Models in Biology: Mathematics, Statistics and Computing.*)

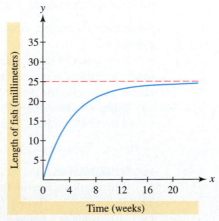

FIGURE 4.96 Size of a Small Fish

Identifying Asymptotes

The graph in Figure 4.95 limits the domain of P to natural numbers because a fraction of a ball is not allowed. However, the implied domain of $f(x) = \dfrac{1}{x}$ is the set of nonzero real numbers.

EXAMPLE 3 *Analyzing the graph of $f(x) = \dfrac{1}{x}$*

The graph of $f(x) = \dfrac{1}{x}$ is shown in Figure 4.97. Numerical values of $f(x)$ are listed in Tables 4.8 and 4.9. Relate these values to the graph of f.

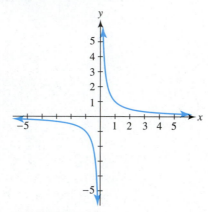

FIGURE 4.97

TABLE 4.8

x	0	0.001	0.01	0.1	1	10	100	1000
$f(x) = \dfrac{1}{x}$	—	1000	100	10	1	0.1	0.01	0.001

TABLE 4.9

x	-1000	-100	-10	-1	-0.1	-0.01	-0.001	0
$f(x) = \dfrac{1}{x}$	-0.001	-0.01	-0.1	-1	-10	-100	-1000	—

SOLUTION Table 4.8 shows positive values of x. As x approaches 0 from the right, the $f(x)$-values increase without bound. This is denoted by $f(x) \to \infty$ as $x \to 0^+$. Slightly right of the y-axis there are large y-values on the graph of $f(x) = \dfrac{1}{x}$. The y-axis ($x = 0$) is a vertical asymptote.

As x assumes large positive values in Table 4.8, the $f(x)$-values decrease and tend toward 0. This is expressed as $f(x) \to 0$ as $x \to \infty$. As a result, the graph of f levels off above the x-axis. The x-axis ($y = 0$) is a horizontal asymptote.

Table 4.9 shows negative values of x. As x approaches 0 from the left, the $f(x)$-values decrease without bound. This is denoted by $f(x) \to -\infty$ as $x \to 0^-$. The graph of $f(x) = \dfrac{1}{x}$ falls rapidly just left of the y-axis, indicating that the y-axis ($x = 0$) is a vertical asymptote.

When x decreases in Table 4.9, the $f(x)$-values are negative and tend toward 0. This is denoted by $f(x) \to 0$ as $x \to -\infty$. The graph of f levels off below the x-axis ($y = 0$), which is a horizontal asymptote.

EXAMPLE 4   *Determining horizontal and vertical asymptotes visually*

Use the graph of each rational function to determine any vertical or horizontal asymptotes.

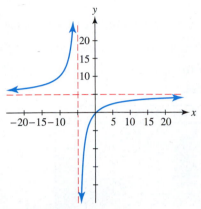

FIGURE 4.98

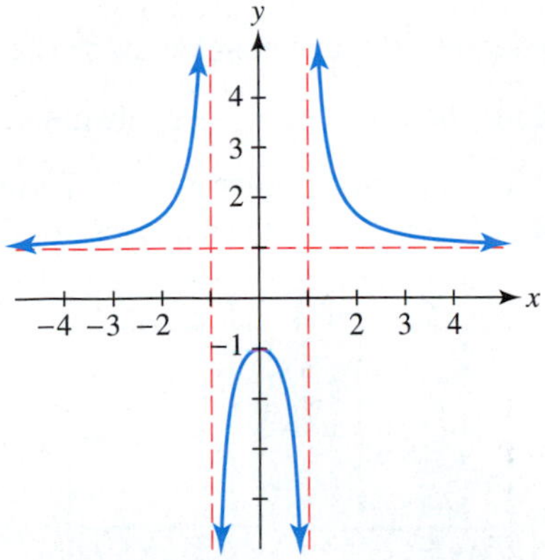

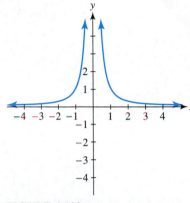

FIGURE 4.99

FIGURE 4.100

SOLUTION   In Figure 4.98, $y = 5$ is a horizontal asymptote and $x = -5$ is a vertical asymptote. In Figure 4.99, $x = \pm 1$ are vertical asymptotes and $y = 1$ is a horizontal asymptote. As x approaches 0 from either side in Figure 4.100, the y-values become large without bound. The y-axis ($x = 0$) is a vertical asymptote. As x approaches ∞ or $-\infty$, the graph of f approaches the x-axis from above. The x-axis ($y = 0$) is a horizontal asymptote.

The following may be used to find vertical and horizontal asymptotes.

Finding vertical and horizontal asymptotes

Let f be a rational function given by $f(x) = \dfrac{p(x)}{q(x)}$ written in lowest terms.

Vertical Asymptote

To find a vertical asymptote, set the denominator, $q(x)$, equal to 0 and solve. If k is a zero of $q(x)$, then $x = k$ is a vertical asymptote. *Caution:* If k is a zero of both $q(x)$ *and* $p(x)$, then $f(x)$ is *not* written in lowest terms, and $x - k$ is a common factor that should be canceled.

Horizontal Asymptote

(a) If the degree of the numerator is less than the degree of the denominator, then $y = 0$ (the x-axis) is a horizontal asymptote.

(b) If the degree of the numerator equals the degree of the denominator, then

$y = \dfrac{a}{b}$ is a horizontal asymptote, where a is the leading coefficient of the numerator, and b is the leading coefficient of the denominator.

(c) If the degree of the numerator is greater than the degree of the denominator, then there are no horizontal asymptotes.

EXAMPLE 5   *Finding asymptotes*

For each rational function, determine any horizontal or vertical asymptotes.

(a) $f(x) = \dfrac{6x - 1}{3x + 3}$ **(b)** $g(x) = \dfrac{x + 1}{x^2 - 4}$ **(c)** $h(x) = \dfrac{x^2 - 1}{x + 1}$

SOLUTION   **(a)** The degrees of the numerator and the denominator are both 1. Since the ratio of the leading coefficients is $\dfrac{6}{3} = 2$, the graph of f has a horizontal asymptote of $y = 2$.

This is supported numerically in Figures 4.101 and 4.102, where the y-values approach 2 as the x-values increase or decrease.

Graphing Calculator Help

To make a table of values, see Appendix B (page AP-9).

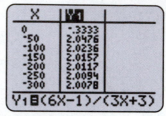

FIGURE 4.101   **FIGURE 4.102**

When $x = -1$ the denominator, $3x + 3$, equals 0 and the numerator, $6x + 1$, does not equal 0. Thus, $x = -1$ is a vertical asymptote. A graph of f is shown in Figure 4.103.

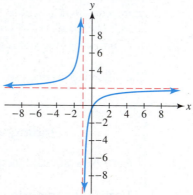

FIGURE 4.103

(b) The degree of the numerator is one less than the degree of the denominator, so the x-axis, or $y = 0$, is a horizontal asymptote. The denominator, $x^2 - 4$, equals 0 and the numerator, $x + 1$, does not equal 0 when $x = \pm 2$. Thus, $x = \pm 2$ are vertical asymptotes. See Figure 4.104.

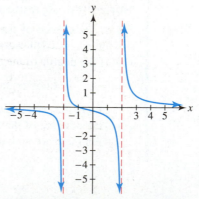

FIGURE 4.104

(c) The degree of the numerator is greater than the degree of the denominator so there are no horizontal asymptotes. When $x = -1$, both numerator and denominator equal 0. We can simplify $h(x)$ as follows.

$$h(x) = \frac{x^2 - 1}{x + 1} = \frac{(x + 1)(x - 1)}{x + 1} = x - 1, \qquad x \neq -1$$

The graph of $h(x)$ is the line $y = x - 1$ with the point $(-1, -2)$ missing. There are no vertical asymptotes. See Figure 4.105.

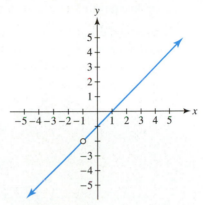

FIGURE 4.105

Note: Calculators typically graph in connected or dot mode. (See Example 6.) If connected mode is used when graphing a rational function, it may appear as though the calculator is graphing vertical asymptotes automatically. However, in most instances the calculator is connecting points inappropriately.

Sometimes rational functions can be graphed in connected mode using a *decimal* or *friendly* viewing rectangle. Decimal windows are used in Examples 7 and 8.

EXAMPLE 6   *Analyzing a rational function with technology*

Let $f(x) = \dfrac{2x^2 + 1}{x^2 - 4}$.

(a) Graph f. Find the domain of f.
(b) Identify any vertical or horizontal asymptotes.
(c) Sketch a graph of f that includes the asymptotes.

SOLUTION   (a) A graph of f is shown in Figure 4.106 using dot mode. The function is undefined when $x^2 - 4 = 0$, or when $x = \pm 2$. The domain of f is $D = \{x | x \neq \pm 2\}$.

$[-6, 6, 1]$ by $[-6, 6, 1]$

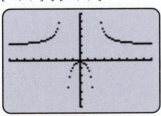

FIGURE 4.106

(b) When $x = \pm 2$ the denominator, $x^2 - 4$, equals 0 and the numerator, $2x^2 + 1$, does not equal 0. Therefore, $x = \pm 2$ are vertical asymptotes. The degree of the numerator equals the degree of the denominator, and the ratio of the leading co-efficients is $\dfrac{2}{1} = 2$. A horizontal asymptote of the graph of f is $y = 2$.

(c) A graph of f and its asymptotes is shown in Figure 4.107.

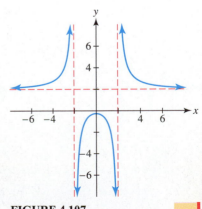

FIGURE 4.107

A third type of asymptote, which is neither vertical nor horizontal, occurs when the numerator of a rational function has degree *one more* than the degree of the denominator. For example, let $f(x) = \dfrac{x^2 + 2}{x - 1}$. If $x - 1$ is divided into $x^2 + 2$, the quotient is $x + 1$ with remainder 3. Thus

Algebra Review
To review division of
polynomials, see Section 4.3.

$$f(x) = x + 1 + \frac{3}{x - 1}$$

is an equivalent representation of f. For large values of $|x|$ the ratio $\dfrac{3}{x - 1}$ approaches zero and the graph of f approaches $y = x + 1$. The line $y = x + 1$ is called a **slant asymptote** or **oblique asymptote** of the graph of f. A graph of f with vertical asymptote $x = 1$ and slant asymptote $y = x + 1$ is shown in Figure 4.108.

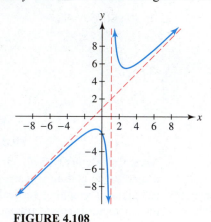

FIGURE 4.108

Rational Equations

A rational equation can be written in the form $f(x) = 0$, where f is a rational function. Rational equations often contain rational expressions. Examples of rational equations include

$$\frac{x^2 - 1}{x^2 + x + 3} = 0, \qquad \frac{3x}{x^3 + x} = 5, \qquad \text{and} \qquad \frac{1}{x - 1} + \frac{1}{x} = 5.$$

Rational equations can be solved symbolically, graphically, and numerically.

EXAMPLE 7   *Solving a rational equation*

Solve $\dfrac{4x}{x - 1} = 6$ symbolically, graphically, and numerically.

SOLUTION   ***Symbolic Solution***

$$\frac{4x}{x - 1} = 6 \qquad \text{Given equation}$$

$$4x = 6(x - 1) \qquad \text{Cross multiply.}$$

$$4x = 6x - 6 \qquad \text{Distributive property}$$

$$-2x = -6 \qquad \text{Subtract } 6x.$$

$$x = 3 \qquad \text{Divide by } -2.$$

Graphical Solution   Graph $Y_1 = 4X/(X - 1)$ and $Y_2 = 6$. Their graphs intersect at $(3, 6)$, so the solution is 3. See Figure 4.109.

[−9.4, 9.4, 1] by [−9.4, 9.4, 1]

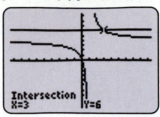

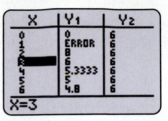

FIGURE 4.109   **FIGURE 4.110**

Numerical Solution   In Figure 4.110, $Y_1 = Y_2$ when $x = 3$.

A common approach to solve rational equations is to multiply both sides of the equation by a common denominator. This technique clears fractions from an equation and is used in Examples 8 and 9.

EXAMPLE 8

Solving a rational equation

Solve $\dfrac{6}{x^2} - \dfrac{5}{x} = 1$ symbolically and graphically.

SOLUTION

Symbolic Solution

$$\frac{6}{x^2} - \frac{5}{x} = 1 \qquad \text{Given equation}$$

$$\frac{6}{x^2} \cdot x^2 - \frac{5}{x} \cdot x^2 = 1 \cdot x^2 \qquad \text{Multiply each term by } x^2.$$

$$6 - 5x = x^2 \qquad \text{Simplify.}$$

$$0 = x^2 + 5x - 6 \qquad \text{Add } 5x \text{ and subtract 6.}$$

$$0 = (x + 6)(x - 1) \qquad \text{Factor.}$$

$$x + 6 = 0 \qquad \text{or} \qquad x - 1 = 0 \qquad \text{Zero-product property}$$

$$x = -6 \qquad \text{or} \qquad x = 1 \qquad \text{Solve.}$$

Algebra Review
To review clearing fractions, see Chapter R (page R-38).

Graphical Solution   Graph $Y_1 = 6/X^2 - 5/X$ and $Y_2 = 1$. The graphs intersect at $(-6, 1)$ and $(1, 1)$. See Figures 4.111 and 4.112. The solutions are -6 and 1.

[−9.4, 9.4, 1] by [−6.2, 6.2, 1]   [−9.4, 9.4, 1] by [−6.2, 6.2, 1]

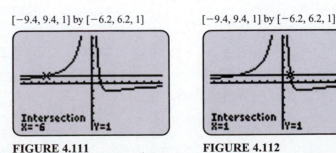

FIGURE 4.111   **FIGURE 4.112**

The next example illustrates the importance of checking possible solutions when solving rational equations.

EXAMPLE 9 *Solving a rational equation*

Solve $\dfrac{1}{x+3} + \dfrac{1}{x-3} = \dfrac{6}{x^2-9}$ symbolically. Check the result.

SOLUTION

$$\frac{1}{x+3} + \frac{1}{x-3} = \frac{6}{x^2-9}$$ Given equation

$$\frac{(x+3)(x-3)}{x+3} + \frac{(x+3)(x-3)}{x-3} = \frac{6(x+3)(x-3)}{x^2-9}$$ Multiply by $(x+3)(x-3)$.

$$(x-3) + (x+3) = 6$$ Reduce.

$$2x = 6$$ Combine terms.

$$x = 3$$ Divide by 2.

When 3 is substituted for *x*, two expressions in the given equation are undefined. There are no solutions.

Rational equations are used in real-world applications such as in construction, which is discussed in the next example. Steps for solving application problems (pages 98–99) have been used to structure our solution.

EXAMPLE 10 *Designing a box*

A box with rectangular sides and a top is being designed to hold 324 cubic inches and to have a surface area of 342 square inches. If the length of the box is four times longer than the height, find possible dimensions of the box.

SOLUTION   **STEP 1:** We are asked to find the dimensions of a box. If *x* is the height of the box, and *y* is the width, then the length of the box is four times longer than the height, or $4x$.

Geometry Review

To review formulas related to box shapes, see Chapter R (page R-4).

     x: Height of the box

     y: Width of the box

     $4x$:   Length of the box

STEP 2: To relate these variables to an equation, sketch a box as illustrated in Figure 4.113. The volume *V* of the box is height times width times length.

$$V = xy(4x)$$
$$= 4x^2 y$$

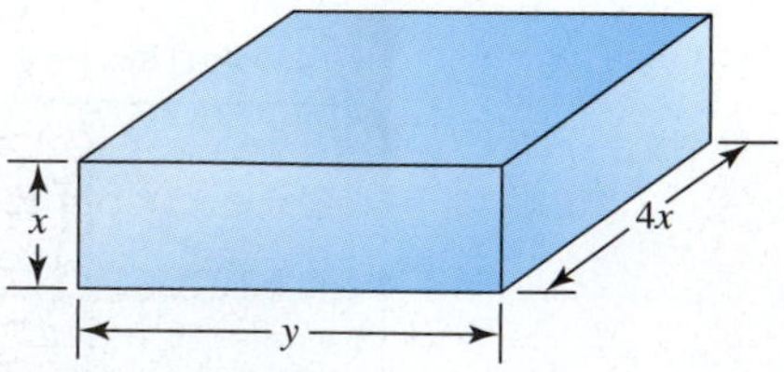

FIGURE 4.113

The surface area *A* of this box is determined by finding the area of the 6 rectangular sides: left and right sides, front and back, and top and bottom.

$$A = 2(4x \cdot x) + 2(x \cdot y) + 2(4x \cdot y)$$
$$= 8x^2 + 10xy$$

If we solve $V = 4x^2y$ for y, and let $V = 324$, we obtain

$$y = \frac{V}{4x^2} = \frac{324}{4x^2} = \frac{81}{x^2}.$$

Substituting $y = \frac{81}{x^2}$ into the formula for A eliminates the y variable.

$$
\begin{aligned}
A &= 8x^2 + 10xy && \text{Area formula} \\
&= 8x^2 + 10x \cdot \frac{81}{x^2} && \text{Let } y = \frac{81}{x^2}. \\
&= 8x^2 + \frac{810}{x} && \text{Simplify.}
\end{aligned}
$$

Since the surface area is $A = 342$ square inches, the height x can be determined by solving the rational equation

$$8x^2 + \frac{810}{x} = 342.$$

STEP 3: Figures 4.114 and 4.115 show the graphs of $Y_1 = 8X^2 + 810/X$ and $Y_2 = 342$. There are two solutions, $x = 3$ and $x = 4.5$.

Graphing Calculator Help

To find a point of intersection, see Appendix B (page AP-11).

[0, 6, 1] by [300, 400, 20]

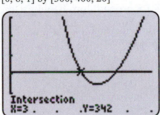

FIGURE 1.114

[0, 6, 1] by [300, 400, 20]

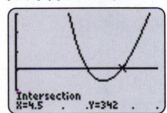

FIGURE 1.115

If the height is $x = 3$ inches, then the length is $4 \cdot 3 = 12$ inches, and the width is $y = \frac{81}{3^2} = 9$ inches. (Note that $y = \frac{81}{x^2}$.) If the height is 4.5 inches, then the length is $4 \cdot 4.5 = 18$ inches, and the width is $y = \frac{81}{4.5^2} = 4$ inches. Thus the dimensions of the box in inches can be either $3 \times 9 \times 12$ or $4.5 \times 4 \times 18$.

STEP 4: We can check our results directly. If the dimensions are $3 \times 9 \times 12$, then

$$V = 3 \cdot 9 \cdot 12 = 324 \qquad \text{and}$$
$$S = 2(3 \cdot 9) + 2(3 \cdot 12) + 2(9 \cdot 12) = 342.$$

If the dimensions are $4.5 \times 4 \times 18$, then

$$V = 4.5 \cdot 4 \cdot 18 = 324 \qquad \text{and}$$
$$S = 2(4.5 \cdot 4) + 2(4.5 \cdot 18) + 2(4 \cdot 18) = 342.$$

In both cases our results check.

Variation

In Section 2.2 direct variation was discussed. Sometimes a quantity y varies directly as a power of a variable. For example, the area A of a circle varies directly as the second power of the radius r. That is, $A = \pi r^2$.

Direct variation as the *n*th power

Let x and y denote two quantities and n be a positive number. Then y is **directly proportional to the *n*th power** of x, or y **varies directly as the *n*th power** of x, if there exists a nonzero number k such that

$$y = kx^n.$$

The number k is called the *constant of variation* or the *constant of proportionality*. In the formula $A = \pi r^2$, $k = \pi$.

EXAMPLE 11   *Modeling a pendulum*

The time T required for a pendulum to swing back and forth once is called its *period*. See Figure 4.116. The length L of a pendulum is directly proportional to the *n*th power of T, where n is a positive integer. Table 4.10 lists the period T for various lengths L.
(a) Find the constant of proportionality k and the value of n.
(b) Predict T for a pendulum having a length of 5 feet.

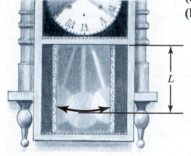

FIGURE 4.116

TABLE 4.10

L (ft)	1.0	1.5	2.0	2.5	3.0	3.5	4.0
T (sec)	1.11	1.36	1.57	1.76	1.92	2.08	2.22

SOLUTION   (a) The equation $L = kT^n$ models the pendulum. It follows that $k = \dfrac{L}{T^n}$. The ratio $\dfrac{L}{T^n}$ equals the constant k for some n. To find k and n see Table 4.11. The ratio is constant when $n = 2$. In this case $k \approx 0.81$ and $L = 0.81T^2$.

TABLE 4.11

L	T	L/T	L/T^2	L/T^3
1.0	1.11	0.90	0.81	0.73
1.5	1.36	1.10	0.81	0.60
2.0	1.57	1.27	0.81	0.52
2.5	1.76	1.42	0.81	0.46
3.0	1.92	1.56	0.81	0.42
3.5	2.08	1.68	0.81	0.39
4.0	2.22	1.80	0.81	0.37

(b) If $L = 5$, then $5 = 0.81T^2$. It follows that $T = \sqrt{5/0.81} \approx 2.48$ seconds.

When two quantities vary inversely, an increase in one quantity results in a decrease in the second quantity. For example, it takes 4 hours to travel 100 miles at 25 miles per hour and 2 hours to travel 100 miles at 50 miles per hour. Greater speed results in less travel time. If s represents the average speed of a car and t is the time to travel 100 miles, then $s \cdot t = 100$ or $t = \dfrac{100}{s}$. Doubling the speed cuts the time in half, tripling the speed reduces the time by one-third. The quantities t and s are said to *vary inversely*. The constant of variation is 100.

Inverse variation as the *n*th power

Let x and y denote two quantities and n be a positive number. Then y is **inversely proportional to the *n*th power** of x, or y **varies inversely as the *n*th power** of x, if there exists a nonzero number k such that

$$y = \frac{k}{x^n}.$$

If $y = \dfrac{k}{x}$, then y is **inversely proportional** to x or y **varies inversely** as x.

Inverse variation occurs when measuring the intensity of light. If we increase our distance from a lightbulb, the intensity of the light decreases. Intensity I is inversely proportional to the second power of the distance d. The equation $I = \dfrac{k}{d^2}$ models this phenomenon.

EXAMPLE 12   *Modeling the intensity of light*

At a distance of 3 meters, a 100-watt bulb produces an intensity of 0.88 watt per square meter.   (Source: R. Weidner and R. Sells, *Elementary Classical Physics,* Volume 2.)
(a) Find the constant of proportionality k.
(b) Determine the intensity at a distance of 2 meters.

SOLUTION   **(a)** Substitute $d = 3$ and $I = 0.88$ into the equation $I = \dfrac{k}{d^2}$. Solve for k.

$$0.88 = \frac{k}{3^2} \quad \text{or} \quad k = 7.92$$

(b) Since $I = \dfrac{7.92}{d^2}$, let $d = 2$ and find I.

$$I = \frac{7.92}{2^2} = 1.98$$

The intensity at 2 meters is 1.98 watts per square meter.

Putting it all Together
4.5

The following table summarizes some important concepts about rational functions and equations.

Concept	Explanation	Examples
Rational function	$f(x) = \dfrac{p(x)}{q(x)}$, where $p(x)$ and $q(x)$ are polynomials with $q(x) \neq 0$.	$f(x) = \dfrac{x - 1}{x^2 + 2x + 1}$ $g(x) = 1 + \dfrac{1}{x}$ \qquad (*Note*: $1 + \dfrac{1}{x} = \dfrac{x + 1}{x}$.)
Graph of a rational function	The graph of a rational function is continuous, except at x-values where the denominator equals zero.	The graph of $f(x) = \dfrac{3x^2 + 1}{x^2 - 4}$ is discontinuous at $x = \pm 2$. It has vertical asymptotes of $x = \pm 2$ and a horizontal asymptote of $y = 3$.
Vertical asymptote	If k is a zero of the denominator, but not of the numerator, then $x = k$ is a vertical asymptote.	The graph of $f(x) = \dfrac{2x + 1}{x - 2}$ has vertical asymptotes at $x = 2$ because 2 is a zero of $x - 2$, but not a zero of $2x + 1$.
Horizontal asymptote	A horizontal asymptote occurs when the degree of the numerator is less than or equal to the degree of the denominator.	$f(x) = \dfrac{-4x^2 + 1}{3x^2 - x}$ Horizontal asymptote: $y = -\dfrac{4}{3}$. $f(x) = \dfrac{x}{4x^2 + 2x}$ Horizontal asymptote: $y = 0$.

Concept	Explanation	Examples
Rational equation	A rational equation contains rational expressions in the form $\dfrac{p(x)}{q(x)}$, where $p(x)$ and $q(x)$ are polynomials.	To solve the rational equation $$\frac{4}{2x-1}=8,$$ multiply both sides by $2x-1$. $$4 = 8(2x-1)$$ $$4 = 16x - 8$$ $$12 = 16x$$ $$x = \frac{3}{4}$$ Be sure to check your solutions.

The following table summarizes some concepts about variation.

Concept	Equation	Examples
Varies directly as the nth power of x	$y = kx^n$	Let y vary directly as the third power of x. If the constant of variation is $k = 5$, then $y = 5x^3$.
Varies inversely as the nth power of x	$y = \dfrac{k}{x^n}$	Let y vary inversely as the square of x. If the constant of variation is $k = 3$, then $y = \dfrac{3}{x^2}$.

4.5 EXERCISES

Rational Functions

Exercises 1–6: Determine whether f is a rational function and state its domain.

1. $f(x) = \dfrac{x^3 - 5x + 1}{4x - 5}$

2. $f(x) = \dfrac{6}{x^2}$

3. $f(x) = x^2 - x - 2$

4. $f(x) = \dfrac{x^2 + 1}{\sqrt{x - 8}}$

5. $f(x) = \dfrac{|x - 1|}{x + 1}$

6. $f(x) = \dfrac{4}{x} + 1$

Asymptotes

Exercises 7–10: Identify any horizontal or vertical asymptotes in the graph. State the domain of f.

7.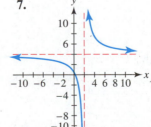

8.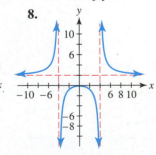

9.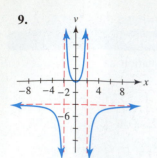

10.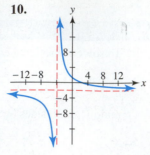

c.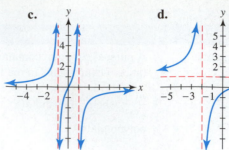

d.

Exercises 11–20: Find any vertical or horizontal asymptotes.

11. $f(x) = \dfrac{4x + 1}{2x - 6}$

12. $f(x) = \dfrac{x + 6}{5 - 2x}$

13. $f(x) = \dfrac{3}{x^2 - 5}$

14. $f(x) = \dfrac{3x^2}{x^2 - 9}$

15. $f(x) = \dfrac{x^4 + 1}{x^2 + 3x - 10}$

16. $f(x) = \dfrac{6x^2 - x - 2}{2x^2 + x - 6}$

17. $f(x) = \dfrac{3x(x + 2)}{(x + 2)(x - 1)}$

18. $f(x) = \dfrac{x}{x^3 - x}$

19. $f(x) = \dfrac{x^2 - 9}{x + 3}$

20. $f(x) = \dfrac{2x^2 - 3x + 1}{2x - 1}$

Exercises 21–24: Let a be a positive constant. Match f(x) with its graph (a–d) without using a calculator.

21. $f(x) = \dfrac{a}{x - 1}$

22. $f(x) = \dfrac{2x + a}{x - 1}$

23. $f(x) = \dfrac{x - a}{x + 2}$

24. $f(x) = \dfrac{-2x}{x^2 - a}$

a.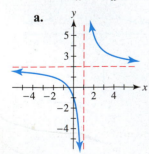

b.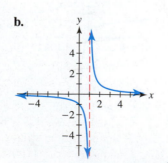

Exercises 25–32: Complete the following.

(a) *Find the domain of f.*
(b) *Graph f in an appropriate viewing rectangle.*
(c) *Find any vertical or horizontal asymptotes.*
(d) *Sketch a graph of f that includes any asymptotes.*

25. $f(x) = \dfrac{x + 3}{x - 2}$

26. $f(x) = \dfrac{6 - 2x}{x + 3}$

27. $f(x) = \dfrac{4x + 1}{x^2 - 4}$

28. $f(x) = \dfrac{0.5x^2 + 1}{x^2 - 9}$

29. $f(x) = \dfrac{4}{1 - 0.25x^2}$

30. $f(x) = \dfrac{x^2}{1 + 0.25x^2}$

31. $f(x) = \dfrac{x^2 - 4}{x - 2}$

32. $f(x) = \dfrac{4(x - 1)}{x^2 - x - 6}$

Exercises 33 and 34: In the accompanying table, Y_1 represents a rational function. Give a possible equation for a horizontal asymptote of the graph of Y_1.

33.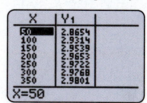

34.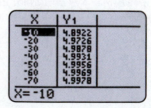

Exercises 35–38: Write a symbolic representation of a rational function f that satisfies the conditions.

35. Vertical asymptote $x = -3$,
horizontal asymptote $y = 1$

36. Vertical asymptote $x = 4$,
horizontal asymptote $y = -3$

37. Vertical asymptotes $x = \pm 3$, horizontal asymptote $y = 0$

38. Vertical asymptotes $x = -2$ and $x = 4$, horizontal asymptote $y = 5$

Exercises 39–44: Find any vertical or slant asymptotes of the graph of f. Use a graphing calculator to help sketch a graph of f. Show all asymptotes.

39. $f(x) = \dfrac{x^2 + 1}{x + 1}$

40. $f(x) = \dfrac{2x^2 - 5x - 2}{x - 2}$

41. $f(x) = \dfrac{0.5x^2 - 2x + 2}{x + 2}$

42. $f(x) = \dfrac{0.5x^2 - 5}{x - 3}$

43. $f(x) = \dfrac{x^2 + 2x + 1}{x - 1}$

44. $f(x) = \dfrac{2x^2 + 3x + 1}{x - 2}$

45. *Interpreting an Asymptote* Suppose that an insect population in millions is modeled by $f(x) = \dfrac{10x + 1}{x + 1}$, where $x \geq 0$ is in months.

    (a) Graph f in $[0, 14, 1]$ by $[0, 14, 1]$. Find the equation of the horizontal asymptote.

    (b) Determine the initial insect population.

    (c) What happens to the population after several months?

    (d) Interpret the horizontal asymptote.

46. *Interpreting an Asymptote* Suppose that the population of a species of fish in thousands is modeled by $f(x) = \dfrac{x + 10}{0.5x^2 + 1}$, where $x \geq 0$ is in years.

    (a) Graph f in $[0, 12, 1]$ by $[0, 12, 1]$. What is the horizontal asymptote?

    (b) Determine the initial population.

    (c) What happens to the population of this fish after many years?

    (d) Interpret the horizontal asymptote.

Rational Equations

Exercises 47–52: Solve the rational equation

    (a) symbolically,

    (b) graphically, and

    (c) numerically.

47. $\dfrac{2x}{x + 2} = 6$

48. $\dfrac{3x}{2x - 1} = 3$

49. $2 - \dfrac{5}{x} + \dfrac{2}{x^2} = 0$

50. $\dfrac{1}{x^2} + \dfrac{1}{x} = 2$

51. $\dfrac{1}{x + 1} + \dfrac{1}{x - 1} = \dfrac{1}{x^2 - 1}$

52. $\dfrac{4}{x - 2} = \dfrac{3}{x - 1}$

Exercises 53–62: Find all real solutions to the equation. Check your results.

53. $\dfrac{x + 1}{x - 5} = 0$

54. $\dfrac{x - 2}{x + 3} = 1$

55. $\dfrac{6(1 - 2x)}{x - 5} = 4$

56. $\dfrac{2}{5(2x + 5)} + 3 = -1$

57. $\dfrac{1}{x + 2} + \dfrac{1}{x} = 1$

58. $\dfrac{2x}{x - 1} = 5 + \dfrac{2}{x - 1}$

59. $\dfrac{1}{x} - \dfrac{2}{x^2} = 5$

60. $\dfrac{1}{x^2 - 2} = \dfrac{1}{x}$

61. $\dfrac{x^3 - 4x}{x^2 + 1} = 0$

62. $\dfrac{1}{x + 2} + \dfrac{1}{x + 3} = \dfrac{2}{x^2 + 5x + 6}$

Applications of Rational Functions

63. *Time Spent in Line* Suppose the average number of vehicles arriving at the main gate of an amusement park is equal to 10 per minute, while the average number of vehicles being admitted through the gate per minute is equal to x. Then the average waiting time in minutes for each vehicle at the gate can be computed by $f(x) = \dfrac{x - 5}{x^2 - 10x}$, where $x > 10$.

    (Source: F. Mannering.)

    (a) Estimate the admittance rate x that results in an average wait of 15 seconds.

    (b) If one attendant can serve 5 vehicles per minute, how many attendants are needed to keep the average wait to 15 seconds or less?

64. *Length of Lines* (Refer to Example 2.) Determine the traffic intensity x when the average number of vehicles in line equals 3.

65. *Construction* (Refer to Example 10.) Find possible dimensions for a box with a volume of 196 cubic inches, a surface area of 280 square inches, and a length that is twice the width.

66. *Minimizing Surface Area* An aluminum can is being designed to hold a volume of 100π cubic centimeters. Estimate the radius r and height h that result in the least amount of aluminum used in the construction of the can. See the accompanying figure.

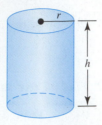

Geometry Review To review formulas for cylinders, see Chapter R (page R-4).

67. *Train Curves* When curves are designed for trains, sometimes the outer rail is elevated or banked, so that a locomotive can safely negotiate the curve at a higher speed. See the accompanying figure. Suppose a circular curve is being designed for 60 miles per hour. The rational function given by $f(x) = \dfrac{2540}{x}$ computes the elevation y in inches of the outer track for a curve with a radius of x feet, where $y = f(x)$. (Source: L. Haefner, *Introduction to Transportation Systems.*)

(a) Evaluate $f(400)$ and interpret its meaning.
(b) Graph f in [0, 600, 100] by [0, 50, 5]. Discuss how the elevation of the outer rail changes with the radius x.
(c) Interpret the horizontal asymptote.
(d) What radius is associated with an elevation of 12.7 inches?

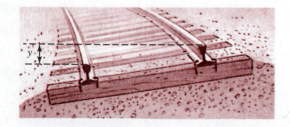

68. *Cost-Benefit* A cost-benefit function C computes the cost in millions of dollars of implementing a city recycling project when x percent of the citizens participate, where $C(x) = \dfrac{1.2x}{100 - x}$.

(a) Graph C in [0, 100, 10] by [0, 10, 1]. Interpret the graph as x approaches 100.
(b) If 75% participation is expected, determine the cost for the city.

(c) The city plans to spend $5 million on this recycling project. Estimate graphically the percentage of participation that can be expected.
(d) Solve part (c) symbolically.

69. *Mortality Rates* The table contains incidence ratios by age for deaths from coronary heart disease (CHD) and lung cancer (LC) when comparing smokers (21–39 cigarettes per day) to nonsmokers. For example, the incidence ratio of 10 means that smokers are 10 times more likely than nonsmokers to die from lung cancer between the ages of 55 and 64. If the incidence ratio is x, then the percentage P of deaths caused by smoking can be calculated using the rational function $P(x) = \dfrac{x - 1}{x}$, where the output is in decimal form. (Source: A. Walker, *Observation and Inference.*)

Age	CHD	LC
55–64	1.9	10
65–74	1.7	9

(a) Calculate the percentage of lung cancer deaths in the age group 65–74 that can be attributed to smoking.
(b) Determine the percentage of deaths from coronary heart disease in the age group from 55–64 that can be attributed to smoking.

70. *Probability* A container holds x balls numbered 1 through x. Only one ball has the winning number.
(a) Determine a rational function f that computes the probability or likelihood of *not* drawing the winning ball.
(b) What is the domain of f?
(c) What happens to the probability of *not* drawing the winning ball as the number of balls increases?
(d) Interpret the horizontal asymptote of the graph of f.

71. *Braking Distance* The **grade** x of a hill is a measure of its steepness. For example, if a road rises 10 feet for every 100 feet of horizontal distance, then it has an uphill grade of $x = \dfrac{10}{100}$ or 10%. See the accompanying figure. Grades are typically kept quite small—usually less than 10%. The braking (or stopping) distance D for a car traveling at 50 miles per hour on a wet, uphill grade is given by $D(x) = \dfrac{2500}{30(0.3 + x)}$. (Source: L. Haefner.)

(a) Evaluate $D(0.05)$ and interpret the result.
(b) Describe what happens to the braking distance as the hill becomes steeper. Does this agree with your driving experience?
(c) Estimate the grade associated with a braking distance of 220 feet.

72. *Braking Distance* (Refer to the previous exercise.) If a car is traveling 50 miles per hour downhill, then its braking distance on a wet pavement is given by $D(x) = \dfrac{2500}{30(0.3 + x)}$, where $x < 0$ for a downhill grade.
(a) Evaluate $D(-0.1)$ and interpret the result.
(b) What happens to the braking distance as the downhill grade becomes steeper? Does this agree with your driving experience?
(c) The graph of D has a vertical asymptote at $x = -0.3$. Give the physical significance of this asymptote.
(d) Estimate the grade associated with a braking distance of 350 feet.

73. *Slippery Roads* If a car is moving at 50 miles per hour on a level highway, then its braking (or stopping) distance depends on the road conditions. This distance in feet can be computed by $D(x) = \dfrac{2500}{30x}$, where x is the coefficient of friction between the tires and the road and $0 < x \le 1$. A smaller value of x indicates that the road is more slippery.
(a) Identify and interpret the vertical asymptote.
(b) Estimate the coefficient of friction associated with a braking distance of 340 feet.

74. *Concentration of a Drug* The concentration of a drug in a medical patient's bloodstream is given by $f(t) = \dfrac{5}{t^2 + 1}$, where the input $t \ge 0$ is in hours and the output is in milligrams per liter.
(a) Does the concentration of the drug increase or decrease? Explain.
(b) The patient should not take a second dose until the concentration is below 1.5 milligrams per liter. How long should the patient wait before taking a second dose?

Variation

Exercises 75–78: Find the constant of proportionality k for the given conditions.

75. $y = \dfrac{k}{x}$, and $y = 2$ when $x = 3$

76. $y = \dfrac{k}{x^2}$, and $y = \dfrac{1}{4}$ when $x = 8$

77. $y = kx^3$, and $y = 64$ when $x = 2$

78. $y = kx^{3/2}$, and $y = 96$ when $x = 16$

Exercises 79–82: Solve the variation problem.

79. Suppose T varies directly as the $\dfrac{3}{2}$ power of x. When $x = 4$, $T = 20$. Find T when $x = 16$.

80. Suppose y varies directly as the second power of x. When $x = 3$, $y = 10.8$. Find y when $x = 1.5$.

81. Let y be inversely proportional to x. When $x = 6$, $y = 5$. Find y when $x = 15$.

82. Let z be inversely proportional to the third power of t. When $t = 5$, $z = 0.08$. Find z when $t = 2$.

Exercises 83–86: Assume that the constant of proportionality is positive.

83. Let y be inversely proportional to x. If x doubles, what happens to y?

84. Let y vary inversely as the second power of x. If x doubles, what happens to y?

85. Suppose y varies directly as the third power of x. If x triples, what happens to y?

86. Suppose y is directly proportional to the second power of x. If x is halved, what happens to y?

Exercises 87 and 88: The data in the table satisfy the equation $y = kx^n$, where n is a positive integer. Determine k and n.

87.

x	2	3	4	5
y	2	4.5	8	12.5

88.

x	3	5	7	9
y	32.4	150	411.6	874.8

Exercises 89 and 90: The data in the table satisfy the equation $y = \dfrac{k}{x^n}$, where n is a positive integer. Determine k and n.

89.

x	2	3	4	5
y	1.5	1	0.75	0.6

90.

x	2	4	6	8
y	9	2.25	1	0.5625

91. *Allometric Growth*   The weight y of a fiddler crab is directly proportional to the 1.25 power of the weight x of its claws. A crab with a body weight of 1.9 grams has claws weighing 1.1 grams. Estimate the weight of a fiddler crab with claws weighing 0.75 gram.   (Source: D. Brown.)

92. *Gravity*   The weight of an object varies inversely as the second power of the distance from the center of Earth. The radius of Earth is approximately 4000 miles. If a person weighs 160 pounds on Earth's surface, what would this individual weigh 8000 miles above the surface of Earth?

93. *Hubble Telescope*   The brightness or intensity of starlight varies inversely as the square of its distance from Earth. The Hubble Telescope can see stars whose intensities are $\frac{1}{50}$ of the faintest star now seen by ground-based telescopes. Determine how much farther the Hubble Telescope can see into space than ground-based telescopes.   (Source: National Aeronautics and Space Administration.)

94. *Volume*   The volume V of a cylinder with a fixed height is directly proportional to the square of its radius r. If a cylinder with a radius of 10 inches has a volume of 200 cubic inches, what is the volume of a cylinder with the same height and a radius of 5 inches?

95. *Electrical Resistance*   The electrical resistance R of a wire varies inversely as the square of its diameter d. If a 25-foot wire with a diameter of 2 millimeters has a resistance of 0.5 ohm, find the resistance of a wire having the same length and a diameter of 3 millimeters.

96. *Strength of a Beam*   The strength of a rectangular wood beam varies directly as the square of the depth of its cross section. If a beam with a depth of 3.5 inches can support 1000 pounds, how much weight can the same type of beam hold if its depth is 10.5 inches?

Writing about Mathematics

97. Suppose that the symbolic representation of a rational function f is given.
   (a) Explain how to find any vertical or horizontal asymptotes of the graph of f.
   (b) Discuss what a horizontal asymptote represents.

98. Discuss how to find the domain of a rational function symbolically and graphically.

4.6 POLYNOMIAL AND RATIONAL INEQUALITIES

Polynomial Inequalities • Rational Inequalities

Introduction

Waiting in line has become part of almost everyone's life. When people arrive randomly at a line, rational functions can be used to estimate the average number of people standing in line. For example, if an attendant at a ticket booth can wait on 30 customers per hour, and if customers arrive at an average rate of x per hour, then the average number of customers waiting in line is computed by

$$f(x) = \frac{x^2}{900 - 30x},$$

where $0 \le x < 30$. For example, $f(28) \approx 13$ indicates that if customers arrive, on average, at 28 per hour, then the average number of people in line is 13. If a line length of 8 customers or fewer is acceptable, then $f(x)$ can be used to estimate customer arrival rates x that one attendant can accommodate by solving the rational inequality

$$\frac{x^2}{900 - 30x} \leq 8.$$

In this section we discuss solving polynomial inequalities, and then use similar techniques to solve rational inequalities.   (Source: N. Garber and L. Hoel, *Traffic and Highway Engineering.*)

Polynomial Inequalities

In Section 3.3 a strategy for solving quadratic inequalities was presented. This strategy involves first finding boundary numbers (x-values) where equality holds. Once the boundary numbers are known, a graph or a table of test values can be used to determine the intervals where inequality holds. This strategy can be applied to other types of inequalities. For example, consider the graph of $p(x) = -x^4 + 5x^2 - 4$ with boundary numbers -2, -1, 1, and 2 shown in Figure 4.117. The solution set to the inequality $p(x) > 0$ can be written in interval notation as $(-2, -1) \cup (1, 2)$ and corresponds to where the graph is above the x-axis. (The symbol $\cup$ means union and indicates that x can be in either interval.) The intervals $(-2, -1)$ and $(1, 2)$ are shaded on the x-axis in Figure 4.118. Similarly, the solution set to the inequality $p(x) < 0$ is $(-\infty, -2) \cup (-1, 1) \cup (2, \infty)$ and corresponds to where the graph of $y = p(x)$ is below the x-axis. These intervals are shaded on the x-axis in Figure 4.119.

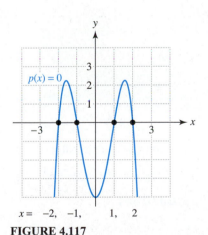

$x = -2, -1, \quad 1, \quad 2$

FIGURE 4.117

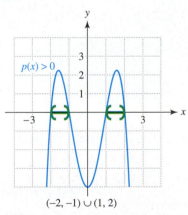

$(-2, -1) \cup (1, 2)$

FIGURE 4.118

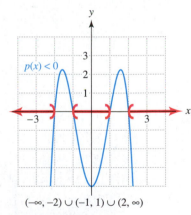

$(-\infty, -2) \cup (-1, 1) \cup (2, \infty)$

FIGURE 4.119

Polynomial inequalities can also be solved symbolically. For example, to solve $-x^4 + 5x^2 - 4 > 0$, begin by finding the boundary numbers.

$$-x^4 + 5x^2 - 4 = 0 \qquad \text{Replace} > \text{with} =.$$
$$x^4 - 5x^2 + 4 = 0 \qquad \text{Multiply by } -1.$$
$$(x^2 - 4)(x^2 - 1) = 0 \qquad \text{Factor.}$$
$$x^2 - 4 = 0 \quad \text{or} \quad x^2 - 1 = 0 \qquad \text{Zero-product property}$$
$$x = \pm 2 \quad \text{or} \quad x = \pm 1 \qquad \text{Square root property}$$

The boundary numbers $-2, -1, 1$, and 2 separate the number line into five intervals:

$$(-\infty, -2), (-2, -1), (-1, 1), (1, 2), \text{ and } (2, \infty),$$

as illustrated in Figure 4.120.

Algebra Review
To review interval notation, see
Section 2.4.

FIGURE 4.120

The polynomial $p(x) = -x^4 + 5x^2 - 4$ is either always positive or always negative on each of these intervals. To determine which is the case, we can choose one test value (x-value) from each interval and substitute this test value into $p(x)$. From Table 4.12 we see that the test value $x = -1.5$ results in

$$p(-1.5) = -(1.5)^4 + 5(1.5)^2 - 4 = 2.1875 > 0.$$

Since $x = -1.5$ is in the interval $(-2, -1)$, it follows that $p(x)$ is positive on this interval. The sign of $p(x)$ on the other intervals is determined similarly. Thus the solution set to $p(x) = -x^4 + 5x^2 - 4 > 0$ is $(-2, -1) \cup (1, 2)$, which is consistent with the graphical solution in Figure 4.118.

TABLE 4.12

Interval	Test Value x	$-x^4 + 5x^2 - 4$	Positive or Negative?
$(-\infty, -2)$	-3	-40	Negative
$(-2, -1)$	-1.5	2.1875	Positive
$(-1, 1)$	0	-4	Negative
$(1, 2)$	1.5	2.1875	Positive
$(2, \infty)$	3	-40	Negative

These symbolic and graphical procedures are now summarized verbally.

Solving polynomial inequalities

STEP 1: If necessary, write the inequality as $p(x) < 0$, where $p(x)$ is a polynomial and $<$ may be replaced by $>$, $\leq$, or $\geq$.

STEP 2: Solve $p(x) = 0$ either symbolically or graphically. The solutions are called boundary numbers.

STEP 3: Use the boundary numbers to separate the number line into disjoint intervals. On each interval, $p(x)$ is either always positive or always negative.

STEP 4: To solve the inequality, either make a table of test values for $p(x)$ or use a graph of $p(x)$. For example, the solution set for $p(x) < 0$ corresponds to intervals where test values result in negative outputs or to intervals where the graph of $p(x)$ is below the x-axis.

EXAMPLE 1 *Solving a polynomial inequality*

Solve $x^3 \geq 2x^2 + 3x$ symbolically and graphically.

SOLUTION *Symbolic Solution*

STEP 1: Begin by writing the inequality as $x^3 - 2x^2 - 3x \geq 0$.

STEP 2: Replace the $\geq$ symbol with an equals sign and solve the resulting equation.

$$x^3 - 2x^2 - 3x = 0 \qquad \text{Replace } \geq \text{ with } =.$$
$$x(x^2 - 2x - 3) = 0 \qquad \text{Factor out } x.$$
$$x(x + 1)(x - 3) = 0 \qquad \text{Factor the trinomial.}$$
$$x = 0 \quad \text{or} \quad x = -1 \quad \text{or} \quad x = 3 \qquad \text{Zero-product property}$$

Algebra Review

To review factoring, see Chapter R, (page R-27).

The boundary numbers are -1, 0, and 3.

STEP 3: The boundary numbers separate the number line into four disjoint intervals:

$$(-\infty, -1), (-1, 0), (0, 3), \text{ and } (3, \infty),$$

which is illustrated in Figure 4.121.

Graphing Calculator Help

To make a table like Figure 4.122, see Appendix B (page AP-11).

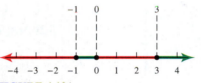

FIGURE 4.121

STEP 4: In Table 4.13 the expression $x^3 - 2x^2 - 3x$ is evaluated at test values from each interval. The solution set is $[-1, 0] \cup [3, \infty)$. In Figure 4.122 a graphing calculator has been used to evaluate the same test values. (*Note*: The boundary numbers are included in the solution set because the inequality involves $\geq$ rather than $>$.)

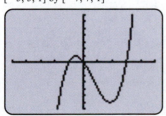

FIGURE 4.122

$[-5, 5, 1]$ by $[-7, 7, 1]$

TABLE 4.13

Interval	Test Value x	$x^3 - 2x^2 - 3x$	Positive or Negative?
$(-\infty, -1)$	-2	-10	Negative
$(-1, 0)$	-0.5	0.875	Positive
$(0, 3)$	1	-4	Negative
$(3, \infty)$	4	20	Positive

FIGURE 4.123

Graphical Solution Graph $Y_1 = X^3 - 2X^2 - 3X$, as shown in Figure 4.123. The zeros or x-intercepts are located at -1, 0, and 3. The graph of Y_1 is positive (or above the x-axis) when $-1 < x < 0$ or when $3 < x < \infty$. If we include the boundary numbers, this result agrees with the symbolic solution.

Rational Inequalities

In the introduction we discussed how a rational inequality can be used to estimate the number of people standing in line at a ticket booth. In the next example, this inequality is solved graphically.

EXAMPLE 2 *Modeling customers in a line*

A ticket booth attendant can wait on 30 customers per hour. To keep the time waiting in line reasonable, the line length should not exceed 8 customers. Solve the inequality $\dfrac{x^2}{900 - 30x} \le 8$ to determine the rates x at which customers can arrive before a second attendant is needed. Note that the x-values are limited to $0 \le x < 30$.

SOLUTION

[0, 30, 5] by [0, 10, 2]

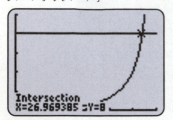

Intersection
X=26.969385 Y=8

FIGURE 4.124

Graphing Calculator Help

To find a point of intersection, see Appendix B (page AP-11).

Graph $Y_1 = X^2/(900 - 30X)$ and $Y_2 = 8$ for $0 \le x \le 30$, as shown in Figure 4.124. The only point of intersection on this interval is near $(26.97, 8)$. The graph of Y_1 is below the graph of Y_2 for x-values to the left of this point. We conclude that if the arrival rate is about 27 customers per hour or less, then the line length does not exceed 8 customers. If the arrival rate is more than 27 customers per hour, a second ticket booth attendant is needed.

To solve rational inequalities symbolically, we can use the same basic techniques that we used to solve polynomial inequalities, with one important modification: boundary numbers also occur at x-values where the denominator of any rational expression in the equation equals 0. The following steps can be used to solve a rational inequality symbolically.

Solving rational inequalities symbolically

STEP 1: If necessary, write the inequality in the form $\dfrac{p(x)}{q(x)} > 0$, where $p(x)$ and $q(x)$ are polynomials. Note that $>$ may be replaced by $<$, $\le$, or $\ge$.

STEP 2: Solve the equations $p(x) = 0$ and $q(x) = 0$. The solutions are boundary numbers.

STEP 3: Use the boundary numbers to separate the number line into disjoint intervals. On each interval, $\dfrac{p(x)}{q(x)}$ is either always positive or always negative.

STEP 4: Use a table of test values to solve the inequality in **STEP 1**.

EXAMPLE 3 *Solving a rational inequality*

Solve $\dfrac{2 - x}{2x} > 0$ symbolically. Support your answer graphically.

SOLUTION

Symbolic Solution Since the inequality is already in the form $\dfrac{p(x)}{q(x)} > 0$, **STEP 1** is unnecessary.

STEP 2: Set the numerator and the denominator equal to 0 and solve.

Numerator	Denominator
$2 - x = 0$	$2x = 0$
$x = 2$	$x = 0$

STEP 3: The boundary numbers are 0 and 2, which separate the number line into three disjoint intervals: $(-\infty, 0)$, $(0, 2)$, and $(2, \infty)$.

STEP 4: Table 4.14 shows a table of test values. We can see that the expression is positive between the two boundary numbers or when $0 < x < 2$. In interval notation the solution set is $(0, 2)$.

TABLE 4.14

Interval	Test Value x	$(2 - x)/(2x)$	Positive or Negative?
$(-\infty, 0)$	-1	-1.5	Negative
$(0, 2)$	1	0.5	Positive
$(2, \infty)$	4	-0.25	Negative

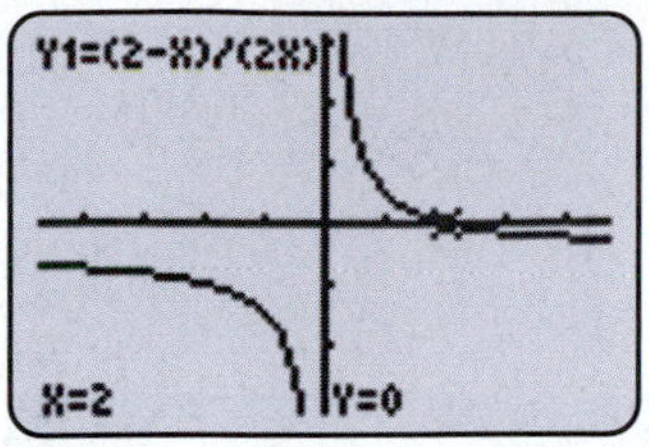

$[-4.7, 4.7, 1]$ by $[-3.1, 3.1, 1]$

FIGURE 4.125

Graphical Solution   Graph $Y_1 = (2 - X)/(2X)$, as shown in Figure 4.125. The graph has a vertical asymptote at $x = 0$ and an x-intercept at $x = 2$. Between these boundary numbers the graph of Y_1 is positive (or above the x-axis.) This agrees with our symbolic solution.

EXAMPLE 4   *Solving a rational inequality symbolically*

Solve $\dfrac{1}{x} \le \dfrac{2}{x + 1}$.

SOLUTION   **STEP 1:**   Begin by writing the inequality in the form $\dfrac{p(x)}{q(x)} \le 0$.

Algebra Review
To review subtraction of rational expressions, see Chapter R (page R-36).

$$\frac{1}{x} - \frac{2}{x + 1} \le 0 \qquad \text{Subtract } \frac{2}{x + 1}.$$

$$\frac{1}{x} \cdot \frac{(x + 1)}{(x + 1)} - \frac{2}{x + 1} \cdot \frac{x}{x} \le 0 \qquad \text{Common denominator is } x(x + 1).$$

$$\frac{x + 1}{x(x + 1)} - \frac{2x}{x(x + 1)} \le 0 \qquad \text{Multiply.}$$

$$\frac{1 - x}{x(x + 1)} \le 0 \qquad \text{Subtract and combine terms.}$$

STEP 2:   Find the zeros of the numerator and the denominator.

Numerator	*Denominator*
$1 - x = 0$	$x(x + 1) = 0$
$x = 1$	$x = 0$ or $x = -1$

STEP 3:   The boundary numbers are -1, 0, and 1, which separate the number line into four disjoint intervals: $(-\infty, -1)$, $(-1, 0)$, $(0, 1)$, and $(1, \infty)$.

STEP 4:   Table 4.15 can be used to solve the inequality $\dfrac{1 - x}{x(x + 1)} \le 0$. The solution set is $(-1, 0) \cup [1, \infty)$. (*Note:* The boundary numbers -1 and 0 are not included in the solution set because the given inequality is undefined when $x = -1$ or $x = 0$.)

TABLE 4.15

Interval	Test Value x	$(1 - x)/(x(x + 1))$	Positive or Negative?
$(-\infty, -1)$	-2	1.5	Positive
$(-1, 0)$	-0.5	-6	Negative
$(0, 1)$	0.5	$0.\overline{6}$	Positive
$(1, \infty)$	2	$-0.1\overline{6}$	Negative

Putting it all Together

4.6

*T*he following table outlines basic concepts for solving polynomial and rational inequalities.

Concept	Description
Polynomial inequality	Can be written as $p(x) < 0$, where $p(x)$ is a polynomial and $<$ may be replaced by $>$, $\leq$, or $\geq$. **Examples:** $x^3 - x \leq 0$ $\qquad\qquad 2x^4 - 3x^2 \geq 5x + 1$
Solving a polynomial inequality	Follow the steps to solve a polynomial inequality presented in this section. Either graphical or symbolic methods can be used. **Example:** A graph of $x^3 - 2x^2 - 5x + 6$ is shown. The boundary numbers are -2, 1, and 3. The solution set to $x^3 - 2x^2 - 5x + 6 > 0$ is $(-2, 1) \cup (3, \infty)$. $[-5, 5, 1]$ by $[-10, 10, 2]$
Rational inequality	Can be written as $\dfrac{p(x)}{q(x)} < 0$, where $p(x)$ and $q(x) \neq 0$ are polynomials and $<$ may be replaced by $>$, $\leq$, or $\geq$. **Examples:** $\dfrac{x - 3}{x + 2} \geq 0$ $\qquad\qquad 2x - \dfrac{2}{x^2 - 1} > 5$
Solving a rational inequality	Follow the steps to solve a rational inequality presented in this section. Either graphical or symbolic methods can be used. **Example:** A graph of $\dfrac{x - 1}{2x + 1}$ is shown. Boundary numbers occur where either the numerator or denominator equals zero: $x = 1$ or $x = -\dfrac{1}{2}$. The solution set to $\dfrac{x - 1}{2x + 1} < 0$ is $\left(-\dfrac{1}{2}, 1\right)$. $[-4.7, 4.7, 1]$ by $[-3.1, 3.1, 1]$

4.6 EXERCISES

Graphical Solutions

Exercises 1–6: Use the graph of f to solve the equation and inequalities.

    (a) $f(x) = 0$
    (b) $f(x) > 0$
    (c) $f(x) < 0$

1.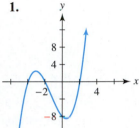

2.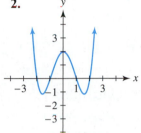

3.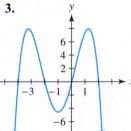

4.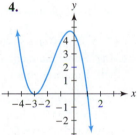

5.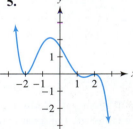

6.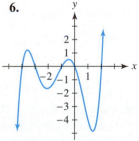

Exercises 7–12: Use the graph of the rational function f to complete the following.

    (a) Identify x-values where either $f(x)$ is undefined or $f(x) = 0$.
    (b) Solve $f(x) > 0$.
    (c) Solve $f(x) < 0$.

7.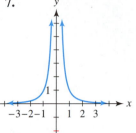

8.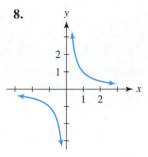

9.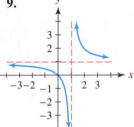

10.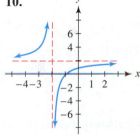

11.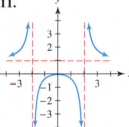

12.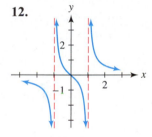

Polynomial Inequalities

Exercises 13–18: Solve the polynomial inequality

    (a) symbolically, and
    (b) graphically.

13. $x^3 - x > 0$

14. $8x^3 < 27$

15. $x^3 + x^2 \geq 2x$

16. $2x^3 \leq 3x^2 + 5x$

17. $x^4 - 13x^2 + 36 < 0$

18. $4x^4 - 5x^2 - 9 \geq 0$

Exercises 19–24: Solve the polynomial inequality.

19. $7x^4 \geq 14x^2$

20. $3x^4 - 4x^2 < 7$

21. $(x - 1)(x - 2)(x + 2) \geq 0$

22. $-(x + 1)^2(x - 2) \geq 0$

23. $2x^4 + 2x^3 \leq 12x^2$

24. $x^3 + 6x^2 + 9x > 0$

Exercises 25–28: Solve the polynomial inequality graphically.

25. $x^3 - 7x^2 + 14x \leq 8$

26. $2x^3 + 3x^2 - 3x < 2$

27. $3x^4 - 7x^3 - 2x^2 + 8x > 0$

28. $x^4 - 5x^3 \leq 5x^2 + 45x + 36$

Rational Inequalities

Exercises 29–34: Solve the rational inequality
  (a) symbolically, and
  (b) graphically.

29. $\dfrac{1}{x} < 0$

30. $\dfrac{1}{x^2} > 0$

31. $\dfrac{4}{x + 3} \geq 0$

32. $\dfrac{x - 1}{x + 1} < 0$

33. $\dfrac{5}{x^2 - 4} < 0$

34. $\dfrac{x}{x^2 - 1} \geq 0$

Exercises 35–42: Solve the rational inequality.

35. $\dfrac{(x + 1)^2}{x - 2} \leq 0$

36. $\dfrac{2x}{(x - 2)^2} > 0$

37. $\dfrac{3 - 2x}{1 + x} < 0$

38. $\dfrac{x + 1}{4 - 2x} \geq 1$

39. $\dfrac{1}{x - 3} \leq \dfrac{5}{x - 3}$

40. $\dfrac{3}{2 - x} > \dfrac{x}{2 + x}$

41. $2 - \dfrac{5}{x} + \dfrac{2}{x^2} \geq 0$

42. $\dfrac{1}{x - 1} + \dfrac{1}{x + 1} > \dfrac{3}{4}$

Applications

43. *Waiting in Line* (Refer to Example 2.) A parking lot attendant can wait on 40 cars per hour. If cars arrive randomly at a rate of x cars per hour, then the average line length to enter the ramp is given by $f(x) = \dfrac{x^2}{1600 - 40x}$, where $0 \leq x < 40$.
  (a) Solve the inequality $f(x) \leq 8$.
  (b) Interpret your answer from part (a).

44. *Time Spent in Line* If a parking ramp attendant can wait on 3 vehicles per minute, and vehicles are leaving the ramp at x vehicles per minute, then the average wait in minutes for a car trying to exit is given by $f(x) = \dfrac{1}{3 - x}$.
  (a) Solve the three-part inequality
$$5 \leq \dfrac{1}{3 - x} \leq 10.$$
  (b) Interpret your results from part (a).

45. *Slippery Roads* The coefficient of friction x measures the friction between the tires of a car and the road, where $0 < x \leq 1$. A smaller value of x indicates that the road is more slippery. If a car is traveling at 60 miles per hour, then the braking distance D in feet is given by $D(x) = \dfrac{120}{x}$.
  (a) What happens to the braking distance as the coefficient of friction becomes smaller?
  (b) Find values for the coefficient of friction x that correspond to a braking distance of 400 feet or more.

46. *Average Temperature*   The monthly average high temperature in degrees Fahrenheit at Daytona Beach, Florida, can be approximated by

$$f(x) = 0.0145x^4 - 0.426x^3 + 3.53x^2 - 6.22x + 72.0,$$

where $x = 1$ corresponds to January, $x = 2$ to February, and so on. Estimate graphically when the monthly average high temperature is 75°F or more.

47. *Geometry*   A cubical box is being manufactured to hold 213 cubic inches. If this measurement can vary between 212.8 cubic inches and 213.2 cubic inches, inclusively, by how much can the length x of a side of the cube vary?

48. *Construction*   A cylindrical aluminum can is being manufactured so that its height h is 8 centimeters more than its radius r. Estimate values for the radius to the nearest hundredth that result in the can having a volume between 1000 and 1500 cubic centimeters.

Geometry Review   To review formulas for cylinders, see Chapter R (page R-5).

Writing about Mathematics

49. Describe the steps to graphically solve a polynomial inequality in the form $p(x) > 0$.

50. Describe the steps to symbolically solve a rational inequality in the form $f(x) > 0$.

CHECKING BASIC CONCEPTS FOR SECTIONS 4.5 AND 4.6

1. Graph $f(x) = \dfrac{2x - 1}{x + 2}$ using a graphing calculator.

   (a) Find the domain of f.

   (b) Identify any vertical or horizontal asymptotes.

   (c) Sketch a graph of f that includes all asymptotes.

2. Solve $\dfrac{3x - 1}{1 - x} = 1$.

3. Solve $2x^3 + x^2 - 6x < 0$. Write the solution set in interval notation.

4. Solve $\dfrac{x^2 - 1}{x + 2} \geq 0$. Write the solution set in interval notation.

5. *Time Spent in Line*   Suppose a parking ramp attendant can wait on 4 vehicles per minute and vehicles are leaving the ramp randomly at an average rate of x vehicles per minute. Then the average time T in minutes spent waiting in line and paying the attendant is given by

$$T(x) = \frac{1}{4 - x},$$

where $x < 4$.   (Source: N. Garber and L. Hoel, *Traffic and Highway Engineering.*)

   (a) Evaluate $T(2)$ and interpret the result.

   (b) Graph T in [0, 4, 1] by [0, 4, 1].

   (c) What happens to the waiting time as x increases from 0 to 4?

   (d) Find x if the waiting time is 5 minutes.

4.7   POWER FUNCTIONS AND RADICAL EQUATIONS

Rational Exponents and Radical Notation • Power Functions and Models • Equations Involving Radicals • Power Regression

Introduction

Johannes Kepler (1571–1630) was the first to recognize that the orbits of planets are elliptical, rather than circular. He also found that a power function models the relationship between a planet's distance from the sun and its period of revolution. Table 4.16 lists the average distance x from the sun and the time y in years for several planets to orbit the sun. The distance x has been normalized so that Earth is one unit away from the sun. For example, Jupiter is 5.2 times farther from the sun than Earth and requires 11.9 years to orbit the sun. (Source: C. Ronan, *The Natural History of the Universe.*)

TABLE 4.16

Planet	x (distance)	y (period)
Mercury	0.387	0.241
Venus	0.723	0.615
Earth	1.00	1.00
Mars	1.52	1.88
Jupiter	5.20	11.9
Saturn	9.54	29.5

A scatterplot of the data in Table 4.16 is shown in Figure 4.126. To model this data we might try a polynomial, such as $f(x) = x$ or $g(x) = x^2$. Figure 4.127 shows that $f(x) = x$ increases too slowly, and $g(x) = x^2$ increases too fast. To model this data, a new type of function is required. That is, we need a function in the form $h(x) = x^b$, where $1 < b < 2$. Polynomials only allow the exponent b to be a nonnegative integer, whereas *power functions* allow b to be any real number. See Figure 4.128.

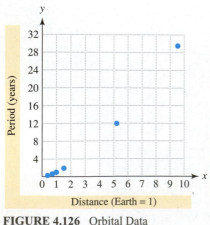

FIGURE 4.126   Orbital Data

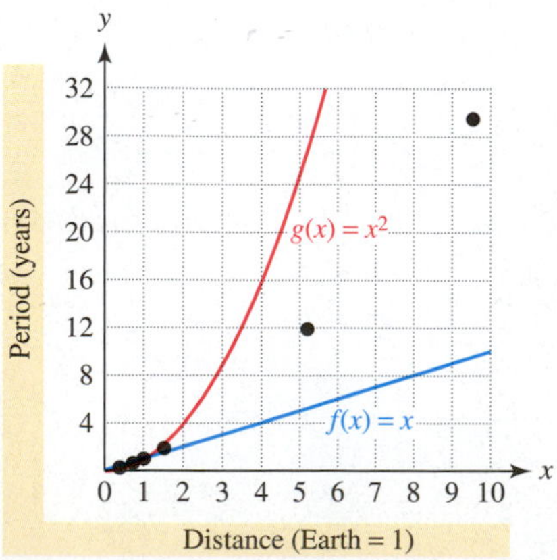

FIGURE 4.127   Polynomial Model

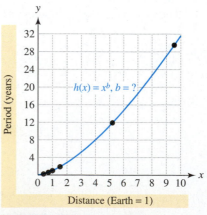

FIGURE 4.128   Power Model

In this section we discuss power functions and use a power function in Example 6 to model Kepler's data. Because power functions use rational numbers as exponents, we begin by reviewing properties of exponents and radical notation.

Rational Exponents and Radical Notation

The following properties can be used to simplify expressions with rational exponents.

Properties of exponents

Let m and n be positive integers, and r and p be real numbers. Asume that b is a real number and that $\sqrt[n]{b}$ is defined.

	Property	**Example**
1.	$b^{m/n} = (b^m)^{1/n} = (b^{1/n})^m$	$4^{3/2} = (4^3)^{1/2} = (4^{1/2})^3 = 2^3 = 8$
2.	$b^{m/n} = \sqrt[n]{b^m} = \left(\sqrt[n]{b}\right)^m$	$8^{2/3} = \sqrt[3]{8^2} = \left(\sqrt[3]{8}\right)^2 = 2^2 = 4$
3.	$(b^r)^p = b^{rp}$	$(2^{3/2})^4 = 2^6 = 64$
4.	$b^{-r} = \dfrac{1}{b^r}$	$4^{-1/2} = \dfrac{1}{4^{1/2}} = \dfrac{1}{2}$
5.	$b^r b^p = b^{r+p}$	$3^{5/2} \cdot 3^{3/2} = 3^{(5/2)+(3/2)} = 3^4 = 81$
6.	$\dfrac{b^r}{b^p} = b^{r-p}$	$\dfrac{5^{5/4}}{5^{3/4}} = 5^{(5/4)-(3/4)} = 5^{1/2}$

Algebra Review

To review integer exponents, see Chapter R (page R-14).

EXAMPLE 1 *Applying properties of exponents*

Simplify each expression by hand.

(a) $16^{3/4}$ **(b)** $\dfrac{-4^{1/3}}{4^{5/6}}$ **(c)** $27^{-2/3} \cdot 27^{1/3}$ **(d)** $(5^{3/4})^{2/3}$ **(e)** $(-125)^{-4/3}$

SOLUTION **(a)** $16^{3/4} = (\sqrt[4]{16})^3 = (2)^3 = 8$

(b) $\dfrac{-4^{1/3}}{4^{5/6}} = -4^{(1/3)-(5/6)} = -4^{-1/2} = -\dfrac{1}{\sqrt{4}} = -\dfrac{1}{2}$

(c) $27^{-2/3} \cdot 27^{1/3} = 27^{(-2/3)+(1/3)} = 27^{-1/3} = \dfrac{1}{\sqrt[3]{27}} = \dfrac{1}{3}$

(d) $(5^{3/4})^{2/3} = 5^{(3/4)(2/3)} = 5^{1/2}$ or $\sqrt{5}$

(e) $(-125)^{-4/3} = \dfrac{1}{(\sqrt[3]{-125})^4} = \dfrac{1}{(-5)^4} = \dfrac{1}{625}$

EXAMPLE 2 *Writing radicals with rational exponents*

Use positive rational exponents to write each expression.
(a) $\sqrt{x}$ **(b)** $\sqrt[3]{x^2}$ **(c)** $(\sqrt[4]{z})^{-5}$ **(d)** $\sqrt{\sqrt[3]{y} \cdot \sqrt[4]{y}}$

SOLUTION **(a)** $\sqrt{x} = x^{1/2}$ **(b)** $\sqrt[3]{x^2} = (x^2)^{1/3} = x^{2/3}$

(c) $(\sqrt[4]{z})^{-5} = (z^{1/4})^{-5} = z^{-5/4} = \dfrac{1}{z^{5/4}}$

(d) $\sqrt{\sqrt[3]{y} \cdot \sqrt[4]{y}} = \left(y^{1/3} \cdot y^{1/4}\right)^{1/2} = \left(y^{(1/3)+(1/4)}\right)^{1/2} = \left(y^{7/12}\right)^{1/2} = y^{7/24}$

■ **MAKING CONNECTIONS**

Rational Exponents and Radical Notation Any expression that is written with (noninteger) rational exponents can be written in radical notation, and any expression written in radical notation can be written with rational exponents. Normally we choose the form that is most convenient for a given situation.

Power Functions and Models

Functions with real number exponents are often used to model physical characteristics of living organisms. For example, taller animals tend to be heavier than shorter animals. There is a relationship between an animal's height and weight. (See also Example 4.) This area of study in biology is called *allometry,* which attempts to model the relative sizes of different characteristics of an organism.

Power functions have real number exponents, and a special type of power function is a root function. These functions are defined in the following.

Power function

A function f given by $f(x) = x^b$, where b is a constant, is a **power function**. If $b = \dfrac{1}{n}$ for some integer $n \geq 2$, then f is a **root function** given by $f(x) = x^{1/n}$, or equivalently, $f(x) = \sqrt[n]{x}$.

Symbolic representations of power functions include

$$f_1(x) = x^2, \qquad f_2(x) = x^\pi, \qquad f_3(x) = x^{0.4}, \qquad \text{and} \qquad f_4(x) = \sqrt[3]{x^2}.$$

Frequently, the domain of a power function f is restricted to nonnegative numbers. Suppose the rational number $\dfrac{p}{q}$ is written in lowest terms. Then the domain of $f(x) = x^{p/q}$ is all real numbers whenever q is odd and all nonnegative real numbers whenever q is even. If b is an irrational number, the domain of $f(x) = x^b$ is all nonnegative real numbers. For example, the domain of $f(x) = x^{1/3}$ $\left(f(x) = \sqrt[3]{x}\right)$ is all real numbers, whereas the domains of $g(x) = x^{1/2}$ $\left(g(x) = \sqrt{x}\right)$ and $h(x) = x^{\sqrt{2}}$ are all nonnegative numbers.

Graphs of three common power functions are shown in Figures 4.129–4.131.

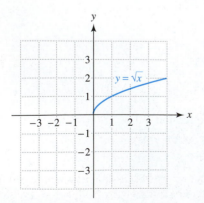

FIGURE 4.129 Square Root

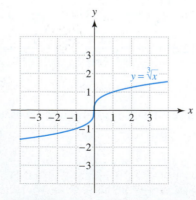

FIGURE 4.130 Cube Root

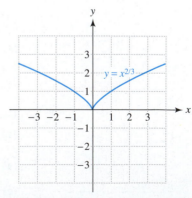

FIGURE 4.131 Cube Root Squared

EXAMPLE 3 *Graphing power functions*

Graph $f(x) = x^b$, $b = 0.3$, 1, and 1.7, for $x \geq 0$. Discuss the effect that b has on the graph of f.

SOLUTION The graphs of $y = x^{0.3}$, $y = x^1$, and $y = x^{1.7}$ are shown in Figure 4.132. Larger values of b cause the graph of f to increase faster.

Graphing Calculator Help

To enter $x^{0.3}$ use X^0.3.

[0, 9, 1] by [0, 6, 1]

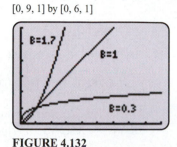

FIGURE 4.132

Note: Graphs of power functions with negative exponents are investigated in the Extended and Discovery Exercises from this section.

In the next example we model the size of the wings of birds.

EXAMPLE 4 *Modeling wing size of a bird*

Heavier birds have larger wings with more surface areas than do lighter birds. For some species of birds, this relationship can be modeled by $S(w) = 0.2w^{2/3}$, where w is the weight of the bird in kilograms and S is the surface area of the wings in square meters. (Source: C. Pennycuick, *Newton Rules Biology.*)

(a) Approximate $S(0.5)$ and interpret the result.

(b) What weight corresponds to a surface area of 0.25 square meter?

SOLUTION (a) $S(0.5) = 0.2(0.5)^{2/3} \approx 0.126$. The wings of a bird that weighs 0.5 kilogram have a surface area of about 0.126 square meter.

(b) To answer this question we must solve the equation $0.2w^{2/3} = 0.25$.

$$0.2w^{2/3} = 0.25$$

$$w^{2/3} = \frac{0.25}{0.2} \qquad \text{Divide by 0.2.}$$

$$(w^{2/3})^3 = \left(\frac{0.25}{0.2}\right)^3 \qquad \text{Cube both sides.}$$

$$w^2 = \left(\frac{0.25}{0.2}\right)^3 \qquad \text{Simplify.}$$

$$w = \pm\sqrt{\left(\frac{0.25}{0.2}\right)^3} \qquad \text{Square root property.}$$

$$w \approx \pm 1.4 \qquad \text{Approximate.}$$

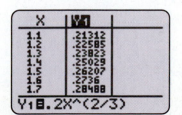

FIGURE 4.133

Since w must be positive, this means that the wings of a 1.4-kilogram bird have a surface area of about 0.25 square meter. Numerical support is shown in Figure 4.133, where a table of $Y_1 = 0.2X^{(2/3)}$ is approximately 0.25 when $x = 1.4$.

Solving an equation with rational exponents

Solve $2x^{5/2} - 7 = 23$. Approximate the answer to the nearest hundredth, and give graphical support.

SOLUTION *Symbolic Solution* Start by adding 7 to both sides.

$$2x^{5/2} = 30 \qquad \text{Add 7 to both sides.}$$
$$x^{5/2} = 15 \qquad \text{Divide by 2.}$$
$$(x^{5/2})^2 = 15^2 \qquad \text{Square both sides.}$$
$$x^5 = 225 \qquad \text{Properties of exponents}$$
$$x = 225^{1/5} \qquad \text{Take the fifth root.}$$
$$x \approx 2.95 \qquad \text{Approximate.}$$

Graphical Solution Graphical support is shown in Figure 4.134, where the graphs of $Y_1 = 2X^{(5/2)} - 7$ and $Y_2 = 23$ intersect near $(2.95, 23)$.

$[-5, 5, 1]$ by $[-40, 40, 10]$

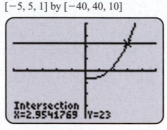

FIGURE 4.134

Graphing Calculator Help

When entering $X^{(5/2)}$, be sure to put parentheses around the fraction $5/2$.

In the next example we use a power function to model the planetary data presented in the introduction to this section.

Modeling the period of planetary orbits

Use the data in Table 4.16 on page 312 to complete the following.
(a) Make a scatterplot of the data. Estimate graphically a value for b so that $f(x) = x^b$ models the data.
(b) Numerically check the accuracy of f.
(c) The average distances of Uranus, Neptune, and Pluto from the sun are 19.2, 30.1, and 39.5, respectively. Use f to estimate the periods of revolution for these planets. Compare these answers to the actual values of 84.0, 164.8, and 248.5 years.

SOLUTION **(a)** Make a scatterplot of the data and then graph $y = x^b$ for different values of b. By viewing the graphs of $y = x^{1.4}$, $y = x^{1.5}$, and $y = x^{1.6}$ in Figures 4.135–4.137, it can be seen that $b \approx 1.5$.

$[0, 10, 1]$ by $[0, 30, 10]$

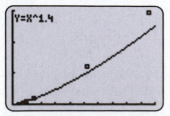

FIGURE 4.135

$[0, 10, 1]$ by $[0, 30, 10]$

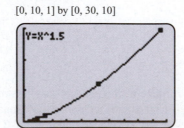

FIGURE 4.136

$[0, 10, 1]$ by $[0, 30, 10]$

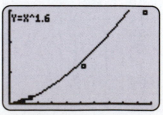

FIGURE 4.137

(b) Let $f(x) = x^{1.5}$ and table $Y_1 = X^{\wedge}1.5$. The values shown in Figure 4.138 model the data in Table 4.16 remarkably well.

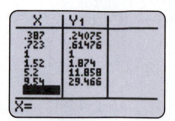

FIGURE 4.138

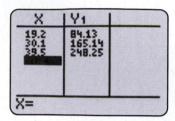

FIGURE 4.139

Graphing Calculator Help

To make a scatterplot, see Appendix B (page AP-6). To make a table like Figure 4.138, see Appendix B (page AP-11).

(c) To approximate the number of years for Uranus, Neptune, and Pluto to orbit the sun, evaluate $f(x) = x^{1.5}$ at $x = 19.2$, 30.1, and 39.5, as shown in Figure 4.139. These values are close to the actual values. ▮

■ **CLASS DISCUSSION**

Suppose in Example 6 we let $f(x) = x^b$ and require that $f(9.54) = 29.5$. Can we symbolically solve the resulting equation, $9.54^b = 29.5$? Explain. ■

Equations Involving Radicals

When solving equations that contain square roots, it is common to square both sides of an equation. This is done in the next example.

EXAMPLE 7   *Solving an equation containing a square root*

Solve $x = \sqrt{15 - 2x}$. Check your solutions.

SOLUTION   Begin by squaring both sides of the equation.

$$x = \sqrt{15 - 2x} \qquad \text{Given equation}$$
$$x^2 = (\sqrt{15 - 2x})^2 \qquad \text{Square both sides.}$$
$$x^2 = 15 - 2x \qquad \text{Simplify.}$$
$$x^2 + 2x - 15 = 0 \qquad \text{Add } 2x \text{ and subtract } 15.$$
$$(x + 5)(x - 3) = 0 \qquad \text{Factor.}$$
$$x = -5 \quad \text{or} \quad x = 3 \qquad \text{Solve.}$$

Now substitute these values in the given equation $x = \sqrt{15 - 2x}$.

$$-5 \neq \sqrt{15 - 2(-5)} = 5, \qquad 3 = \sqrt{15 - 2(3)}$$

Graphing Calculator Help

To find a point of intersection, see Appendix B (page AP-11).

$[-9, 9, 1]$ by $[-6, 6, 1]$

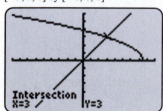

FIGURE 4.140

Thus 3 is the only solution. This result is supported graphically in Figure 4.140, where $Y_1 = X$ and $Y_2 = \sqrt{(15 - 2X)}$. Notice that no point of intersection occurs when $x = -5$. ▮

The value -5 in Example 7 is called an **extraneous solution** because it does not satisfy the given equation. It is important to check results whenever *squaring* has been used to solve an equation.

Sometimes it is necessary to cube both sides of an equation.

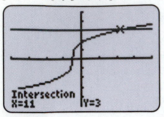

[−20, 20, 5] by [−5, 5, 1]

FIGURE 4.141

| **EXAMPLE 8** | *Solving an equation containing a cube root* |

Solve $\sqrt[3]{2x + 5} - 2 = 1$. Support your result graphically.

SOLUTION   Start by adding 2 to both sides. Then cube both sides.

$$\sqrt[3]{2x + 5} = 3 \qquad \text{Add 2 to both sides.}$$
$$(\sqrt[3]{2x + 5})^3 = 3^3 \qquad \text{Cube both sides.}$$
$$2x + 5 = 27 \qquad \text{Simplify.}$$
$$2x = 22 \qquad \text{Subtract 5 from both sides.}$$
$$x = 11 \qquad \text{Divide by 2.}$$

The only solution is $x = 11$. Graphical support is shown in Figure 4.141, where the graphs of $Y_1 = \sqrt[3]{(2X + 5)}$ and $Y_2 = 3$ intersect at $(11, 3)$.

Power Regression

Rather than visually fit a curve to data, as was done in Example 6, we can use least-squares regression to fit the data. Least-squares regression was introduced in Section 2.6. In the next example we apply this technique to data from biology.

| **EXAMPLE 9** | *Modeling the length of a bird's wing* |

Table 4.17 lists the weight W and the wingspan L for birds of a particular species.

TABLE 4.17   Weights and Wingspans

W (kilograms)	0.5	1.5	2.0	2.5	3.0
L (meters)	0.77	1.10	1.22	1.31	1.40

Source: C. Pennycuick.

(a) Use power regression to model the data with $L = aW^b$. Graph the data and the equation.

(b) Approximate the wingspan for a bird weighing 3.2 kilograms.

SOLUTION   **(a)** Let x be the weight W and y be the length L. Enter the data, and then select power regression (PwrReg), as shown in Figures 4.142 and 4.143. The results are shown in Figure 4.144. Let

$$y = 0.9674x^{0.3326} \qquad \text{or} \qquad L = 0.9674W^{0.3326}.$$

Graphing Calculator Help

To find an equation of least-squares fit, see Appendix B (page AP-15).

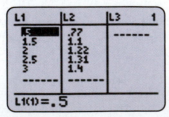

FIGURE 4.142

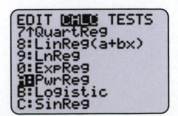

FIGURE 4.143

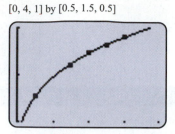

[0, 4, 1] by [0.5, 1.5, 0.5]

FIGURE 4.144　　　　　**FIGURE 4.145**

The data and equation are graphed in Figure 4.145.

(b) If a bird weighs 3.2 kilograms, this model predicts the wingspan to be

$$L = 0.9674(3.2)^{0.3326} \approx 1.42 \text{ meters.}$$

Putting it all Together 4.7

*T*he following table outlines important concepts in this section.

Concept	Symbolic Representation	Examples
Rational exponents	$x^{m/n} = (x^m)^{1/n}$ $= (x^{1/n})^m$	$9^{3/2} = (9^3)^{1/2} = (729)^{1/2} = 27$ $9^{3/2} = (9^{1/2})^3 = (3)^3 = 27$
Radical notation	$x^{1/2} = \sqrt{x}$ $x^{1/3} = \sqrt[3]{x}$ $x^{m/n} = \sqrt[n]{x^m}$ $= (\sqrt[n]{x})^m$	$25^{1/2} = \sqrt{25} = 5$ $27^{1/3} = \sqrt[3]{27} = 3$ $8^{2/3} = \sqrt[3]{8^2} = \sqrt[3]{64} = 4$ $\sqrt{4^3} = (\sqrt{4})^3 = (2)^3 = 8$
Power function	$f(x) = x^b$, where b is a constant.	$f(x) = x^{5/4}$ $g(x) = x^{-\pi}$
Root function	$f(x) = x^{1/n}$, where $n \geq 2$ is an integer.	$f(x) = x^{1/2}$ or $f(x) = \sqrt{x}$ $g(x) = x^{1/5}$ or $g(x) = \sqrt[5]{x}$

In this section we studied power functions and root functions. In the first four chapters, we studied several types of functions. The following summary of functions lists several types of functions. This summary may be used as a reference for future work.

Type of Function	Examples	Graphs
Linear function $f(x) = ax + b$	$f(x) = 0.5x - 1$ $g(x) = -3x + 2$ $h(x) = 2$	
Polynomial function $f(x) = a_n x^n + \cdots + a_2 x^2 + a_1 x + a_0$	$f(x) = x^2 - 1$ $g(x) = x^3 - 4x - 1$ $h(x) = -x^4 + 4x^2 - 2$	
Rational function $f(x) = \dfrac{p(x)}{q(x)}$, where $p(x)$ and $q(x)$ are polynomials.	$f(x) = \dfrac{1}{x}$ $g(x) = \dfrac{2x - 1}{x + 2}$ $h(x) = \dfrac{1}{x^2 - 1}$	
Root function $f(x) = x^{1/n}$, where $n \geq 2$ is an integer.	$f(x) = x^{1/2} = \sqrt{x}$ $g(x) = x^{1/3} = \sqrt[3]{x}$ $h(x) = x^{1/4} = \sqrt[4]{x}$	
Power function $f(x) = x^b$, where b is a constant.	$f(x) = x^{2/3}$ $g(x) = x^{\sqrt{2}}$ $h(x) = x^3$	

4.7 EXERCISES

Properties of Exponents

Exercises 1–16: Evaluate the expression by hand. Then check your result with a calculator.

1. $8^{2/3}$

2. $-16^{3/2}$

3. $16^{-3/4}$

4. $25^{-3/2}$

5. $-81^{0.5}$

6. $32^{1/5}$

7. $64^{1/6}$

8. $16^{-0.25}$

9. $(-9^{3/4})^2$

10. $(4^{-1/2})^{-4}$

11. $(0.5^{-2})^2$

12. $\left(\dfrac{1}{4}\right)^{-3/2}$

13. $\left(\dfrac{2}{3}\right)^{-2}$

14. $\left(\dfrac{1}{9}\right)^{-1/2} + \left(\dfrac{1}{16}\right)^{-1/2}$

15. $1^{-1} + 2^{-1} + 3^{-1}$

16. $(8^{-1/3} + 27^{-1/3})^2$

Exercises 17–26: Use positive rational exponents to rewrite the expression.

17. $\sqrt[3]{2x}$

18. $\sqrt{x+1}$

19. $\sqrt[3]{z^5}$

20. $\sqrt[5]{x^2}$

21. $(\sqrt[4]{y})^{-3}$

22. $(\sqrt[3]{y^2})^{-5}$

23. $\sqrt{x} \cdot \sqrt[3]{x}$

24. $(\sqrt[5]{z})^{-3}$

25. $\sqrt{y} \cdot \sqrt{y}$

26. $\dfrac{\sqrt[3]{x}}{\sqrt{x}}$

Power Functions

Exercises 27–30: Evaluate f(x) at the given x. Approximate each result to the nearest hundredth.

27. $f(x) = x^{1.62}$, $x = 1.2$

28. $f(x) = x^{-0.71}$, $x = 3.8$

29. $f(x) = x^{3/2} - x^{1/2}$, $x = 50$

30. $f(x) = x^{5/4} - x^{-3/4}$, $x = 7$

Exercises 31 and 32: Match f(x) with its graph. Assume that a and b are constants with $0 < a < 1 < b$.

31. $f(x) = x^a$

32. $f(x) = x^b$

a.

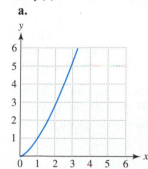

b.

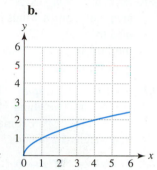

Exercises 33–42: Solve the equation. Check each solution.

33. $x^3 = 8$

34. $x^4 = \dfrac{1}{81}$

35. $x^{1/4} = 3$

36. $x^{1/3} = \dfrac{1}{5}$

37. $x^{2/5} = 4$

38. $x^{2/3} = 16$

39. $2(x^{1/5} - 2) = 0$

40. $x^{1/2} + x^{1/2} = 8$

41. $2x^{1/3} - 5 = 1$

42. $4x^{3/2} + 5 = 21$

Exercises 43–48: Use translations of graphs to help sketch a graph of f.

43. $f(x) = \sqrt{x} + 1$

44. $f(x) = \sqrt[3]{x} - 1$

45. $f(x) = x^{2/3} - 1$

46. $f(x) = \sqrt{x-1}$

47. $f(x) = \sqrt{x+2} - 1$

48. $f(x) = (x-1)^{2/3}$

Equations Involving Radicals

Exercises 49–60: Solve each equation. Check your answers.

49. $\sqrt{x+2} = x - 4$

50. $\sqrt{2x+1} = 13$

51. $\sqrt{3x+7} = 3x + 5$

52. $\sqrt{1-x} = x + 5$

53. $\sqrt{5x-6} = x$

54. $x - 5 = \sqrt{5x-1}$

55. $\sqrt{x+1} + 3 = \sqrt{3x+4}$ (*Hint*: You will need to square the equation twice.)

56. $\sqrt{x} = \sqrt{x-5} + 1$

57. $\sqrt[3]{x+1} = -3$

58. $\sqrt[3]{z} + 5 = 4$

59. $\sqrt[3]{x + 1} = \sqrt[3]{2x - 1}$

60. $\sqrt[3]{2x^2 + 1} = \sqrt[3]{1 - x}$

Applications

61. *Modeling Wing Size* Suppose that the surface area S of a bird's wings in square feet can be modeled by $S(w) = 1.27w^{2/3}$, where w is the weight of the bird in pounds. Estimate the weight of a bird with wings having a surface area of 3 square feet.

62. *Modeling Wingspan* The wingspan L in feet of a bird weighing W pounds is given by $L = 2.43W^{0.3326}$. Estimate the wingspan of a bird that weighs 5.2 pounds.

63. *Modeling Planetary Orbits* The formula $f(x) = x^{1.5}$ calculates the number of years it would take for a planet to orbit the sun if its average distance from the sun is x times farther than Earth. If there were a planet located 15 times farther from the sun than Earth, how many years would it take for the planet to orbit the sun?

64. *Modeling Planetary Orbits* (Refer to Exercise 63.) If there were a planet that took 200 years to orbit the sun, what would be its average distance x from the sun compared to Earth?

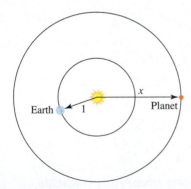

65. *Trout and Pollution* Rainbow trout are sensitive to zinc ions in the water. High concentrations are lethal. The average survival times x in minutes for trout in various concentrations of zinc ions y in milligrams per liter (mg/l) are listed in the table. (Source: C. Mason, *Biology of Freshwater Pollution.*)

x (min)	0.5	1	2	3
y (mg/l)	4500	1960	850	525

(a) This data can be modeled by $f(x) = ax^b$, where a and b are constants. Determine a. (*Hint:* Let $f(1) = 1960$.)

(b) Estimate b.

(c) Evaluate $f(4)$ and interpret the result.

66. *Asbestos and Cancer* Insulation workers who were exposed to asbestos and employed before 1960 experienced an increased likelihood of lung cancer. If a group of insulation workers have a cumulative total of 100,000 years of work experience with their first date of employment x years ago, then the number of lung cancer cases occurring within the group can be modeled by $N(x) = 0.00437x^{3.2}$. (Source: A. Walker, *Observation and Inference.*)

(a) Calculate $N(x)$ when $x = 5$, 10, and 20. What happens to the likelihood of cancer as x increases?

(b) If x doubles, does the number of cancer cases also double?

(c) Solve the equation $N(x) = 100$, and interpret your answer.

67. *Fiddler Crab Size* Allometric relations often can be modeled by $f(x) = ax^b$, where a and b are constants. One study of the male fiddler crab showed a connection between the weight of the large claws and its total body weight. For a crab weighing over 0.75 gram, the weight of its claws can be estimated by $f(x) = 0.445x^{1.25}$. The input x is the weight of the crab in grams and the output $f(x)$ is the weight of the claws in grams. (Sources: J. Huxley, *Problems of Relative Growth*; D. Brown and P. Rothery, *Models in Biology: Mathematics, Statistics and Computing.*)

(a) Predict the weight of the claws for a 2-gram crab.

(b) Approximate graphically the weight of a crab that has 0.5-gram claws.

(c) Solve part (b) symbolically.

68. *Weight and Height* Allometry can be applied to height-weight relationships for humans. The average weight in pounds for men and women can be estimated by $f(x) = ax^{1.7}$, where x represents a person's height in inches and a is a constant that depends on the sex of the individual.

(a) If an average weight for a 68-inch tall man is 152 pounds, approximate a. Use f to estimate the average weight of a 66-inch tall man.

(b) If an average weight for a 68-inch tall woman is 137 pounds, approximate a. Use f to estimate the average weight of a 70-inch tall woman.

Exercises 69 and 70: Power Regression The table contains data that can be modeled by a function of the form $f(x) = ax^b$. Use regression to find the constants a and b to the nearest hundredth. Graph f and the data.

69.

x	2	4	6	8
$f(x)$	3.7	4.2	4.6	4.9

70.

x	3	6	9	12
$f(x)$	23.8	58.5	99.2	144

71. *Pulse Rate and Weight* According to one model, the rate at which an animal's heart beats varies with its weight. Smaller animals tend to have faster pulses, whereas larger animals tend to have slower pulses. The accompanying table lists average pulse rates in beats per minute (bpm) for animals with various weights in pounds (lb). Use regression (or some other method) to find values for a and b so that $f(x) = ax^b$ models these data.

Weight (lb)	40	150	400	1000	2000
Pulse (bpm)	140	72	44	28	20

Source: C. Pennycuick.

72. *Pulse Rate and Weight* (Continuation of Exercise 71) Use the results in the previous exercise to calculate the pulse rates for a 60-pound dog and a 2-ton whale.

Writing about Mathematics

73. Can a function be both a polynomial function and a power function? Explain.

74. Explain the basic steps needed to solve equations that contain square roots of variables.

Extended and Discovery Exercises

1. *Odd Root Functions* Graph $y = \sqrt[n]{x}$ for $n = 3, 5$, and 7. State some generalizations about a graph of an odd root function.

2. *Even Root Functions* Graph $y = \sqrt[n]{x}$ for $n = 2, 4$, and 6. State some generalizations about a graph of an even root function.

3. *Power Functions* Graph $y = x^b$ for $n = -1, -3$, and -5. State some generalizations about a graph of a power function, where b is a negative odd integer.

4. *Power Functions* Graph $y = x^b$ for $n = -2, -4$, and -6. State some generalizations about a graph of a power function, where b is a negative even integer.

CHECKING BASIC CONCEPTS FOR SECTION 4.7

1. Simplify each expression without a calculator.
   (a) $-4^{3/2}$ **(b)** $(8^{-2})^{1/3}$ **(c)** $\sqrt[3]{27^2}$

2. Solve the equation $4x^{3/2} - 3 = 29$.

3. Solve the equation $\sqrt{5x - 4} = x - 2$.

4. Find constants a and b so that $f(x) = ax^b$ models the data.

x	1	2	3	4
$f(x)$	2	2.83	3.46	4

Chapter 4 *Summary*

CONCEPT	EXPLANATION AND EXAMPLES

Section 4.1

ABSOLUTE AND LOCAL EXTREMA

The accompanying graph has the following extrema.

Absolute maximum: none   Absolute minimum: -14.8

Local maximum: 1   Local minimums: $-4.3, -14.8$

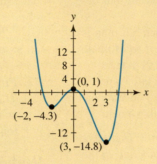

EVEN FUNCTION

$f(-x) = f(x)$

Graph is symmetric with respect to the y-axis.

Example: $f(x) = x^4 - 3x^2$

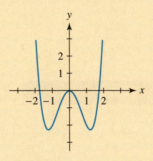

ODD FUNCTION

$f(-x) = -f(x)$

Graph is symmetric with respect to the origin.

Example: $f(x) = x - x^3$

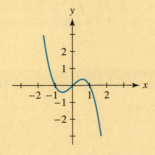

CONCEPT	EXPLANATION AND EXAMPLES

Section 4.2

GRAPH OF POLYNOMIALS

$f(x) = a_n x^n + \cdots + a_2 x^2 + a_1 x + a_0$

Degree: n Leading coefficient: a_n

Polynomials have continuous graphs with no breaks or corners.

A polynomial of degree $n \geq 1$ has at most n x-intercepts and at most $n - 1$ turning points. Its domain is all real numbers.

REFLECTIONS OF GRAPHS

The graph of $y = -f(x)$ is a reflection of the graph of $y = f(x)$ across the x-axis.

Example: $y = -x^2 - 1$ is a reflection of $y = x^2 + 1$ across the x-axis.

The graph of $y = f(-x)$ is a reflection of the graph of $y = f(x)$ across the y-axis.

Example: $y = (-x)^3 + 2(-x)$ is a reflection of $y = x^3 + 2x$ across the y-axis.

PIECEWISE–POLYNOMIAL FUNCTION

Example:

$f(x) = \begin{cases} x^2 - 2 & \text{if } x < 0 \\ 1 - 2x & \text{if } x \geq 0 \end{cases}$

$f(-2) = (-2)^2 - 2 = 2$

$f(0) = 1 - 2(0) = 1$

$f(2) = 1 - 2(2) = -3$

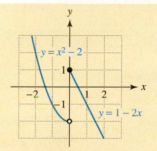

Section 4.3

FACTOR THEOREM

A polynomial $f(x)$ has a factor $x - k$ if and only if $f(k) = 0$.

Example: $f(x) = x^2 - 3x + 2$;
$\qquad f(1) = 0$ implies $(x - 1)$ is a factor of $x^2 - 3x + 2$.

COMPLETE FACTORED FORM

$f(x) = a_n(x - c_1)(x - c_2) \cdots (x - c_n)$, where a_n is the leading coefficient of the polynomial $f(x)$ and the c_k are its zeros.

Example: $f(x) = -2x^3 + 8x$ has zeros of $-2, 0, 2$.
$\qquad f(x) = -2(x + 2)(x - 0)(x - 2)$

MULTIPLE ZEROS

Odd multiplicity: graph crosses the x-axis.
Even multiplicity: graph touches but does not cross the x-axis.

Example: Let $f(x) = -2(x + 1)^3(x - 4)^2$;
$\qquad f(x)$ has a zero of -1 with odd multiplicity 3 and a zero of 4 with even multiplicity 2.

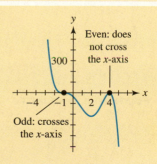

CONCEPT	EXPLANATION AND EXAMPLES

Section 4.4

IMAGINARY UNIT

$i = \sqrt{-1}, \quad i^2 = -1$

COMPLEX NUMBER

$a + bi$, where a and b are real numbers (standard form)

Complex numbers include all real numbers. We can add, subtract, multiply, and divide complex numbers.

FUNDAMENTAL THEOREM OF ALGEBRA

A polynomial $f(x)$ of degree $n \geq 1$ has at least one complex zero.

Explanation: With complex numbers, any polynomial may be written in complete factored form.

Example: $x^2 + 1 = (x + i)(x - i)$
$3x^2 - 3x - 6 = 3(x + 1)(x - 2)$

NUMBER OF ZEROS THEOREM

A polynomial of degree n has at most n distinct zeros.

Example: A cubic polynomial has at most 3 zeros.

Section 4.5

RATIONAL FUNCTION

$f(x) = \dfrac{p(x)}{q(x)}$, where $p(x)$ and $q(x) \neq 0$ are polynomials.

Example: $f(x) = \dfrac{2x - 3}{x - 1}$

Horizontal asymptote: $y = 2$
Vertical asymptote: $x = 1$

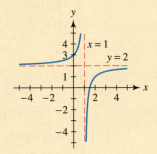

SOLVING RATIONAL EQUATIONS

Multiply both sides of the equation by a common denominator.

Example: $\dfrac{-24}{x - 3} - 4 = x + 3$ (Multiply by $x - 3$.)

$-24 - 4(x - 3) = (x + 3)(x - 3)$
$-24 - 4x + 12 = x^2 - 9$
$0 = x^2 + 4x + 3$
$0 = (x + 3)(x + 1)$
$x = -3 \quad$ or $\quad x = -1 \quad$ (Both solutions check.)

Section 4.6

POLYNOMIAL INEQUALITY

Write the inequality as $p(x) > 0$, where $>$ may be replaced by $\geq$, $<$, or $\leq$. Replace the inequality sign with an equals sign, and solve this equation. The solutions are called boundary numbers. Then use a graph or table to find the solution set to the given inequality.

CONCEPT	EXPLANATION AND EXAMPLES
RATIONAL INEQUALITY	As with polynomial inequalities, find the boundary numbers, including x-values where any expressions are undefined.

Example: $\dfrac{(x+2)(x-3)}{x} \geq 0$; Boundary numbers: $-2, 0, 3$

Solution set: $[-2, 0) \cup [3, \infty)$

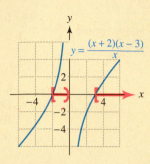

Section 4.7

RATIONAL EXPONENTS	$x^{m/n} = \sqrt[n]{x^m} = (\sqrt[n]{x})^m$

Examples: $25^{3/2} = \sqrt{25^3} = (\sqrt{25})^3 = 5^3 = 125$

POWER FUNCTION	$f(x) = x^b$, where b is a constant.

Examples: $f(x) = x^{4/3}$, $g(x) = x^{\sqrt{3}}$

SOLVING RADICAL EQUATIONS	When an equation contains a square root, isolate the square root and then square both sides. Be sure to check your answers.

Example: $x + \sqrt{3x - 3} = 1$

$$\sqrt{3x - 3} = 1 - x \qquad \text{Subtract } x.$$
$$3x - 3 = (1 - x)^2 \qquad \text{Square both sides.}$$
$$3x - 3 = 1 - 2x + x^2 \qquad \text{Expand.}$$
$$0 = x^2 - 5x + 4 \qquad \text{Combine terms.}$$
$$0 = (x - 1)(x - 4) \qquad \text{Factor.}$$
$$x = 1 \quad \text{or} \quad x = 4 \qquad \text{Solve.}$$

Check: $x = 1$ is a solution, but $x = 4$ is not.

$$1 + \sqrt{3(1) - 3} = 1 \qquad 4 + \sqrt{3(4) - 3} \neq 1$$

Review Exercises

Exercises 1 and 2: Complete the following.

(a) *Graph f.*

(b) *Estimate the intervals on which f is increasing or decreasing.*

1. $f(x) = x^2 - x$ **2.** $f(x) = x^3 - 4x$

3. Use the complete numerical representations for the functions *f*, *g*, and *h* to classify each as increasing, decreasing, or neither.

x	f(x)	g(x)	h(x)
1	−1	−3	−8
7	10	−7	−2
4	3	−5	4
−2	−6	−1	5

4. Use the graph of *f* to estimate the intervals where *f* is increasing or decreasing.

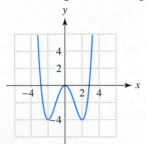

Exercises 5 and 6: The table is a complete representation of f. Decide if f is even, odd, or neither.

5.

x	−4	−2	0	2	4
f(x)	13	7	0	−7	−13

6.

x	−5	−3	−1	1	3	5
f(x)	−6	2	7	7	2	−6

Exercises 7–10: Determine if f is even, odd, or neither.

7. $f(x) = 2x^6 - 5x^4 - x^2$

8. $f(x) = -5x^3 - 18$

9. $f(x) = 7x^5 + 3x^3 - x$

10. $f(x) = \dfrac{1}{1 + x^2}$

11. The polynomial $f(x) = \dfrac{1}{2}x^3 - 3x^2 + \dfrac{11}{2}x - 3$ has zeros 1, 2, and 3. Write its complete factored form.

12. Find the complete factored form of $f(x) = x^4 + 2x^3 - 13x^2 - 14x + 24$.

13. Find the complete factored form of $f(x) = 2x^3 + 3x^2 - 18x + 8$ graphically.

14. Let $f(x)$ be given by

$$f(x) = \begin{cases} 2x & \text{if } 0 \le x < 2 \\ 8 - x^2 & \text{if } 2 \le x \le 4. \end{cases}$$

(a) Sketch a graph of *f*. Is *f* continuous?

(b) Evaluate $f(1)$ and $f(3)$.

(c) Solve the equation $f(x) = 2$.

Exercises 15–18: Evaluate the expression by hand. Then check your result with a calculator.

15. $(36^{3/4})^2$ **16.** $(9^{-3/2})^{-2}$

17. $(2^{-3/2} \cdot 2^{1/2})^{-3}$ **18.** $\left(\dfrac{4}{9}\right)^{-3/2}$

Exercises 19–22: Write the expression using positive exponents.

19. $\sqrt[3]{x^4}$ **20.** $(\sqrt[4]{z})^{-1/2}$

21. $\sqrt[3]{y} \cdot \sqrt{y}$ **22.** $\sqrt{x} \cdot \sqrt[3]{x^2} \cdot \sqrt[4]{x^3}$

Exercises 23–24: Give the domain of the power function. Approximate f(3) to the hundredth.

23. $f(x) = x^{5/2}$ **24.** $f(x) = x^{-2/3}$

Exercises 25 and 26: Use the graph of f to estimate any

(a) *local extrema, and*

(b) *absolute extrema.*

25.   **26.**

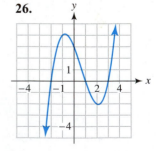

27. Graph $f(x) = -0.25x^4 + 0.67x^3 + 9.5x^2 - 20x - 50$.

(a) Approximate any local extrema.

(b) Approximate any absolute extrema.

(c) Determine intervals where *f* is increasing or decreasing.

28. Let $f(x) = \dfrac{2x^2}{x^2 - 4}$.

    (a) Find the domain of f.

    (b) Identify any vertical or horizontal asymptotes.

    (c) Graph f with a graphing calculator.

    (d) Sketch a graph of f that includes all asymptotes.

29. Graph each quartic (degree 4) polynomial. Count the local maxima, local minima, and x-intercepts.

    (a) $f(x) = x^4 + 2x^3 - 9x^2 - 2x + 20$

    (b) $g(x) = -0.4x^4 + 3.6x^2$

    (c) $h(x) = x^4 + 2x^3 - 21x^2 - 22x + 40$

30. Graph each quintic (degree 5) polynomial. Estimate any turning points.

    (a) $f(x) = 0.03x^5 - 0.21x^4 + 0.21x^3 + 0.57x^2 - 0.48x - 0.6$

    (b) $g(x) = -0.02x^5 + 1$

    (c) $h(x) = 0.01x^5 - 0.09x^4 - 0.22x^3 + 2.72x^2 - 0.96x - 12.8$

31. Sketch a graph of a function f with vertical asymptote $x = -2$ and horizontal asymptote $y = 2$.

32. Find the quotient and remainder when $2x^3 - x^2 - 4x + 1$ is divided by $x - 2$.

33. Solve $9x = 3x^3$.

34. Solve $x^4 - 3x^2 + 2 = 0$.

35. Write the complete factored form of a quintic (degree 5) polynomial $f(x)$ that has zeros -2 and 2 with multiplicities 2 and 3, respectively.

36. Find the complete factored form of $f(x) = x^3 - 2x^2 - 5x + 6$, given $c = 3$ is a zero.

Exercises 37 and 38: Sketch a graph of a polynomial that satisfies the given conditions.

37. Cubic, two x-intercepts, and a positive leading coefficient

38. Quartic (degree 4) with no x-intercepts

39. What is the maximum number of times that a horizontal line can intersect the graph of each type of polynomial?

    (a) linear (degree 1)   **(b)** quadratic

    (c) cubic

40. (a) Give the maximum number of times that a horizontal line can intersect the graph of a polynomial of degree $n \geq 1$.

    (b) Discuss the maximum number of real solutions to the polynomial equation $a_n x^n + \cdots + a_2 x^2 + a_1 x + a_0 = k$, where k is a constant and $n \geq 1$.

Exercises 41–44: Write the following in standard form.

41. $(2 - 3i) + (-3 + 3i)$

42. $(-5 + 3i) - (-3 - 5i)$

43. $(3 + 2i)(-4 - i)$

44. $\dfrac{3 + 2i}{2 - i}$

Exercises 45–48: Find all real and imaginary solutions.

45. $4x^2 + 9 = 0$ **46.** $2x^2 + 3 = 2x$

47. $x^3 + x = 0$ **48.** $x^4 + 3x^2 + 2 = 0$

49. Determine graphically the number of real zeros and the number of imaginary zeros of $f(x) = x^3 - 3x^2 + 3x - 9$.

50. Write a polynomial $f(x)$ in complete factored form that has degree 3, leading coefficient 4, and zeros $1, 3i$, and $-3i$. Then write $f(x)$ in expanded form.

51. Find the complete factorization of $f(x) = 2x^2 + 4$.

52. The graph of a semicircle with a radius of 5 and center $(0, 0)$ is given by $y = \sqrt{25 - x^2}$. Use reflections to graph the entire circle in $[-9, 9, 1]$ by $[-6, 6, 1]$.

Exercises 53 and 54: Use the rational zero test to determine any rational zeros of $f(x)$. Then, find the complete factored form of $f(x)$.

53. $f(x) = 2x^3 + x^2 - 13x + 6$

54. $f(x) = x^3 + x^2 - 11x - 11$

Exercises 55–62: Solve the equation. Check your results.

55. $x^5 = 1024$

56. $x^{1/3} = 4$

57. $\sqrt{x - 2} = x - 4$

58. $x^{3/2} = 27$

59. $2x^{1/4} + 3 = 6$

60. $\sqrt{x - 2} = 14 - x$

61. $\sqrt[3]{2x - 3} + 1 = 4$

62. $x^{1/3} + 3x^{1/3} = -2$

Exercises 63–66: Solve the rational equation. Support your results graphically or numerically.

63. $\dfrac{5x + 1}{x + 3} = 3$ **64.** $\dfrac{1}{x} - \dfrac{1}{x^2} + 2 = 0$

65. $\dfrac{1}{x + 2} + \dfrac{1}{x - 2} = \dfrac{1}{x^2 - 4}$

66. $x - \dfrac{1}{x} = 4$

Exercises 67–72: Solve the inequality either graphically or symbolically.

67. $x^3 + x^2 - 6x > 0$

68. $2x^3 + x^2 \le 10x$

69. $x^4 + 4 < 5x^2$

70. $\dfrac{1 - x}{x} < 0$

71. $\dfrac{2x - 1}{x + 2} > 0$

72. $\dfrac{1}{x} + \dfrac{1}{x + 2} \le \dfrac{4}{3}$

Applications

73. *Allometry* One of the earliest studies in allometry was performed by Bryan Robinson during the eighteenth century. He found that the pulse rate of an animal could be approximated by $f(x) = 1607x^{-0.75}$. The input x is the length of the animal in inches, and the output $f(x)$ is the approximate number of heartbeats per minute. (Source: H. Lancaster, *Quantitative Methods in Biology and Medical Sciences*.)

  (a) Use f to estimate the pulse rates of a 2-foot dog and a 5.5-foot person.

  (b) What length corresponds to a pulse rate of 400 beats per minute?

74. *Corporate Mergers* The number of corporate mergers changed during the past decade. The table lists the approximate numbers of mergers from 1987 to 1995. (Source: Securities Data Company.)

Year	1987	1988	1989	1990	1991
Mergers	3100	3800	5300	5600	5100

Year	1992	1993	1994	1995
Mergers	5300	6200	7500	8800

  (a) Make a line graph of the data.

  (b) Suggest a possible degree for a polynomial that models these data. Explain your reasoning.

  (c) Use least-squares regression to find a polynomial that models these data.

75. *Daylight Hours* The accompanying graph shows the deviation in the number of daylight hours from the average of 12 hours at a latitude of 40°N. In this graph x represents the month, where $x = 0$ corresponds to the spring equinox on March 21. For example, $x = 2$ represents May 21 and $x = -3$ represents December 21. At the spring equinox, day and night are of equal length so the deviation is 0. (Source: J. Williams, *The Weather Almanac 1995*.)

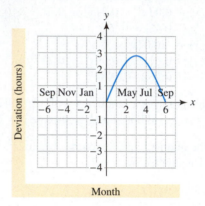

Month

  (a) Let f model this deviation in daylight. Is f an odd or even function? Explain.

  (b) Sketch the left side of the graph for $-6 \le x \le 0$.

76. *Falling Object* If an object is dropped from a height h, then the time t for the object to strike the ground is directly proportional to the square root of h. If it requires 1 second for an object to fall 16 feet, how long does it take for an object to fall 256 feet?

77. *Animals and Trotting Speeds* Taller animals tend to take longer, but fewer steps per second than shorter animals. The relationship between the shoulder height h in meters of an animal and an animal's stepping frequency F in steps per second, while *trotting*, is shown in the table. (Source: C. Pennycuick, *Newton Rules Biology*.)

h	0.5	1.0	1.5	2.0	2.5
F	2.6	1.8	1.5	1.3	1.2

  (a) Find values for constants a and b so that $f(x) = ax^b$ models the data.

(b) Estimate the stepping frequency for an elephant with a 3-meter shoulder height.

78. *Electrical Circuits*   The figure shows the voltage y in a household electrical circuit over a time interval of $\dfrac{1}{60}$ second.

(a) What is the maximum voltage?

(b) Determine the times when the voltage is decreasing.

(c) Use your results from part (b) to determine when the voltage is increasing. Let $0 \le x \le \dfrac{1}{60}$.

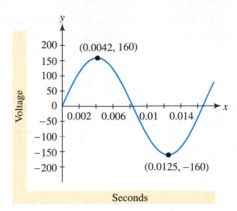

Seconds

Extended and Discovery Exercises

Exercises 1–2: *Velocity*   *Suppose that a person is riding a bicycle on a straight road, and that $f(t)$ computes the total distance in feet that the rider has traveled after t seconds. To calculate the person's average velocity between time t_1 and time t_2, we can evaluate the **difference quotient***

$$\frac{f(t_2) - f(t_1)}{t_2 - t_1}.$$

For example, if $f(t) = t^2$ then $f(10) = 100$ indicates that the person has traveled 100 feet after 10 seconds, and $f(15) = 225$ indicates that the person has traveled 225 feet after 15 seconds. Therefore the average velocity between 10 seconds and 15 seconds is

$$\frac{225 - 100}{15 - 10} = 25 \text{ feet per second.}$$

(a) *For the given $f(t)$ and the indicated values of t_1 and t_2, calculate the average velocity of the bike rider. Make a table to organize your work.*

(b) *Make a conjecture about the velocity of the bike rider precisely at time t_1.*

1. $f(t) = t^2$
   (i) $t_1 = 10, t_2 = 11$
   (ii) $t_1 = 10, t_2 = 10.1$
   (iii) $t_1 = 10, t_2 = 10.01$
   (iv) $t_1 = 10, t_2 = 10.001$

2. $f(t) = \sqrt{t}$
   (i) $t_1 = 4, t_2 = 5$
   (ii) $t_1 = 4, t_2 = 4.1$
   (iii) $t_1 = 4, t_2 = 4.01$
   (iv) $t_1 = 4, t_2 = 4.001$

Exercises 3–6: *Instantaneous Velocity*   *In Exercises 1 and 2 we calculated the average velocity of a bike rider over smaller and smaller time intervals. If we calculate the average velocity as the time interval approaches 0, we say that we are calculating the **instantaneous velocity**. To find the instantaneous velocity, we can calculate the average velocity between times $t_1 = a$ to $t_2 = a + h$. The difference quotient becomes*

$$\frac{f(t_2) - f(t_1)}{t_2 - t_1} = \frac{f(a + h) - f(a)}{h}.$$

For example, if $f(t) = t^2$, the difference quotient can be simplified as follows.

$$\frac{f(a + h) - f(a)}{h} = \frac{(a + h)^2 - a^2}{h}$$

$$= \frac{a^2 + 2ah + h^2 - a^2}{h}$$

$$= \frac{2ah + h^2}{h}$$

$$= 2a + h$$

If we let h approach 0 ($h \to 0$), then the instantaneous velocity becomes $2a + 0 = 2a$ at time $t_1 = a$. For example, if $a = 10$, then $v = 2(10) = 20$. That is, after 10 seconds the bike rider is traveling at 20 feet per second. Does this result agree with your findings from Exercise 1?

(a) *For the given $f(t)$, calculate the difference quotient $\dfrac{f(a + h) - f(a)}{h}$. Be sure to simplify completely.*

(b) Let h approach 0 and determine a formula for the instantaneous velocity at time $t_1 = a$.

(c) Calculate the instantaneous velocity at times $a = 5, 10,$ and 15.

3. $f(t) = 5t$

4. $f(t) = t^2 + 1$

5. $f(t) = t^2 + 2t$

6. $f(t) = \dfrac{200}{t}$

7. $f(t) = t^3$

8. $f(t) = \sqrt{t}$ (*Hint*: To simplify the ratio, multiply the numerator and denominator by $\sqrt{a + h} + \sqrt{a}$. This procedure is called **rationalizing** the numerator.)

*Exercises 9–12: **Average Rates of Change*** *These exercises investigate the relationship between polynomial functions and their average rates of change. For example, the average rate of change of $f(x) = x^2$ from x to $x + 0.001$ for any x can be calculated and graphed as shown in the accompanying figures. The graph of f is a parabola, and the graph of its average rate of change is a line. There is an interesting relationship between a graph of a polynomial function and a graph of its average rate of change. Try to discover what it is by completing the following.*

(a) *Graph each function and its average rate of change from x to $x + 0.001$.*

(b) *Compare the graph of each function to the graph of its average rate of change. How are turning points on the graph of a function related to its average rate of change?*

(c) *Make a generalization about these results. Test your generalization.*

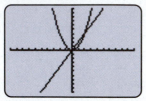

$[-10, 10, 1]$ by $[-10, 10, 1]$

9. Linear Functions
$f_1(x) = 3x + 1$ \qquad $f_2(x) = -2x + 6$
$f_3(x) = 1.5x - 5$ \qquad $f_4(x) = -4x - 2.5$

10. Quadratic Functions
$f_1(x) = 2x^2 - 3x + 1$ \qquad $f_2(x) = -0.5x^2 + 2x + 2$
$f_3(x) = x^2 + x - 2$ \qquad $f_4(x) = -1.5x^2 - 4x + 6$

11. Cubic Functions
$f_1(x) = 0.5x^3 - x^2 - 2x + 1$
$f_2(x) = -x^3 + x^2 + 3x - 5$
$f_3(x) = 2x^3 - 5x^2 + x - 3$
$f_4(x) = -x^3 + 3x - 4$

12. Quartic Functions
$f_1(x) = 0.05x^4 + 0.2x^3 - x^2 - 2.4x$
$f_2(x) = -0.1x^4 + 0.1x^3 + 1.3x^2 - 0.1x - 1.2$
$f_3(x) = 0.1x^4 + 0.4x^3 - 0.2x^2 - 2.4x - 2.4$

*The important thing
is not to
stop questioning.*

—

Albert Einstein

—

Chapter

5

Exponential and Logarithmic Functions

In 1900 the population of the world was approximately 1.6 billion. At that time the Swedish scientist Svante Arrhenius first predicted a greenhouse effect resulting from emissions of carbon dioxide by the industrialized countries. His classic calculation made use of logarithms and predicted that a doubling of the carbon dioxide concentration in the atmosphere would raise the average global temperature by 7°F to 11°F. Since then, the world population has increased dramatically to approximately 6 billion. With this increase in population, there has been a corresponding increase in emissions of greenhouse gases such as carbon dioxide, methane, and chlorofluorocarbons. These emissions have created the potential to alter the earth's climate and destroy portions of the ozone layer.

When quantities such as population and pollution increase rapidly, nonlinear functions and equations are used to model their growth. In this chapter, exponential and logarithmic functions are introduced. They occur in a wide variety of applications. Examples include acid rain, the decline of the bluefin tuna, air pollution, global warming, salinity of the ocean, demand for liver transplants, diversity of bird species, the relationship between caloric intake and land ownership in developing countries, hurricanes, earthquakes, and increased skin cancer due to a decrease in the ozone layer. Understanding these issues and discovering solutions will require creativity, innovation, and mathematics. This chapter presents some of the mathematical tools necessary for modeling these trends.

Reference: M. Kraljic, *The Greenhouse Effect.*

5.1 COMBINING FUNCTIONS

Arithmetic Operations on Functions • Composition of Functions

Introduction

Addition, subtraction, multiplication, and division can be performed on numbers and variables. Arithmetic operations also can be used to combine functions. For example, to model the stopping distance of a car, we compute two quantities. The first quantity is the *reaction distance,* which is the distance that a car travels between the time when a driver first recognizes a hazard and the time when the brakes are applied. The second quantity is *braking distance,* which is the distance that a car travels after the brakes have been applied. *Stopping distance* is equal to the sum of the reaction distance and the braking distance. One way to determine stopping distance is to find one function that models the reaction distance and a second function that calculates the braking distance. The stopping distance is then found by adding the two functions.

Arithmetic Operations on Functions

Highway engineers frequently assume that drivers have a reaction time of 2.5 seconds or less. During this time, a car continues to travel at a constant speed, until a driver is able to move his or her foot from the accelerator to the brake pedal. If a car travels at x miles per hour, then $r(x) = (11/3)x$ computes the distance in feet that a car travels in 2.5 seconds. For example, a driver traveling at 55 miles per hour might have a reaction distance of $r(55) = (11/3)(55) \approx 201.7$ feet.

The braking distance in feet for a car traveling on wet, level pavement at x miles per hour can be approximated by $b(x) = x^2/9$. For instance, a car traveling at 55 miles per hour would need $b(55) = 55^2/9 \approx 336.1$ feet to stop after the brakes have been applied. The estimated stopping distance at 55 miles per hour is the sum of these distances: $201.7 + 336.1 = 537.8$ feet. See the accompanying figure.   (Source: L. Haefner, *Introduction to Transportation Systems.*)

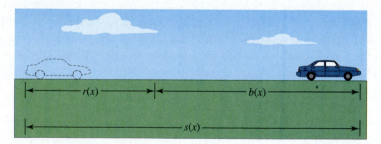

The concept of finding the sum of two functions can be represented symbolically, graphically, and numerically, as illustrated in the next example.

EXAMPLE 1   *Representing stopping distance symbolically, graphically, and numerically*

A driver traveling at 60 miles per hour with a reaction time of 2.5 seconds attempts to stop on wet, level pavement in order to avoid an accident.
(a) Write a symbolic representation for a function s using r and b that computes the stopping distance for a car traveling at x miles per hour. Evaluate $s(60)$.

(b) Graph *r, b,* and *s* in [45, 70, 5] by [0, 800, 100]. Interpret the graph.

(c) Illustrate the relationship between *r, b,* and *s* numerically.

SOLUTION   **(a)** *Symbolic Representation*   Let *s(x)* be the sum of the two formulas for *r* and *b*.

$$s(x) = r(x) + b(x) = \frac{11}{3}x + \frac{1}{9}x^2$$

The stopping distance for a car traveling at 60 miles per hour is

$$s(60) = \frac{11}{3}(60) + \frac{1}{9}(60)^2 = 620 \text{ feet.}$$

[45, 70, 5] by [0, 800, 100]

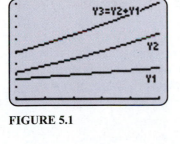

FIGURE 5.1

(b) *Graphical Representation*   Let $Y_1 = (11/3)X$, $Y_2 = (X\text{^}2)/9$, and $Y_3 = (11/3)X + (X\text{^}2)/9$ or $Y_3 = Y_1 + Y_2$, as shown in Figure 5.1. For any *x*-value in the graph, the sum of Y_1 and Y_2 equals Y_3. Figure 5.2 illustrates that

$$s(60) = r(60) + b(60) = 220 + 400 = 620.$$

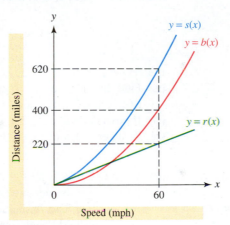

FIGURE 5.2

(c) *Numerical Representation*   Table 5.1 shows numerical representations of *r(x)*, *b(x)*, and *s(x) = r(x) + b(x)*. Notice that values for *s(x)* can be found by adding *r(x)* and *b(x)*. For example, when *x* = 60, *r*(60) = 220, *b*(60) = 400, and *s*(60) = 220 + 400 = 620.

TABLE 5.1

x	0	12	24	36	48	60
r(x)	0	44	88	132	176	220
b(x)	0	16	64	144	256	400
s(x)	0	60	152	276	432	620

We now formally define arithmetic operations on functions.

> ## Operations on functions
>
> If $f(x)$ and $g(x)$ both exist, the sum, difference, product, and quotient of two functions f and g are defined by
>
> $$(f + g)(x) = f(x) + g(x),$$
> $$(f - g)(x) = f(x) - g(x),$$
> $$(fg)(x) = f(x) \cdot g(x), \quad \text{and}$$
> $$\left(\frac{f}{g}\right)(x) = \frac{f(x)}{g(x)}, \quad \text{where } g(x) \neq 0.$$

The domains of the sum, difference, and product of f and g include x-values that are in both the domain of f and the domain of g. The domain of the quotient f/g includes all x-values in both the domain of f and the domain of g, where $g(x) \neq 0$.

EXAMPLE 2 *Performing arithmetic operations on functions symbolically*

Let $f(x) = 2 + \sqrt{x - 1}$ and $g(x) = x^2 - 4$.
(a) Find the domain of $f(x)$ and $g(x)$. Then find the domains of $(f + g)(x)$, $(f - g)(x)$, $(fg)(x)$, and $(f/g)(x)$.
(b) If possible, evaluate $(f + g)(5)$, $(f - g)(1)$, $(fg)(0)$, and $(f/g)(3)$.
(c) Write expressions for $(f + g)(x)$, $(f - g)(x)$, $(fg)(x)$, and $(f/g)(x)$.

SOLUTION (a) Whenever $x \geq 1$, $f(x) = 2 + \sqrt{x - 1}$ is defined. Therefore the domain of f is $\{x \mid x \geq 1\}$. The domain of $g(x) = x^2 - 4$ is all real numbers. The domains of $f + g$, $f - g$, and fg include all x-values in *both* the domain of f and the domain of g. Thus their domains are $\{x \mid x \geq 1\}$.

To determine the domain of f/g, we must also exclude x-values where $g(x) = x^2 - 4 = 0$. This occurs when $x = \pm 2$. Thus the domain of f/g is $\{x \mid x \geq 1, x \neq 2\}$. (Note that $x \neq -2$ is satisfied if $x \geq 1$.)

(b) The expressions can be evaluated as follows.

$$(f + g)(5) = f(5) + g(5) = \left(2 + \sqrt{5 - 1}\right) + (5^2 - 4) = 4 + 21 = 25$$
$$(f - g)(1) = f(1) - g(1) = \left(2 + \sqrt{1 - 1}\right) - (1^2 - 4) = 2 - (-3) = 5$$

$(fg)(0)$ is undefined since 0 is not in the domain of $f(x)$.

$$\left(\frac{f}{g}\right)(3) = \frac{f(3)}{g(3)} = \frac{2 + \sqrt{3 - 1}}{3^2 - 4} = \frac{2 + \sqrt{2}}{5}$$

(c) The sum, difference, product, and quotient of f and g are calculated as follows.

$$(f + g)(x) = f(x) + g(x) = \left(2 + \sqrt{x - 1}\right) + (x^2 - 4) = \sqrt{x - 1} + x^2 - 2$$
$$(f - g)(x) = f(x) - g(x) = \left(2 + \sqrt{x - 1}\right) - (x^2 - 4) = \sqrt{x - 1} - x^2 + 6$$
$$(fg)(x) = f(x) \cdot g(x) = \left(2 + \sqrt{x - 1}\right)(x^2 - 4)$$
$$\left(\frac{f}{g}\right)(x) = \frac{f(x)}{g(x)} = \frac{2 + \sqrt{x - 1}}{x^2 - 4}$$

EXAMPLE 3 *Evaluating combinations of functions graphically, numerically, and symbolically*

If possible, use the given representations of the functions f and g to evaluate $(f + g)(4)$, $(f - g)(-2)$, $(fg)(1)$, and $(f/g)(0)$.

(a)

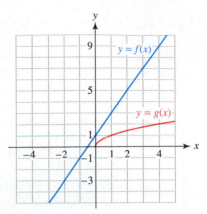

FIGURE 5.3

(b)

TABLE 5.2

x	-2	0	1	4
$f(x)$	-3	1	3	9
$g(x)$	Error	0	1	2

(c) $f(x) = 2x + 1$, $g(x) = \sqrt{x}$

SOLUTION **(a)** *Graphical Evaluation* In Figure 5.3, $f(4) = 9$ and $g(4) = 2$. Thus

$$(f + g)(4) = f(4) + g(4) = 9 + 2 = 11.$$

Algebra Review
To review the domain of a function, see Section 1.3.

Although $f(-2) = -3$, $g(-2)$ is undefined because -2 is not in the domain of g. Thus $(f - g)(-2)$ is undefined. The domains of f and g include 1, and $(fg)(1) = f(1)g(1) = 3(1) = 3$. The graph of g intersects the origin, so $g(0) = 0$. Thus $(f/g)(0)$ is undefined.

(b) *Numerical Evaluation* Numerical evaluation of f and g is performed using Table 5.2. From the table $f(4) = 9$ and $g(4) = 2$. As in part (a),

$$(f + g)(4) = f(4) + g(4) = 9 + 2 = 11.$$

In the table $g(-2)$ is undefined (error), so $(f - g)(-2)$ is also undefined. The calculations of $(fg)(1)$ and $(f/g)(0)$ are done in a similar manner.

(c) *Symbolic Evaluation* Use the formulas $f(x) = 2x + 1$ and $g(x) = \sqrt{x}$.

$$(f + g)(4) = f(4) + g(4) = (2 \cdot 4 + 1) + \sqrt{4} = 9 + 2 = 11$$
$$(f - g)(-2) = f(-2) - g(-2) = (2 \cdot (-2) + 1) - \sqrt{-2} \text{ is undefined.}$$
$$(fg)(1) = f(1)g(1) = (2 \cdot 1 + 1)\sqrt{1} = 3(1) = 3$$
$$\left(\frac{f}{g}\right)(0) = \frac{f(0)}{g(0)} \text{ is undefined, since } g(0) = 0.$$

The next example involves an application from business where the difference between two functions occurs.

EXAMPLE 4 *Finding the difference of two functions*

Once the sound track for a compact disc (CD) has been recorded, the cost of producing the master disc can be significant. After this disc has been made, additional discs can be produced inexpensively. A typical cost for producing the master disc is $2000, while additional discs cost approximately $2 each.   (Source: Windcrest Productions.)

(a) Assuming no other expenses, find a function C that outputs the cost of producing the master disc plus x additional compact discs. Find the cost of making the master disc and 1500 additional compact discs.

(b) Suppose that each CD is sold for $12. Find a function R that computes the revenue received from selling x compact discs. Find the revenue from selling 1500 compact discs.

(c) Assuming that the master CD is not sold, determine a function P that outputs the profit from selling x compact discs. How much profit will there be from selling 1500 discs?

SOLUTION **(a)** The cost of producing the master disc for $2000 plus x additional discs at $2 each is given by $C(x) = 2x + 2000$. The $2000 cost is sometimes called a *fixed cost*. The cost of manufacturing the master disc and 1500 additional compact discs is

$$C(1500) = 2(1500) + 2000 = \$5000.$$

(b) The revenue from x discs at $12 each is computed by $R(x) = 12x$. The revenue from selling 1500 compact discs is $R(1500) = 12(1500) = \$18,000$.

(c) Profit P is equal to revenue minus cost. This can be written using function notation.

$$
\begin{aligned}
P(x) &= R(x) \; - \;\; C(x) \\
&= 12x - (2x + 2000) \\
&= 12x - 2x - 2000 \\
&= 10x - 2000
\end{aligned}
$$

Algebra Review

To review arithmetic operations on polynomials, see Chapter R (page R-21).

If $x = 1500$, $P(1500) = 10(1500) - 2000 = \$13,000$.

Composition of Functions

Many tasks in life are performed in sequence. For example, to go to a movie we might get into a car, drive to the movie theater, and get out of the car. The order in which these tasks are performed is important.

A similar situation occurs with functions. For example, to convert miles to inches we might first convert miles to feet and then feet to inches. Since there are 5280 feet in a mile, $f(x) = 5280x$ converts x miles to an equivalent number of feet. Then $g(x) = 12x$ changes feet to inches. To convert x miles to inches, we combine the functions f and g in sequence. Figure 5.4 illustrates how to convert 5 miles to inches. First, $f(5) = 5280(5) = 26,400$. Then the output of 26,400 feet from f is used as input for g. The number of inches in 26,400 feet is $g(26,400) = 12(26,400) = 316,800$. This computation is called the *composition* of g and f.

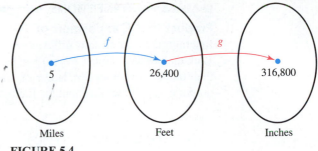

FIGURE 5.4

The composition of g and f shown in Figure 5.4 can be expressed symbolically. The symbol $\circ$ is used to denote composition of two functions.

$$(g \circ f)(5) = g(f(5)) \qquad \text{First compute } f(5).$$
$$= g(5280 \cdot 5) \qquad f(x) = 5280x$$
$$= g(26{,}400) \qquad \text{Simplify.}$$
$$= 12(26{,}400) \qquad g(x) = 12x$$
$$= 316{,}800 \qquad \text{Simplify.}$$

A distance of 5 miles is equivalent to 316,800 inches. The concept of composition of two functions is now defined formally.

Composition of functions

If f and g are functions, then the **composite function** $g \circ f$, or **composition** of g and f is defined by

$$(g \circ f)(x) = g(f(x)).$$

The domain of $g \circ f$ is all x in the domain of f such that $f(x)$ is in the domain of g.

Note: We read $g(f(x))$ as "g of f of x."

EXAMPLE 5   *Finding a symbolic representation of a composite function*

Find a symbolic representation for the composite function $g \circ f$ that converts x miles into inches.

SOLUTION   Let $f(x) = 5280x$ and $g(x) = 12x$.

$$(g \circ f)(x) = g(f(x)) \qquad \text{Definition of composition.}$$
$$= g(5280x) \qquad f(x) = 5280x \text{ is the input for } g.$$
$$= 12(5280x) \qquad g \text{ multiplies the input by 12.}$$
$$= 63{,}360x \qquad \text{Simplify.}$$

Thus $(g \circ f)(x) = 63{,}360x$ converts x miles to inches.

■ **MAKING CONNECTIONS**

Product and Composition of Two Functions Computing a product of two functions is fundamentally different from computing a composition of two functions. With the product $(fg)(x) = f(x) \cdot g(x)$, both f and g receive the *same* input x. Then their outputs are multiplied. However, with the composition $(f \circ g)(x) = f(g(x))$, the output $g(x)$ provides the input for f. These concepts are illustrated in Figure 5.5.

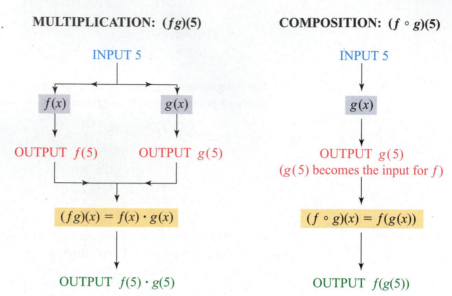

FIGURE 5.5 Comparison of Multiplication and Composition

■ **EXAMPLE 6** *Evaluating a composite function symbolically*

Let $f(x) = x^2 + 3x + 2$ and $g(x) = \dfrac{1}{x}$.

(a) Evaluate $(f \circ g)(2)$ and $(g \circ f)(2)$. How do they compare?
(b) Find symbolic expressions for $(f \circ g)(x)$ and $(g \circ f)(x)$. Are they equivalent expressions?
(c) Find the domains of $(f \circ g)(x)$ and $(g \circ f)(x)$.

SOLUTION

(a) $(f \circ g)(2) = f(g(2)) = f\left(\dfrac{1}{2}\right) = \left(\dfrac{1}{2}\right)^2 + 3\left(\dfrac{1}{2}\right) + 2 = 3.75.$

$(g \circ f)(2) = g(f(2)) = g(2^2 + 3 \cdot 2 + 2) = g(12) = \dfrac{1}{12} \approx 0.0833.$

The results are not equal.

(b) Symbolic representations for $(f \circ g)(x)$ and $(g \circ f)(x)$ also can be found.

$$(f \circ g)(x) = f(g(x)) = f\left(\dfrac{1}{x}\right) = \left(\dfrac{1}{x}\right)^2 + 3\left(\dfrac{1}{x}\right) + 2 = \dfrac{1}{x^2} + \dfrac{3}{x} + 2$$

$$(g \circ f)(x) = g(f(x)) = g(x^2 + 3x + 2) = \dfrac{1}{x^2 + 3x + 2}$$

The expressions for $(f \circ g)(x)$ and $(g \circ f)(x)$ are not equivalent.

(c) The domain of f is all real numbers and the domain of g is $\{x \mid x \neq 0\}$. The domain of $(f \circ g)(x) = f(g(x))$ consists of all x in the domain of g such that $g(x)$ is in the domain of f. Thus the domain of $(f \circ g)(x) = \dfrac{1}{x^2} + \dfrac{3}{x} + 2$ is $\{x \mid x \neq 0\}$.

The domain of $(g \circ f)(x) = g(f(x))$ consists of all x in the domain of f such that $f(x)$ is in the domain of g. Since $x^2 + 3x + 2 = 0$ when $x = -1$ or -2, the domain of $(g \circ f)(x) = \dfrac{1}{x^2 + 3x + 2}$ is $\{x \mid x \neq -1 \text{ or } x \neq -2\}$.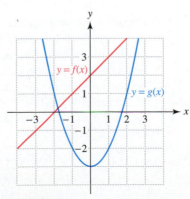

EXAMPLE 7 *Finding symbolic representations for composite functions*

Find $(f \circ g)(x)$ and $(g \circ f)(x)$.

(a) $f(x) = x + 2$, $\quad g(x) = x^3 - 2x^2 - 1$

(b) $f(x) = \sqrt{2x}$, $\quad g(x) = \dfrac{1}{x + 1}$

(c) $f(x) = 2x - 3$, $\quad g(x) = x^2 + 5$

SOLUTION **(a)** Begin by writing $(f \circ g)(x) = f(g(x)) = f(x^3 - 2x^2 - 1)$. Function f adds 2 to the input. That is, $f(\text{input}) = (\text{input}) + 2$ because $f(x) = x + 2$. Thus

$$f(x^3 - 2x - 1) = (x^3 - 2x^2 - 1) + 2 = x^3 - 2x^2 + 1.$$

To find $(g \circ f)(x)$, begin by writing $(g \circ f)(x) = g(f(x)) = g(x + 2)$. Function g does the following to its input: $g(\text{input}) = (\text{input})^3 - 2(\text{input})^2 - 1$. Thus

$$g(x + 2) = (x + 2)^3 - 2(x + 2)^2 - 1.$$

(b) $(f \circ g)(x) = f(g(x)) = f\!\left(\dfrac{1}{x + 1}\right) = \sqrt{2 \cdot \dfrac{1}{x + 1}} = \sqrt{\dfrac{2}{x + 1}}$

$(g \circ f)(x) = g(f(x)) = g(\sqrt{2x}) = \dfrac{1}{\sqrt{2x} + 1}$

(c) $(f \circ g)(x) = f(g(x)) = f(x^2 + 5) = 2(x^2 + 5) - 3 = 2x^2 + 7$

$(g \circ f)(x) = g(f(x)) = g(2x - 3) = (2x - 3)^2 + 5$

The next example shows how to evaluate a composition of functions graphically.

EXAMPLE 8 *Evaluating a composite function graphically*

Use the graphs of f and g shown in Figure 5.6 to evaluate each expression.

(a) $(f \circ g)(2)$ **(b)** $(g \circ f)(-3)$

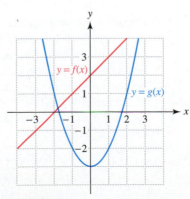

FIGURE 5.6

SOLUTION   **(a)** Because $(f \circ g)(2) = f(g(2))$, first evaluate $g(2)$. From Figure 5.7, $g(2) = 1$ and

$$(f \circ g)(2) = f(g(2)) = f(1).$$

To complete the evaluation, use Figure 5.8 to determine that $f(1) = 3$. Thus $(f \circ g)(2) = 3$.

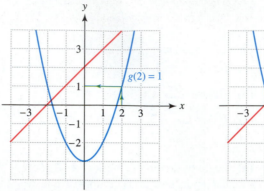

FIGURE 5.7   **FIGURE 5.8**

(b) Because $(g \circ f)(-3) = g(f(-3))$, first evaluate $f(-3)$. From Figure 5.9, $f(-3) = -1$ and

$$(g \circ f)(-3) = g(f(-3)) = g(-1).$$

From Figure 5.10, $g(-1) = -2$. Thus $(g \circ f)(-3) = -2$.

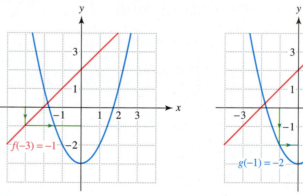

FIGURE 5.9   **FIGURE 5.10**

EXAMPLE 9   *Evaluating a composite function numerically*

Ozone in the stratosphere filters out approximately 90% of the harmful ultraviolet (UV) rays from the sun. Depletion of the ozone layer has caused an increase in the amount of UV radiation reaching the surface of the earth. An increase in UV radiation is associated with skin cancer. In Table 5.3 the function f computes the approximate percent *increase* in UV radiation resulting from an x percent *decrease* in the thickness of the ozone layer. The function g shown in Table 5.4 computes the expected percent increase in cases of skin cancer resulting from an x percent increase in UV radiation. (Source: R. Turner, D. Pearce, and I. Bateman, *Environmental Economics.*)

TABLE 5.3   Percent Increase in UV Radiation

x	0	1	2	3	4	5	6
$f(x)$	0	1.5	3.0	4.5	6.0	7.5	9.0

TABLE 5.4   Percent Increase in Skin Cancer

x	0	1.5	3.0	4.5	6.0	7.5	9.0
$g(x)$	0	5.25	10.5	15.75	21.0	26.25	31.5

(a) Find $(g \circ f)(2)$ and interpret this calculation.

(b) Determine a tabular representation for $g \circ f$. Describe what $(g \circ f)(x)$ computes.

SOLUTION   (a) $(g \circ f)(2) = g(f(2)) = g(3.0) = 10.5$. This means that a 2% decrease in the thickness of the ozone layer results in a 3% increase in UV radiation. This could cause a 10.5% increase in skin cancer.

(b) The values for $(g \circ f)(x)$ can be found in a manner similar to part (a). See Table 5.5.

TABLE 5.5

x	0	1	2	3	4	5	6
$(g \circ f)(x)$	0	5.25	10.5	15.75	21.0	26.25	31.5

The composition $(g \circ f)(x)$ computes the percent increase in cases of skin cancer resulting from an x percent decrease in the ozone layer.

EXAMPLE 10   *Modeling the urban heat island phenomena*

Cities are made up of large amounts of concrete and asphalt that heat up in the daytime from sunlight but do not cool off completely at night. As a result, urban areas tend to be warmer than the surrounding rural areas. This effect is called the *urban heat island* and has been documented in cities throughout the world. In Figure 5.11 function f computes the average increase in nighttime summer temperatures in degrees Celsius at Sky Harbor Airport in Phoenix from 1948 to 1990. In this graph 1948 is the base year with a zero temperature increase. This rise in urban temperature increased peak demand for electricity. In Figure 5.12 function g computes the percent increase in electrical demand for an average nighttime temperature increase of x degrees Celsius.   (Source: W. Cotton and R. Pielke, *Human Impacts on Weather and Climate.*)

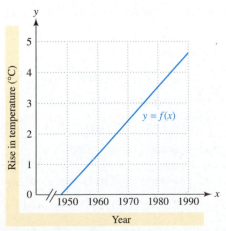

FIGURE 5.11 Nighttime Temperature Increase

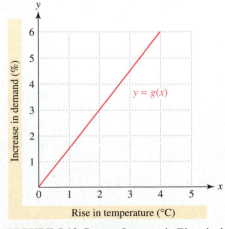

FIGURE 5.12 Percent Increase in Electrical Demand

(a) Evaluate $(g \circ f)(1975)$ graphically.   **(b)** Interpret $(g \circ f)(x)$.

SOLUTION **(a)** To evaluate $(g \circ f)(1975) = g(f(1975))$ graphically, first find $f(1975)$. In Figure 5.13, $f(1975) \approx 3$. Next let 3 be input for $g(x)$. In Figure 5.14, $g(3) \approx 4.5$. Thus $(g \circ f)(1975) = g(f(1975)) \approx g(3) \approx 4.5$. In 1975 the average nighttime temperature had risen about 3°C since 1948. This resulted in a 4.5% increase in peak demand for electricity.

(b) $(g \circ f)(x)$ computes the percent increase in peak electrical demand during year x.

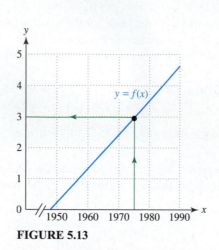

FIGURE 5.13

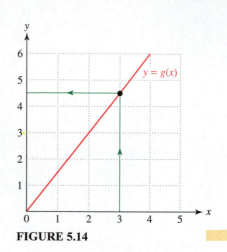

FIGURE 5.14

When you are solving problems, it is sometimes helpful to recognize a function as the composition of two simpler functions. For example, $h(x) = \sqrt[3]{x^2}$ can be thought of as the composition of the cube root function, $g(x) = \sqrt[3]{x}$, and the squaring function, $f(x) = x^2$. Then function h can be written as $h(x) = g(f(x)) = \sqrt[3]{x^2}$. This concept is demonstrated in the next example.

EXAMPLE 11   *Writing a function as a composition of two functions*

Find functions f and g so that $h(x) = (g \circ f)(x)$.

(a) $h(x) = (x + 3)^2$ **(b)** $h(x) = \sqrt{2x - 7}$ **(c)** $h(x) = \dfrac{1}{x^2 + 2x}$

SOLUTION **(a)** Let $f(x) = x + 3$ and $g(x) = x^2$. Then,

$$(g \circ f)(x) = g(f(x)) = g(x + 3) = (x + 3)^2.$$

(b) Let $f(x) = 2x - 7$ and $g(x) = \sqrt{x}$. Then,

$$(g \circ f)(x) = g(f(x)) = g(2x - 7) = \sqrt{2x - 7}.$$

(c) Let $f(x) = x^2 + 2x$ and $g(x) = \dfrac{1}{x}$. Then,

$$(g \circ f)(x) = g(f(x)) = g(x^2 + 2x) = \dfrac{1}{x^2 + 2x}.$$

Putting it all Together

5.1

Addition, subtraction, multiplication, and division can be used to combine functions. Composition of functions is another way to combine functions, and is fundamentally different from arithmetic operations on functions. In the composition $(g \circ f)(x)$, the *output* $f(x)$ is used as *input* for g.

The following table summarizes some concepts involved with combining functions.

Concept	Notation	Examples
Sum of two functions	$(f + g)(x) = f(x) + g(x)$	$f(x) = x^2, g(x) = 2x + 1$ $(f + g)(3) = f(3) + g(3)$ $= 9 + 7 = 16$ $(f + g)(x) = f(x) + g(x)$ $= x^2 + 2x + 1$
Difference of two functions	$(f - g)(x) = f(x) - g(x)$	$f(x) = 3x, g(x) = 2x + 1$ $(f - g)(1) = f(1) - g(1)$ $= 3 - 3 = 0$ $(f - g)(x) = f(x) - g(x)$ $= 3x - (2x + 1)$ $= x - 1$
Product of two functions	$(fg)(x) = f(x) \cdot g(x)$	$f(x) = x^3, g(x) = 1 - 3x$ $(fg)(-2) = f(-2) \cdot g(-2)$ $= (-8)(7) = -56$ $(fg)(x) = f(x) \cdot g(x)$ $= x^3(1 - 3x)$ $= x^3 - 3x^4$
Quotient of two functions	$\left(\dfrac{f}{g}\right)(x) = \dfrac{f(x)}{g(x)}, g(x) \neq 0$	$f(x) = x^2 - 1, g(x) = x + 2$ $\left(\dfrac{f}{g}\right)(2) = \dfrac{f(2)}{g(2)}$ $= \dfrac{3}{4}$ $\left(\dfrac{f}{g}\right)(x) = \dfrac{f(x)}{g(x)}$ $= \dfrac{x^2 - 1}{x + 2}, x \neq -2$
Composition of two functions	$(g \circ f)(x) = g(f(x))$	$f(x) = x^3, g(x) = x^2 - 2x + 1$ $(g \circ f)(2) = g(f(2)) = g(8)$ $= 64 - 16 + 1 = 49$ $(g \circ f)(x) = g(f(x))$ $= g(x^3)$ $= (x^3)^2 - 2(x^3) + 1$ $= x^6 - 2x^3 + 1$

5.1 EXERCISES

Arithmetic Operations on Functions

Exercises 1 and 2: Use the table to evaluate each expression, if possible.

(a) $(f + g)(2)$
(b) $(f - g)(4)$
(c) $(fg)(-2)$
(d) $(f/g)(0)$

1.

x	-2	0	2	4
$f(x)$	0	5	7	10
$g(x)$	6	0	-2	5

2.

x	-2	0	2	4
$f(x)$	-4	8	5	0
$g(x)$	2	-1	4	0

3. Use the table in Exercise 1 to complete the following table.

x	-2	0	2	4
$(f + g)(x)$				
$(f - g)(x)$				
$(fg)(x)$				
$(f/g)(x)$				

4. Use the table in Exercise 2 to complete the table in Exercise 3.

Exercises 5–8: Use the graph to evaluate each expression.

5. (a) $(f + g)(2)$
   (b) $(f - g)(1)$
   (c) $(fg)(0)$
   (d) $(f/g)(1)$

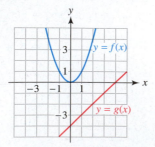

6. (a) $(f + g)(1)$
   (b) $(f - g)(0)$
   (c) $(fg)(-1)$
   (d) $(f/g)(1)$

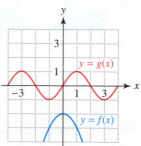

7. (a) $(f + g)(0)$
   (b) $(f - g)(-1)$
   (c) $(fg)(1)$
   (d) $(f/g)(2)$

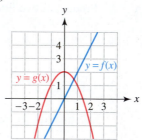

8. (a) $(f + g)(-1)$
   (b) $(f - g)(-2)$
   (c) $(fg)(0)$
   (d) $(f/g)(2)$

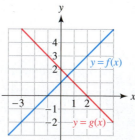

Exercises 9–12: Use $f(x)$ and $g(x)$ to evaluate each expression symbolically.

9. $f(x) = 2x - 3, g(x) = 1 - x^2$
   (a) $(f + g)(3)$ (b) $(f - g)(-1)$
   (c) $(fg)(0)$ (d) $(f/g)(2)$

10. $f(x) = 4x - x^3$, $g(x) = x + 3$
   (a) $(g + g)(-2)$ (b) $(f - g)(0)$
   (c) $(gf)(1)$ (d) $(g/f)(-3)$

11. $f(x) = 2x + 1$, $g(x) = \dfrac{1}{x}$

   (a) $(f + g)(2)$ (b) $(f - g)\left(\dfrac{1}{2}\right)$
   (c) $(fg)(4)$ (d) $(f/g)(0)$

12. $f(x) = \sqrt[3]{x^2}$, $g(x) = |x - 3|$
   (a) $(f + g)(-8)$ (b) $(f - g)(-1)$
   (c) $(fg)(0)$ (d) $(f/g)(27)$

Exercises 13–24: Use $f(x)$ and $g(x)$ to find each expression symbolically. Identify its domain.
   (a) $(f + g)(x)$
   (b) $(f - g)(x)$
   (c) $(fg)(x)$
   (d) $(f/g)(x)$

13. $f(x) = 2x$, $g(x) = x^2$

14. $f(x) = 1 - 4x$, $g(x) = 3x + 1$

15. $f(x) = x - \sqrt{x - 1}$, $g(x) = x + \sqrt{x - 1}$

16. $f(x) = \sqrt{1 - x}$, $g(x) = x^3$

17. $f(x) = \dfrac{1}{x + 1}$, $g(x) = \dfrac{3}{x + 1}$

18. $f(x) = 1 - x^2$, $g(x) = 3x^2 + 5$

19. $f(x) = 2x$, $g(x) = 3 - 4x$

20. $f(x) = x^{1/2}$, $g(x) = 3$

21. $f(x) = \dfrac{1}{2x - 4}$, $g(x) = \dfrac{x}{2x - 4}$

22. $f(x) = \dfrac{1}{x}$, $g(x) = x^3$

23. $f(x) = x + 2$, $g(x) = x^2 + 2x$

24. $f(x) = 6 - x$, $g(x) = \sqrt{x - 4}$

25. Let $f(x) = \sqrt{x}$ and $g(x) = x + 1$.
   (a) Graph $f(x)$, $g(x)$, and $(f + g)(x)$ in $[0, 9, 1]$ by $[0, 15, 1]$.
   (b) Explain how the graph of $(f + g)(x)$ can be found using the graphs of $f(x)$ and $g(x)$.

26. Let $f(x) = 0.2x + 5$ and $g(x) = 0.1x + 1$.
   (a) Graph $f(x)$, $g(x)$, and $(f - g)(x)$ in $[0, 10, 1]$ by $[0, 10, 1]$.
   (b) Explain how the graph of $(f - g)(x)$ can be found using the graphs of $f(x)$ and $g(x)$.

Composition of Functions

Exercises 27 and 28: Numerical representations for the functions f and g are given. Evaluate the expression, if possible.
   (a) $(g \circ f)(1)$
   (b) $(f \circ g)(4)$
   (c) $(f \circ f)(3)$

27.

x	1	2	3	4
$f(x)$	4	3	1	2

x	1	2	3	4
$g(x)$	2	3	4	5

28.

x	1	3	4	6
$f(x)$	2	6	5	7

x	2	3	5	7
$g(x)$	4	2	6	0

29. Use the tables for $f(x)$ and $g(x)$ in Exercise 27 to complete the composition shown in the diagram.

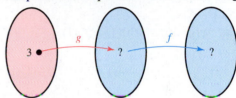

30. Use the tables for $f(x)$ and $g(x)$ in Exercise 28 to complete the composition shown in the diagram.

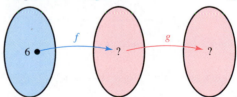

Exercises 31–34: Use the graph to evaluate each expression.

31. (a) $(f \circ g)(4)$ (b) $(g \circ f)(3)$ (c) $(f \circ f)(2)$

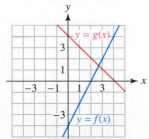

32. (a) $(f \circ g)(2)$
(b) $(g \circ g)(0)$
(c) $(g \circ f)(4)$

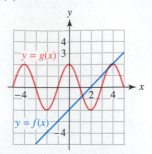

33. (a) $(f \circ g)(1)$ **(b)** $(g \circ f)(-2)$
(c) $(g \circ g)(-2)$

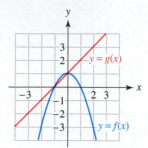

34. (a) $(f \circ g)(-2)$ **(b)** $(g \circ f)(1)$

(c) $(f \circ f)(0)$

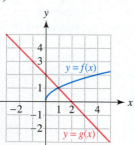

Exercises 35–38: Use $f(x)$ and $g(x)$ to evaluate each expression.

35. $f(x) = \sqrt{x + 5}, g(x) = x^2$
(a) $(f \circ g)(2)$ **(b)** $(g \circ f)(-1)$

36. $f(x) = |x^2 - 4|, g(x) = 2x^2 + x + 1$
(a) $(f \circ g)(1)$ **(b)** $(g \circ f)(-3)$

37. $f(x) = 5x - 2, g(x) = |x|$
(a) $(f \circ g)(-4)$ **(b)** $(g \circ f)(5)$

38. $f(x) = \dfrac{1}{x - 4}, g(x) = 5$
(a) $(f \circ g)(3)$ **(b)** $(g \circ f)(8)$

Exercises 39–52: Use $f(x)$ and $g(x)$ to find each of the following. Identify its domain.

    (a) $(f \circ g)(x)$
    (b) $(g \circ f)(x)$
    (c) $(f \circ f)(x)$

39. $f(x) = x^3$, $g(x) = x^2 + 3x - 1$

40. $f(x) = 2 - x$, $g(x) = \dfrac{1}{x^2}$

41. $f(x) = x^2$, $g(x) = \sqrt{1 - x}$

42. $f(x) = x + 2$, $g(x) = x^4 + x^2 - 3x - 4$

43. $f(x) = 2 - 3x$, $g(x) = x^3$

44. $f(x) = \sqrt{x}$, $g(x) = 1 - x^2$

45. $f(x) = x - 4$, $g(x) = 5 - x^2$

46. $f(x) = \sqrt{4 - x}$, $g(x) = x^2$

47. $f(x) = \dfrac{1}{x + 1}$, $g(x) = 5x$

48. $f(x) = \dfrac{1}{3x}$, $g(x) = \dfrac{2}{x - 1}$

49. $f(x) = x + 4$, $g(x) = \sqrt{4 - x^2}$

50. $f(x) = 2x + 1$, $g(x) = 4x^3 - 5x^2$

51. $f(x) = \sqrt{x - 1}$, $g(x) = 3x$

52. $f(x) = x^2 - 1$, $g(x) = 2 - 3x^2$

Exercises 53–60: (Refer to Example 11.) Find functions f and g so that $h(x) = (g \circ f)(x)$.

53. $h(x) = \sqrt{x - 2}$ **54.** $h(x) = (x + 2)^4$

55. $h(x) = 5(x + 2)^2 - 4$ **56.** $h(x) = \dfrac{1}{x + 2}$

57. $h(x) = 4(2x + 1)^3$ **58.** $h(x) = \sqrt[3]{x^2 + 1}$

59. $h(x) = (x^3 - 1)^2$ **60.** $h(x) = 4(x - 5)^{-2}$

Applications

61. *Revenue, Cost, and Profit* Suppose that it costs $150,000 to produce a master disc for a music video and $1.50 to produce each copy.
(a) Write a cost function C that outputs the cost of producing the master disc and x copies.
(b) If the music video is sold for $6.50 each, find a function R that outputs the revenue received from selling x music videos. What is the revenue from selling 8000 compact discs?
(c) Assuming that the master disc is not sold, find a function P that outputs the profit from sell-

ing x music videos. What is the profit from selling 40,000 videos?

(d) How many videos must be sold to break even? That is, how many videos must be sold for the revenue to equal the cost?

62. *Profit* (Refer to Example 4.) Determine a profit function P that results if the compact discs are sold for $15 each. Find the profit from selling 3000 compact discs.

63. *Skin Cancer* (Refer to Example 9 and Tables 5.3 and 5.4.) If possible, calculate the composition and interpret the result.
(a) $(g \circ f)(1)$ (b) $(f \circ g)(21)$

64. *Skin Cancer* In Example 9 the functions f and g are both linear.
(a) Find symbolic representations for f and g.
(b) Determine $(g \circ f)(x)$.
(c) Evaluate $(g \circ f)(3.5)$ and interpret the result.

65. *Urban Heat Island* (Refer to Example 10.) The functions f and g are given by

$$f(x) = 0.11(x - 1948) \quad \text{and} \quad g(x) = 1.5x.$$

(a) Evaluate $(g \circ f)(1960)$.
(b) Find $(g \circ f)(x)$.
(c) What type of functions are f, g, and $g \circ f$?

66. *Urban Heat Island* (Refer to Example 10 and Figures 5.11 and 5.12.) If possible, calculate the composition and interpret the result.
(a) $(g \circ f)(1980)$
(b) $(f \circ g)(3)$

67. *Circular Wave* A marble is dropped into a lake and it results in a circular wave, whose radius increases at a rate of 6 inches per second. Write a formula for C that gives the circumference of the circular wave in inches after t seconds.

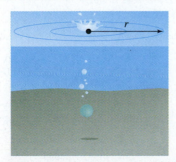

Geometry Review To review formulas related to circles, see Chapter R (page R-2).

68. *Circular Wave* (Refer to Exercise 67.) Write a function A that gives the area contained inside the circular wave in square inches after t seconds.

69. *Geometry* The surface area of a cone (excluding the bottom) is given by $S = \pi r \sqrt{r^2 + h^2}$, where r is its radius and h is its height, as shown in the accompanying figure. If the height is twice the radius, write a formula for S in terms of r.

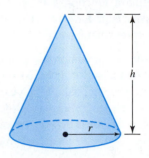

70. *Sphere* The volume V of a sphere with radius r is given by $V = \dfrac{4}{3}\pi r^3$, and the surface area S is given by $S = 4\pi r^2$. Show that $V = \dfrac{4}{3}\pi \left(\dfrac{S}{4\pi}\right)^{3/2}$.

71. *Hollywood Movies* From 1991 to 1996 the cost to produce and market Hollywood movies increased. In the table, f computes the average cost to produce a movie, and g computes the average cost to market a movie. (Source: Motion Picture Association of America.)

x (yr)	1991	1992	1993
$f(x)$ ($ millions)	26	28	30
$g(x)$ ($ millions)	12	14	14

x (yr)	1994	1995	1996
$f(x)$ ($ millions)	33	37	41
$g(x)$ ($ millions)	17	17	20

(a) Make a table for a function h that computes the average cost to produce *and* market a movie in the year x.
(b) Write an equation that relates $f(x)$, $g(x)$, and $h(x)$.

72. *Petroleum Spillage* Large amounts of petroleum products enter the oceans each year. In the table on the next page, f computes the total amount of petroleum that enters the world's oceans. The function g computes petroleum spillage into the

oceans caused by oil tankers. Amounts are given in thousands of tons. (Source: B. Freedman, *Environmental Ecology.*)

Year	1973	1979	1981	1983	1989
$f(x)$	6110	4670	3570	3200	570
$g(x)$	1380	900	1050	1100	—

(a) Define a function h by $h(x) = f(x) - g(x)$. Make a table for $h(x)$. What is the domain of h?

(b) Interpret what h computes.

73. *Acid Rain* A common air pollutant responsible for acid rain is sulfur dioxide (SO_2). Emissions of SO_2 from burning coal during year x are computed by $f(x)$ in the table. Emissions of SO_2 from burning oil are computed by $g(x)$. Amounts are given in millions of tons. (Source: B. Freedman.)

x	1860	1900	1940	1970	2000
$f(x)$	2.4	12.6	24.2	32.4	55.0
$g(x)$	0.0	0.2	2.3	17.6	23.0

(a) Evaluate $(f + g)(1970)$.

(b) Interpret $(f + g)(x)$.

(c) Make a table for $(f + g)(x)$.

74. (Refer to Exercise 73.) Make a table for a function h defined by $h(x) = (f/g)(x)$. Round values for $h(x)$ to the nearest hundredth. What information does h give regarding the use of coal and oil in the United States?

75. *Methane Emissions* Methane is a greenhouse gas. It lets sunlight into the atmosphere but blocks heat from escaping the earth's atmosphere. Methane is a by-product of burning fossil fuels. In the table, f models the predicted methane emissions in millions of tons produced by developed countries. The function g models the same emissions for developing countries. (Source: A. Nilsson, *Greenhouse Earth.*)

x	1990	2000	2010	2020	2030
$f(x)$	27	28	29	30	31
$g(x)$	5	7.5	10	12.5	15

(a) Make a table for a function h that models the total predicted methane emissions for developed *and* developing countries.

(b) Write an equation that relates $f(x)$, $g(x)$, and $h(x)$.

76. (Refer to the previous exercise.) The accompanying figure shows graphical representations of the func-

tions f and g that model methane emissions. Use their graphs to sketch a graph of the function h.

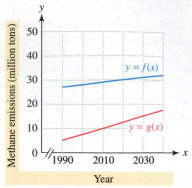

77. (Refer to Exercises 75 and 76.) Symbolic representations for f and g are $f(x) = 0.1x - 172$ and $g(x) = 0.25x - 492.5$, where x is the year. Find a symbolic representation for h.

78. *Stopping Distance* (Refer to Example 1.) A car is traveling at 60 miles per hour. The driver has a reaction time of 1.25 seconds.

(a) Determine a function r that computes the reaction distance for this driver.

(b) Find a symbolic representation for a function s that computes the stopping distance for this driver traveling at x miles per hour.

(c) Evaluate $s(60)$ and interpret the result.

79. *China's Energy Production* Predicted energy production in China is shown in the figure. The function f computes total coal production, and the function g computes total coal *and* oil production. Energy units are in million metric tons of oil equivalent (Mtoe). Let the function h compute China's oil production. (Source: A. Pascal, *Global Energy: The Changing Outlook.*)

(a) Write an equation that relates $f(x)$, $g(x)$, and $h(x)$.

(b) Evaluate $h(1995)$ and $h(2000)$.

(c) Determine a symbolic representation for h. (*Hint:* h is linear.)

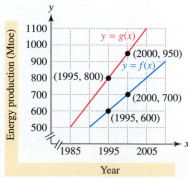

80. *AIDS* The accompanying table lists both the cumulative numbers of AIDS cases and cumulative deaths reported for selected years x. The numbers of AIDS cases are modeled by $f(x) = 3200(x - 1982)^2 + 1586$ and the numbers of deaths are modeled by $g(x) = 1900(x - 1982)^2 + 619$. (Source: Department of Health and Human Services.)

Year	Cases	Deaths
1982	1586	619
1984	10,927	5605
1986	41,910	24,593
1988	106,304	61,911
1990	196,576	120,811
1992	329,205	196,283
1994	441,528	270,533

(a) Graph $h(x) = \dfrac{g(x)}{f(x)}$ in [1982, 1994, 2] by [0, 1, 0.1]. Interpret the graph.

(b) Compute the ratio $\dfrac{\text{Deaths}}{\text{Cases}}$ for each year. Compare these ratios with the results from part (a).

81. *Swimming Pools* In the accompanying figures, f computes the cubic feet of water in a pool after x days, and g converts cubic feet to gallons.
(a) Estimate the gallons of water in the pool when $x = 2$.
(b) Interpret $(g \circ f)(x)$.

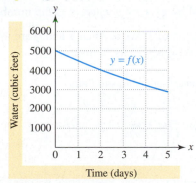

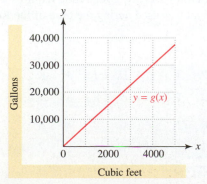

82. *Temperature* The figures show graphs of f and g. The function f computes the temperature on a summer day after x hours, and g converts Fahrenheit temperature to Celsius temperature.
(a) Evaluate $(g \circ f)(2)$.
(b) Interpret $(g \circ f)(x)$.

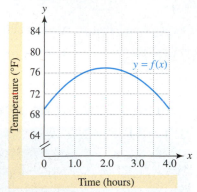

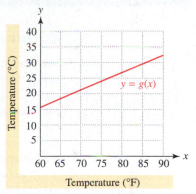

83. Let $f(x) = k$ and $g(x) = ax + b$, where k, a, and b are constants.
(a) Find $(f \circ g)(x)$. What type of function is $f \circ g$?
(b) Find $(g \circ f)(x)$. What type of function is $g \circ f$?

84. Show that if $f(x) = ax + b$ and $g(x) = cx + d$, then $(g \circ f)(x)$ also represents a linear function. Find the slope of the graph of $(g \circ f)(x)$.

Writing about Mathematics

85. Describe differences between $(fg)(x)$ and $(f \circ g)(x)$. Give examples.

86. Describe differences between $(f \circ g)(x)$ and $(g \circ f)(x)$. Give examples.

Extended and Discovery Exercises

Exercises 1–4:   **Combining Polynomials**   *Suppose that $f(x)$ is a polynomial with degree n and $g(x)$ is a polynomial with degree m and $m > n \geq 1$. What can be said about the degree D of each expression?*

1. $(gf)(x)$

2. $(g/f)(x)$ (Assume that $f(x)$ divides into $g(x)$ with no remainder.)

3. $(g \circ f)(x)$ **4.** $(f + g)(x)$

5.2   INVERSE FUNCTIONS AND THEIR REPRESENTATIONS

Inverse Operations • One-to-One Functions • Symbolic Representations of Inverse Functions • Other Representations of Inverse Functions

Introduction

Many actions are reversible. A closed door can be opened—an open door can be closed. One hundred dollars can be withdrawn from or deposited into a savings account. These actions undo or cancel each other. But not all actions are reversible. Explosions and weather are two examples. In mathematics this concept of reversing a calculation and arriving at the original result is associated with an inverse.

To introduce the basic concept of an inverse function, refer to Table 5.6, which can be used to convert gallons to pints. It is apparent that there are 8 pints in a gallon, and $f(x) = 8x$ converts x gallons to an equivalent number of pints. However, if we want to reverse the computation and convert pints to gallons, a different function g is required. This conversion is calculated by $g(x) = \dfrac{x}{8}$, where x is the number of pints.

Function g performs the inverse operation of f. We say that f and g are inverse functions and write this as $g(x) = f^{-1}(x)$. We read f^{-1} as "f inverse."

TABLE 5.6   Converting Gallons to Pints

x (gallons)	1	2	3	4	5
$f(x)$ (pints)	8	16	24	32	40

Inverse Operations

Throughout time, codes have been used for secrecy. Every code requires a consistent method to decode it. A simple code that was used during the time of the Roman Empire is shown in Table 5.7.

TABLE 5.7

Letter	A	B	C	D	E	F	G	H	I	J	K	L	M
Code	C	D	E	F	G	H	I	J	K	L	M	N	O

Letter	N	O	P	Q	R	S	T	U	V	W	X	Y	Z
Code	P	Q	R	S	T	U	V	W	X	Y	Z	A	B

Source: A. Sinkov, *Elementary Cryptanalysis: A Mathematical Approach.*

Table 5.7 can be used to code the word HELP as JGNR and decode it back to HELP. Coding and decoding are inverse actions or operations. It is essential that both the coding and decoding procedures give exactly one output for each input.

Actions and their inverses occur in everyday life. Suppose a person opens a car door, gets in, and starts the engine. What are the inverse actions? They are turn off the engine, get out, and close the car door. Notice that the order must be reversed as well as applying the inverse operation at each step.

In mathematics there are basic operations that can be considered inverse operations of each other. For example, if we begin with 10 and add 5, the result is 15. To undo this operation, subtract 5 from 15 to obtain 10. Addition and subtraction are inverse operations. The same is true for multiplication and division. If we multiply a number by 2 and then divide by 2, the original number is obtained. Division and multiplication are inverse operations.

EXAMPLE 1 *Finding inverse actions and operations*

For each of the following, state the inverse actions or operations.
(a) Put on a coat and go outside.
(b) Subtract 7 from x and divide the result by 2.

SOLUTION **(a)** To find the inverse actions, reverse the order and apply the inverse action at each step. The inverse actions would be to come inside and take off the coat.
(b) We must reverse the order and apply the inverse operation at each step. The inverse operations would be to multiply x by 2 and add 7. The original operations could be expressed as $\dfrac{x-7}{2}$, and the inverse operations could be written as $2x + 7$.

Inverse operations can be described by functions. If $f(x) = x + 5$, then the *inverse function* of f is given by $f^{-1}(x) = x - 5$. For example, $f(5) = 10$, and $f^{-1}(10) = 5$. If input x produces output y with function f, input y produces output x with function f^{-1}. This can be seen numerically in Table 5.8.

TABLE 5.8

x	0	5	10	15
$f(x)$	5	10	15	20

x	5	10	15	20
$f^{-1}(x)$	0	5	10	15

From Table 5.8, note that $f(0) = 5$ and $f^{-1}(5) = 0$, and that $f(15) = 20$ and $f^{-1}(20) = 15$. In general, if $f(a) = b$ then $f^{-1}(b) = a$. That is, if f outputs b with input a, then f^{-1} must output a with input b. *Outputs and inputs are interchanged for inverse functions.* This statement is illustrated in Figure 5.15.

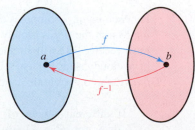

FIGURE 5.15 $f(a) = b; f^{-1}(b) = a$

When $f(x) = x + 5$ and $f^{-1}(x) = x - 5$ are applied in sequence, the output of f is used as input for f^{-1}. This is composition of functions.

$$(f^{-1} \circ f)(x) = f^{-1}(f(x)) \qquad \text{Definition of composition}$$
$$= f^{-1}(x + 5) \qquad f(x) = x + 5$$
$$= (x + 5) - 5 \qquad f^{-1} \text{ subtracts 5 from its input.}$$
$$= x \qquad \text{Simplify.}$$

The composition $f^{-1} \circ f$ with input x produces output x. The same action occurs for $f \circ f^{-1}$.

$$(f \circ f^{-1})(x) = f(f^{-1}(x)) \qquad \text{Definition of composition}$$
$$= f(x - 5) \qquad f^{-1}(x) = x - 5$$
$$= (x - 5) + 5 \qquad f \text{ adds 5 to its input.}$$
$$= x \qquad \text{Simplify.}$$

Another pair of inverse functions is given by $f(x) = x^3$ and $f^{-1}(x) = \sqrt[3]{x}$. Before a formal definition of inverse functions is given, we must discuss the concept of one-to-one functions.

■ **MAKING CONNECTIONS**

The Notation f^{-1} and Negative Exponents If a represents a real number, then $a^{-1} = \dfrac{1}{a}$. For example, $4^{-1} = \dfrac{1}{4}$. On the other hand, if f represents a function, then $f^{-1}(x) \neq \dfrac{1}{f(x)}$. Instead, $f^{-1}(x)$ represents the inverse function of f. For example, if $f(x) = 5x$, then $f^{-1}(x) = \dfrac{x}{5} \neq \dfrac{1}{5x}$.

One-to-One Functions

Does every function have an inverse function? The next example answers this question.

EXAMPLE 2 *Determining if a function has an inverse function*

Table 5.9 represents a function f that computes the percentage of the time that the sky is cloudy in Augusta, Georgia, where x corresponds to the standard numbers for the months. Determine if f has an inverse function.

TABLE 5.9 Cloudy Skies in Augusta

x (month)	1	2	3	4	5	6	7	8	9	10	11	12
$f(x)$ (%)	43	40	39	29	28	26	27	25	30	26	31	39

Source: J. Williams, *The Weather Almanac 1995.*

SOLUTION For each input, f computes exactly one output. For example, $f(3) = 39$ means that during March the sky is cloudy 39% of the time. If f has an inverse function, the inverse must receive 39 as input and produce exactly one output. Both March and December have cloudy skies 39% of the time. Given an input of 39, it is impossible for an inverse *function* to output both 3 and 12. Therefore f does not have an inverse function.

If different inputs of f produce the same output, then an inverse function of f does not exist. However, if different inputs always produce different outputs, f is a *one-to-one function*. Every one-to-one function has an inverse function. For example, $f(x) = x^2$ is not one-to-one because $f(-2) = 4$ and $f(2) = 4$. Therefore $f(x) = x^2$ does not have an inverse function.

> ## One-to-one function
>
> A function f is a **one-to-one function** if, for elements c and d in the domain of f,
>
> $$c \neq d \qquad \text{implies} \qquad f(c) \neq f(d).$$
>
> That is, different inputs always result in different outputs.

EXAMPLE 3   *Determining if a function is one-to-one graphically*

Use the graph of f to determine if f is a one-to-one function in Figures 5.16 and 5.17.

(a)

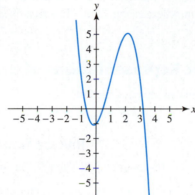

FIGURE 5.16

(b)

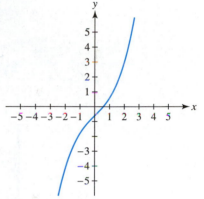

FIGURE 5.17

SOLUTION   **(a)** To decide if the graph represents a one-to-one function, determine if different inputs (x-values) always correspond to different outputs (y-values). In Figure 5.18 the horizontal line $y = 2$ intersects the graph of f at $(-1, 2)$, $(1, 2)$, and $(3, 2)$. This means that $f(-1) = f(1) = f(3) = 2$. Three distinct inputs, -1, 1, and 3, produce the same output, 2. Therefore f is not one-to-one and does not have an inverse function.

(b) Any horizontal line will intersect the graph of f at most once. For example, if the line $y = 3$ is graphed, it intersects the graph of f at one point, $(2, 3)$. See Figure 5.19. Only input 2 results in output 3. This is true in general. Therefore f is one-to-one and has an inverse function.

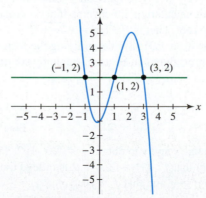

FIGURE 5.18

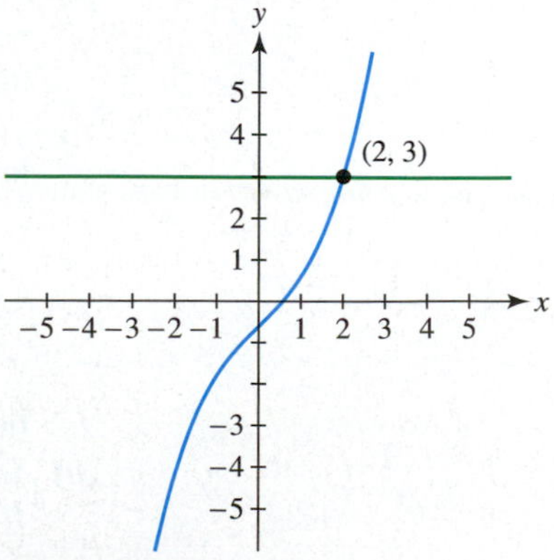

FIGURE 5.19

Note: To show that f is not one-to-one, it was not necessary to find the actual points of intersection—we have to show only that a horizontal line can intersect the graph of f more than once.

The technique of visualizing horizontal lines to determine if a graph represents a one-to-one function is called the *horizontal line test*.

Horizontal line test

If every horizontal line intersects the graph of a function f at most once, then f is a one-to-one function.

■ **CLASS DISCUSSION**

Use the horizontal line test to explain why a nonconstant, linear function has an inverse function, whereas a quadratic function does not. ■

Symbolic Representations of Inverse Functions

We now define an inverse function.

Inverse function

Let f be a one-to-one function. Then f^{-1} is the **inverse function** of f, if

$$(f \circ f^{-1})(x) = f(f^{-1}(x)) = x \qquad \text{for every } x \text{ in the domain of } f^{-1}, \text{ and}$$
$$(f^{-1} \circ f)(x) = f^{-1}(f(x)) = x \qquad \text{for every } x \text{ in the domain of } f.$$

In the next example, we find an inverse function and verify that it is correct.

EXAMPLE 4   *Finding and verifying an inverse function*

Let f be the one-to-one function given by $f(x) = x^3 - 2$.
(a) Find $f^{-1}(x)$.
(b) Verify that your result from part (b) is correct.

SOLUTION   **(a)** Since $f(x) = x^3 - 2$, function f cubes the input x and then subtracts 2. To reverse this calculation, the inverse function must add 2 to the input x and then take the cube root. That is, $f^{-1}(x) = \sqrt[3]{x + 2}$.

If we have difficulty determining the formula for an inverse mentally, the following procedure can be used. Begin by solving $y = f(x)$ for x.

$$\begin{aligned} y &= x^3 - 2 &\qquad & y = f(x) \\ y + 2 &= x^3 &\qquad & \text{Add 2.} \\ \sqrt[3]{y + 2} &= x &\qquad & \text{Take the cube root.} \end{aligned}$$

Interchange x and y to obtain $y = \sqrt[3]{x + 2}$. This gives us the formula for $f^{-1}(x)$.
(b) To verify that $f^{-1}(x) = \sqrt[3]{x + 2}$ is indeed the inverse of $f(x)$, we must show that $f(f^{-1}(x)) = x$ and that $f^{-1}(f(x)) = x$.

$$f(f^{-1}(x)) = f\left(\sqrt[3]{x+2}\right) \qquad f^{-1}(x) = \sqrt[3]{x+2}$$
$$= \left(\sqrt[3]{x+2}\right)^3 - 2 \qquad f(x) = x^3 - 2$$
$$= (x+2) - 2 \qquad \text{Cube the expression.}$$
$$= x \qquad \text{Combine terms.}$$
$$f^{-1}(f(x)) = f^{-1}(x^3 - 2) \qquad f(x) = x^3 - 2$$
$$= \sqrt[3]{(x^3 - 2) + 2} \qquad f^{-1}(x) = \sqrt[3]{x+2}$$
$$= \sqrt[3]{x^3} \qquad \text{Combine terms.}$$
$$= x \qquad \text{Simplify.}$$

Algebra Review
To review fractional exponents and radical notation, see Section 4.7.

These calculations verify that our results are correct.

This symbolic technique is now summarized verbally.

Finding a symbolic representation for f^{-1}

To find a formula for f^{-1}, perform the following steps.

STEP 1: Verify that f is a one-to-one function.

STEP 2: Solve the equation $y = f(x)$ for x, resulting in the equation $x = f^{-1}(y)$.

STEP 3: Interchange x and y to obtain $y = f^{-1}(x)$.

To verify $f^{-1}(x)$, show that $(f^{-1} \circ f)(x) = x$ and $(f \circ f^{-1})(x) = x$.

EXAMPLE 5 *Finding and verifying an inverse function*

Suppose that a person x inches tall sprains an ankle. An approximation for the length of a crutch (in inches) that this person might need is calculated by $f(x) = \frac{18}{25}x + 2$.
(a) Explain why f is a one-to-one function.
(b) Find a symbolic representation for f^{-1}.
(c) Verify that this is the inverse function of f.
(d) Interpret the meaning of $f^{-1}(56)$.

SOLUTION **(a)** Since f is a linear function, its graph is a line with a slope of $\frac{18}{25}$.

Every horizontal line intersects it at most once. See Figure 5.20. By the horizontal line test, f is one-to-one.

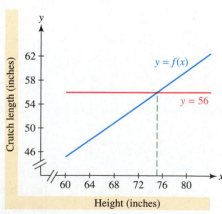

FIGURE 5.20

(b) To find $f^{-1}(x)$, solve the equation $y = f(x)$ for x.

$$y = \frac{18}{25}x + 2 \qquad y = f(x)$$

$$y - 2 = \frac{18}{25}x \qquad \text{Subtract 2.}$$

$$\frac{25}{18}(y - 2) = x \qquad \text{Multiply by } \frac{25}{18}, \text{ the reciprocal of } \frac{18}{25}.$$

Now interchange x and y to obtain $y = \frac{25}{18}(x - 2)$. The formula for the inverse is, therefore,

$$f^{-1}(x) = \frac{25}{18}(x - 2).$$

(c) We must show that $(f^{-1} \circ f)(x) = x$ and $(f \circ f^{-1})(x) = x$.

$$(f^{-1} \circ f)(x) = f^{-1}(f(x)) \qquad \text{Definition of composition}$$

$$= f^{-1}\left(\frac{18}{25}x + 2\right) \qquad f(x) = \frac{18}{25}x + 2$$

$$= \frac{25}{18}\left(\left(\frac{18}{25}x + 2\right) - 2\right) \qquad f^{-1}(x) = \frac{25}{18}(x - 2)$$

$$= \frac{25}{18}\left(\frac{18}{25}x\right) \qquad \text{Simplify.}$$

$$= x \qquad \text{It checks.}$$

Similarly,

$$(f \circ f^{-1})(x) = f(f^{-1}(x)) \qquad \text{Definition of composition}$$

$$= f\left(\frac{25}{18}(x - 2)\right) \qquad f^{-1}(x) = \frac{25}{18}(x - 2)$$

$$= \frac{18}{25}\left(\frac{25}{18}(x - 2)\right) + 2 \qquad f(x) = \frac{18}{25}x + 2$$

$$= (x - 2) + 2 \qquad \text{Simplify.}$$

$$= x \qquad \text{It checks.}$$

(d) The expression $f(x)$ calculates the proper crutch length for a person x inches tall, and $f^{-1}(x)$ computes the height of a person requiring a crutch x inches long. Therefore $f^{-1}(56) = \frac{25}{18}(56 - 2) = 75$ means that a 56-inch crutch would be appropriate for a person 75 inches tall. See Figure 5.20.

EXAMPLE 6   *Restricting the domain of a function*

Let $f(x) = (x - 1)^2$.
(a) Does f have an inverse function? Explain.
(b) Restrict the domain of f so that f^{-1} exists.
(c) Find $f^{-1}(x)$ for the restricted domain.

SOLUTION   **(a)** The graph of $f(x) = (x - 1)^2$, shown in Figure 5.21, does not pass the horizontal line test. Therefore f is not one-to-one and does not have an inverse function.

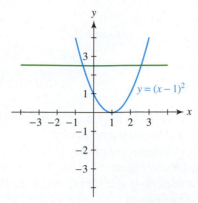

FIGURE 5.21 Not One-to-One   **FIGURE 5.22** Restricting the Domain

(b) If we restrict the domain of f to $D = \{x \mid x \geq 1\}$, then f becomes a one-to-one function. To illustrate this, the graph of $y = (x - 1)^2$ for $x \geq 1$ is shown in Figure 5.22. This graph passes the horizontal line test and now f^{-1} exists.

(c) Assume that $x \geq 1$ and solve the equation $y = f(x)$ for y.

$$y = (x - 1)^2 \qquad y = f(x)$$

$$\sqrt{y} = x - 1 \qquad \text{Take the positive square root.}$$
$$\qquad\qquad\qquad \textit{Note: } x \geq 1 \text{ implies that } x - 1 \geq 0.$$

$$\sqrt{y} + 1 = x \qquad \text{Add 1.}$$

Thus $f^{-1}(x) = \sqrt{x} + 1$.

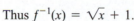

FIGURE 5.23

The graph of $f(x) = x^5 + 3x^3 - x^2 + 5x - 3$ is shown in Figure 5.23. By the horizontal line test, f is a one-to-one function and f^{-1} exists. However, it would be difficult to obtain a symbolic representation for f because we would need to solve the equation $y = x^5 + 3x^3 - x^2 + 5x - 3$ for x. It is important to realize that many functions have inverses, but their inverses can be either difficult or impossible to find symbolically.

Other Representations of Inverse Functions

Given a symbolic representation of a one-to-one function f, we can often find its inverse f^{-1} by using the techniques discussed in the previous subsection. Numerical and graphical representations of a one-to-one function can also be used to find its inverse.

Numerical Representations.   In Table 5.10, f has domain $D = \{1940, 1970, 1995\}$. It computes the percentage of the U.S. population with 4 or more years of college in year x.

TABLE 5.10

x	1940	1970	1995
$f(x)$	5	11	23

Source: Bureau of the Census.

Function f is one-to-one because different inputs produce different outputs. Therefore f^{-1} exists. Since $f(1940) = 5$, it follows that $f^{-1}(5) = 1940$. In a similar manner, $f^{-1}(11) = 1970$ and $f^{-1}(23) = 1995$. Table 5.11 shows a numerical representation of f^{-1}.

TABLE 5.11

x	5	11	23
$f^{-1}(x)$	1940	1970	1995

The domain of f is $\{1940, 1970, 1995\}$ and the range is $\{5, 11, 23\}$. The domain of f^{-1} is $\{5, 11, 23\}$ and the range of f^{-1} is $\{1940, 1970, 1995\}$. The functions f and f^{-1} interchange domains and ranges. This result is true in general for inverse functions.

The diagrams in Figures 5.24 and 5.25 demonstrate this property. To obtain f^{-1} from f, the arrows for f are simply reversed. This reversal causes the domains and ranges to be interchanged.

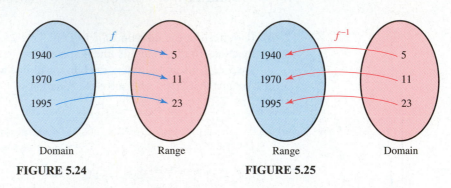

FIGURE 5.24   **FIGURE 5.25**

Figure 5.26 shows a function f that is not one-to-one. In Figure 5.27 the arrows defining f have been reversed. This is a relation. However, this relation does not represent the inverse *function*, since input 4 produces two outputs, 1 and 2.

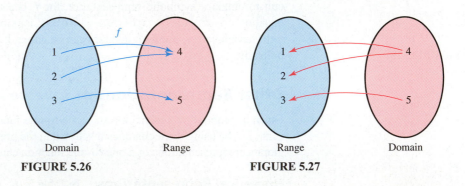

FIGURE 5.26   **FIGURE 5.27**

The relationship between domains and ranges is summarized in the following.

Domains and ranges of inverse functions

The domain of f equals the range of f^{-1}.
The range of f equals the domain of f^{-1}.

Graphical Representations. If the point $(2, 5)$ lies on the graph of f, then $f(2) = 5$ and $f^{-1}(5) = 2$. Therefore the point $(5, 2)$ must lie on the graph of f^{-1}. In general, if the point (a, b) lies on the graph of f, then the point (b, a) lies on the graph of f^{-1}. Refer to Figure 5.28. If a line segment is drawn between the points (a, b) and (b, a), the line $y = x$ is a *perpendicular bisector* of this line segment. Figure 5.29 shows pairs of points in the form (a, b) and (b, a). Figure 5.30 contains continuous graphs of f and f^{-1} passing through these points. The graph of f^{-1} is a *reflection* of the graph of f across the line $y = x$.

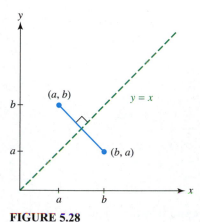

FIGURE 5.28

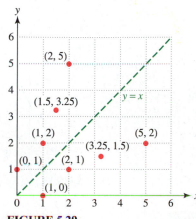

FIGURE 5.29

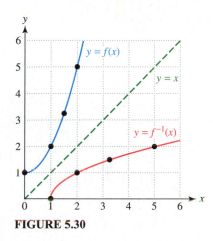

FIGURE 5.30

> ## Graphs of functions and their inverses
>
> The graph of f^{-1} is a reflection of the graph of f across the line $y = x$.

EXAMPLE 7   *Representing an inverse function graphically*

Let $f(x) = x^3 + 2$. Graph f. Then sketch a graph of f^{-1}.

SOLUTION   Figure 5.31 shows a graph of f. To sketch a graph of f^{-1}, reflect the graph of f across the line $y = x$. The graph of f^{-1} appears as though it were the image of the graph of f in a mirror located along $y = x$. See Figure 5.32.

$[-5, 5, 1]$ by $[-5, 5, 1]$

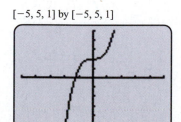

FIGURE 5.31

Graphing Calculator Help

To graph an inverse function, see Appendix B, (page AP-17).

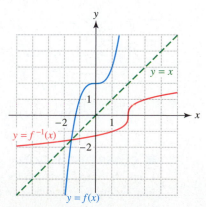

FIGURE 5.32

EXAMPLE 8   *Graphically evaluating f and f⁻¹*

Use the graph of f in Figure 5.33 to evaluate each expression.
(a) $f(2)$
(b) $f^{-1}(3)$
(c) $f^{-1}(-3)$

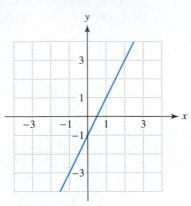

FIGURE 5.33

SOLUTION   **(a)** To evaluate $f(2)$, find 2 on the x-axis, move upward to the graph of f, and then move across to the y-axis to obtain $f(2) = 3$, as shown in Figure 5.34.

(b) Evaluating $f^{-1}(3)$ is a slightly different process because we are given a graph of f, not f^{-1}. We must reverse the method used in part (a). Start by finding 3 on the y-axis, move across to the graph of f, and then downward to the x-axis to obtain $f^{-1}(3) = 2$, as shown in Figure 5.35. Notice from part (a) that $f(2) = 3$ and $f^{-1}(3) = 2$ in part (b).

(c) Find -3 on the y-axis, move left to the graph of f, and then upward to the x-axis. We can see from Figure 5.36 that $f^{-1}(-3) = -1$.

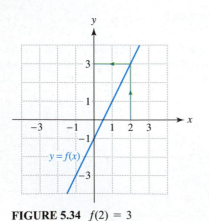

FIGURE 5.34 $f(2) = 3$

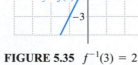

FIGURE 5.35 $f^{-1}(3) = 2$

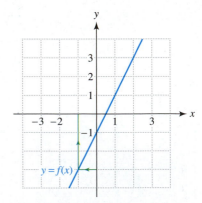

FIGURE 5.36 $f^{-1}(-3) = -1$

■ **CLASS DISCUSSION**

Does an inverse function for $f(x) = |2x - 1|$ exist? Explain. What difficulties would you encounter if you tried to evaluate $f^{-1}(3)$ graphically? ■

Putting it all Together 5.2

The following table summarizes some important concepts about inverse functions.

Concept	Comments	Examples
One-to-one function	f is one-to-one if different inputs always result in different outputs. That is, $a \neq b$ implies $f(a) \neq f(b)$.	$f(x) = x^2 - 4x$ is not one-to-one because $f(0) = 0$ *and* $f(4) = 0$. With this function, *different* inputs can result in the *same* output.
Horizontal line test	If every horizontal line intersects the graph of f at most once, then f is one-to-one.	 f is not one-to-one because a horizontal line can intersect it more than once.
Inverse function	If a function is one-to-one, it has an inverse function f^{-1} that satisfies both $$f^{-1}(f(x)) = x$$ and $$f(f^{-1}(x)) = x.$$	$f(x) = 3x - 1$ is one-to-one and has inverse function $f^{-1}(x) = \dfrac{x+1}{3}$. $$\begin{aligned} f^{-1}(f(x)) &= f^{-1}(3x - 1) \\ &= \frac{(3x - 1) + 1}{3} \\ &= x \end{aligned}$$ $$\begin{aligned} f(f^{-1}(x)) &= f\!\left(\frac{x+1}{3}\right) \\ &= 3\!\left(\frac{x+1}{3}\right) - 1 \\ &= x \end{aligned}$$
Domains and ranges of inverse functions	The domain of f equals the range of f^{-1}. The range of f equals the domain of f^{-1}.	Let $f(x) = (x + 2)^2$ with domain $x \geq -2$. Then, $f^{-1}(x) = \sqrt{x} - 2$. Domain of $f = \{x \mid x \geq -2\}$ = range of f^{-1}. Range of $f = \{y \mid y \geq 0\}$ = domain of f^{-1}.

Both functions and their inverses have verbal, symbolic, numerical, and graphical representations. The following table summarizes these representations.

Concept	Examples
Verbal representations	Given a verbal representation of f, apply the inverse operations in reverse order to find f^{-1}. *Example:* The function f multiplies 2 times x and then adds 25. The function f^{-1} subtracts 25 from x and divides the result by 2.
Symbolic representations	If possible, solve the equation $y = f(x)$ for x. For example, let $f(x) = 3x - 5$. $y = 3x - 5$ is equivalent to $\dfrac{y + 5}{3} = x$. Interchange x and y to obtain $f^{-1}(x) = \dfrac{x + 5}{3}$.
Numerical representations	<table><tr><td>x</td><td>1</td><td>2</td><td>3</td></tr><tr><td>$f(x)$</td><td>0</td><td>5</td><td>7</td></tr></table> <table><tr><td>x</td><td>0</td><td>5</td><td>7</td></tr><tr><td>$f^{-1}(x)$</td><td>1</td><td>2</td><td>3</td></tr></table>
Graphical representations	The graph of f^{-1} can be obtained by reflecting the graph of f across the line $y = x$.

5.2 EXERCISES

Inverse Operations

Exercises 1–4: State the inverse action or actions.

1. Opening a window

2. Climbing up a ladder

3. Walking into a classroom, sitting down, and opening a book

4. Opening the door and turning on the lights

Exercises 5–12: Describe verbally the inverse of the statement. Then, express both the statement and its inverse symbolically.

5. Add 2 to x.

6. Multiply x by 5.

7. Subtract 2 from x and multiply the result by 3.

8. Divide x by 20 and add 10.

9. Take the cube root of x and add 1.

10. Multiply x by -2 and add 3.

11. Take the reciprocal of a nonzero number x.

12. Take the square root of a positive number x.

One-to-One Functions

Exercises 13–18: Use the graph of f to determine if f is one-to-one.

13.
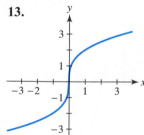

14.
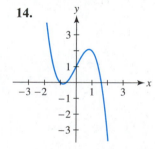

15.
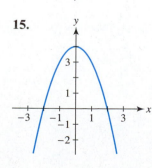

16.
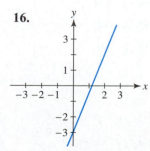

17.   18.
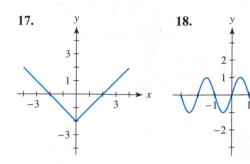

Exercises 19–22: Use the table to determine if f is one-to-one.

19.

x	1	2	3	4
$f(x)$	4	3	3	5

20.

x	-2	0	2	4
$f(x)$	4	2	0	-2

21.

x	0	2	4	6	8
$f(x)$	-1	0	4	1	-3

22.

x	-2	-1	0	1	2
$f(x)$	4	1	0	1	4

Exercises 23–30: Determine if f is one-to-one. Justify your answer graphically.

23. $f(x) = 2x - 7$

24. $f(x) = x^2 - 1$

25. $f(x) = |x - 1|$

26. $f(x) = x^3$

27. $f(x) = \dfrac{1}{1 + x^2}$

28. $f(x) = \dfrac{1}{x}$

29. $f(x) = 3x - x^3$

30. $f(x) = x^{2/3}$

Exercises 31–34: Decide if the situation could be modeled by a one-to-one function.

31. The distance between the ground and a person who is riding a Ferris wheel after x seconds

32. The cumulative numbers of AIDS cases from 1980 to present

33. The population of the United States from 1980 to 2000

34. The height y of a stone thrown upward after x seconds

Symbolic Representations of Inverse Functions

Exercises 35–46: Find a symbolic representation for $f^{-1}(x)$.

35. $f(x) = \sqrt[3]{x}$ **36.** $f(x) = 2x$

37. $f(x) = -2x + 10$ **38.** $f(x) = x^3 + 2$

39. $f(x) = 3x - 1$ **40.** $f(x) = \dfrac{x - 1}{2}$

41. $f(x) = 2x^3 - 5$ **42.** $f(x) = 1 - \dfrac{1}{2}x^3$

43. $f(x) = x^2 - 1, \quad x \geq 0$

44. $f(x) = (x + 2)^2, \quad x \leq -2$

45. $f(x) = \dfrac{1}{2x}$ **46.** $f(x) = \dfrac{2}{\sqrt{x}}$

Exercises 47–50: Restrict the domain of $f(x)$ so that f is one-to-one. Then find $f^{-1}(x)$.

47. $f(x) = 4 - x^2$ **48.** $f(x) = 2(x + 3)^2$

49. $f(x) = (x - 2)^2 + 4$ **50.** $f(x) = x^4 - 1$

Exercises 51–58: Find a symbolic representation for $f^{-1}(x)$. Identify the domain and range of f^{-1}. Verify that f and f^{-1} are inverses.

51. $f(x) = 5x - 15$

52. $f(x) = (x + 3)^2, x \geq -3$

53. $f(x) = \dfrac{1}{x + 3}$ **54.** $f(x) = \dfrac{2}{x - 1}$

55. $f(x) = 2x^3$ **56.** $f(x) = 1 - 4x^3$

57. $f(x) = x^2, x \geq 0$ **58.** $f(x) = \sqrt[3]{1 - x}$

Numerical Representations of Inverse Functions

Exercises 59–62: Use the table for $f(x)$ to find a table for $f^{-1}(x)$. Identify the domains and ranges of f and f^{-1}.

59.

x	1	2	3
$f(x)$	5	7	9

60.

x	1	10	100
$f(x)$	0	1	2

61.

x	0	1	2	3	4
$f(x)$	0	1	4	9	16

62.

x	-2	-1	0	1	2
$f(x)$	$\dfrac{1}{4}$	$\dfrac{1}{2}$	1	2	4

Exercises 63–66: Use $f(x)$ to complete the table for $f^{-1}(x)$.

63. $f(x) = x + 5$

x	-3	0	3	6
$f^{-1}(x)$				

64. $f(x) = 4x$

x	0	2	4	6
$f^{-1}(x)$				

65. $f(x) = x^3$

x	-8	-1	8	27
$f^{-1}(x)$				

66. $f(x) = \dfrac{1}{x}$

x	-3	-1	4	5
$f^{-1}(x)$				

Exercises 67–74: Use the tables to evaluate the following.

x	0	1	2	3	4
$f(x)$	1	3	5	4	2

x	-1	1	2	3	4
$g(x)$	0	2	1	4	5

67. $f^{-1}(3)$

68. $f^{-1}(5)$

69. $g^{-1}(4)$

70. $g^{-1}(0)$

71. $(f \circ g^{-1})(1)$

72. $(g^{-1} \circ g^{-1})(2)$

73. $(g \circ f^{-1})(5)$

74. $(f^{-1} \circ g)(4)$

Graphical Representations of Inverse Functions

75. *Interpreting an Inverse* The graph of f computes the number of dollars in a savings account after x years. Estimate each of the following. Interpret what $f^{-1}(x)$ computes.

(a) $f(1)$ **(b)** $f^{-1}(110)$ **(c)** $f^{-1}(160)$

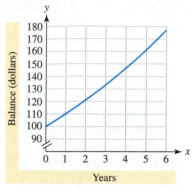

76. *Interpreting an Inverse* The graph of f computes the Celsius temperature of a pan of water after x minutes. Estimate each of the following. Interpret what $f^{-1}(x)$ computes.

(a) $f(4)$ **(b)** $f^{-1}(90)$ **(c)** $f^{-1}(80)$

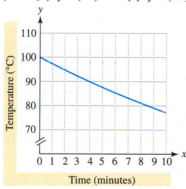

Exercises 77–78: Use the graph of f to evaluate each expression.

77. (a) $f(-1)$ **(b)** $f^{-1}(-2)$
     (c) $f^{-1}(0)$ **(d)** $(f^{-1} \circ f)(3)$

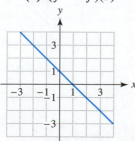

78. (a) $f(1)$ **(b)** $f^{-1}(1)$
     (c) $f^{-1}(4)$ **(d)** $(f \circ f^{-1})(2.5)$

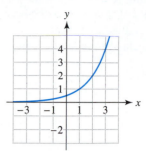

Exercises 79–82: Use the graph of f to sketch a graph of f^{-1}.

79.

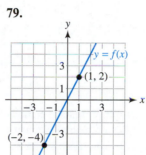

80.

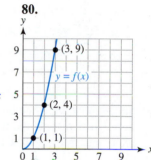

81.

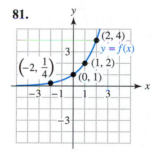

82.

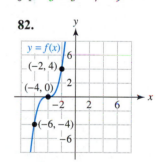

Exercises 83–86: Graph $y = f(x)$, $y = f^{-1}(x)$, and $y = x$ in a square viewing rectangle such as $[-4.7, 4.7, 1]$ by $[-3.1, 3.1, 1]$.

83. $f(x) = 3x - 1$

84. $f(x) = \dfrac{3 - x}{2}$

85. $f(x) = \dfrac{1}{3}x^3 - 1$

86. $f(x) = \sqrt[3]{x - 1}$

Graphing Calculator Help To learn about a square viewing rectangle, see Appendix B (page AP-9).

Applications

87. *Advertising Costs* The line graph represents a function f that computes the cost of a 30-second commercial during a Super Bowl telecast. Perform each calculation and interpret the results. (Source: *USA Today.*)

(a) Evaluate $f(1995)$.

(b) Solve $f(x) = 1$ for x.

(c) Evaluate $f^{-1}(1)$.

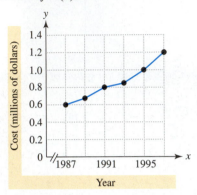

88. *Social Security* The line graph represents a function f that models the number of Social Security recipients from 1990 to 2030. (Source: Social Security Administration.)

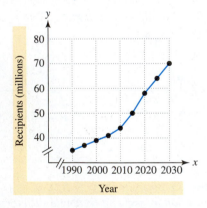

(a) Solve the equation $f(x) = 50$ for x.

(b) Evaluate $f^{-1}(50)$.

89. *Volume* The volume V of a sphere with radius r is $V = \frac{4}{3}\pi r^3$.

(a) Does V represent a one-to-one function?

(b) What does the inverse of V compute?

(c) Find a formula for the inverse.

(d) Normally we interchange x and y to find the inverse of a function. Does it make sense to interchange V and r in part (c) of this exercise? Explain.

90. *Temperature* The formula $f(x) = \frac{9}{5}x + 32$ converts a Celsius temperature to an equivalent Fahrenheit temperature.

(a) Find a symbolic representation for f^{-1} and interpret the result.

(b) What Celsius temperature is equivalent to 68°F?

91. *Codes* Use the code in Table 5.7 on page 352 to code MATH, then decode HWPEVKQPU. If a function f is used to code a word x, why is it essential that f is one-to-one?

92. *Codes* Consider the following method to code a word. First let A = 1, B = 2, C = 3, . . . , Z = 26. For a given word, add the numbers associated with each letter in the word. The sum represents the code for the word. For example, since M = 13, A = 1, T = 20, and H = 8, the word MATH is coded as $13 + 1 + 20 + 8 = 42$. Is this a good way to code words? Explain.

93. *Planetary Orbits* The formula $f(x) = x^{3/2}$ calculates the time in years that it takes a planet to orbit the sun if the planet is x times farther from the sun than Earth.

(a) Find the inverse of f.

(b) What does the inverse of f calculate?

94. *Height and Weight* The formula $W = \frac{25}{7}h - \frac{800}{7}$ approximates the recommended minimum weight for a person h inches tall, where $h \geq 62$.

(a) What is the recommended minimum weight for someone 70 inches tall?

(b) Does W represent a one-to-one function?

(c) Find a formula for the inverse of W.

(d) What does W^{-1} compute?

95. *Converting Units* The accompanying tables represent a function f that converts yards to feet and a function g that converts miles to yards. Evaluate each expression and interpret the results.

x (yd)	1760	3520	5280	7040	8800
$f(x)$ (ft)	5280	10,560	15,840	21,120	26,400

x (mi)	1	2	3	4	5
$g(x)$ (yd)	1760	3520	5280	7040	8800

(a) $(f \circ g)(2)$ **(b)** $f^{-1}(26{,}400)$

(c) $(g^{-1} \circ f^{-1})(21{,}120)$

96. (Refer to the previous exercise.)
(a) Find symbolic representations for $f(x)$, $g(x)$, and $(f \circ g)(x)$.
(b) Express $(g^{-1} \circ f^{-1})(x)$ symbolically. What does this function compute?

97. *Converting Units* The accompanying tables represent a function f that converts tablespoons to cups and a function g that converts cups to quarts. Evaluate each expression and interpret the results.

x (tbsp)	32	64	96	128
$f(x)$ (c)	2	4	6	8

x (c)	2	4	6	8
$g(x)$ (qt)	0.5	1	1.5	2

(a) $(g \circ f)(96)$
(b) $g^{-1}(2)$
(c) $(f^{-1} \circ g^{-1})(1.5)$

98. (Refer to the previous exercise.)
(a) Find symbolic representations for $f(x)$, $g(x)$, and $(g \circ f)(x)$.
(b) Express $(f^{-1} \circ g^{-1})(x)$ symbolically. What does this function compute?

99. *Air Pollution* Tiny particles suspended in the air are necessary for clouds to form. Because of this fact, experts believe that air pollutants may cause an increase in cloud cover. From 1930 to 1980 the percentage of cloud cover over the world's oceans was monitored. The linear function given by $f(x) = 0.06(x - 1930) + 62.5$, where $1930 \leq x \leq 1980$, approximates this percentage. (Source: W. Cotton and R. Pielke, *Human Impacts on Weather and Climate.*)
(a) Evaluate $f(1930)$ and $f(1980)$. How did the amount of cloud cover over the oceans change during this 50-year period?
(b) What does $f^{-1}(x)$ compute?
(c) Use your results from part (a) to evaluate $f^{-1}(62.5)$ and $f^{-1}(65.5)$.
(d) Find $f^{-1}(x)$ symbolically.

100. *Rise in Sea Level* Due to the greenhouse effect, the global sea level could rise through thermal expansion and partial melting of the polar ice caps. The table represents a function f that models this expected rise in sea level in centimeters for the year x. (This model assumes no changes in current trends.) (Source: A. Nilsson, *Greenhouse Earth.*)

x (yr)	1990	2000	2030	2070	2100
$f(x)$ (cm)	0	1	18	44	66

(a) Is f a one-to-one function? Explain.
(b) Use $f(x)$ to find a table for $f^{-1}(x)$. Interpret f^{-1}.

Writing about Mathematics

101. Explain how to find verbal, numerical, graphical, and symbolic representations of an inverse function. Give examples.

102. Can a one-to-one function have more than one x-intercept or more than one y-intercept? Explain.

103. If the graphs of $y = f(x)$ and $y = f^{-1}(x)$ intersect at a point (x, y), what can be said about this point? Explain your reasoning.

104. If $f(x) = ax^2 + bx + c$ with $a \neq 0$, does $f^{-1}(x)$ exist? Explain.

Extended and Discovery Exercises

1. *Interpreting an Inverse* Let $f(x)$ compute the distance traveled in miles after x hours by a car with a velocity of 60 miles per hour.
(a) Explain what $f^{-1}(x)$ computes.
(b) Interpret the solution to the equation $f(x) = 200$.
(c) Explain how to solve the equation in part (b) using $f^{-1}(x)$.

2. *Interpreting an Inverse* Let $f(x)$ compute the height in feet of a rocket after x seconds of upward flight.
(a) Explain what $f^{-1}(x)$ computes.
(b) Interpret the solution to the equation $f(x) = 5000$.
(c) Explain how to solve the equation in part (b) using $f^{-1}(x)$.

CHECKING BASIC CONCEPTS FOR SECTIONS 5.1 AND 5.2

1. Use the table to evaluate each expression, if possible.

x	-2	-1	0	1	2
$f(x)$	0	1	-2	-1	2
$g(x)$	1	-2	-1	2	0

(a) $(f + g)(1)$ **(b)** $(f - g)(-1)$
(c) $(fg)(0)$ **(d)** $(f/g)(2)$
(e) $(f \circ g)(2)$ **(f)** $(g \circ f)(-2)$

2. Let $f(x) = x^2 + 3x - 2$ and $g(x) = 3x - 1$. Find each expression.
(a) $(f + g)(x)$ **(b)** $(f/g)(x)$
(c) $(f \circ g)(x)$

3. If $f(x) = 5 - 2x$, find $f^{-1}(x)$.

5.3 EXPONENTIAL FUNCTIONS AND MODELS

Linear and Exponential Growth • Exponential Models • Compound Interest • The Natural Exponential Function

Introduction

Even though modern exponential notation was not developed until 1637 by the great French mathematician René Descartes, some of the earliest applications involving exponential functions occurred in the calculation of interest. The custom of charging interest dates back to at least 2000 B.C. in ancient Babylon, where interest rates ran as high as 33%. Today, exponential functions are used not only to calculate interest, but also to model a wide variety of phenomena in business, biology, medicine, engineering, and education. This section discusses exponential functions and their representations. (Source: *Historical Topics for the Mathematics Classroom, Thirty-first Yearbook,* NCTM.)

Linear and Exponential Growth

A linear function g can be written as $g(x) = mx + b$, where m represents the rate of change in $g(x)$ for each unit increase in x. For example, Table 5.12 shows a numerical representation of a linear function g. Each time x increases by 1 unit, $g(x)$ increases by 2 units. We can write a formula $g(x) = 2x + 3$ because the rate of change is 2 and $g(0) = 3$.

TABLE 5.12   A Linear Function

x	0	1	2	3	4	5
$y = g(x)$	3	5	7	9	11	13

An exponential function is fundamentally different from a linear function. Rather than *adding* a fixed amount to the previous y-value for each unit increase in x, an exponential function *multiplies* the previous y-value by a fixed amount for each unit increase in x. Table 5.13 shows an exponential function f. Note that in Table 5.13 consecutive y-values are found by multiplying the previous y-value by 2.

TABLE 5.13 An Exponential Function

x	0	1	2	3	4	5
$y = f(x)$	3	6	12	24	48	96

Compare the following patterns for calculating the linear function g and the exponential function f.

$g(0) = 3$

$g(1) = \underbrace{3}_{g(0)} + 2 = 3 + 2 \cdot 1 = 5$

$g(2) = \underbrace{3 + 2}_{g(1)} + 2 = 3 + 2 \cdot 2 = 7$

$g(3) = \underbrace{3 + 2 + 2}_{g(2)} + 2 = 3 + 2 \cdot 3 = 9$

$g(4) = \underbrace{3 + 2 + 2 + 2}_{g(3)} + 2 = 3 + 2 \cdot 4 = 11$

$g(5) = \underbrace{3 + 2 + 2 + 2 + 2}_{g(4)} + 2 = 3 + 2 \cdot 5 = 13$

$f(0) = 3$

$f(1) = \underbrace{3}_{f(0)} \cdot 2 = 3 \cdot 2^1 = 6$

$f(2) = \underbrace{3 \cdot 2}_{f(1)} \cdot 2 = 3 \cdot 2^2 = 12$

$f(3) = \underbrace{3 \cdot 2 \cdot 2}_{f(2)} \cdot 2 = 3 \cdot 2^3 = 24$

$f(4) = \underbrace{3 \cdot 2 \cdot 2 \cdot 2}_{f(3)} \cdot 2 = 3 \cdot 2^4 = 48$

$f(5) = \underbrace{3 \cdot 2 \cdot 2 \cdot 2 \cdot 2}_{f(4)} \cdot 2 = 3 \cdot 2^5 = 96$

Notice that if x is a positive integer then

$$g(x) = 3 + \underbrace{(2 + 2 + \cdots + 2)}_{x \text{ terms}} \quad \text{and} \quad f(x) = 3 \cdot \underbrace{(2 \cdot 2 \cdot \cdots \cdot 2)}_{x \text{ factors}}.$$

Using these patterns we can write formulas for $g(x)$ and $f(x)$ as follows.

$$g(x) = 3 + 2x \qquad f(x) = 3 \cdot 2^x$$

This discussion gives motivation for the following definition.

Exponential function

A function f represented by

$$f(x) = Ca^x, \qquad a > 0, \quad a \neq 1, \quad \text{and} \quad C > 0,$$

is an **exponential function with base a**.

Note: Some definitions for an exponential function require that $C = 1$.

Examples of exponential functions include

$$f(x) = 3^x, \qquad g(x) = 5(1.7)^x, \quad \text{and} \quad h(x) = 4\left(\frac{1}{2}\right)^x.$$

Their bases are 3, 1.7, and $\frac{1}{2}$, respectively. When evaluating exponential functions, remember that any nonzero number raised to the 0 power equals 1. Thus, $f(0) = 3^0 = 1$, $g(0) = 5(1.7)^0 = 5$, and $h(0) = 4\left(\frac{1}{2}\right)^0 = 4$.

Note: If $f(x) = Ca^x$, then $f(0) = Ca^0 = C(1) = C$. That is, C equals the value of the function at $x = 0$. If x represents time, then C often equals the *initial value* of the quantity being modeled by f.

For large values of x, an exponential function with $a > 1$ grows faster than any linear function. Figures 5.37 and 5.38 show graphs of $f(x) = 2^x$ and $g(x) = 2x$, respectively. The points shown in the table have also been plotted. Notice that the graph of the exponential function f increases more rapidly than the graph of the linear function g for large values of x.

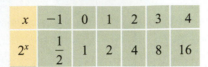

x	-1	0	1	2	3	4
2^x	$\frac{1}{2}$	1	2	4	8	16

x	-1	0	1	2	3	4
$2x$	-2	0	2	4	6	8

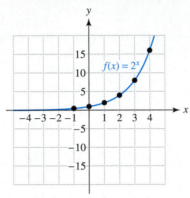

FIGURE 5.37 An Exponential Function

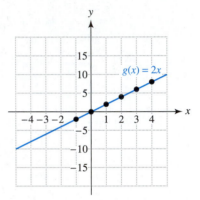

FIGURE 5.38 A Linear Function

EXAMPLE 1   *Recognizing linear and exponential data*

For each table of data, find either a linear or exponential function that models the data.

(a)

x	0	1	2	3
y	-3	-1.5	0	1.5

(b)

x	0	1	2	3	4
y	16	4	1	$\frac{1}{4}$	$\frac{1}{16}$

(c)

x	0	1	2	3
y	3	4.5	6.75	10.125

(d)

x	0	1	2	3	4
y	16	12	8	4	0

SOLUTION

(a) For each unit increase in x, the y-values increase by 1.5, so the data are linear. Because $y = -3$ when $x = 0$, it follows that the data can be modeled by $f(x) = 1.5x - 3$.

(b) For each unit increase in x, the y-values are multiplied by $\frac{1}{4}$. This is an exponential function with $C = 16$ and $a = \frac{1}{4}$, so $f(x) = 16\left(\frac{1}{4}\right)^x$.

(c) Since the data do not change by a fixed amount for each unit increase in x, the data are not linear. To determine if the data are exponential, calculate ratios of consecutive y-values.

$$\frac{4.5}{3} = 1.5, \qquad \frac{6.75}{4.5} = 1.5, \qquad \frac{10.125}{6.75} = 1.5$$

For each unit increase in x, the next y-value in the table can be found by multiplying the previous y-value by 1.5, so let $a = 1.5$. Since $y = 3$ when $x = 0$, let $C = 3$. Thus $f(x) = 3(1.5)^x$.

(d) For each unit increase in x, the next y-value is found by adding -4 to the previous y-value, so the data can be modeled by $f(x) = -4x + 16$.

EXAMPLE 2 *Finding linear and exponential functions*

A college graduate signs a contract for an annual salary of $50,000 and can choose either Option A: a fixed annual raise of $6000 each year, or Option B: a 10% raise each year. However, the graduate must stay with the same option.
(a) Write a formula that gives the annual salary during the nth year for each option.
(b) Discuss which option is better.

SOLUTION (a) **Option A:** The salary for the first year is $50,000, for the second year is $56,000, for the third year is $62,000, and so on. Next year's salary is found by adding $6000 to this year's salary. These salaries can be modeled by a linear function g with $m = 6000$, passing through the point $(1, 50,000)$. Using the point-slope for a line, we obtain

$$g(n) = 6000(n - 1) + 50,000 \quad \text{or} \quad g(n) = 6000n + 44,000.$$

Algebra Review

To review point-slope form of a line, see Section 2.2

Option B: The salary for the first year is $50,000. Since 10% of $50,000 is $5000, the salary for the second year is $50,000 + $5000 = $55,000. This is equivalent to multiplying $50,000 by $1 + 0.10 = 1.10$. In general, to calculate next year's salary we can multiply this year's salary by 1.10.

$$f(1) = 50,000 = 50,000(1.10)^0 = \$50,000$$

$$f(2) = 50,000(1.1) = 50,000(1.10)^1 = \$55,000$$

$$f(3) = 50,000(1.1)(1.1) = 50,000(1.10)^2 = \$60,500$$

$$f(4) = 50,000(1.1)(1.1)(1.1) = 50,000(1.10)^3 = \$66,550$$

From the pattern above, we can write an exponential function f that models this situation

$$f(n) = 50,000(1.10)^{n-1}.$$

Note: The exponent for function f must be $n - 1$ rather than n, because during the first year ($n = 1$) the salary is $50,000, not $55,000.

(b) Table 5.14 shows the salaries under each option. During the 5th year the two options have comparable salaries. If the graduate is planning on changing jobs after a few years, Option A is better. However, if the graduate plans to stay for a long time, Option B is better.

TABLE 5.14 Salaries with Linear and Exponential Growth

x (year)	1	3	5	7	9	11	13
Option A	$50,000	$62,000	$74,000	$86,000	$98,000	$110,000	$122,000
Option B	$50,000	$60,500	$73,205	$88,578	$107,179	$129,687	$156,921

If an exponential function is written as $f(x) = Ca^x$ with $a > 1$, then $f(x)$ experiences **exponential growth**, as illustrated in Figure 5.39. In Example 2, the salary described by Option B experienced exponential growth, with a **growth factor** of $a = 1.10$. If $0 < a < 1$, then the y-values decrease by a factor of a for each unit increase in x. In this case $f(x) = Ca^x$ experiences **exponential decay** with **decay factor** a. See Figure 5.40. Exponential decay was modeled in Example 1(b), where the decay factor was $a = \dfrac{1}{4}$.

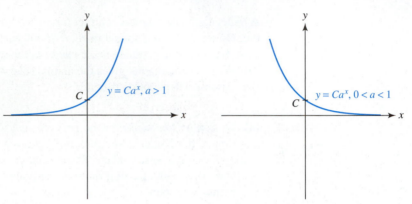

FIGURE 5.39 Exponential Growth   **FIGURE 5.40** Exponential Decay

To investigate further the effect that a has on the graph of an exponential function, we can let $C = 1$ in $f(x) = Ca^x$, and graph $y = 1^x$, $y = 1.3^x$, $y = 1.7^x$, and $y = 2.5^x$, as shown in Figure 5.41. (Note that $f(x) = 1^x$ does *not* represent an exponential function. However, the graph of $y = 1^x$ represents the boundary between exponential growth and decay.) As a increases, the graph of $y = a^x$ increases at a faster rate. On the other hand, the graphs of $y = 0.7^x$, $y = 0.5^x$, and $y = 0.15^x$ decrease faster as a decreases. See Figure 5.42. The graph of $y = a^x$ is increasing when $a > 1$ and decreasing when $0 < a < 1$. Note that the graph of $y = a^x$, $a > 0$, always passes through the point $(0, 1)$.

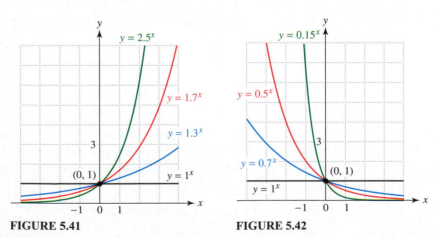

FIGURE 5.41   **FIGURE 5.42**

In Section 3.4 we learned that the graph of $y = f(-x)$ is a reflection of the graph of $y = f(x)$ across the y-axis. As a result, the graph of $y = a^{-x}$ is a reflection of $y = a^x$ across the y-axis. For example, the graphs of $y = 2^x$ and $y = 2^{-x}$ are reflections across the y-axis. See Figure 5.43. Note that by properties of exponents, $2^{-x} = \left(\dfrac{1}{2}\right)^x$.

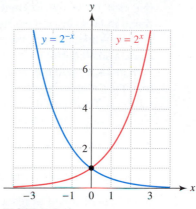

FIGURE 5.43

Algebra Review
To review properties of exponents, see Chapter R (page R-14) and Section 4.7.

■ **MAKING CONNECTIONS**

Exponential Functions and Polynomial Functions   An exponential function has a *variable* for an exponent, whereas a polynomial function has a *constant* exponent. For example, $f(x) = 3^x$ represents an exponential function, and $g(x) = x^3$ represents a polynomial function.

EXAMPLE 3 *Comparing exponential and polynomial functions*

Compare $f(x) = 3^x$ and $g(x) = x^3$ graphically and numerically for $x \geq 0$.

SOLUTION   *Graphical Comparison*   The graphs $Y_1 = 3^\wedge X$ and $Y_2 = X^\wedge 3$ are shown in Figure 5.44. For $x \geq 6$, the graph of the exponential function Y_1 increases significantly faster than the graph of the polynomial function Y_2.

Numerical Comparison   Table $Y_1 = 3^\wedge X$ and $Y_2 = X^\wedge 3$, as shown in Figure 5.45. The values for Y_1 increase faster than the values for Y_2.

[0, 12, 1] by [0, 10,000, 1000]

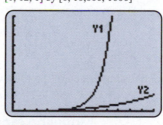

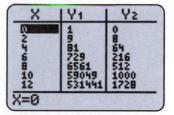

FIGURE 5.44   **FIGURE 5.45**

The results of Example 3 are true in general. For large enough inputs, exponential functions with $a > 1$ eventually become greater than any polynomial function. This is an example of *exponential growth*.

Exponential Models

Radioactivity is an application of exponential decay. When an element such as uranium undergoes radioactive decay, atoms change from one element to another. The time it takes for half of the atoms to decay into a different element is called the **half-life**, and this time varies for different elements. For example, radioactive carbon-14, which is found in all living things, has a half-life of about 5700 years and can be used to date fossils. While animals are alive, they breathe both carbon dioxide and oxygen. Because a small portion of normal atmospheric carbon dioxide is made up of radioactive carbon-14, a fixed percentage of their bodies is composed of carbon-14. When an animal dies, it quits breathing and the carbon-14 continues to disintegrate without being replaced. To determine the time since an animal died, scientists measure the percentage of carbon-14 remaining in its bones. If initially there was 1 gram of carbon-14 present, then after 5700 years there would be $\frac{1}{2}$ gram, after $2 \cdot 5700 = 11{,}400$ years there would be $\frac{1}{4}$ gram, and in general after each 5700-year period, the amount of carbon-14 would be reduced by half. Because the amount of carbon-14 is being reduced by a constant factor of $\frac{1}{2}$ every 5700 years, we can model the amount of carbon-14 with an exponential function.

If the initial amount of carbon-14 equals C grams, then $f(x) = Ca^x$ models the amount of carbon-14 present after x years, where a needs to be determined for carbon-14. After 5700 years, there will be $\frac{1}{2}C$ grams, so $f(5700) = \frac{1}{2}C$ implies that

$$Ca^{5700} = \frac{1}{2}C.$$

To solve this equation for a, begin by dividing by C.

$$a^{5700} = \frac{1}{2} \qquad \text{Divide by } C.$$

Algebra Review
To review fractional exponents, see Section 4.7.

$$\left(a^{5700}\right)^{1/5700} = \left(\frac{1}{2}\right)^{1/5700} \qquad \text{Raise to the } \frac{1}{5700}\text{th power.}$$

$$a = \left(\frac{1}{2}\right)^{1/5700} \qquad \text{Properties of exponents}$$

Using properties of exponents, we write $f(x) = Ca^x$ as

$$f(x) = C\left(\left(\frac{1}{2}\right)^{1/5700}\right)^x = C\left(\frac{1}{2}\right)^{x/5700}.$$

This discussion is summarized in the following.

Modeling radioactive decay

If a radioactive sample containing C units has a half-life of k years, then the amount A remaining after x years is given by

$$A(x) = C\left(\frac{1}{2}\right)^{x/k}.$$

For example, the half-life of radium-226 is about 1600 years. After 9600 years a 2-gram sample decays to

$$A(9600) = 2\left(\frac{1}{2}\right)^{9600/1600} = 0.03125 \text{ gram.}$$

EXAMPLE 4   *Finding the age of a fossil*

A fossil contains 5% of the amount of the carbon-14 that it contained when it was alive. Graphically estimate its age. This situation is illustrated in Figure 5.46.

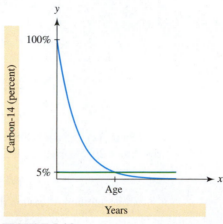

FIGURE 5.46

SOLUTION   The initial amount of carbon-14 is 100% or 1, the final amount is 5% or 0.05, and the half-life is 5700 years, so $A = 0.5$, $C = 1$, and $k = 5700$ in $A(x) = C\left(\dfrac{1}{2}\right)^{x/k}$. To determine the age of the fossil, solve

$$0.05 = 1\left(\frac{1}{2}\right)^{x/5700}$$

for x. Graph $Y_1 = 0.05$ and $Y_2 = 0.5\char`\^(X/5700)$, as shown in Figure 5.47. Their graphs intersect near $(24{,}635, 0.05)$, so the fossil is about 24,635 years old.

Graphing Calculator Help

To find a point of intersection, see Appendix B (page AP-11).

[0, 50,000, 10,000] by [0, 0.1, 0.01]

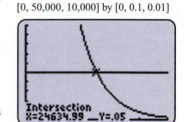

FIGURE 5.47

EXAMPLE 5   *Modeling atmospheric CO_2 concentrations*

Future concentrations of atmospheric carbon dioxide (CO_2) in parts per million (ppm) are shown in Table 5.15. (These concentrations assume that current trends continue.) The CO_2 levels in the year 2000 were greater than they have been in any time in the previous 160,000 years. The increase in concentrations of CO_2 has been accelerated by the burning of fossil fuels and by deforestation.   (Source: R. Turner, *Environmental Economics.*)

TABLE 5.15   Concentrations of Atmospheric CO_2

Year	2000	2050	2100	2150	2200
CO_2 (ppm)	364	467	600	769	987

(a) Let $x = 0$ correspond to 2000 and $x = 200$ to 2200. Find values for C and a so that $f(x) = Ca^x$ models this data.

(b) Estimate CO_2 concentrations for the year 2025.

SOLUTION   **(a)** The concentration is 364 when $x = 0$, so $C = 364$. This gives $f(x) = 364a^x$. Next, we estimate a value for a. One possibility for determining a is to require that the graph of f pass through the point (2200, 987). It then follows that $f(200) = 987$.

$$364 \cdot a^{200} = 987 \qquad \text{\textcolor{blue}{$f(200) = 987$}}$$

$$a^{200} = \frac{987}{364} \qquad \text{\textcolor{blue}{Divide by 364.}}$$

$$(a^{200})^{1/200} = \left(\frac{987}{364}\right)^{1/200} \qquad \text{\textcolor{blue}{Take the $\frac{1}{200}$th power.}}$$

$$a = \left(\frac{987}{364}\right)^{1/200} \qquad \text{\textcolor{blue}{Properties of exponents}}$$

$$a \approx 1.005 \qquad \text{\textcolor{blue}{Approximate.}}$$

Thus $f(x) = 364(1.005)^x$. Answers for $f(x)$ may vary slightly.

(b) Since 2025 corresponds to $x = 25$, evaluate $f(25)$.

$$f(25) = 364(1.005)^{25} \approx 412.$$

Concentration of carbon dioxide could reach 412 ppm by 2025.

Compound Interest

Suppose $1000 is deposited in an account paying 10% annual interest. At the end of 1 year, the account will contain $1000 plus 10% of $1000, or $100, in interest. Let A_0 represent the **principal** or initial amount deposited. Then, the amount A_1 in the account after 1 year can be computed as follows.

$$A_1 = A_0 + (0.10)A_0 \qquad \text{\textcolor{blue}{Principal plus interest}}$$

$$= A_0(1 + 0.10) \qquad \text{\textcolor{blue}{Factor.}}$$

$$= 1000(1.10) \qquad \text{\textcolor{blue}{$A_0 = \$1000$}}$$

$$= 1100 \qquad \text{\textcolor{blue}{Simplify.}}$$

The sum of the principal and interest is $1100.

During the second year, the account earns interest on $1100. The amount A_2 after the second year will be equal to A_1 plus 10% of A_1.

$$A_2 = A_1 + (0.10)A_1$$

$$= A_1(1 + 0.10) \qquad \text{\textcolor{blue}{Factor.}}$$

$$= A_0(1 + 0.10)(1 + 0.10) \qquad \text{\textcolor{blue}{$A_1 = A_0(1 + 0.10)$}}$$

$$= A_0(1 + 0.10)^2$$

$$= 1000(1.10)^2$$

$$= 1210$$

After 2 years there will be $1210.

We would like to determine a general formula for A_n, the amount in the account after n years. To do this, compute A_3 and observe a pattern.

$$A_3 = A_2 + (0.10)A_2$$

$$= A_2(1 + 0.10)$$

$$= A_0(1 + 0.10)^2(1 + 0.10) \qquad \text{\textcolor{blue}{$A_2 = A_0(1 + 0.10)^2$}}$$

$$= A_0(1 + 0.10)^3$$

$$= 1000(1.10)^3$$

$$= 1331$$

The amount is $1331. We can see that in general, $A_n = A_0(1 + 0.10)^n$. This type of interest is said to be *compounded annually,* because it is paid once a year.

If the interest rate had been r, expressed in decimal form, then $A_n = A_0(1 + r)^n$. Notice that in the expression $A_0(1 + r)^n$, the variable n occurs as an exponent.

EXAMPLE 6   *Calculating an account balance*

If the principal is $2000 and the interest rate is 8% compounded annually, calculate the account balance after 4 years.

SOLUTION   The initial amount is $A_0 = 2000$, the interest rate is $r = 0.08$, and the number of years is $n = 4$.

$$A_4 = 2000(1 + 0.08)^4 \approx 2720.98$$

After 4 years the account contains $2720.98.

In most savings accounts, interest is paid more often than once a year. In this case a smaller amount of interest is paid more frequently. For example, suppose $1000 is deposited in an account paying 10% annual interest, compounded quarterly. After 3 months the interest would amount to one-fourth of 10% or 2.5% of $1000. The account balance would be $1000(1 + 0.025) = \$1025$. After the next 3-month period, interest would be paid on the $1025. In a manner similar to annual compounding, the balance would be $1000(1 + 0.025)^2 \approx \$1050.63$ after 6 months, $1000(1 + 0.025)^3 \approx \$1076.89$ after 9 months, and $1000(1 + 0.025)^4 \approx \$1103.81$ after a year. With annual compounding the amount is $1100. The difference of $3.81 is due to compounding quarterly. Although this amount is small after 1 year, compounding more frequently can have a dramatic effect over a long time.

> **Compound interest**
>
> If A_0 dollars is deposited in an account paying an annual rate of interest r, compounded (paid) m times per year, then after n years the account will contain A_n dollars, where
>
> $$A_n = A_0\left(1 + \frac{r}{m}\right)^{mn}. \qquad P_e{}^{rt}$$

EXAMPLE 7   *Comparing compound interest*

Suppose $1000 is deposited by a 20-year-old worker in an Individual Retirement Account (IRA) that pays an annual interest rate of 12%. Describe the effect on the balance at age 65, if interest were compounded annually and quarterly.

SOLUTION   *Compounded Annually*   Let $A_0 = 1000$, $r = 0.12$, $m = 1$, and $n = 65 - 20 = 45$.

$$A_{45} = 1000(1 + 0.12)^{45} \approx \$163{,}987.60$$

Compounded Quarterly   Let $A_0 = 1000$, $r = 0.12$, $m = 4$, and $n = 45$.

$$A_{45} = 1000\left(1 + \frac{0.12}{4}\right)^{4(45)} = 1000(1 + 0.03)^{180} \approx \$204{,}503.36$$

Quarterly compounding results in an increase of $40,515.76!

■ **CLASS DISCUSSION**

In Example 7, make a conjecture about the effect on the IRA balance after 45 years if the interest rate were 6% instead of 12%. Test your conjecture. ■

The Natural Exponential Function

In Example 7, compounding interest quarterly rather than annually made a significant difference in the balance after 45 years. What would happen if interest were compounded daily or even hourly? Would there be a limit to the amount of interest that could be earned? To answer these questions, suppose \$1 was deposited in an account at the very high interest rate of 100%. Table 5.16 shows the amount of money after 1 year, compounding m times during the year. The first column represents the time interval between compound interest payments. In Table 5.16, the formula $A_n = A_0\left(1 + \dfrac{r}{m}\right)^{mn}$ reduces to $A_1 = \left(1 + \dfrac{1}{m}\right)^m$, since $A_0 = n = r = 1$.

TABLE 5.16

Time	m	A_1
Year	1	2.000000
Month	12	2.613035
Day	365	2.714567
Hour	8760	2.718127
Minute	525,600	2.718279
Second	31,536,000	2.718282

Notice that A_1 levels off near \$2.72. If compounding were done more frequently, by letting m become large without bound, it would be called *continuous compounding*. The expression $\left(1 + \dfrac{1}{m}\right)^m$ reaches a limit of approximately 2.718281828. This numeric value is so important in mathematics that it has been given its own symbol, e, sometimes called **Euler's number**. The number e has many of the same characteristics as π. Its decimal expansion never terminates or repeats in a pattern. It is an irrational number.

Value of e

To nine decimal places, $e \approx 2.718281828$.

Continuous compounding can be applied to population growth. Compounding annually would mean that all births and deaths occur on December 31. Similarly, compounding quarterly would mean that births and deaths occur at the end of March, June, September, and December. In large populations, births and deaths occur *continuously* throughout the year. Compounding continuously is a *natural* way to model large populations.

The natural exponential function

The function f, represented by $f(x) = e^x$, is the **natural exponential function**.

EXAMPLE 8 *Evaluating the natural exponential function*

Use a calculator to approximate to four decimal places $f(x) = e^x$ when $x = 1, 0.5$, and -2.56.

SOLUTION Figure 5.48 shows that $f(1) = e^1 \approx 2.7183$, $f(0.5) = e^{0.5} \approx 1.6487$, and $f(-2.56) = e^{-2.56} \approx 0.0773$.

Graphing Calculator Help

When evaluating e^x, be sure to use the built-in key for e rather than use an approximation, such as 2.72.

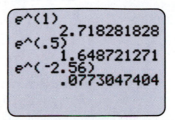

FIGURE 5.48

■ **CLASS DISCUSSION**

Graph $y = 2^x$ and $y = 3^x$ on the same coordinate axes, using the viewing rectangle $[-3, 3, 1]$ by $[0, 4, 1]$. Make a conjecture about how the graph of $y = e^x$ will appear. Test your conjecture. ■

When interest is compounded continuously, the following can be used to compute the amount of money in an account.

Continuously compounded interest

If A_0 dollars is deposited in an account paying an annual rate of interest r, compounded continuously, then after n years the account will contain A_n dollars, where

$$A_n = A_0 e^{rn}.$$

EXAMPLE 9 *Calculating continuously compounded interest*

Suppose $1000 is deposited into an IRA with an interest rate of 12%, compounded continuously.
(a) How much money will there be after 45 years?
(b) Determine graphically and numerically the 10-year period when the account balance increases the most.

SOLUTION **(a)** Let $A_0 = 1000$, $r = 0.12$, and $n = 45$. Then $A_{45} = 1000e^{(0.12)45} \approx \$221,406.42$. This is more than the $204,503.36 that resulted from compounding quarterly in Example 7.
(b) Graph $Y_1 = 1000e^{\wedge}(.12X)$, as shown in Figure 5.49 on the next page. From the graph we can see that the account balance increases the most during the last 10 years. Numerical support for this conclusion is shown on the next page in Figure 5.50.

Graphing Calculator Help

To make a table like Figure 5.50, see Appendix B (page AP-11).

[0, 45, 5] by [0, 250,000, 100,000]

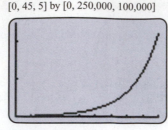

FIGURE 5.49   **FIGURE 5.50**

EXAMPLE 10   *Modeling traffic flow*

At an intersection cars arrive randomly with an average rate of 30 cars per hour. Highway engineers estimate the likelihood or probability that at least one car will enter the intersection within a period of x minutes with $f(x) = 1 - e^{-0.5x}$.   (Source: F. Mannering and W. Kilareski, *Principles of Highway Engineering and Traffic Analysis.*)

(a) Evaluate $f(2)$ and interpret the answer.

(b) Graph f for $0 \le x \le 60$. What happens to the likelihood that at least one car enters the intersection during a 60-minute period?

SOLUTION   **(a)** $f(2) = 1 - e^{-0.5(2)} = 1 - e^{-1} \approx 0.63$. There is a 63% chance that at least one car will enter the intersection during a 2-minute period.

(b) Graph $Y_1 = 1 - e^{\wedge}(-0.5X)$, as shown in Figure 5.51. As time progresses, the probability increases and begins to approach 1. That is, it is almost certain that at least one car will enter the intersection during a 60-minute period.

[0, 60, 10] by [0, 1.2, 0.2]

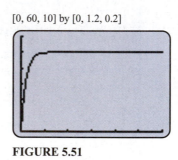

FIGURE 5.51

EXAMPLE 11   *Modeling the growth of E. coli bacteria*

A type of bacteria that inhabits the intestines of animals is named *E. coli* (*Escherichia coli*). These bacteria are capable of rapid growth and can be dangerous to humans—especially children. In one study, *E. coli* bacteria were capable of doubling in number every 49.5 minutes. Their number N after x minutes could be modeled by $N(x) = N_0 e^{0.014x}$. Suppose $N_0 = 500,000$ is the initial number of bacteria per milliliter. (Source: G. S. Stent, *Molecular Biology of Bacterial Viruses.*)

(a) Make a conjecture about the number of bacteria per milliliter after 99 minutes. Verify your conjecture.

(b) Determine graphically the elapsed time when there were 25 million bacteria per milliliter.

SOLUTION **(a)** Since the bacteria double every 49.5 minutes, there would be 1,000,000 per milliliter after 49.5 minutes and 2,000,000 after 99 minutes. This is verified by evaluating

$$N(99) = 500,000e^{0.014(99)} \approx 2,000,000.$$

(b) *Graphical Solution* Solve $N(x) = 25,000,000$ by graphing $Y_1 = 500000e^\wedge(0.014X)$ and $Y_2 = 25000000$. Their graphs intersect near (279.4, 25,000,000), as shown in Figure 5.52. Thus in a 1 milliliter sample, half a million *E. coli* bacteria could increase to 25 million in approximately 279 minutes, or 4 hours and 39 minutes.

$[0, 400, 100]$ by $[0, 3 \times 10^7, 1 \times 10^7]$

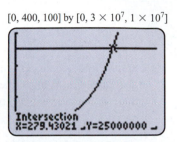

Intersection
X=279.43021 Y=25000000

FIGURE 5.52

Putting it all Together
5.3

*T*he following table summarizes some important concepts involving exponential functions.

Concept	Comments	Examples
Exponential function $f(x) = Ca^x,$ where $a > 0, a \neq 1,$ and $C > 0.$	Exponential growth occurs when $a > 1$, and exponential decay occurs when $0 < a < 1$. C often represents the initial amount present.	$f(x) = 5(0.8^x)$ (Decay) $g(x) = 3^x$ (Growth)  Decay: $0 < a < 1$ Growth: $a > 1$
Linear growth $y = mx + b$	If data increase by a fixed amount m for each unit increase in x, they can be modeled by a linear function.	The following data can be modeled by $f(x) = 3x + 2$ because the data increase 3 units for each unit increase in x, and because $y = 2$ when $x = 0$.

x	0	1	2	3
y	2	5	8	11

+3 +3 +3

Concept	Comment	Example
Exponential growth $$y = Ca^x$$	If data increase by a constant factor a for each unit increase in x, they can be modeled by an exponential function.	The following data can be modeled by $f(x) = 2(3)^x$ because the data increase by a factor of 3 for each unit increase in x and because $y = 2$ when $x = 0$. 	x	0	1	2	3	 	y	2	6	18	54	 $\times 3$ $\times 3$ $\times 3$
Radioactive decay	If a radioactive sample has a half-life of k years and contains C units, then the amount A remaining after x years is given by $$A(x) = C\left(\frac{1}{2}\right)^{x/k}.$$	A 5-gram sample of radioactive material with a half-life of 300 years is modeled by $A(x) = 5\left(\frac{1}{2}\right)^{x/300}$.
Interest compounded m times per year $$A_n = A_0\left(1 + \frac{r}{m}\right)^{mn}$$	A_0 is the principal, r is the interest rate, m is the number of times interest is paid each year, n is the number of years, and A_n is the amount after n years.	\$500 at 8%, compounded monthly for 3 years yields: $$500\left(1 + \frac{0.08}{12}\right)^{12(3)}$$ $$\approx \$635.12.$$
The number e	The number e is an irrational number and important in mathematics, much like π.	$e \approx 2.718282$
Natural exponential function	This function is an exponential function with base e.	$f(x) = e^x$
Interest compounded continuously $$A_n = A_0 e^{rn}$$	A_0 is the principal, r is the interest rate, n is the number of years, and A_n is the amount after n years.	\$500 at 8% compounded continuously for 3 years yields: $500e^{0.08(3)} \approx \$635.62.$

5.3   EXERCISES

Exponents

Exercises 1–16: Simplify the expression without a calculator.

1. 2^{-3}

2. $(-3)^{-2}$

3. $4^{1/2}$

4. $\left(\frac{1}{2}\right)^{-3}$

5. $27^{2/3}$

6. $8^{-2/3}$

7. $\left(\frac{1}{8}\right)^{-1}$

8. $a^{0.5}a^{1.5}$

9. $4^{1/6}4^{1/3}$

10. $\dfrac{9^{5/6}}{9^{1/3}}$

11. $e^x e^x$

12. $16^{-1/2}27^{1/3}$

13. 3^0

14. $5\left(\frac{3}{4}\right)^0$

15. $(5^{101})^{1/101}$

16. $(8^{27})^{1/27}$

Exercises 17–18: Use a calculator to approximate the expression to the nearest hundredth.

17. (a) $3^{5/2}$ **(b)** 2^{π} **(c)** $-e^{\sqrt{3}}$

18. (a) π^2 **(b)** e^{-1} **(c)** $2^{0.145}$

Linear and Exponential Growth

Exercises 19–24: (Refer to Example 1.) Find either a linear or an exponential function that models the data in the table.

19.

x	0	1	2	3	4
y	2	0.75	−0.50	−1.75	−3

20.

x	0	1	2	3	4
y	2	8	32	128	512

21.

x	−3	−2	−1	0	1
y	64	32	16	8	4

22.

x	−2	−1	0	1	2
y	3	5.5	8	10.5	13

23.

x	−4	−2	0	2	4
y	0.3125	1.25	5	20	80

24.

x	−15	−5	5	15	25
y	22	24	26	28	30

25. *Job Offer* A company offers a college graduate $40,000 per year with a guaranteed 8% raise each year. Is this an example of linear or exponential growth? Find a function f that computes the salary during the nth year.

26. *Job Offer* A new employee is offered $35,000 per year with a guaranteed $5000 raise each year. Is this an example of linear or exponential growth? Find a function f that computes the salary during the nth year.

27. *Comparing Growth* Which function becomes larger for $0 \le x \le 10$: $f(x) = 2^x$ or $g(x) = x^2$?

28. *Salaries* If you were offered 1¢ for the first week of work, 3¢ for the second week, 5¢ for the third week, 7¢ for the fourth week, and so on for a year, would you accept the offer? Would you accept an offer that pays 1¢ for the first week of work, 2¢ for

the second week, 4¢ for the third week, 8¢ for the fourth week, and so on for a year? Explain your answers.

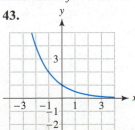

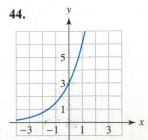

Exponential Functions

Exercises 29–34: Find C and a so that $f(x) = Ca^x$ satisfies the given conditions.

29. $f(0) = 5$ and for each unit increase in x, the output is multiplied by 1.5.

30. $f(1) = 3$ and for each unit increase in x, the output is multiplied by $\dfrac{3}{4}$.

31. $f(0) = 10$ and $f(1) = 20$

32. $f(0) = 7$ and $f(-1) = 1$

33. $f(1) = 9$ and $f(2) = 27$

34. $f(-1) = \dfrac{1}{4}$ and $f(1) = 4$

35. *Tire Pressure* The pressure in a tire with a leak is initially 30 pounds per square inch and can be modeled by $f(x) = 30(0.9)^x$ after x minutes. What is the tire's pressure after 9.5 minutes?

36. *Population* The population of California in 1998 was about 33 million and increased by a factor of 1.012 each year. Estimate the population of California in 2000.

Exercises 37–42: Sketch a graph of $y = f(x)$.

37. $f(x) = 2^x$ **38.** $f(x) = 4^x$

39. $f(x) = 3^{-x}$ **40.** $f(x) = 3(2^{-x})$

41. $f(x) = 2\left(\dfrac{1}{3}\right)^x$ **42.** $f(x) = 2(3^x)$

Exercises 43–46: Use the graph of $y = Ca^x$ to determine values for C and a.

43.

44.

45.

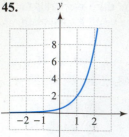

46.

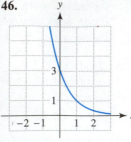

b.

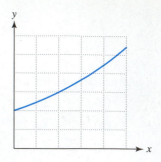

47. Match the symbolic representation of f with its graphical representation (a–d). Do not use a calculator.

  i. $f(x) = e^x$ ii. $f(x) = 3^{-x}$

 iii. $f(x) = 1.5^x$ iv. $f(x) = 0.99^x$

c.

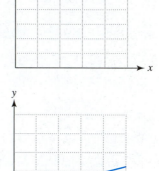

a.

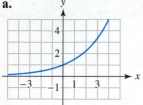

b.

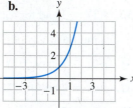

c.

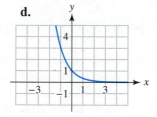

d.

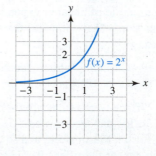

d.

48. *Modeling Phenomena*   Match the situation with the graph (a–d) that models it best.

  i.   Balance of an account after x years earning 10% interest compounded continuously

  ii.   Balance of an account after x years earning 5% interest compounded annually

 iii.   Air pressure in a car tire with a large hole in it after x minutes

 iv.   Air pressure in a car tire with a pinhole in it after x minutes

a.

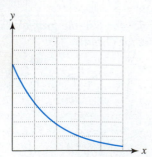

Exercises 49 and 50: The graph of $y = f(x)$ is shown in the accompanying figure. Sketch a graph of each equation using translations of graphs and reflections. Do not use a graphing calculator.

Algebra Review   To review translations of graphs, see Section 3.4.

49. $f(x) = 2^x$

  (a) $y = 2^x - 2$ **(b)** $y = 2^{x-1}$

  (c) $y = 2^{-x}$ **(d)** $y = -2^x$

50. $f(x) = e^{-0.5x}$

   (a) $y = -e^{-0.5x}$ **(b)** $y = e^{-0.5x} - 3$

   (c) $y = e^{-0.5(x-2)}$ **(d)** $y = e^{0.5x}$

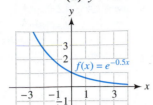

Exercises 51–56: Approximate $f(x)$ to four decimal places.

51. $f(x) = e^x$, $x = 3.1$

52. $f(x) = e^{2x}$, $x = -0.43$

53. $f(x) = 4e^{-1.2x}$, $x = -2.4$

54. $f(x) = -2.1e^{-0.71x}$, $x = 1.9$

55. $f(x) = \dfrac{e^x - e^{-x}}{2}$, $x = -0.7$

56. $f(x) = 4\left(e^{-0.3x} - e^{-0.6x}\right)$, $x = 1.6$

Compound Interest

Exercises 57–64: Use the compound interest formula to determine the final value of each amount.

57. $600 at 7% compounded annually for 5 years

58. $2300 at 11% compounded semiannually for 10 years

59. $950 at 3% compounded daily for 20 years (*Hint:* Let $m = 365$.)

60. $3300 at 8% compounded quarterly for 2 years

61. $2000 at 10% compounded continuously for 8 years

62. $100 at 19% compounded continuously for 50 years

63. $1600 at 10.4% compounded monthly for 2.5 years

64. $2000 at 8.7% compounded annually for 5 years

65. *Investments* Compare investing $2000 at 10% compounded monthly for 20 years with investing $2000 at 13% compounded monthly for 20 years.

66. *Lake Property* In some states, lake shore property is increasing in value by 15% per year. Determine the value of a $90,000 lake lot after 5 years.

67. *College Tuition* If college tuition is currently $8000 per year, inflating at 6% per year, what will be the cost of tuition in 10 years?

68. *Doubling Time* How long does it take for an investment to double its value if the interest is 12% compounded annually? 6% compounded annually? (*Hint:* Use trial and error or graphing.)

69. *Interest* A principal of $100 is deposited into an account paying 5% interest compounded monthly. In a similar account, $200 is deposited. Make a conjecture about how the amounts in each account will compare after 10 years. Explain your reasoning.

70. *Interest* A principal of $2000 is deposited into two different accounts. One account pays 10% interest compounded annually, and the other pays 5% interest compounded annually.

   (a) Make a conjecture whether the 10% account will accrue twice the interest as the 5% account after 5 years.

   (b) Test your conjecture by computing the actual amounts.

71. *Federal Debt* In fiscal year 1996 the federal budget deficit was $107.3 billion. At the same time, 30-year treasury bonds were paying 6.83% interest. Suppose the American taxpayer loaned $107.3 billion to the federal government at 6.83% compounded annually. If the federal government waited 30 years to pay the entire amount back, including the interest, how much would this be? (Source: Department of the Treasury.)

72. In the previous exercise suppose that interest rates were 2% higher. How much would the federal government owe after 30 years? Is the national debt sensitive to interest rates?

73. *Annuity* If x dollars is deposited every 2 weeks (26 times per year) into an account paying an annual interest rate r, expressed in decimal form, then the amount A_n in the account after n years can be approximated by the formula

$$A_n = x\left[\frac{(1 + r/26)^{26n} - 1}{(r/26)}\right].$$

If $50 is deposited every 2 weeks into an account paying 8% interest, approximate the amount in the account after 10 years.

74. (Refer to the previous exercise.) Suppose a retirement account pays 10% annual interest. Determine how much a 20-year-old worker should deposit in this account every 2 weeks in order to have 1 million dollars at age 65.

Applications

75. *Radioactive Radium-226* The half-life of radium-226 is about 1600 years. After 3000 years what percentage P of a sample of radium remains?

(*Hint*: Let $P = 1\left(\dfrac{1}{2}\right)^{x/1600}$.)

76. *Radioactive Strontium-90* Radioactive strontium-90 has a half-life of about 28 years and sometimes contaminates the soil. After 50 years, what percentage of a sample of radioactive strontium would remain?

77. *Radioactive Carbon-14* (Refer to Example 4.) A fossil contains 10% of the carbon-14 it contained when it was alive. Graphically estimate its age.

78. *Radioactive Carbon-14* (Refer to Example 4.) If a fossil contained 0.01 milligram of carbon-14 when it was alive and now contains 0.002 milligram of carbon-14, estimate its age.

79. *Radioactive Cesium-137* Radioactive cesium-137 was emitted in large amounts in the Chernobyl nuclear power station accident in Russia on April 26, 1986. The amount of cesium remaining after x years in an initial sample of 100 milligrams can be described by $A(x) = 100e^{-0.02295x}$. (Source: C. Mason, *Biology of Freshwater Pollution.*)

(a) How much is remaining after 50 years? Is the half-life of cesium more or less than 50 years?

(b) Estimate graphically the half-life of cesium-137.

80. *Radioactive Polonium-210* The accompanying table shows the amount y of polonium in grams remaining after x days from an initial sample of 2 milligrams.

x (days)	0	100	200	300
y (milligrams)	2	1.22	0.743	0.453

(a) Use the table to determine mentally if the half-life of polonium is more or less than 200 days.

(b) Find a formula that models the amount A of polonium in the table after x days.

(c) Estimate graphically the half-life of polonium.

81. *Greenhouse Gases* (Refer to Example 5.) Chlorofluorocarbons (CFCs) are gases created by people that increase the greenhouse effect and damage the ozone layer. CFC-12 is one type of chlorofluorocarbon used in refrigeration, air conditioning, and foam insulation. The following table lists future concentrations of CFC-12 in parts per billion (ppb), if current trends continue.

Year	2000	2005	2010	2015	2020
CFC-12 (ppb)	0.72	0.88	1.07	1.31	1.60

(a) Let $x = 0$ correspond to 2000 and $x = 20$ to 2020. Find values for C and a so that $f(x) = Ca^x$ models this data.

(b) Estimate the CFC-12 concentration in 2013.

82. *Modeling Traffic Flow* At an intersection cars arrive randomly with an average rate of 50 cars per hour. The likelihood or probability that at least one car will enter the intersection within a period of x minutes can be estimated by $f(x) = 1 - e^{-5x/6}$.

(a) Find the likelihood that at least one car enters the intersection during a 3-minute period.

(b) Graphically determine the value of x that gives a 50–50 chance of at least one car entering the intersection during an interval of x-minutes.

83. *Swimming Pool Maintenance* Chlorine is frequently used to disinfect swimming pools. The chlorine concentration should remain between 1.5 and 2.5 parts per million. On warm, sunny days with many swimmers agitating the water, 30% of the chlorine can dissipate into the air or combine with other chemicals. (Source: D. Thomas, *Swimming Pool Operators Handbook.*)

(a) Let $f(x) = 2.5(0.7)^x$ model the amount of chlorine in a pool after x days. What is the initial concentration of chlorine in the pool?

(b) If no chlorine is added, estimate graphically or numerically the number of days before chlorine should be added.

84. *Filters* Impurities in water are frequently removed using filters. Suppose that a 1-inch filter allows 10% of the impurities to pass through it. The other 90% is trapped in the filter.

(a) Find a formula in form $f(x) = 100a^x$, that calculates the percentage of impurities passing through x inches of this type of filter.

(b) Use $f(x)$ to estimate the percentage of impurities passing through 2.3 inches of the filter.

85. *Thickness of Runways* Heavier aircraft require runways with thicker pavement for landings and takeoffs. A pavement 6 inches thick can accommodate an aircraft weighing 80,000 pounds, whereas a 12-inch pavement is necessary for a 350,000-pound plane. The relation between pavement thickness x in inches and gross weight y in thousands of pounds can be modeled by $y = 18.29(1.279)^x$. Complete the table. Round values to the nearest thousand pounds. (Source: Federal Aviation Administration.)

x (inches)	6	7.5	9	10.5	12
y (lb × 1000)	80	?	?	?	350

86. *Survival of Reindeer* For all types of animals, the percentage that survive into the next year decreases. In one study, the survival rate of a sample of reindeer was modeled by $S(t) = 100(0.999993)^{t^5}$. The function S outputs the percentage of reindeer that survive t years. (Source: D. Brown.)

(a) Evaluate $S(4)$ and $S(15)$. Interpret the results.
(b) Graph S in [0, 15, 5] by [0, 110, 10]. Interpret the graph.

87. *Trains and Horsepower* The faster a locomotive travels, the more horsepower is needed. The function given by $H(x) = 0.157(1.033)^x$ calculates this horsepower for a level track. The input x is in miles per hour and the output $H(x)$ is the horsepower required per ton of cargo. (Source: L. Haefner, *Introduction to Transportation Systems.*)

(a) Graph H in [0, 100, 10] by [0, 5, 1]. Is H increasing or decreasing?
(b) Evaluate $H(30)$ and interpret the result.
(c) Determine the horsepower needed to move a 5000-ton train 30 miles per hour.
(d) Some types of locomotives are rated for 1350 horsepower. How many locomotives of this type would be needed in part (c)?

88. *Drug Concentrations* Sometimes after a patient takes a drug, the amount of medication A in the bloodstream can be modeled by $A = Ce^{-rt}$, where

C is the initial concentration in milligrams per liter, r is the hourly percentage decrease (in decimal) of the drug in the bloodstream, and t is the elapsed time in hours. Suppose that a drug's concentration is initially 2 milligrams per liter and that $r = 0.2$. Find the drug concentration after 3.5 hours.

89. *Computer Growth* Between 1971 and 1995 the numbers of transistors that could be placed on a single chip is shown in the table and modeled by $f(x) = 0.00168(1.39)^{x-1970}$. The input x is the year and the output $f(x)$ is the number of transistors in millions. (Source: Intel.)

Year	Chip	Transistors (millions)
1971	4004	0.0023
1986	386DX	0.275
1989	486DX	1.2
1993	Pentium	3.3
1995	P6	5.5

(a) Graph f and the data pairs (year, transistors) in [1968, 1998, 2] by [−1, 6, 1].
(b) Use $f(x)$ to estimate an average annual percentage increase in the number of transistors between 1971 and 1995.

90. *Incubation Time of AIDS* With some diseases such as AIDS, there is a delay time between the initial infection and the time when symptoms begin. This delay time can vary greatly from one individual to the next, but it appears to have an overall pattern. In one study $N(x) = N_0 e^{-0.1x^2/2}$ models the delay time for AIDS patients receiving infected blood transfusions between 1978 and 1986. The formula $N(x)$ computes the number of people who have not developed AIDS x years after their initial infection. (Source: G. Medley, "Incubation period of AIDS in patients infected via blood transfusions.")

(a) Evaluate $N(0)$ and interpret N_0.
(b) Let $N_0 = 100$. Graphically approximate the number of years before 50% of the patients have developed symptoms of AIDS.

91. *Tree Density* Ecologists studied the spacing between individual trees in a forest in British Columbia. This lodgepole pine forest was 40 to 50 years old and had approximately 1600 trees per acre that were randomly spaced. The probability or likelihood that there is at least one tree located in a circle with a radius of x feet is estimated by $P(x) = 1 - e^{-0.1144x}$. For example, $P(7) \approx 0.55$ means that if a person picks a point at random in the forest, there is a 55% chance that at least one tree will be located within 7 feet. See the accompanying figure. (Source: E. Pielou, *Populations and Community Ecology.*)

(a) Evaluate $P(2)$ and $P(20)$, and interpret the results.

(b) Graph P. Explain verbally why it is logical for P to be an increasing function.

(c) Solve $P(x) = 0.5$ and interpret the result.

92. *E. Coli Growth*   (Refer to Example 11.)
   (a) Approximate the number of *E. coli* after 3 hours.
   (b) Estimate graphically the elapsed time before there are 10 million bacteria per milliliter.

Writing about Mathematics

93. Explain how a linear function and an exponential function differ.

94. Discuss the domain and range of an exponential function f. Is f one-to-one? Explain.

Extended and Discovery Exercises

Exercises 1–3: **Present Value**   *In the compound interest formula $A = P(1 + r/m)^{mn}$, we can think of P as the present value of an investment and A as the future value of an investment. For example, if we are saving for college and need a future value of A dollars, then P would represent the amount needed in an account today to reach our goal in n years at an interest rate of r, compounded m times per year. If we solve the equation for P, it results in*

$$P = A(1 + r/m)^{-mn}.$$

1. Verify that the two formulas are equivalent by transforming the first equation into the second.

2. What should the present value of a savings account be to cover $30,000 of college expenses in 12.5 years, if the account pays 7.5% interest compounded quarterly?

3. If you need $15,000 to buy a car in 3 years, what should the present value of a savings account be to reach this goal, if the account pays 5% compounded monthly?

5.4 LOGARITHMIC FUNCTIONS AND MODELS

The Common Logarithmic Function • Solving Equations • Logarithms with Other Bases • General Logarithmic Equations

Introduction

In Chapter 1, numbers were introduced as a way to measure and quantify data. Then, functions were developed to model and describe data. Applications involving functions frequently result in equations. Although we have solved many equations, one equation that we have not solved *symbolically* is $a^x = k$. This exponential equation occurs frequently in applications. In this section we discuss logarithmic functions and how they can be used to solve exponential equations.

The Common Logarithmic Function

Exponential functions are capable of modeling data that exhibit rapid growth. However, a new type of nonlinear function is needed to model data that grows much slower. Table 5.17 lists the growth of a bacteria colony at 1-day intervals and illustrates exponential growth: each time x increases by 1 day, the number of bacteria y increases by a factor of 10. This growth is modeled by $f(x) = 10^x$, which calculates the number of bacteria on day x.

TABLE 5.17   Growth of Bacteria

x (day)	0	1	2	3	4
y (bacteria in thousands)	1	10	100	1000	10,000

Now consider Table 5.18, where the roles of x and y have been interchanged. Table 5.18 can be used to calculate the day when there were x bacteria. Notice that the function represented by this table grows very slowly: each time x increases by a factor of 10, y increases by 1. This function is called the *common (or base-10) logarithmic function,* and is written as $g(x) = \log x$.

TABLE 5.18   Growth of a Bacteria Colony

x (bacteria in thousands)	1	10	100	1000	10,000
y (day)	0	1	2	3	4

The common exponential function and the common logarithmic function are represented numerically in Tables 5.19 and 5.20.

TABLE 5.19   Common Exponential Function

x	-4	-3	-2	-1	0	0.5	1	2	π
10^x	10^{-4}	10^{-3}	10^{-2}	10^{-1}	10^0	$10^{0.5}$	10^1	10^2	10^π

TABLE 5.20   Common Logarithmic Function

x	10^{-4}	10^{-3}	10^{-2}	10^{-1}	10^0	$10^{0.5}$	10^1	10^2	10^π
$\log x$	-4	-3	-2	-1	0	0.5	1	2	π

Note: The common (or base-10) logarithmic function denoted by $\log x$ outputs k if the input x can be written as $x = 10^k$ for some real number k. That is, $\log 10^k = k$ for all real numbers k. This concept is illustrated visually in the accompanying figure.

Input   Output

$$x = 10^k \longrightarrow \boxed{\log 10^k} \longrightarrow k$$

Common Logarithm

The points $(10^{-1}, -1)$, $(10^0, 0)$, $(10^{0.5}, 0.5)$, and $(10^1, 1)$ are located on the graph of $y = \log x$. They are plotted in Figure 5.53 on the next page. Any *positive* real number x can be expressed as $x = 10^k$ for some real number k. In this case $\log x = \log 10^k = k$. The graph of $y = \log x$ is a continuous curve, as shown in Figure 5.54. The y-axis is a vertical asymptote. The common logarithmic function is one-to-one and always increasing. Its domain is all positive real numbers, and its range is all real numbers.

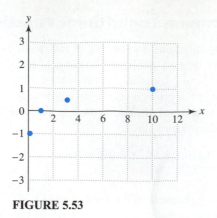

FIGURE 5.53

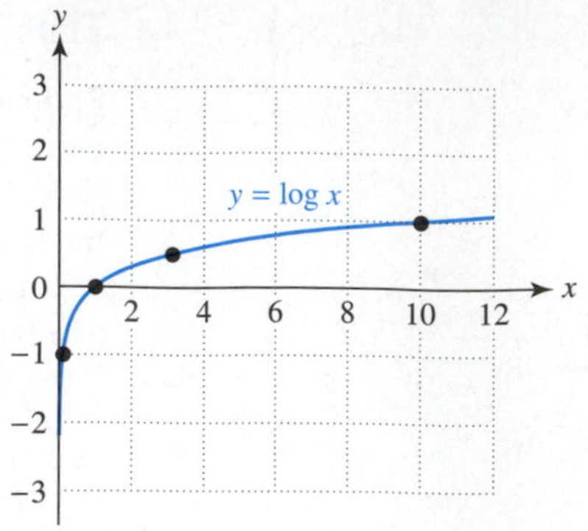

FIGURE 5.54 The Common Logarithmic Function

We have shown *verbal, numerical, graphical,* and *symbolic* representations of the common logarithmic function. A formal definition of the common logarithm is now given.

> ### Common logarithm
>
> The **common logarithm of a positive number** x, denoted $\log x$, is defined by
>
> $$\log x = k \quad \text{if and only if} \quad x = 10^k,$$
>
> where k is a real number. The function given by
>
> $$f(x) = \log x$$
>
> is called the **common logarithmic function**.

The common logarithmic function outputs an *exponent k,* which may be positive, negative, or zero. However, a valid input must be positive.

EXAMPLE 1 *Evaluating common logarithms*

Simplify each logarithm by hand.

(a) $\log 1$ **(b)** $\log \dfrac{1}{1000}$ **(c)** $\log \sqrt{10}$

SOLUTION

(a) Since $1 = 10^0$, $\log 1 = \log 10^0 = 0$.

Algebra Review

To review properties of exponents, see Section 4.7, and Chapter R (page R-15).

(b) $\log \dfrac{1}{1000} = \log 10^{-3} = -3.$

(c) $\log \sqrt{10} = \log (10^{1/2}) = \dfrac{1}{2}.$

FIGURE 5.55

Much like the square root function, the common logarithmic function does not have an easy-to-evaluate formula. For example, $\sqrt{4} = 2$ and $\sqrt{100} = 10$ can be calculated mentally, but for $\sqrt{2}$ we usually rely on a calculator. Similarly, $\log 100 = 2$ can be found mentally since $100 = 10^2$, whereas $\log 12$ can be approximated using a calculator. See Figure 5.55. To check that $\log 12 \approx 1.0791812$, evaluate $10^{1.0791812} \approx 12$. Another similarity between the square root function and the common logarithmic function is that their domains do not include negative numbers. If outputs are restricted to real numbers, both $\sqrt{-3}$ and $\log(-3)$ are undefined expressions.

The human ear is extremely sensitive and is able to detect pressures on the eardrum that range from 10^{-16} w/cm^2 (watts per square centimeter) to 10^{-4} w/cm^2. Sound with an intensity of 10^{-4} w/cm^2 is painful to the human eardrum. Because of this wide range of intensities, scientists use logarithms to measure sound in decibels.

EXAMPLE 2   *Applying common logarithms to sound*

Sound levels in decibels (db) can be computed by $f(x) = 10 \log (10^{16}x)$, where x is the intensity of the sound in watts per square centimeter.   (Source: R. Weidner and R. Sells, *Elementary Classical Physics.*)

(a) At what decibel level does the threshold for pain occur?
(b) Make a table of f for the intensities $x = 10^{-13}, 10^{-12}, 10^{-11}, \ldots, 10^{-4}$.
(c) If the intensity x increases by a factor of 10, what is the corresponding increase in decibels?

SOLUTION   (a) The threshold for pain occurs when $x = 10^{-4}$.

$$\begin{aligned} f(10^{-4}) &= 10 \log (10^{16} \cdot 10^{-4}) && \text{Substitute } x = 10^{-4}. \\ &= 10 \log (10^{12}) && \text{Add exponents when multiplying like bases.} \\ &= 10(12) && \text{Evaluate the logarithm.} \\ &= 120 && \text{Multiply.} \end{aligned}$$

The human eardrum begins to hurt at 120 decibels.

(b) Other values for x in Table 5.21 are computed in a similar manner.

TABLE 5.21

x (w/cm^2)	10^{-13}	10^{-12}	10^{-11}	10^{-10}	10^{-9}	10^{-8}	10^{-7}	10^{-6}	10^{-5}	10^{-4}
$f(x)$ (db)	30	40	50	60	70	80	90	100	110	120

(c) From Table 5.21, a tenfold increase in x results in an increase of 10 decibels.

■ CLASS DISCUSSION

If the sound level increases by 60 db, by what factor does the intensity increase? ■

Solving Equations

The graphs of $y = 10^x$ and $y = \log x$ are shown in Figure 5.56. Notice that they are reflections of each other across the line $y = x$. Both functions are one-to-one.

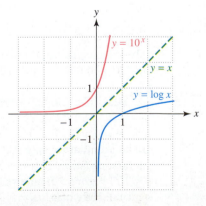

FIGURE 5.56

If $f(x) = 10^x$, then its inverse function is given by $f^{-1}(x) = \log x$. To see this numerically, consider Table 5.22. Notice that $\log 10^x = x$ for each input x. In a similar manner, $10^{\log x} = x$ for any positive x. For example, $\log 10^2 = 2$ and $10^{\log 100} = 100$.

TABLE 5.22

x	-2	-1	0	1	2
10^x	10^{-2}	10^{-1}	10^0	10^1	10^2

x	10^{-2}	10^{-1}	10^0	10^1	10^2
$\log x$	-2	-1	0	1	2

Inverse properties are summarized in the following.

Inverse properties of the common logarithm

The following inverse properties hold for the common logarithm.

$$\log 10^x = x \qquad \text{for any real number } x, \text{ and}$$

$$10^{\log x} = x \qquad \text{for any positive number } x.$$

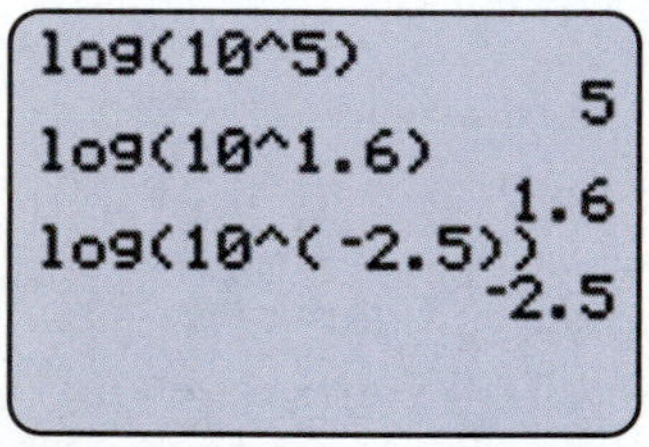

```
log(10^5)
              5
log(10^1.6)
            1.6
log(10^(-2.5))
           -2.5
```

FIGURE 5.57

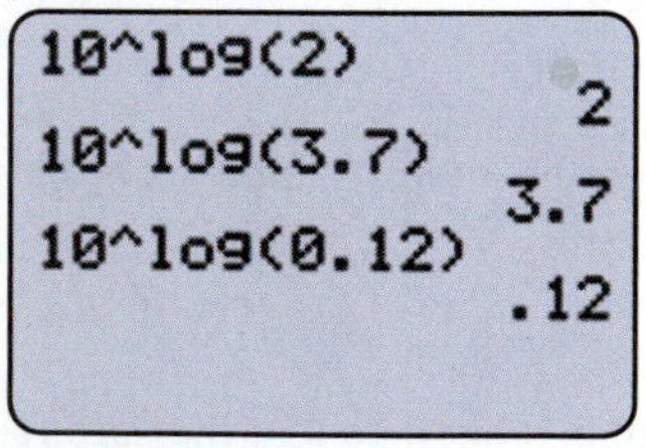

```
10^log(2)
              2
10^log(3.7)
            3.7
10^log(0.12)
            .12
```

FIGURE 5.58

A graphing calculator has been used to illustrate these properties in Figures 5.57 and 5.58.

An exponential equation has a variable that occurs as an exponent in an expression. To solve the exponential equation $10^x = k$, we take the common logarithm of both sides and then apply the inverse property $\log 10^x = x$ for any real number x. The following example illustrates this method.

$$10^x = 5 \qquad \textcolor{blue}{\text{Given exponential equation}}$$

$$\log 10^x = \log 5 \qquad \textcolor{blue}{\text{Take the common logarithm.}}$$

$$x = \log 5 \qquad \textcolor{blue}{\text{Inverse property: } \log 10^x = x}$$

Converting $10^x = 5$ to the equivalent equation $x = \log 5$ is sometimes referred to as changing **exponential form** to **logarithmic form**. In the second step, the common logarithm is taken of both sides of the equation. We include this second step to emphasize the fact that inverse properties are being used to solve these equations.

EXAMPLE 3   *Solving equations of the form $10^x = k$*

Solve each equation, if possible.
(a) $10^x = 0.001$ **(b)** $10^x = 55$ **(c)** $10^x = -5$

SOLUTION   **(a)** Take the common logarithm of both sides of the equation $10^x = 0.001$. Then,

$$\log 10^x = \log 0.001 \qquad \text{or} \qquad x = \log 10^{-3} = -3.$$

(b) In a similar manner, $10^x = 55$ is equivalent to $x = \log 55 \approx 1.7404$.
(c) The equation $10^x = -5$ has no real solution, because -5 is not in the range of 10^x. Figure 5.59 shows that the graphs of $Y_1 = 10\text{^}X$ and $Y_2 = -5$ do not intersect. Note that $\log(-5)$ is undefined.

$[-6, 6, 1]$ by $[-6, 6, 1]$

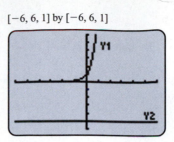

FIGURE 5.59

EXAMPLE 4   *Solving an exponential equation*

Solve $4(10^{3x}) = 244$.

SOLUTION   Begin by dividing both sides by 4.

$$4(10^{3x}) = 244 \qquad \text{Given equation}$$
$$10^{3x} = 61 \qquad \text{Divide by 4.}$$
$$\log 10^{3x} = \log 61 \qquad \text{Take the common logarithm.}$$
$$3x = \log 61 \qquad \text{Inverse property: } \log 10^k = k.$$
$$x = \frac{\log 61}{3} \qquad \text{Divide by 3.}$$
$$x \approx 0.595 \qquad \text{Approximate.}$$

A logarithmic equation contains logarithms. To solve logarithmic equations we *exponentiate* both sides of the equation and then apply the inverse property $10^{\log x} = x$.

$$\log x = 2.5 \qquad \text{Given logarithmic equation}$$
$$10^{\log x} = 10^{2.5} \qquad \text{Exponentiate both sides; base 10.}$$
$$x = 10^{2.5} \qquad \text{Inverse property: } \log 10^x = x$$
$$x \approx 316.23 \qquad \text{Approximate.}$$

Converting the equation $\log x = k$ to the equivalent equation $x = 10^k$ is called changing *logarithmic form* to *exponential form*.

EXAMPLE 5   *Solving equations of the form log x = k*

Solve each equation.
(a) $\log x = 3$ **(b)** $\log x = -2$ **(c)** $\log x = 2.7$

SOLUTION   **(a)**
$$\log x = 3 \qquad \text{Given equation}$$
$$10^{\log x} = 10^3 \qquad \text{Exponentiate both sides; base 10.}$$
$$x = 10^3 \qquad \text{Inverse property: } 10^{\log x} = x$$
$$x = 1000 \qquad \text{Simplify.}$$

$[0, 700, 100]$ by $[0, 4, 1]$

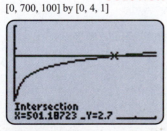

Intersection
X=501.18723 Y=2.7

FIGURE 5.60

(b) Similarly, $\log x = -2$ is equivalent to $x = 10^{-2} = 0.01$.

(c) $\log x = 2.7$ is equivalent to $x = 10^{2.7} \approx 501.2$. Graphical support is shown in Figure 5.60, where the graphs of $Y_1 = \log X$ and $Y_2 = 2.7$ intersect near $(501.2, 2.7)$.

EXAMPLE 6 *Solving a logarithmic equation*

Solve $5 \log 2x = 16$.

SOLUTION Begin by dividing both sides by 5.

$$5 \log 2x = 16 \qquad \text{Given equation}$$

$$\log 2x = 3.2 \qquad \text{Divide both sides by 5.}$$

$$10^{\log 2x} = 10^{3.2} \qquad \text{Exponentiate both sides; base 10.}$$

$$2x = 10^{3.2} \qquad \text{Inverse property: } 10^{\log k} = k$$

$$x = \frac{10^{3.2}}{2} \qquad \text{Divide both sides by 2.}$$

$$x \approx 792.4 \qquad \text{Approximate.}$$

Some types of data grow slowly and can be modeled by $f(x) = a + b \log x$. For example, a larger area of land tends to have a wider variety of birds. However, if the land area doubles, the number of species of birds does not double. In actuality, the land area has to more than double before the number of species doubles.

EXAMPLE 7 *Modeling data with logarithms*

Table 5.23 lists the number of species of birds on islands of different sizes near New Guinea.
(a) Find values for a and b so that $f(x) = a + b \log x$ models the data.
(b) Determine symbolically and graphically the island size that might have 50 different species of birds.

TABLE 5.23

Area (mi²)	0.1	1	10	100	1000
Species of Birds	26	39	52	65	78

Source: B. Freedman, *Environmental Ecology.*

SOLUTION **(a)** Begin by substituting $x = 1$ into the formula to determine a.

$$a + b \log 1 = 39 \qquad f(1) = 39$$

$$a + b(0) = 39 \qquad \log 1 = \log 10^0 = 0$$

$$a = 39 \qquad \text{Simplify.}$$

Thus $f(x) = 39 + b \log x$. Although we can use any data point in Table 5.22 to find b, we let $x = 10$.

$$39 + b \log 10 = 52 \qquad f(10) = 52$$

$$b \log 10 = 13 \qquad \text{Subtract 39.}$$

$$b = 13 \qquad \log 10^1 = 1$$

The data is modeled by $f(x) = 39 + 13 \log x$ and verified in Figure 5.61.

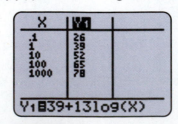

FIGURE 5.61

(b) *Symbolic Solution*   We must solve the equation $f(x) = 50$.

$$39 + 13 \log x = 50 \qquad \textcolor{blue}{f(x) = 50}$$
$$13 \log x = 11 \qquad \textcolor{blue}{\text{Subtract 39.}}$$
$$\log x = \frac{11}{13} \qquad \textcolor{blue}{\text{Divide by 13.}}$$
$$10^{\log x} = 10^{11/13} \qquad \textcolor{blue}{\text{Exponentiate both sides; base 10.}}$$
$$x = 10^{11/13} \qquad \textcolor{blue}{\text{Inverse property: } \log 10^k = k}$$
$$x \approx 7 \qquad \textcolor{blue}{\text{Approximate.}}$$

To have 50 species of birds, an island should be about 7 square miles.

Graphical Solution   The graphs of $Y_1 = 39 + 13 \log (X)$ and $Y_2 = 50$ in Figure 5.62 intersect near the point $(7.02, 50)$. This result agrees with our symbolic solution.

[0, 15, 5] by [0, 70, 10]

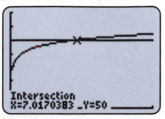

FIGURE 5.62

Air pollutants frequently cause acid rain. A measure of the acidity is pH, which ranges between 1 and 14. Pure water is neutral and has a pH of 7. Acidic solutions have a pH less than 7, whereas alkaline solutions have a pH greater than 7. A pH value measures the concentration of the hydrogen ions in a solution. It can be computed by $f(x) = -\log x$, where x represents the hydrogen ion concentration in moles per liter. Pure water exposed to normal carbon dioxide in the atmosphere has a pH of 5.6. If the pH of a lake drops below this level, it is indicative of an *acid lake*.

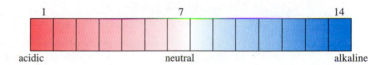

1	7	14
acidic	neutral	alkaline

EXAMPLE 8   *Analyzing acid rain*

In rural areas of Europe, rainwater typically has a hydrogen ion concentration of $x = 10^{-4.7}$.   (Source: G. Howells, *Acid Rain and Acid Water.*)
(a) Find its pH. What effect might this have on a lake with a pH of 5.6?
(b) Seawater has a pH of 8.2. Compared to seawater, how many times greater is the hydrogen ion concentration in rainwater from rural Europe?

SOLUTION   **(a)** $f(10^{-4.7}) = -\log 10^{-4.7} = -(-4.7) = 4.7$. The pH is 4.7, which could cause the lake to become more acidic.

(b) To find the hydrogen ion concentration in seawater, solve $f(x) = 8.2$ for x.

$$-\log x = 8.2 \qquad \textcolor{blue}{f(x) = 8.2}$$
$$\log x = -8.2 \qquad \textcolor{blue}{\text{Multiply by } -1.}$$
$$10^{\log x} = 10^{-8.2} \qquad \textcolor{blue}{\text{Exponentiate both sides; base 10.}}$$
$$x = 10^{-8.2} \qquad \textcolor{blue}{\text{Inverse property}}$$

Since $\dfrac{10^{-4.7}}{10^{-8.2}} = 10^{3.5}$, the hydrogen ion concentration in the rainwater is $10^{3.5} \approx$ 3162 times greater than it is in seawater.

Logarithms with Other Bases

The development of base-a logarithms can be done with any positive base $a \neq 1$. For example, in computer science base-2 logarithms are frequently used. A numerical representation of the base-2 logarithmic function, denoted $f(x) = \log_2 x$, is shown in Table 5.24. If x can be expressed in the form $x = 2^k$ for some k, then $\log_2 x = k$. A graph of $y = \log_2 x$ is shown in Figure 5.63.

TABLE 5.24   Base-2 Logarithm

x	$2^{-3.1}$	2^{-2}	$2^{-0.5}$	2^0	$2^{0.5}$	2^2	$2^{3.1}$
$\log_2 x$	-3.1	-2	-0.5	0	0.5	2	3.1

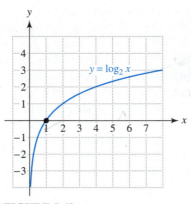

FIGURE 5.63

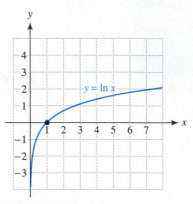

FIGURE 5.64

In a similar manner, a numerical representation for the base-e logarithm is shown in Table 5.25. The base-e logarithm is referred to as the **natural logarithm** and denoted either $\log_e x$ or $\ln x$. Natural logarithms are used in mathematics, science, economics, and technology. A graph of $y = \ln x$ is shown in Figure 5.64.

TABLE 5.25   Natural Logarithm

x	$e^{-3.1}$	e^{-2}	$e^{-0.5}$	e^0	$e^{0.5}$	e^2	$e^{3.1}$
$\ln x$	-3.1	-2	-0.5	0	0.5	2	3.1

■ CLASS DISCUSSION
Make a numerical representation for a base-4 logarithm. Evaluate $\log_4 16$. ■

A base-a logarithm is now defined.

Logarithm

The **logarithm with base a of a positive number x**, denoted by $\log_a x$, is defined by

$$\log_a x = k \quad \text{if and only if} \quad x = a^k,$$

where $a > 0$, $a \neq 1$, and k is a real number. The function, given by

$$f(x) = \log_a x,$$

is called the **logarithmic function with base a**.

Remember that *a logarithm is an exponent*. The expression $\log_a x$ is the exponent k such that $a^k = x$. Logarithms with base a satisfy inverse properties similar to those for common logarithms.

> ### Inverse properties
> The following inverse properties hold for logarithms with base a.
>
> $$\log_a a^x = x \quad \text{for any real number } x, \text{ and}$$
> $$a^{\log_a x} = x \quad \text{for any positive number } x.$$

The inverse function of $f(x) = a^x$ is $f^{-1}(x) = \log_a x$. Therefore the graph of $y = \log_a x$ is a reflection of the graph of $y = a^x$ across the line $y = x$. Figure 5.65 shows graphs of $f(x) = 2^x$ and $f^{-1}(x) = \log_2 x$. The graph of f^{-1} is a reflection of f in the line $y = x$. The rapid growth in f is called *exponential growth*. On the other hand, the graph of f^{-1} begins to level off. This slower rate of growth is called **logarithmic growth**. Since the range of f is all positive numbers, the domain of f^{-1} is all positive numbers. Notice that when $0 < x < 1$, $f^{-1}(x) = \log_2 x$ outputs negative numbers, and when $x > 1$, $f^{-1}(x)$ outputs positive numbers.

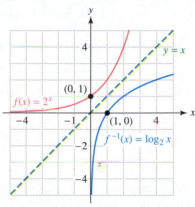

FIGURE 5.65

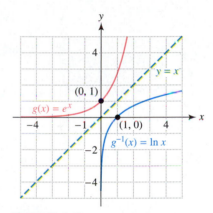

FIGURE 5.66

Graphs of $g(x) = e^x$ and $g^{-1}(x) = \ln x$ are shown in Figure 5.66. Since $e \approx 2.7183 > 2$, the graph of g increases faster than f, and the graph of g^{-1} levels off faster than the graph of f^{-1}.

■ MAKING CONNECTIONS

Exponential and Logarithmic Functions The inverse of an exponential function is a logarithmic function, and the inverse of a logarithmic function is an exponential function. For example,

$$\text{if } f(x) = 10^x \quad \text{then} \quad f^{-1}(x) = \log x,$$
$$\text{if } g(x) = \ln x \quad \text{then} \quad g^{-1}(x) = e^x, \quad \text{and}$$
$$\text{if } h(x) = 2^x \quad \text{then} \quad h^{-1}(x) = \log_2 x.$$

EXAMPLE 9 *Evaluating logarithms*

Evaluate each logarithm.

(a) $\log_2 8$ **(b)** $\log_5 \dfrac{1}{25}$ **(c)** $\log_7 49$ **(d)** $\ln e^{-7}$

SOLUTION   **(a)** To determine $\log_2 8$, express 8 as 2^k for some k. Since $8 = 2^3$, $\log_2 8 = \log_2 2^3 = 3$.

(b) $\log_5 \dfrac{1}{25} = \log_5 \dfrac{1}{5^2} = \log_5 5^{-2} = -2$ **(c)** $\log_7 49 = \log_7 7^2 = 2$

(d) $\ln e^{-7} = \log_e e^{-7} = -7$

General Logarithmic Equations

To solve the equation $a^x = k$, take the base-a logarithm of both sides.

EXAMPLE 10   *Solving equations of the form $a^x = k$*

Solve each equation.

(a) $3^x = \dfrac{1}{27}$ **(b)** $e^x = 5$

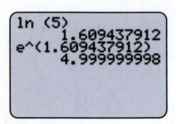

FIGURE 5.67

SOLUTION   **(a)** $3^x = \dfrac{1}{27}$ Given equation

$\log_3 3^x = \log_3 \dfrac{1}{27}$ Take the base-3 logarithm of both sides.

$\log_3 3^x = \log_3 3^{-3}$ Properties of exponents

$x = -3$ Inverse property: $\log_a a^k = k$

(b) Take the natural logarithm of both sides. Then,

$$\ln e^x = \ln 5 \quad \text{is equivalent to} \quad x = \ln 5 \approx 1.6094.$$

Many calculators are able to compute natural logarithms. The evaluation of $\ln 5$ is shown in Figure 5.67. Notice that $e^{1.609437912} \approx 5$.

To solve the equation $\log_a x = k$, exponentiate both sides of the equation by using base a. This is illustrated in the next example.

EXAMPLE 11   *Solving equations of the form $\log_a x = k$*

Solve each equation.
(a) $\log_2 x = 5$ **(b)** $\log_5 x = -2$ **(c)** $\ln x = 4.3$

SOLUTION   **(a)** $\log_2 x = 5$ Given equation

$2^{\log_2 x} = 2^5$ Exponentiate both sides; base 2.

$x = 2^5$ Inverse property: $a^{\log_a x} = x$

$x = 32$ Simplify.

(b) In a similar manner, $\log_5 x = -2$ is equivalent to $x = 5^{-2} = \dfrac{1}{25}$.
(c) $\ln x = 4.3$ is equivalent to $x = e^{4.3} \approx 73.7$.

EXAMPLE 12   *Solving exponential and logarithmic equations*

Solve each equation symbolically. Support your results graphically.
(a) $5e^x - 8 = 37$ **(b)** $5 \ln 2x + 3 = 10$

SOLUTION **(a)** *Symbolic Solution* Begin by adding 8 to both sides.

$$5e^x - 8 = 37 \qquad \text{Given equation}$$
$$5e^x = 45 \qquad \text{Add 8.}$$
$$e^x = 9 \qquad \text{Divide by 5.}$$
$$\ln e^x = \ln 9 \qquad \text{Take the natural logarithm.}$$
$$x = \ln 9 \qquad \text{Inverse property: } \log_a a^x = x$$
$$x \approx 2.197 \qquad \text{Approximate.}$$

Graphical Solution The graphs of $Y_1 = 5*e^\wedge(X) - 8$ and $Y_2 = 37$ intersect near the point $(2.197, 37)$, as shown in Figure 5.68.

[0, 5, 1] by [0, 50, 10]

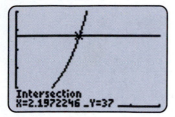

FIGURE 5.68

(b) *Symbolic Solution* Start by subtracting 3 from both sides.

$$5 \ln 2x + 3 = 10 \qquad \text{Given equation}$$
$$5 \ln 2x = 7 \qquad \text{Subtract 3.}$$
$$\ln 2x = \frac{7}{5} \qquad \text{Divide by 5.}$$
$$e^{\ln 2x} = e^{7/5} \qquad \text{Exponentiate both sides; base } e.$$
$$2x = e^{7/5} \qquad \text{Inverse property: } a^{\log_a x} = x$$
$$x = \frac{e^{7/5}}{2} \qquad \text{Divide by 2.}$$
$$x \approx 2.028 \qquad \text{Approximate.}$$

Graphical Solution The graphs of $Y_1 = 5 \ln(2X) + 3$ and $Y_2 = 10$ in Figure 5.69 intersect near the point $(2.028, 10)$. This result agrees with the symbolic solution.

Graphing Calculator Help

To find a point of intersection, see Appendix B (page AP-11).

[−5, 5, 1] by [−20, 20, 5]

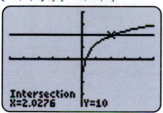

FIGURE 5.69

Putting it all Together *5.4*

*T*he following table summarizes some important concepts about base-*a* logarithms. Common and natural logarithms satisfy the same properties.

Concept	Explanation	Examples
Base-*a* logarithm	The base-*a* logarithm of a positive number x is $$\log_a x = k$$ if and only if $$x = a^k.$$ That is, a logarithm is an exponent k.	$\log 100 = \log 10^2 = 2$ $\log_2 8 = \log_2 2^3 = 3$ $\log_3 \sqrt[3]{3} = \log_3 3^{1/3} = \frac{1}{3}$ $\ln 5 \approx 1.609$ (using a calculator)

Concept	Explanation	Examples
Graph of $y = \log_a x$	The graph of the base-a logarithm *always* passes through the point $(1, 0)$ because $\log_a 1 = \log_a a^0 = 0$. The y-axis is a vertical asymptote.	
Inverse properties	a^x and $\log_a x$ calculate inverse operations. That is, $$\log_a a^x = x$$ and $$a^{\log_a x} = x.$$	$\log 10^{-3} = -3$ $\log_4 4^6 = 6$ $10^{\log x} = x$ $4^{\log_4 2x} = 2x$ $e^{\ln 5} = 5$
Inverse functions	The inverse function of $f(x) = a^x$ is $f^{-1}(x) = \log_a x$	$f(x) = 10^x \qquad f^{-1}(x) = \log x$ $f(x) = e^x \qquad f^{-1}(x) = \ln x$ $f(x) = \log_2 x \qquad f^{-1}(x) = 2^x$
Exponential equations	To solve $a^x = k$, take the base-a logarithm of both sides.	$10^x = 15 \qquad\qquad e^x = 20$ $\log 10^x = \log 15 \qquad \ln e^x = \ln 20$ $x = \log 15 \qquad\qquad x = \ln 20$
Logarithmic equations	To solve $\log_a x = k$, exponentiate both sides; base a.	$\log x = 3 \qquad\qquad \ln x = 5$ $10^{\log x} = 10^3 \qquad e^{\ln x} = e^5$ $x = 1000 \qquad\qquad x = e^5$

5.4 EXERCISES

Common Logarithms

Exercises 1 and 2: Complete the table for the common logarithm.

1.

x	10^0	10^4	10^8	$10^{1.2}$
$\log x$	?	4	?	?

2.

x	10^{-20}	$10^{-\pi}$	10^{55}	$10^{7.89}$
$\log x$	?	?	55	?

Exercises 3–6: Evaluate the following expressions by hand, if possible. Check your results with a calculator.

3. (a) $\log 10$ (b) $\log 10{,}000$
    (c) $20 \log 0.1$ (d) $\log 10 + \log 0.001$

4. (a) $\log 100$ (b) $\log 1{,}000{,}000$
    (c) $5 \log 0.01$ (d) $\log 0.1 - \log 1000$

5. (a) $2 \log 0.1 + 4$ (b) $\log 10^{1/2}$
    (c) $3 \log 100 - \log 1000$ (d) $\log(-10)$

6. (a) $\log(-4)$ **(b)** $\log 1$
(c) $\log 0$ **(d)** $-6\log 100$

Exercises 7 and 8: Determine mentally an integer n so that the logarithm is between n and n + 1. Check your result with a calculator.

7. (a) $\log 79$ **(b)** $\log 500$
(c) $\log 5$ **(d)** $\log 0.5$

8. (a) $\log 63$ **(b)** $\log 5000$
(c) $\log 9$ **(d)** $\log 0.04$

9. Solve each equation.
(a) $10^x = 1000$ **(b)** $10^x = 1{,}000{,}000$
(c) $10^x = 0.01$ **(d)** $10^x = 0.0001$

10. Solve each equation.
(a) $10^x = 250$ **(b)** $10^x = 4$ **(c)** $10^x = 0.5$

Exercises 11 and 12: Find the exact value of each expression. Use a calculator to support your results.

11. (a) $\log\sqrt{1000}$ **(b)** $\log\sqrt[3]{10}$
(c) $\log\sqrt{10} + \log\sqrt[5]{0.1}$
(d) $\log\sqrt{0.01} - \log\sqrt{1}$

12. (a) $\log\sqrt{100{,}000}$ **(b)** $\log\sqrt[3]{100}$
(c) $2\log\sqrt{0.1}$ **(d)** $10\log\sqrt[5]{10}$

General Logarithms

Exercises 13–26: Evaluate the logarithm.

13. $\log_2 64$

14. $\log_2 \dfrac{1}{4}$

15. $\log_4 2$

16. $\log_3 9$

17. $\ln 1$

18. $\ln e$

19. $\log_a \dfrac{1}{a}$

20. $\log_a (a^2 \cdot a^3)$

21. $\log_5 5^0$

22. $\ln\sqrt{e}$

23. $\log_2 \dfrac{1}{16}$

24. $\log_8 8^k$

25. $2^{\log_2 k}$

26. $\ln e^4 + \ln e^{-4}$

Exercises 27–30: Complete the table without the aid of a calculator.

27. $f(x) = \log_2 x$

x	$\frac{1}{16}$	1	8	32
$f(x)$				

28. $f(x) = \log_4 x$

x	$\frac{1}{4}$	4	16	64
$f(x)$				

29. $f(x) = 2\log_2 (x - 5)$

x	6	7	21
$f(x)$			

30. $f(x) = 2\log_3 (2x)$

x	$\frac{1}{18}$	$\frac{3}{2}$	$\frac{9}{2}$
$f(x)$			

Solving Equations

Exercises 31–38: Solve the equation symbolically.

31. $2^x = 72$ **32.** $3^x = 101$

33. $5^x = 0.25$ **34.** $4^x = 11$

35. $e^x = 25$ **36.** $e^x = 0.23$

37. $e^{-x} = 3$ **38.** $e^{-x} = \dfrac{1}{2}$

Exercises 39–46: Solve the equation.

39. $10^x - 5 = 95$ **40.** $2 \cdot 10^x = 66$

41. $10^{3x} = 100$ **42.** $4 \cdot 10^{2x} + 1 = 21$

43. $e^x + 1 = 24$ **44.** $1 - 2e^x = -5$

45. $2^x + 1 = 15$ **46.** $3 \cdot 5^x = 125$

Exercises 47–52: Solve the equation symbolically. Approximate answers to 4 decimal places.

47. $\log x = 2.3$ **48.** $\log_2 x = -3$

49. $\log_2 x = 1.2$ **50.** $\log_4 x = 3.7$

51. $\ln x = -2$ **52.** $\ln x = 2$

Exercises 53–62: Solve the equation. Approximate answers to 4 decimal places.

53. $2\log x = 6$ **54.** $\log 4x = 2$

55. $2\log 5x = 4$ **56.** $6 - \log x = 3$

57. $4\ln x = 3$ **58.** $\ln 5x = 8$

59. $5\ln x - 1 = 6$ **60.** $2\ln 3x = 8$

61. $4\log_2 x = 16$

62. $\log_3 5x = 10$

Exercises 63 and 64: Find values for a and b so that f(x) model the data exactly.

63. $f(x) = a + b \log x$

x	1	10	100	1000
y	5	7	9	11

64. $f(x) = a + b \log_2 x$

x	1	2	4	8
y	3.1	6	8.9	11.8

Graphs of Logarithmic Equations

Exercises 65 and 66: Use the graph of f to sketch a graph of f^{-1}. Give a symbolic representation of f^{-1}.

65. $f(x) = e^x$

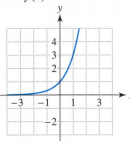

66. $f(x) = \log_4 x$

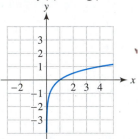

Exercises 67–70: Match the function f with its graph (a–d).

67. $f(x) = \log_2 x$

68. $f(x) = \log x$

69. $f(x) = \log (x + 2)$

70. $f(x) = 2 + \log_2 x$

(a)

(b)

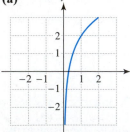

(c)

(d)

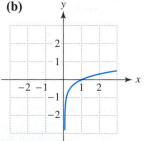

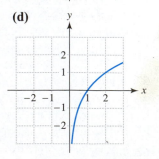

Exercises 71–74: Graph f and state its domain.

71. $f(x) = \log (x + 1)$

72. $f(x) = \log (x - 3)$

73. $f(x) = \ln (-x)$

74. $f(x) = \ln (x^2 + 1)$

Applications

75. *Diversity of Birds* (Refer to Example 7.) The accompanying table lists the number of species of birds on islands of various sizes. Find values for a and b so that $f(x) = a + b \log x$ models this data. Then use f to estimate the size of an island that might have 16 species of birds.

Area (km²)	0.1	1	10	100	1000
Species of Birds	3	7	11	15	19

76. *Diversity of Insects* The accompanying table lists the number of types of insects found in wooded regions with various acreages. Find values for a and b so that $f(x) = a + b \log x$ models this data. Then use f to estimate an acreage that might have 1200 types of insects.

Area (acres)	10	100	1000	10,000	100,000
Types of Insects	500	800	1100	1400	1700

77. *Growth of Bacteria* The accompanying table lists the growth of a colony of bacteria in millions after an elapsed time of x days.

x (days)	0	1	2	3	4
y (millions of bacteria)	3	6	12	24	48

(a) Find values for C and a so that $f(x) = Ca^x$ models the data.

(b) Estimate when there were 16 million bacteria in the colony.

78. *Growth of an Investment* The growth of an investment is shown in the accompanying table.

x (years)	0	5	10	15	20
y (dollars)	100	300	900	2700	8100

(a) Find values for C and a so that $f(x) = Ca^x$ models the data.

(b) Estimate when the account contained $2000.

79. *Runway Length* There is a mathematical relation between an airplane's weight x and the runway length required at takeoff. For some airplanes the minimum runway length in thousands of feet may be modeled by $L(x) = 3 \log x$, where x is measured in thousands of pounds. (Source: L. Haefner, *Introduction to Transportation Systems.*)

(a) Graph L in $[0, 50, 10]$ by $[0, 6, 1]$. Interpret the graph.

(b) If the weight of an airplane increases tenfold from 10,000 to 100,000 pounds, does the length of the required runway also increase by a factor of 10? Explain.

(c) Generalize your answer from part (b).

80. *Runway Length* (Refer to the previous exercise.) Estimate the maximum weight of a plane that can take off from a runway that is 10 thousand feet long.

81. *Acid Rain* (Refer to Example 8.) Find the hydrogen ion concentration for the following pH levels of acid rain. (Source: G. Howells.)

(a) 4.92 (pH of rain at Amsterdam Islands in the Indian Ocean)

(b) 3.9 (pH of rain at some locations in the eastern United States)

82. *Growth in Salary* Suppose that a person's salary is initially \$30,000 and is modeled by $f(x)$, where x represents the number of years of experience. Use $f(x)$ to approximate the years of experience when the salary exceeds \$60,000.

(a) $f(x) = 30,000(1.1)^x$

(b) $f(x) = 30,000 \log (10 + x)$

Would most people prefer that their salaries increase exponentially or logarithmically?

83. *Earthquakes* The Richter scale is used to measure the intensity of earthquakes. Intensity corresponds to the amount of energy released by an earthquake. If an earthquake has an intensity of x then its *magnitude*, as computed by the Richter scale, is given by $R(x) = \log \dfrac{x}{I_0}$, where I_0 is the intensity of a small, measurable earthquake.

(a) On July 26, 1963, an earthquake in Yugoslavia had a magnitude of 6.0 on the Richter scale, and on August 19, 1977, an earthquake in Indonesia measured 8.0. Find the intensity x for each of these earthquakes if $I_0 = 1$.

(b) How many times more intense was the Indonesian earthquake than the Yugoslavian earthquake?

84. *Modeling Algae Growth* When sewage was accidentally dumped into Lake Tahoe, the concentration of the algae *Selenastrum* increased from 1000 cells per milliliter to approximately 1,000,000 cells per milliliter within 6 days. Let f model the algae concentration after x days have elapsed. (Source: A. Payne, "Responses of the three test algae of the algal assay procedure: bottle test.")

(a) Compute the average rate of change in f from 0 to 6 days.

(b) The specific growth rate r is defined by $r = \dfrac{\log N_2 - \log N_1}{x_2 - x_1}$, where N_1 is the algae concentration at time x_1 and N_2 is the algae concentration at time x_2. Compute r in this example.

(c) Discuss why environmental scientists might use the specific growth rate r, rather than the average rate of change, to describe algae growth.

85. *Hurricanes* Hurricanes are some of the largest storms on earth. They are very low pressure areas with diameters of over 500 miles. The barometric air pressure in inches of mercury at a distance of x miles from the eye of a typical hurricane is modeled by $f(x) = 0.48 \ln(x + 1) + 27$. (Source: A. Miller and R. Anthes, *Meteorology.*)

(a) Evaluate $f(0)$ and $f(100)$. Interpret the results.

(b) Graph f in $[0, 250, 50]$ by $[25, 30, 1]$. Describe how air pressure changes as one moves away from the eye of the hurricane.

(c) At what distance from the eye of the hurricane is the air pressure 28 inches of mercury?

86. *Predicting Wind Speed* Wind speed typically varies in the first 20 meters above the ground. Close to the ground wind speed is often less than it is at 20 meters above the ground. For this reason, the National Weather Service usually measures wind speeds at heights between 5 and 10 meters. For a particular day, let $f(x) = 1.2 \ln x + 2.3$ compute the wind speed in meters per second at a height x meters above the ground for $x \geq 1$. (Source: A. Miller.)

(a) Find the wind speed at a height of 5 meters.

(b) Graph f in [0, 20, 5] by [0, 7, 1]. Interpret the graph.

(c) Estimate the height where the wind speed is 5 meters per second.

87. *Cooling an Object* A pot of boiling water with a temperature of 100°C is set in a room with a temperature of 20°C. The temperature T of the water after x hours is given by $T(x) = 20 + 80e^{-x}$.

(a) Estimate the temperature of the water after 1 hour.

(b) How long did it take the water to cool to 60°C?

88. *Warming an Object* A can of soda with a temperature of 5°C is set in a room with a temperature of 20°C. The temperature T of the soda after x minutes is given by $T(x) = 20 - 15(10)^{-0.05x}$.

(a) Estimate the temperature of the soda after 5 minutes.

(b) After how many minutes was the temperature of the soda 15°C?

89. *Modeling Traffic Flow* (Refer to Example 10, Section 5.3.) At an intersection cars arrive randomly with an average volume of 20 cars per hour. The likelihood or probability that no car enters the intersection within a period of x minutes can be estimated by $f(x) = e^{-x/3}$.

(a) What is the probability that no car enters the intersection during a 5-minute period?

(b) Determine the value of x that gives a 30% chance that no car enters the intersection during an interval of x-minutes.

90. *Population Growth* The population of Tennessee in millions is given by $P(x) = 4.88e^{0.0133x}$, where $x = 0$ corresponds to 1990, $x = 1$ to 1991, and so on. (Source: Bureau of the Census.)

(a) Determine symbolically the year when the population of Tennessee was 5.4 million.

(b) Solve part (a) graphically.

Writing about Mathematics

91. Describe the relationship between an exponential function and a logarithmic function with the same base. Explain why logarithms are needed to solve exponential equations.

92. Give verbal, numerical, graphical, and symbolic representations of a base-5 logarithmic function.

Extended and Discovery Exercise

1. *Greenhouse Effect* According to one model, the future increases in average global temperatures (due to carbon dioxide levels exceeding 280 parts per million) can be estimated using $T = 6.5 \ln (C/280)$, where C is the concentration of atmospheric carbon dioxide in parts per million (ppm) and T is in degrees Fahrenheit. Let future amounts of carbon dioxide be modeled by $C(x) = 364(1.005)^x$, where $x = 0$ corresponds to the year 2000 and $x = 100$ to 2100, and so on. (Source: W. Clime, *The Economics of Global Warming*.)

(a) Use composition of functions to write T as a function of x. Evaluate T when $x = 100$ and interpret the result.

(b) Graph $C(x)$ in [0, 200, 50] by [0, 1000, 100] and $T(x)$ in [0, 200, 50] by [0, 10, 1]. Describe each graph.

(c) How does an exponential growth in carbon dioxide concentrations affect the rise in global temperature?

CHECKING BASIC CONCEPTS FOR SECTIONS 5.3 AND 5.4

1. If the principal is $1200 and the interest rate is 9.5% compounded monthly, calculate the account balance after 4 years. Determine the balance if the interest is compounded continuously.

2. Find values for C and a so that $f(x) = Ca^x$ models the data in the table.

x	0	1	2	3
y	4	2	1	0.5

3. Explain verbally what $\log_2 15$ represents. Is it equal to an integer? (Do not use a calculator.)

4. Evaluate each of the following logarithms by hand.
(a) $\log_6 36$ **(b)** $\log\sqrt{10} + \log 0.01$
(c) $\ln \dfrac{1}{e^2}$

5. Solve each equation.
(a) $e^x = 5$ **(b)** $10^x = 25$ **(c)** $\log x = 1.5$

6. Solve the equation symbolically. Support your results graphically.
(a) $2e^x + 1 = 25$ **(b)** $\log 2x = 2.3$
(c) $\log x^2 = 1$

7. *Population*   In July 1994 the population of New York state in millions was modeled by $f(x) = 18.2e^{0.001x}$, and the population of Florida in millions was modeled by $g(x) = 14e^{0.0168x}$. In both formulas x is the year, where $x = 0$ corresponds to July 1994.   (Source: Bureau of the Census.)
(a) Find the population of New York and Florida in July 1994.
(b) Assuming these trends continue, estimate graphically the year when the population of Florida will equal the population of New York. What will their populations be at this time?

5.5 PROPERTIES OF LOGARITHMS

Basic Properties of Logarithms • Change of Base Formula

Introduction

The discovery of logarithms by John Napier (1550–1617) played an important role in the history of science. Logarithms were instrumental for Johannes Kepler (1571–1630) to calculate the positions of the planet Mars, which led to his discovery of the laws of planetary motion. Kepler's laws were used by Isaac Newton (1642–1727) to discover the universal laws of gravitation. Although calculators and computers have made tables of logarithms obsolete, applications involving logarithms still play an important role in modern-day computation. One reason for this is that logarithms possess several important properties. For example, the loudness of a sound can be measured in decibels by the formula $f(x) = 10 \log(10^{16}x)$, where x is the intensity of the sound in watts per square centimeter. In Example 4, we use properties of logarithms to simplify this formula.

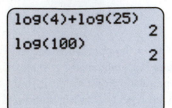

FIGURE 5.70

FIGURE 5.71

Basic Properties of Logarithms

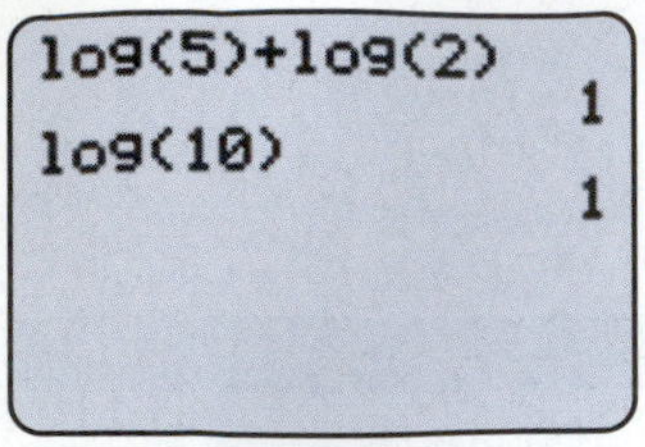

Logarithms possess several important properties. One property of logarithms states that the sum of the logarithms of two numbers equals the logarithm of their product. For example, we see in Figure 5.70 that

$$\log 5 + \log 2 = \log 10 \qquad 5 \cdot 2 = 10$$

and in Figure 5.71 that

$$\log 4 + \log 25 = \log 100. \qquad 4 \cdot 25 = 100$$

These calculations illustrate a basic property of logarithms: $\log_a m + \log_a n = \log_a (mn)$.

Four properties of logarithms are as follows.

Properties of logarithms

For positive numbers m, n, and $a \neq 1$, and any real number r:

1. $\log_a 1 = 0$
2. $\log_a m + \log_a n = \log_a (mn)$

3. $\log_a m - \log_a n = \log_a \left(\dfrac{m}{n}\right)$

4. $\log_a (m^r) = r \log_a m$

Algebra Review

To review properties of exponents, see Section 4.7, and Chapter R (page R-14).

The properties of logarithms are a direct result of the properties of exponents and the inverse property $\log_a a^k = k$, as shown in the following.

Property 1:   Since $a^0 = 1$, it follows that $\log_a 1 = \log_a a^0 = 0$.

Examples: $\log 1 = \log 10^0 = 0$ and $\ln 1 = \ln e^0 = 0$.

Property 2:   If m and n are positive numbers, then we can write $m = a^c$ and $n = a^d$ for some real numbers c and d.

$$\log_a m + \log_a n = \log_a a^c + \log_a a^d = c + d$$
$$\log_a (mn) = \log_a (a^c a^d) = \log_a (a^{c+d}) = c + d$$

Thus $\log_a m + \log_a n = \log_a (mn)$.

Example: Let $m = 100$ and $n = 1000$.

$$\log m + \log n = \log 100 + \log 1000 = \log 10^2 + \log 10^3 = 2 + 3 = 5$$
$$\log (mn) = \log (100 \cdot 1000) = \log 100{,}000 = \log 10^5 = 5$$

Property 3:   Let $m = a^c$ and $n = a^d$ for some real numbers c and d.

$$\log_a m - \log_a n = \log_a a^c - \log_a a^d = c - d$$
$$\log_a \left(\frac{m}{n}\right) = \log_a \left(\frac{a^c}{a^d}\right) = \log_a (a^{c-d}) = c - d$$

Thus $\log_a m - \log_a n = \log_a \left(\dfrac{m}{n}\right)$.

Example: Let $m = 100$ and $n = 1000$.

$$\log m - \log n = \log 100 - \log 1000 = \log 10^2 - \log 10^3 = 2 - 3 = -1$$

$$\log\left(\frac{m}{n}\right) = \log\left(\frac{100}{1000}\right) = \log\left(\frac{1}{10}\right) = \log\left(10^{-1}\right) = -1$$

Property 4: Let $m = a^c$ and r be any real number.

$$\log_a m^r = \log_a (a^c)^r = \log_a (a^{cr}) = cr$$

$$r \log_a m = r \log_a a^c = rc$$

Thus $\log_a (m^r) = r \log_a m$.

Example: Let $m = 100$ and $r = 3$.

$$\log m^r = \log 100^3 = \log 1{,}000{,}000 = \log 10^6 = 6$$

$$r \log m = 3 \log 100 = 3 \log 10^2 = 3 \cdot 2 = 6$$

Caution: $\log_a (m + n) \neq \log_a m + \log_a n$; $\log_a (m - n) \neq \log_a m - \log_a n$

EXAMPLE 1	*Recognizing properties of logarithms*

Use a calculator to evaluate each pair of expressions. Then state which property of logarithms this calculation illustrates.

(a) $\ln 5 + \ln 4, \ln 20$ **(b)** $\log 10 - \log 5, \log 2$ **(c)** $\log 5^2, 2 \log 5$

SOLUTION   **(a)** From Figure 5.72, we see that the two expressions are equal. These calculations illustrate Property 2 because $\ln 5 + \ln 4 = \ln (5 \cdot 4) = \ln 20$.

(b) The two expressions are equal in Figure 5.73, and these calculations illustrate Property 3 because

$$\log 10 - \log 5 = \log \frac{10}{5} = \log 2.$$

(c) The two expressions are equal in Figure 5.74, and these calculations illustrate Property 4 because

$$\log 5^2 = 2 \log 5.$$

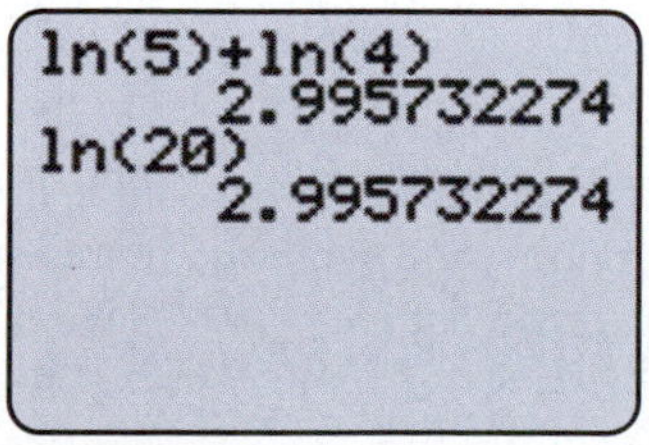

```
ln(5)+ln(4)
          2.995732274
ln(20)
          2.995732274
```

```
log(10)-log(5)
          .3010299957
log(2)
          .3010299957
```

```
log(5²)
          1.397940009
2log(5)
          1.397940009
```

FIGURE 5.72    **FIGURE 5.73**    **FIGURE 5.74**

EXAMPLE 2	*Expanding logarithmic expression*

Use properties of logarithms to expand each expression.

(a) $\log xy$ **(b)** $\ln \dfrac{6}{z}$ **(c)** $\log_4 \dfrac{\sqrt[3]{x}}{\sqrt{k}}$

SOLUTION   (a) By Property 2, $\log xy = \log x + \log y$.

(b) By Property 3, $\ln \dfrac{6}{z} = \ln 6 - \ln z$.

(c) Begin by using Property 3.

$$\log_4 \frac{\sqrt[3]{x}}{\sqrt{k}} = \log_4 \sqrt[3]{x} - \log_4 \sqrt{k} \qquad \text{Property 3}$$

$$= \log_4 x^{1/3} - \log_4 k^{1/2} \qquad \text{Properties of exponents}$$

$$= \frac{1}{3} \log_4 x - \frac{1}{2} \log_4 k \qquad \text{Property 4}$$

■ EXAMPLE 3   *Applying properties of logarithms*

Expand each expression.

(a) $\log 2x^4$ (b) $\ln \dfrac{7x^3}{k}$ (c) $\log \dfrac{\sqrt{x+1}}{(x-2)^3}$

SOLUTION   (a)

$$\log 2x^4 = \log 2 + \log x^4 \qquad \text{Property 2}$$
$$= \log 2 + 4 \log x \qquad \text{Property 4}$$

(b)

$$\ln \frac{7x^3}{k} = \ln 7x^3 - \ln k \qquad \text{Property 3}$$
$$= \ln 7 + \ln x^3 - \ln k \qquad \text{Property 2}$$
$$= \ln 7 + 3 \ln x - \ln k \qquad \text{Property 4}$$

(c)

$$\log \frac{\sqrt{x+1}}{(x-2)^3} = \log \sqrt{x+1} - \log (x-2)^3 \qquad \text{Property 3}$$

$$= \log (x+1)^{1/2} - \log (x-2)^3 \qquad \text{Properties of exponents}$$

$$= \frac{1}{2} \log (x+1) - 3 \log (x-2) \qquad \text{Property 4}$$

Sometimes properties of logarithms are used in applications to simplify a formula. This is illustrated in the next example.

■ EXAMPLE 4   *Analyzing sound with decibels*

Sound levels in decibels (db) can be computed by $f(x) = 10 \log (10^{16}x)$.
(a) Use properties of logarithms to simplify the symbolic representation for f.
(b) Ordinary conversation has an intensity of $x = 10^{-10}$ w/cm^2. Find the decibel level.

SOLUTION   (a) To simplify the formula use Property 2.

$$f(x) = 10 \log (10^{16}x)$$

$$= 10(\log 10^{16} + \log x) \qquad \text{Property 2}$$

$$= 10(16 + \log x) \qquad \text{Evaluate the logarithm.}$$

$$= 160 + 10 \log x \qquad \text{Distributive property}$$

(b) $f(10^{-10}) = 160 + 10 \log (10^{-10}) = 160 + 10(-10) = 160 - 100 = 60$
Ordinary conversation occurs at about 60 decibels.

The next two examples demonstrate how properties of logarithms can be used to combine logarithmic expressions.

EXAMPLE 5 *Applying properties of logarithms*

Write each expression as one logarithm.

(a) $\ln 2e + \ln \dfrac{1}{e}$ **(b)** $\log_2 27 + \log_2 x^3$ **(c)** $\log x^3 - \log x^2$

SOLUTION **(a)** By Property 2, $\ln 2e + \ln \dfrac{1}{e} = \ln \left(2e \cdot \dfrac{1}{e} \right) = \ln 2$.

(b) By Property 2, $\log_2 27 + \log_2 x^3 = \log_2 (27x^3)$.

(c) By Property 3, $\log x^3 - \log x^2 = \log \dfrac{x^3}{x^2} = \log x$.

EXAMPLE 6 *Collecting terms in logarithmic expressions*

Write each expression as one logarithm.

(a) $\log 5 + \log 15 - \log 3$ **(b)** $2 \ln x - \dfrac{1}{2} \ln y - 3 \ln z$

(c) $5 \log_3 x + \log_3 2x - \log_3 y$

SOLUTION

(a)
$$\begin{aligned}
\log 5 + \log 15 - \log 3 &= \log (5 \cdot 15) - \log 3 && \text{Property 2} \\
&= \log \left(\frac{5 \cdot 15}{3} \right) && \text{Property 3} \\
&= \log 25 && \text{Simplify.}
\end{aligned}$$

(b)
$$\begin{aligned}
2 \ln x - \frac{1}{2} \ln y - 3 \ln z &= \ln x^2 - \ln y^{1/2} - \ln z^3 && \text{Property 4} \\
&= \ln \left(\frac{x^2}{y^{1/2}} \right) - \ln z^3 && \text{Property 3} \\
&= \ln \frac{x^2}{y^{1/2} z^3} && \text{Property 3} \\
&= \ln \frac{x^2}{z^3 \sqrt{y}} && \text{Properties of exponents}
\end{aligned}$$

(c)
$$\begin{aligned}
5 \log_3 x + \log_3 2x - \log_3 y &= \log_3 x^5 + \log_3 2x - \log_3 y && \text{Property 4} \\
&= \log_3 (x^5 \cdot 2x) - \log_3 y && \text{Property 2} \\
&= \log_3 \frac{2x^6}{y} && \text{Property 3}
\end{aligned}$$

Change of Base Formula

Occasionally it is necessary to evaluate a logarithmic function with a base other than 10 or e. This can be done using a change of base formula.

> ### Change of base formula
>
> Let x, $a \neq 1$, and $b \neq 1$ be positive real numbers. Then,
>
> $$\log_a x = \frac{\log_b x}{\log_b a}.$$

The change of base formula can be derived as follows.

$$y = \log_a x$$

$a^y = a^{\log_a x}$	Exponentiate both sides; base a.
$a^y = x$	Inverse property
$\log_b a^y = \log_b x$	Take base-b logarithm of both sides.
$y \log_b a = \log_b x$	Property 4
$y = \dfrac{\log_b x}{\log_b a}$	Divide by $\log_b a$.
$\log_a x = \dfrac{\log_b x}{\log_b a}$	Substitute $\log_a x$ for y.

To calculate $\log_2 5$, evaluate $\dfrac{\log 5}{\log 2} \approx 2.322$. The change of base formula was used with $x = 5$, $a = 2$, and $b = 10$. We could have also evaluated $\dfrac{\ln 5}{\ln 2} \approx 2.322$.

EXAMPLE 7   *Applying the change of base formula*

Use a calculator to approximate each expression to the nearest thousandth.
(a) $\log_4 20$ **(b)** $\log_2 125 + \log_7 39$

SOLUTION   **(a)** Using the change of base formula, we have $\log_4 20 = \dfrac{\log 20}{\log 4} \approx 2.161$. We could also evaluate $\dfrac{\ln 20}{\ln 4}$ to obtain the same result, as shown in Figure 5.75.

(b) $\log_2 125 + \log_7 39 = \dfrac{\log 125}{\log 2} + \dfrac{\log 39}{\log 7} \approx 8.848$. See Figure 5.76.

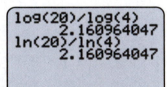

FIGURE 5.75   FIGURE 5.76

The change of base formula can be used to graph logarithmic functions with bases other than 10 or e.

EXAMPLE 8   *Using the change of base formula*

Estimate graphically the solution to the equation $\log_2 (x^3 + x - 1) = 5$.

SOLUTION   Graph $Y_1 = \log (X^3 + X - 1)/\log (2)$ and $Y_2 = 5$. See Figure 5.77. Their graphs intersect near the point $(3.104, 5)$. The solution is $x \approx 3.104$.

Graphing Calculator Help

To find a point of intersection, see Appendix B (page AP-11).

$[-10, 10, 1]$ by $[-10, 10, 1]$

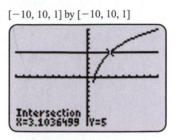

FIGURE 5.77

Putting it all Together

5.5

*T*he following table summarizes some properties of logarithms.

Concept	Explanation	Examples
Properties of logarithms	1. $\log_a 1 = 0$ 2. $\log_a m + \log_a n = \log_a (mn)$ 3. $\log_a m - \log_a n = \log_a \left(\dfrac{m}{n}\right)$ 4. $\log_a (m^r) = r \log_a m$	$\ln 1 = 0$ $\log 3 + \log 6 = \log (3 \cdot 6) = \log 18$ $\log_3 8 - \log_3 2 = \log_3 \dfrac{8}{2} = \log_3 4$ $\log 6^7 = 7 \log 6$
Change of base formula	Let x, $a \neq 1$, and $b \neq 1$ be positive real numbers. Then, $$\log_a x = \frac{\log_b x}{\log_b a}.$$	$\log_3 6 = \dfrac{\log 6}{\log 3} = \dfrac{\ln 6}{\ln 3} \approx 1.631$
Graphing logarithmic functions	Use the change of base formula to graph $y = \log_a x$, whenever $\log_a x \neq \log x$ or $\log_a x \neq \ln x$.	To graph $\log_2 x$, let $Y_1 = \log (X)/\log (2)$ or $Y_1 = \ln (X)/\ln (2)$.

5.5 EXERCISES

Properties of Logarithms

Exercises 1–6: (Refer to Example 1.) Use a calculator to evaluate each pair of expressions. Then state which property of logarithms this calculation illustrates.

1. $\log 4 + \log 7$, $\quad \log 28$ 2. $\ln 12 + \ln 5$, $\quad \ln 60$

3. $\ln 72 - \ln 8$, $\quad \ln 9$ 4. $3 \log 4$, $\quad \log 4^3$

5. $10 \log 2$, $\quad \log 1024$

6. $\log_2 100 - \log_2 20$, $\quad \log_2 5$

Exercises 7–20: (Refer to Examples 2 and 3.) Expand each expression.

7. $\log \dfrac{6}{z}$ 8. $\ln \dfrac{xy}{z}$

9. $\log \dfrac{x^2}{3}$ 10. $\log 3x^6$

11. $\ln \dfrac{2x^7}{3k}$ 12. $\ln \dfrac{kx^3}{5}$

13. $\log_2 4k^2x^3$ 14. $\log \dfrac{5kx^2}{11}$

15. $\log_5 \dfrac{x^3}{(x-4)^4}$ 16. $\log_8 \dfrac{(3x-2)^2}{x^2+1}$

17. $\log_2 \dfrac{\sqrt{x}}{z^2}$ 18. $\log \sqrt{\dfrac{xy^2}{z}}$

19. $\ln \sqrt[3]{\dfrac{2x+6}{(x+1)^5}}$ 20. $\log \dfrac{\sqrt{x^2+4}}{\sqrt[3]{x-1}}$

Exercises 21–34: (Refer to Examples 5 and 6.) Write the expression as one logarithm.

21. $\log 2 + \log 3$ 22. $\log \sqrt{2} + \log \sqrt[3]{2}$

23. $\ln \sqrt{5} - \ln 25$ 24. $\ln 33 - \ln 11$

25. $\log \sqrt{x} + \log x^2 - \log x$

26. $\log \sqrt[4]{x} + \log x^4 - \log x^2$

27. $\ln \dfrac{1}{e^2} + \ln 2e$ 28. $\ln 4e^3 - \ln 2e^2$

29. $2 \ln x - 4 \ln y + \dfrac{1}{2} \ln z$

30. $\dfrac{1}{3} \log_5 (x+1) + \dfrac{1}{3} \log_5 (x-1)$

31. $\log 4 - \log x + 7 \log \sqrt{x}$

32. $\ln 3e - \ln \dfrac{1}{4e}$

33. $2 \log (x^2 - 1) + 4 \log (x-2) - \dfrac{1}{2} \log y$

34. $\log_3 x + \log_3 \sqrt{x+3} - \dfrac{1}{3} \log_3 (x-4)$

Exercises 35–42: Complete the following.

(a) *Make a table of $f(x)$ and $g(x)$ starting at $x = 1$, incrementing by 1. Determine whether $f(x) = g(x)$.*

(b) *If possible, use properties of logarithms to show that $f(x) = g(x)$.*

35. $f(x) = \log 3x + \log 2x$, $\quad g(x) = \log 6x^2$

36. $f(x) = \log 2 + \log x$, $\quad g(x) = \log 2x$

37. $f(x) = \ln 3x - \ln 2x$, $\quad g(x) = \ln x$

38. $f(x) = \ln 2x^2 - \ln x$, $\quad g(x) = \ln 2x$

39. $f(x) = \log x^4$, $\quad g(x) = 4 \log x$

40. $f(x) = \log x^2 + \log x^3$, $\quad g(x) = 5 \log x$

41. $f(x) = \ln x^4 - \ln x^2$, $\quad g(x) = 2 \ln x$

42. $f(x) = (\ln x)^2$, $\quad g(x) = 2 \ln x$

Change of Base Formula

Exercises 43–52: Use the change of base formula to approximate each logarithm to the nearest thousandth.

43. $\log_2 25$ 44. $\log_3 67$

45. $\log_5 130$ 46. $\log_6 0.77$

47. $\log_2 5 + \log_2 7$ 48. $\log_9 85 + \log_7 17$

49. $\sqrt{\log_4 46}$

50. $2 \log_5 15 + \sqrt[3]{\log_3 67}$

51. $\dfrac{\log_2 12}{\log_2 3}$ 52. $\dfrac{\log_7 125}{\log_7 25}$

Exercises 53–56: Solve the equation graphically. Express any solutions to the nearest thousandth.

53. $\log_2 (x^3 + x^2 + 1) = 7$

54. $\log_3 (1 + x^2 + 2x^4) = 4$

55. $\log_2 (x^2 + 1) = 5 - \log_3 (x^4 + 1)$

56. $\ln (x^2 + 2) = \log_2 (10 - x^2)$

Applications

57. *Runway Length* (Refer to Exercise 79, Section 5.4.) Use a natural logarithm to write the formula $L(x) = 3 \log x$. Evaluate $L(50)$ for each formula. Do your answers agree?

58. *Allometry* The equation $y = bx^a$ is used in applications involving allometry. Another form of this equation is $\log y = \log b + a \log x$. Use properties of logarithms to obtain this second equation from the first. (Source: H. Lancaster, *Quantitative Methods in Biological and Medical Sciences.*)

59. *Decibels* (Refer to Example 4.) Use a natural logarithm to write the formula

$$f(x) = 160 + 10 \log x.$$

Evaluate $f(5 \times 10^{-8})$ for each formula. Do your answers agree?

60. *Decibels* If the intensity x of a sound increases by a factor of 10, then the decibel level increases by how much? (*Hint:* Let $f(x) = 160 + 10 \log x$, and evaluate $f(10x)$.)

61. *Light Absorption* When sunlight passes through lake water, its initial intensity I_0 decreases to a weaker intensity I at a depth of x feet according to the formula

$$\ln I - \ln I_0 = -kx,$$

where k is a positive constant. Solve this equation for I.

62. *Dissolving Salt* If C grams of salt are added to a sample of water, the amount A of undissolved salt is modeled by $A = Ca^x$, where x is time. Solve the equation for x.

Writing about Mathematics

63. A student insists that

$$\log (x + y) \quad \text{and} \quad \log x + \log y$$

are equal. How could you convince the student otherwise?

64. A student insists that

$$\log \left(\frac{x}{y}\right) \quad \text{and} \quad \frac{\log x}{\log y}$$

are equal. How could you convince the student otherwise?

5.6 EXPONENTIAL AND LOGARITHMIC EQUATIONS

Exponential Equations • Logarithmic Equations

Introduction

The population of the world has grown rapidly during the past century. As a result, heavy demands have been made on the world's resources. Exponential functions and equations are often used to model this rapid growth. Logarithms are used to model slower growth.

Exponential Equations

The population of the world was 3 billion in 1960, 6 billion in 1999, and can be modeled by $f(x) = 3(1.018)^{x-1960}$, where x is the year. We can use f to estimate the year when world population reached 5 billion by solving the exponential equation

$$3(1.018)^{x-1960} = 5.$$

An equation where one or more variables occur in the exponent of an expression is called an **exponential equation**. In the next example we use Property 4 of logarithms, $\log_a (m)^r = r \log_a m$, to solve this equation.

■ CLASS DISCUSSION

What is the growth factor for $f(x) = 3(1.018)^{x-1960}$? By what percentage has world population grown, on average, each year from 1960 to 1999? ■

EXAMPLE 1   *Modeling world population*

World population in billions during year x can be modeled by $f(x) = 3(1.018)^{x-1960}$ and is shown in Figure 5.78. Solve the equation $3(1.018)^{x-1960} = 5$ symbolically to estimate the year when world population reached 5 billion.

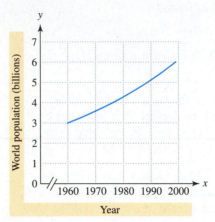

FIGURE 5.78

SOLUTION   First, divide both sides by 3, and then take the common logarithm of both sides. (The natural logarithm could also be used.)

$$3(1.018)^{x-1960} = 5 \qquad \text{Given equation}$$

$$(1.018)^{x-1960} = \frac{5}{3} \qquad \text{Divide by 3.}$$

$$\log (1.018)^{x-1960} = \log \frac{5}{3} \qquad \text{Take the common logarithm.}$$

$$(x - 1960) \log (1.018) = \log \frac{5}{3} \qquad \text{Property 4: } \log (m)^r = r \log m$$

$$x - 1960 = \frac{\log (5/3)}{\log (1.018)} \qquad \text{Divide by } \log (1.018).$$

$$x = 1960 + \frac{\log (5/3)}{\log (1.018)} \qquad \text{Add 1960.}$$

$$x \approx 1988.6 \qquad \text{Approximate.}$$

This model predicts world population reached 5 billion during 1988. ■

Exponential equations can be solved symbolically, graphically, or numerically. These methods are demonstrated in the next two examples, where the rapid growth in gambling during the early 1990s and the rapid decline of the bluefin tuna are modeled.

EXAMPLE 2   *Solving an exponential equation*

Revenues in the United States from all forms of legal gambling increased between 1991 and 1995. The function represented by $f(x) = 26.6e^{0.131x}$ models these revenues in billions of dollars. In this formula x represents the year, where $x = 0$ corresponds to 1991.   (Source: *International Gaming & Wagering Business.*)

(a) Estimate gambling revenues in 1995 to the nearest billion dollars.
(b) Determine symbolically the year when these revenues reached $30 billion.
(c) Support your result in part (b) graphically and numerically.

SOLUTION   **(a)** Since $x = 4$ corresponds to 1995, evaluate $f(4)$.

$$f(4) = 26.6e^{0.131(4)} \approx 44.9$$

In 1995 gambling revenues were approximately $45 billion.

(b) *Symbolic Solution*   Solve the exponential equation $f(x) = 30$.

$26.6e^{0.131x} = 30$	$f(x) = 30$
$e^{0.131x} = \dfrac{30}{26.6}$	Divide by 26.6.
$\ln e^{0.131x} = \ln \dfrac{30}{26.6}$	Take the natural logarithm of both sides.
$0.131x = \ln \dfrac{30}{26.6}$	Inverse property: $\ln e^k = k$
$x = \dfrac{\ln (30/26.6)}{0.131}$	Divide by 0.131.
$x \approx 0.92$	Approximate x.

Gambling revenues reached $30 billion when $x \approx 1$, or in 1992.

(c) *Graphical Solution*   Graph $Y_1 = 26.6e^{\wedge}(.131X)$ and $Y_2 = 30$. In Figure 5.79 their graphs intersect near (0.92, 30).

Numerical Solution   Numerical support is shown in Figure 5.80, where $Y_1 \approx Y_2 = 30$ when $x = 1$.

Graphing Calculator Help

To find a point of intersection, see Appendix B (page AP-11).

[0, 5, 1] by [0, 50, 10]

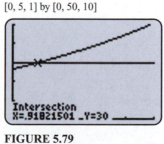

X	Y1	Y2
0	26.6	30
1	30.323	30
2	34.567	30
3	39.406	30
4	44.921	30
5	51.209	30
6	58.376	30

X=1

Intersection
X=.91821501   Y=30

FIGURE 5.79   **FIGURE 5.80**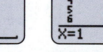

To solve an exponential equation with a base other than 10 or e, Property 4 of logarithms may be used. This technique was demonstrated in Example 1 and is also used in the next example.

EXAMPLE 3   *Modeling the decline of bluefin tuna*

Bluefin tuna are large fish that can weigh 1500 pounds and swim at speeds of 55 miles per hour. Because they are used for sushi, a prime fish can be worth over $30,000. As a result, the western Atlantic bluefin tuna have been exploited, and their numbers have declined exponentially. Their numbers in thousands from 1974 to 1991 can be modeled by $f(x) = 230(0.881)^x$, where x is the year and $x = 0$ corresponds to 1974. (Source: B. Freedman, *Environmental Ecology.*)

(a) Estimate the number of bluefin tuna in 1974 and 1991.
(b) Determine symbolically the year when their numbers were 50 thousand.
(c) Support your result in part (b) graphically.

SOLUTION (a) To determine their numbers in 1974 and 1991, evaluate $f(0)$ and $f(17)$.

$$f(0) = 230(0.881)^0 = 230(1) = 230$$
$$f(17) = 230(0.881)^{17} \approx 26.7$$

Bluefin tuna decreased from 230 thousand in 1974 to fewer than 27 thousand in 1991.

(b) *Symbolic Solution* Solve the equation $f(x) = 50$.

$$230(0.881)^x = 50 \qquad \text{\color{blue}{$f(x) = 50$}}$$

$$0.881^x = \frac{5}{23} \qquad \text{\color{blue}{Divide by 230.}}$$

$$\ln 0.881^x = \ln \frac{5}{23} \qquad \text{\color{blue}{Take the natural logarithm of both sides. (The common logarithm could also be used.)}}$$

$$x \ln 0.881 = \ln \frac{5}{23} \qquad \text{\color{blue}{Property 4: $\ln m^r = r \ln m$}}$$

$$x = \frac{\ln (5/23)}{\ln 0.881} \qquad \text{\color{blue}{Divide by $\ln 0.881$.}}$$

$$x \approx 12.04 \qquad \text{\color{blue}{Approximate x.}}$$

The population of the bluefin tuna was about 50 thousand in 1974 + 12.04 ≈ 1986.

[0, 17, 1] by [0, 250, 50]

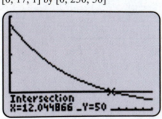

FIGURE 5.81

(c) *Graphical Solution* Graph $Y_1 = 230*.881\text{\textasciicircum}X$ and $Y_2 = 50$. In Figure 5.81 their graphs intersect near (12.04, 50).

In real applications, a modeling function is seldom provided with the data. However, it is not uncommon to be given the general form of a function that describes a data set.

EXAMPLE 4   *Finding a modeling function*

The gap between available organs for liver transplants and people who need them has widened. The number of individuals waiting for liver transplants in 1988 was 2329, and in 1995 it increased to 10,532. These numbers can be modeled by $f(x) = Ca^{(x-1988)}$, where C and a are constants.   (Source: United Network for Organ Sharing.)

(a) Approximate C and a.

(b) Use f to estimate the number of individuals waiting to receive a liver transplant in 1998.

SOLUTION   (a) In 1988 there were 2329 waiting, so $f(1988) = 2329$. This equation can be used to find C.

$$f(1988) = Ca^{(1988-1988)} = Ca^0 = C(1) = 2329$$

Thus $f(x) = 2329a^{(x-1988)}$. In 1995 the number waiting was 10,532.

$$
\begin{aligned}
f(1995) &= 10{,}532 \\
2329a^{(1995-1988)} &= 10{,}532 &&\text{Substitute.} \\
2329a^7 &= 10{,}532 &&\text{Simplify.} \\
a^7 &= \frac{10{,}532}{2329} &&\text{Divide by 2329.} \\
a &= \left(\frac{10{,}532}{2329}\right)^{1/7} &&\text{Take the seventh root of both sides.} \\
a &\approx 1.2406 &&\text{Approximate } a.
\end{aligned}
$$

Thus let $f(x) = 2329(1.2406)^{(x-1988)}$.

(b) To estimate this number in 1998, evaluate $f(1998)$.

$$f(1998) = 2329(1.2406)^{(1998-1988)} \approx 20{,}113$$

In 1998 more than 20,000 people may have needed liver transplants.

Exponential equations can occur in many forms. Although some types of exponential equations cannot be solved symbolically, Example 5 shows four equations that can.

EXAMPLE 5   *Solving exponential equations symbolically*

Solve each equation.

(a) $10^{x+2} = 10^{3x}$ (b) $5(1.2)^x + 1 = 26$ (c) $\left(\frac{1}{4}\right)^{x-1} = \frac{1}{10}$

(d) $5^{x-3} = e^{2x}$

SOLUTION   (a) Start by taking the common logarithm of both sides.

$$
\begin{aligned}
10^{x+2} &= 10^{3x} \\
\log 10^{x+2} &= \log 10^{3x} &&\text{Take the common logarithm.} \\
x + 2 &= 3x &&\text{Inverse property: } \log 10^k = k \\
2 &= 2x &&\text{Subtract } x. \\
x &= 1 &&\text{Divide by 2.}
\end{aligned}
$$

(b) Begin by subtracting 1 from both sides of the equation.

$$5(1.2)^x = 25 \qquad \text{Subtract 1.}$$

$$(1.2)^x = 5 \qquad \text{Divide by 5.}$$

$$\log (1.2)^x = \log 5 \qquad \text{Take the common logarithm.}$$

$$x \log (1.2) = \log 5 \qquad \text{Property 4: } \log m^r = r \log m$$

$$x = \frac{\log 5}{\log (1.2)} \qquad \text{Divide by } \log (1.2).$$

$$x \approx 8.827 \qquad \text{Approximate.}$$

(c) Begin by taking the common logarithm of both sides.

$$\left(\frac{1}{4}\right)^{x-1} = \frac{1}{10}$$

$$\log \left(\frac{1}{4}\right)^{x-1} = \log \frac{1}{10} \qquad\qquad\qquad \text{Take the common logarithm.}$$

$$(x - 1) \log \left(\frac{1}{4}\right) = \log \frac{1}{10} \qquad\qquad\qquad \text{Property 4: } \log m^r = r \log m$$

$$(x - 1) = \frac{\log (1/10)}{\log (1/4)} \qquad\qquad\qquad \text{Divide by } \log \frac{1}{4}.$$

$$x = 1 + \frac{\log (1/10)}{\log (1/4)} \approx 2.661 \qquad \text{Add 1 and approximate.}$$

This also could have been solved by taking the natural logarithm of both sides.

(d) Begin by taking the natural logarithm of both sides. (The common logarithm could also be used.)

$$5^{x-3} = e^{2x} \qquad \text{Given equation}$$

$$\ln 5^{x-3} = \ln e^{2x} \qquad \text{Take natural logarithm.}$$

$$(x - 3) \ln 5 = 2x \qquad \text{Property 4; inverse property}$$

$$x \ln 5 - 3 \ln 5 = 2x \qquad \text{Distributive property}$$

$$x \ln 5 - 2x = 3 \ln 5 \qquad \text{Subtract } 2x; \text{ add } 3 \ln 5.$$

$$x(\ln 5 - 2) = 3 \ln 5 \qquad \text{Factor out } x.$$

$$x = \frac{3 \ln 5}{\ln 5 - 2} \qquad \text{Divide by } \ln 5 - 2.$$

$$x \approx -12.36 \qquad \text{Approximate.}$$

If a hot object is put in a room with temperature T, then according to **Newton's law of cooling**, the temperature of the object after time t is modeled by

$$f(t) = T + Da^t,$$

where $0 < a < 1$ and D is the initial temperature difference between the object and the room. This phenomenon is discussed in the next example.

EXAMPLE 6   *Modeling coffee cooling*

A pot of coffee with a temperature of 100°C is set down in a room with a temperature of 20°C. The coffee cools to 60°C after 1 hour.
(a) Find values for *T, D,* and *a* so that $f(t) = T + Da^t$ models the data.
(b) Find the temperature of the coffee after half an hour.
(c) How long did it take for the coffee to reach 50°C? Support your result graphically.

SOLUTION   **(a)** The room has temperature $T = 20°C$ and the initial temperature difference between the coffee and the room is $D = 100 - 20 = 80°C$. It follows that $f(t) = 20 + 80a^t$. We can use the fact that the temperature of the coffee after 1 hour was 60°C to determine *a*.

$$f(1) = 60$$

$$20 + 80a^1 = 60 \qquad \text{Let } t = 1 \text{ in } f(t) = 10 + 80a^t.$$

$$80a = 40 \qquad \text{Subtract 20.}$$

$$a = \frac{1}{2} \qquad \text{Divide by 80.}$$

Thus $f(t) = 20 + 80\left(\frac{1}{2}\right)^t$.

(b) After half an hour the temperature is

$$f\left(\frac{1}{2}\right) = 20 + 80\left(\frac{1}{2}\right)^{1/2} \approx 76.6°C.$$

(c) *Symbolic Solution*   To determine when the coffee reached 50°C, solve the equation $f(t) = 50$.

$$20 + 80\left(\frac{1}{2}\right)^t = 50 \qquad \textcolor{blue}{f(t) = 50}$$

$$80\left(\frac{1}{2}\right)^t = 30 \qquad \textcolor{blue}{\text{Subtract 20.}}$$

$$\left(\frac{1}{2}\right)^t = \frac{3}{8} \qquad \textcolor{blue}{\text{Divide by 80.}}$$

$$\log\left(\frac{1}{2}\right)^t = \log\frac{3}{8} \qquad \textcolor{blue}{\text{Take the common logarithm.}}$$

$$t\,\log\left(\frac{1}{2}\right) = \log\frac{3}{8} \qquad \textcolor{blue}{\text{Property 4}}$$

$$t = \frac{\log(3/8)}{\log(1/2)} \qquad \textcolor{blue}{\text{Divide by } \log\frac{1}{2}.}$$

$$t \approx 1.415 \qquad \textcolor{blue}{\text{Approximate.}}$$

[0, 3, 1] by [0, 100, 10]

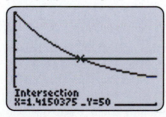

FIGURE 5.82

The temperature of the coffee reaches 50°C after about 1.415 hours or after about 1 hour and 25 minutes.

Graphical Solution   The graphs of $Y_1 = 20 + 80(1/2)^X$ and $Y_2 = 50$ intersect near (1.415, 50), as shown in Figure 5.82. This result agrees with the symbolic solution.

Some exponential equations cannot be solved symbolically, but they can be solved graphically. This is demonstrated in the next example.

EXAMPLE 7   *Solving an exponential equation graphically*

Solve $e^{-x} + 2x = 3$ graphically. Approximate all solutions to the nearest hundredth.

SOLUTION   The graphs of $Y_1 = e^{\wedge}(-X) + 2X$ and $Y_1 = 3$ intersect near the points $(-1.92, 3)$ and $(1.37, 3)$, as shown in Figures 5.83 and 5.84. Thus the solutions are $x \approx -1.92$ and $x \approx 1.37$.

$[-6, 6, 1]$ by $[-4, 4, 1]$ $[-6, 6, 1]$ by $[-4, 4, 1]$

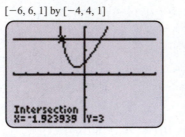   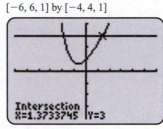

FIGURE 5.83   **FIGURE 5.84**

Logarithmic Equations

Logarithmic equations contain logarithms. Like exponential equations, logarithmic equations also occur in applications. To solve a logarithmic equation, we use the inverse property $a^{\log_a x} = x$. This technique is illustrated in the next example.

EXAMPLE 8   *Solving a logarithmic equation*

Solve $3 \log_3 x = 12$.

SOLUTION   Begin by dividing both sides by 3.

$$\log_3 x = 4 \qquad \text{Divide by 3.}$$
$$3^{\log_3 x} = 3^4 \qquad \text{Exponentiate both sides; base 3.}$$
$$x = 81 \qquad \text{Inverse property: } a^{\log_a x} = x.$$

EXAMPLE 9   *Solving a logarithmic equation symbolically*

In developing countries there is a relationship between the amount of land a person owns and the average daily calories consumed. This relationship is modeled by $f(x) = 280 \ln (x + 1) + 1925$, where x is the amount of land owned in acres and $0 \le x \le 4$. (Source: D. Grigg, *The World Food Problem.*)

(a) Find the average caloric intake for a person in a developing country who owns no land.

(b) Graph f in $[0, 4, 1]$ by $[1800, 2500, 100]$. Interpret the graph.

(c) Determine symbolically the number of acres owned by someone whose average intake is 2000 calories per day.

SOLUTION   **(a)** Since $f(0) = 280 \ln(0 + 1) + 1925 = 1925$, a person without land consumes an average of 1925 calories per day.

[0, 4, 1] by [1800, 2500, 100]

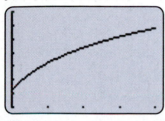

FIGURE 5.85

(b) Graph $Y_1 = 280 \ln(X + 1) + 1925$, as shown in Figure 5.85. As the amount of land x increases, the caloric intake y also increases. However, it begins to level off. This would be expected because there is a limit to the number of calories an average person would eat, regardless of his or her economic status.

(c) Solve the equation $f(x) = 2000$.

$$280 \ln(x + 1) + 1925 = 2000$$

$$280 \ln(x + 1) = 75 \qquad \text{Subtract 1925.}$$

$$\ln(x + 1) = \frac{75}{280} \qquad \text{Divide by 280.}$$

$$e^{\ln(x+1)} = e^{75/280} \qquad \text{Exponentiate both sides; base } e.$$

$$x + 1 = e^{75/280} \qquad \text{Inverse property: } e^{\ln k} = k$$

$$x = e^{75/280} - 1 \qquad \text{Subtract 1.}$$

$$x \approx 0.307 \qquad \text{Approximate.}$$

A person who owns about 0.3 acre has an average intake of 2000 calories per day.

Like exponential equations, logarithmic equations can occur in many forms. The next example illustrates three equations that can be solved symbolically.

EXAMPLE 10   *Solving logarithmic equations symbolically*

Solve each equation.
(a) $\log(2x + 1) = 2$ **(b)** $\log_2 4x = 2 - \log_2 x$
(c) $2 \ln(x + 1) = \ln(1 - 2x)$

SOLUTION   **(a)** To solve the equation, exponentiate both sides of the equation using base 10.

$$\log(2x + 1) = 2$$

$$10^{\log(2x+1)} = 10^2 \qquad \text{Exponentiate both sides; base 10.}$$

$$2x + 1 = 100 \qquad \text{Inverse property: } 10^{\log k} = k$$

$$x = 49.5 \qquad \text{Solve for } x.$$

(b) To solve this equation, we apply properties of logarithms.

$$\log_2 4x = 2 - \log_2 x$$

$$\log_2 4x + \log_2 x = 2 \qquad \text{Add } \log_2 x.$$

$$\log_2 4x^2 = 2 \qquad \text{Property 2: } \log_a m + \log_a n = \log_a(mn)$$

$$2^{\log_2 4x^2} = 2^2 \qquad \text{Exponentiate both sides; base 2.}$$

$$4x^2 = 4 \qquad \text{Inverse property: } 2^{\log_2 k} = k$$

$$x = \pm 1 \qquad \text{Solve for } x.$$

However, $x = -1$ is not a solution since $\log_2 x$ is undefined for negative values of x. Thus the only solution is $x = 1$.

(c) Start by applying Property 4 to the given equation.

$2 \ln (x + 1) = \ln (1 - 2x)$	Given equation
$\ln (x + 1)^2 = \ln (1 - 2x)$	Property 4
$e^{\ln (x+1)^2} = e^{\ln (1-2x)}$	Exponentiate; base e.
$(x + 1)^2 = 1 - 2x$	Inverse property: $a^{\log_a x} = x$
$x^2 + 2x + 1 = 1 - 2x$	Expand binomial.
$x^2 + 4x = 0$	Combine terms.
$x(x + 4) = 0$	Factor.
$x = 0$ or $x = -4$	Zero-product property

Substituting $x = 0$ and $x = -4$ into the given equation shows that $x = 0$ is a solution, but $x = -4$ is not a solution.

■ **CLASS DISCUSSION**
Compare the first step for solving each pair of equations.

i. $10^x = 5$ and $\log x = 5$
ii. $e^x = 5$ and $\ln x = 5$
iii. $4^x = 5$ and $\log_4 x = 5$ ■

EXAMPLE 11   *Modeling the life span of a robin*

The life spans of 129 robins were monitored over a 4-year period in one study. The equation $y = \dfrac{2 - \log (100 - x)}{0.42}$ can be used to calculate the number of years y for x percent of the robin population to die. For example, to find the time when 40% of the robins had died, substitute $x = 40$ into the equation. The result is $y \approx 0.53$, or about half a year.   (Source: D. Lack, *The Life of a Robin*.)
(a) Graph y in $[0, 100, 10]$ by $[0, 5, 1]$. Interpret the graph.
(b) Estimate graphically the percentage of the robins that died after 2 years.

SOLUTION   (a) The graph of $Y_1 = (2 - \log (100 - X))/.42$ is shown in Figure 5.86. At first, the graph increases slowly between 0% and 60%. Since years are listed on the y-axis and percentage on the x-axis, this slow increase in f indicates that not many years pass by before a large percentage of the robins die. When $x \geq 60$, the graph begins to increase dramatically. This indicates that a small percentage has a comparatively long life span.

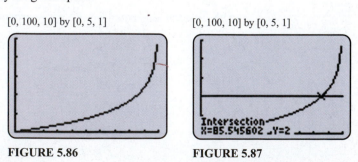

[0, 100, 10] by [0, 5, 1]   [0, 100, 10] by [0, 5, 1]

FIGURE 5.86   **FIGURE 5.87**

(b) To determine the percentage that died after 2 years, graph Y_1 and $Y_2 = 2$. Their graphs intersect near $(85.5, 2)$, as shown in Figure 5.87. After 2 years, approximately 85.5% of the robins had died—a surprisingly high percentage.

■ MAKING CONNECTIONS

Solving Exponential and Logarithmic Equations   At some point in the process of solving an exponential equation, we often take a logarithm of both sides of the equation. Similarly, when solving a logarithmic equation, we often exponentiate both sides of the equation. ■

The following table summarizes techniques that can be used to solve some types of exponential and logarithmic equations symbolically.

Concept	Explanation	Example
Exponential equations	Typical form: $Ca^x = k$ Solve for a^x. Then take a logarithm of both sides. Use the inverse property: $$\log_a a^x = x$$	Solve $4e^x = 24$. $$e^x = 6$$ $$\ln e^x = \ln 6$$ $$x = \ln 6$$
Logarithmic equations	*Equation*: $C \log_a x = k$ Solve for $\log_a x$. Then exponentiate both sides with base a. Use $a^{\log_a x} = x$.	Solve $4 \log x = 10$. $$\log x = 2.5$$ $$10^{\log x} = 10^{2.5}$$ $$x = 10^{2.5} \approx 316$$
	Equation: $\log_a bx \pm \log_a cx = k$ When more than one logarithm with the same base occurs, use properties of logarithms to combine logarithms. Be sure to check any solutions.	Solve $\log x + \log 4x = 2$. $$\log 4x^2 = 2$$ $$4x^2 = 10^2$$ $$x^2 = 25$$ $$x = \pm 5$$ The only solution is $x = 5$.

Putting it all Together

5.6

5.6 EXERCISES

Solving Exponential Equations

Exercises 1–4: The graphical and symbolic representations of f and g are given.

   (a) *Use the graph to estimate the solution of $f(x) = g(x)$.*

   (b) *Solve $f(x) = g(x)$ symbolically.*

1. $f(x) = e^x$, $g(x) = 7.5$

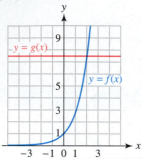

2. $f(x) = 0.1(10^x)$, $g(x) = 0.5$

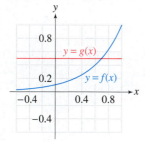

3. $f(x) = 10^{0.2x}$, $g(x) = 2.5$

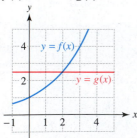

4. $f(x) = e^{-0.7x}$, $g(x) = 2$

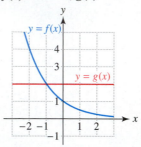

Exercises 5–24: Use common or natural logarithms to solve the exponential equation symbolically. Support your results graphically or numerically.

5. $4e^x = 5$ **6.** $2e^{-x} = 8$

7. $2(10^x) + 5 = 45$ **8.** $100 - 5(10^x) = 7$

9. $2.5e^{-1.2x} = 1$ **10.** $9.5e^{0.005x} = 19$

11. $1.2(0.9^x) = 0.6$ **12.** $0.05(1.15)^x = 5$

13. $4(1.1^{x-1}) = 16$ **14.** $3(2^{x-2}) = 99$

15. $10^{(x^2)} = 10^{3x-2}$ **16.** $e^{2x} = e^{5x-3}$

17. $\left(\dfrac{1}{5}\right)^x = -5$ **18.** $2^x = -4$

19. $4^{x-1} = 3^{2x}$ **20.** $3^{1-2x} = e^{0.5x}$

21. $3(1.4)^x - 4 = 60$ **22.** $2(1.05)^x + 3 = 10$

23. $5(1.015)^{x-1980} = 8$

24. $30 - 3(0.75)^{x-1} = 29$

Solving Logarithmic Equations

Exercises 25–44: Solve the logarithmic equation symbolically. Support your answer graphically or numerically.

25. $3 \log x = 2$ **26.** $5 \ln x = 10$

27. $\ln 2x = 5$ **28.** $\ln 4x = 1.5$

29. $\log 2x^2 = 2$ **30.** $\log (2 - x) = 0.5$

31. $160 + 10 \log x = 50$

32. $160 + 10 \log x = 120$

33. $\ln x + \ln x^2 = 3$ **34.** $\log x^5 = 4 + 3 \log x$

35. $2 \log_2 x = 4.2$ **36.** $3 \log_2 3x = 1$

37. $\log x + \log 2x = 2$ **38.** $\ln 2x + \ln 3x = \ln 6$

39. $\ln(x - 1) = 1$

40. $\log x + \log(2x + 5) = \log 7$

41. $2 \ln x = \ln(2x + 1)$

42. $\log(x^2 + 3) = 2 \log(x + 1)$

43. $\log(x + 1) + \log(x - 1) = \log 3$

44. $\ln(x^2 - 4) - \ln(x + 2) = \ln(3 - x)$

Solving Equations Graphically

Exercises 45–50: The following equations cannot be solved symbolically. Solve these equations graphically and round your answers to the nearest hundredth.

45. $2x + e^x = 2$ **46.** $xe^x - 1 = 0$

47. $x^2 - x \ln x = 2$ **48.** $x \ln|x| = -2$

49. $xe^{-x} + \ln x = 1$ **50.** $2^{x-2} = \log x^4$

Applications

51. *Credit Cards*   From 1987 to 1996 the number of Visa cards and MasterCards was up 80% to 376 million. The function given by $f(x) = 36.2e^{0.14x}$ models the amount of credit card spending from Thanksgiving to Christmas in billions of dollars. In this formula $x = 0$ corresponds to 1987 and $x = 9$ to 1996. (Source: National Credit Counseling Services.)

   (a) Determine symbolically the year when this amount reached $55 billion.

   (b) Solve part (a) graphically or numerically.

52. *Gambling*   U.S. gambling revenues in billions of dollars from 1991 to 1995 can be modeled by $f(x) = 26.6e^{0.131x}$, where x is the year and $x = 0$ corresponds to 1991. (See Example 2.) Determine symbolically the year when revenues reached $45 billion. Support your answer graphically.

53. *Liver Transplants*   (Refer to Example 4.) Determine when the number of individuals waiting for liver transplants might reach 8500.

54. *Bluefin Tuna*   The number of Atlantic bluefin tuna in thousands can be modeled by $f(x) = 230(0.881)^x$, where x is the year and $x = 0$ corresponds to 1974. (See Example 3.) Estimate the year when the number of bluefin tuna reached 95 thousand.

55. *Population Growth*   In 2000 India's population reached 1 billion and in 2025 it is projected to be 1.4 billion. (Source: Bureau of the Census.)

   (a) Find values for C and a so that $f(x) = Ca^{x-2000}$ models the population of India in year x.

   (b) Estimate India's population in 2010.

   (c) Use f to determine the year when India's population might reach 1.5 billion.

56. *Midair Near Collisions*   The table shows the number of airliner near collisions y in year x. (Source: Federal Aviation Administration.)

x	1989	1991	1993	1995
y	131	78	44	34

   (a) Approximate constants C and a so that $f(x) = Ca^{(x-1989)}$ models the data.

   (b) Support your answer graphically, by graphing f and the data.

Exercises 57 and 58: Continuous Compounding   Suppose that A_0 dollars is deposited in a savings account paying 9% interest compounded continuously. After x years the account will contain $A(x) = A_0 e^{0.09x}$ dollars.

   (a) Solve the equation $A(x) = k$ for the given values of A_0 and k.

   (b) Interpret your results.

57. $A_0 = 500$ and $k = 750$

58. $A_0 = 1000$ and $k = 2000$

59. *Doubling Time*   How long does it take for $1000 to double in value if the interest rate is 8.5% compounded quarterly? $\left(Hint: A_n = A_0\left(1 + \dfrac{r}{m}\right)^{mn}. \right)$

60. *Doubling Time*   How long does it take for $1000 to double in value if the interest rate is 12% compounded continuously? (*Hint:* $A_n = A_0 e^{rn}$.)

61. *Radioactive Carbon-14*   The percentage P of radioactive carbon-14 remaining in a fossil after t years is given by $P = 100\left(\dfrac{1}{2}\right)^{t/5700}$. Suppose a fossil contains 35% of the carbon-14 it contained when it was alive. Estimate the age of the fossil. (*Hint:* Let $P = 35$ and solve for t.)

62. *Radioactive Radium-226* The amount A of radium in milligrams remaining in a sample after t years is given by $A(t) = 0.02\left(\dfrac{1}{2}\right)^{t/1600}$. How many years will it take for the sample to decay to 0.004 milligrams?

63. *Modeling Traffic Flow* Cars arrive randomly at an intersection with an average traffic volume of 1 car per minute. The likelihood or probability that at least one car enters the intersection during a period of x minutes can be estimated by $f(x) = 1 - e^{-x}$.
 (a) What is the probability that at least one car enters the intersection during a 5-minute period?
 (b) Determine the value of x that gives a 40% chance that at least one car enters the intersection during an interval of x-minutes.

64. *Drug Concentrations* The concentration of a drug in a patient's bloodstream after t hours is modeled by $C(t) = 11(0.72)^t$, where C is measured in milligrams per liter.
 (a) What is the initial concentration of the drug?
 (b) How long does it take for the concentration to decrease to 50% of its initial level?

65. *Diversity* In 1995 the U.S. racial mix was 75.3% white, 9.0% Hispanic, 12.0% African-American, 2.9% Asian/Pacific Islanders, and 0.8% Native American. Their respective percentages of executives, managers, and administrators in private-industry communication firms were 84.0%, 9.5%, 3.4%, 2.0%, and 0.0%. If the percentage P of African-American executives, managers, and administrators increases at an annual rate of 7% per year, determine the year when their representation will reach 12.0% of the number of executives, managers, and administrators in communications. (*Hint:* Let $P(x) = 3.4(1.07)^x$, where $x = 0$ corresponds to 1995, and solve $P(x) = 12$.) (Source: Labor Department's Glass Ceiling Commission.)

66. *Atmospheric Pressure* As altitude increases, air pressure decreases. The atmospheric pressure in millibars (mb) at a given altitude x in meters is listed in the table. (Source: A. Miller and J. Thompson, *Elements of Meteorology.*)

Altitude (m)	0	5000	10,000	15,000
Pressure (mb)	1013	541	265	121

Altitude (m)	20,000	25,000	30,000
Pressure (mb)	55	26	12

 (a) The data can be modeled by $f(x) = 1013e^{kx}$. Approximate the constant k.
 (b) Graph f and the data in the same viewing rectangle. Comment on the fit.
 (c) Estimate the atmospheric pressure at an altitude of 23,000 feet.

67. *Cooling a Soda Can* A soda can at 80°F is put into a cooler containing ice at 32°F. The temperature of the soda can after t minutes is modeled by $f(t) = 32 + 48(0.9)^t$.
 (a) Evaluate $f(30)$ and interpret your results.
 (b) How long did it take for the soda can to cool to 50°F?

68. *Warming a Soda Can* (Refer to Example 6.) According to Newton's law of cooling, if a cold object is brought into a warm room, the object's temperature after time t can be modeled by $f(t) = T + Da^t$. In this formula T is the air temperature of the room, and D represents the temperature difference between the object and the room. Suppose that a can of soda, initially at 5°C, warms to 18°C after 2 hours in a room that has a temperature of 20°C.
 (a) Find values for T, D, and a so that $f(t)$ models the situation. (*Hint:* If the object is cooler than T, then $D < 0$.)
 (b) Find the temperature of the soda can after 1.5 hours.
 (c) How long did it take for the soda can to warm to 15°C?

69. *Life Span of Robins* Solve Example 11, part (b) on page 424 symbolically.

70. *Salinity* The salinity of the oceans changes with latitude and with depth. In the tropics, the salinity increases on the surface of the ocean due to rapid evaporation. In the higher latitudes, there is less evaporation and rainfall causes the salinity to be less on the surface than at lower depths. The function given by $f(x) = 31.5 + 1.1 \log (x + 1)$ models salinity to depths of 1000 meters at a latitude of 57.5°N. The input x is the depth in meters and the output $f(x)$ is in grams of salt per kilogram of seawater. (Source: D. Hartman, *Global Physical Climatology.*)
 (a) Table f starting at $x = 0$, incrementing by 100. Discuss any trends.
 (b) Numerically approximate the depth where the salinity equals 33.
 (c) Solve part (b) symbolically.

71. *Caloric Intake* The formula

$$f(x) = 280 \ln(x + 1) + 1925$$

models the number of calories consumed daily by a person owning x acres of land in a developing country. (See Example 9.) Estimate the number of acres owned for the average intake to be 2300 calories per day.

72. *Modeling Data* Use the data table to complete the following.

x	1	2	5	10
y	2.5	2.1	1.6	1.2

(a) Find values for a and b so that the formula $f(x) = a + b \log x$ models the data.

(b) Solve the equation $f(x) = 1.8$.

73. *Reducing Carbon Emissions* When fossil fuels are burned, carbon is released into the atmosphere. Governments could reduce carbon emissions by placing a tax on fossil fuels. For example, a higher tax might lead people to use less gasoline. The **cost-benefit** equation

$$\ln(1 - P) = -0.0034 - 0.0053x$$

estimates the relationship between a tax of x dollars per ton of carbon and the percent P reduction in emissions of carbon, where P is in decimal

form. Determine P when $x = 60$. Interpret the result. (Source: W. Clime, *The Economics of Global Warming.*)

74. The cost-benefit function from the previous exercise can be represented as $P(x) = 1 - e^{-0.0034 - 0.0053x}$.

(a) Graph P in $[0, 1000, 100]$ by $[0, 1, 0.1]$. Discuss the benefit of continuing to raise taxes on fossil fuels.

(b) According to this model, what tax would result in a 50% reduction in carbon emissions? Solve this problem either symbolically or graphically.

(c) Verify that

$$\ln(1 - P) = -0.0034 - 0.0053x$$

and

$$P = 1 - e^{-0.0034 - 0.0053x}$$

are equivalent by solving the first equation for P.

75. *Investments* The formula $A_n = A_0\left(1 + \dfrac{r}{m}\right)^{nm}$ can be used to calculate the future value of an investment. Solve the equation for n.

76. *Decibels* The formula $D = 160 + 10 \log x$ can be used to calculate loudness of a sound. Solve the equation for x.

Writing about Mathematics

77. Explain how to solve the equation $Ca^x = k$ symbolically, where C and k are constants. Demonstrate your method.

78. Explain how to solve the equation $b \log_a x = k$ symbolically, where b and k are constants. Demonstrate your method.

CHECKING BASIC CONCEPTS FOR SECTIONS 5.5 AND 5.6

1. Use properties of logarithms to expand $\log \dfrac{x^2 y^3}{\sqrt[3]{z}}$.

2. Use properties of logarithms to write $\dfrac{1}{2} \ln x - 3 \ln y + \ln z$ as one logarithm.

3. Solve each equation.
(a) $5(1.4)^x - 4 = 25$
(b) $4^{2-x} = 4^{2x+1}$

4. Solve each equation.
(a) $5 \log_2 2x = 25$

(b) $\ln(x + 1) + \ln(x - 1) = \ln 3$

5. The temperature T of a cooling object in degrees Fahrenheit after x minutes is given by

$$T = 80 + 120(0.9)^x.$$

(a) What happens to the temperature of the object after a long time?

(b) After how long is the object's temperature $100°F$?

5.7 CONSTRUCTING NONLINEAR MODELS

Exponential Model • Logarithmic Model • Logistic Model

Introduction

If data change at a constant rate, then they can be modeled with a linear function. However, real-life data usually change at a nonconstant rate. For example, a tree grows slowly when it is small and then gradually grows faster as it becomes larger. Finally, when the tree is mature, its height begins to level off. This type of growth is nonlinear.

Three types of data are shown in Figures 5.88–5.90, where t represents time. In Figure 5.88 the data increase rapidly, and an exponential function might be an appropriate modeling function. In Figure 5.89 we see data that are growing, but at a slower rate than in Figure 5.88. This data could be modeled by a logarithmic function. Finally in Figure 5.90 the data increase slowly, then faster, and finally level off. These data might represent the height of a tree over a 50-year period. To model these data, we need a new type of function called a *logistic function*.

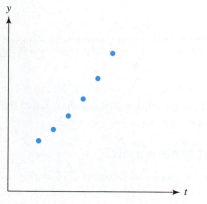

FIGURE 5.88 Exponential Data

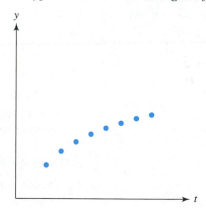

FIGURE 5.89 Logarithmic Data

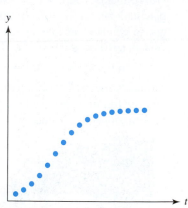

FIGURE 5.90 Logistic Data

Exponential Model

The numbers of certified female automotive technicians changed dramatically during the early 1990s. In the next example we find an exponential function that models these numbers.

EXAMPLE 1   *Using exponential regression*

The National Institute for Automotive Service Excellence (ASE) reported that the number of females working in automotive repair is increasing. Table 5.26 shows the net total of female ASE-certified technicians.

TABLE 5.26

Year	1988	1989	1990	1991	1992	1993	1994	1995
Total	556	614	654	737	849	1086	1329	1592

Source: National Institute for Automotive Service Excellence.

(a) What type of function might model this data?

(b) Use least-squares regression to find an exponential function given by $f(x) = ab^x$ that models the data.

(c) Use f to estimate the number of certified female technicians in 2000. Round the result to the nearest hundred.

SOLUTION   **(a)** Let y be the number of female technicians and x be the year, where $x = 0$ corresponds to 1988, $x = 1$ to 1989, and so on, until $x = 7$ corresponds to 1995. A scatterplot of the data is shown in Figure 5.91. The data is increasing rapidly, and an exponential function might model this data.

$[-1, 8, 1]$ by $[400, 1700, 100]$

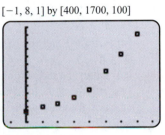

FIGURE 5.91

Graphing Calculator Help

To make a scatterplot, see Appendix B (page AP-6).

(b) The formula $f(x) = 507.1(1.166)^x$ for f is found by using least-squares regression in Figures 5.92 and 5.93. A scatterplot of the data, together with a graph of f, is shown in Figure 5.94.

$[-1, 8, 1]$ by $[400, 1700, 100]$

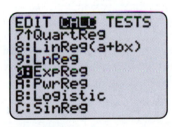

FIGURE 5.92

FIGURE 5.93

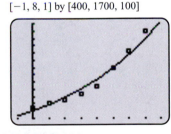

FIGURE 5.94

Graphing Calculator Help

To find an equation of least-squares fit, see Appendix B (page AP-15).

(c) Since $x = 12$ corresponds to the year 2000,

$$f(12) = 507.1(1.166)^{12} \approx 3202.$$

According to this model, the number of certified female automotive technicians in 2000 was about 3200.

Logarithmic Model

When buying a certificate of deposit (CD), a higher interest rate is usually given if the money is deposited for a longer period of time. However, one does not usually receive twice the interest rate for a 2-year CD than for a 1-year CD. Instead the rate of interest gradually increases with time. In the next example we model real interest rates with a logarithmic function.

■ **EXAMPLE 2** *Modeling interest rates*

Table 5.27 lists the interest rates for certificates of deposit during January 1997. Use the data to complete the following.

TABLE 5.27 Yield on Certificates of Deposit

Time	6 mo	1 yr	2.5 yr	5 yr
Yield	4.75%	5.03%	5.25%	5.54%

Source: *USA Today.*

(a) Make a scatterplot of the data. What type of function might model these data?
(b) Use least-squares regression to obtain a formula, $f(x) = a + b \ln x$, that models the data.
(c) Graph f and the data in the same viewing rectangle.

SOLUTION **(a)** Enter the data points (0.5, 4.75), (1, 5.03), (2.5, 5.25), and (5, 5.54) into your calculator. A scatterplot of the data is shown in Figure 5.95. The data increase but are gradually leveling off. A logarithmic modeling function may be appropriate.

[0, 6, 1] by [4, 6, 1]

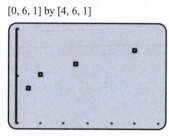

FIGURE 5.95

(b) In Figures 5.96 and 5.97 least-square regression has been used to find a logarithmic function f given (approximately) by $f(x) = 5 + 0.33 \ln x$.
(c) A graph of f and the data are shown in Figure 5.98.

| FIGURE 5.96 | FIGURE 5.97 |

[0, 6, 1] by [4, 6, 1]

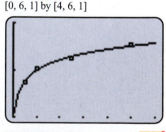

FIGURE 5.98

Logistic Model

In real life, populations of bacteria, insects, and animals do not continue to grow indefinitely. Initially, population growth may be slow. Then, as their numbers increase, so does the rate of growth. After a region has become heavily populated or saturated, the population usually levels off because of limited resources.

This type of growth may be modeled by a **logistic function** represented by $f(x) = \dfrac{c}{1 + ae^{-bx}}$, where *a, b,* and *c* are positive constants. A typical graph of a logistic function f is shown in Figure 5.99. The graph of f is referred to as a **sigmoidal curve**. The next example demonstrates how a logistic function can be used to describe the growth of a yeast culture.

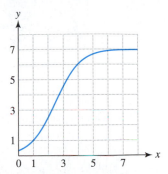

FIGURE 5.99 A Sigmoidal Curve

EXAMPLE 3   *Modeling logistic growth*

One of the earliest studies about population growth was done using yeast plants in 1913. A small amount of yeast was placed in a container with a fixed amount of nourishment. The units of yeast were recorded every 2 hours. The data is listed in Table 5.28.   (Source: T. Carlson, *Biochem.;* D. Brown, *Models in Biology.*)

TABLE 5.28

Time	0	2	4	6	8	10	12	14	16	18
Yeast	9.6	29.0	71.1	174.6	350.7	513.3	594.8	640.8	655.9	661.8

(a) Make a scatterplot of the data. Describe the growth.
(b) Use least-squares regression to find a logistic function f that models the data.
(c) Graph f and the data in the same viewing rectangle.
(d) Approximate graphically the time when the amount of yeast was 200 units.

SOLUTION   **(a)** A scatterplot of the data is shown in Figure 5.100. The yeast increase slowly at first. Then they grow more rapidly, until the amount of yeast gradually levels off. The limited amount of nourishment causes this leveling off.

$[-2, 20, 1]$ by $[-100, 800, 100]$

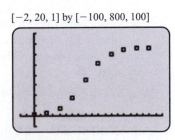

FIGURE 5.100

(b) In Figures 5.101 and 5.102 on the next page we see least-squares regression being used to find a logistics function f given by

$$f(x) = \frac{661.8}{1 + 74.46e^{-0.552x}}.$$

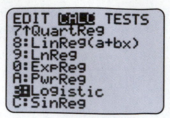

FIGURE 5.101

FIGURE 5.102

(c) In Figure 5.103 the data and f are graphed in the same viewing rectangle. The fit for this real data is remarkably good.

$[-2, 20, 1]$ by $[-100, 800, 100]$

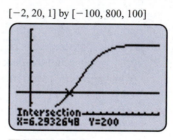

FIGURE 5.103

Graphing Calculator Help

To find an equation of least-squares fit, see Appendix B (page AP-15). To copy a regression equation into Y_1, see page AP-17.

(d) The graphs of $Y_1 = f(x)$ and $Y_2 = 200$ intersect near (6.29, 200), as shown in Figure 5.104. The amount of yeast reached 200 units after about 6.29 hours.

$[-2, 20, 1]$ by $[-100, 800, 100]$

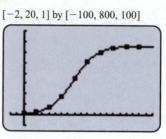

FIGURE 5.104

Graphing Calculator Help

To find a point of intersection, see Appendix B (page AP-11).

■ **CLASS DISCUSSION**

In Example 3, suppose that after 18 hours the experiment had been extended and more nourishment were provided for the yeast plants. Sketch a possible graph of the amount of yeast. ■

■ **MAKING CONNECTIONS**

Logistic Functions and Horizontal Asymptotes   If a logistic function is given by $f(x) = \dfrac{c}{1 + ae^{-bx}}$, where a, b, and c are positive constants, then the graph of f has a horizontal asymptote of $y = c$. (Try to explain why this is true.) In Example 3, the value of c was 661.8. This means that the amount of yeast leveled off at 661.8 units.

Putting it all Together

5.7

The following table summarizes the basics of exponential, logarithmic, and logistic models. Least-squares regression can be used to determine the constants a, b, c, and C.

Concept	Formula	Comments
Exponential model	$f(x) = Ca^x$, $f(x) = ab^x$, or $f(x) = Ce^{kx}$	Exponential functions can be used to model data that increase or decrease rapidly over time.
Logarithmic model	$f(x) = a + b \log x$ or $f(x) = a + b \ln x$	Logarithmic functions can be used to model data that increase gradually over time.
Logistic model	$f(x) = \dfrac{c}{1 + ae^{-bx}}$	Logistic functions can be used to model data that at first increase slowly, then increase rapidly, and finally level off.

5.7 EXERCISES

Selecting a Model

Exercises 1–4: Select an appropriate type of modeling function for the data shown in the graph. Choose from the following.

    i. *Exponential*
   ii. *Logarithmic*
  iii. *Logistic*

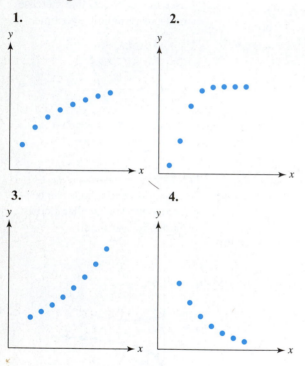

1.

2.

3.

4.

Exercises 5–8: Make a scatterplot of the data. Then find an exponential, logarithmic, or logistic function f that best models the data.

5.

x	1	2	3	4
y	2.04	3.47	5.90	10.02

6.

x	1	2	3	4	5
y	1.98	2.35	2.55	2.69	2.80

7.

x	1	2	3	4	5
y	1.1	3.1	4.3	5.2	5.8

8.

x	1	2	3	4	5	6
y	1	2	4	7	9	10

Applications

9. *Heart Disease Death Rates* The accompanying table contains heart disease death rates per 100,000 people in 1996 for selected ages.

Age	30	40	50	60	70
Death Rate	30.5	108.2	315	776	2010

Source: Department of Health and Human Services.

  (a) Make a scatterplot of the data in [25, 75, 5] by [−100, 2100, 200].
  (b) Find a function f that models the data.
  (c) Estimate the heart disease death rate for people who are 80 years old.

10. *Hurricanes* The accompanying table shows the air pressure y in inches of mercury x miles from the eye of a hurricane.

x (mi)	2	4	8
y (in. of mercury)	27.3	27.7	28.04

x (mi)	15	30	100
y (in. of mercury)	28.3	28.7	29.3

Source: A. Miller and R. Anthes, *Meteorology.*

  (a) Make a scatterplot of the data in [0, 110, 10] by [27, 30, 1].
  (b) Find a function f that models the data.
  (c) Estimate the air pressure at 50 miles.

11. *Atmospheric Density* The table lists the atmospheric density y in kilograms per cubic meter (kg/m^3) at an altitude of x meters. (Source: A. Miller.)

x (m)	0	5000	10,000	15,000
y (kg/m^3)	1.2250	0.7364	0.4140	0.1948

x (m)	20,000	25,000	30,000
y (kg/m^3)	0.0889	0.0401	0.0184

  (a) Use exponential regression to estimate the constants a and b so that $f(x) = ab^x$ models the data.
  (b) Predict the density at 7000 meters. (The actual value is 0.59 kg/m^3.)

12. *Bird Populations*   Near New Guinea there is a relationship between the number of bird species found on an island and the size of the island. The accompanying table lists species of birds y found on an island with an area of x square kilometers. (Source: B. Freedman, *Environmental Ecology.*)

x (km^2)	0.1	1	10	100	1000
y (species)	10	15	20	25	30

(a) Find a function f that models the data.
(b) Predict the number of bird species on an island of 5000 square kilometers.

13. *Fertilizer Usage*   Between 1950 and 1980 the use of chemical fertilizers increased. The table lists worldwide average usage y in kilograms per hectare of cropland during year x, where $x = 0$ corresponds to 1950. (*Note:* 1 hectare $\approx$ 2.47 acres.)   (Source: D. Grigg.)

x	0	13	22	29
y	12.4	27.9	54.3	77.1

(a) Graph the data. Is the data linear?
(b) Find a function f that models the data.
(c) Predict the chemical fertilizer usage in 1989. The actual value was 98.7 kilograms per hectare. What does this indicate about usage of fertilizer during the 1980s?

14. *Social Security*   If major reform occurs in the Social Security system, individuals may be able to invest some of their contributions into individual accounts. Many of these accounts would be managed by financial firms that charge fees. The following table lists the amount in billions of dollars that may be collected if fees are 0.93% of the assets each year.   (Source: Social Security Advisory Council.)

Year	2005	2010	2015	2020
Fees ($ billions)	20	41	80	136

(a) Use exponential regression to find a and b so that $f(x) = ab^x$ models the data. Let $x = 5$ correspond to 2005, $x = 10$ to 2010, and so on.
(b) Graph f and the data.
(c) Estimate the fees in 2013.

15. *Insect Population*   The accompanying table shows the density y of a species of insect measured in thousands per acre during a period of x days.

x (days)	2	4	6	8
y ($\times$ 1000/acre)	0.38	1.24	2.86	4.22

x (days)	10	12	14
y ($\times$ 1000/acre)	4.78	4.94	4.98

(a) Find a function f that models the data.
(b) Use f to estimate the insect density after a long time.

16. *Telecommuting*   Some workers use technology such as fax machines, e-mail, computers, and multiple phone lines to work at home, rather than in the office. However, because of the need for teamwork and collaboration in the workplace, fewer employees are telecommuting than expected. The table lists the expected telecommuters in millions from 1997 to 2006.   (Source: USA Today.)

Year	1997	1998	1999	2000
Telecommuters	9.2	9.6	10.0	10.4

Year	2001	2002	2003	2004
Telecommuters	10.6	11.0	11.1	11.2

Year	2005	2006
Telecommuters	11.3	11.4

Find values for a and b so that $f(x) = a + b \ln x$ models the data, where $x = 1$ corresponds to 1997, $x = 2$ to 1998, and so on.

17. *Heart Disease*   As age increases, so does the likelihood of coronary heart disease (CHD). The fraction of people x years old with some CHD is modeled by $f(x) = \dfrac{0.9}{1 + 271e^{-0.122x}}$.   (Source: D. Hosmer and S. Lemeshow, *Applied Logistic Regression.*)

(a) Evaluate $f(25)$ and $f(65)$. Interpret the results.
(b) At what age does this likelihood equal 50%?

18. *Modeling Tree Growth*   The height of a tree in feet after x years is modeled by $f(x) = \dfrac{50}{1 + 47.5e^{-0.22x}}$.

(a) Make a table of f starting at $x = 10$, incrementing by 10. What appears to be the maximum height of the tree?
(b) Graph f and identify the horizontal asymptote. Explain its significance.
(c) After how long was the tree 30 feet tall?

Writing About Mathematics

19. How can you distinguish between data that illustrate exponential growth and data that illustrate logarithmic growth?

20. Give an example of data that could be modeled by a logistic function and explain why.

Extended and Discovery Exercise

1. For medical reasons, dyes may be injected into the bloodstream to determine the health of internal organs. In one study involving animals, the dye BSP was injected to assess the blood flow in the liver. The results are listed in the accompanying table, where x represents the elapsed time in minutes and y is the concentration of the dye in the bloodstream in milligrams per milliliter (mg/ml). (Sources: F. Harrison, "The measurement of liver blood flow in conscious calves"; D. Brown and P. Rothery, *Models in Biology: Mathematics, Statistics and Computing.*)

x (min)	1	2	3	4	5	7
y (mg/ml)	0.102	0.077	0.057	0.045	0.036	0.023

x (min)	9	13	16	19	22
y (mg/ml)	0.015	0.008	0.005	0.004	0.003

  (a) Find a function that models the data.

  (b) Estimate the elapsed time when the concentration of the dye reaches 30% of its initial concentration of 0.133 mg/ml.

CHECKING BASIC CONCEPTS FOR SECTION 5.7

Exercises 1–3: Find a function that models the data in the table. Choose from exponential, logarithmic, or logistic functions.

1.

x	2	3	4	5	6	7
y	0.72	0.86	1.04	1.24	1.49	1.79

2.

x	2	3	4	5	6	7
y	0.08	1.30	2.16	2.83	3.38	3.84

3.

x	2	3	4	5	6	7
y	0.25	0.86	2.19	3.57	4.23	4.43

Chapter 5 Summary

CONCEPT	EXPLANATION AND EXAMPLES

Section 5.1

| ARITHMETIC OPERATIONS ON FUNCTIONS | Addition: $(f + g)(x) = f(x) + g(x)$
 Subtraction: $(f - g)(x) = f(x) - g(x)$
 Multiplication: $(fg)(x) = f(x) \cdot g(x)$
 Division: $(f/g)(x) = \dfrac{f(x)}{g(x)}, \ g(x) \neq 0$ |

CONCEPT	EXPLANATION AND EXAMPLES

Examples: Let $f(x) = x^2 - 5$, $g(x) = x^2 - 4$.

$$(f + g)(x) = (x^2 - 5) + (x^2 - 4) = 2x^2 - 9$$

$$(f - g)(x) = (x^2 - 5) - (x^2 - 4) = -1$$

$$(fg)(x) = (x^2 - 5)(x^2 - 4) = x^4 - 9x^2 + 20$$

$$(f/g)(x) = \frac{x^2 - 5}{x^2 - 4}, \; x \neq \pm 2$$

COMPOSITION OF FUNCTIONS

Composition: $(g \circ f)(x) = g(f(x))$

$$(f \circ g)(x) = f(g(x))$$

Examples: Let $f(x) = 3x + 2$, $g(x) = 2x^2 - 4x + 1$.

$$g(f(x)) = g(3x + 2)$$
$$= 2(3x + 2)^2 - 4(3x + 2) + 1$$

$$f(g(x)) = f(2x^2 - 4x + 1)$$
$$= 3(2x^2 - 4x + 1) + 2$$
$$= 6x^2 - 12x + 5$$

Section 5.2

INVERSE FUNCTION

The inverse function of f is f^{-1} if
$f^{-1}(f(x)) = x$　for every x in the domain of f, and
$f(f^{-1}(x)) = x$　for every x in the domain of f^{-1}.
Note: If $f(a) = b$, then $f^{-1}(b) = a$.

Example:　Find the inverse function of $f(x) = 4x - 5$.

$$y = 4x - 5 \text{ is equivalent to } \frac{y + 5}{4} = x. \text{ (Solve for } x.)$$

Thus $f^{-1}(x) = \dfrac{x + 5}{4}$.

Check:　$f^{-1}(f(x)) = f^{-1}(4x - 5)$

$$= \frac{(4x - 5) + 5}{4}$$

$$= x$$

$$f(f^{-1}(x)) = f\left(\frac{x + 5}{4}\right)$$

$$= 4\left(\frac{x + 5}{4}\right) - 5$$

$$= x \qquad \text{(It checks.)}$$

ONE-TO-ONE FUNCTION

If different inputs always result in different outputs, then f is one-to-one. That is, $a \neq b$ implies $f(a) \neq f(b)$.
Note: If f is one-to-one, then f has an inverse denoted f^{-1}.

Example:　$f(x) = x^2 + 1$ is not one-to-one because $f(2) = f(-2) = 5$.

CONCEPT	EXPLANATION AND EXAMPLES
HORIZONTAL LINE TEST	If every horizontal line intersects the graph of a function f at most once, then f is a one-to-one function.

Section 5.3

CONCEPT	EXPLANATION AND EXAMPLES
EXPONENTIAL FUNCTION	$f(x) = Ca^x$, $a > 0$, $a \neq 1$, and $C > 0$. Exponential growth: $a > 1$; Exponential decay: $0 < a < 1$ **Example:** $f(x) = 3(2^x)$ (growth); $f(x) = 1.2(0.5)^x$ (decay)
NATURAL EXPONENTIAL FUNCTION	$f(x) = e^x$, where $e \approx 2.718282$.
COMPOUND INTEREST	$A_n = A_0\left(1 + \dfrac{r}{m}\right)^{mn}$, where A_0 is the initial principal, r is the interest expressed as a decimal, m is the number of times interest is paid each year, and n is the number of years. **Example:** $A_4 = 2000\left(1 + \dfrac{0.10}{12}\right)^{12(4)} \approx \$2978.71$ calculates the future value of \$2000 invested at 10% compounded monthly for 4 years.
CONTINUOUSLY COMPOUNDED INTEREST	$A_n = A_0 e^{rn}$, where A_0 is the initial principal, r is the interest expressed as a decimal, n is the number of years. **Example:** $A_4 = 2000e^{0.10(4)} \approx \$2983.65$ calculates the future value of \$2000 invested at 10% compounded continuously for 4 years.
EXPONENTIAL DATA	For each unit increase in x, the y-values increase (or decrease) by a constant factor a. **Example:** The data in the table are modeled by $y = 5(2)^x$. <table><tr><td>x</td><td>0</td><td>1</td><td>2</td><td>3</td></tr><tr><td>y</td><td>5</td><td>10</td><td>20</td><td>40</td></tr></table>

Section 5.4

CONCEPT	EXPLANATION AND EXAMPLES
COMMON LOGARITHM **NATURAL LOGARITHM** **GENERAL LOGARITHM**	$\log x = k$ if and only if $x = 10^k$ $\ln x = k$ if and only if $x = e^k$ $\log_a x = k$ if and only if $x = a^k$ **Examples:** $\log 100 = 2$ because $100 = 10^2$. $\ln \sqrt{e} = \dfrac{1}{2}$ because $\sqrt{e} = e^{1/2}$. $\log_2 \dfrac{1}{8} = -3$ because $\dfrac{1}{8} = 2^{-3}$.

CONCEPT	EXPLANATION AND EXAMPLES
INVERSE PROPERTIES	$\log 10^k = k, \qquad 10^{\log k} = k$ $\ln e^k = k, \qquad e^{\ln k} = k$ $\log_a a^k = k, \qquad a^{\log_a k} = k$ **Examples:** $10^{\log 100} = 100$ $e^{\ln 23} = 23$ $\log_4 64 = \log_4 4^3 = 3$
INVERSE FUNCTIONS	The inverse function of $f(x) = a^x$ is $f^{-1}(x) = \log_a x$. **Examples:** If $f(x) = 10^x$, then $f^{-1}(x) = \log x$. If $f(x) = \ln x$, then $f^{-1}(x) = e^x$. If $f(x) = \log_5 x$, then $f^{-1}(x) = 5^x$.

Section 5.5

CONCEPT	EXPLANATION AND EXAMPLES
PROPERTIES OF LOGARITHMS	1. $\log_a 1 = 0$ 2. $\log_a m + \log_a n = \log_a (mn)$ 3. $\log_a m - \log_a n = \log_a \left(\dfrac{m}{n}\right)$ 4. $\log_a (m^r) = r \log_a m$ **Examples:** 1. $\log_4 1 = 0$ 2. $\log 2 + \log 5 = \log (2 \cdot 5) = \log 10 = 1$ 3. $\log 500 - \log 5 = \log (500/5) = \log 100 = 2$ 4. $3 \log_2 2 = \log_2 2^3 = 3$
CHANGE OF BASE FORMULA	$\log_a x = \dfrac{\log_b x}{\log_b a}$ **Example:** $\log_3 23 = \dfrac{\log 23}{\log 3} \approx 2.854$

Section 5.6

CONCEPT	EXPLANATION AND EXAMPLES
SOLVING EXPONENTIAL EQUATIONS	To solve an exponential equation we typically take a logarithm of both sides. **Example:** $4e^x = 48$ Given equation $e^x = 12$ Divide by 4. $\ln e^x = \ln 12$ Take the natural logarithm. $x = \ln 12$ Inverse property
SOLVING LOGARITHMIC EQUATIONS	To solve a logarithmic equation we typically need to exponentiate both sides. **Example:** $5 \log_3 x = 10$ Given equation $\log_3 x = 2$ Divide by 5. $3^{\log_3 x} = 3^2$ Exponentiate; base 3. $x = 9$ Inverse property

CONCEPT	EXPLANATION AND EXAMPLES

Section 5.7

EXPONENTIAL MODEL	$f(x) = Ca^x$, $f(x) = ab^x$, or $f(x) = Ce^{kx}$ Models data that can increase or decrease rapidly
LOGARITHMIC MODEL	$f(x) = a + b \ln x$ or $f(x) = a + b \log x$ Models data that increase slowly
LOGISTIC MODEL	$f(x) = \dfrac{c}{1 + ae^{-bx}}$, a, b, and c are positive constants. Models data that increase slowly at first, then increase more rapidly, and finally level off near the value of c.

Review Exercises

1. Use the table to evaluate each expression, if possible.

x	-1	0	1	3
$f(x)$	3	5	7	9
$g(x)$	-2	0	1	9

  (a) $(f + g)(1)$ **(b)** $(f - g)(3)$
  (c) $(fg)(-1)$ **(d)** $(f/g)(0)$

2. Use the graph to evaluate each expression.
  (a) $(f - g)(2)$ **(b)** $(fg)(0)$

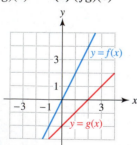

3. Use $f(x) = x^2$ and $g(x) = 1 - x$ to evaluate each expression.
  (a) $(f + g)(3)$ **(b)** $(f - g)(-2)$
  (c) $(fg)(1)$ **(d)** $(f/g)(3)$

4. Use $f(x) = x^2 + 3x$ and $g(x) = x^2 - 1$ to find each expression. Identify its domain.
  (a) $(f + g)(x)$ **(b)** $(f - g)(x)$
  (c) $(fg)(x)$ **(d)** $(f/g)(x)$

5. Numerical representations for f and g are given. Evaluate the expressions.
  (a) $(g \circ f)(-2)$ **(b)** $(f \circ g)(3)$

x	-2	0	2	4
$f(x)$	1	4	3	2

x	1	2	3	4
$g(x)$	2	4	-2	0

6. Use the graph to evaluate each expression.
  (a) $(f \circ g)(2)$ **(b)** $(g \circ f)(0)$

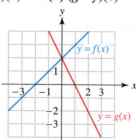

7. Use $f(x) = \sqrt{x}$ and $g(x) = x^2 + x$ to evaluate each expression.
  (a) $(f \circ g)(2)$ **(b)** $(g \circ f)(9)$

8. Use $f(x) = x^2 + 1$ and $g(x) = x^3 - x^2 + 2x + 1$ to find the following.
  (a) $(f \circ g)(x)$ **(b)** $(g \circ f)(x)$

Exercises 9 and 10: Describe the inverse operations of the given statement. Then, express both the statement and its inverse symbolically.

9. Divide x by 10 and add 6.

10. Subtract 5 from x and take the cube root.

Exercises 11 and 12: Determine if f is one-to-one.

11. $f(x) = 3x - 1$

12. $f(x) = 3x^2 - 2x + 1$

13. Use the table to determine if f is one-to-one.

x	−1	−2	0	4
$f(x)$	−5	4	3	4

14. Use the table of f to determine a table for f^{-1}. Identify the domains and ranges of f and f^{-1}.

x	−1	0	4	6
$f(x)$	6	4	3	1

15. The function f computes the dollars in a savings account after x years. Use the graph to evaluate the following. Interpret f^{-1}.

   (a) $f(1)$ **(b)** $f^{-1}(1200)$

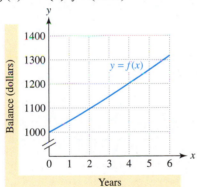

Years

16. Use the graph of f to sketch a graph of f^{-1}.

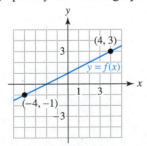

Exercises 17 and 18: Find $f^{-1}(x)$.

17. $f(x) = 3x - 5$

18. $f(x) = \sqrt[3]{x} + 1$

19. If $f(x) = \dfrac{1}{x + 7}$, find $f^{-1}(x)$.

20. Verify that $f(x) = 2x - 1$ and $f^{-1}(x) = \dfrac{x + 1}{2}$ are inverses.

21. Determine the final value of $1200 invested at 9% compounded semiannually for 3 years.

22. Determine the final value of $500 invested at 6.5% compounded continuously for 8 years.

Exercises 23 and 24: Evaluate $f(x)$ for the given x. Approximate the result to the nearest thousandth.

23. $f(x) = e^{-2x}, x = -1.2$

24. $f(x) = 4e^{0.23x}, x = 5.7$

25. Solve $e^x = 19$ symbolically. Support your result graphically or numerically.

26. Solve $2^x - x^2 = x$ graphically. Round each solution to the nearest thousandth.

Exercises 27–30: Evaluate the expression without a calculator.

27. $\log 1000$

28. $\log 0.001$

29. $10 \log 0.01 + \log \dfrac{1}{10}$

30. $\log 100 + \log \sqrt[3]{10}$

Exercises 31–34: Evaluate the logarithm without a calculator.

31. $\log_3 9$

32. $\log_5 \dfrac{1}{25}$

33. $\ln e$

34. $\log_2 32$

Exercises 35 and 36: Approximate the logarithm to the nearest thousandth.

35. $\log_3 18$

36. $\log_2 173$

Exercises 37–44: Solve the equation using common or natural logarithms. Support your results graphically or numerically.

37. $10^x = 125$

38. $1.5^x = 55$

39. $e^{0.1x} = 5.2$

40. $4e^{2x} - 5 = 3$

41. $5^{-x} = 10$

42. $3(10^{-x}) = 6$

43. $50 - 3(0.78)^{x-10} = 21$

44. $5(1.3)^x + 4 = 104$

Exercises 45–46: Find either a linear or an exponential function that models the data in the table.

45.

x	0	1	2	3	4
y	1.5	3	6	12	24

46.

x	0	1	2	3	4
y	3	4.5	6	7.5	9

Exercises 47–52: Solve the equation.

47. $\log x = 1.5$

48. $\log_3 x = 4$

49. $\ln x = 3.4$

50. $\log x = 2.2$

51. $2 \log_4 (x + 2) + 5 = 12$

52. $4 - \ln (5 - x) = \dfrac{5}{2}$

Exercises 53 and 54: Use properties of logarithms to write each expression as one logarithm.

53. $\log 6 + \log 5x$ **54.** $\log \sqrt{3} - \log \sqrt[3]{3}$

55. Expand $\ln \dfrac{4}{x^2}$.   **56.** Expand $\log \dfrac{4x^3}{k}$.

Exercises 57–62: Solve the logarithmic equation symbolically. Support your results graphically or numerically.

57. $8 \log x = 2$

58. $\ln 2x = 2$

59. $2 \log 3x + 5 = 15$

60. $5 \log_2 x = 25$

61. $2 \log (x + 2) = \log (x + 8)$

62. $\ln (5 - x) - \ln (5 + x) = -\ln 9$

63. Suppose that b is the y-intercept on the graph of a one-to-one function f. What is the x-intercept on the graph of f^{-1}? Explain your reasoning.

64. Let f be a linear function given by $f(x) = ax + b$ with $a \neq 0$.
   (a) Show that f^{-1} is also a linear function by finding $f^{-1}(x)$.
   (b) How is the slope of the graph of f related to the slope of the graph of f^{-1}?

65. *Diversity*   In 1995 the U.S. racial mix was 75.3% white, 9.0% Hispanic, 12.0% African-American, 2.9% Asian/Pacific Islanders, and 0.8% Native American. Their respective percentages of execu-

tives, managers, and administrators in private-industry retail trade were 80.8%, 4.8%, 4.9%, 5.2%, and 0.0%. If the percentage P of Hispanic executives, managers, and administrators increases at a rate of 5% per year, estimate the year when their representation will reach 9.0% of the total number in retail trade. (*Hint:* Let $P(x) = 4.8(1.05)^x$, where $x = 0$ corresponds to 1995, and solve $P(x) = 9$.)   (Source: Labor Department's Glass Ceiling Commission.)

66. *Combining Functions*   The total number of gallons of water passing through a pipe after x seconds is computed by $f(x) = 10x$. Another pipe delivers $g(x) = 5x$ gallons per second. Find a function h that computes the total volume of water passing through both pipes in x seconds.

67. *Converting Units*   The accompanying figures show graphs of a function f that converts fluid ounces to pints and a function g that converts pints to quarts. Evaluate each expression. Interpret the results.
   (a) $(g \circ f)(32)$ **(b)** $f^{-1}(1)$
   (c) $(f^{-1} \circ g^{-1})(1)$

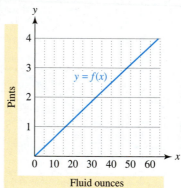

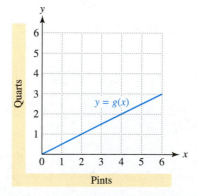

68. *Population and Tax Revenue*   The accompanying graph of f shows the population in thousands of a city from 1980 until 2000. The graph of g computes the expected tax revenue in millions of dollars that can be raised from a population of x thousand. Evaluate and interpret $(g \circ f)(1988)$.

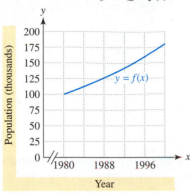

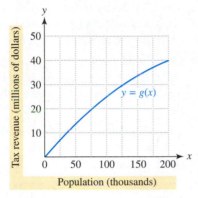

69. **Modeling Growth** The function given by

$$W(x) = 175.6(1 - 0.66e^{-0.24x})^3$$

models the weight in milligrams of a small fish called the *Lebistes reticulatus* after x weeks, where $0 \le x \le 14$. (Source: D. Brown and P. Rothery, *Models in Biology: Mathematics, Statistics and Computing.*)
(a) Evaluate $W(1)$.
(b) Solve the equation $W(x) = 50$ graphically. Interpret the result.
(c) Solve part (b) symbolically.

70. **Test Scores** Let scores on a standardized test be modeled by $f(x) = 36e^{-(x-20)^2/49}$. The function f computes the number in thousands of people that received score x.
(a) Graph f in [0, 40, 10] by [0, 40, 10]. What score did the largest number of people receive?
(b) Solve the equation $f(x) = 30$ symbolically. Interpret your result.
(c) Solve part (b) graphically.

71. **Tire Pressure** A car tire has a small leak, and the tire pressure in pounds per square inch after t minutes is given by $P(t) = 32e^{-0.2t}$. After how many minutes was the pressure 15 pounds per square inch?

72. **Radioactive Decay** After 23 days a 10-milligram sample of a radioactive material decays to 5 milligrams. After how many days will there be 1 milligram of the material? $\left(Hint: A(t) = 10\left(\dfrac{1}{2}\right)^{t/23}.\right)$

73. **Modeling Epidemics** In 1666 the village of Eyam, located in England, experienced an outbreak of the Great Plague. Out of 261 people in the community, only 83 people survived. The accompanying table shows a function f that computes the number of people who had not (yet) been infected after x days. (Source: G. Raggett, "Modeling the Eyam plague.")

x	0	15	30	45
$f(x)$	254	240	204	150

x	60	75	90	125
$f(x)$	125	103	97	83

(a) Use a table to represent a function g that computes the number of people in Eyam that were infected after x days.
(b) Write an equation that shows the relationship between $f(x)$ and $g(x)$.
(c) Use graphing to decide which equation represents $g(x)$ better,

$$y_1 = \frac{171}{1 + 18.6e^{-0.0747x}} \quad \text{or} \quad y_2 = 18.3(1.024)^x.$$

(d) Use your results from parts (b) and (c) to find a formula for $f(x)$.

74. **Greenhouse Gases** Methane is a greenhouse gas that is produced when fossil fuels are burned. In 1600 methane had an atmospheric concentration of 700 parts per billion (ppb), whereas in 2000 its concentration is about 1700 ppb. (Source: D. Wuebbles and J. Edmonds, *Primer on Greenhouse Gases.*)
(a) Let $x = 0$ correspond to 1600 and $x = 400$ to 2000. Then, $f(x) = 700(10^{0.000964x})$ models the methane concentrations. Graph f for $0 \le x \le 400$ and explain why f is one-to-one.
(b) Find $f^{-1}(x)$. Evaluate $f^{-1}(1000)$ and interpret the result.

75. **Exponential Regression** The data in the table can be modeled by $f(x) = ab^x$. Use regression to estimate the constants a and b. Graph f and the data.

x	1	2	3	4
y	2.59	1.92	1.42	1.05

76. *Logarithmic Regression*   The data in the table can be modeled by $f(x) = a + b \ln x$. Use regression to estimate the constants a and b. Graph f and the data.

x	2	3	4	5
y	2.93	3.42	3.76	4.03

Extended and Discovery Exercises

1. *Lunar Orbits*   The table lists the orbital distances and periods of several moons of Jupiter. Let x represent the distance and y the period. Find a function f that models this data. Try more than one type of function such as linear, quadratic, power, and exponential. Graph each function and the data. Discuss which function models the data best. (Source: M. Zeilik, *Introductory Astronomy and Astrophysics.*)

Moons of Jupiter	Distance (10^3km)	Period (days)
Metis	128	0.29
Almathea	181	0.50
Thebe	222	0.67
Io	422	1.77
Europa	671	3.55
Ganymede	1070	7.16
Callisto	1883	16.69

2. *Modeling Data*   There is a procedure to determine whether data can be modeled by $y = ax^b$, where a and b are constants. That is, the following procedure can be used to model data with a power function. Start by taking the natural logarithm of both sides of this equation.

$\ln y = \ln (ax^b)$

$\ln y = \ln a + \ln x^b$ Property 2:
$\ln (mn) = \ln m + \ln n$

$\ln y = \ln a + b \ln x$ Property 4:
$\ln (m^r) = r \ln m$

If we let $z = \ln y$, $d = \ln a$, and $w = \ln x$, then $\ln y = \ln a + b \ln x$ becomes $z = d + bw$. Thus the data points $(w, z) = (\ln x, \ln y)$ lie on the line having a slope of b and y-intercept $d = \ln a$. The following steps provide a procedure for finding the constants a and b.

Modeling data with the equation $y = ax^b$

If a data set (x, y) can be modeled by the equation $y = ax^b$, then the following procedure can be applied to determine the constants a and b.

STEP 1: Let $w = \ln x$ and $z = \ln y$ for each data point. Graph the points (w, z). If this data is not linear, then do *not* use this procedure.

STEP 2: Find an equation of a line in the form $z = bw + d$ that models the data points (w, z). (Linear regression may be used.)

STEP 3: The slope of the line equals the constant b. The value of a is given by $a = e^d$.

Apply this procedure to the data in Exercise 1.

3. *Global Warming*   Greenhouse gases such as carbon dioxide trap heat from the sun. Presently, the net incoming solar radiation reaching Earth's surface is approximately 240 watts per square meter (w/m^2). Any portion of this amount that is due to greenhouse gases is called *radiative forcing*. The accompanying table lists the estimated increase in radiative forcing R over the levels in 1750. (Source: A. Nilsson, *Greenhouse Earth.*)

x (year)	1800	1850	1900	1950	2000
$R(x)$ (w/m^2)	0.2	0.4	0.6	1.2	2.4

(a) Estimate constants C and k so that $R(x) = Ce^{kx}$ models the data. Let $x = 0$ correspond to 1800.

(b) Estimate the year when the additional radiative forcing could reach 3 w/m^2.

4. *Global Warming*   (Refer to the previous exercise.) The relationship between radiative forcing R and the increase in average global temperature T in degrees

Fahrenheit can be modeled by $T(R) = 1.03R$. For example, $T(2) = 1.03(2) = 2.06$ means that if the earth's atmosphere traps an additional 2 watts per square meter, then the average global temperature may increase by 2.06°F, *if* all other factors remained constant.   (Source: W. Clime, *The Economics of Global Warming.*)

(a) Use $R(x)$ from the previous exercise to express $(T \circ R)(x)$ symbolically, where x is the year and $x = 0$ corresponds to the year 1800.

(b) Evaluate $(T \circ R)(100)$ and interpret its meaning.

*Exercises 5 and 6: Evaluate the expression as accurately as possible. Try to decide if it is precisely an integer. (*Hint: *Use computer software capable of calculating a large number of decimal places.)*

5. $\left(\dfrac{1}{\pi} \ln (640{,}320^3 + 744) \right)^2$

   (Source: I. J. Good, "What is the most amazing approximate integer in the universe?")

6. $e^{\pi\sqrt{163}}$ (Source: W. Cheney and D. Kincaid, *Numerical Mathematics and Computing.*)

The future belongs to those who believe in the beauty of their dreams.

—

Eleanor Roosevelt

—

Systems of Equations and Inequalities

One of the most important inventions of the twentieth century was the electronic digital computer. In 1940 John Atanasoff, a physicist at Iowa State University, needed to find the solution to a system of 29 equations. Since this task was too difficult to solve by hand, it led him to invent the first fully electronic digital computer. Today modern supercomputers compute trillions of arithmetic operations in a single second and solve over 600,000 equations simultaneously. The need to solve a *mathematical problem* resulted in one of the most profound inventions of our time.

A recent breakthrough in visualizing our world is the invention of digital photography. By the year 2006, TV stations may be required by the Federal Communications Commission (FCC) to transmit digital signals. Digital pictures are represented by num-

bers rather than film. They are crystal clear without interference. One of the first digital pictures was done by James Blinn of NASA in 1981. It was a simulation of the Voyager-Saturn flyby and required substantial mathematics to create.

Mathematics plays a central role in new technology. In this chapter mathematics is used to solve systems of equations, represent digital pictures, compute movement in computer graphics, and even represent color on computer monitors. Throughout history, many important discoveries and inventions have been based on the creative insights of a few people. These insights have had a profound impact on our society.

References: A. Tucker, *Fundamentals of Computing I; USA Today;* NASA/JPL.

6.1 FUNCTIONS AND EQUATIONS OF TWO VARIABLES

Functions of Two Variables • Systems of Equations • The Method of Substitution • Graphical and Numerical Methods • Joint Variation

Introduction

Many quantities in everyday life depend on more than one variable. Calculating the area of a rectangular room requires both its length and width. The heat index is a function of temperature and humidity, and windchill is determined using temperature and wind speed. Information about grades and credit hours is necessary to compute a grade point average.

In earlier chapters we discussed functions of one input or one variable. Quantities determined by more than one variable often are computed by a function of more than one variable. The mathematical concepts concerning functions of one input also apply to functions of more than one input. One unifying concept about all functions is that each produces *at most one output* each time it is evaluated.

Functions of Two Variables

The arithmetic operations of addition, subtraction, multiplication, and division are computed by functions of two inputs. In order to perform addition, two numbers must be provided. The addition of x and y results in one output z. The addition function f can be represented symbolically by $f(x, y) = x + y$, where $z = f(x, y)$. For example, the addition of 3 and 4 can be written as $z = f(3, 4) = 3 + 4 = 7$. In this case, $f(x, y)$ is a **function of two inputs** or a **function of two variables**. The **independent variables** are x and y, and z is the **dependent variable**. The output z depends on the inputs x and y. In a similar manner, a division function can be defined by $g(x, y) = \dfrac{x}{y}$, where $z = g(x, y)$.

■ **MAKING CONNECTIONS**

Independent and Dependent Variables In Chapters 1 through 5 the input for a function f was usually represented by x and the output by y. This was expressed as $y = f(x)$. For functions of two inputs, it is common to use x and y as inputs and z as the output. This is expressed as $z = f(x, y)$. ■

EXAMPLE 1 *Evaluating a function of more than one input*

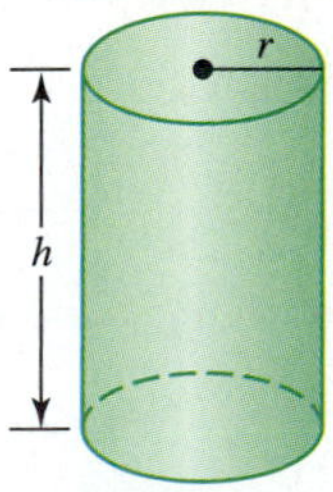

For each function, evaluate the expression and interpret the result.

(a) $f(3, -4)$, where $f(x, y) = xy$ represents the multiplication function

(b) $f(120, 5)$, where $f(m, g) = \dfrac{m}{g}$ computes the gas mileage when traveling m miles on g gallons of gasoline

(c) $f(0.5, 2)$, where $f(r, h) = \pi r^2 h$ calculates the volume of a cylindrical barrel with a radius r and height h (See Figure 6.1.)

FIGURE 6.1

SOLUTION **(a)** $f(3, -4) = (3)(-4) = -12$. The product of 3 and -4 is -12.

(b) $f(120, 5) = \dfrac{120}{5} = 24$. If a car travels 120 miles on 5 gallons of gasoline, its mileage is 24 miles per gallon.

(c) $f(0.5, 2) = \pi(0.5^2)(2) = 0.5\pi \approx 1.57$. If a barrel has a radius of 0.5 foot and a height of 2 feet, it holds about 1.57 cubic feet of liquid.

The surface area of the skin covering the human body is a function of more than one variable. A taller person tends to have a larger surface area, as does a heavier person. Both height and weight influence the surface area of a person's body. The following example demonstrates a function that can be used to estimate this surface area.

EXAMPLE 2   *Estimating the surface area of the human body*

A formula to determine the surface area of a person's body in square meters is computed by $S(w, h) = 0.007184(w^{0.425})(h^{0.725})$, where w is weight in kilograms and h is height in centimeters. Use S to estimate the surface area of a human body that is 65 inches tall and weighs 154 pounds.   (Source: H. Lancaster, *Quantitative Methods in Biological and Medical Sciences.*)

SOLUTION   First, convert 65 inches to centimeters. There are approximately 2.54 centimeters in 1 inch, so this person's height is $65 \cdot 2.54 = 165.1$ centimeters. Next, convert pounds to kilograms. There are about 2.2 pounds in 1 kilogram, so the individual weighs $\frac{154}{2.2} = 70$ kilograms.

$$S(70, 165.1) = 0.007184(70^{0.425})(165.1^{0.725}) \approx 1.77$$

Thus a person who weighs 70 kilograms (154 pounds) and is 165.1 centimeters (65 inches) tall has a surface area of approximately 1.77 square meters.

We often use formulas with more than one variable. Sometimes it is necessary to solve an equation for a variable. For example, to find the radius r of a circle that has an area A of 50 square inches, we might first solve the equation $A = \pi r^2$ for r.

$$A = \pi r^2 \qquad \text{Area of a circle}$$

$$\frac{A}{\pi} = r^2 \qquad \text{Divide by } \pi.$$

$$\pm\sqrt{\frac{A}{\pi}} = r \qquad \text{Square root property}$$

$$r = \sqrt{\frac{A}{\pi}} \qquad \text{For this problem, } r > 0.$$

If a circle has an area of 50 square inches, then its radius is

$$r = \sqrt{\frac{50}{\pi}} \approx 3.99 \text{ inches.}$$

EXAMPLE 3   *Solving for a variable*

The equation $P = 2L + 2W$ calculates the perimeter of a rectangle with length L and width W, as illustrated in Figure 6.2.
(a) Solve $P = 2L + 2W$ for L.
(b) Find L for a rectangle with $P = 21$ feet and $W = 3.5$ feet.

Geometry Review
To review formulas related to
rectangles, see Chapter R
(page R-1).

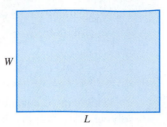

FIGURE 6.2

SOLUTION   **(a)**

$$P = 2L + 2W \qquad \text{Given formula}$$

$$P - 2W = 2L \qquad \text{Subtract } 2W.$$

$$\frac{P - 2W}{2} = L \qquad \text{Divide by 2.}$$

Given the perimeter P and the width W, the length L can be computed with this formula.

(b) $L = \dfrac{P - 2W}{2} = \dfrac{21 - 2(3.5)}{2} = 7$ feet.

Systems of Equations

A **linear equation in two variables** can be written in the form

$$ax + by = k,$$

where a, b, and k are constants, and a and b are not equal to 0. Examples of linear equations in two variables include

$$2x - 3y = 4, \qquad -x - 5y = 0, \qquad \text{and} \qquad 5x - y = 10.$$

The average of two numbers, x and y, can be found with the formula $\dfrac{x + y}{2}$. To find two numbers whose average is 10, we solve the linear equation

$$\frac{x + y}{2} = 10.$$

(Note that $\dfrac{x + y}{2} = 10$ is a linear equation because it can be written as $\dfrac{1}{2}x + \dfrac{1}{2}y = 10$.)

The numbers 5 and 15 average to 10, as do the numbers 0 and 20. In fact, any pair of numbers whose sum is 20 will satisfy this equation. There are infinitely many solutions to this linear equation. For a unique solution, another restriction must be placed on the variables x and y.

Many situations involving two variables result in the need to determine values for x and y that satisfy two different equations. For example, suppose that we would like to find a pair of numbers whose average is 10, and whose difference is 2. The function $f(x, y) = \dfrac{x + y}{2}$ calculates the average of two numbers and $g(x, y) = x - y$ computes their difference. The solution could be found by solving two linear equations.

$$f(x, y) = 10$$

$$g(x, y) = 2$$

These equations can be written as follows.

$$\frac{x + y}{2} = 10$$

$$x - y = 2$$

This pair of equations is called a **system of linear equations** because we are solving more than one linear equation at once. A **solution** to a system of equations consists of an x-value and a y-value that satisfy *both* equations simultaneously. The set of all solutions is called the **solution set**. Using trial and error, we see that $x = 11$ and $y = 9$ satisfy both equations. This is the only solution and can be expressed as the ordered pair $(11, 9)$.

EXAMPLE 4 *Determining a system of equations*

In 1996 hamburger sales totaled $23.9 billion at McDonald's and Burger King. McDonald's sales exceeded Burger King's by $8.9 billion.   (Source: Technomic.)
(a) Write a system of equations involving two variables, whose solution gives the individual sales of each company.
(b) Is the resulting system of equations linear or nonlinear?

SOLUTION **(a)** When setting up a system of equations, it is essential that we identify what each variable represents. Let x represent the hamburger sales of McDonald's and y the hamburger sales of Burger King. Their combined sales are $23.9 billion, so $x + y = 23.9$. McDonald's sales exceed Burger King's by $8.9 billion, so $x - y = 8.9$. Thus the system of equations is as follows.

$$x + y = 23.9$$

$$x - y = 8.9$$

(b) Both equations are written in the form $ax + by = k$. Therefore it is a system of linear equations.

Systems of equations that are not linear are called **systems of nonlinear equations**. The following are two examples of systems of nonlinear equations.

$$
\begin{array}{ll}
x^2 - 5y = 8 & x^2 + y^2 = 5 \\
2x + 3y = 3 & \sqrt{x} - y = -2
\end{array}
$$

The Method of Substitution

The next example describes the **method of substitution**. This procedure is often used to solve systems of equations involving two variables.

EXAMPLE 5 *Solving a linear system of equations*

Use the method of substitution to solve the system of equations in Example 4. Interpret the result, and check the solution.

SOLUTION In this system x represents 1996 hamburger sales for McDonald's in billions of dollars, and y represents the corresponding sales for Burger King.

$$x + y = 23.9$$

$$x - y = 8.9$$

Algebra Review

To review how to solve a linear equation, see Section 2.3.

Begin by solving an equation for a convenient variable. Then substitute the result into the other equation. For example, if we solve the first equation for y, the result is $y = 23.9 - x$. Then, substitute $(23.9 - x)$ for y into the second equation.

$$x - (23.9 - x) = 8.9$$

This equation can be solved for x.

$$x - (23.9 - x) = 8.9$$
$$x - 23.9 + x = 8.9 \qquad \text{Distributive property}$$
$$2x = 32.8 \qquad \text{Combine } x\text{-terms and add 23.9.}$$
$$x = 16.4 \qquad \text{Divide by 2.}$$

Thus McDonald's sold $16.4 billion in hamburgers in 1996. Since $y = 23.9 - x$, it follows that $y = 23.9 - 16.4 = 7.5$. Burger King's share of the market was $7.5 billion.

To check if $(16.4, 7.5)$ is a solution, substitute $x = 16.4$ and $y = 7.5$ into both equations.

$$16.4 + 7.5 \stackrel{?}{=} 23.9 \qquad \text{True}$$
$$16.4 - 7.5 \stackrel{?}{=} 8.9 \qquad \text{True}$$

Since *both* equations are satisfied, $(16.4, 7.5)$ is a solution.

The method of substitution is summarized verbally in the following.

The method of substitution

To use the method of substitution to solve a system of two equations with two variables, perform the following steps.

STEP 1: Choose a variable in one of the two equations. Solve the equation for that variable.

STEP 2: Substitute the result into the other equation and solve this equation.

STEP 3: Use the result from **STEP 2** to determine the value of the other variable. To do this, you may want to use the equation you found in **STEP 1**.

EXAMPLE 6 *Using the method of substitution*

Solve the system symbolically. Give graphical support.

$$5x - 2y = -16$$
$$x + 4y = -1$$

SOLUTION *Symbolic Solution*

STEP 1: Begin by solving one of the equations for one of the variables. One possibility is to solve the second equation for x

$$x + 4y = -1 \qquad \text{Second equation}$$
$$x = -4y - 1 \qquad \text{Subtract } 4y \text{ from both sides.}$$

STEP 2: Next, substitute $(-4y - 1)$ for x into the first equation and solve the resulting equation for y.

$$5x - 2y = -16 \qquad \text{First equation}$$
$$5(-4y - 1) - 2y = -16 \qquad \text{Let } x = -4y - 1.$$
$$-20y - 5 - 2y = -16 \qquad \text{Distributive property}$$
$$-5 - 22y = -16 \qquad \text{Combine like terms.}$$
$$-22y = -11 \qquad \text{Add 5 to both sides.}$$
$$y = \frac{1}{2} \qquad \text{Divide both sides by } -22.$$

STEP 3: Now find the value of x by using the equation $x = -4y - 1$, which was found in **STEP 1**. Since $y = \frac{1}{2}$, it follows that $x = -4\left(\frac{1}{2}\right) - 1 = -3$. The solution can be written as an ordered pair: $\left(-3, \frac{1}{2}\right)$.

Graphical Solution   We can use the intersection-of-graphs method to solve this problem. Start by solving each equation for y.

$$5x - 2y = -16 \quad \text{is equivalent to} \quad y = \frac{5x + 16}{2}.$$

$$x + 4y = -1 \quad \text{is equivalent to} \quad y = \frac{-x - 1}{4}.$$

The graphs of $Y_1 = (5X + 16)/2$ and $Y_2 = (-X - 1)/4$ intersect at $(-3, 0.5)$ in Figure 6.3. The graphical solution supports our symbolic solution. Notice that the graph of each linear equation is a line.

Graphing Calculator Help

To find a point of intersection, see Appendix B (page AP-11).

$[-6, 6, 1]$ by $[-4, 4, 1]$

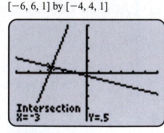

Intersection
X=-3   Y=.5

FIGURE 6.3

The method of substitution can also be used to solve systems of nonlinear equations. In the next example we solve a system of nonlinear equations with two solutions. In general, a system of nonlinear equations can have any number of solutions.

EXAMPLE 7   *Solving a system of nonlinear equations*

Solve the system symbolically.

$$6x + 2y = 10$$
$$2x^2 - 3y = 11$$

SOLUTION   **STEP 1:** Begin by solving one of the equations for one of the variables. One possibility is to solve the first equation for y.

$$6x + 2y = 10 \qquad \text{First equation}$$
$$2y = 10 - 6x \qquad \text{Subtract } 6x \text{ from both sides.}$$
$$y = 5 - 3x \qquad \text{Divide both sides by 2.}$$

STEP 2: Next, substitute $(5 - 3x)$ for y into the second equation and solve the resulting quadratic equation for x.

Algebra Review
To review factoring, see Chapter R (page R-27).

$$2x^2 - 3y = 11 \qquad \text{Second equation}$$
$$2x^2 - 3(5 - 3x) = 11 \qquad \text{Let } y = 5 - 3x.$$
$$2x^2 - 15 + 9x = 11 \qquad \text{Distributive property}$$
$$2x^2 + 9x - 26 = 0 \qquad \text{Subtract 11 and simplify.}$$
$$(2x + 13)(x - 2) = 0 \qquad \text{Factor.}$$
$$x = -\frac{13}{2} \text{ or } x = 2 \qquad \text{Zero-product property}$$

STEP 3: Now find the corresponding y-values for each x-value. From **STEP 1**, we know that $y = 5 - 3x$, so it follows that $y = 5 - 3\left(-\frac{13}{2}\right) = \frac{49}{2}$ or $y = 5 - 3(2) = -1$. Thus the solutions are $\left(-\frac{13}{2}, \frac{49}{2}\right)$ and $(2, -1)$. ◼

| EXAMPLE 8 | *Solving a system of nonlinear equations*

A circle with radius r, centered at the origin, has an equation $x^2 + y^2 = r^2$. Use the method of substitution to determine the points where the graph of $y = 2x$ intersects this circle when $r = \sqrt{5}$. Sketch a graph that illustrates the solutions.

SOLUTION   Since $r^2 = 5$, solve the following system.

$$x^2 + y^2 = 5$$
$$y = 2x$$

Substitute $(2x)$ for y into the first equation.

Geometry Review
To review equations of circles, see Chapter R (page R-10).

$$x^2 + y^2 = 5 \qquad \text{First equation}$$
$$x^2 + (2x)^2 = 5 \qquad y = 2x$$
$$x^2 + 4x^2 = 5 \qquad \text{Square the expression.}$$
$$5x^2 = 5 \qquad \text{Add like terms.}$$
$$x^2 = 1 \qquad \text{Divide by 5.}$$
$$x = \pm 1 \qquad \text{Square root property}$$

Since $y = 2x$ we see that when $x = 1, y = 2$ and when $x = -1, y = -2$. The graphs of $x^2 + y^2 = 5$ and $y = 2x$ intersect at the points $(1, 2)$ and $(-1, -2)$. This nonlinear system has two solutions, which are shown in Figure 6.4 on the next page.

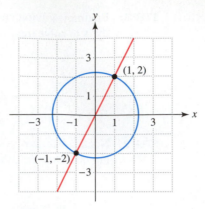

FIGURE 6.4

Graphical and Numerical Methods

The next example illustrates how a system with two variables can be solved graphically and numerically.

EXAMPLE 9 *Modeling roof trusses*

Linear systems occur in the design of roof trusses for homes and buildings. See Figure 6.5. One of the simplest types of roof trusses is an equilateral triangle. If a 200-pound force is applied to the peak of a truss, as shown in Figure 6.6, then the weights W_1 and W_2 exerted on each rafter of the truss are determined by the following system of linear equations. (Source: R. Hibbeler, *Structural Analysis.*)

$$W_1 - W_2 = 0$$

$$\frac{\sqrt{3}}{2}(W_1 + W_2) = 200$$

Estimate the solution graphically and numerically.

FIGURE 6.5

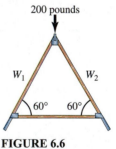

FIGURE 6.6

SOLUTION *Graphical Solution* Begin by solving each equation for the variable W_2.

$$W_2 = W_1$$

$$W_2 = \frac{400}{\sqrt{3}} - W_1$$

Graph the equations $Y_1 = X$ and $Y_2 = (400/\sqrt{3}) - X$. Their graphs intersect near the point (115.47, 115.47), as shown in Figure 6.7. This means that each rafter supports a weight of approximately 115 pounds. (*Remark:* W_1 and W_2 represent the forces parallel to the rafters. Most of the weight is pushing downward, while a smaller portion is attempting to spread the two rafters apart.)

Numerical Solution Numerical support is shown in Figure 6.8, where $Y_1 \approx Y_2$ when $x = 115$.

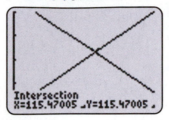

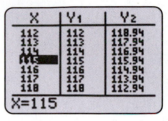

FIGURE 6.7   **FIGURE 6.8**

EXAMPLE 10   *Determining the dimensions of a cylinder*

The volume V of a cylindrical container with a radius r and height h is computed by $V(r, h) = \pi r^2 h$. See Figure 6.9. The lateral surface area S of the container, excluding the circular top and bottom, is computed by $S(r, h) = 2\pi rh$.

Geometry Review
To review formulas related to cylinders, see Chapter R (page R-4).

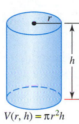
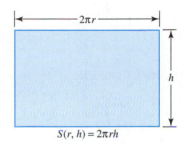

$V(r, h) = \pi r^2 h$ $S(r, h) = 2\pi rh$

FIGURE 6.9

(a) Write a system of equations, whose solution is the dimensions for a cylinder with a volume of 38 cubic inches and a lateral surface area of 63 square inches.
(b) Solve the system of equations graphically and numerically.

SOLUTION   **(a)** The equations $V(r, h) = 38$ and $S(r, h) = 63$ must be satisfied. This results in the following system of nonlinear equations.

$$\pi r^2 h = 38$$
$$2\pi rh = 63$$

(b) ***Graphical Solution*** To find the solution graphically, we can solve each equation for h, and then apply the intersection-of-graphs method.

$$h = \frac{38}{\pi r^2}$$
$$h = \frac{63}{2\pi r}$$

Let r correspond to x and h to y. Graph $Y_1 = 38/(\pi X^2)$ and $Y_2 = 63/(2\pi X)$. Their graphs intersect near the point $(1.206, 8.312)$, as shown in Figure 6.10 on the next page. Therefore a cylinder with a radius of $r \approx 1.206$ inches and height of $h \approx 8.312$ inches has a volume of 38 cubic inches and lateral surface area of 63 square inches.

Numerical Solution Numerical support is shown in Figure 6.11, where $Y_1 \approx Y_2$ when $x = 1.2$.

[0, 4, 1] by [0, 20, 5]

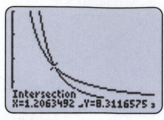

FIGURE 6.10

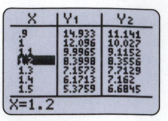

FIGURE 6.11

Sometimes it is either difficult or impossible to solve a system of nonlinear equations symbolically. However, it might be possible to solve such a system graphically.

EXAMPLE 11   *Solving a system of nonlinear equations graphically*

Solve the system graphically to the nearest thousandth.

$$2x^3 - y = 2$$
$$\ln x^2 - 3y = -1$$

[−6, 6, 1] by [−4, 4, 1]

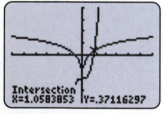

FIGURE 6.12

SOLUTION   Begin by solving both equations for y. The first equation becomes $y = 2x^3 - 2$. Solving the second equation for y gives the following results.

$$\ln x^2 - 3y = -1 \qquad \text{Second equation}$$
$$\ln x^2 + 1 = 3y \qquad \text{Add } 3y \text{ and 1 to both sides.}$$
$$\frac{\ln x^2 + 1}{3} = y \qquad \text{Divide both sides by 3.}$$

The graphs of $Y_1 = 2X^3 - 2$ and $Y_2 = (\ln (X^2) + 1)/3$ in Figure 6.12 intersect at one point. To the nearest thousandth, the solution is (1.058, 0.371).

Joint Variation

A quantity may depend on more than one variable. For example, the volume V of a cylinder is given by $V = \pi r^2 h$. We say that V *varies jointly* as h and the square of r. The *constant of variation* is π.

Joint variation

Let m and n be real numbers. Then z **varies jointly** as the nth power of x and the mth power of y if a nonzero real number k exists such that

$$z = kx^n y^m.$$

EXAMPLE 12   *Modeling the amount of wood in a tree*

In forestry it is common to estimate the volume of timber in a given area of forest. To do this, formulas are developed to find the amount of wood contained in a tree with height h and diameter d. See Figure 6.13. One study concluded that the volume V of

wood in a tree varies jointly as the 1.12 power of h and the 1.98 power of d. (The diameter is measured 4.5 feet above the ground.)   (Source: B. Ryan, B. Joiner, and T. Ryan, *Minitab Handbook.*)

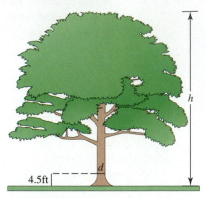

h

d

4.5ft

FIGURE 6.13

(a) Write an equation that relates V, h, and d.
(b) A tree with a 13.8-inch diameter and a 64-foot height has a volume of 25.14 cubic feet. Estimate the constant of variation k.
(c) Estimate the volume of wood in a tree with $d = 11$ inches and $h = 47$ feet.

SOLUTION   **(a)** $V = kh^{1.12}d^{1.98}$, where k is the constant of variation.
(b) Substitute $d = 13.8$, $h = 64$, and $V = 25.14$ into the equation and solve for k.

$$25.14 = k(64^{1.12})(13.8^{1.98})$$

$$k = \frac{25.14}{(64^{1.12})(13.8^{1.98})} \approx 0.00132$$

Thus let $V = 0.00132h^{1.12}d^{1.98}$.
(c) $V = 0.00132(47^{1.12})(11^{1.98}) \approx 11.4$ cubic feet.

Putting it all Together 6.1

*M*any mathematical concepts that were presented in earlier chapters can be applied to functions and equations of two variables. Functions of two variables often lead to equations involving two variables. Systems of equations can be solved symbolically, graphically, or numerically.

The following table summarizes some mathematical concepts involved with functions and equations of two variables.

Concept	Comments	Example
Function of two inputs or variables	$z = f(x, y)$, where x and y are inputs and z is the output.	$f(x, y) = x^2 + 5y$ $f(2, 3) = 2^2 + 5(3) = 19$
System of two linear equations	Equations written as $ax + by = k$.	$2x - 3y = 6$ $5x + 4y = -8$
System of two nonlinear equations	A system of equations that is not linear is nonlinear.	$5x^2 - 4xy = 20$ $\dfrac{5}{x} - 2y = 18$

Concept	Comments	Example
Method of substitution	Solve one equation for a variable and substitute the result into the second equation and solve.	$x - y = 1$ $x + y = 5$ If $x - y = 1$, then $x = 1 + y$. Substitute this into the second equation, $(1 + y) + y = 5$. This results in $y = 2$ and $x = 3$.
Graphical method for two equations	Solve both equations for the same variable. Then apply the intersection-of-graphs method.	If $x + y = 3$ then $y = 3 - x$. If $4x - y = 2$ then $y = 4x - 2$. Graph and locate the point of intersection at $(1, 2)$.

6.1 EXERCISES

Functions of More than One Input

Exercises 1 and 2: Evaluate f for the indicated inputs and interpret the result.

1. $f(5, 8)$, where $f(b, h) = \dfrac{1}{2}bh$ (f computes the area of a triangle with base b and height h.)

2. $f(20, 35)$, where $f(w, l) = wl$ (f computes the area of a rectangle with width w and length l.)

Exercises 3–6: (Refer to Example 2.) Estimate, to the nearest tenth, the surface area of a human body with weight w and height h.

3. $w = 86$ kilograms, $h = 185$ centimeters

4. $w = 68$ kilograms, $h = 165$ centimeters

5. $w = 132$ pounds, $h = 62$ inches

6. $w = 220$ pounds, $h = 75$ inches

Exercises 7–10: Write the symbolic representation for $f(x, y)$, if the function f computes the following quantity.

7. The sum of y and twice x

8. The product of x^2 and y^2

9. The product of x and y divided by $1 + x$

10. The square root of the sum of x and y

Exercises 11 and 12: **Life Expectancy** *The table lists life expectancy by birth year and sex. Let this table be a numerical representation of a function f, where $f(x, y)$ computes the life expectancy at birth of someone born in the year x whose sex is y.* (Source: Department of Health and Human Services.)

Year	Male	Female
1920	53.6	54.6
1940	60.8	65.2
1960	66.6	73.1
1980	70.0	77.4
1997	73.6	79.2

11. Evaluate $f(1960, \text{Male})$. Interpret the r

12. Evaluate $f(1997, \text{Female}) - f(1997$ pret the result.

Exercises 13–16: Solve the given equation for the indicated variable.

13. $A = \dfrac{1}{2}bh$

(a) b
(b) h

14. $V = \dfrac{1}{3}\pi r^2 h$

(a) h
(b) r

15. $z = \dfrac{2x^2}{y+1}$

(a) x
(b) y

16. $z = 2x + 4y$

(a) x
(b) y

Systems of Equations

Exercises 17–20: Determine which ordered pairs are solutions to the given system of equations.

17. $(2, 1), (-2, 1), (1, 0)$
$2x + y = 5$
$x + y = 3$

18. $(3, 2), (3, -4), (5, 0)$
$x - y = 5$
$2x + y = 10$

19. $(4, -3), (0, 5), (4, 3)$
$x^2 + y^2 = 25$
$2x + 3y = -1$

20. $(4, 8), (8, 4), (-4, -8)$
$xy = 32$
$x + y = 12$

Exercises 21 and 22: The figure shows the graph of a system of two linear equations. Use the graph to estimate the solution to this system of equations. Then solve the system symbolically.

21.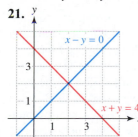

22.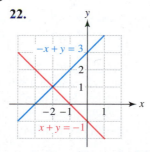

Exercises 23–26: Solve the system of linear equations

(a) *graphically,*
(b) *numerically, and*
(c) *symbolically.*

23. $2x + y = 1$
$x - 2y = 3$

24. $3x + 2y = -2$
$2x - y = -6$

25. $-2x + y = 0$
$7x - 2y = 3$

26. $x - 4y = 15$
$3x - 2y = 15$

Exercises 27–38: Solve the system of linear equations.

27. $x + 2y = 0$
$3x + 7y = 1$

28. $-2x - y = -2$
$3x + 4y = -7$

29. $3x - 5y = -38$
$-4x + y = 28$

30. $7x - 2y = 5$
$x + 9y = 10$

31. $2x - 9y = -17$
$8x + 5y = 14$

32. $3x + 6y = 0$
$4x - 2y = -5$

33. $\dfrac{1}{2}x - y = -5$
$x + \dfrac{1}{2}y = 10$

34. $-x - \dfrac{1}{3}y = -4$
$\dfrac{1}{3}x + 2y = 7$

35. $0.2x - 0.1y = 0.5$
$0.4x + 0.3y = 2.5$

36. $0.3x + 0.2y = 9$
$0.5x - 0.1y = 2$

37. $20x - 10y = 30$
$-5x - 20y = -30$

38. $100x + 200y = 300$
$200x + 100y = 0$

Exercises 39–42: Solve the system of nonlinear equations

(a) *graphically,*
(b) *symbolically.*

39. $x^2 + y^2 = 16$
$x - y = 0$

40. $x^2 - y = 1$
$3x + y = -1$

41. $xy = 63$
$x - y = 2$

42. $x^2 + y^2 = 2$
$x^2 - y = 0$

Exercises 43–54: Solve the system of nonlinear equations.

43. $x^2 - y = 0$
$2x + y = 0$

44. $x^2 - y = 3$
$x + y = 3$

45. $xy = 8$
$x + y = 6$

46. $2x - y = 0$
$2xy = 4$

47. $x^2 + y^2 = 20$
$y = 2x$

48. $x^2 + y^2 = 9$
$x + y = 3$

49. $x^2 + y^2 = 6$
$\sqrt{x} - y = 0$

50. $x^2 + y^2 = 2$
$2x - y = 1$

51. $x^2 - y = 4$
$x^2 + y = 4$

52. $x^2 + x = y$
$2x^2 - y = 2$

53. $x^3 - x = 3y$
$x - y = 0$

54. $x^4 + y = 4$
$3x^2 - y = 0$

Exercises 55 and 56: The area of a rectangle with length l and width w is computed by $A(l, w) = lw$, and its perimeter is calculated by $P(l, w) = 2l + 2w$. Assume that $l > w$ and use the method of substitution to solve the system of equations for l and w. Interpret the solution.

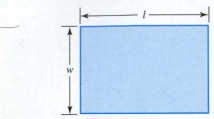

Geometry Review To review formulas related to rectangles, see Chapter R (page R-1).

55. $A(l, w) = 35$
$P(l, w) = 24$

56. $A(l, w) = 300$
$P(l, w) = 70$

Exercises 57–62: Approximate, to the nearest thousandth, any solutions to the system of nonlinear equations graphically.

57. $x^3 - 3x + y = 1$
$x^2 + 2y = 3$

58. $x^2 + y = 5$
$x + y^2 = 6$

59. $2x^3 - x^2 = 5y$
$2^{-x} - y = 0$

60. $x^4 - 3x^3 = y$
$\log x^2 - y = 0$

61. $e^{2x} + y = 4$
$\ln x - 2y = 0$

62. $3x^2 + y = 3$
$(0.3)^x + 4y = 1$

Applications

63. *Bank Robberies* The total number of bank robberies in 1994 and 1995 was 13,787. From 1994 to 1995 the number of bank robberies declined by 271. (Source: Federal Bureau of Investigation.)
 (a) Write a system of equations whose solution represents the number of robberies committed in each of these years.
 (b) Solve the system graphically.
 (c) Solve the system symbolically.

64. *Medical Waste* In 1996 hospital incinerators caused significant toxic pollution. The two states that produced the largest medical waste pollution were New York and California. Together they accounted for 185,000 tons. New York's total exceeded California's by 32,900 tons. (Source: Environmental Working Group.)
 (a) Set up a system of linear equations whose solution represents the incinerator pollution from each state.
 (b) Solve the system graphically.
 (c) Solve the system symbolically.

65. *Tourists* In 1996 a total of 6.2 million tourists visited Fort Lauderdale, Florida. There were four times as many American tourists as foreign tourists. Determine the number of foreign and American tourists that visited Fort Lauderdale in 1996. Round your answers to the nearest tenth of a million. (Source: Greater Fort Lauderdale Convention & Visitor Bureau.)

66. *Card Catalogs* Libraries are moving toward computer catalogs. From 1968 to 1996 the number of 3×5-inch cards sold to libraries by the Library of Congress declined by 78.19 million. The number of cards sold in 1996 was only 0.72% of the 1968 number. How many cards were sold in 1968 and in 1996? Round your answers to the nearest hundredth of a million. (Source: Library of Congress.)

67. *Geometry* (Refer to Example 10.) Approximate graphically the radius and height of a cylindrical container with a volume of 50 cubic inches and a lateral surface area of 65 square inches.

68. *Geometry* (Refer to Example 10.) Determine graphically if it is possible to construct a cylindrical container, *including* the top and bottom, with a volume of 38 cubic inches and a surface area of 38 square inches.

69. *Heart Rate* In one study the maximum heart rates of conditioned athletes were examined. A group of athletes was exercised to exhaustion. Let x represent an athlete's heart rate 5 seconds after stopping exercise and y this rate after 10 seconds. It was found that the maximum heart rate H for these athletes satisfied the following two equations.

$$H = 0.491x + 0.468y + 11.2$$
$$H = -0.981x + 1.872y + 26.4$$

If an athlete had a maximum heart rate of $H = 180$, determine x and y graphically. Interpret your answer. (Source: V. Thomas, *Science and Sport.*)

70. *Heart Rate* Repeat the previous exercise for an athlete with a maximum heart rate of 195.

71. *Roof Truss* (Refer to Example 9.) The forces or weights W_1 and W_2 exerted on each rafter for the roof truss shown in the figure are determined by the system of linear equations. Solve the system.

$$W_1 + \sqrt{2}W_2 = 300$$
$$\sqrt{3}W_1 - \sqrt{2}W_2 = 0$$

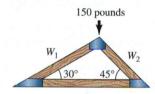

72. *Height and Weight* The relationship between a professional basketball player's height h in inches and weight w in pounds was modeled using two samples of players. The resulting equations that modeled each sample were $w = 7.46h - 374$ and $w = 7.93h - 405$. Assume that $65 \le h \le 85$.
  (a) Use each equation to predict the weight of a professional basketball player who is 6'11".
  (b) Determine graphically the height where the two models give the same weight.
  (c) For each model, what change in weight is associated with a 1-inch increase in height?

Joint Variation

Exercises 73 and 74: Approximate the constant of variation to the nearest hundredth.

73. The variable z varies jointly as the second power of x and the third power of y. When $x = 2$ and $y = 2.5$, $z = 31.9$.

74. The variable z varies jointly as the 1.5 power of x and the 2.1 power of y. When $x = 4$ and $y = 3.5$, $z = 397$.

75. *Wind Power* The electrical power generated by a windmill varies jointly as the square of the diameter of the blades and the cube of the wind velocity. If a windmill with 8-foot blades and a 10 mile-per-hour wind generates 2405 watts, how much power would be generated if the blades were 6 feet and the wind were 20 miles per hour?

76. *Strength of a Beam* The strength of a rectangular beam varies jointly as its width and the square of its thickness. If a beam 5.5 inches wide and 2.5 inches thick supports 600 pounds, how much can a similar beam that is 4 inches wide and 1.5 inches thick support?

77. *Volume of Wood* (Refer to Example 12.) One cord of wood contains 128 cubic feet. Estimate the number of cords in a tree that is 105 feet tall and has a diameter of 38 inches.

78. *Carpeting* The cost of carpet for a rectangular room varies jointly as its width and length. If a room 10 feet wide and 12 feet long costs $1560 to carpet, find the cost to carpet a room that is 11 feet by 23 feet. Interpret the constant of variation.

79. *Surface Area* Use the results of Example 2 to find a formula for $S(w, h)$ that calculates the surface area of a person's body if w is given in pounds and h is given in inches.

80. *Surface Area* Use the results of the previous exercise to solve Exercises 5 and 6.

Writing about Mathematics

81. Give an example of a quantity occurring in everyday life that can be computed by a function of more than one input. Identify the inputs and the output.

82. Give an example of a system of linear equations with two variables. Explain how to solve the system graphically and symbolically.

Extended and Discovery Exercises

Exercises 1–6: Colors on Computer Monitors   *Although it might not seem possible to describe color using mathematics, it is done every day on computer screens. Colored light can be broken down into three basic colors: red, green, and blue. Some computer monitors are capable of creating 256 intensities for each of these three colors, which may be numbered from 0 to 255. The number 0 indicates the absence of a color, while 255 represents the brightest intensity of a color. The color displayed on a screen can be computed by $f(r, g, b)$, where r represents the brightness of red light, g of green light, and b of blue light. The expression $f(255, 0, 0)$ indicates the brightest intensity of red with no green or blue light. Therefore, $f(255, 0, 0) = red$. In a similar manner, $f(0, 255, 0) = green$ and $f(0, 0, 255) = blue$. By mixing intensities of these three basic colors, thousands of colors can be generated. For example, $f(255, 255, 0) = yellow$, since red and green light combine to make yellow light.*
(Source: I. Kerlow, *The Art of 3-D Computer Animation and Imaging.*)

1. Make a conjecture about what color $f(255, 0, 255)$ represents.

2. Make a conjecture about what inputs create white light.

3. Turquoise is a greenish-blue color. Make a conjecture about inputs that create turquoise light.

4. Make a conjecture about whether $f(141, 43, 17)$ represents rust or cream color.

5. If there are 256 intensities for each basic color, how many colors can be created on a computer screen by mixing different intensities of these colors?

6. Some computers are capable of generating 65,536 intensities for each of the three basic colors. Approximate the number of colors that could be generated by this computer.

6.2 LINEAR SYSTEMS OF EQUATIONS AND INEQUALITIES IN TWO VARIABLES

Types of Linear Systems in Two Variables • The Elimination Method • Systems of Inequalities • Linear Programming

Introduction

Large systems of equations and inequalities involving thousands of variables are solved by economists, scientists, engineers, and mathematicians every day. Computers and sophisticated numerical methods are essential for finding their solutions. In this section we focus on systems involving two variables. Graphical, numerical, and symbolic techniques can be applied to equations in two variables. Many of the concepts presented in this section apply to larger systems of linear equations.

Types of Linear Systems in Two Variables

A system of linear equations in two variables can be written in the form

$$a_1 x + b_1 y = c_1$$
$$a_2 x + b_2 y = c_2,$$

where a_1, b_1, c_1, a_2, b_2, and c_2 are constants. The graph of this system consists of *two* lines in the xy-plane. The lines can intersect at a point, coincide, or be parallel, as illustrated in Figures 6.14–6.16. **Coincident lines** are identical lines and indicate that the two equations are equivalent and have the same graph.

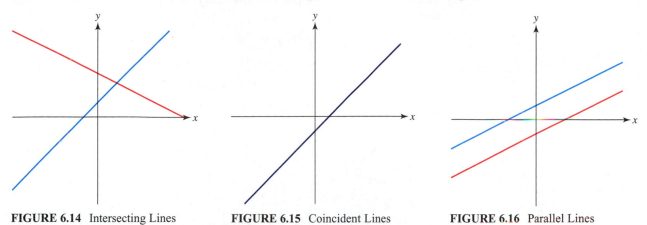

FIGURE 6.14 Intersecting Lines   **FIGURE 6.15** Coincident Lines   **FIGURE 6.16** Parallel Lines

A **consistent** system of linear equations has either one solution or infinitely many solutions. If the system has one solution, then the equations are **independent**, and they can be represented by lines that intersect at one point, as shown in Figure 6.14. If the system has infinitely many solutions, then the equations are **dependent**, and they can be represented by coincident lines, as illustrated in Figure 6.15. An **inconsistent** system has no solutions and can be represented by parallel lines, as shown in Figure 6.16.

■ **CLASS DISCUSSION**

Explain why a system of linear equations cannot have two or three solutions. ■

This discussion is summarized by the following.

> ### Possible graphs of a system of two linear equations in two variables
>
> 1. The graphs of the two equations are intersecting lines. The system is *consistent*. There is one solution, which is given by the coordinates of the point of intersection. In this case the equations are *independent*.
> 2. The graphs of the two equations are the same line. The system is *consistent*. There are infinitely many solutions, and the equations are *dependent*.
> 3. The graphs of the two equations consist of distinct parallel lines. The system is *inconsistent*. There are no solutions.

The Elimination Method

The substitution method, presented in Section 6.1, is a symbolic method for solving small systems of equations. A second symbolic method is the **elimination method**. The next example demonstrates this technique.

EXAMPLE 1   *Using elimination to solve a system*

Use elimination to solve each system of equations, if possible. Identify the system as consistent or inconsistent. If the system is consistent, state whether the equations are dependent or independent. Support your results graphically.

(a) $2x - y = -4$
$3x + y = -1$

(b) $4x - y = 10$
$-4x + y = -10$

(c) $x - y = 6$
$x - y = 3$

SOLUTION (a) *Symbolic Solution* We can eliminate the y-variable by adding the two equations.

$$
\begin{array}{l}
2x - y = -4 \\
\underline{3x + y = -1} \\
5x \quad\quad = -5 \quad \text{or} \quad x = -1
\end{array}
$$

Now the y-variable can be determined by substituting $x = -1$ into either equation.

$$2x - y = -4 \qquad \text{First equation}$$
$$2(-1) - y = -4 \qquad \text{Let } x = -1.$$
$$-y = -2 \qquad \text{Add 2.}$$
$$y = 2 \qquad \text{Multiply by } -1.$$

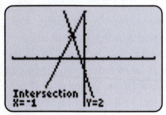

The solution is $(-1, 2)$. There is a unique solution so the system is consistent and the equations are independent.

Graphical Solution For a graphical solution, start by solving each equation for y.

$$2x - y = -4 \quad \text{is equivalent to} \quad y = \quad 2x + 4.$$
$$3x + y = -1 \quad \text{is equivalent to} \quad y = -3x - 1.$$

The graphs of $Y_1 = 2X + 4$ and $Y_2 = -3X - 1$ intersect at $(-1, 2)$, as shown in Figure 6.17.

(b) *Symbolic Solution* If we add the equations, we obtain the following result.

$$
\begin{array}{l}
4x - y = \quad 10 \\
\underline{-4x + y = -10} \\
\quad\quad 0 = 0
\end{array}
$$

The equation $0 = 0$ is always true. The two equations are equivalent: if we multiply the first equation by -1, we obtain the second equation. Thus there are infinitely many solutions, and we can write the solution set in set builder notation.

$$\{(x, y) \mid 4x - y = 10\}$$

Some examples of solutions are $(3, 2)$, $(4, 6)$, and $(1, -6)$. The system is consistent and the equations are dependent.

Graphical Solution For a graphical solution, start by solving each equation for y.

$$4x - y = 10 \quad \text{is equivalent to} \quad y = 4x - 10.$$
$$-4x + y = -10 \quad \text{is equivalent to} \quad y = 4x - 10.$$

Clearly, the graphs of $Y_1 = 4X - 10$ and $Y_2 = 4x - 10$, shown in Figure 6.18, are identical. Their graphs are coincident.

(c) *Symbolic Solution* If we subtract the equations, we obtain the following result.

$$
\begin{array}{l}
x - y = 6 \\
\underline{x - y = 3} \\
\quad 0 = 3
\end{array}
$$

Graphing Calculator Help

To find a point of intersection, see Appendix B (page AP-11).

$[-6, 6, 1]$ by $[-4, 4, 1]$

FIGURE 6.17

$[-30, 30, 10]$ by $[-20, 20, 10]$

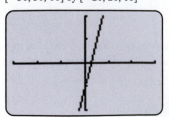

FIGURE 6.18

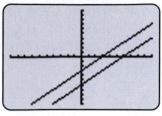

FIGURE 6.19

The equation $0 = 3$ is never true. Therefore there are no solutions, and the system is inconsistent.

Graphical Solution   For a graphical solution start by solving each equation for y.

$$x - y = 6 \quad \text{is equivalent to} \quad y = x - 6.$$
$$x - y = 3 \quad \text{is equivalent to} \quad y = x - 3.$$

The graphs of $Y_1 = X - 6$ and $Y_2 = X - 3$, shown in Figure 6.19, are parallel. The lines are parallel and never intersect, so there are no solutions.

The next example illustrates how a linear system can occur in an application involving real data. We solve this system symbolically, graphically, and numerically.

EXAMPLE 2   *Using the elimination method*

Title IX is landmark legislation that prohibits sex discrimination in sports programs. In 1997 the national average spent on two varsity athletes, one female and one male, was $6050 for Division I-A schools. However, average expenditures for a male athlete exceeded those for a female athlete by $3900. Determine how much was spent per varsity athlete for each gender. Give graphical and numerical support.   (Source: *USA Today.*)

SOLUTION   *Symbolic Solution*   Let x represent average expenditures per male athlete and y average expenditures per female athlete in 1997. Since the average amount spent on one female and one male athlete was $6050, the equation $(x + y)/2 = 6050$ must be satisfied. The expenditures for a male athlete exceeded those for a female athlete by $3900. Thus $x - y = 3900$.

$$\frac{x + y}{2} = 6050$$
$$x - y = 3900$$

Mutiply the first equation by 2 and then add the resulting equations.

$$
\begin{aligned}
x + y &= 12{,}100 \\
\underline{x - y} &= \underline{3900} \\
2x &= 16{,}000
\end{aligned}
$$

By adding the equations, we have *eliminated* the y-variable. The solution to $2x = 16{,}000$ is $x = 8000$. Thus the average expenditure per male athlete in 1997 was $8000. To determine y, we can substitute $x = 8000$ into either equation.

$$8000 - y = 3900 \quad \text{implies} \quad y = 4100.$$

The average amount spent on each female athlete was $4100.

Graphical Solution Solve each equation for y and graph $Y_1 = 12100 - X$ and $Y_2 = X - 3900$. Their graphs intersect at $(8000, 4100)$. See Figure 6.20. This agrees with the symbolic solution.

Numerical Solution Make a table of Y_1 and Y_2. When $x = 8000$, $Y_1 = Y_2 = 4100$. See Figure 6.21.

[0, 15,000, 5000] by [0, 15,000, 5000]

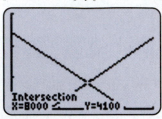

FIGURE 6.20　　　　**FIGURE 6.21**

Sometimes multiplication is performed before elimination is used, as illustrated in the next example.

EXAMPLE 3 *Multiplying before using elimination*

Solve each system of equations using elimination. Support your results graphically.

(a) $2x - 3y = 18$ 　　(b) $5x + 10y = 10$
　　$5x + 2y = 7$ 　　　　$x + 2y = 2$

SOLUTION (a) *Symbolic Solution* If we multiply the first equation by 2 and the second equation by 3, then the y-coefficients become -6 and 6. Addition eliminates the y-variable.

$$4x - 6y = 36$$
$$\underline{15x + 6y = 21}$$
$$19x \qquad = 57 \quad \text{or} \quad x = 3$$

Substituting $x = 3$ into $2x - 3y = 18$ results in

$$2(3) - 3y = 18 \quad \text{or} \quad y = -4.$$

The solution is $(3, -4)$.

Graphical Solution To show graphical support let $Y_1 = (2X - 18)/3$ and $Y_2 = (7 - 5X)/2$. Their graphs intersect at $(3, -4)$. See Figure 6.22.

(b) *Symbolic Solution* If the second equation is multiplied by 5, subtraction eliminates both variables.

$$5x + 10y = 10$$
$$\underline{5x + 10y = 10}$$
$$0 = 0$$

[−10, 10, 1] by [−10, 10, 1]

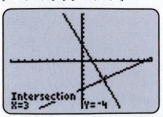

FIGURE 6.22

The statement $0 = 0$ is always true. This means that the equations are dependent. There are infinitely many solutions. The solution set is $\{(x, y) \mid x + 2y = 2\}$.

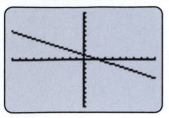

FIGURE 6.23

Graphical Solution   To give graphical support, let $Y_1 = (10 - 5X)/10$ and $Y_2 = (2 - X)/2$. Their graphs are identical. See Figure 6.23.

Systems of Inequalities

A linear inequality in two variables can be written as

$$ax + by \leq c,$$

where a, b, and c are constants with a and b not equal to zero. (The symbol $\leq$ can be replaced by $\geq$, $<$, or $>$.) If an ordered pair (x, y) makes the inequality a true statement, then (x, y) is a solution. The set of all solutions is called the *solution set.* The graph of an inequality includes all points (x, y) in the solution set.

The graph of a linear inequality is a **half-plane**, which may include the boundary. For example, to graph the linear inequality $3x - 2y \leq 6$, solve the inequality for y. This results in $y \geq \frac{3}{2}x - 3$. Then graph the line $y = \frac{3}{2}x - 3$, which is the boundary. The inequality includes the line and the half-plane above the line, as shown in Figure 6.24.

A second approach is to write the equation $3x - 2y = 6$ as $y = \frac{3}{2}x - 3$ and graph. To determine which half-plane to shade, select a **test point**. For example, the point $(0, 0)$ satisfies the inequality $3x - 2y \leq 6$, so shade the region containing $(0, 0)$.

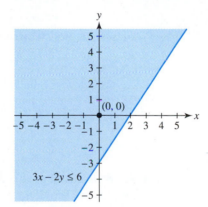

FIGURE 6.24

EXAMPLE 4   *Modeling plant growth*

Two factors that have a critical effect on plant growth are temperature and precipitation. If a region has too little precipitation, it will be a desert. Forests tend to grow in regions where trees can exist at relatively low temperatures and there is sufficient rainfall. At other levels of precipitation and temperature, grasslands may prevail. Figure 6.25 illustrates the relationship among forests, grasslands, and deserts, as suggested by annual average temperature T in degrees Fahrenheit and precipitation P in inches.   (Source: A. Miller and J. Thompson, *Elements of Meteorology.*)
(a) Determine a system of linear inequalities that describes where grasslands are likely to occur.
(b) Bismarck, North Dakota, has annual average temperature of 40°F and precipitation of 15 inches. According to the graph, what type of plant growth would you expect near Bismarck? Do these values satisfy the system of inequalities from part (a)?

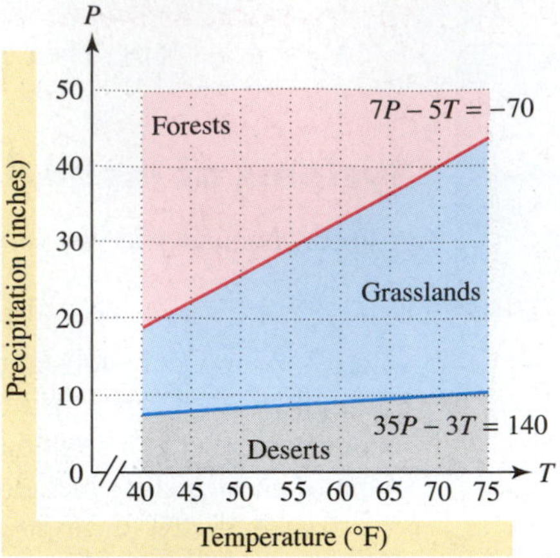

FIGURE 6.25 Effect of Temperature and Precipitation on Plant Growth

SOLUTION   **(a)** Grasslands occur for ordered pairs (T, P) lying between the two lines in Figure 6.25. The boundary between deserts and grassland is determined by the equation $35P - 3T = 140$. Solving for P results in

$$P = \frac{3}{35}T + \frac{140}{35}.$$

Grasslands grow where values of P are above the line. This region is described by $P > \frac{3}{35}T + \frac{140}{35}$, or equivalently, $35P - 3T > 140$. In a similar manner, the region below the boundary between grasslands and forests is represented by $7P - 5T < -70$. Thus grasslands satisfy the following system of inequalities.

$$35P - 3T > 140$$
$$7P - 5T < -70$$

(b) For Bismarck, $T = 40$ and $P = 15$. From Figure 6.25, it appears that the point $(40, 15)$ lies between the two lines, so the graph predicts that grasslands will exist around Bismarck. Substituting these values for T and P into the system of inequalities results in the following true statements.

$$35(15) - 3(40) = 405 > 140$$
$$7(15) - 5(40) = -95 < -70$$

The temperature and precipitation for Bismarck satisfies the system of inequalities for grasslands.

■ **CLASS DISCUSSION**

Use Example 4 to find an inequality that describes levels of temperature and precipitation in deserts. ■

In Section 6.1, we learned that systems of equations could be either linear or nonlinear. In an analogous manner, systems of inequalities can be linear or nonlinear. The next example illustrates a system of each type. Both are solved graphically.

EXAMPLE 5   *Solving systems of inequalities graphically*

Solve each system of inequalities by shading the solution set. Use the graph to identify one solution.

(a) $y > x^2$
    $x + y < 4$

(b) $x + 3y \leq 9$
    $2x - y \leq -1$

SOLUTION   **(a)** This is a nonlinear system. Graph the parabola $y = x^2$ and the line $y = 4 - x$. Since $y > x^2$ and $y < 4 - x$, the region satisfying the system lies above the parabola and below the line. It does not include the boundary, which is shown using a dashed line and curve. See Figure 6.26.

Any point in the shaded region represents a solution. For example, $(0, 2)$ lies in the shaded region and is a solution since $x = 0$ and $y = 2$ satisfy both inequalities.

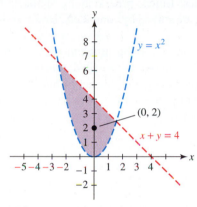

FIGURE 6.26

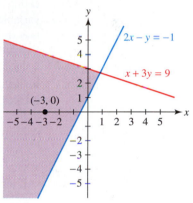

FIGURE 6.27

Graphing Calculator Help

To shade a graph, see Appendix B (page AP-18).

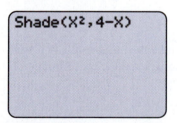

FIGURE 6.28

$[-5, 5, 1]$ by $[-2, 8, 1]$

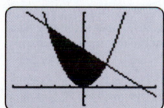

FIGURE 6.29

(b) Begin by solving each linear inequality for y.

$$y \leq \frac{9 - x}{3}$$

$$y \geq 2x + 1$$

Graph $y = \dfrac{9 - x}{3}$ $\left(\text{or } y = -\dfrac{1}{3}x + 3\right)$ and $y = 2x + 1$. The region satisfying the system is below the line $x + 3y = 9$ and above the line $2x - y = -1$. Because equality is included, the boundaries are part of the region, which are shown as solid lines in Figure 6.27. The point $(-3, 0)$ is a solution since it lies in the shaded region and satisfies both inequalities.

Graphing calculators can be used to shade regions in the *xy*-plane. See Figure 6.28. The solution set shown in Figure 6.26 is also shown in Figure 6.29, where a graphing calculator has been used. However, the boundary is not dashed.

Figures 6.30 and 6.31 on the following page show a different method of shading a solution set; we have shaded the area below $Y_1 = (9 - X)/3$ and above $Y_2 = 2X + 1$. The solution set is the region shaded with both vertical *and* horizontal lines and corresponds to the shaded region in Figure 6.27.

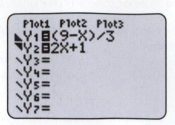

FIGURE 6.30

$[-6, 6, 1]$ by $[-6, 6, 1]$

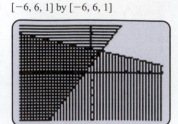

FIGURE 6.31

Linear Programming

An important application of mathematics to business and social sciences is linear programming. **Linear programming** is a procedure used to optimize quantities such as cost and profit. It was developed during World War II as a method of efficiently allocating supplies. Linear programming applications frequently contain thousands of variables and are solved by computers. However, here we focus on problems involving two variables.

A linear programming problem consists of a linear **objective function** and a system of linear inequalities called **constraints**. The solution set for the system of linear inequalities is called the set of **feasible solutions**. The objective function describes a quantity that is to be optimized. For example, linear programming is often used to maximize profit or minimize cost. The following example illustrates these concepts.

EXAMPLE 6   *Finding maximum profit*

Suppose a small company manufacturers two products—radios and CD players. Each radio results in a profit of $15 and each CD player provides a profit of $35. Due to demand, the company must produce at least 5 and not more than 25 radios per day. The number of radios cannot exceed the number of CD players, and the number of CD players cannot exceed 30. How many of each should the company manufacture to obtain maximum profit?

SOLUTION   Let x represent the number of radios produced daily and y the number of CD players produced daily. Since the profit from x radios is $15x$ dollars and the profit from y CD players is $35y$ dollars, the total daily profit P is given by

$$P = 15x + 35y.$$

The company produces from 5 to 25 radios per day, so the inequalities

$$x \geq 5 \quad \text{and} \quad x \leq 25$$

must be satisfied. The requirements that the number of radios cannot exceed the number of CD players and the number of CD players cannot exceed 30 indicate that

$$x \leq y \quad \text{and} \quad y \leq 30.$$

Since the number of radios and CD players cannot be negative, we have

$$x \geq 0 \quad \text{and} \quad y \geq 0.$$

Listing all the constraints on production gives

$$x \geq 5, \quad x \leq 25, \quad y \leq 30, \quad x \leq y, \quad x \geq 0, \quad \text{and} \quad y \geq 0.$$

Graphing these constraints results in the shaded region shown in Figure 6.32. This shaded region is the set of feasible solutions. The vertices (or corners) of this region are (5, 5), (25, 25), (25, 30), and (5, 30).

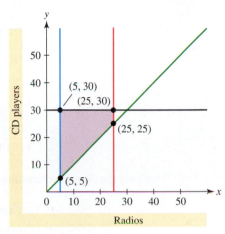

FIGURE 6.32

It can be shown that maximum profit occurs at a vertex of the region of feasible solutions. Thus we evaluate P at each vertex, as shown in Table 6.1.

TABLE 6.1

Vertex	$P = 15x + 35y$	
(5, 5)	$15(5) + 35(5) = 250$	
(25, 25)	$15(25) + 35(25) = 1250$	
(25, 30)	$15(25) + 35(30) = 1425$	⟵ Maximum Profit
(5, 30)	$15(5) + 35(30) = 1125$	

The maximum value of P is 1425 at vertex (25, 30). Thus the maximum profit is $1425 and it occurs when 25 radios and 30 CD players are manufactured.

The following holds for linear programming problems.

Fundamental theorem of linear programming

If the optimal value for a linear programming problem exists, then it occurs at a vertex of the region of feasible solutions.

EXAMPLE 7   *Finding the minimum of an objective function*

Find the minimum value of $C = 2x + 3y$ subject to the following constraints.

$$x \geq 0 \quad \text{and} \quad y \geq 0$$
$$x + y \geq 4 \quad \text{and} \quad 2x + y \leq 8$$

SOLUTION   Sketch the region determined by the constraints and find all vertices. See Figure 6.33.

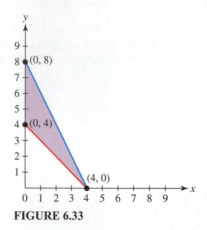

FIGURE 6.33

Evaluate the objective function C at each vertex, as shown in Table 6.2.

TABLE 6.2

Vertex	$C = 2x + 3y$	
$(4, 0)$	$2(4) + 3(0) = 8$	⟵ Minimum Value
$(0, 8)$	$2(0) + 3(8) = 24$	
$(0, 4)$	$2(0) + 3(4) = 12$	

The minimum value for C is 8 and it occurs at vertex $(4, 0)$, or when $x = 4$ and $y = 0$.

The following procedure describes how to solve a linear programming problem.

Solving a linear programming problem

STEP 1:   Read the problem carefully. Consider making a table to display the information given.

STEP 2:   Use the table to write the objective function and all the constraints.

STEP 3:   Sketch a graph of the region of feasible solutions. Identify all vertices or corner points.

STEP 4:   Evaluate the objective function at each vertex. A maximum (or a minimum) occurs at a vertex. *Note:* If the region is unbounded, a maximum (or minimum) may not exist.

EXAMPLE 8   *Minimizing cost*

A breeder is buying two brands of food, A and B, for her animals. Each serving is a mixture of the two foods and should contain at least 40 grams of protein and 30 grams of fat. Brand A costs 90 cents per unit and Brand B costs 60 cents per unit. Each unit of Brand A contains 20 grams of protein and 10 grams of fat, whereas each unit of Brand B contains 10 grams of protein and 10 grams of fat. Determine how much of each brand should be bought to obtain a minimum cost per serving.

SOLUTION **STEP 1:** After reading the problem carefully, begin by listing the information, as illustrated in Table 6.3.

TABLE 6.3

Brand	Units	Protein	Fat	Cost
A	x	20	10	90¢
B	y	10	10	60¢
Minimum		40	30	

STEP 2: If x units of Brand A are purchased at 90¢ per unit and if y units of Brand B are purchased at 60¢ per unit, then the cost C is given by $C = 90x + 60y$. Each serving requires at least 40 grams of protein. If x units of Brand A are bought, each containing 20 grams of protein, if y units of Brand B are bought, each containing 10 grams of protein, and if each serving requires at least 40 grams of protein, then we can write $20x + 10y \geq 40$. Similarly, since each serving requires at least 30 grams of fat, we can write $10x + 10y \geq 30$. The linear programming problem can be written as follows.

Minimize:	$C = 90x + 60y$	Cost
Subject to:	$20x + 10y \geq 40$	Protein
	$10x + 10y \geq 30$	Fat
	$x \geq 0$	
	$y \geq 0$	

STEP 3: The region that contains the feasible solutions is shown in Figure 6.34. The vertices for this region are $(0, 4)$, $(1, 2)$, and $(3, 0)$.

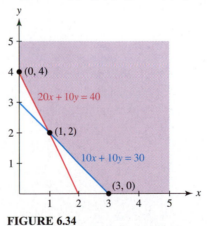

FIGURE 6.34

STEP 4: Evaluate the objective function C at each vertex, as shown in Table 6.4.

TABLE 6.4

Vertex	$C = 90x + 60y$	
$(0, 4)$	240	
$(1, 2)$	210	⟵ Minimum Cost
$(3, 0)$	270	

The minimum cost occurs when 1 unit of Brand A and 2 units of Brand B are mixed, at a cost of $2.10 per serving.

Putting it all Together

6.2

Systems of linear equations can have no solutions, one solution, or an infinite number of solutions. The substitution and elimination methods may be used to solve systems of equations. Graphical and numerical methods can also be applied to systems of equations and inequalities involving two variables.

The following table summarizes some important mathematical concepts from this section.

Concept	Comments	Example
Consistent system of linear equations in two variables.	A consistent linear system has either one or infinitely many solutions. Its graph is either intersecting lines or identical lines.	$$\begin{aligned} x + y &= 10 \\ x - y &= 4 \\ \hline 2x &= 14 \end{aligned}$$ Solution is $x = 7, y = 3$. Since the equations have exactly one solution, they are independent.
Dependent system of linear equations in two variables	A dependent linear system has infinitely many solutions. The graph consists of two identical lines.	$2x + 2y = 2$ and $x + y = 1$ are equivalent (dependent) equations. The solution set is $\{(x, y) \mid x + y = 1\}$.
Inconsistent system of linear equations in two variables	An inconsistent linear system has no solutions. The graph is two parallel lines.	$$\begin{aligned} x + y &= 1 \\ x + y &= 2 \quad \text{(Subtract)} \\ \hline 0 &= -1 \quad \text{(Always false)} \end{aligned}$$ The solution set is empty.
Elimination method	By performing arithmetic operations on a system, a variable is eliminated.	$$\begin{aligned} 2x + y &= 5 \\ x - y &= 1 \quad \text{(Add.)} \\ \hline 3x &= 6 \quad \text{so} \quad x = 2 \end{aligned}$$ and $y = 1$.
Linear inequality in two variables	$ax + by \le c$ ($\le$ may be replaced by $<$, $>$, or $\ge$.) The solution set is typically a shaded region in the xy-plane.	$$\begin{aligned} 2x - 3y &\le 12 \\ -2x + y &\le 4 \end{aligned}$$

Concept	Comments	Example
Linear programming	In a linear programming problem, the maximum or minimum of an objective function is found, subject to a set of constraints. If a solution exists, it occurs at a vertex in the region of feasible solutions.	Maximize the objective function $$P = 2x + 3y,$$ subject to the following constraints. $$2x + y \leq 6$$ $$x + 2y \leq 6$$ $$x \geq 0, \qquad y \geq 0$$ The maximum of $P = 10$ occurs at vertex $(2, 2)$.

6.2 EXERCISES

Consistent and Inconsistent Linear Systems

Exercises 1–4: The figure represents a system of linear equations. Classify each system as consistent or inconsistent. Solve the system graphically and symbolically, if possible.

1.
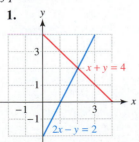

2.
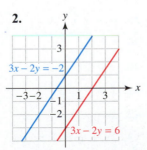

3.
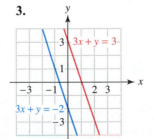

4.
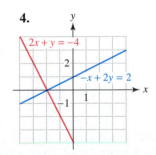

The Elimination Method

Exercises 5–16: Use elimination to solve each system of equations, if possible. Identify the system as consistent or inconsistent. If the system is consistent, state whether the equations are dependent or independent. Support your results graphically or numerically.

5. $x + y = 20$
     $x - y = 8$

6. $2x + y = 15$
     $x - y = 0$

7. $x + 3y = 10$
     $x - 2y = -5$

8. $4x + 2y = 10$
     $-2x - y = 10$

9. $x + y = 500$
     $0.05x + 0.05y = 25$

10. $2x + 3y = 4$
     $x - 2y = -5$

11. $5x + y = -5$
     $7x - 3y = -29$

12. $2x + 3y = 5$
     $5x - 2y = 3$

13. $2x + 4y = 7$
     $-x - 2y = 5$

14. $4x - 3y = 5$
     $3x + 4y = 2$

15. $2x + 3y = 2$
     $x - 2y = -5$

16. $x - 3y = 1$
     $2x - 6y = 2$

Exercises 17–28: Solve the system, if possible.

17. $\frac{1}{2}x - y = 5$

     $x - \frac{1}{2}y = 4$

18. $\frac{1}{2}x - \frac{1}{3}y = 1$

     $\frac{1}{3}x - \frac{1}{2}y = 1$

19. $2x - 3y = \frac{5}{3}$

     $6x - 2y = \frac{8}{3}$

20. $4x + 3y = 8$

     $-2x + 6y = 1$

21. $7x - 3y = -17$

     $-21x + 9y = 51$

22. $-\frac{1}{3}x + \frac{1}{6}y = -1$

     $2x - y = 6$

23. $\frac{2}{3}x + \frac{4}{3}y = \frac{1}{3}$

     $-2x - 4y = 5$

24. $5x - 2y = 7$

     $10x - 4y = 6$

25. $-5x + 2y = 15$
     $x - 6y = -17$

26. $2x - 3y = 1$
     $3x - 2y = 2$

27. $0.2x + 0.3y = 8$
     $-0.4x + 0.2y = 0$

28. $\frac{1}{2}x - \frac{3}{2}y = 9$

     $-\frac{3}{2}x + \frac{5}{2}y = -17$

Exercises 29–32: Write a system of linear equations with two variables whose solution satisfies the problem. State what each variable represents. Then solve the system

   (a) *graphically, and*
   (b) *symbolically using elimination.*

29. The screen of a rectangular television set is 2 inches wider than it is high. If the perimeter of the screen is 38 inches, find its dimensions.

30. The sum of two numbers is 300 and their difference is 8. Find the two numbers.

31. Admission prices to a movie are \$4 for children and \$7 for adults. If 75 tickets were sold for \$456, how many of each type of ticket were sold?

32. A sample of 16 dimes and quarters has a value of \$2.65. How many of each type of coin are there?

Inequalities

Exercises 33–36: Match the system of inequalities with the appropriate graph (a–d). Use the graph to identify a solution.

33. $x + y \geq 2$
     $x - y \leq 1$

34. $2x - y > 0$
     $x - 2y \leq 1$

35. $\frac{1}{2}x^3 - y > 0$

     $2x - y \leq 1$

36. $x^2 + y \leq 4$

     $x^2 - y \leq 2$

a.

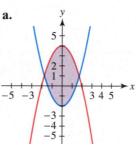

b.

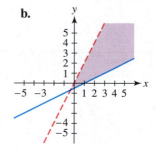

c.

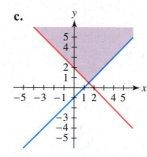

d.

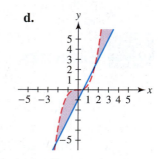

Exercises 37–44: Graph the inequality.

37. $x \geq y$

38. $y > -3$

39. $x < 1$

40. $y > 2x$

41. $x + y \leq 2$

42. $x + y > -3$

43. $2x + y > 4$

44. $2x + 3y \leq 6$

Exercises 45–50: Graph each system of inequalities. Use the graph to identify a solution.

45. $y \geq x^2$
$x + y \leq 6$

46. $y \leq \sqrt{x}$
$y \geq 1$

47. $x + 2y > -2$
$x + 2y < 5$

48. $x - y \leq 3$
$x + y \leq 3$

49. $x^2 + y^2 \leq 16$
$x + y < 2$

50. $x^2 + y \leq 4$
$x^2 - y \leq 3$

Exercises 51–58: Graph the system of inequalities.

51. $x + 2y \leq 4$
$2x - y \geq 6$

52. $3x - y \leq 3$
$x + 2y \leq 2$

53. $3x + 2y < 6$
$x + 3y \leq 6$

54. $4x + 3y \geq 12$
$2x + 6y \geq 4$

55. $x - 2y \geq 0$
$x - 3y \leq 3$

56. $2x - 4y \geq 4$
$x + y \leq 0$

57. $x^2 + y^2 \leq 4$
$y \geq 1$

58. $x^2 - y \leq 0$
$x^2 + y^2 \leq 6$

Applications

59. *Student Loans* A student takes out two loans totaling $3000 to help pay for college expenses. One loan is at 8% interest, and the other is at 10%. Interest for both loans is compounded annually.
(a) If the first-year interest is $264, determine a system of linear equations whose solution is the amount of each loan.
(b) Find the amount of each loan.

60. *Student Loans* (Refer to the previous exercise.) Suppose that both loans have an interest rate of 10% and the total first-year interest is $300. If possible, determine the amount of each loan. Interpret your results.

61. *Student Loans* (Refer to the previous two exercises.) Suppose that both loans are at 10% and the total annual interest is $264. If possible, determine the amount of each loan. Interpret your results.

62. *Maximizing Area* Suppose a rectangular pen for a pet is to be made using 40 feet of fence. Let *l* represent its length and *w* its width with $l \geq w$.
(a) Find the dimensions that result in an area of 91 square feet.
(b) Write a formula for the area *A* in terms of *w*. Graph *A*.

(c) What is the maximum area possible for the pen? Interpret this result.

63. *Television Sets* In 1999 there were 32 million television sets sold. For every 10 television sets sold with stereo sound, 19 television sets were sold without stereo sound. (Source: Consumer Electronics.)
(a) Write a linear system whose solution gives the number of television sets sold with stereo sound and the number of television sets sold without stereo sound.
(b) Solve the system and interpret the result.

64. *Charitable Giving* In 1995 corporations and individuals gave $123.6 billion to charitable organizations. Individuals gave 15.7 times more than corporations. Use elimination to determine how much individuals gave and how much corporations gave. Round answers to the nearest tenth of a billion dollars. (Source: Giving USA '95 AAFRC Trust for Philanthropy.)

65. *Meat Consumption* In 1994 the average person in the United States ate 4.4 times as much beef as turkey. The average consumption of beef and turkey was 77.8 pounds per person. Use elimination to determine the average annual consumption per person for each type of meat. Round your answers to the nearest tenth of a pound.

66. In 1996 and 1997 total world production of motor vehicles reached 104.8 million. There were 2.1 million more vehicles produced in 1997 than in 1996. Determine how many motor vehicles were produced in each year. (Source: American Automobile Manufacturers Association.)

67. *Air Speed* A jet airliner travels 1680 miles in 3 hours with a tail wind. The return trip, into the wind, takes 3.5 hours. Find both the air speed of the jet and the wind speed. (*Hint:* First find the ground speed of the airplane in each direction.)

68. *River Current* A tugboat can push a barge 60 miles upstream in 15 hours. The same tugboat and barge can make the return trip downstream in 6 hours. Determine the speed of the current in the river.

69. *Traffic Control* The figure shows two intersections labeled A and B that involve one-way streets. The numbers and variables represent the average traffic flow rates measured in vehicles per hour. For example, an average of 500 vehicles per hour enter intersection A from the west, whereas 150 vehicles per hour enter this intersection from the north. A stoplight will control the unknown traffic flow denoted by the variables x and y. Use the fact that the number of vehicles entering an intersection must equal the number leaving to determine x and y.

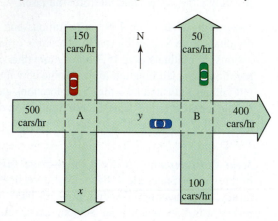

70. *Traffic Control* (Refer to the previous exercise and accompanying figure.) Suppose that the number of vehicles entering intersection A from the west varies between 400 and 600. If all other traffic flows remain the same as in the figure, what effect does this have on the ranges of the values for x and y?

Exercises 71–74: **Weight and Height** *The following graph shows a weight and height chart. The weight w is listed in pounds and the height h in inches. The shaded area is a recommended region.* (Source: Department of Agriculture.)

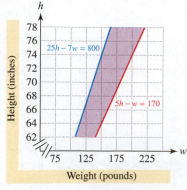

71. What does this chart indicate about an individual who weighs 125 pounds and is 70 inches tall?

72. If a person is 74 inches tall, use the graph to estimate the recommended weight range.

73. Use the graph to find a system of linear inequalities that describes the recommended region.

74. Explain why inequalities, rather than equalities, are more appropriate to describe recommended weight and height combinations.

Linear Programming

Exercises 75 and 76: The graph shows a region of feasible solutions. Find the maximum and minimum values of P over this region.

75. $P = 3x + 5y$

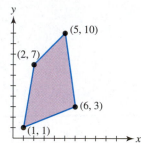

76. $P = 6x + y$

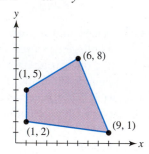

Exercises 77–80: The graph shows a region of feasible solutions. Find the maximum and minimum of C over this region.

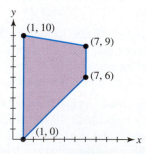

77. $C = 3x + 5y$ **78.** $C = 5x + 5y$

79. $C = 10y$ **80.** $C = 3x - y$

Exercises 81–82: Write a system of linear inequalities that describes the shaded region.

81.   **82.**

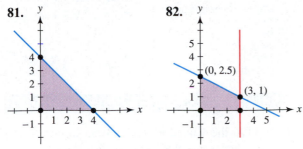

83. Find the minimum value of $C = 4x + 2y$ subject to the following constraints.

$$x \geq 0$$
$$y \geq 0$$
$$x + y \geq 3$$
$$2x + 3y \leq 12$$

84. Find the maximum value of $P = 3x + 5y$ subject to the following constraints.

$$x \geq 0$$
$$y \geq 0$$
$$3x + y \leq 8$$
$$x + 3y \leq 8$$

Exercises 85–86: If possible, maximize and minimize z subject to the given constraints.

85. $z = 7x + 6y$

$$x \geq 0$$
$$y \geq 0$$
$$x + y \leq 8$$
$$x + y \geq 4$$

86. $z = 8x + 3y$

$$x \geq 0$$
$$y \geq 0$$
$$4x + y \geq 12$$
$$x + 2y \geq 6$$

87. *Maximizing Profit*   Rework Example 6 if the profit from each radio is $20 and the profit from each CD player is $15.

88. *Minimizing Cost*   (Refer to Example 8.) A breeder is mixing Brand A and Brand B. Each serving should contain at least 60 grams of protein and 30 grams of fat. Brand A costs 80 cents per unit and Brand B costs 50 cents per unit. Each unit of Brand A contains 15 grams of protein and 10 grams of fat, whereas each unit of Brand B contains 20 grams of protein and 5 grams of fat. Determine how much of each food should be bought to achieve a minimum cost per serving.

89. *Maximizing Revenue*   A refinery produces both gasoline and fuel oil, and sells gasoline for $1.00 per gallon and fuel oil for $0.90 per gallon. The refinery can produce at most 600,000 gallons a day, but must produce at least 2 gallons of fuel oil for every gallon of gasoline. Furthermore, at least 150,000 gallons of fuel oil must be produced each day for the coming winter. Determine how much of each type of fuel should be produced to maximize revenue.

90. *Maximizing Storage*   An office manager wants to buy filing cabinets. Cabinet X costs $100, requires 6 square feet of floor space, and holds 8 cubic feet. Cabinet Y costs $200, requires 8 square feet of floor space, and holds 12 cubic feet. No more than $1400 can be spent, and the office has room for no more than 72 square feet of cabinets. The office manager wants the maximum storage capacity within the limits imposed by funds and space. How many of each type of cabinet should be bought?

91. *Maximizing Profit*   A business manufactures two parts, X and Y. Machines A and B are needed to make each part. To make part X, machine A is needed 4 hours and machine B is needed 2 hours. To make part Y, machine A is needed 1 hour and machine B is needed 3 hours. Machine A is available for 40 hours each week and machine B is available for 30 hours. The profit from part X is $500 and the profit from part Y is $600. How many parts of each type should be made to maximize weekly profit?

92. *Minimizing Cost*   Two substances, X and Y, are found in pet food. Each substance contains the ingredients A and B. Substance X is 20% ingredient A and 50% ingredient B. Substance Y is 50% ingredient A and 30% ingredient B. The cost of substance X is $2 per pound and the cost of substance Y is $3 per pound. The pet store needs at least 251 pounds of ingredient A and at least 200 pounds of ingredient B. If cost is to be minimal, how many pounds of each substance should be ordered? Find the minimum cost.

Writing about Mathematics

93. Give the general form of a system of linear equations in two variables. Discuss what distinguishes a system of linear equations from a system of nonlinear equations.

94. Discuss what the words *consistent* and *inconsistent* mean with regards to linear systems. Give an example of each type of system.

CHECKING BASIC CONCEPTS FOR SECTIONS 6.1 AND 6.2

1. Evaluate $f(13, 18)$ if

$$f(x, y) = \sqrt{(x - 1)^2 + (y - 2)^2}.$$

(The function f computes the distance that the point (x, y) is from the point $(1, 2)$.)

2. Solve the system of nonlinear equations using the method of substitution.

$$2x^2 - y = 0$$
$$3x + 2y = 7$$

3. Solve the system of inequalities graphically. Use the graph to identify one solution.

$$x^2 - y < 3$$
$$x - y \geq 1$$

4. In 1996 the United States and the former Soviet Union had a combined total of 13,820 strategic nuclear warheads. The United States had 480 more warheads than the former Soviet Union. (Sources: Arms Control Association, Natural Resources Defense Council.) Find a system of linear equations whose solution gives the number of nuclear warheads for each country. Solve the system.

6.3 SOLUTIONS OF LINEAR SYSTEMS USING MATRICES

Representing Systems of Linear Equations with Matrices • Row-Echelon Form • Gaussian Elimination • Solving Systems of Linear Equations with Technology

Introduction

The methods of substitution and elimination can be applied to linear systems that involve more than two variables. As the number of variables increases, these methods become more cumbersome and less efficient. However, by combining these two methods, they provide a state-of-the-art numerical method, capable of solving systems of linear equations that contain thousands of variables. This numerical method is called *Gaussian elimination with backward substitution*. Even though this method dates back to Carl Friedrich Gauss (1777–1855), it continues to be one of the most efficient methods for solving systems of linear equations.

Representing Systems of Linear Equations with Matrices

Arrays of numbers occur frequently in many different situations. Spreadsheets often make use of arrays, where data are displayed in a tabular format. A **matrix** is a rectangular array of elements. The following are examples of matrices whose elements are real numbers.

$$
\begin{bmatrix} 4 & -7 \\ -2 & 9 \end{bmatrix}
\quad
\begin{bmatrix} -1 & -5 & 3 \\ 1.2 & 0 & -1.3 \\ 4.1 & 5 & 7 \end{bmatrix}
\quad
\begin{bmatrix} -3 & -6 & 9 & 5 \\ \sqrt{2} & -8 & -8 & 0 \\ 3 & 0 & 19 & -7 \\ -11 & -3 & 7 & 8 \end{bmatrix}
\quad
\begin{bmatrix} 5 & -2 \\ -2 & \pi \\ 1 & -1 \end{bmatrix}
\quad
\begin{bmatrix} 1 & -0.5 & 9 \\ 5 & 0.4 & -3 \end{bmatrix}
$$

2×2 3×3 4×4 3×2 2×3

The dimension of a matrix is given much like the dimensions of a rectangular room. We might say a room is n feet long and m feet wide. The **dimension** of a matrix is $n \times m$ (n by m), if it has n rows and m columns. For example, the last matrix has a dimension of 2×3, because it has 2 rows and 3 columns. If the number of rows and columns are equal, the matrix is a **square matrix**. The first three matrices are square matrices.

Matrices are frequently used to represent systems of linear equations. A linear system with three equations and three variables can be written as follows.

$$
\begin{aligned}
a_1 x + b_1 y + c_1 z &= d_1 \\
a_2 x + b_2 y + c_2 z &= d_2 \\
a_3 x + b_3 y + c_3 z &= d_3
\end{aligned}
$$

The a_k, b_k, c_k, and d_k are constants and x, y, and z are variables. The coefficients of the variables can be represented using a 3×3 matrix. This matrix is called the **coefficient matrix** of the linear system.

$$
\begin{bmatrix} a_1 & b_1 & c_1 \\ a_2 & b_2 & c_2 \\ a_3 & b_3 & c_3 \end{bmatrix}
$$

If we enlarge the matrix to include the constants d_k, the system may be written as

$$
\left[\begin{array}{ccc|c} a_1 & b_1 & c_1 & d_1 \\ a_2 & b_2 & c_2 & d_2 \\ a_3 & b_3 & c_3 & d_3 \end{array} \right].
$$

Because this matrix is an enlargement of the coefficient matrix, it is commonly called an **augmented matrix**. The vertical line between the third and fourth columns corresponds to where the equals sign occurs in each equation.

■ **CLASS DISCUSSION**
Give a general form of a system of linear equations with four equations and four variables. Write its augmented matrix. ■

EXAMPLE 1   *Representing a linear system with an augmented matrix*

Express each linear system with an augmented matrix. State the dimension of the matrix.
(a) $\begin{aligned} 3x - 4y &= 6 \\ -5x + y &= -5 \end{aligned}$
(b) $\begin{aligned} 2x - 5y + 6z &= -3 \\ 3x + 7y - 3z &= 8 \\ x + 7y &= 5 \end{aligned}$

SOLUTION (a) This system has two equations with two variables. It can be represented by an augmented matrix having dimension 2×3.

$$\begin{bmatrix} 3 & -4 & | & 6 \\ -5 & 1 & | & -5 \end{bmatrix}$$

(b) This system has three equations with three variables. Note that variable z does not appear in the third equation. A 0 is inserted for its coefficient.

$$\begin{bmatrix} 2 & -5 & 6 & | & -3 \\ 3 & 7 & -3 & | & 8 \\ 1 & 7 & 0 & | & 5 \end{bmatrix}$$

Since this matrix has 3 rows and 4 columns, its dimension is 3×4.

EXAMPLE 2 *Converting an augmented matrix into a linear system*

Write the linear system represented by the augmented matrix. Let the variables be x, y, and z.

(a) $\begin{bmatrix} 1 & 0 & 2 & | & -3 \\ 2 & 2 & 10 & | & 3 \\ -1 & 2 & 3 & | & 5 \end{bmatrix}$
(b) $\begin{bmatrix} 1 & 2 & 3 & | & -4 \\ 0 & 1 & -6 & | & 7 \\ 0 & 0 & 1 & | & 8 \end{bmatrix}$

SOLUTION (a) The first column corresponds to x, the second to y, and the third to z. When a zero appears, the variable for that column does not appear in the equation. The vertical line corresponds to the location of the equals sign. The last column represents the constant terms.

$$\begin{array}{rcrcr} x & & + & 2z & = & -3 \\ 2x & + & 2y & + & 10z & = & 3 \\ -x & + & 2y & + & 3z & = & 5 \end{array}$$

(b) The augmented matrix represents the following linear system.

$$\begin{array}{rcr} x + 2y + 3z & = & -4 \\ y - 6z & = & 7 \\ z & = & 8 \end{array}$$

Row-Echelon Form

To solve linear systems with augmented matrices, a convenient form is row-echelon form. The following matrices are in row-echelon form.

$$\begin{bmatrix} 1 & 3 & 0 & -1 \\ 0 & 1 & -6 & 1 \\ 0 & 0 & 1 & -2 \end{bmatrix} \begin{bmatrix} 1 & 2 & 0 \\ 0 & 1 & 4 \end{bmatrix} \begin{bmatrix} 1 & 3 & -1 & 5 \\ 0 & 1 & -1 & 3 \\ 0 & 0 & 1 & 0 \end{bmatrix} \begin{bmatrix} 1 & 3 & -1 & 5 \\ 0 & 0 & 1 & 3 \\ 0 & 0 & 0 & 0 \end{bmatrix} \begin{bmatrix} 1 & 3 & 5 \\ 0 & 0 & 1 \end{bmatrix}$$

The elements of the **main diagonal** are blue in each matrix. Scanning down the main diagonal of a matrix in row-echelon form, we see that it first contains only 1's, and then possibly 0's. The first nonzero element in each nonzero row is 1. This 1 is called a **leading 1**. If two rows contain a leading 1, then the row with the left-most leading 1 is listed first in the matrix. Rows containing only 0's occur last in the matrix. All elements below the main diagonal are 0.

The next example demonstrates a technique called **backward substitution**. It can be used to solve linear systems represented by an augmented matrix in row-echelon form.

EXAMPLE 3   *Solving a linear system with backward substitution*

Solve the system of linear equations.

(a) $\begin{bmatrix} 1 & 1 & 3 & | & 12 \\ 0 & 1 & -2 & | & -4 \\ 0 & 0 & 1 & | & 3 \end{bmatrix}$ (b) $\begin{bmatrix} 1 & -1 & 5 & | & 5 \\ 0 & 1 & 3 & | & 3 \\ 0 & 0 & 0 & | & 0 \end{bmatrix}$

SOLUTION   (a) The linear system is represented by

$$\begin{aligned} x + y + 3z &= 12 \\ y - 2z &= -4 \\ z &= 3. \end{aligned}$$

Since $z = 3$, substitute this value into the second equation to find y.

$$y - 2(3) = -4 \quad \text{or} \quad y = 2$$

Then $y = 2$ and $z = 3$ can be substituted into the first equation to determine x.

$$x + 2 + 3(3) = 12 \quad \text{or} \quad x = 1$$

The solution is $x = 1$, $y = 2$, and $z = 3$, and can be expressed as the **ordered triple** $(1, 2, 3)$.

(b) The linear system is represented by

$$\begin{aligned} x - y + 5z &= 5 \\ y + 3z &= 3 \\ 0 &= 0. \end{aligned}$$

The last equation $0 = 0$ is always true. Its presence usually indicates infinitely many solutions. Use the second equation to write y in terms of z.

$$y = 3 - 3z$$

Next, substitute $(3 - 3z)$ for y in the first equation and write x in terms of z.

$$\begin{aligned} x - (3 - 3z) + 5z &= 5 \\ x - 3 + 3z + 5z &= 5 \qquad \text{Distributive property} \\ x &= 8 - 8z \qquad \text{Solve for } x. \end{aligned}$$

The solution can be written as the ordered triple $(8 - 8z, 3 - 3z, z)$, where z is any real number. There are infinitely many solutions. For example, if we let $z = 1$, then $y = 3 - 3(1) = 0$ and $x = 8 - 8(1) = 0$. Thus one solution is $(0, 0, 1)$.

Gaussian Elimination

If an augmented matrix is not in row-echelon form, it can be transformed into row-echelon form using *Gaussian elimination*. This method uses three basic matrix row transformations.

> ## Matrix row transformations
>
> For any augmented matrix representing a system of linear equations, the following row transformations result in an equivalent system of linear equations.
>
> 1. Any two rows may be interchanged.
> 2. The elements of any row may be multiplied by a nonzero constant.
> 3. Any row may be changed by adding to (or subtracting from) its elements a multiple of the corresponding elements of another row.

When we transform a matrix into row-echelon form, we also are transforming a system of linear equations. The next two examples illustrate how Gaussian elimination with backward substitution is performed.

EXAMPLE 4 *Transforming a matrix into row-echelon form*

Use Gaussian elimination with backward substitution to solve the linear system of equations.

$$\begin{aligned} x + y + z &= 1 \\ -x + y + z &= 5 \\ y + 2z &= 5 \end{aligned}$$

SOLUTION The linear system is written to the right and illustrates how each row transformation affects the corresponding system of linear equations. Note that it is *not* necessary to write the system of equations to the right of the augmented matrix.

Augmented Matrix	*Linear System*
$\begin{bmatrix} 1 & 1 & 1 & \vert & 1 \\ -1 & 1 & 1 & \vert & 5 \\ 0 & 1 & 2 & \vert & 5 \end{bmatrix}$	$\begin{aligned} x + y + z &= 1 \\ -x + y + z &= 5 \\ y + 2z &= 5 \end{aligned}$

The first step is to add the first equation to the second equation to obtain a 0 where the coefficient of x in the second row is highlighted. This row operation is denoted $R_2 + R_1$, and the result becomes the new row 2. It is important to write down each row operation so that we can check our work easily.

$$R_2 + R_1 \rightarrow \begin{bmatrix} 1 & 1 & 1 & \vert & 1 \\ 0 & 2 & 2 & \vert & 6 \\ 0 & 1 & 2 & \vert & 5 \end{bmatrix} \qquad \begin{aligned} x + y + z &= 1 \\ 2y + 2z &= 6 \\ y + 2z &= 5 \end{aligned}$$

To have the matrix in row-echelon form, we need the highlighted 2 in the second row to be a 1. Divide each element in row 2 by 2 and denote it $\dfrac{R_2}{2}$.

$$\frac{R_2}{2} \rightarrow \begin{bmatrix} 1 & 1 & 1 & \vert & 1 \\ 0 & 1 & 1 & \vert & 3 \\ 0 & 1 & 2 & \vert & 5 \end{bmatrix} \qquad \begin{aligned} x + y + z &= 1 \\ y + z &= 3 \\ y + 2z &= 5 \end{aligned}$$

Finally, for the matrix to be in row-echelon form, we need a 0 where the 1 is highlighted in row 3. Subtract row 2 from row 3 and denote it $R_3 - R_2$.

$$R_3 - R_2 \rightarrow \begin{bmatrix} 1 & 1 & 1 & | & 1 \\ 0 & 1 & 1 & | & 3 \\ 0 & 0 & 1 & | & 2 \end{bmatrix} \qquad \begin{aligned} x + y + z &= 1 \\ y + z &= 3 \\ z &= 2 \end{aligned}$$

The matrix is now in row-echelon form, and we see that $z = 2$. Backward substitution may be applied to find the solution. Substituting $z = 2$ into the second equation gives

$$y + 2 = 3 \quad \text{or} \quad y = 1.$$

Finally, let $y = 1$ and $z = 2$ in the first equation to determine x.

$$x + 1 + 2 = 1 \quad \text{or} \quad x = -2$$

The solution of the system is $x = -2$, $y = 1$, and $z = 2$, or $(-2, 1, 2)$.

EXAMPLE 5   *Transforming a matrix into row-echelon form*

Use Gaussian elimination with backward substitution to solve the linear system of equations.

$$\begin{aligned} 2x + 4y + 4z &= 4 \\ x + 3y + z &= 4 \\ -x + 3y + 2z &= -1 \end{aligned}$$

SOLUTION   The initial linear system and augmented matrix are written first.

$$\qquad \textbf{\textit{Augmented Matrix}} \qquad\qquad \textbf{\textit{Linear System}}$$

$$\begin{bmatrix} 2 & 4 & 4 & | & 4 \\ 1 & 3 & 1 & | & 4 \\ -1 & 3 & 2 & | & -1 \end{bmatrix} \qquad \begin{aligned} 2x + 4y + 4z &= 4 \\ x + 3y + z &= 4 \\ -x + 3y + 2z &= -1 \end{aligned}$$

First we obtain a 1 where the coefficient of x in the first row is highlighted. This can be accomplished by either multiplying the first equation by $\frac{1}{2}$, or interchanging rows 1 and 2. We multiply row 1 by $\frac{1}{2}$. This operation is denoted $\frac{1}{2}R_1$, to indicate that row 1 in the previous augmented matrix is multiplied by $\frac{1}{2}$.

$$\frac{1}{2}R_1 \rightarrow \begin{bmatrix} 1 & 2 & 2 & | & 2 \\ 1 & 3 & 1 & | & 4 \\ -1 & 3 & 2 & | & -1 \end{bmatrix} \qquad \begin{aligned} x + 2y + 2z &= 2 \\ x + 3y + z &= 4 \\ -x + 3y + 2z &= -1 \end{aligned}$$

The next step is to eliminate the x-variable in rows 2 and 3 by obtaining zeros in the highlighted positions. To do this, subtract row 1 from row 2, and add row 1 to row 3.

$$\begin{aligned} R_2 - R_1 &\rightarrow \\ R_3 + R_1 &\rightarrow \end{aligned} \begin{bmatrix} 1 & 2 & 2 & | & 2 \\ 0 & 1 & -1 & | & 2 \\ 0 & 5 & 4 & | & 1 \end{bmatrix} \qquad \begin{aligned} x + 2y + 2z &= 2 \\ y - z &= 2 \\ 5y + 4z &= 1 \end{aligned}$$

Since we have a 1 for the coefficient of y in the second row, the next step is to eliminate the y-variable in row 3, and obtain a zero where the y-coefficient of 5 is highlighted. Multiply row 2 by -5, and add the result to row 3.

$$R_3 + (-5)R_2 \rightarrow \begin{bmatrix} 1 & 2 & 2 & | & 2 \\ 0 & 1 & -1 & | & 2 \\ 0 & 0 & 9 & | & -9 \end{bmatrix} \qquad \begin{aligned} x + 2y + 2z &= 2 \\ y - z &= 2 \\ 9z &= -9 \end{aligned}$$

Finally, make the coefficient of z in the third row equal 1 by multiplying row 3 by $\frac{1}{9}$.

$$\frac{1}{9}R_3 \rightarrow \begin{bmatrix} 1 & 2 & 2 & | & 2 \\ 0 & 1 & -1 & | & 2 \\ 0 & 0 & 1 & | & -1 \end{bmatrix} \qquad \begin{aligned} x + 2y + 2z &= 2 \\ y - z &= 2 \\ z &= -1 \end{aligned}$$

The final matrix is in row-echelon form. Backward substitution may be applied to find the solution. Substituting $z = -1$ into the second equation gives

$$y - (-1) = 2 \quad \text{or} \quad y = 1.$$

Next, substitute $y = 1$ and $z = -1$ into the first equation to determine x.

$$x + 2(1) + 2(-1) = 2 \quad \text{or} \quad x = 2$$

The solution of the system is $x = 2$, $y = 1$, and $z = -1$, or $(2, 1, -1)$.

■ MAKING CONNECTIONS

A Geometric Interpretation of Systems of Linear Equations   Solving a linear equation in one variable is reduced to finding the x-value on the number line that satisfies the equation. Solving two linear equations in two variables is often equivalent to finding the xy-coordinates where two lines intersect.

Solving linear equations in three variables also has a geometric interpretation. The graph of a linear equation in the three variables x, y, and z is a flat plane. Finding a unique solution is equivalent to locating a point where three planes in space intersect, as illustrated in Figure 6.35.

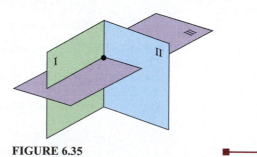

FIGURE 6.35

Sometimes it is convenient to express a matrix in reduced row-echelon form. A matrix in row-echelon form is in **reduced row-echelon form** if every element above and below a leading 1 in a column is 0. The following matrices are examples of reduced row-echelon form.

$$\begin{bmatrix} 1 & 0 \\ 0 & 1 \end{bmatrix} \quad \begin{bmatrix} 1 & 0 \\ 0 & 0 \end{bmatrix} \quad \begin{bmatrix} 1 & 0 & 0 \\ 0 & 1 & 0 \\ 0 & 0 & 1 \end{bmatrix} \quad \begin{bmatrix} 1 & 0 & 3 \\ 0 & 1 & -2 \end{bmatrix} \quad \begin{bmatrix} 1 & 0 & 0 & 3 \\ 0 & 1 & 0 & 1 \\ 0 & 0 & 1 & -1 \end{bmatrix} \quad \begin{bmatrix} 1 & 0 & 4 & 8 \\ 0 & 1 & -1 & 2 \\ 0 & 0 & 0 & 0 \end{bmatrix}$$

If an augmented matrix is in reduced row-echelon form, it usually is straight-forward to solve the system of linear equations.

EXAMPLE 6 *Determining a solution from a matrix in reduced row-echelon form*

Each matrix represents a system of linear equations. Find the solution.

(a) $\begin{bmatrix} 1 & 0 & 0 & | & 3 \\ 0 & 1 & 0 & | & -1 \\ 0 & 0 & 1 & | & 2 \end{bmatrix}$
(b) $\begin{bmatrix} 1 & 0 & | & 6 \\ 0 & 1 & | & -5 \end{bmatrix}$

SOLUTION
(a) The top row represents $1x + 0y + 0z = 3$ or $x = 3$. Using similar reasoning for the second and third rows: $y = -1$ and $z = 2$. The solution is $(3, -1, 2)$.
(b) The system involves two equations and two unknowns. The solution is $(6, -5)$.

Gaussian elimination may be used to transform an augmented matrix into re-duced row-echelon form. It requires more effort than transforming a matrix into row-echelon form, but often eliminates the need for backward substitution.

EXAMPLE 7 *Transforming a matrix into reduced row-echelon form*

Use Gaussian elimination to transform the augmented matrix of the linear system into reduced row-echelon form. State the solution.

$$\begin{aligned} 2x + \ y + 2z &= 10 \\ x \quad\quad + 2z &= 5 \\ x - 2y + 2z &= 1 \end{aligned}$$

SOLUTION The linear system has been written to the right for illustrative purposes.

Augmented Matrix	*Linear System*

$\begin{bmatrix} 2 & 1 & 2 & | & 10 \\ 1 & 0 & 2 & | & 5 \\ 1 & -2 & 2 & | & 1 \end{bmatrix}$
$\begin{aligned} 2x + \ y + 2z &= 10 \\ x \quad\quad + 2z &= 5 \\ x - 2y + 2z &= 1 \end{aligned}$

Start by obtaining a leading 1 in the first row by interchanging rows 1 and 2.

$\begin{array}{c} R_2 \rightarrow \\ R_1 \rightarrow \\ \ \end{array} \begin{bmatrix} 1 & 0 & 2 & | & 5 \\ 2 & 1 & 2 & | & 10 \\ 1 & -2 & 2 & | & 1 \end{bmatrix}$
$\begin{aligned} x \quad\quad + 2z &= 5 \\ 2x + \ y + 2z &= 10 \\ x - 2y + 2z &= 1 \end{aligned}$

Next subtract 2 times row 1 from row 2. Then, subtract row 1 from row 3. This eliminates the x-variable from the second and third equations.

$\begin{array}{c} \ \\ R_2 - 2R_1 \rightarrow \\ R_3 - \ R_1 \rightarrow \end{array} \begin{bmatrix} 1 & 0 & 2 & | & 5 \\ 0 & 1 & -2 & | & 0 \\ 0 & -2 & 0 & | & -4 \end{bmatrix}$
$\begin{aligned} x \quad\quad + 2z &= 5 \\ y - 2z &= 0 \\ -2y \quad\quad &= -4 \end{aligned}$

To eliminate the y-variable in row 3, add two times row 2 to row 3.

$\begin{array}{c} \ \\ \ \\ R_3 + 2R_2 \rightarrow \end{array} \begin{bmatrix} 1 & 0 & 2 & | & 5 \\ 0 & 1 & -2 & | & 0 \\ 0 & 0 & -4 & | & -4 \end{bmatrix}$
$\begin{aligned} x \quad\quad + 2z &= 5 \\ y - 2z &= 0 \\ -4z &= -4 \end{aligned}$

To obtain a leading 1 in row 3, multiply by $-\frac{1}{4}$.

$$-\frac{1}{4}R_3 \to \begin{bmatrix} 1 & 0 & 2 & | & 5 \\ 0 & 1 & -2 & | & 0 \\ 0 & 0 & 1 & | & 1 \end{bmatrix} \qquad \begin{array}{rcl} x & + \ 2z & = 5 \\ y & - \ 2z & = 0 \\ & z & = 1 \end{array}$$

The matrix is in row-echelon form. It can be transformed into reduced row-echelon form by subtracting 2 times row 3 from row 1, and adding 2 times row 3 to row 2.

$$\begin{array}{c} R_1 - 2R_3 \to \\ R_2 + 2R_3 \to \end{array} \begin{bmatrix} 1 & 0 & 0 & | & 3 \\ 0 & 1 & 0 & | & 2 \\ 0 & 0 & 1 & | & 1 \end{bmatrix} \qquad \begin{array}{rcl} x & = 3 \\ y & = 2 \\ z & = 1 \end{array}$$

This matrix is in reduced row-echelon form. The solution is $(3, 2, 1)$.

Solving Systems of Linear Equations with Technology

Gaussian elimination is a numerical method. If the arithmetic at each step is done exactly, then it may be thought of as an exact symbolic procedure. However, in real applications the augmented matrix is usually quite large and its elements are not all integers. As a result, calculators and computers often are used to solve systems of equations. Their solutions usually are approximate. The next three examples use a graphing calculator to solve systems of linear equations.

EXAMPLE 8 *Solving a system of equations using technology*

Use a graphing calculator to solve the system of linear equations in Example 7.

SOLUTION To solve the system

$$\begin{array}{rcrcrcr} 2x & + & y & + & 2z & = & 10 \\ x & & & + & 2z & = & 5 \\ x & - & 2y & + & 2z & = & 1 \end{array}$$

enter the augmented matrix

Graphing Calculator Help

To enter the elements of a matrix, see Appendix B (page AP-18).

$$A = \begin{bmatrix} 2 & 1 & 2 & | & 10 \\ 1 & 0 & 2 & | & 5 \\ 1 & -2 & 2 & | & 1 \end{bmatrix},$$

as shown in Figures 6.36–6.38.

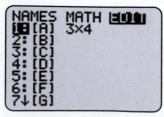

FIGURE 6.36

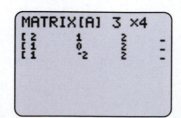

FIGURE 6.37

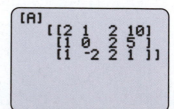

FIGURE 6.38

Now we can use a graphing calculator to transform matrix A into reduced row-echelon form, as illustrated in Figures 6.39 and 6.40. Notice that the reduced row-echelon form obtained from the graphing calculator agrees with our results from Example 7. The solution is $(3, 2, 1)$.

Graphing Calculator Help

To transform a matrix into reduced row-echelon form, see Appendix B (page AP-19).

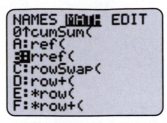

FIGURE 6.39

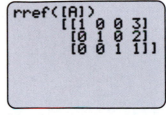

FIGURE 6.40

EXAMPLE 9 *Transforming a matrix into reduced row-echelon form*

Three food shelters are operated by a charitable organization. Three different quantities are computed, which include monthly food costs F in dollars, number of people served per month N, and monthly charitable receipts R in dollars. The data are shown in Table 6.5. (Source: D. Sanders, *Statistics: A First Course.*)

TABLE 6.5

Food Costs (F)	Number Served (N)	Charitable Receipts (R)
3000	2400	8000
4000	2600	10,000
8000	5900	14,000

(a) Model this data using the equation $F = aN + bR + c$, where a, b, and c are constants.

(b) Predict the food costs for a shelter that serves 4000 people and receives charitable receipts of $12,000. Round your answer to the nearest hundred dollars.

SOLUTION **(a)** Since $F = aN + bR + c$, the constants a, b, and c satisfy the following three equations.

$$3000 = a(2400) + b(8000) + c$$
$$4000 = a(2600) + b(10{,}000) + c$$
$$8000 = a(5900) + b(14{,}000) + c$$

This system can be rewritten as

$$2400a + 8000b + c = 3000$$
$$2600a + 10{,}000b + c = 4000$$
$$5900a + 14{,}000b + c = 8000.$$

The resulting augmented matrix is

$$A = \begin{bmatrix} 2400 & 8000 & 1 & 3000 \\ 2600 & 10{,}000 & 1 & 4000 \\ 5900 & 14{,}000 & 1 & 8000 \end{bmatrix}.$$

Figure 6.41 shows the matrix A. The fourth column of A may be viewed using the arrow keys. In Figure 6.42, A has been transformed into reduced row-echelon form. From Figure 6.42, we see that $a \approx 0.6897$, $b \approx 0.4310$, and $c \approx -2103$. Thus let $F = 0.6897N + 0.431R - 2103$.

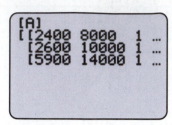

| FIGURE 6.41 | FIGURE 6.42 |

(b) To predict the food costs for a shelter that serves 4000 people and receives charitable receipts of \$12,000, let $N = 4000$ and $R = 12,000$ and evaluate F.

$$F = 0.6897(4000) + 0.431(12,000) - 2103 = 5827.8.$$

This model predicts monthly food costs of approximately \$5800.

The next example shows how a system of linear equations can be used to determine a quadratic function.

EXAMPLE 10   *Determining a quadratic function*

More than half of private-sector employees cannot carry vacation days into a new year. The average number y of paid days off for full-time workers at medium to large companies after x years is listed in Table 6.6.   (Source: Bureau of Labor Statistics.)

TABLE 6.6

x (year)	1	15	30
y (days)	9.4	18.8	21.9

(a) Determine the coefficients for $f(x) = ax^2 + bx + c$ so that f models these data.
(b) Graph f with the data in $[-4, 32, 5]$ by $[8, 23, 2]$.
(c) Estimate the number of paid days off after 3 years of experience. Compare it to the actual value of 11.2 days.

SOLUTION   **(a)** For f to model the data, the three equations $f(1) = 9.4$, $f(15) = 18.8$, and $f(30) = 21.9$ must be satisfied.

$$f(1) = a(1)^2 + b(1) + c = 9.4$$
$$f(15) = a(15)^2 + b(15) + c = 18.8$$
$$f(30) = a(30)^2 + b(30) + c = 21.9$$

The associated augmented matrix is

$$\begin{bmatrix} 1^2 & 1 & 1 & | & 9.4 \\ 15^2 & 15 & 1 & | & 18.8 \\ 30^2 & 30 & 1 & | & 21.9 \end{bmatrix}.$$

Figure 6.43 shows a portion of the matrix represented in reduced row-echelon form.

Graphing Calculator Help

To plot data and to graph an equation, see Appendix B (page AP-10).

$[-4, 32, 5]$ by $[8, 23, 2]$

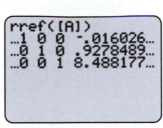

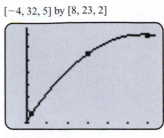

FIGURE 6.43 **FIGURE 6.44**

The solution is $a \approx -0.016026$, $b \approx 0.92785$, and $c \approx 8.4882$.

(b) Graph $Y_1 = -0.016026X^2 + 0.92785X + 8.4882$ along with the points $(1, 9.4)$, $(15, 18.8)$, and $(30, 21.9)$. The graph of f passes through the points as expected. See Figure 6.44.

(c) To estimate the number of paid days off after 3 years, evaluate $f(3)$.

$$f(3) = -0.016026(3)^2 + 0.92785(3) + 8.4882 \approx 11.1$$

This is quite close to the actual value of 11.2 days.

Putting it all Together
6.3

A matrix is a rectangular array of elements. The elements are often real numbers. An augmented matrix may be used to represent a system of linear equations. One of the most common methods for solving a system of linear equations is Gaussian elimination with backward substitution. Through a sequence of matrix row operations, an augmented matrix is transformed into row-echelon form or reduced row-echelon form. Backward substitution is frequently used to find the solution when a matrix is in row-echelon form. Technology can be used to solve systems of linear equations.

The following table summarizes some mathematical concepts involved with solving systems of linear equations in three variables.

Augmented Matrix

A linear system can be represented by an augmented matrix.

$$\begin{bmatrix} 2 & 0 & -3 & | & 2 \\ -1 & 2 & -2 & | & -5 \\ 1 & -2 & -1 & | & 7 \end{bmatrix} \qquad \begin{aligned} 2x - 3z &= 2 \\ -x + 2y - 2z &= -5 \\ x - 2y - z &= 7 \end{aligned}$$

Row-Echelon Form

The following matrices are in row-echelon form. They represent the three situations: no solutions, one solution, and infinitely many solutions.

$$\begin{bmatrix} 1 & -2 & 1 & | & 0 \\ 0 & 1 & 2 & | & 3 \\ 0 & 0 & 0 & | & 1 \end{bmatrix} \qquad \begin{bmatrix} 1 & -2 & 1 & | & 0 \\ 0 & 1 & 2 & | & 3 \\ 0 & 0 & 1 & | & 1 \end{bmatrix} \qquad \begin{bmatrix} 1 & -2 & 1 & | & 0 \\ 0 & 1 & 2 & | & 3 \\ 0 & 0 & 0 & | & 0 \end{bmatrix}$$

No solutions   One solution   Infinitely many solutions
                     $(1, 1, 1)$ $(6 - 5z, 3 - 2z, z)$

Backward Substitution

Backward substitution can be used to solve a system of linear equations represented by an augmented matrix in row-echelon form.

$$\begin{bmatrix} 1 & -2 & 1 & | & 3 \\ 0 & 1 & -2 & | & -3 \\ 0 & 0 & 1 & | & 2 \end{bmatrix}$$

From the last row, $z = 2$.

Substitute $z = 2$ into the second row, $y - 2(2) = -3$ or $y = 1$.

Let $z = 2$ and $y = 1$ in the first row, $x - 2(1) + 2 = 3$ or $x = 3$.

The solution is $(3, 1, 2)$.

6.3 EXERCISES

Dimensions of Matrices and Augmented Matrices

Exercises 1 and 2: State the dimension of each matrix.

1. (a) $\begin{bmatrix} 1 \\ 2 \\ 3 \end{bmatrix}$ **(b)** $\begin{bmatrix} a & b & c \\ d & e & b \end{bmatrix}$ **(c)** $\begin{bmatrix} 3 & 0 \\ 1 & -4 \end{bmatrix}$

2. (a) $\begin{bmatrix} -1 & 1 \end{bmatrix}$ **(b)** $\begin{bmatrix} 1 & -1 \\ 7 & 5 \\ -4 & 0 \end{bmatrix}$

(c) $\begin{bmatrix} 1 & 3 & 8 & -3 \\ 1 & -1 & 1 & -2 \\ 4 & 5 & 0 & -1 \end{bmatrix}$

Exercises 3–6: Represent the linear system by an augmented matrix, and state the dimension of the matrix.

3. $\begin{aligned} 5x - 2y &= 3 \\ -x + 3y &= -1 \end{aligned}$ **4.** $\begin{aligned} 3x + y &= 4 \\ -x + 4y &= 5 \end{aligned}$

5. $\begin{aligned} -3x + 2y + z &= -4 \\ 5x \quad\quad - z &= 9 \\ x - 3y - 6z &= -9 \end{aligned}$

6. $\begin{aligned} x + 2y - z &= 2 \\ -2x + y - 2z &= -3 \\ 7x + y - z &= 7 \end{aligned}$

Exercises 7–10: Write the system of linear equations that the augmented matrix represents.

7. $\begin{bmatrix} 3 & 2 & | & 4 \\ 0 & 1 & | & 5 \end{bmatrix}$ **8.** $\begin{bmatrix} -2 & 1 & | & 5 \\ 7 & 9 & | & 2 \end{bmatrix}$

9. $\begin{bmatrix} 3 & 1 & 4 & | & 0 \\ 0 & 5 & 8 & | & -1 \\ 0 & 0 & -7 & | & 1 \end{bmatrix}$

10. $\begin{bmatrix} 1 & -1 & 3 & | & 2 \\ -2 & 1 & 1 & | & -2 \\ -1 & 0 & -2 & | & 1 \end{bmatrix}$

Row-Echelon Form

Exercises 11 and 12: Determine if the matrix is in row-echelon form.

11. (a) $\begin{bmatrix} 1 & 3 & | & 2 \\ 0 & 1 & | & -1 \end{bmatrix}$ **(b)** $\begin{bmatrix} 1 & 4 & -1 & | & 0 \\ 0 & -1 & 1 & | & 3 \\ 0 & 2 & 1 & | & 7 \end{bmatrix}$

(c) $\begin{bmatrix} 1 & 6 & -8 & | & 5 \\ 0 & 1 & 7 & | & 9 \\ 0 & 0 & 1 & | & 11 \end{bmatrix}$

12. (a) $\begin{bmatrix} 1 & 3 & | & 2 \\ 0 & -1 & | & -1 \end{bmatrix}$ **(b)** $\begin{bmatrix} 1 & 3 & -1 & | & 8 \\ 0 & 1 & 5 & | & 3 \\ 0 & 0 & 0 & | & 0 \end{bmatrix}$

(c) $\begin{bmatrix} 0 & 0 & 1 & | & 1 \\ 0 & 1 & 7 & | & 9 \\ 1 & 2 & -1 & | & 11 \end{bmatrix}$

Exercises 13–20: The augmented matrix is in row-echelon form and represents a linear system. Solve the system using backward substitution, if possible. Write the solution as either an ordered pair or an ordered triple.

13. $\begin{bmatrix} 1 & 2 & | & 3 \\ 0 & 1 & | & -1 \end{bmatrix}$ **14.** $\begin{bmatrix} 1 & -5 & | & 6 \\ 0 & 0 & | & 1 \end{bmatrix}$

15. $\begin{bmatrix} 1 & -1 & | & 2 \\ 0 & 1 & | & 0 \end{bmatrix}$ **16.** $\begin{bmatrix} 1 & 4 & | & -2 \\ 0 & 1 & | & 3 \end{bmatrix}$

17. $\begin{bmatrix} 1 & 1 & -1 & | & 4 \\ 0 & 1 & -1 & | & 2 \\ 0 & 0 & 1 & | & 1 \end{bmatrix}$ **18.** $\begin{bmatrix} 1 & -2 & -1 & | & 0 \\ 0 & 1 & -3 & | & 1 \\ 0 & 0 & 1 & | & 2 \end{bmatrix}$

19. $\begin{bmatrix} 1 & 2 & -1 & | & 5 \\ 0 & 1 & -2 & | & 1 \\ 0 & 0 & 0 & | & 0 \end{bmatrix}$ **20.** $\begin{bmatrix} 1 & -1 & 2 & | & 8 \\ 0 & 1 & -4 & | & 2 \\ 0 & 0 & 1 & | & -1 \end{bmatrix}$

Solving Systems with Gaussian Elimination

Exercises 21–24: Perform each row operation on the given matrix by completing the matrix at the right.

21. $\begin{bmatrix} 2 & -4 & 6 & | & 10 \\ -3 & 5 & 3 & | & 2 \\ 4 & 8 & 4 & | & -8 \end{bmatrix}$ $\begin{matrix} \frac{1}{2}R_1 \rightarrow \\ \\ \frac{1}{4}R_3 \rightarrow \end{matrix}$ $\begin{bmatrix} 1 & & & | & \\ -3 & 5 & 3 & | & 2 \\ & & 1 & | & \end{bmatrix}$

22. $\begin{bmatrix} 1 & -2 & 1 & | & 3 \\ 1 & 4 & 0 & | & -1 \\ 2 & 0 & 1 & | & 5 \end{bmatrix}$ $\begin{matrix} \\ R_2 - R_1 \rightarrow \\ R_3 - 2R_1 \rightarrow \end{matrix}$ $\begin{bmatrix} 1 & -2 & 1 & | & 3 \\ & 6 & & | & \\ & & & | & -1 \end{bmatrix}$

23. $\begin{bmatrix} 1 & -1 & 1 & | & 2 \\ -1 & 2 & -2 & | & 0 \\ 1 & 7 & 0 & | & 5 \end{bmatrix}$ $\begin{matrix} \\ R_2 + R_1 \rightarrow \\ R_3 - R_1 \rightarrow \end{matrix}$ $\begin{bmatrix} 1 & -1 & 1 & | & 2 \\ & & & | & \\ & & & | & \end{bmatrix}$

24. $\begin{bmatrix} 1 & -2 & 3 & | & 6 \\ 2 & 1 & 4 & | & 5 \\ -3 & 5 & 3 & | & 2 \end{bmatrix}$ $\begin{matrix} \\ R_2 - 2R_1 \rightarrow \\ R_3 + 3R_1 \rightarrow \end{matrix}$ $\begin{bmatrix} 1 & -2 & 3 & | & 6 \\ & & & | & \\ & & & | & \end{bmatrix}$

Exercises 25–32: Use Gaussian elimination with backward substitution to solve the system of linear equations. Write the solution as an ordered pair or an ordered triple, whenever possible.

25. $\begin{aligned} x + 2y &= 3 \\ -x - y &= 7 \end{aligned}$ **26.** $\begin{aligned} 2x + 4y &= 10 \\ x - 2y &= -3 \end{aligned}$

27. $\begin{aligned} x + 2y + z &= 3 \\ x + y - z &= 3 \\ -x - 2y + z &= -5 \end{aligned}$ **28.** $\begin{aligned} x + y + z &= 6 \\ 2x + 3y - z &= 3 \\ x + y + 2z &= 10 \end{aligned}$

29. $\begin{aligned} 3x + y + 3z &= 14 \\ x + y + z &= 6 \\ -2x - 2y + 3z &= -7 \end{aligned}$

30. $\begin{aligned} x + 3y - 2z &= 3 \\ -x - 2y + z &= -2 \\ 2x - 7y + z &= 1 \end{aligned}$

31. $\begin{aligned} x + 2y - z &= 2 \\ 2x + 5y + z &= 8 \\ 3x + 7y &= 5 \end{aligned}$ **32.** $\begin{aligned} x + y + z &= 3 \\ x + y + 2z &= 4 \\ 2x + 2y + 3z &= 7 \end{aligned}$

Exercises 33–40: Solve the system, if possible.

33. $\begin{aligned} x - y + z &= 1 \\ x + 2y - z &= 2 \\ y - z &= 0 \end{aligned}$

34. $\begin{aligned} x - y - 2z &= -11 \\ x - 2y - z &= -11 \\ -x + y + 3z &= 14 \end{aligned}$

35. $\begin{aligned} 2x - 4y + 2z &= 11 \\ x + 3y - 2z &= -9 \\ 4x - 2y + z &= 7 \end{aligned}$

36. $\begin{aligned} x - 4y + z &= 9 \\ 3y - 2z &= -7 \\ -x + z &= 0 \end{aligned}$

37. $\begin{aligned} 3x - 2y + 2z &= -18 \\ -x + 2y - 4z &= 16 \\ 4x - 3y - 2z &= -21 \end{aligned}$

38. $\begin{aligned} 2x - y - z &= 0 \\ x - y - z &= -2 \\ 3x - 2y - 2z &= -2 \end{aligned}$

39. $\begin{aligned} x - 4y + 3z &= 26 \\ -x + 3y - 2z &= -19 \\ -y + z &= 10 \end{aligned}$ **40.** $\begin{aligned} 4x - y - z &= 0 \\ 4x - 2y &= 0 \\ 2x + z &= 1 \end{aligned}$

*Exercises 41–46: **Reduced Row-Echelon Form** Find the solution by transforming the augmented matrix of the linear system into reduced row-echelon form.*

41. $\begin{aligned} x - y &= 1 \\ x + y &= 5 \end{aligned}$ **42.** $\begin{aligned} 2x + 3y &= 1 \\ x - 2y &= -3 \end{aligned}$

43. $\begin{aligned} x + 2y + z &= 3 \\ y - z &= -2 \\ -x - 2y + 2z &= 6 \end{aligned}$

44. $\begin{aligned} x + z &= 2 \\ x - y - z &= 0 \\ -2x + y &= -2 \end{aligned}$

45. $\begin{aligned} x - y + 2z &= 7 \\ 2x + y - 4z &= -27 \\ -x + y - z &= 0 \end{aligned}$

46. $\begin{aligned} 2x - 4y - 6z &= 2 \\ x - 3y + z &= 12 \\ 2x + y + 3z &= 5 \end{aligned}$

*Exercises 47–52: **Technology** Use technology to find the solution. Approximate values to the nearest thousandth.*

47. $\begin{aligned} 5x - 7y + 9z &= 40 \\ -7x + 3y - 7z &= 20 \\ 5x - 8y - 5z &= 15 \end{aligned}$

48. $12x - 4y - 7z = 8$
$-8x - 6y + 9z = 7$
$34x + 6y - 2z = 5$

49. $2.1x + 0.5y + 1.7z = 4.9$
$-2x + 1.5y - 1.7z = 3.1$
$5.8x - 4.6y + 0.8z = 9.3$

50. $53x + 95y + 12z = 108$
$81x - 57y - 24z = -92$
$-9x + 11y - 78z = 21$

51. $0.1x + 0.3y + 1.7z = 0.6$
$0.6x + 0.1y - 3.1z = 6.2$
$2.4y + 0.9z = 3.5$

52. $103x - 886y + 431z = 1200$
$-55x + 981y = 1108$
$-327x + 421y + 337z = 99$

Applications

53. *Food Shelters* (Refer to Example 9.) Three food shelters have monthly food costs F in dollars, number of people served per month N, and monthly charitable receipts R in dollars, as shown in the table.

Food Costs (F)	Number Served (N)	Charitable Receipts (R)
1300	1800	5000
5300	3200	12,000
6500	4500	13,000

(a) Model this data using $F = aN + bR + c$, where a, b, and c are constants.

(b) Predict the food costs for a shelter that serves 3500 people and receives charitable receipts of $12,500. Round your answer to the nearest hundred dollars.

54. *Computing Time* When computers are programmed to solve large linear systems involved in applications, such as designing aircraft or large electrical circuits, they frequently use Gaussian elimination with backward substitution. Solving a linear system with n equations and n variables requires a computer to perform a total of $T(n) = \frac{2}{3}n^3 + \frac{3}{2}n^2 - \frac{7}{6}n$ arithmetic operations (additions, subtractions, multiplications, and divisions). (Source: R. Burden and J. Faires, *Numerical Analysis.*)

(a) John Atanasoff, the inventor of the modern digital computer, needed to solve a system of 29 linear equations. Evaluate $T(29)$ to find the number of arithmetic operations this would require. Would it be too many to do by hand?

(b) Compute T for $n = 10, 100, 1000, 10,000$, and $100,000$. List the results in a table.

(c) If the number of equations and variables increases by a factor of 10, does the number of arithmetic operations also increase by a factor of 10? Explain.

(d) Discuss why supercomputers are needed to solve large systems of linear equations.

55. *Pumping Water* Three pumps are being used to empty a small swimming pool. The first pump is twice as fast as the second pump. The first two pumps can empty the pool in 8 hours, while all three pumps can empty it in 6 hours. How long would it take each pump to empty the pool individually? (*Hint:* Let x represent the fraction of the pool that the first pump can empty in 1 hour. Let y and z represent this fraction for the second and third pumps, respectively.)

56. *Pumping Water* Suppose in the previous exercise the first pump is three times as fast as the third pump, the first and second pumps can empty the pool in 6 hours, and all three pumps can empty the pool in 8 hours.

(a) Are these data realistic? Explain your reasoning.

(b) Make a conjecture about a mathematical solution to these data.

(c) Test your conjecture by solving the problem.

57. *Investment* A sum of $5000 is invested in three mutual funds that pay 8%, 11%, and 14% interest rates. The amount of money invested in the fund paying 14% equals the total amount of money invested in the other two funds, and the total annual interest from all three funds is $595.

(a) Write a system of equations whose solution gives the amount invested in each mutual fund. Be sure to state what each variable represents.

(b) Solve the system of equations.

58. *Investment* A sum of $10,000 is invested in three accounts that pay 6%, 8%, and 10% interest. Twice as much money is invested in the account paying 10% as the account paying 6%, and the total annual interest from all three accounts is $842.

(a) Write a system of equations whose solution gives the amount invested in each account. Be sure to state what each variable represents.

(b) Solve the system of equations.

Exercises 59 and 60: **Traffic Flow**   *The accompanying figure shows three one-way streets with intersections A, B, and C. Numbers indicate the average traffic flow in vehicles per minute. The variables x, y, and z denote unknown traffic flows that need to be determined for timing of stoplights.*

(a) *If the number of vehicles per minute entering an intersection must equal the number exiting an intersection, verify that the accompanying system of linear equations describes the traffic flow.*
(b) *Rewrite the system and solve.*
(c) *Interpret your solution.*

59. A: $x + 5 = y + 7$
    B: $z + 6 = x + 3$
    C: $y + 3 = z + 4$

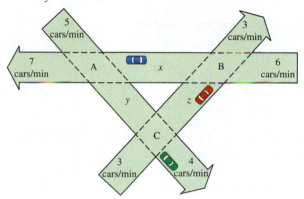

60. A: $x + 7 = y + 4$
    B: $4 + 5 = x + z$
    C: $y + 8 = 9 + 4$

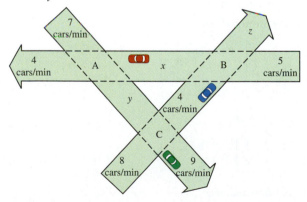

Exercises 61–64: Each set of data can be modeled by $f(x) = ax^2 + bx + c$, *where x represents the year.*

(a) *Find a linear system whose solution represents values of a, b, and c.*
(b) *Use technology to find the solution.*
(c) *Graph f and the data in the same viewing rectangle.*
(d) *Make your own prediction using f.*

61. **Chronic Health Care**   A large percentage of the U.S. population will require chronic health care in the coming decades. The average caregiving age is 50–64, while the typical person needing chronic care is 85 or older. The ratio of potential caregivers to those needing chronic health care will shrink in the coming years, as shown in the table.   (Source: Robert Wood Johnson Foundation, *Chronic Care in America: A 21st Century Challenge.*)

Year	1990	2010	2030
Ratio	11	10	6

62. **Home Health Care**   The table shows the cost of Medicare home health care y in billions of dollars during the year x. In this table $x = 0$ corresponds to 1990, and $x = 6$ corresponds to 1996.   (Source: Health Care Financing Administration.)

x	0	3	6
y	3.9	10.5	18.1

63. **Women in the Military**   The table shows the percentage y of the enlisted people in the military who are women. In this table $x = 3$ corresponds to 1973 and $x = 26$ to 1996.   (Source: Department of Defense.)

x	3	18	26
y	2.2	10.4	12.8

64. **Carbon Dioxide Levels**   Carbon dioxide (CO_2) is a greenhouse gas. Its concentration in parts per million (ppm) have been measured at Mauna Loa, Hawaii, during past years. The table lists measurements for three selected years.   (Source: A. Nilsson, *Greenhouse Earth.*)

Year	1958	1973	1992
CO_2 (ppm)	315	325	354

Writing about Mathematics

65. A linear equation in three variables can be represented by a flat plane. Describe geometrically possible situations that can occur when a system of three linear equations has either no solution or an infinite number of solutions.

66. Give an example of an augmented matrix in row-echelon form that represents a system of linear equations that has no solution. Explain your reasoning.

6.4 PROPERTIES AND APPLICATIONS OF MATRICES

Matrix Notation • Sums, Differences, and Scalar Multiples of Matrices • Matrix Products • Technology and Matrices

Introduction

Matrices occur in many fields of study and have a wide variety of applications. New technologies, such as digital photography and computer graphics, frequently utilize matrices to achieve their goals. Many of these technologies have become specialized fields that require substantial interdisciplinary skills, including mathematics. In this section we discuss properties of matrices and some of their applications.

Matrix Notation

The following notation is used to denote elements in a matrix.

$$\begin{bmatrix} a_{11} & a_{12} \\ a_{21} & a_{22} \end{bmatrix} \quad \begin{bmatrix} a_{11} & a_{12} & a_{13} \\ a_{21} & a_{22} & a_{23} \\ a_{31} & a_{32} & a_{33} \end{bmatrix} \quad \begin{bmatrix} a_{11} & a_{12} & a_{13} & a_{14} \\ a_{21} & a_{22} & a_{23} & a_{24} \\ a_{31} & a_{32} & a_{33} & a_{34} \\ a_{41} & a_{42} & a_{43} & a_{44} \end{bmatrix} \quad \begin{bmatrix} a_{11} & a_{12} \\ a_{21} & a_{22} \\ a_{31} & a_{32} \end{bmatrix} \quad \begin{bmatrix} a_{11} & a_{12} & a_{13} \\ a_{21} & a_{22} & a_{23} \end{bmatrix}$$

A general element in matrix A is denoted by a_{ij}. This refers to the element in the ith row, jth column. For example, a_{23} would be the element of A located in the second row, third column. Two m by n matrices A and B are **equal** if corresponding elements are equal. If A and B have different dimensions, they cannot be equal. For example,

$$\begin{bmatrix} 3 & -3 & 7 \\ 2 & 6 & -2 \\ 4 & 2 & 5 \end{bmatrix} = \begin{bmatrix} 3 & -3 & 7 \\ 2 & 6 & -2 \\ 4 & 2 & 5 \end{bmatrix}$$

because all corresponding elements are equal. However,

$$\begin{bmatrix} 1 & 4 \\ -3 & 2 \\ 4 & -7 \end{bmatrix} \neq \begin{bmatrix} 1 & 4 \\ -3 & 2 \\ 5 & -7 \end{bmatrix}$$

because $4 \neq 5$ in row 3 and column 1, and

$$\begin{bmatrix} 1 & 2 & 3 \\ 4 & 5 & 6 \end{bmatrix} \neq \begin{bmatrix} 1 & 2 \\ 4 & 5 \end{bmatrix}$$

because the matrices have different dimensions.

EXAMPLE 1 *Determining matrix elements*

Let a_{ij} denote an element in A and b_{ij} an element in B, where

$$A = \begin{bmatrix} 3 & -3 & 7 \\ 1 & 6 & -2 \\ 4 & 2 & 5 \end{bmatrix} \quad \text{and} \quad B = \begin{bmatrix} 3 & x & 7 \\ 1 & 6 & -2 \\ 4 & 5 & 2 \end{bmatrix}.$$

(a) Identify a_{12}, b_{32}, and a_{13}.

(b) Compute $a_{31}b_{13} + a_{32}b_{23} + a_{33}b_{33}$.

(c) Is there a value for x that will make the statement $A = B$ true?

SOLUTION   (a) The element a_{12} is located in the first row, second column of A. Thus, $a_{12} = -3$. In a similar manner, $b_{32} = 5$ and $a_{13} = 7$.

(b) $a_{31}b_{13} + a_{32}b_{23} + a_{33}b_{33} = (4)(7) + (2)(-2) + (5)(2) = 34$

(c) No, since $a_{32} = 2 \neq 5 = b_{32}$ and $a_{33} = 5 \neq 2 = b_{33}$. Even if we let $x = -3$, there are other corresponding elements in A and B that are not equal.

Sums, Differences, and Scalar Multiples of Matrices

The FCC has mandated all television stations to transmit digital signals by the year 2006. Digital photography is a new technology in which matrices play an important role. Figure 6.45 shows a black-and-white photograph. In Figure 6.46 a grid has been laid over this photograph. Each square in the grid represents a pixel in a digitized photograph. A gray scale is illustrated in Figure 6.47 on the following page. In this scale 0 corresponds to white, and 11 represents black. As the scale increases from 0 to 11, the shade of gray darkens.

FIGURE 6.45 Photograph to Be Digitized

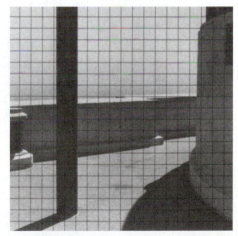

FIGURE 6.46 Placing a Grid over the Photograph

Source: *Visualization* by Friedhoff and Benzon © 1991 by W. H. Freeman and Company.

In Figure 6.48 on the next page the shades of gray are stored as numbers in a matrix with dimension 20×20. Notice how the dark pole in the picture is represented by the fifth and sixth columns in the matrix. Both these columns contain the numbers 10 and 11, which indicates a dark portion in the photograph. The matrix is a numerical or digital representation of the photograph, which can be stored and transmitted over the Internet. In real applications, low-resolution, digital photographs use a grid with 512 rows and 512 columns. The gray scale typically contains 64 levels instead of the 12 in our example. High-resolution photography uses 2048 rows and columns in the grid with 256 levels of gray.

1	1	1	1	10	11	1	1	1	1	1	1	1	1	1	1	1	1	9	8
1	1	1	1	10	11	1	1	1	1	1	1	1	1	1	1	1	2	8	9
1	1	1	1	10	11	1	1	1	1	1	1	1	1	1	1	1	4	8	10
1	1	1	1	10	11	1	1	1	1	1	1	1	1	1	1	1	4	8	10
1	1	1	1	10	11	1	1	1	1	1	1	1	1	1	1	1	5	8	10
1	1	1	1	10	11	1	1	1	1	1	1	1	1	1	1	1	5	8	8
1	0	0	0	10	11	0	0	0	0	0	0	1	1	0	0	0	5	8	8
0	2	2	2	10	11	2	3	3	4	4	5	5	7	7	7	6	8	8	
3	5	9	8	10	11	7	7	7	7	8	8	8	7	6	6	6	6	8	8
11	9	9	6	10	11	7	7	8	8	8	8	8	6	5	5	5	6	8	8
10	9	9	6	10	11	7	8	8	8	8	8	8	3	5	6	4	6	8	8
10	7	6	6	10	11	6	7	7	6	6	6	6	4	3	2	2	5	8	8
4	3	8	7	10	11	5	4	3	2	2	2	2	2	2	2	4	8	8	8
6	4	2	2	10	11	2	2	2	2	2	3	2	2	2	7	9	8	8	8
2	2	2	2	10	11	2	2	2	2	2	2	2	5	7	9	9	7	8	8
2	2	2	2	10	11	2	2	2	2	4	2	8	9	9	10	9	8	7	8
2	2	2	2	10	11	2	2	2	2	2	5	7	9	10	11	9	10	9	7
2	2	2	2	10	11	2	2	2	2	2	7	9	9	9	10	11	10	9	7
2	2	5	7	9	8	2	2	2	2	7	9	9	9	10	11	11	11	10	7
7	9	7	5	2	2	2	2	2	2	8	9	9	10	11	11	11	11	11	10

0	1	2	3	4	5	6	7	8	9	10	11

FIGURE 6.47 Gray Levels   **FIGURE 6.48** Digitized Photograph

Source: *Visualization* by Friedhoff and Benzon © 1991 by W. H. Freeman and Company. Used with permission.

Matrix Addition and Subtraction.   To simplify the concept of digital photography, we reduce the grid to 3×3 and have four gray levels, where 0 represents white, 1 light gray, 2 dark gray, and 3 black. Suppose that we would like to digitize the letter T shown in Figure 6.49. The gray levels are shown in Figure 6.50.

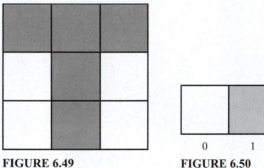

FIGURE 6.49

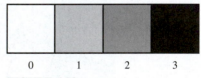

     0   1   2   3

FIGURE 6.50

Since the T is dark gray and the background is white, Figure 6.49 can be represented by

$$A = \begin{bmatrix} 2 & 2 & 2 \\ 0 & 2 & 0 \\ 0 & 2 & 0 \end{bmatrix}.$$

Suppose that we want to make the entire picture darker. If we change every element in A to 3, the entire picture would be black. A more acceptable solution would be to darken each pixel by one gray level. This corresponds to adding 1 to each element in the matrix A and can be accomplished efficiently using matrix notation.

$$\begin{bmatrix} 2 & 2 & 2 \\ 0 & 2 & 0 \\ 0 & 2 & 0 \end{bmatrix} + \begin{bmatrix} 1 & 1 & 1 \\ 1 & 1 & 1 \\ 1 & 1 & 1 \end{bmatrix} = \begin{bmatrix} 2+1 & 2+1 & 2+1 \\ 0+1 & 2+1 & 0+1 \\ 0+1 & 2+1 & 0+1 \end{bmatrix} = \begin{bmatrix} 3 & 3 & 3 \\ 1 & 3 & 1 \\ 1 & 3 & 1 \end{bmatrix}$$

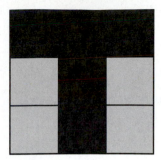

FIGURE 6.51

To add two matrices of equal dimension, add corresponding elements. The result is shown as a picture in Figure 6.51. Notice that the background is now light gray and the T is black. The entire picture is darker.

To lighten the picture in Figure 6.51, subtract 1 from each element. To subtract two matrices of equal dimension, subtract corresponding elements.

$$\begin{bmatrix} 3 & 3 & 3 \\ 1 & 3 & 1 \\ 1 & 3 & 1 \end{bmatrix} - \begin{bmatrix} 1 & 1 & 1 \\ 1 & 1 & 1 \\ 1 & 1 & 1 \end{bmatrix} = \begin{bmatrix} 3-1 & 3-1 & 3-1 \\ 1-1 & 3-1 & 1-1 \\ 1-1 & 3-1 & 1-1 \end{bmatrix} = \begin{bmatrix} 2 & 2 & 2 \\ 0 & 2 & 0 \\ 0 & 2 & 0 \end{bmatrix}$$

Increasing the contrast in a picture causes a light area to become lighter and a dark area to become darker. As a result, there are fewer pixels with intermediate gray levels. Changing contrast is different from making the entire picture lighter or darker.

The digital picture in Figure 6.52 was taken by *Voyager 1* and faintly shows the rings of Saturn. By increasing the contrast in the picture, as shown in Figure 6.53, the rings are much clearer. Notice that light gray features have become lighter, while dark gray areas have become darker. With digital photography, contrast enhancement could be performed on Earth after *Voyager 1* sent back the original picture.

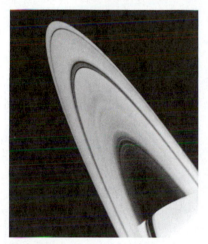

FIGURE 6.52 Rings of Saturn
Source: NASA.

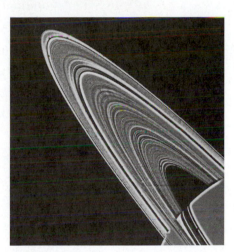

FIGURE 6.53 Enhanced Photograph

EXAMPLE 2 *Applying addition of matrices to digital photography*

Increase the contrast of the + sign in Figure 6.54 by changing light gray to white and dark gray to black. Use matrices to represent this computation.

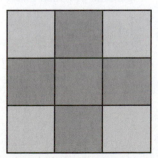

FIGURE 6.54

SOLUTION Figure 6.54 can be represented by the matrix A.

$$A = \begin{bmatrix} 1 & 2 & 1 \\ 2 & 2 & 2 \\ 1 & 2 & 1 \end{bmatrix}$$

To change the contrast, we can reduce each 1 in the matrix A to 0 and increase each 2 to 3. The addition of the matrix B can accomplish this task.

$$A + B = \begin{bmatrix} 1 & 2 & 1 \\ 2 & 2 & 2 \\ 1 & 2 & 1 \end{bmatrix} + \begin{bmatrix} -1 & 1 & -1 \\ 1 & 1 & 1 \\ -1 & 1 & -1 \end{bmatrix} = \begin{bmatrix} 0 & 3 & 0 \\ 3 & 3 & 3 \\ 0 & 3 & 0 \end{bmatrix}$$

FIGURE 6.55 The picture corresponding to $A + B$ is shown in Figure 6.55.

■ **CLASS DISCUSSION**
Discuss how the negative image of a digital picture might be represented. ■

■ **EXAMPLE 3** *Adding and subtracting matrices*

If $A = \begin{bmatrix} 7 & 8 & -1 \\ 0 & -1 & 6 \end{bmatrix}$ and $B = \begin{bmatrix} 5 & -2 & 10 \\ -3 & 2 & 4 \end{bmatrix}$, find the following.

(a) $A + B$ **(b)** $B + A$ **(c)** $A - B$

SOLUTION **(a)** $A + B = \begin{bmatrix} 7 & 8 & -1 \\ 0 & -1 & 6 \end{bmatrix} + \begin{bmatrix} 5 & -2 & 10 \\ -3 & 2 & 4 \end{bmatrix}$

$$= \begin{bmatrix} 7+5 & 8+(-2) & -1+10 \\ 0+(-3) & -1+2 & 6+4 \end{bmatrix}$$

$$= \begin{bmatrix} 12 & 6 & 9 \\ -3 & 1 & 10 \end{bmatrix}$$

(b) $B + A = \begin{bmatrix} 5 & -2 & 10 \\ -3 & 2 & 4 \end{bmatrix} + \begin{bmatrix} 7 & 8 & -1 \\ 0 & -1 & 6 \end{bmatrix}$

$$= \begin{bmatrix} 5+7 & -2+8 & 10+(-1) \\ -3+0 & 2+(-1) & 4+6 \end{bmatrix}$$

$$= \begin{bmatrix} 12 & 6 & 9 \\ -3 & 1 & 10 \end{bmatrix}$$

Notice that $A + B = B + A$. The commutative property for matrix addition holds in general, provided that A and B have the same dimension.

(c) $A - B = \begin{bmatrix} 7 & 8 & -1 \\ 0 & -1 & 6 \end{bmatrix} - \begin{bmatrix} 5 & -2 & 10 \\ -3 & 2 & 4 \end{bmatrix}$

$$= \begin{bmatrix} 7-5 & 8-(-2) & -1-10 \\ 0-(-3) & -1-2 & 6-4 \end{bmatrix}$$

$$= \begin{bmatrix} 2 & 10 & -11 \\ 3 & -3 & 2 \end{bmatrix}$$

■ **CLASS DISCUSSION**
If matrices A and B have the same dimension, does $A - B = B - A$? ■

Multiplication of a Matrix by a Scalar. The matrix

$$B = \begin{bmatrix} 1 & 1 & 1 \\ 1 & 1 & 1 \\ 1 & 1 & 1 \end{bmatrix}$$

can be used to darken a digital picture. Suppose that a photograph is represented by a matrix A with gray levels 0 through 11. Every time the matrix B is added to A, the picture becomes slightly darker. For example, if

$$A = \begin{bmatrix} 0 & 5 & 0 \\ 5 & 5 & 5 \\ 0 & 5 & 0 \end{bmatrix}$$

then the addition of $A + B + B$ would darken the picture by two gray levels, and could be computed by

$$A + B + B = \begin{bmatrix} 0 & 5 & 0 \\ 5 & 5 & 5 \\ 0 & 5 & 0 \end{bmatrix} + \begin{bmatrix} 1 & 1 & 1 \\ 1 & 1 & 1 \\ 1 & 1 & 1 \end{bmatrix} + \begin{bmatrix} 1 & 1 & 1 \\ 1 & 1 & 1 \\ 1 & 1 & 1 \end{bmatrix} = \begin{bmatrix} 2 & 7 & 2 \\ 7 & 7 & 7 \\ 2 & 7 & 2 \end{bmatrix}.$$

A simpler way to write the expression $A + B + B$ is $A + 2B$. Multiplying B by 2 to obtain $2B$ is called **scalar multiplication**.

$$2B = 2\begin{bmatrix} 1 & 1 & 1 \\ 1 & 1 & 1 \\ 1 & 1 & 1 \end{bmatrix} = \begin{bmatrix} 2(1) & 2(1) & 2(1) \\ 2(1) & 2(1) & 2(1) \\ 2(1) & 2(1) & 2(1) \end{bmatrix} = \begin{bmatrix} 2 & 2 & 2 \\ 2 & 2 & 2 \\ 2 & 2 & 2 \end{bmatrix}$$

Each element of B is multiplied by the real number 2.

Sometimes a matrix B is denoted $B = [b_{ij}]$, where b_{ij} represents the element in the ith row, jth column. In this way, we could write $2B$ as $2[b_{ij}] = [2b_{ij}]$. This indicates that to calculate $2B$, multiply each b_{ij} by 2. In a similar manner, a matrix A is sometimes denoted by $[a_{ij}]$.

Some operations on matrices are now summarized.

Operations on matrices

Matrix Addition

The sum of two $m \times n$ matrices A and B is the $m \times n$ matrix $A + B$, in which each element is the sum of the corresponding elements of A and B. This is written as $A + B = [a_{ij}] + [b_{ij}] = [a_{ij} + b_{ij}]$. If A and B have different dimensions, then $A + B$ is undefined.

Matrix Subtraction

The difference of two $m \times n$ matrices A and B is the $m \times n$ matrix $A - B$, in which each element is the difference of the corresponding elements of A and B. This is written as $A - B = [a_{ij}] - [b_{ij}] = [a_{ij} - b_{ij}]$. If A and B have different dimensions, then $A - B$ is undefined.

Multiplication of a Matrix by a Scalar

The product of a scalar (real number) k and an $m \times n$ matrix A is the $m \times n$ matrix kA, in which each element is k times the corresponding element of A. This is written as $kA = k[a_{ij}] = [ka_{ij}]$.

EXAMPLE 4 *Performing scalar multiplication*

If $A = \begin{bmatrix} 2 & 7 & 11 \\ -1 & 3 & -5 \\ 0 & 9 & -12 \end{bmatrix}$, find $-4A$.

SOLUTION
$$-4A = -4\begin{bmatrix} 2 & 7 & 11 \\ -1 & 3 & -5 \\ 0 & 9 & -12 \end{bmatrix} = \begin{bmatrix} -4(2) & -4(7) & -4(11) \\ -4(-1) & -4(3) & -4(-5) \\ -4(0) & -4(9) & -4(-12) \end{bmatrix} = \begin{bmatrix} -8 & -28 & -44 \\ 4 & -12 & 20 \\ 0 & -36 & 48 \end{bmatrix}$$

EXAMPLE 5 *Performing operations on matrices*

If possible, perform the indicated operations using

$$A = \begin{bmatrix} 4 & -2 \\ 3 & 5 \end{bmatrix}, B = \begin{bmatrix} 0 & 1 \\ -2 & 3 \end{bmatrix}, C = \begin{bmatrix} 1 & -1 \\ 0 & 7 \\ -4 & 2 \end{bmatrix}, \text{ and } D = \begin{bmatrix} -1 & -3 \\ 9 & -7 \\ 1 & 8 \end{bmatrix}.$$

(a) $A + 3B$ **(b)** $A - C$ **(c)** $-2C - 3D$

SOLUTION **(a)** $A + 3B = \begin{bmatrix} 4 & -2 \\ 3 & 5 \end{bmatrix} + 3\begin{bmatrix} 0 & 1 \\ -2 & 3 \end{bmatrix} = \begin{bmatrix} 4 & -2 \\ 3 & 5 \end{bmatrix} + \begin{bmatrix} 0 & 3 \\ -6 & 9 \end{bmatrix} = \begin{bmatrix} 4 & 1 \\ -3 & 14 \end{bmatrix}$

(b) $A - C$ is undefined because the dimension of A is 2×2 and unequal to the dimension of C, which is 3×2.

(c) $-2C - 3D = -2\begin{bmatrix} 1 & -1 \\ 0 & 7 \\ -4 & 2 \end{bmatrix} - 3\begin{bmatrix} -1 & -3 \\ 9 & -7 \\ 1 & 8 \end{bmatrix} =$

$$\begin{bmatrix} -2 & 2 \\ 0 & -14 \\ 8 & -4 \end{bmatrix} - \begin{bmatrix} -3 & -9 \\ 27 & -21 \\ 3 & 24 \end{bmatrix} = \begin{bmatrix} 1 & 11 \\ -27 & 7 \\ 5 & -28 \end{bmatrix}$$

Matrix Products

Addition, subtraction, and multiplication can be performed on numbers, variables, and functions. The same operations apply to matrices. Matrix multiplication is different from scalar multiplication.

Suppose two students are taking day classes at one college and night classes at another, in order to graduate on time. Tables 6.7 and 6.8 list the number of credits taken by the students and the cost per credit at each college.

TABLE 6.7 Credits

	College A	College B
Student 1	10	7
Student 2	11	4

TABLE 6.8

	Cost per credit
College A	$60
College B	$80

The cost of tuition is computed by multiplying the number of credits times the cost of each credit. Student 1 is taking 10 credits at $60 each and 7 credits at $80 each. The total tuition for Student 1 is $10(\$60) + 7(\$80) = \$1160$. In a similar manner, the tuition for Student 2 is given by $11(\$60) + 4(\$80) = \$980$.

The information in these tables can be represented by matrices. Let A represent Table 6.7 and B represent Table 6.8.

$$A = \begin{bmatrix} 10 & 7 \\ 11 & 4 \end{bmatrix} \quad \text{and} \quad B = \begin{bmatrix} 60 \\ 80 \end{bmatrix}$$

The matrix product AB calculates total tuition for each student.

$$AB = \begin{bmatrix} 10 & 7 \\ 11 & 4 \end{bmatrix}\begin{bmatrix} 60 \\ 80 \end{bmatrix} = \begin{bmatrix} 10(60) + 7(80) \\ 11(60) + 4(80) \end{bmatrix} = \begin{bmatrix} 1160 \\ 980 \end{bmatrix}$$

Generalizing from this example provides the following definition of matrix multiplication.

Matrix multiplication

The **product** of an $m \times n$ matrix A and an $n \times k$ matrix B is the $m \times k$ matrix AB, which is computed as follows. To find the element of AB in the ith row and jth column, multiply each element in the ith row of A by the corresponding element in the jth column of B. The sum of these products will give the element of row i, column j in AB.

Note: In order to compute the product of two matrices, the number of columns in the first matrix must equal the number of rows in the second matrix, as illustrated in Figure 6.56.

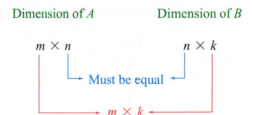

Dimension of A Dimension of B

$m \times n$ $n \times k$

Must be equal

$m \times k$

Dimension of AB

FIGURE 6.56

EXAMPLE 6   *Multiplying matrices*

If possible, compute each product using

$$A = \begin{bmatrix} 1 & -1 \\ 0 & 3 \\ 4 & -2 \end{bmatrix}, B = \begin{bmatrix} -1 \\ -2 \end{bmatrix}, C = \begin{bmatrix} 1 & 2 & 3 \\ 4 & 5 & 6 \end{bmatrix}, \text{and } D = \begin{bmatrix} 1 & -1 & 2 \\ 0 & 3 & -2 \\ -3 & 4 & 5 \end{bmatrix}.$$

(a) AB **(b)** CA **(c)** DC **(d)** CD

SOLUTION   **(a)** The dimension of A is 3×2 and the dimension of B is 2×1. The dimension of AB is 3×1 and can be found as follows.

Dimension of A Dimension of B

$$3 \times 2 \qquad\qquad\qquad 2 \times 1$$

Must be equal

$$3 \times 1$$

Dimension of AB

$$AB = \begin{bmatrix} 1 & -1 \\ 0 & 3 \\ 4 & -2 \end{bmatrix} \begin{bmatrix} -1 \\ -2 \end{bmatrix} = \begin{bmatrix} (1)(-1) + (-1)(-2) \\ (0)(-1) + (3)(-2) \\ (4)(-1) + (-2)(-2) \end{bmatrix} = \begin{bmatrix} 1 \\ -6 \\ 0 \end{bmatrix}$$

(b) The dimension of C is 2×3, and the dimension of A is 3×2. Thus CA has dimension 2×2.

$$CA = \begin{bmatrix} 1 & 2 & 3 \\ 4 & 5 & 6 \end{bmatrix} \begin{bmatrix} 1 & -1 \\ 0 & 3 \\ 4 & -2 \end{bmatrix}$$

$$= \begin{bmatrix} 1(1) + 2(0) + 3(4) & 1(-1) + 2(3) + 3(-2) \\ 4(1) + 5(0) + 6(4) & 4(-1) + 5(3) + 6(-2) \end{bmatrix}$$

$$= \begin{bmatrix} 13 & -1 \\ 28 & -1 \end{bmatrix}$$

(c) The dimension of D is 3×3 and the dimension of C is 2×3. Therefore DC is undefined. Note that D has 3 columns and C has only 2 rows.

(d) The dimension of C is 2×3 and the dimension of D is 3×3. Thus CD has dimension 2×3.

$$CD = \begin{bmatrix} 1 & 2 & 3 \\ 4 & 5 & 6 \end{bmatrix} \begin{bmatrix} 1 & -1 & 2 \\ 0 & 3 & -2 \\ -3 & 4 & 5 \end{bmatrix}$$

$$= \begin{bmatrix} 1(1) + 2(0) + 3(-3) & 1(-1) + 2(3) + 3(4) & 1(2) + 2(-2) + 3(5) \\ 4(1) + 5(0) + 6(-3) & 4(-1) + 5(3) + 6(4) & 4(2) + 5(-2) + 6(5) \end{bmatrix}$$

$$= \begin{bmatrix} -8 & 17 & 13 \\ -14 & 35 & 28 \end{bmatrix}$$

■ **MAKING CONNECTIONS**

The Commutative Property and Matrix Multiplication In Example 6 it was shown that $CD \neq DC$. Unlike multiplication of numbers, variables, and functions, matrix multiplication is *not* commutative. Instead, matrix multiplication is similar to function composition, where for a general pair of functions $f \circ g \neq g \circ f$.

Square matrices have the same number of rows as columns and have dimension $n \times n$ for some natural number n. When we multiply two square matrices, both having dimension $n \times n$, the resulting matrix also has dimension $n \times n$, as illustrated in the next example.

EXAMPLE 7   *Multiplying square matrices*

If $A = \begin{bmatrix} 1 & 0 & 7 \\ 3 & 2 & -1 \\ -5 & -2 & 5 \end{bmatrix}$ and $B = \begin{bmatrix} 4 & -6 & 7 \\ 8 & 9 & 10 \\ 0 & 1 & -3 \end{bmatrix}$, find AB.

SOLUTION $AB = \begin{bmatrix} 1 & 0 & 7 \\ 3 & 2 & -1 \\ -5 & -2 & 5 \end{bmatrix} \begin{bmatrix} 4 & -6 & 7 \\ 8 & 9 & 10 \\ 0 & 1 & -3 \end{bmatrix}$

$$= \begin{bmatrix} 1(4) + 0(8) + 7(0) & 1(-6) + 0(9) + 7(1) & 1(7) + 0(10) + 7(-3) \\ 3(4) + 2(8) - 1(0) & 3(-6) + 2(9) - 1(1) & 3(7) + 2(10) - 1(-3) \\ -5(4) - 2(8) + 5(0) & -5(-6) - 2(9) + 5(1) & -5(7) - 2(10) + 5(-3) \end{bmatrix}$$

$$= \begin{bmatrix} 4 & 1 & -14 \\ 28 & -1 & 44 \\ -36 & 17 & -70 \end{bmatrix}$$

Technology and Matrices

Computing arithmetic operations on large matrices by hand can be a difficult task, prone to errors. Many graphing calculators have the capability to perform addition, subtraction, multiplication, and scalar multiplication with matrices, as the next two examples demonstrate.

EXAMPLE 8   *Multiplying matrices with technology*

Use a graphing calculator to find the product AB from Example 7.

SOLUTION   First enter the matrices A and B into your calculator, as illustrated in Figures 6.57 and 6.58. Then find their product on the home screen, as shown in Figure 6.59. Notice that the answer agrees with our results from Example 7.

Graphing Calculator Help

To enter the elements of a matrix, see Appendix B (page AP-18). To multiply two matrices, see Appendix B (page AP-19).

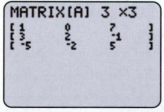

FIGURE 6.57

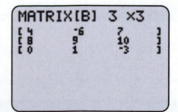

FIGURE 6.58

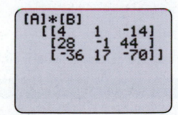

FIGURE 6.59

EXAMPLE 9   *Using technology to evaluate a matrix expression*

Evaluate the expression $2A + 3B^3$, where

$$A = \begin{bmatrix} 3 & -1 & 2 \\ -1 & 6 & -1 \\ 2 & -1 & 9 \end{bmatrix} \quad \text{and} \quad B = \begin{bmatrix} 1 & -2 & 5 \\ 3 & 1 & -1 \\ 5 & 2 & 1 \end{bmatrix}.$$

SOLUTION   In the expression $2A + 3B^3$, B^3 is equal to BBB. Enter each matrix into a calculator and evaluate the expression. Figure 6.60 shows the result of this computation.

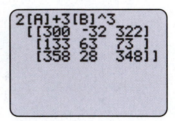

FIGURE 6.60

Putting it all Together

6.4

*A*ddition, subtraction, and multiplication can be performed on numbers, variables, and functions. In this section we learned how these operations also apply to matrices. The following table provides examples of these operations.

Matrix Addition

$$\begin{bmatrix} 1 & 2 & 3 \\ 5 & 6 & 7 \end{bmatrix} + \begin{bmatrix} -1 & 0 & 8 \\ 9 & -2 & 10 \end{bmatrix} = \begin{bmatrix} 1 + (-1) & 2 + 0 & 3 + 8 \\ 5 + 9 & 6 + (-2) & 7 + 10 \end{bmatrix} = \begin{bmatrix} 0 & 2 & 11 \\ 14 & 4 & 17 \end{bmatrix}$$

Both matrices must have the same dimension for their sum to be defined.

Matrix Subtraction

$$\begin{bmatrix} 1 & -4 \\ -3 & 4 \\ 2 & 7 \end{bmatrix} - \begin{bmatrix} 5 & 1 \\ 3 & 6 \\ 8 & -9 \end{bmatrix} = \begin{bmatrix} 1 - 5 & -4 - 1 \\ -3 - 3 & 4 - 6 \\ 2 - 8 & 7 - (-9) \end{bmatrix} = \begin{bmatrix} -4 & -5 \\ -6 & -2 \\ -6 & 16 \end{bmatrix}$$

Both matrices must have the same dimension for their difference to be defined.

Scalar Multiplication

$$3 \begin{bmatrix} 3 & -2 \\ 0 & 1 \end{bmatrix} = \begin{bmatrix} 3(3) & 3(-2) \\ 3(0) & 3(1) \end{bmatrix} = \begin{bmatrix} 9 & -6 \\ 0 & 3 \end{bmatrix}$$

Matrix Multiplication

$$\begin{bmatrix} 0 & 1 \\ 2 & -3 \end{bmatrix}\begin{bmatrix} 3 & -5 \\ 4 & 6 \end{bmatrix} = \begin{bmatrix} 0(3) + 1(4) & 0(-5) + 1(6) \\ 2(3) + (-3)(4) & 2(-5) + (-3)(6) \end{bmatrix} = \begin{bmatrix} 4 & 6 \\ -6 & -28 \end{bmatrix}$$

For a matrix product to be defined, the number of columns in the first matrix must equal the number of rows in the second matrix. Matrix multiplication is not commutative. That is, $AB \neq BA$ in general.

6.4 EXERCISES

Elements of Matrices

Exercises 1–4: Determine the following for the matrix A, if possible.

(a) $a_{12}, a_{21},$ and a_{32}

(b) $a_{11} a_{22} + 3a_{23}$

1. $A = \begin{bmatrix} 1 & 2 & 3 \\ 4 & 5 & 6 \end{bmatrix}$

2. $A = \begin{bmatrix} 1 & 2 & 3 & 4 \\ 5 & 6 & 7 & 8 \\ 9 & 10 & 11 & 12 \end{bmatrix}$

3. $A = \begin{bmatrix} 1 & -1 & 4 \\ 3 & -2 & 5 \\ 7 & 0 & -6 \end{bmatrix}$

4. $A = \begin{bmatrix} 1 & -2 \\ -4 & 5 \end{bmatrix}$

Addition, Subtraction, and Scalar Multiples of Matrices

Exercises 5–10: If possible, find each of the following.

(a) $A + B$ (b) $3A$ (c) $2A - 3B$

5. $A = \begin{bmatrix} 2 & -6 \\ 3 & 1 \end{bmatrix}, \quad B = \begin{bmatrix} -1 & 0 \\ -2 & 3 \end{bmatrix}$

6. $A = \begin{bmatrix} 1 & -2 & 5 \\ 3 & -4 & -1 \end{bmatrix}, \quad B = \begin{bmatrix} 0 & -1 & -5 \\ -3 & 1 & 2 \end{bmatrix}$

7. $A = \begin{bmatrix} 1 & -1 & 0 \\ 1 & 5 & 9 \\ -4 & 8 & -5 \end{bmatrix}, \quad B = \begin{bmatrix} 2 & 8 & -1 \\ 6 & -1 & 3 \end{bmatrix}$

8. $A = \begin{bmatrix} 6 & 2 & 9 \\ 3 & -2 & 0 \\ -1 & 4 & 8 \end{bmatrix}, \quad B = \begin{bmatrix} 1 & 0 & -1 \\ 3 & 0 & 7 \\ 0 & -2 & -5 \end{bmatrix}$

9. $A = \begin{bmatrix} -2 & -1 \\ -5 & 1 \\ 2 & -3 \end{bmatrix}, \quad B = \begin{bmatrix} 2 & -1 \\ 3 & 1 \\ 7 & -5 \end{bmatrix}$

10. $A = \begin{bmatrix} 0 & 1 \\ 3 & 2 \\ 4 & -9 \end{bmatrix}, \quad B = \begin{bmatrix} 5 & 2 & -7 \\ 8 & -2 & 0 \end{bmatrix}$

Exercises 11–16: Evaluate the matrix expression.

11. $2\begin{bmatrix} 2 & -1 \\ 5 & 1 \\ 0 & 3 \end{bmatrix} + \begin{bmatrix} 5 & 0 \\ 7 & -3 \\ 1 & 1 \end{bmatrix} - \begin{bmatrix} 9 & -4 \\ 4 & 4 \\ 1 & 6 \end{bmatrix}$

12. $-3\begin{bmatrix} 3 & 8 \\ -1 & -9 \end{bmatrix} + 5\begin{bmatrix} 4 & -8 \\ 1 & 6 \end{bmatrix}$

13. $\begin{bmatrix} 4 & 6 \\ 3 & -7 \end{bmatrix} - 2\begin{bmatrix} 1 & 0 \\ -4 & 1 \end{bmatrix}$

14. $\begin{bmatrix} 5 & -1 & 6 \\ -2 & 10 & 12 \\ 5 & 2 & 9 \end{bmatrix} - \begin{bmatrix} -1 & 2 & 2 \\ 2 & -1 & 2 \\ 2 & 2 & -1 \end{bmatrix}$

15. $2\begin{bmatrix} 2 & -1 & -1 \\ -1 & 2 & -1 \\ -1 & -1 & 2 \end{bmatrix} + 3\begin{bmatrix} 1 & 2 & 3 \\ 2 & 1 & 3 \\ 2 & 3 & 1 \end{bmatrix}$

16. $3\begin{bmatrix} 1 & 0 & 3 & -1 \\ 0 & 1 & 2 & -1 \\ 1 & 0 & -3 & 1 \end{bmatrix} - 4\begin{bmatrix} -1 & 0 & 0 & 4 \\ 0 & -1 & 3 & 2 \\ 2 & 0 & 1 & -1 \end{bmatrix}$

Matrices and Digital Photography

Exercises 17–20: **Digital Photography**   *(Refer to the discussion of digital photography in this section.) Consider the following simplified digital photograph that has a 3 × 3 grid with four gray levels numbered from 0 to 3. It shows the number 1 in dark gray on a light gray background. Let A be the 3 × 3 matrix that represents this figure digitally.*

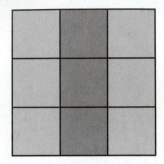

17. Find the matrix A.

18. Find a matrix B such that $A + B$ represents the entire picture becoming one gray level darker. Evaluate $A + B$.

19. Find matrix B such that $A - B$ represents the entire picure becoming lighter by one gray level. Evaluate $A - B$.

20. (Refer to Example 2.) Find a matrix B such that $A + B$ will result in a contrast enhancement of A by one gray level. Evaluate $A + B$.

21. *Negative Image*   The negative image of a picture interchanges black and white. The number 1 is represented by the matrix A. Determine a matrix B such that $B - A$ represents the negative image of the picture presented by A. Evaluate $B - A$.

$$A = \begin{bmatrix} 0 & 3 & 0 \\ 0 & 3 & 0 \\ 0 & 3 & 0 \end{bmatrix}$$

22. *Negative Image*   (Refer to the previous exercise.) Consider the matrix A representing a digital photograph. Find a matrix B that represents the negative image of this picture.

$$A = \begin{bmatrix} 0 & 3 & 0 \\ 1 & 3 & 1 \\ 2 & 3 & 2 \end{bmatrix}$$

Exercises 23 and 24: **Digital Photography**   *The accompanying digital photograph represents the letter F*

using 20 pixels in a 5 × 4 grid. Assume that there are four gray levels from 0 to 3.

23. Find a matrix A that represents a digital photograph of this letter F.

24. (Continuation of the previous exercise)
    (a) Find a matrix B such that $B - A$ represents the negative image of A.
    (b) Find a matrix C where $A + C$ represents a decrease in the contrast of A by one gray level.

Exercises 25–28: **Digitizing Letters**   *Complete the following.*

    (a) *Design a matrix A with a dimension 4 × 4 that represents a digital photograph of the given letter. Assume that there are four gray levels from 0 to 3.*
    (b) *Find a matrix B such that B − A represents the negative image of the picture represented by the matrix A from part (a).*

25. Z   26. N

27. L   28. O

Matrix Multiplication

Exercises 29–32: **Tuition Costs**   *(Refer to the discussion before Example 6.)*

    (a) *Find a matrix A and a column matrix B that describe the following tables involving credits and tuition costs.*
    (b) *Find the matrix product AB, and interpret the result.*

29.

	College A	College B
Student 1	12	4
Student 2	8	7

	Cost per credit
College A	$55
College B	$70

30.

	College A	College B
Student 1	15	2
Student 2	12	4

	Cost per credit
College A	$90
College B	$75

31.

	College A	College B
Student 1	10	5
Student 2	9	8
Student 3	11	3

	Cost per credit
College A	$60
College B	$70

32.

	College A	College B	College C
Student 1	6	0	3
Student 2	11	3	0
Student 3	0	12	3

	Cost per credit
College A	$50
College B	$65
College C	$60

Exercises 33–46: If possible, find the matrix products AB and BA.

33. $A = \begin{bmatrix} 1 & -1 \\ 2 & 0 \end{bmatrix}, \qquad B = \begin{bmatrix} -2 & 3 \\ 1 & 2 \end{bmatrix}$

34. $A = \begin{bmatrix} -3 & 5 \\ 2 & 7 \end{bmatrix}, \qquad B = \begin{bmatrix} -1 & 2 \\ 0 & 7 \end{bmatrix}$

35. $A = \begin{bmatrix} 5 & -7 & 2 \\ 0 & 1 & 5 \end{bmatrix}, \quad B = \begin{bmatrix} 9 & 8 & 7 \\ 1 & -1 & -2 \end{bmatrix}$

36. $A = \begin{bmatrix} 2 & 1 & -1 \\ 0 & 2 & 1 \\ 3 & 2 & -1 \end{bmatrix}, \quad B = \begin{bmatrix} 1 & 0 \\ 2 & -1 \\ 3 & 1 \end{bmatrix}$

37. $A = \begin{bmatrix} 2 & -3 \\ 5 & 3 \end{bmatrix}, \qquad B = \begin{bmatrix} -3 \\ 4 \\ 1 \end{bmatrix}$

38. $A = \begin{bmatrix} 3 & -1 \\ 2 & -2 \\ 0 & 4 \end{bmatrix}, \qquad B = \begin{bmatrix} 1 & -4 & 0 \\ -1 & 3 & 2 \end{bmatrix}$

39. $A = \begin{bmatrix} 2 & -1 & 3 \\ 0 & 1 & 0 \\ 2 & -2 & 3 \end{bmatrix}, \qquad B = \begin{bmatrix} 1 & 5 & -1 \\ 0 & 1 & 3 \\ -1 & 2 & 1 \end{bmatrix}$

40. $A = \begin{bmatrix} 1 & -2 & 5 \\ 1 & 0 & -2 \\ 1 & 3 & 2 \end{bmatrix}, \qquad B = \begin{bmatrix} -1 & 4 & 2 \\ -3 & 0 & 1 \\ 5 & 1 & 0 \end{bmatrix}$

41. $A = \begin{bmatrix} 2 & -1 \\ 3 & 1 \end{bmatrix}, \qquad B = \begin{bmatrix} 1 \\ 3 \end{bmatrix}$

42. $A = [5 \quad -3], \qquad B = \begin{bmatrix} 1 \\ 3 \end{bmatrix}$

43. $A = \begin{bmatrix} -3 & 1 \\ 2 & -4 \end{bmatrix}, \qquad B = \begin{bmatrix} 1 & 0 & -2 \\ -4 & 8 & 1 \end{bmatrix}$

44. $A = \begin{bmatrix} 6 & 1 & 0 \\ -2 & 5 & 1 \\ 4 & -7 & 10 \end{bmatrix}, \qquad B = \begin{bmatrix} 10 \\ 20 \\ 30 \end{bmatrix}$

45. $A = \begin{bmatrix} 1 & 0 & -2 \\ 3 & -4 & 1 \\ 2 & 0 & 5 \end{bmatrix}, \qquad B = \begin{bmatrix} 1 \\ -1 \\ 3 \end{bmatrix}$

46. $A = \begin{bmatrix} 1 & -1 & 3 & -2 \\ 1 & 0 & 3 & 4 \\ 2 & -2 & 0 & 8 \end{bmatrix}, \quad B = \begin{bmatrix} 1 & -1 \\ 0 & 5 \\ 2 & 3 \\ -5 & 4 \end{bmatrix}$

47. *Auto Parts*   A store owner makes two separate orders for three types of auto parts: I, II, and III. The number of parts ordered is represented by the matrix A.

$$A = \begin{matrix} & \begin{matrix} \text{I} & \text{II} & \text{III} \end{matrix} & \\ \begin{bmatrix} 3 & 4 & 8 \\ 5 & 6 & 2 \end{bmatrix} & \begin{matrix} \text{Order 1} \\ \text{Order 2} \end{matrix} \end{matrix}$$

For example, in order 1 there were 4 parts of type II ordered. The cost in dollars of each part can be represented by the matrix B.

$$B = \begin{matrix} & \text{Cost} & \\ \begin{bmatrix} 10 \\ 20 \\ 30 \end{bmatrix} & \begin{matrix} \text{Part I} \\ \text{Part II} \\ \text{Part III} \end{matrix} \end{matrix}$$

Find AB and interpret the result.

48. *Car Sales* Two car dealers buy four different makes of cars: I, II, III, and IV. The number and make of automobiles bought by each dealer is represented by the matrix *A*.

$$A = \begin{bmatrix} \overset{\text{I}}{1} & \overset{\text{II}}{3} & \overset{\text{III}}{8} & \overset{\text{IV}}{4} \\ 3 & 5 & 7 & 0 \end{bmatrix} \begin{matrix} \text{Dealer 1} \\ \text{Dealer 2} \end{matrix}$$

For example, dealer 2 bought 7 cars of type III. The cost in thousands of dollars of each type of car can be represented by the matrix *B*.

$$B = \begin{bmatrix} \overset{\text{Cost}}{15} \\ 21 \\ 28 \\ 38 \end{bmatrix} \begin{matrix} \text{Make I} \\ \text{Make II} \\ \text{Make III} \\ \text{Make IV} \end{matrix}$$

Find *AB* and interpret the result.

*Exercises 49–52: **Properties of Matrices** Use a graphing calculator to evaluate the expression with the given matrices A, B, and C. Compare your answers for parts (a) and (b). Then, interpret the results.*

$$A = \begin{bmatrix} 2 & -1 & 3 \\ 1 & 3 & -5 \\ 0 & -2 & 1 \end{bmatrix}, \quad B = \begin{bmatrix} 6 & 2 & 7 \\ 3 & -4 & -5 \\ 7 & 1 & 0 \end{bmatrix},$$

$$\text{and } C = \begin{bmatrix} 1 & 4 & -3 \\ 8 & 1 & -1 \\ 4 & 6 & -2 \end{bmatrix}$$

49. (a) $A(B + C)$ **(b)** $AB + AC$

50. (a) $(A - B)C$ **(b)** $AC - BC$

51. (a) $(A - B)^2$ **(b)** $A^2 - AB - BA + B^2$

52. (a) $(AB)C$ **(b)** $A(BC)$

Writing about Mathematics

53. Discuss whether matrix multiplication is more like multiplication of functions or composition of functions. Explain your reasoning.

54. Describe one application of matrices.

Extended and Discovery Exercises

*Exercises 1–4: **Representing Colors** Colors for computer monitors are often described using ordered triples. One model, called the RGB system, uses red, green, and blue to generate all colors. The accompanying figure describes the relationships of these colors in this system. For example, red is $(1, 0, 0)$, green is $(0, 1, 0)$, and blue is $(0, 0, 1)$. Since equal amounts of red and green combine to form yellow, yellow is represented by $(1, 1, 0)$. Similarly, magenta (a deep reddish purple) is a mixture of blue and red and represented by $(1, 0, 1)$. Cyan is $(0, 1, 1)$, since it is a mixture of blue and green.*

Another color model uses cyan, magenta, and yellow. It is referred to as the CMY model and is used in the four-color printing process for textbooks. In this system, cyan is $(1, 0, 0)$, magenta is $(0, 1, 0)$, and yellow is $(0, 0, 1)$. In the CMY model, red is created by mixing magenta and yellow. Thus, red is $(0, 1, 1)$ in this system. To convert ordered triples in the RGB model to ordered triples in the CMY model, we can use the following matrix equation. In both of these systems, color intensities vary between 0 and 1. (Sources: I. Kerlow, *The Art of 3-D Computer Animation and Imaging;* R. Wolff.)

$$\begin{bmatrix} C \\ M \\ Y \end{bmatrix} = \begin{bmatrix} 1 \\ 1 \\ 1 \end{bmatrix} - \begin{bmatrix} R \\ G \\ B \end{bmatrix}$$

Blue = (0, 0, 1)
Magenta = (1, 0, 1)
Cyan = (0, 1, 1)
White = (1, 1, 1)
Black = (0, 0, 0)
Green = (0, 1, 0)
Red = (1, 0, 0)
Yellow = (1, 1, 0)

1. In the RGB model, aquamarine is (0.631, 1, 0.933). Use the matrix equation to determine the mixture of cyan, magenta, and yellow that makes aquamarine in the CMY model.

2. In the RGB model, rust is (0.552, 0.168, 0.066). Use the matrix equation to determine the mixture of cyan, magenta, and yellow that makes rust in the CMY model. Explain your results.

3. Use the given matrix equation to find a matrix equation that changes colors represented by ordered triples in the CMY model into ordered triples in the RGB model.

4. In the CMY model, (0.012, 0, 0.597) is a cream color. Use the matrix equation from the previous exercise to determine the mixture of red, green, and blue that makes a cream color in the RGB model.

CHECKING BASIC CONCEPTS FOR SECTIONS 6.3 AND 6.4

1. Solve the system of linear equations using Gaussian elimination and backward substitution.

$$
\begin{aligned}
x \qquad\ + z &= 2 \\
x + \ y - z &= 1 \\
-x - 2y - z &= 0
\end{aligned}
$$

2. Solve the system of linear equations in the previous exercise using technology.

3. Perform the following operations on the given matrices A and B.

$$
A = \begin{bmatrix} 1 & 0 & 1 \\ -1 & 1 & 2 \\ 1 & 3 & 0 \end{bmatrix}, \quad B = \begin{bmatrix} -1 & 1 & 2 \\ 0 & 4 & 1 \\ 1 & -2 & 0 \end{bmatrix}
$$

(a) $A + B$ (b) $2A - B$ (c) AB

6.5 INVERSES OF MATRICES

Matrix Inverses • Representing Linear Systems with Matrix Equations • Solving Linear Systems with Inverses • Finding Inverses Symbolically

Introduction

In Section 5.2 we discussed how the inverse function f^{-1} will undo or cancel the computation performed by the function f. Like functions, some matrices have inverses. The inverse of a matrix A will undo or cancel the computation performed by A. For example, matrices play an important role in computer graphics. If a matrix A is capable of rotating a figure on a screen 90° clockwise, then the inverse matrix would cause the figure to rotate 90° counterclockwise. This section discusses matrix inverses and some of their applications.

Matrix Inverses

In computer graphics the matrix

$$
A = \begin{bmatrix} 1 & 0 & h \\ 0 & 1 & k \\ 0 & 0 & 1 \end{bmatrix}
$$

is used to translate a point (x, y), horizontally h units and vertically k units. The translation is to the right if $h > 0$ and to the left if $h < 0$. Similarly, the translation is upward if $k > 0$ and downward if $k < 0$. A point (x, y) is represented using the 3×1 **column matrix**

$$X = \begin{bmatrix} x \\ y \\ 1 \end{bmatrix}.$$

(Source: C. Pokorny and C. Gerald, *Computer Graphics.*)

The third element in X is always equal to 1. For example, the point $(-1, 2)$ could be translated 3 units right and 4 units downward by computing the matrix product.

$$AX = \begin{bmatrix} 1 & 0 & 3 \\ 0 & 1 & -4 \\ 0 & 0 & 1 \end{bmatrix} \begin{bmatrix} -1 \\ 2 \\ 1 \end{bmatrix} = \begin{bmatrix} 2 \\ -2 \\ 1 \end{bmatrix} = Y.$$

Its new location is $(2, -2)$. In the matrix A, $h = 3$ and $k = -4$. See Figure 6.61.

If A translates a point 3 units right and 4 units downward, then the inverse matrix translates a point 3 units left and 4 units upward. This would return a point to its original position after being translated by A. Therefore the inverse matrix of A, denoted A^{-1}, is given by

$$A^{-1} = \begin{bmatrix} 1 & 0 & -3 \\ 0 & 1 & 4 \\ 0 & 0 & 1 \end{bmatrix}.$$

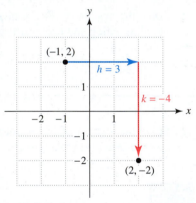

FIGURE 6.61 Translation of a Point

In A^{-1}, $h = -3$ and $k = 4$. The matrix product $A^{-1}Y$ results in

$$A^{-1}Y = \begin{bmatrix} 1 & 0 & -3 \\ 0 & 1 & 4 \\ 0 & 0 & 1 \end{bmatrix} \begin{bmatrix} 2 \\ -2 \\ 1 \end{bmatrix} = \begin{bmatrix} -1 \\ 2 \\ 1 \end{bmatrix} = X.$$

The matrix A^{-1} translates the point located at $(2, -2)$ to its original coordinates of $(-1, 2)$. The two translations acting on the point $(-1, 2)$ can be represented by the following computation.

$$A^{-1}AX = \begin{bmatrix} 1 & 0 & -3 \\ 0 & 1 & 4 \\ 0 & 0 & 1 \end{bmatrix} \begin{bmatrix} 1 & 0 & 3 \\ 0 & 1 & -4 \\ 0 & 0 & 1 \end{bmatrix} \begin{bmatrix} -1 \\ 2 \\ 1 \end{bmatrix}$$

$$= \begin{bmatrix} 1 & 0 & 0 \\ 0 & 1 & 0 \\ 0 & 0 & 1 \end{bmatrix} \begin{bmatrix} -1 \\ 2 \\ 1 \end{bmatrix}$$

$$= \begin{bmatrix} -1 \\ 2 \\ 1 \end{bmatrix} = X$$

That is, the action of A followed by A^{-1} on the point $(-1, 2)$ results in $(-1, 2)$. In a similar manner, if we reverse the order of A^{-1} and A to compute $AA^{-1}X$, the result is again X.

$$AA^{-1}X = \begin{bmatrix} 1 & 0 & 3 \\ 0 & 1 & -4 \\ 0 & 0 & 1 \end{bmatrix} \begin{bmatrix} 1 & 0 & -3 \\ 0 & 1 & 4 \\ 0 & 0 & 1 \end{bmatrix} \begin{bmatrix} -1 \\ 2 \\ 1 \end{bmatrix}$$

$$= \begin{bmatrix} 1 & 0 & 0 \\ 0 & 1 & 0 \\ 0 & 0 & 1 \end{bmatrix} \begin{bmatrix} -1 \\ 2 \\ 1 \end{bmatrix}$$

$$= \begin{bmatrix} -1 \\ 2 \\ 1 \end{bmatrix} = X$$

Notice that the matrix product of $A^{-1}A$ and AA^{-1} resulted in a matrix with 1's on its main diagonal and 0's elsewhere. This matrix is called the **identity matrix**.

The $n \times n$ identity matrix

The **$n \times n$ identity matrix**, denoted by I_n, has only 1's on its main diagonal and 0's elsewhere.

Some examples of identity matrices are

$$I_2 = \begin{bmatrix} 1 & 0 \\ 0 & 1 \end{bmatrix}, \qquad I_3 = \begin{bmatrix} 1 & 0 & 0 \\ 0 & 1 & 0 \\ 0 & 0 & 1 \end{bmatrix}, \qquad \text{and} \qquad I_4 = \begin{bmatrix} 1 & 0 & 0 & 0 \\ 0 & 1 & 0 & 0 \\ 0 & 0 & 1 & 0 \\ 0 & 0 & 0 & 1 \end{bmatrix}.$$

If A is any $n \times n$ matrix, then $I_nA = A$ and $AI_n = A$. For instance, if

$$A = \begin{bmatrix} 2 & 3 \\ 4 & 5 \end{bmatrix}$$

then

$$I_2A = \begin{bmatrix} 1 & 0 \\ 0 & 1 \end{bmatrix} \begin{bmatrix} 2 & 3 \\ 4 & 5 \end{bmatrix} = \begin{bmatrix} 2 & 3 \\ 4 & 5 \end{bmatrix} = A, \qquad \text{and}$$

$$AI_2 = \begin{bmatrix} 2 & 3 \\ 4 & 5 \end{bmatrix} \begin{bmatrix} 1 & 0 \\ 0 & 1 \end{bmatrix} = \begin{bmatrix} 2 & 3 \\ 4 & 5 \end{bmatrix} = A.$$

Next we formally define the inverse of an $n \times n$ matrix A, whenever it exists.

Inverse of a square matrix

Let A be an $n \times n$ matrix. If there exists an $n \times n$ matrix, denoted A^{-1}, that satisfies

$$A^{-1}A = I_n \qquad \text{and} \qquad AA^{-1} = I_n,$$

then A^{-1} is the **inverse** of A.

If A^{-1} exists, then A is **invertible** or **nonsingular**. On the other hand, if a matrix A is not invertible then it is **singular**. Not every matrix has an inverse. For example, the **zero matrix** with dimension 3×3 is given by

$$O_3 = \begin{bmatrix} 0 & 0 & 0 \\ 0 & 0 & 0 \\ 0 & 0 & 0 \end{bmatrix}.$$

The matrix O_3 does not have an inverse. The product of O_3 with any 3×3 matrix B would again be O_3, rather than the identity matrix I_3.

EXAMPLE 1   *Verifying an inverse*

Determine if B is the inverse of A, where

$$A = \begin{bmatrix} 5 & 3 \\ -3 & -2 \end{bmatrix} \quad \text{and} \quad B = \begin{bmatrix} 2 & 3 \\ -3 & -5 \end{bmatrix}.$$

SOLUTION   For B to be the inverse of A, it must satisfy the equations $AB = I_2$ and $BA = I_2$.

$$AB = \begin{bmatrix} 5 & 3 \\ -3 & -2 \end{bmatrix} \begin{bmatrix} 2 & 3 \\ -3 & -5 \end{bmatrix} = \begin{bmatrix} 1 & 0 \\ 0 & 1 \end{bmatrix} = I_2$$

$$BA = \begin{bmatrix} 2 & 3 \\ -3 & -5 \end{bmatrix} \begin{bmatrix} 5 & 3 \\ -3 & -2 \end{bmatrix} = \begin{bmatrix} 1 & 0 \\ 0 & 1 \end{bmatrix} = I_2$$

Thus B is the inverse of A. That is, $B = A^{-1}$.

The next example discusses the significance of an inverse matrix in computer graphics.

EXAMPLE 2   *Interpreting an inverse matrix*

The matrix A can be used to rotate a point $90°$ clockwise about the origin, where

$$A = \begin{bmatrix} 0 & 1 & 0 \\ -1 & 0 & 0 \\ 0 & 0 & 1 \end{bmatrix} \quad \text{and} \quad A^{-1} = \begin{bmatrix} 0 & -1 & 0 \\ 1 & 0 & 0 \\ 0 & 0 & 1 \end{bmatrix}.$$

(a) Use A to rotate the point $(-2, 0)$ clockwise $90°$ about the origin.
(b) Make a conjecture about the effect of A^{-1} on the resulting point.
(c) Test this conjecture.

SOLUTION   **(a)** First, let the point $(-2, 0)$ be represented by the column matrix

$$X = \begin{bmatrix} -2 \\ 0 \\ 1 \end{bmatrix}.$$

Then compute

$$AX = \begin{bmatrix} 0 & 1 & 0 \\ -1 & 0 & 0 \\ 0 & 0 & 1 \end{bmatrix} \begin{bmatrix} -2 \\ 0 \\ 1 \end{bmatrix} = \begin{bmatrix} 0 \\ 2 \\ 1 \end{bmatrix} = Y.$$

If the point $(-2, 0)$ is rotated 90° clockwise about the origin, its new location is $(0, 2)$. See Figure 6.62.

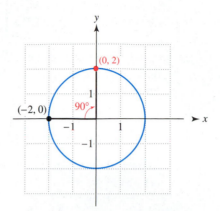

FIGURE 6.62 Rotating a Point about the Origin

(b) Since A^{-1} represents the inverse operation of A, A^{-1} will rotate the point located at $(0, 2)$ counterclockwise 90°, back to $(-2, 0)$.

(c) This conjecture is correct since

$$A^{-1}Y = \begin{bmatrix} 0 & -1 & 0 \\ 1 & 0 & 0 \\ 0 & 0 & 1 \end{bmatrix} \begin{bmatrix} 0 \\ 2 \\ 1 \end{bmatrix} = \begin{bmatrix} -2 \\ 0 \\ 1 \end{bmatrix} = X.$$

■ CLASS DISCUSSION

What will the results be of the computations AAX and $A^{-1}A^{-1}X$? ■

Representing Linear Systems with Matrix Equations

In Section 6.3 linear systems were solved using Gaussian elimination with backward substitution. This method used an augmented matrix to represent a system of linear equations. A system of linear equations can also be represented by a matrix equation.

$$3x - 2y + 4z = 5$$
$$2x + y + 3z = 9$$
$$-x + 5y - 2z = 5$$

Let A, X, and B be matrices defined as

Coefficient Matrix	Variable Matrix		Constant Matrix

$$A = \begin{bmatrix} 3 & -2 & 4 \\ 2 & 1 & 3 \\ -1 & 5 & -2 \end{bmatrix}, \quad X = \begin{bmatrix} x \\ y \\ z \end{bmatrix}, \quad \text{and} \quad B = \begin{bmatrix} 5 \\ 9 \\ 5 \end{bmatrix}.$$

The matrix product AX is given by

$$AX = \begin{bmatrix} 3 & -2 & 4 \\ 2 & 1 & 3 \\ -1 & 5 & -2 \end{bmatrix} \begin{bmatrix} x \\ y \\ z \end{bmatrix} = \begin{bmatrix} 3x + (-2)y + 4z \\ 2x + 1y + 3z \\ (-1)x + 5y + (-2)z \end{bmatrix} = \begin{bmatrix} 3x - 2y + 4z \\ 2x + y + 3z \\ -x + 5y - 2z \end{bmatrix}.$$

Thus the matrix equation $AX = B$ simplifies to

$$\begin{bmatrix} 3x - 2y + 4z \\ 2x + y + 3z \\ -x + 5y - 2z \end{bmatrix} = \begin{bmatrix} 5 \\ 9 \\ 5 \end{bmatrix}.$$

This matrix equation $AX = B$ is equivalent to the original system of linear equations. Any system of linear equations can be represented by a matrix equation in the form $AX = B$.

EXAMPLE 3 *Representing linear systems with matrix equations*

Represent each system of linear equations in the form $AX = B$.
(a) $3x - 4y = 7$
        $-x + 6y = -3$

(b) $x - 5y \qquad = 2$
        $-3x + 2y + z = -7$
        $4x + 5y + 6z = 10$

SOLUTION **(a)** This linear system comprises two equations and two variables. The equivalent matrix equation is

$$AX = \begin{bmatrix} 3 & -4 \\ -1 & 6 \end{bmatrix} \begin{bmatrix} x \\ y \end{bmatrix} = \begin{bmatrix} 7 \\ -3 \end{bmatrix} = B.$$

(b) The equivalent matrix equation is

$$AX = \begin{bmatrix} 1 & -5 & 0 \\ -3 & 2 & 1 \\ 4 & 5 & 6 \end{bmatrix} \begin{bmatrix} x \\ y \\ z \end{bmatrix} = \begin{bmatrix} 2 \\ -7 \\ 10 \end{bmatrix} = B.$$

Solving Linear Systems with Inverses

The matrix equation $AX = B$ can be solved using A^{-1}, if it exists.

$AX = B$	Linear system
$A^{-1}AX = A^{-1}B$	Multiply both sides by A^{-1}.
$I_nX = A^{-1}B$	$A^{-1}A = I_n$
$X = A^{-1}B$	$I_nX = X$ for any $n \times 1$ matrix X.

To solve a linear system, multiply both sides of the matrix equation $AX = B$ by A^{-1}, if it exists. The solution to the system can be written as $X = A^{-1}B$.

Note: Since matrix multiplication is not commutative, it is essential to multiply both sides of the equation on the *left* by A^{-1}. That is, $X = A^{-1}B \neq BA^{-1}$ in general.

EXAMPLE 4 *Solving a linear system using the inverse of a 2 × 2 matrix*

Write the linear system as a matrix equation in the form $AX = B$. Find A^{-1} and solve for X.

$$4x - 5y = 8.1$$
$$7x - 9y = -4.7$$

SOLUTION The linear system can be written as

$$AX = \begin{bmatrix} 4 & -5 \\ 7 & -9 \end{bmatrix} \begin{bmatrix} x \\ y \end{bmatrix} = \begin{bmatrix} 8.1 \\ -4.7 \end{bmatrix} = B.$$

The matrix A^{-1} can be found with a calculator, as shown in Figure 6.63. The solution to the system, given by the product $A^{-1}B$, is $x = 96.4$ and $y = 75.5$. See Figure 6.64.

Graphing Calculator Help

To find the inverse of a matrix, see Appendix B (page AP-20). To solve a linear system with a matrix inverse, see Appendix B (page AP-21).

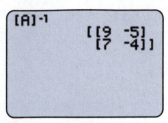

FIGURE 6.63

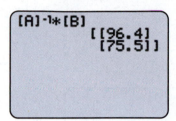

FIGURE 6.64

EXAMPLE 5 *Solving a linear system using the inverse of a 3 × 3 matrix*

Write the linear system as a matrix equation in the form $AX = B$. Find A^{-1} and solve for X.

$$\begin{aligned} x + 3y - z &= 6 \\ -2y + z &= -2 \\ -x + y - 3z &= 4 \end{aligned}$$

SOLUTION The linear system can be written as

$$AX = \begin{bmatrix} 1 & 3 & -1 \\ 0 & -2 & 1 \\ -1 & 1 & -3 \end{bmatrix} \begin{bmatrix} x \\ y \\ z \end{bmatrix} = \begin{bmatrix} 6 \\ -2 \\ 4 \end{bmatrix} = B.$$

The matrix A^{-1} can be found with a graphing calculator, as shown in Figure 6.65. The solution to the system is $x = 4.5$, $y = -0.5$, and $z = -3$. See Figure 6.66.

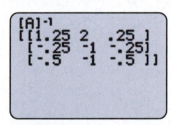

FIGURE 6.65

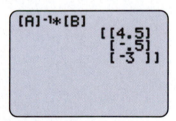

FIGURE 6.66

EXAMPLE 6 *Modeling blood pressure*

In one study of adult males, it was believed that systolic blood pressure P was affected by both age A in years and weight W in pounds. This was modeled by $P(A, W) = a + bA + cW$, where a, b, and c are constants. Table 6.9 lists three individuals with representative blood pressures for the group. (Reference: C. H. Brase and C. P. Brase, *Understandable Statistics*.)

(a) Use Table 6.9 to approximate values for the constants a, b, and c.

(b) Estimate a typical systolic blood pressure for an individual who is 55 years old and weighs 175 pounds.

TABLE 6.9

P	A	W
113	39	142
138	53	181
152	65	191

SOLUTION   **(a)** Determine the constants a, b, and c in $P(A, W) = a + bA + cW$ by solving the following three equations.

$$P(39, 142) = a + b(39) + c(142) = 113$$
$$P(53, 181) = a + b(53) + c(181) = 138$$
$$P(65, 191) = a + b(65) + c(191) = 152$$

These three equations can be rewritten as follows.

$$a + 39b + 142c = 113$$
$$a + 53b + 181c = 138$$
$$a + 65b + 191c = 152$$

FIGURE 6.67

This system can be represented by the matrix equation $AX = B$.

$$AX = \begin{bmatrix} 1 & 39 & 142 \\ 1 & 53 & 181 \\ 1 & 65 & 191 \end{bmatrix} \begin{bmatrix} a \\ b \\ c \end{bmatrix} = \begin{bmatrix} 113 \\ 138 \\ 152 \end{bmatrix} = B$$

The solution, $X = A^{-1}B$, is shown in Figure 6.67. The values for the constants are $a \approx 32.78$, $b \approx 0.9024$, and $c \approx 0.3171$. Thus P is given by $P(A, W) = 32.78 + 0.9024A + 0.3171W$.

(b) Evaluate $P(55, 175) = 32.78 + 0.9024(55) + 0.3171(175) \approx 137.9$. This model predicts that a typical (male) individual 55 years old, weighing 175 pounds, has a systolic blood pressure of approximately 138. Clearly, this could vary greatly among individuals.

Finding Inverses Symbolically

In Examples 4–6, technology was used to find A^{-1}. The inverse matrix can be found symbolically by first forming the augmented matrix $[A \mid I_n]$, and then performing matrix row operations, until the left side of the augmented matrix becomes the identity matrix. The resulting augmented matrix can be written as $[I_n \mid A^{-1}]$, where the right side of the matrix is A^{-1}. The next two examples illustrate this method.

EXAMPLE 7   *Finding the inverse of a 2 × 2 matrix symbolically*

Find A^{-1} if

$$A = \begin{bmatrix} 1 & 4 \\ 2 & 9 \end{bmatrix}.$$

SOLUTION   Begin by forming a 2 × 4 augmented matrix. Perform matrix row operations to obtain the identity matrix on the left side, and perform the same operation on the right side of this matrix.

$$\begin{bmatrix} 1 & 4 & | & 1 & 0 \\ 2 & 9 & | & 0 & 1 \end{bmatrix}$$

$$R_2 - 2R_1 \rightarrow \begin{bmatrix} 1 & 4 & | & 1 & 0 \\ 0 & 1 & | & -2 & 1 \end{bmatrix}$$

$$R_1 - 4R_1 \rightarrow \begin{bmatrix} 1 & 0 & | & 9 & -4 \\ 0 & 1 & | & -2 & 1 \end{bmatrix}$$

Since the 2×2 identity matrix appears on the left side, it follows that the right side equals A^{-1}. That is,

$$A^{-1} = \begin{bmatrix} 9 & -4 \\ -2 & 1 \end{bmatrix}.$$

Furthermore, it can be verified that $A^{-1}A = I_2 = AA^{-1}$.

EXAMPLE 8   *Finding the inverse of a 3 × 3 matrix symbolically*

Find A^{-1} if

$$A = \begin{bmatrix} 1 & 0 & 1 \\ 2 & 1 & 3 \\ -1 & 1 & 1 \end{bmatrix}.$$

SOLUTION   Begin by forming the following 3×6 augmented matrix. Perform matrix row operations to obtain the identity matrix on the left side, and perform the same operation on the right side of this matrix.

$$\left[\begin{array}{ccc|ccc} 1 & 0 & 1 & 1 & 0 & 0 \\ 2 & 1 & 3 & 0 & 1 & 0 \\ -1 & 1 & 1 & 0 & 0 & 1 \end{array} \right]$$

$$\begin{matrix} \\ R_2 - 2R_1 \rightarrow \\ R_3 + R_1 \rightarrow \end{matrix} \left[\begin{array}{ccc|ccc} 1 & 0 & 1 & 1 & 0 & 0 \\ 0 & 1 & 1 & -2 & 1 & 0 \\ 0 & 1 & 2 & 1 & 0 & 1 \end{array} \right]$$

$$\begin{matrix} \\ \\ R_3 - R_2 \rightarrow \end{matrix} \left[\begin{array}{ccc|ccc} 1 & 0 & 1 & 1 & 0 & 0 \\ 0 & 1 & 1 & -2 & 1 & 0 \\ 0 & 0 & 1 & 3 & -1 & 1 \end{array} \right]$$

$$\begin{matrix} R_1 - R_3 \rightarrow \\ R_2 - R_3 \rightarrow \\ \end{matrix} \left[\begin{array}{ccc|ccc} 1 & 0 & 0 & -2 & 1 & -1 \\ 0 & 1 & 0 & -5 & 2 & -1 \\ 0 & 0 & 1 & 3 & -1 & 1 \end{array} \right]$$

The right side is equal to A^{-1}. That is,

$$A^{-1} = \begin{bmatrix} -2 & 1 & -1 \\ -5 & 2 & -1 \\ 3 & -1 & 1 \end{bmatrix}.$$

It can be verified that $A^{-1}A = I_3 = AA^{-1}$.

Note: If it is not possible to obtain the identity matrix on the left side of the augmented matrix by using matrix row operations, then A^{-1} does not exist.

*Putting
it all
Together*

6.5

The inverse of a matrix A is denoted A^{-1}, if it exists. The inverse matrix will undo or cancel the operations performed by A. Inverse matrices frequently are used in computer graphics. They also can be used to solve systems of linear equations.

The following table summarizes some of the mathematical concepts presented in this section.

Concept	Comments	Examples
Identity matrix	The $n \times n$ identity matrix I_n has only 1's on the main diagonal and 0's elsewhere. When it is multiplied by any $n \times n$ matrix A, the result is A.	$\begin{bmatrix} 1 & 0 \\ 0 & 1 \end{bmatrix}\begin{bmatrix} 2 & 3 \\ 4 & 5 \end{bmatrix} = \begin{bmatrix} 2 & 3 \\ 4 & 5 \end{bmatrix}$ and $\begin{bmatrix} 2 & 3 \\ 4 & 5 \end{bmatrix}\begin{bmatrix} 1 & 0 \\ 0 & 1 \end{bmatrix} = \begin{bmatrix} 2 & 3 \\ 4 & 5 \end{bmatrix},$ where $I_2 = \begin{bmatrix} 1 & 0 \\ 0 & 1 \end{bmatrix}.$
Matrix inverse	If an $n \times n$ matrix A has an inverse it is unique, is denoted A^{-1}, and satisfies the equations $AA^{-1} = I_n$ and $A^{-1}A = I_n$. Matrix inverses can be found using technology. They can also be found using pencil and paper by performing matrix row operations on the augmented matrix $[A \mid I_n]$ until it is transformed to $[I_n \mid A^{-1}]$.	If $A = \begin{bmatrix} 2 & 3 \\ 3 & 5 \end{bmatrix}$, then $A^{-1} = \begin{bmatrix} 5 & -3 \\ -3 & 2 \end{bmatrix}$ because $AA^{-1} = \begin{bmatrix} 2 & 3 \\ 3 & 5 \end{bmatrix}\begin{bmatrix} 5 & -3 \\ -3 & 2 \end{bmatrix} = \begin{bmatrix} 1 & 0 \\ 0 & 1 \end{bmatrix}$ and $A^{-1}A = \begin{bmatrix} 5 & -3 \\ -3 & 2 \end{bmatrix}\begin{bmatrix} 2 & 3 \\ 3 & 5 \end{bmatrix} = \begin{bmatrix} 1 & 0 \\ 0 & 1 \end{bmatrix}.$
Systems of linear equations	Systems of linear equations can be written using the matrix equation $AX = B$. If A is invertible, then there will be a unique solution given by $X = A^{-1}B$. If A is not invertible, then there could be either no solution or an infinite number of solutions. In the latter case, Gaussian elimination should be applied.	The linear system $\quad\begin{aligned} 2x - y &= 3 \\ x + 2y &= 4 \end{aligned}$ can be written as $AX = B$, where $A = \begin{bmatrix} 2 & -1 \\ 1 & 2 \end{bmatrix}, X = \begin{bmatrix} x \\ y \end{bmatrix},$ and $B = \begin{bmatrix} 3 \\ 4 \end{bmatrix}.$ The solution to the system is given by $X = A^{-1}B = \begin{bmatrix} 0.4 & 0.2 \\ -0.2 & 0.4 \end{bmatrix}\begin{bmatrix} 3 \\ 4 \end{bmatrix} = \begin{bmatrix} 2 \\ 1 \end{bmatrix}.$ The solution is $x = 2$ and $y = 1$.

6.5 EXERCISES

Identifying Inverse Matrices

Exercises 1–4: Determine if B is the inverse matrix of A by calculating AB and BA.

1. $A = \begin{bmatrix} 4 & 3 \\ 5 & 4 \end{bmatrix}$, $B = \begin{bmatrix} 4 & -3 \\ -5 & 4 \end{bmatrix}$

2. $A = \begin{bmatrix} -1 & 2 \\ -3 & 8 \end{bmatrix}$, $B = \begin{bmatrix} -4 & 1 \\ -2 & 0.5 \end{bmatrix}$

3. $A = \begin{bmatrix} 1 & -1 & 2 \\ 0 & 1 & -1 \\ 1 & 0 & 2 \end{bmatrix}$, $B = \begin{bmatrix} 2 & 2 & -1 \\ -1 & 0 & 1 \\ -1 & -1 & 1 \end{bmatrix}$

4. $A = \begin{bmatrix} 2 & 1 & 1 \\ -1 & 0 & -1 \\ 0 & 2 & -1 \end{bmatrix}$, $B = \begin{bmatrix} 2 & 3 & -1 \\ -1 & -2 & 1 \\ -2 & -4 & 1 \end{bmatrix}$

Exercises 5–8: Determine the value of the constant k in the matrix B so that $B = A^{-1}$.

5. $A = \begin{bmatrix} 1 & 1 \\ 1 & 2 \end{bmatrix}$, $B = \begin{bmatrix} 2 & -1 \\ -1 & k \end{bmatrix}$

6. $A = \begin{bmatrix} -2 & 2 \\ 1 & -2 \end{bmatrix}$, $B = \begin{bmatrix} -1 & k \\ -0.5 & -1 \end{bmatrix}$

7. $A = \begin{bmatrix} 1 & 3 \\ -1 & -5 \end{bmatrix}$, $B = \begin{bmatrix} k & 1.5 \\ -0.5 & -0.5 \end{bmatrix}$

8. $A = \begin{bmatrix} -2 & 5 \\ -3 & 4 \end{bmatrix}$, $B = \begin{bmatrix} \dfrac{4}{7} & -\dfrac{5}{7} \\ k & -\dfrac{2}{7} \end{bmatrix}$

Exercises 9–12: Predict the results of $I_n A$ and $A I_n$. Then, verify your prediction.

9. $I_2 = \begin{bmatrix} 1 & 0 \\ 0 & 1 \end{bmatrix}$, $A = \begin{bmatrix} 1 & -2 \\ 4 & 3 \end{bmatrix}$

10. $I_3 = \begin{bmatrix} 1 & 0 & 0 \\ 0 & 1 & 0 \\ 0 & 0 & 1 \end{bmatrix}$, $A = \begin{bmatrix} 1 & -4 & 3 \\ 1 & 9 & 5 \\ 3 & -5 & 0 \end{bmatrix}$

11. $I_3 = \begin{bmatrix} 1 & 0 & 0 \\ 0 & 1 & 0 \\ 0 & 0 & 1 \end{bmatrix}$, $A = \begin{bmatrix} 0 & 0 & 0 \\ 0 & 0 & 0 \\ 0 & 0 & 0 \end{bmatrix}$

12. $I_4 = \begin{bmatrix} 1 & 0 & 0 & 0 \\ 0 & 1 & 0 & 0 \\ 0 & 0 & 1 & 0 \\ 0 & 0 & 0 & 1 \end{bmatrix}$, $A = \begin{bmatrix} 5 & -2 & 6 & -3 \\ 0 & 1 & 4 & -1 \\ -5 & 7 & 9 & 8 \\ 0 & 0 & 3 & 1 \end{bmatrix}$

Interpreting Inverses

Exercises 13 and 14: **Translations**   *(Refer to the discussion in this section about translating a point.) The matrix product AX performs a translation on the point (x, y), where*

$$A = \begin{bmatrix} 1 & 0 & h \\ 0 & 1 & k \\ 0 & 0 & 1 \end{bmatrix} \quad \text{and} \quad X = \begin{bmatrix} x \\ y \\ 1 \end{bmatrix}.$$

(a) *Predict the new location of the point (x, y), when it is translated by A. Compute $Y = AX$ to verify your prediction.*

(b) *Make a conjecture as to what $A^{-1}Y$ represents. Find A^{-1} and calculate $A^{-1}Y$ to test your conjecture.*

(c) *What will AA^{-1} and $A^{-1}A$ equal?*

13. $A = \begin{bmatrix} 1 & 0 & 2 \\ 0 & 1 & 3 \\ 0 & 0 & 1 \end{bmatrix}$, $(x, y) = (0, 1)$, and $X = \begin{bmatrix} 0 \\ 1 \\ 1 \end{bmatrix}$

14. $A = \begin{bmatrix} 1 & 0 & -4 \\ 0 & 1 & 5 \\ 0 & 0 & 1 \end{bmatrix}$, $(x, y) = (4, 2)$, and $X = \begin{bmatrix} 4 \\ 2 \\ 1 \end{bmatrix}$

Exercises 15 and 16: **Translations**   *(Refer to the discussion in this section about translating a point.) Find a 3×3 matrix A that performs the following translation of a point (x, y) represented by X. Find A^{-1} and describe what it computes.*

15. 3 units to the left and 5 units downward

16. 6 units to the right and 1 unit upward

17. *Rotation* (Refer to Example 2.) The matrix B rotates (x, y) clockwise about the origin 45°, where

$$B = \begin{bmatrix} \dfrac{1}{\sqrt{2}} & \dfrac{1}{\sqrt{2}} & 0 \\[2mm] -\dfrac{1}{\sqrt{2}} & \dfrac{1}{\sqrt{2}} & 0 \\[2mm] 0 & 0 & 1 \end{bmatrix} \quad \text{and}$$

$$B^{-1} = \begin{bmatrix} \dfrac{1}{\sqrt{2}} & -\dfrac{1}{\sqrt{2}} & 0 \\[2mm] \dfrac{1}{\sqrt{2}} & \dfrac{1}{\sqrt{2}} & 0 \\[2mm] 0 & 0 & 1 \end{bmatrix}.$$

(a) Let X represent the point $(-\sqrt{2}, -\sqrt{2})$. Compute $Y = BX$.

(b) Find $B^{-1}Y$. Interpret the computation performed by B^{-1}.

18. *Rotation* (Continuation of the previous exercise.) Predict the result of the computations $BB^{-1}X$ and $B^{-1}BX$ for any point (x, y) represented by X. Explain this result geometrically.

19. *Translations* The matrix A translates a point to the right 4 units and downward 2 units, and the matrix B translates a point to the left 3 units and upward 3 units, where

$$A = \begin{bmatrix} 1 & 0 & 4 \\ 0 & 1 & -2 \\ 0 & 0 & 1 \end{bmatrix} \quad \text{and} \quad B = \begin{bmatrix} 1 & 0 & -3 \\ 0 & 1 & 3 \\ 0 & 0 & 1 \end{bmatrix}.$$

(a) Let X represent the point $(1, 1)$. Predict the result of $Y = ABX$. Check your prediction.

(b) Predict the form of the matrix product AB, and then compute AB.

(c) In this exercise, would you expect $AB = BA$? Verify your answer.

(d) Find $(AB)^{-1}$ mentally. Explain your reasoning.

20. *Rotation* (Refer to Exercises 13 and 17 for matrices A and B.) Use a calculator to find any matrix products.

(a) Let X represent the point $(0, \sqrt{2})$. If this point is rotated about the origin 45° clockwise, and then translated 2 units to the right and 3 units upward, determine its new coordinates geometrically.

(b) Compute the matrix product $Y = ABX$, and explain the result.

(c) Is your answer in part (b) equal to BAX? Interpret your answer.

(d) Find a matrix that translates Y back to X. Test your matrix.

Calculating Inverses

Exercises 21–28: Use a graphing calculator to calculate the inverse of A.

21. $A = \begin{bmatrix} 0.5 & -1.5 \\ 0.2 & -0.5 \end{bmatrix}$

22. $A = \begin{bmatrix} -0.5 & 0.5 \\ 3 & 2 \end{bmatrix}$

23. $A = \begin{bmatrix} 1 & 2 & 0 \\ -1 & 4 & -1 \\ 2 & -1 & 0 \end{bmatrix}$

24. $A = \begin{bmatrix} -2 & 0 & 1 \\ 5 & -4 & 1 \\ 1 & -2 & 0 \end{bmatrix}$

25. $A = \begin{bmatrix} 2 & -2 & 1 \\ 0 & 5 & 8 \\ 0 & 0 & -1 \end{bmatrix}$

26. $A = \begin{bmatrix} 2 & 0 & 2 \\ 1 & 5 & 0 \\ -1 & 0 & 2 \end{bmatrix}$

27. $A = \begin{bmatrix} 3 & -1 & -1 \\ -1 & 3 & -1 \\ -1 & -1 & 3 \end{bmatrix}$

28. $A = \begin{bmatrix} 3 & 1 & 0 & 0 \\ 1 & 3 & 1 & 0 \\ 0 & 1 & 3 & 1 \\ 0 & 0 & 1 & 3 \end{bmatrix}$

Exercises 29–36: (Refer to Examples 7 and 8.) Find A^{-1} without a calculator.

29. $A = \begin{bmatrix} 1 & 2 \\ 1 & 3 \end{bmatrix}$

30. $A = \begin{bmatrix} 1 & 0 \\ 1 & -1 \end{bmatrix}$

31. $A = \begin{bmatrix} -1 & 2 \\ 3 & -5 \end{bmatrix}$

32. $A = \begin{bmatrix} 1 & 3 \\ 2 & 5 \end{bmatrix}$

33. $A = \begin{bmatrix} 0 & 0 & 1 \\ 1 & 0 & 0 \\ 0 & 1 & 0 \end{bmatrix}$

34. $A = \begin{bmatrix} 1 & 0 & 0 \\ 1 & 1 & 0 \\ 0 & 1 & 1 \end{bmatrix}$

35. $A = \begin{bmatrix} 1 & 0 & 1 \\ 2 & 1 & 3 \\ -1 & 1 & 1 \end{bmatrix}$

36. $A = \begin{bmatrix} -2 & 1 & 0 \\ 1 & 0 & 1 \\ -1 & 1 & 0 \end{bmatrix}$

Solving Linear Systems

Exercises 37–44: Complete the following for the given system of linear equations.

(a) Write the system in the form $AX = B$.

(b) Solve the linear system by computing $X = A^{-1}B$ with a calculator. Approximate the solution to the nearest hundredth when appropriate.

37. $\begin{aligned} 1.5x + 3.7y &= 0.32 \\ -0.4x - 2.1y &= 0.36 \end{aligned}$

38. $\begin{aligned} 31x + 18y &= 64.1 \\ 5x - 23y &= -59.6 \end{aligned}$

39. $\begin{aligned} 0.08x - 0.7y &= -0.504 \\ 1.1x - 0.05y &= 0.73 \end{aligned}$

40. $-231x + 178y = -439$
$\quad\;\; 525x - 329y = 2282$

41. $3.1x + 1.9y - \;\;\;\; z = \;\;\;\; 1.99$
$\quad\;\, 6.3x \qquad\quad - 9.9z = -3.78$
$\quad\; -x + 1.5y + \;\;\; 7z = \;\;\;\; 5.3$

42. $17x - 22y - \;\;\, 19z = -25.2$
$\quad\;\;\; 3x + 13y - \;\;\;\; 9z = \;\; 105.9$
$\quad\;\;\;\; x - \;\; 2y + 6.1z = -23.55$

43. $\;\;\;\, 3x - \;\;\; y + \;\;\;\; z = \;\;\;\; 4.9$
$\quad\; 5.8x - 2.1y \qquad\quad = -3.8$
$\quad\; -x \qquad\quad + 2.9z = \;\;\;\; 3.8$

44. $\;\;\;\;\, 1.2x - 0.3y - 0.7z = -0.5$
$\quad -0.4x + 1.3y + 0.4z = \;\;\; 0.9$
$\quad\;\;\; 1.7x + 0.6y + 1.1z = \;\;\; 1.3$

Exercises 45 and 46: Use your results from Exercises 35 and 36 to solve the system of equations.

45. $\;\;\; x + \qquad\;\; z = \;\;\; -7$
$\quad\; 2x + y + 3z = -13$
$\quad -x + y + \;\;\; z = \;\;\; -4$

46. $-2x + y \qquad\quad = -5$
$\quad\;\;\;\; x + \qquad z = -5$
$\quad\; -x + y \qquad\quad = -4$

Applications

47. *Cost of CDs* A music store has compact discs that sell for three prices marked A, B, and C. The last column in the table shows the total cost of a purchase. Use this information to determine the cost of one CD of each type by setting up a matrix equation and solving it with an inverse.

A	B	C	Total
2	3	4	$120.91
1	4	0	$62.95
2	1	3	$79.94

48. *Determining Prices* A group of students bought 3 soft drinks and 2 boxes of popcorn at a movie for $8.50. A second group bought 4 soft drinks and 3 boxes of popcorn for $12.
  (a) Find a matrix equation $AX = B$ whose solution gives the individual prices of a soft drink and a box of popcorn. Solve this matrix equation using A^{-1}.
  (b) Could these prices be determined if both groups had bought 3 soft drinks and 2 boxes of popcorn for $8.50? Try to calculate A^{-1} and explain your results.

49. *Traffic Flow* (Refer to Exercises 59 and 60 in Section 6.3.) The accompanying figure shows four one-way streets with intersections A, B, C, and D. Numbers indicate the average traffic flow in vehicles per minute. The variables x_1, x_2, x_3, and x_4 denote unknown traffic flows.
  (a) The number of vehicles per minute entering an intersection equals the number exiting an intersection. Verify that the given system of linear equations describes the traffic flow.
  A: $x_1 + 5 = 4 + 6$
  B: $x_2 + 6 = x_1 + 3$
  C: $x_3 + 4 = x_2 + 7$
  D: $6 + 5 = x_3 + x_4$
  (b) Write the system as $AX = B$ and solve using A^{-1}.
  (c) Interpret your results.

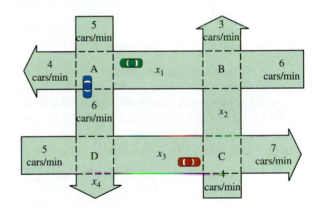

50. *Tire Sales* In one study the relationship among annual tire sales T in thousands, automobile registrations A in millions, and personal disposable income I in millions of dollars was investigated. Representative data for three different years are shown in the table.   (Source: J. Jarrett, *Business Forecasting Methods.*)

T	A	I
10,170	113	308
15,305	133	622
21,289	155	1937

The data were modeled by $T = aA + bI + c$, where a, b, and c are constants.
  (a) Use the data to write a system of linear equations, whose solution gives a, b, and c.
  (b) Solve this linear system. Write a formula for T.
  (c) If $A = 118$ and $I = 311$, predict T. (The actual value for T was 11,314.)

51. *Home Prices* The selling price of a home can depend on several factors. Real estate companies sometimes study how the selling price of a home is related to its size and condition. The accompanying table contains representative data for sales of three homes. Price P is measured in thousands of dollars, home size S in square feet, and condition C is rated on a scale from 1 to 10, where 10 represents excellent condition. The variables were found to be related by the linear equation $P = a + bS + cC$.

P	S	C
122	1500	8
130	2000	5
158	2200	10

(a) Use the table to write a system of linear equations whose solution gives a, b, and c.

(b) Estimate the selling price of a home with 1800 square feet and a condition of 7.

52. *Plate Glass Sales* The amount of plate glass sales G can be affected by the number of new building contracts B issued and automobiles A produced, since plate glass is used in buildings and cars. A plate glass company in California wanted to forecast sales. The table contains sales data for three consecutive years. All units are in millions. (Source: S. Makridakis and S. Wheelwright, *Forecasting Methods for Management.*)

G	A	B
603	5.54	37.1
657	6.93	41.3
779	7.64	45.6

The data were modeled by $G = aA + bB + c$, where a, b, and c are constants.

(a) Write a system of linear equations whose solution gives a, b, and c.

(b) Solve this linear system. Write a formula for G.

(c) For the following year, it was estimated that $A = 7.75$ and $B = 47.4$. Predict G. (The actual value for G was $878 million.)

Writing about Mathematics

53. Discuss how to solve the matrix equation $AX = B$ if A^{-1} exists.

54. Give an example of a 2×2 matrix A that does not have an inverse. Explain what happens if one attempts to find the inverse of A symbolically.

6.6 DETERMINANTS

Definition and Calculation of Determinants • Area of Regions • Cramer's Rule • Limitations on the Method of Cofactors and Cramer's Rule

Introduction

Determinants are used in mathematics for theoretical purposes. However, they also are used to test if a matrix is invertible and to find the area of certain geometric figures. A *determinant* is a real number associated with a square matrix. We begin our discussion by defining a determinant for a 2×2 matrix.

Definition and Calculation of Determinants

The determinant of a matrix with dimension 2×2 is a straightforward arithmetic calculation.

Determinant of a 2 × 2 matrix

The **determinant** of

$$A = \begin{bmatrix} a & b \\ c & d \end{bmatrix}$$

is a real number defined by

$$\det A = ad - cb.$$

Later we define determinants for any $n \times n$ matrix. The following can be used to determine if a matrix is invertible.

Invertible matrix

A square matrix A is invertible if and only if $\det A \neq 0$.

EXAMPLE 1   *Determining if a 2 × 2 matrix is invertible*

Determine if A^{-1} exists by computing the determinant of the matrix A.

(a) $A = \begin{bmatrix} 3 & -4 \\ -5 & 9 \end{bmatrix}$ 　　　　　　　　　 **(b)** $A = \begin{bmatrix} 52 & -32 \\ 65 & -40 \end{bmatrix}$

SOLUTION   **(a)** The determinant of the 2 × 2 matrix A is calculated as follows.

$$\det A = \det \begin{bmatrix} 3 & -4 \\ -5 & 9 \end{bmatrix} = (3)(9) - (-5)(-4) = 7$$

Since $\det A = 7 \neq 0$, the matrix A is invertible and A^{-1} exists.

(b) In a similar manner,

$$\det A = \det \begin{bmatrix} 52 & -32 \\ 65 & -40 \end{bmatrix} = (52)(-40) - (65)(-32) = 0$$

Since $\det A = 0$, A^{-1} does not exist. Try finding A^{-1}. What happens?

We can use determinants of 2 × 2 matrices to find determinants of larger square matrices. In order to do this, we first define the concepts of a *minor* and a *cofactor*.

Minors and cofactors

The **minor**, denoted by M_{ij}, for element a_{ij} in the square matrix A is the real number computed by performing the following steps.

STEP 1: Delete the ith row and jth column from the matrix A.

STEP 2: M_{ij} is equal to the determinant of the resulting matrix.

The **cofactor**, denoted A_{ij}, for a_{ij} is defined by $A_{ij} = (-1)^{i+j} M_{ij}$.

EXAMPLE 2   *Calculating minors and cofactors*

Find the following minors and cofactors for the matrix A.

$$A = \begin{bmatrix} 2 & -3 & 1 \\ -2 & 1 & 0 \\ 0 & -1 & 4 \end{bmatrix}$$

(a) M_{11} and M_{21}

(b) A_{11} and A_{21}

SOLUTION   **(a)** To obtain the minor M_{11}, begin by crossing out the first row and first column of A.

$$A = \begin{bmatrix} 2 & -3 & 1 \\ -2 & 1 & 0 \\ 0 & -1 & 4 \end{bmatrix}$$

The remaining elements form the 2×2 matrix

$$B = \begin{bmatrix} 1 & 0 \\ -1 & 4 \end{bmatrix}.$$

The minor M_{11} is equal to det $B = (1)(4) - (-1)(0) = 4$.

M_{21} is found by crossing out the second row and first column of A.

$$A = \begin{bmatrix} 2 & -3 & 1 \\ -2 & 1 & 0 \\ 0 & -1 & 4 \end{bmatrix}$$

The resulting matrix is

$$B = \begin{bmatrix} -3 & 1 \\ -1 & 4 \end{bmatrix}.$$

Thus $M_{21} = \det B = (-3)(4) - (-1)(1) = -11$.

(b) Since $A_{ij} = (-1)^{i+j} M_{ij}$, A_{11} and A_{21} can be computed as follows.

$$A_{11} = (-1)^{1+1} M_{11} = (-1)^2 (4) = 4$$
$$A_{21} = (-1)^{2+1} M_{21} = (-1)^3 (-11) = 11$$

Using the concept of a cofactor, we can calculate the determinant of any square matrix.

Determinant of a matrix using the method cofactors

Multiply each element in any row or column of the matrix by its cofactor. The sum of the products is equal to the determinant.

To compute the determinant of a 3×3 matrix A, begin by selecting a row or column.

$$A = \begin{bmatrix} a_{11} & a_{12} & a_{13} \\ a_{21} & a_{22} & a_{23} \\ a_{31} & a_{32} & a_{33} \end{bmatrix}$$

For example, if the second row of A is selected, the elements are a_{21}, a_{22}, and a_{23}. Then,

$$\det A = a_{21}A_{21} + a_{22}A_{22} + a_{23}A_{23}.$$

On the other hand, utilizing the elements of a_{11}, a_{21}, and a_{31} in the first column results in

$$\det A = a_{11}A_{11} + a_{21}A_{21} + a_{31}A_{31}.$$

Regardless of the row or column selected, the value of det A is the same. The calculation is easier if some elements in the selected row or column equal 0.

EXAMPLE 3   *Evaluating the determinant of a 3 × 3 matrix*

Find det A, if

$$A = \begin{bmatrix} 2 & -3 & 1 \\ -2 & 1 & 0 \\ 0 & -1 & 4 \end{bmatrix}.$$

SOLUTION   To find the determinant of A, we can select any row or column. If we begin expanding about the first column of A, then

$$\det A = a_{11}A_{11} + a_{21}A_{21} + a_{31}A_{31}.$$

In the first column, $a_{11} = 2$, $a_{21} = -2$, $a_{31} = 0$. In Example 2, the cofactors A_{11} and A_{21} were computed as 4 and 11, respectively. Since A_{31} is multiplied by $a_{31} = 0$, its value is unimportant. Thus

$$\begin{aligned} \det &= a_{11}A_{11} + a_{21}A_{21} + a_{31}A_{31} \\ &= 2(4) + (-2)(11) + (0)A_{31} \\ &= -14. \end{aligned}$$

We could have also expanded about the second row.

$$\begin{aligned} \det A &= a_{21}A_{21} + a_{22}A_{22} + a_{23}A_{23} \\ &= (-2)A_{21} + (1)A_{22} + (0)A_{23} \end{aligned}$$

To complete this computation we need to determine only A_{22}, since A_{21} is known to be 11 and A_{23} is multiplied by 0. To compute A_{22}, delete the second row and column of A to obtain M_{22}.

$$M_{22} = \det \begin{bmatrix} 2 & 1 \\ 0 & 4 \end{bmatrix} = 8 \quad \text{and} \quad A_{22} = (-1)^{2+2}(8) = 8$$

Thus det $A = (-2)(11) + (1)(8) + (0)A_{23} = -14$. The same value for det A is obtained in both calculations.

■ **CLASS DISCUSSION**

If a row or column in matrix A contains only zeros, what is det A? ■

Instead of calculating $(-1)^{i+j}$ for each cofactor, the following sign matrix can be utilized to find determinants of 3×3 matrices. The checkerboard pattern can be expanded to include larger matrices.

$$\begin{bmatrix} + & - & + \\ - & + & - \\ + & - & + \end{bmatrix}$$

For example, if

$$A = \begin{bmatrix} 2 & 3 & 7 \\ -3 & -2 & -1 \\ 4 & 0 & 2 \end{bmatrix},$$

we can compute det A by expanding about the second column to take advantage of the 0. The second column contains $-$, $+$, and $-$ signs. Therefore

$$\det A = -(3) \det \begin{bmatrix} -3 & -1 \\ 4 & 2 \end{bmatrix} + (-2) \det \begin{bmatrix} 2 & 7 \\ 4 & 2 \end{bmatrix} - (0) \det \begin{bmatrix} 2 & 7 \\ -3 & -1 \end{bmatrix}$$

$$= -3(-2) + (-2)(-24) - (0)(19)$$

$$= 54.$$

Many graphing calculators can evaluate the determinant of a matrix, as illustrated in the next example.

EXAMPLE 4   *Using technology to find a determinant*

Find the determinant of A.

(a) $A = \begin{bmatrix} 2 & -3 & 1 \\ -2 & 1 & 0 \\ 0 & -1 & 4 \end{bmatrix}$

(b) $A = \begin{bmatrix} 2 & -3 & 1 & 5 \\ 7 & 1 & -8 & 0 \\ 5 & 4 & 9 & 7 \\ -2 & 3 & 3 & 0 \end{bmatrix}$

SOLUTION   (a) The determinant of this matrix was calculated in Example 3 by hand. To use technology, enter the matrix and evaluate its determinant, as shown in Figure 6.68. The result is det $A = -14$, which agrees with our earlier calculation.

Graphing Calculator Help

To calculate a determinant, see Appendix B (page AP-21).

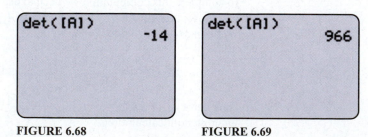

FIGURE 6.68   **FIGURE 6.69**

(b) The determinant of a 4×4 matrix can be computed using cofactors. However, it is considerably easier to use technology. From Figure 6.69 we see that det $A = 966$.

Area of Regions

Determinants may be used to find the area of a triangle. If a triangle has vertices (a_1, b_1), (a_2, b_2), and (a_3, b_3), as shown in Figure 6.70, then its area is equal to the absolute value of D, where

$$D = \frac{1}{2} \det \begin{bmatrix} a_1 & a_2 & a_3 \\ b_1 & b_2 & b_3 \\ 1 & 1 & 1 \end{bmatrix}.$$

If the vertices are entered into the columns of D in a counterclockwise direction, then D will be positive.   (Source: W. Taylor, *The Geometry of Computer Graphics.*)

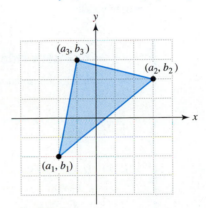

FIGURE 6.70

EXAMPLE 5   *Computing area of a parallelogram*

Calculate the area of the parallelogram in Figure 6.71 with determinants.

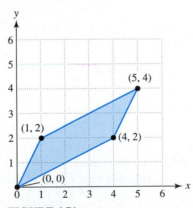

FIGURE 6.71

SOLUTION   To find the area of the parallelogram, we view the parallelogram as comprising two triangles. One possible triangle has vertices at (0, 0), (4, 2), and (1, 2), and the other triangle has vertices at (4, 2), (5, 4), and (1, 2). The area of the parallelogram is equal to the sum of the areas of the two triangles. Since these triangles are congruent, we can calculate the area of one triangle and double it. The area of one triangle is equal to D, where

$$D = \frac{1}{2} \det \begin{bmatrix} 0 & 4 & 1 \\ 0 & 2 & 2 \\ 1 & 1 & 1 \end{bmatrix} = 3.$$

Since the vertices were entered in a counterclockwise direction, D is positive. The area of one triangle is equal to 3. Therefore the area of the parallelogram is twice this value or 6.

■ **CLASS DISCUSSION**

Suppose we are given three distinct vertices and $D = 0$. What must be true about the three points? ■

Cramer's Rule

Determinants can be used to solve linear systems by employing a method called **Cramer's rule**.

> ### Cramer's rule for linear systems in two variables
>
> The solution to the linear system
>
> $$a_1 x + b_1 y = c_1$$
> $$a_2 x + b_2 y = c_2$$
>
> is given by $x = \dfrac{E}{D}$ and $y = \dfrac{F}{D}$, where
>
> $$E = \det \begin{bmatrix} c_1 & b_1 \\ c_2 & b_2 \end{bmatrix}, \quad F = \det \begin{bmatrix} a_1 & c_1 \\ a_2 & c_2 \end{bmatrix}, \quad \text{and} \quad D = \det \begin{bmatrix} a_1 & b_1 \\ a_2 & b_2 \end{bmatrix} \neq 0.$$

EXAMPLE 6 *Using Cramer's rule to solve a linear system in two variables*

Use Cramer's rule to solve the linear system

$$4x + y = 146$$
$$9x + y = 66.$$

SOLUTION In this system $a_1 = 4$, $b_1 = 1$, $c_1 = 146$, $a_2 = 9$, $b_2 = 1$, and $c_2 = 66$. By Cramer's rule, the solution can be found as follows.

$$E = \det \begin{bmatrix} c_1 & b_1 \\ c_2 & b_2 \end{bmatrix} = \det \begin{bmatrix} 146 & 1 \\ 66 & 1 \end{bmatrix} = (146)(1) - (66)(1) = 80$$

$$F = \det \begin{bmatrix} a_1 & c_1 \\ a_2 & c_2 \end{bmatrix} = \det \begin{bmatrix} 4 & 146 \\ 9 & 66 \end{bmatrix} = (4)(66) - (9)(146) = -1050$$

$$D = \det \begin{bmatrix} a_1 & b_1 \\ a_2 & b_2 \end{bmatrix} = \det \begin{bmatrix} 4 & 1 \\ 9 & 1 \end{bmatrix} = (4)(1) - (9)(1) = -5$$

The solution is

$$x = \frac{E}{D} = \frac{80}{-5} = -16 \qquad \text{and} \qquad y = \frac{F}{D} = \frac{-1050}{-5} = 210.$$

Limitations on the Method of Cofactors and Cramer's Rule

Systems of linear equations involving more than two variables can be solved with Cramer's rule. If a linear system has n equations, then Cramer's rule requires the computation of $n + 1$ determinants with dimension $n \times n$. Cramer's rule is seldom employed in real applications because of the substantial number of arithmetic operations needed to compute determinants of large matrices.

It can be shown that the cofactor method of calculating the determinant of an $n \times n$ matrix, $n > 2$, generally involves more than $n!$ multiplication operations. For example, a 4×4 determinant requires over $4! = 4 \cdot 3 \cdot 2 \cdot 1 = 24$ multiplication operations, and a 10×10 determinant would involve over

$$10! = 10 \cdot 9 \cdot 8 \cdot 7 \cdot 6 \cdot 5 \cdot 4 \cdot 3 \cdot 2 \cdot 1 = 3,628,800$$

multiplication operations. Factorials grow rapidly and are discussed in Chapter 8.

In real-life applications, it is not uncommon to solve linear systems that involve thousands of equations. Suppose that we were to solve a modest linear system that involved 20 equations with Cramer's rule, and expand each determinant by the method of cofactors. Then, the calculation of one 20×20 determinant would require over $20!$ multiplication operations. Supercomputers can perform 1 trillion (10^{12}) multiplication operations per second. On this supercomputer, just one 20×20 determinant would require over

$$\frac{20!}{10^{12}} \approx 2,432,902 \text{ seconds} \approx 28.2 \text{ days}$$

to compute. We would need to compute 21 of these determinants. This would require approximately $21 \times 28.2 = 592.2$ days, which is the best part of 2 years. Also, it is not uncommon for the cost of supercomputer time to exceed \$2000 per hour. In addition to the time element, these computations would cost over \$28 million to perform. On a typical personal computer, this task could not be completed in a lifetime.

It becomes obvious why this method is not implemented even on a modest linear system, regardless of the technology available. Modern software packages are more efficient than Cramer's rule on linear systems that involve as few as three variables. (Source: Minnesota Supercomputer Institute.)

Putting it all Together 6.6

*T*he determinant of a square matrix A is a real number, denoted det A. If det $A \neq 0$, then the matrix A is invertible. The computation of a determinant frequently involves a large number of arithmetic operations. Graphing calculators can be a valuable aid in computing determinants. Cramer's rule is a method for solving systems of linear equations that involves determinants.

The following table summarizes the calculation of 2×2 and 3×3 determinants by hand.

Determinants of 2 × 2 Matrices

The determinant of a 2×2 matrix A is given by

$$\det A = \det \begin{bmatrix} a & b \\ c & d \end{bmatrix} = ad - cb.$$

Example: $\det \begin{bmatrix} 6 & -2 \\ 3 & 7 \end{bmatrix} = (6)(7) - (3)(-2) = 48$

Determinants of 3 × 3 Matrices

The determinant of a 3 × 3 matrix A can be reduced to calculating the determinants of three 2 × 2 matrices. This calculation can be performed using cofactors.

$$\det A = \det \begin{bmatrix} a_1 & b_1 & c_1 \\ a_2 & b_2 & c_2 \\ a_3 & b_3 & c_3 \end{bmatrix}$$

$$= a_1 \det \begin{bmatrix} b_2 & c_2 \\ b_3 & c_3 \end{bmatrix} - a_2 \det \begin{bmatrix} b_1 & c_1 \\ b_3 & c_3 \end{bmatrix} + a_3 \det \begin{bmatrix} b_1 & c_1 \\ b_2 & c_2 \end{bmatrix}$$

$$= a_1(b_2c_3 - b_3c_2) - a_2(b_1c_3 - b_3c_1) + a_3(b_1c_2 - b_2c_1)$$

Example: $\det \begin{bmatrix} 1 & -2 & 3 \\ 4 & 5 & -1 \\ -3 & 7 & 8 \end{bmatrix} = (1) \det \begin{bmatrix} 5 & -1 \\ 7 & 8 \end{bmatrix} - (4) \det \begin{bmatrix} -2 & 3 \\ 7 & 8 \end{bmatrix} + (-3) \det \begin{bmatrix} -2 & 3 \\ 5 & -1 \end{bmatrix}$

$$= (1)(47) - 4(-37) - 3(-13) = 234$$

6.6 EXERCISES

Calculating Determinants

Exercises 1–4: Determine if the matrix A is invertible by calculating det A.

1. $A = \begin{bmatrix} 4 & 3 \\ 5 & 4 \end{bmatrix}$

2. $A = \begin{bmatrix} 1 & -3 \\ 2 & 6 \end{bmatrix}$

3. $A = \begin{bmatrix} -4 & 6 \\ -8 & 12 \end{bmatrix}$

4. $A = \begin{bmatrix} 10 & -20 \\ -5 & 10 \end{bmatrix}$

Exercises 5–8: Find the specified minor and cofactor for the matrix A.

5. M_{12} and A_{12} if $A = \begin{bmatrix} 1 & -1 & 3 \\ 2 & 3 & -2 \\ 0 & 1 & 5 \end{bmatrix}$

6. M_{23} and A_{23} if $A = \begin{bmatrix} 1 & 2 & -1 \\ 4 & 6 & -3 \\ 2 & 3 & 9 \end{bmatrix}$

7. M_{22} and A_{22} if $A = \begin{bmatrix} 7 & -8 & 1 \\ 3 & -5 & 2 \\ 1 & 0 & -2 \end{bmatrix}$

8. M_{31} and A_{31} if $A = \begin{bmatrix} 0 & 0 & -1 \\ 6 & -7 & 1 \\ 8 & -9 & -1 \end{bmatrix}$

Exercises 9–12: Find det A by expanding about the first column. State whether A^{-1} exists.

9. $A = \begin{bmatrix} 1 & 4 & -7 \\ 0 & 2 & -3 \\ 0 & -1 & 3 \end{bmatrix}$

10. $A = \begin{bmatrix} 0 & 2 & 8 \\ -1 & 3 & 5 \\ 0 & 4 & 1 \end{bmatrix}$

11. $A = \begin{bmatrix} 5 & 1 & 6 \\ 0 & -2 & 0 \\ 0 & 4 & 0 \end{bmatrix}$

12. $A = \begin{bmatrix} 3 & 2 & 3 \\ 2 & 2 & 2 \\ 1 & 3 & 1 \end{bmatrix}$

Exercises 13–20: Find det A using the method of cofactors.

13. $A = \begin{bmatrix} 2 & 0 & 0 \\ 0 & 3 & 0 \\ 0 & 0 & 5 \end{bmatrix}$

14. $A = \begin{bmatrix} 0 & 0 & 2 \\ 0 & 3 & 0 \\ 5 & 0 & 0 \end{bmatrix}$

15. $A = \begin{bmatrix} 0 & 0 & 0 \\ -8 & 3 & -9 \\ 15 & 5 & 9 \end{bmatrix}$

16. $A = \begin{bmatrix} 1 & 1 & 5 \\ -3 & -3 & 0 \\ 7 & 0 & 0 \end{bmatrix}$

17. $A = \begin{bmatrix} 3 & -1 & 2 \\ 0 & 5 & 7 \\ 1 & 0 & -1 \end{bmatrix}$

18. $A = \begin{bmatrix} 3 & 0 & -1 \\ 2 & 3 & -4 \\ 6 & -5 & 1 \end{bmatrix}$

19. $A = \begin{bmatrix} 1 & -5 & 2 \\ -7 & 1 & 3 \\ 0 & 4 & -2 \end{bmatrix}$

20. $A = \begin{bmatrix} 1 & -1 & 2 \\ -2 & 0 & 1 \\ 1 & 1 & -1 \end{bmatrix}$

Exercises 21–24: Use technology to calculate det A.

21. $A = \begin{bmatrix} 11 & -32 \\ 1.2 & 55 \end{bmatrix}$

22. $A = \begin{bmatrix} 17 & -4 & 3 \\ 11 & 5 & -15 \\ 7 & -9 & 23 \end{bmatrix}$

23. $A = \begin{bmatrix} 2.3 & 5.1 & 2.8 \\ 1.2 & 4.5 & 8.8 \\ -0.4 & -0.8 & -1.2 \end{bmatrix}$

24. $A = \begin{bmatrix} 1 & -1 & 3 & 7 \\ 9 & 2 & -7 & -4 \\ 5 & -7 & 1 & -9 \\ 7 & 1 & 3 & 6 \end{bmatrix}$

Calculating Area

Exercises 25–28: Find the area of the figure.

25.

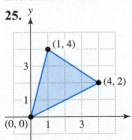

26.

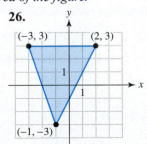

27.

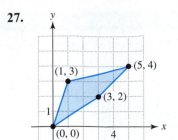

28.

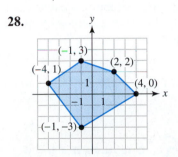

Cramer's Rule

Exercises 29–32: Use Cramer's rule to solve each system of linear equations.

29. $7x + 4y = 23$
$11x - 5y = 70$

30. $-7x + 5y = 8.2$
$6x + 4y = -0.4$

31. $1.7x - 2.5y = -0.91$
$-0.4x + 0.9y = 0.423$

32. $-2.7x + 1.5y = -1.53$
$1.8x - 5.5y = -1.68$

Applying a Concept

Exercises 33–36: Use the concept of the area of a triangle to determine if the three points are collinear.

33. $(1, 3), (-3, 11), (2, 1)$

34. $(3, 6), (-1, -6), (5, 11)$

35. $(-2, -5), (4, 4), (2, 3)$

36. $(4, -5), (-2, 10), (6, -10)$

Writing about Mathematics

37. Choose two matrices A and B with dimension 2×2. Calculate det A, det B, and det (AB). Repeat this process until you are able to discover how these three determinants are related. Summarize your results.

38. Calculate det A and det A^{-1} for different matrices. Compare the determinants. Try to generalize your results.

CHECKING BASIC CONCEPTS FOR SECTIONS 6.5 AND 6.6

1. Find the inverse of the matrix A by hand. Use technology to check your answer.

$$A = \begin{bmatrix} 0 & 0 & 1 \\ 1 & 1 & 0 \\ 1 & 0 & 1 \end{bmatrix}$$

2. Write the following system of linear equations as a matrix equation $AX = B$. Use technology to solve the system utilizing A^{-1}.

$$3.1x - 5.3y = -2.682$$
$$-0.1x + 1.8y = 0.787$$

3. Find the determinant of the matrix A using the method of cofactors. Use technology to check your answer. Is A invertible?

$$A = \begin{bmatrix} 1 & -1 & 2 \\ 2 & 3 & 1 \\ 0 & -2 & 5 \end{bmatrix}$$

Chapter 6 Summary

CONCEPT	EXPLANATION AND EXAMPLES

Section 6.1

FUNCTIONS OF TWO VARIABLES

$z = f(x, y)$, where x and y are inputs to f

Example: $f(x, y) = 2x - 3y$
$f(4, -1) = 2(4) - 3(-1) = 11$

SYSTEM OF LINEAR EQUATIONS WITH TWO VARIABLES

General form
$$a_1 x + b_1 y = c_1$$
$$a_2 x + b_2 y = c_2$$

A linear system can have 0, 1, or infinitely many solutions. A solution can be written as an ordered pair. A linear system may be solved symbolically, graphically, or numerically.

Example: $x - y = 2$
$2x + y = 7$ Solution: $(3, 1)$

METHOD OF SUBSTITUTION FOR TWO EQUATIONS

May be used to solve systems of linear or nonlinear equations

Example: $x - y = -3$
$x + 4y = 17$

Solve the first equation for x to obtain $x = y - 3$. Substitute this result into the second equation and solve for y.

$$(y - 3) + 4y = 17 \quad \text{implies that} \quad y = 4.$$

Then $x = 4 - 3 = 1$ and the solution is $(1, 4)$.

CONCEPT	EXPLANATION AND EXAMPLES

Section 6.2

TYPES OF LINEAR SYSTEMS WITH TWO VARIABLES

Consistent system: Has either one solution (independent equations) or infinitely many solutions (dependent equations).

Inconsistent system: Has no solution

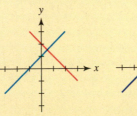 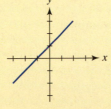 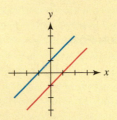

One Solution   Infinitely Many   No Solutions
Solutions

METHOD OF ELIMINATION

Can be used to solve systems of equations.

Example:
$$2x - 3y = 4$$
$$\underline{x + 3y = 11} \quad \text{Add.}$$
$$3x \quad\quad = 15 \quad \text{or} \quad x = 5$$

Substituting $x = 5$ into the first equation gives $y = 2$. The solution is $(5, 2)$.

SYSTEM OF INEQUALITIES WITH TWO VARIABLES

Solution set is often a shaded region in the xy-plane.

Example:
$$x + y \leq 4$$
$$y \geq 0$$
$$x \geq 0$$

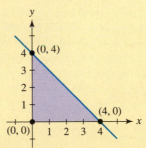

LINEAR PROGRAMMING

Method for maximizing (or minimizing) an objective function subject to a set of constraints.

Example: Maximize $P = 2x + 4y$, subject to
$x + y \leq 4, x \geq 0, y \geq 0$.

The maximum of $P = 16$ occurs at the vertex $(0, 4)$, in the region of feasible solutions. See figure above.

CONCEPT	EXPLANATION AND EXAMPLES

Section 6.3

MATRICES AND SYSTEMS OF LINEAR EQUATIONS

An augmented matrix can be used to represent a system of linear equations.

Augmented Matrix

Example:

$$\begin{array}{rcr} x - 2y + z = & 0 \\ -x + 4y - z = & 4 \\ 2x + y - 3z = & -5 \end{array} \qquad \left[\begin{array}{rrr|r} 1 & -2 & 1 & 0 \\ -1 & 4 & -1 & 4 \\ 2 & 1 & -3 & -5 \end{array}\right]$$

ROW-ECHELON FORM

Examples:

$$\begin{bmatrix} 1 & 2 & -1 \\ 0 & 1 & 2 \end{bmatrix} \qquad \left[\begin{array}{rrr|r} 1 & 3 & -2 & 7 \\ 0 & 1 & 4 & 5 \\ 0 & 0 & 1 & -3 \end{array}\right]$$

GAUSSIAN ELIMINATION WITH BACKWARD SUBSTITUTION

Gaussian elimination can be used to transform a system of linear equations into row-echelon form. Then backward substitution can be used to solve the resulting system of linear equations. Graphing calculators can also be used to solve systems of equations.

Section 6.4

OPERATIONS ON MATRICES

Matrices can be added, subtracted, and multiplied, but there is no division of matrices.

ADDITION

$$\begin{bmatrix} 2 & 4 \\ 5 & 6 \end{bmatrix} + \begin{bmatrix} -2 & 1 \\ 7 & 3 \end{bmatrix} = \begin{bmatrix} 0 & 5 \\ 12 & 9 \end{bmatrix}$$

SUBTRACTION

$$\begin{bmatrix} -3 & 0 \\ 4 & -4 \end{bmatrix} - \begin{bmatrix} 1 & 2 \\ 6 & -7 \end{bmatrix} = \begin{bmatrix} -4 & -2 \\ -2 & 3 \end{bmatrix}$$

SCALAR MULTIPLICATION

$$3\begin{bmatrix} 5 & 1 & 6 & -1 \\ 0 & -2 & 3 & 2 \end{bmatrix} = \begin{bmatrix} 15 & 3 & 18 & -3 \\ 0 & -6 & 9 & 6 \end{bmatrix}$$

MULTIPLICATION

$$\begin{bmatrix} 2 & -1 \\ 0 & 3 \\ -7 & 1 \end{bmatrix} \cdot \begin{bmatrix} 1 & -1 & 0 \\ 3 & -5 & -4 \end{bmatrix} = \begin{bmatrix} -1 & 3 & 4 \\ 9 & -15 & -12 \\ -4 & 2 & -4 \end{bmatrix}$$

Section 6.5

MATRIX INVERSES

The inverse of an $n \times n$ matrix A, denoted A^{-1}, satisfies $A^{-1}A = I_n$ and $AA^{-1} = I_n$, where I_n is the $n \times n$ identity matrix.

CONCEPT	EXPLANATION AND EXAMPLES

Section 6.5 (continued)

Example: $A = \begin{bmatrix} 5 & 2 \\ 2 & 1 \end{bmatrix}$ and $A^{-1} = \begin{bmatrix} 1 & -2 \\ -2 & 5 \end{bmatrix}$

$$\begin{bmatrix} 5 & 2 \\ 2 & 1 \end{bmatrix}\begin{bmatrix} 1 & -2 \\ -2 & 5 \end{bmatrix} = \begin{bmatrix} 1 & 0 \\ 0 & 1 \end{bmatrix} = I_2$$

$$\begin{bmatrix} 1 & -2 \\ -2 & 5 \end{bmatrix}\begin{bmatrix} 5 & 2 \\ 2 & 1 \end{bmatrix} = \begin{bmatrix} 1 & 0 \\ 0 & 1 \end{bmatrix} = I_2$$

MATRIX EQUATIONS

A system of linear equations can be written as a matrix equation.

Example: $2x - 2y = 3$
$-3x + 4y = 2$

$$AX = \begin{bmatrix} 2 & -2 \\ -3 & 4 \end{bmatrix}\begin{bmatrix} x \\ y \end{bmatrix} = \begin{bmatrix} 3 \\ 2 \end{bmatrix} = B$$

The solution can be found as follows.

$$A^{-1}B = \begin{bmatrix} 2 & 1 \\ 1.5 & 1 \end{bmatrix}\begin{bmatrix} 3 \\ 2 \end{bmatrix} = \begin{bmatrix} 8 \\ 6.5 \end{bmatrix}$$

Section 6.6

DETERMINANT OF A 2 × 2 MATRIX

$$\det A = \begin{bmatrix} a & b \\ c & d \end{bmatrix} = ad - cb$$

Example: $\det \begin{bmatrix} 1 & 4 \\ 3 & 5 \end{bmatrix} = (1)(5) - (3)(4) = -7$

DETERMINANT OF A 3 × 3 MATRIX

A 3 × 3 determinant can be reduced to calculating the determinants of 2 × 2 matrices by using cofactors.

Example: $\det \begin{bmatrix} 3 & 1 & -1 \\ 2 & 2 & 0 \\ 0 & 1 & -3 \end{bmatrix}$

$$= 3\begin{bmatrix} 2 & 0 \\ 1 & -3 \end{bmatrix} - 2\begin{bmatrix} 1 & -1 \\ 1 & -3 \end{bmatrix} + 0\begin{bmatrix} 1 & -1 \\ 2 & 0 \end{bmatrix}$$

$$= 3(-6) - 2(-2) + 0(2)$$

$$= -14$$

CRAMER'S RULE

Cramer's rule makes use of determinants to solve systems of linear equations. However, Gaussian elimination with backward substitution is usually more efficient.

Review Exercises

Exercises 1 and 2: Evaluate f for the indicated inputs.

1. $f(3, 6)$, where $f(b, h) = \frac{1}{2}bh$.

2. $f(2, 5)$, where $f(r, h) = \pi r^2 h$

3. *Ultraviolet Light*   There are two types of ultraviolet light: UV-A and UV-B. UV-B is responsible for both skin and eye damage. The table lists the maximum doses of UV-B in the northern hemisphere for selected latitudes L and dates D of the year. (Sources: Orbital Sciences Corporation; J. Williams, *The Weather Almanac 1995.*)

L/D	Mar 21	June 21	Sept 21	Dec 21
0°	325	254	325	272
10°	311	275	280	220
20°	249	292	256	143
30°	179	248	182	80
40°	99	199	127	34
50°	57	143	75	13

(a) A student living at Topeka, Kansas, (39°N) is planning spring break in Honolulu, Hawaii, (21°N) during March 19–23. Approximate how many times stronger the UV-B light is in Hawaii than in Topeka.

(b) Compare the sun's UV-B strength at Glasgow, Montana, (48°N) on June 21 with its strength in Honolulu on December 21.

4. *Wave Height*   The table lists the wave heights produced on the ocean for various wind speeds and durations.   (Source: J. Navarra, *Atmosphere, Weather and Climate.*)

Wind Speed (mph)	Duration of the Wind			
	10 hr	20 hr	30 hr	40 hr
11.5	2 ft	2 ft	2 ft	2 ft
17.3	4 ft	5 ft	5 ft	5 ft
23.0	7 ft	8 ft	9 ft	9 ft
34.5	13 ft	17 ft	18ft	19 ft
46.0	21 ft	28 ft	31 ft	33 ft
57.5	29 ft	40 ft	45 ft	48 ft

(a) What is the expected wave height if a 46 mile-per-hour wind blows for 30 hours?

(b) If the wave height is 5 feet, can the speed of the wind and its duration be determined?

(c) Describe the relationship among wave height, wind speed, and duration of the wind.

(d) If the duration of the wind doubles, does the wave height double?

(e) Estimate the wave height if a 46 mile-per-hour wind blows for 15 hours.

Exercises 5 and 6: Solve the system of equations
(a) graphically, and
(b) symbolically.

5. $3x + y = 1$
$2x - 3y = 8$

6. $x^2 - y = 1$
$x + y = 1$

7. *Area and Perimeter*   Let l represent the length of a rectangle and w its width, where $l \geq w$. Then, its area can be computed by $A(l, w) = lw$ and its perimeter by $P(l, w) = 2l + 2w$. Solve the system of equations determined by $A(l, w) = 77$ and $P(l, w) = 36$.

8. *Cylinder*   Graphically approximate the radius r and height h of a cylindrical container with a volume V of 30 cubic inches and a lateral surface area S of 45 square inches.

Exercises 9 and 10: Use the elimination method to solve each system of linear equations, if possible. Identify the system as consistent or inconsistent.

9. $2x + y = 7$
$x - 2y = -4$

10. $3x + 3y = 15$
$-x - y = -4$

11. *Student Loans*   A student takes out two loans totaling $2000 to help pay for college expenses. One loan is at 7% interest, and the other is at 9%. Interest for both loans is compounded annually.

(a) If the combined total interest for both loans the first year is $156, find the amount of each loan symbolically.

(b) Determine the amount of each loan graphically or numerically.

12. *Dimension of a Screen*   The screen of a rectangular television set is 3 inches wider than it is high. If the perimeter of the screen is 42 inches, find its dimensions by writing a system of linear equations and solving.

Exercises 13 and 14: Graph the system of inequalities. Shade the region that represents the solution set. Use the graph to find a solution.

13. $x^2 + y^2 < 9$
  $x + y > 3$

14. $y \le 4 - x^2$
  $y \ge 2 - x$

Exercises 15 and 16: Each augmented matrix represents a system of linear equations. Solve the system.

15. $\begin{bmatrix} 1 & 5 & | & 6 \\ 0 & 1 & | & 3 \end{bmatrix}$

16. $\begin{bmatrix} 1 & 2 & -2 & | & 8 \\ 0 & 1 & 1 & | & 5 \\ 0 & 0 & 0 & | & 0 \end{bmatrix}$

Exercises 17 and 18: Use Gaussian elimination with backward substitution to solve each system of linear equations.

17. $x + 3y = 8$
  $-x + y = 4$

18. $x + z = 4$
  $x + y - 2z = -3$
  $-x + y + z = 4$

Exercises 19 and 20: Evaluate the following.

  (a) $A + 2B$
  (b) $A - B$

19. $A = \begin{bmatrix} 1 & -3 \\ 2 & -1 \end{bmatrix}$, $B = \begin{bmatrix} 3 & 2 \\ -5 & 1 \end{bmatrix}$

20. $A = \begin{bmatrix} 4 & 0 & 1 \\ -2 & 8 & 9 \end{bmatrix}$, $B = \begin{bmatrix} -5 & 3 & 2 \\ -4 & 0 & 7 \end{bmatrix}$

Exercises 21–24: If possible, find AB and BA.

21. $A = \begin{bmatrix} 2 & 0 \\ -5 & 3 \end{bmatrix}$, $B = \begin{bmatrix} -1 & -2 \\ 4 & 7 \end{bmatrix}$

22. $A = \begin{bmatrix} 1 & -2 \\ 2 & 3 \end{bmatrix}$, $B = \begin{bmatrix} 1 & 0 & 2 \\ -1 & 3 & 4 \end{bmatrix}$

23. $A = \begin{bmatrix} 2 & -1 & 3 \\ 2 & 4 & 0 \end{bmatrix}$, $B = \begin{bmatrix} 1 & 0 \\ -1 & 2 \\ 0 & 3 \end{bmatrix}$

24. $A = \begin{bmatrix} 1 & -1 & 2 \\ 0 & 3 & 4 \\ 1 & 0 & 2 \end{bmatrix}$, $B = \begin{bmatrix} -1 & 0 & 0 \\ 2 & 0 & -1 \\ 1 & 4 & 2 \end{bmatrix}$

Exercises 25 and 26: Use technology to find det A. State if A is invertible.

25. $A = \begin{bmatrix} 13 & 22 \\ 55 & -57 \end{bmatrix}$

26. $A = \begin{bmatrix} 6 & -7 & -1 \\ -7 & 3 & -4 \\ 23 & 54 & 77 \end{bmatrix}$

Exercises 27 and 28: Determine if B is the inverse matrix of A by evaluating AB and BA.

27. $A = \begin{bmatrix} 8 & 5 \\ 6 & 4 \end{bmatrix}$, $B = \begin{bmatrix} 2 & -2.5 \\ -3 & 4 \end{bmatrix}$

28. $A = \begin{bmatrix} -1 & 1 & 2 \\ 1 & 0 & -1 \\ 0 & 1 & 2 \end{bmatrix}$, $B = \begin{bmatrix} -1 & 0 & 1 \\ 2 & 2 & -1 \\ -1 & -1 & -1 \end{bmatrix}$

Exercises 29 and 30: Complete the following.

  (a) Write the system of linear equations in the form $AX = B$.
  (b) Solve the linear system by computing $X = A^{-1}B$.

29. $11x + 31y = -27.6$
  $37x - 19y = 240$

30. $12x + 7y - 3z = 14.6$
  $8x - 11y + 13z = -60.4$
  $-23x + 9z = -14.6$

31. If possible, graphically approximate the solution of each system of equations to the nearest thousandth. Identify each system as consistent or inconsistent. If the system is consistent, determine if the equations are dependent or independent.

  (a) $3.1x + 4.2y = 6.4$
    $1.7x - 9.1y = 1.6$
  (b) $6.3x - 5.1y = 9.3$
    $4.2x - 3.4y = 6.2$
  (c) $0.32x - 0.64y = 0.96$
    $-0.08x + 0.16y = -0.72$

32. *CD Prices*   A music store sells compact discs at two prices marked A and B. Each row in the table represents a purchase. Determine the cost of each type of CD using a matrix inverse.

A	B	Total
1	2	$37.47
2	3	$61.95

Exercises 33 and 34: Find A^{-1}.

33. $A = \begin{bmatrix} 1 & -2 \\ -1 & 1 \end{bmatrix}$

34. $A = \begin{bmatrix} 1 & 0 & 1 \\ 1 & 1 & 1 \\ 0 & 1 & -1 \end{bmatrix}$

Exercises 35 and 36: Find det A using the method of cofactors.

35. $A = \begin{bmatrix} 2 & 1 & 3 \\ 0 & 3 & 4 \\ 1 & 0 & 5 \end{bmatrix}$ **36.** $A = \begin{bmatrix} 3 & 0 & 2 \\ 1 & 3 & 5 \\ -5 & 2 & 0 \end{bmatrix}$

37. *Digital Photography*   Design a 3×3 matrix A that represents a digital photograph of the letter T in black on a white background. Find a matrix B, such that $A + B$ darkens only the white background by one gray level.

38. *Area*   Find the area of the triangle whose vertices are $(0, 0)$, $(5, 2)$, and $(2, 5)$.

39. *Voter Turnout*   The table shows the percent y of voter turnout in the United States for three presidential elections in year x, where $x = 0$ corresponds to 1900. Find a quadratic function defined by $f(x) = ax^2 + bx + c$ that models these data. Graph f together with the data.   (Source: Committee for the Study of the American Electorate.)

x	24	60	96
y	48.9	62.8	48.8

40. *Geometry*   Complete the following.
  (a) Write a system of inequalities that describes possible dimensions for a cylinder with a volume V greater than or equal to 30 cubic inches, and a lateral surface area S less than or equal to 45 square inches.
  (b) Graph and identify the region of solutions.
  (c) Use the graph to estimate one solution to the system of inequalities.

41. *Snowmobile Fatalities*   Through February during the 1996–1997 winter, snowmobile fatalities in Michigan exceeded the average number for an entire winter by 5 fatalities. This was a 19.2% increase over the yearly average, with 2 months still left in the winter. Approximate graphically the average number of snowmobile fatalities in Michigan annually, and the number of fatalities through February during the winter of 1996–1997.   (Source: USA Today.)

42. *Flood Control*   The spillway capacity of a dam is important in flood control. Spillway capacity Q is measured in cubic feet of water per second and depends on the width W in feet of the spillway and the depth D in feet of the water over the spillway. See the figure. Spillway capacity is computed by $Q = f(W, D)$. The table is a numerical representation of f.   (Source: D. Callas, Project Director, *Snapshots of Applications in Mathematics.*)

W\D	1.0	1.5	2.0	2.5	3.0
10	33	61	94	131	173
20	66	122	188	262	345
30	99	183	282	394	518
40	133	244	376	525	690
50	166	305	470	656	863

W\D	3.5	4.0	4.5	5.0	5.5
10	217	266	317	371	428
20	434	531	634	742	856
30	652	797	951	1114	1285
40	870	1062	1268	1485	1713
50	1087	1328	1585	1856	2141

  (a) Evaluate $f(40, 2)$ and interpret the result.
  (b) Solve the equation $f(W, 4) = 1062$. Interpret the result.

(c) If the depth D doubles, does the spillway capacity Q roughly double? Explain.

(d) If the width W doubles, does the spillway capacity Q roughly double? Explain.

43. *Joint Variation*   Suppose P varies jointly as the square of x and the cube of y. If $P = 432$ when $x = 2$ and $y = 3$, find P when $x = 3$ and $y = 5$.

44. *Linear Programming*   Find the maximum value of $P = 3x + 4y$ subject to the following constraints.

$$x \geq 0$$
$$y \geq 0$$
$$x + 3y \leq 12$$
$$3x + y \leq 12$$

Extended and Discovery Exercises

1. To form the **transpose** of a matrix A, denoted A^T, let the first row of A be the first column of A^T, the second row of A be the second column of A^T, and so on, for each row of A. The following are examples of A and A^T. If A has dimension $m \times n$, then A^T has dimension $n \times m$.

$$A = \begin{bmatrix} 3 & -3 & 7 \\ 1 & 6 & -2 \\ 4 & 2 & 5 \end{bmatrix} \quad A^T = \begin{bmatrix} 3 & 1 & 4 \\ -3 & 6 & 2 \\ 7 & -2 & 5 \end{bmatrix}$$

$$A = \begin{bmatrix} 1 & 2 \\ 3 & 4 \\ 5 & 6 \end{bmatrix} \quad A^T = \begin{bmatrix} 1 & 3 & 5 \\ 2 & 4 & 6 \end{bmatrix}$$

Find the transpose of each matrix A.

(a) $A = \begin{bmatrix} 3 & -3 \\ 2 & 6 \\ 4 & 2 \end{bmatrix}$

(b) $A = \begin{bmatrix} 0 & 1 & -2 \\ 2 & 5 & 4 \\ -4 & 3 & 9 \end{bmatrix}$

(c) $A = \begin{bmatrix} 5 & 7 \\ 1 & -7 \\ 6 & 3 \\ -9 & 2 \end{bmatrix}$

Least-Squares Fit and Matrices

The table shows the average cost of tuition and fees y in dollars at 4-year public colleges. In this table x = 0 represents 1980 and x = 15 corresponds to 1995.
(Source: The College Board.)

x	0	5	10	15
y	804	1318	1908	2860

The data are modeled using a line in the accompanying figure.

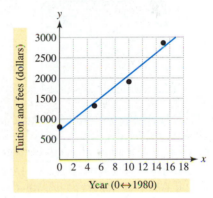

The equation of this line can be found using linear regression based on least squares. (Refer to Section 2.6.) If possible, we would like $f(x) = ax + b$ to satisfy the following four equations.

$$f(0) = a(0) + b = 804$$
$$f(5) = a(5) + b = 1318$$
$$f(10) = a(10) + b = 1908$$
$$f(15) = a(15) + b = 2860$$

Since the data points are not collinear, it is impossible for the graph of f to pass through all four points. These four equations can be written as

$$AX = \begin{bmatrix} 0 & 1 \\ 5 & 1 \\ 10 & 1 \\ 15 & 1 \end{bmatrix} \begin{bmatrix} a \\ b \end{bmatrix} = \begin{bmatrix} 804 \\ 1318 \\ 1908 \\ 2860 \end{bmatrix} = B.$$

The least-squares solution is found by solving the **normal equations**

$$A^{\mathrm{T}}AX = A^{\mathrm{T}}B$$

for X. The solution is $X = (A^{\mathrm{T}}A)^{-1}A^{\mathrm{T}}B$. Using technology, $a = 135.16$ and $b = 708.8$. Thus, f is given by $f(x) = 135.16x + 708.8$. The function f and the data can be graphed. See the accompanying figures.

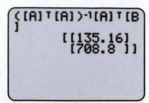

$[-1, 17, 1]$ by $[400, 3200, 200]$

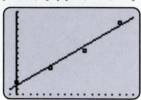

Exercises 2–4: **Least-Square Models** Solve the normal equations to model the data with $f(x) = ax + b$. Plot the data and f in the same viewing rectangle.

2. **Tuition and Fees** The table shows average cost of tuition and fees y in dollars at private 4-year colleges. In this table $x = 0$ corresponds to 1980 and $x = 15$ to 1995. (Source: The College Board.)

x	0	5	10	15
y	3617	6121	9340	12,432

3. **Satellite TV** The table lists the number of satellite television subscribers y in millions. In this table $x = 0$ corresponds to 1995 and $x = 5$ to the year 2000. (Source: USA Today.)

x	0	1	2	3	4	5
y	2.2	4.5	7.9	10.5	13	15

4. **Federal Deficit** The table lists the annual federal deficit y in billions of dollars from 1992 to 1996, where $x = 2$ corresponds to 1992 and $x = 6$ to 1996. (Source: Department of the Treasury.)

x	2	3	4	5	6
y	290	255	203	164	107

Exercises 5 and 6: **Nonlinear Least-Squares** Normal equations can be used to solve problems involving nonlinear regression. Solve the normal equations so that f models the given data.

5. $f(x) = \dfrac{a}{x} + \dfrac{b}{x^2}$

x	1	2	3	4
y	6.30	1.88	0.967	0.619

6. $f(x) = a \log x + bx + c$ (*Hint:* The matrix A has dimension 5×3.)

x	2	3	4	5	6
y	2.45	3.12	3.70	4.25	4.77

Cryptography

7. Businesses and government agencies frequently send classified messages in code. One cryptographic technique that involves matrices is the **polygraphic system**. In this system each letter in the alphabet is associated with a number between 1 and 26. The following table gives a common example. (Source: A. Sinkov, *Elementary Cryptanalysis: A Mathematical Approach.*)

A	B	C	D	E	F	G	H	I	J
1	2	3	4	5	6	7	8	9	10

K	L	M	N	O	P	Q	R	S	T
11	12	13	14	15	16	17	18	19	20

U	V	W	X	Y	Z
21	22	23	24	25	26

For example, the word MATH is coded as 13 1 20 8. Enter these numbers in a matrix B.

$$B = \begin{bmatrix} 13 & 20 \\ 1 & 8 \end{bmatrix}$$

To code these letters a 2×2 matrix, such as

$$A = \begin{bmatrix} 2 & 1 \\ -5 & -2 \end{bmatrix},$$

is multiplied times B to form the product AB.

$$AB = \begin{bmatrix} 2 & 1 \\ -5 & -2 \end{bmatrix} \begin{bmatrix} 13 & 20 \\ 1 & 8 \end{bmatrix} = \begin{bmatrix} 27 & 48 \\ -67 & -116 \end{bmatrix}$$

Since the resulting elements of AB are less than 1 or greater than 26, they may be scaled between 1 and 26 by adding or subtracting multiples of 26.

$$27 - 1(26) = \qquad 48 - 1(26) =$$
$$-67 + 3(26) = \qquad -116 + 5(26) =$$

Thus the word MATH is coded as 1 11 22 14 or AKVN. The advantage of this coding technique is that a particular letter is not always coded the same each time. Use the matrix A to code the following words.

(a) HELP

(b) LETTER (*Hint: B* has dimension 2×3.)

8. (Refer to the previous exercise.) To decode a message, A^{-1} is used. For example, to decode AKVN, write 1 11 22 14. This is stored as

$$C = \begin{bmatrix} 1 & 22 \\ 11 & 14 \end{bmatrix}.$$

To decode the message, evaluate $A^{-1}C$, where

$$A^{-1} = \begin{bmatrix} -2 & -1 \\ 5 & 2 \end{bmatrix}.$$

$$A^{-1}C = \begin{bmatrix} -2 & -1 \\ 5 & 2 \end{bmatrix} \begin{bmatrix} 1 & 22 \\ 11 & 14 \end{bmatrix} = \begin{bmatrix} -13 & -58 \\ 27 & 138 \end{bmatrix}$$

Scaling the matrix elements of $A^{-1}C$ between 1 and 26 results in the following.

$$-13 + (1)26 = \qquad -58 + 3(26) =$$
$$27 - 1(26) = \qquad 138 - 5(26) =$$

The number 13 1 20 8 represents MATH. Use A^{-1} to decode each of the following.

(a) UBNL **(b)** QNABMV

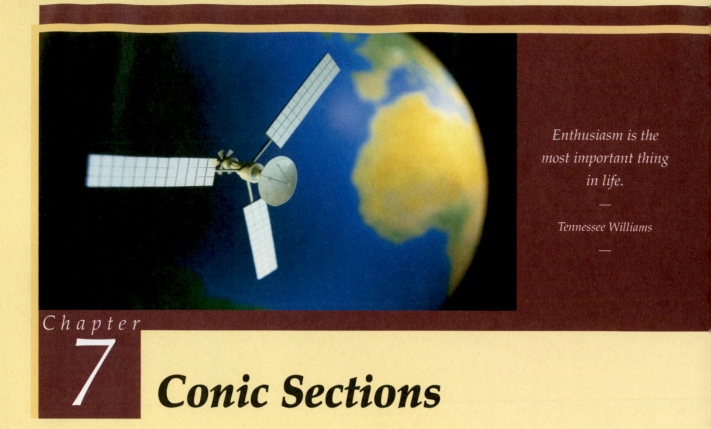

Conic Sections

Throughout history, people have been fascinated by the universe around them and compelled to understand it. Conic sections have played an important role in gaining this understanding. Although conic sections were described and named by the Greek astronomer Apollonius in 200 B.C., it was not until much later that they were used to model motion in the universe. In the sixteenth century Tycho Brahe, the greatest observational astronomer of the age, recorded precise data on planetary movement in the sky. Using Brahe's data in 1619, Johannes Kepler determined that planets move in elliptical orbits around the sun. In 1686 Newton used Kepler's work to show that elliptical orbits are the result of his famous theory of gravitation. We now know that all celestial objects—including planets, comets, asteroids, and satellites—travel in paths described by conic sections. Today scientists search the sky for information about the universe with enormous radio telescopes in the shape of parabolic dishes.

The search to understand the universe has been directed toward both the infinite and the infinitesimal. In 1911 Ernest Rutherford determined the basic structure of the atom. Small atomic particles are capable of traveling in trajectories described by conic sections.

Parabolas, ellipses, and hyperbolas have had a profound influence on our understanding of ourselves and the cosmos around us. In this chapter we learn about these age-old curves.

(Source: *Historical Topics for the Mathematics Classroom, Thirty-first Yearbook,* NCTM.)

7.1 PARABOLAS

Equations and Graphs of Parabolas • Reflective Property of Parabolas • Translations of Parabolas

Introduction

Conic sections are named after the different ways that a plane can intersect a cone. See Figure 7.1. The three basic conic sections are parabolas, ellipses, and hyperbolas. A circle is an example of an ellipse.

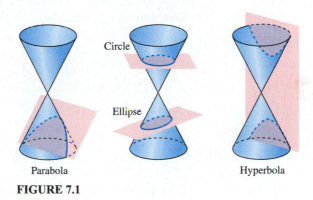

FIGURE 7.1

Figures 7.2–7.4 show examples of conic sections in the *xy*-plane.

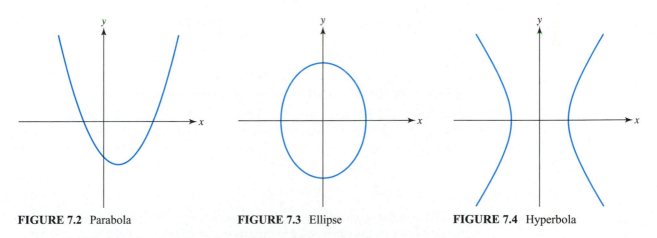

FIGURE 7.2 Parabola   **FIGURE 7.3** Ellipse   **FIGURE 7.4** Hyperbola

In this section we discuss parabolas.

Equations and Graphs of Parabolas

In Chapter 3 we learned that a parabola with vertex $(0, 0)$ can be represented symbolically by the equation $y = ax^2$. With this representation, a parabola can open either upward when $a > 0$ or downward when $a < 0$. The following definition of a parabola allows it to open in any direction.

Parabola

A **parabola** is the set of points in a plane equidistant from a fixed point and a fixed line. The fixed point is called the **focus** and the fixed line is called the **directrix** of the parabola.

Figures 7.5 and 7.6 show two parabolas with vertex V and focus F. The first parabola has a vertical axis (of symmetry) and directrix $y = -p$, and the second parabola has a horizontal axis (of symmetry) and directrix $x = -p$. For any point P located at (x, y) on the parabola, the distance d_1 from F to P is equal to the perpendicular distance d_2 from P to the directrix. By the vertical line test, the parabola in Figure 7.6 cannot be represented by a function.

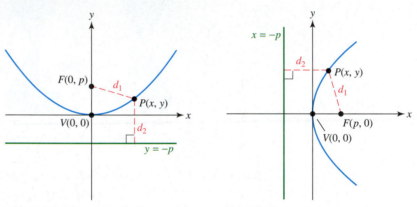

FIGURE 7.5 Vertical Axis   **FIGURE 7.6** Horizontal Axis

■ MAKING CONNECTIONS

Functions and Points   In Figures 7.5 and 7.6, the point P is labeled as $P(x, y)$. This resembles function notation involving two inputs, since the point P is determined by x and y.

We can derive an equation of the parabola shown in Figure 7.5. Since $d_1 = d_2$, the distance formula can be used to express the variables x, y, and p in an equation.

$$d_1 = d_2$$

$\sqrt{(x - 0)^2 + (y - p)^2} = \sqrt{(x - x)^2 + (y - (-p))^2}$	Distance formula
$x^2 + (y - p)^2 = 0^2 + (y + p)^2$	Square both sides.
$x^2 + y^2 - 2py + p^2 = y^2 + 2py + p^2$	Expand binomials.
$x^2 - 2py = 2py$	Subtract y^2 and p^2.
$x^2 = 4py$	Add $2py$.

Algebra Review

To review squaring binomials, see Chapter R (page R-25).

If the value of p is known, then the equation of a parabola with vertex $(0, 0)$ can be found using one of the following equations.

Equation of a parabola with vertex (0, 0)

Vertical Axis

The parabola with a focus at $(0, p)$ and directrix $y = -p$ has equation

$$x^2 = 4py.$$

The parabola opens upward if $p > 0$ and downward if $p < 0$.

Horizontal Axis

The parabola with a focus at $(p, 0)$ and directrix $x = -p$ has equation

$$y^2 = 4px.$$

The parabola opens to the right if $p > 0$ and to the left if $p < 0$.

EXAMPLE 1 *Sketching graphs of parabolas*

Sketch a graph of each parabola. Label the vertex, focus, and directrix.
(a) $x^2 = 8y$ **(b)** $y^2 = -2x$

SOLUTION **(a)** The equation $x^2 = 8y$ is in the form $x^2 = 4py$, where $8 = 4p$. Therefore the parabola has a vertical axis with $p = 2$. Since $p > 0$, the parabola opens upward. The focus is located at $(0, p)$ or $(0, 2)$, and the directrix is $y = -p$ or $y = -2$. The graph of the parabola is shown in Figure 7.7.

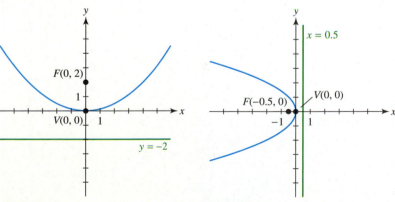

FIGURE 7.7   **FIGURE 7.8**

(b) The equation $y^2 = -2x$ has the form $y^2 = 4px$, where $-2 = 4p$. Therefore the parabola has a horizontal axis with $p = -0.5$. Since $p < 0$, the parabola opens to the left. The focus is located at $(-0.5, 0)$, and the directrix is $x = 0.5$. See Figure 7.8.

In the next example we find the equation of a parabola, given its focus and directrix.

EXAMPLE 2 *Finding the equation of a parabola*

Find the equation of the parabola with focus $(-1.5, 0)$ and directrix $x = 1.5$, as shown in Figure 7.9. Sketch a graph of the parabola.

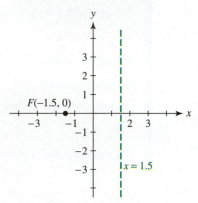

FIGURE 7.9

SOLUTION A parabola always opens toward the focus and away from the directrix. From Figure 7.9 we see that the parabola should open to the left. It follows that $p < 0$ in the equation $y^2 = 4px$. The distance between the focus at $(-1.5, 0)$ and the vertex at $(0, 0)$ is 1.5, and so $p = -1.5 < 0$. (Note that the vertex of the parabola is $(0, 0)$ because the vertex always lies midway between the focus and directrix.) The equation of the parabola is $y^2 = 4(-1.5)x$ or $y^2 = -6x$, and a graph of the parabola is shown in Figure 7.10.

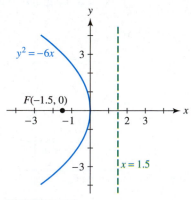

FIGURE 7.10

Reflective Property of Parabolas

When a parabola is rotated about its axis, it sweeps out a shape called a **paraboloid**, as shown in Figure 7.11. Paraboloids have a special reflective property. When incoming rays of light from the sun or distant stars strike the surface of a paraboloid, each ray is reflected toward the focus. See Figure 7.12. If the rays are sunlight, intense heat is produced, which can be used to generate solar heat. Radio signals from distant space also concentrate at the focus. Scientists can analyze these signals by placing a receiver at the focus. This property of a paraboloid can also be used in reverse. If a light source is placed at the focus, then the light is reflected straight ahead, as shown in Figure 7.13. Searchlights, flashlights, and car headlights make use of this property.

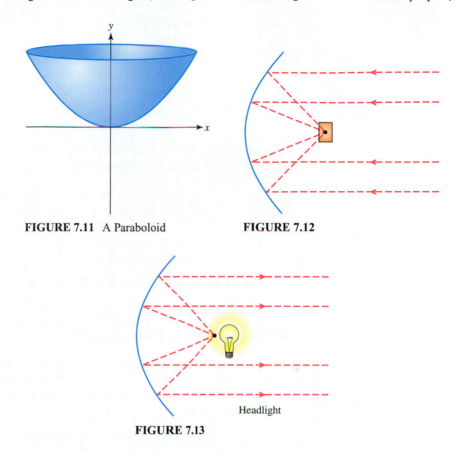

FIGURE 7.11 A Paraboloid   **FIGURE 7.12**

FIGURE 7.13

The next example illustrates this reflective property for radio telescopes.

EXAMPLE 3   *Locating the receiver for a radio telescope*

The U.S. Naval Research Laboratory designed a giant radio telescope weighing 3450 tons. Its parabolic dish has a diameter of 300 feet and a depth of 44 feet. See Figure 7.14.   (Source: J. Mar, *Structure Technology for Large Radio and Radar Telescope Systems.*)
(a) Find an equation in the form $y = ax^2$ that describes a cross section of this dish.
(b) If the receiver is located at the focus, how far should it be from the vertex?

FIGURE 7.14

SOLUTION **(a)** Locate a parabola that passes through $(-150, 44)$ and $(150, 44)$, as in Figure 7.15. Substitute either point into $y = ax^2$.

$$y = ax^2$$
$$44 = a(150)^2$$
$$a = \frac{44}{150^2} = \frac{11}{5625}$$

The equation of the parabola is

$$y = \frac{11}{5625}x^2, \text{ where } -150 \le x \le 150.$$

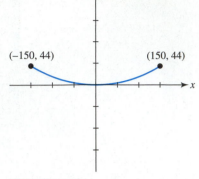

FIGURE 7.15

(b) The value of p represents the distance from the vertex to the focus. To determine p, write the equation in the form $x^2 = 4py$. Then,

$$y = \frac{11}{5625}x^2 \quad \text{is equivalent to} \quad x^2 = \frac{5625}{11}y.$$

It follows that $4p = \dfrac{5625}{11}$ or $p = \dfrac{5625}{44} \approx 127.84$. Therefore the receiver should be located about 127.84 feet from the vertex.

Translations of Parabolas

If the equation of a parabola is given by either $x^2 = 4py$ or $y^2 = 4px$, then its vertex is $(0, 0)$. We can use translations of graphs to find the equation of a parabola with a vertex (h, k) rather than $(0, 0)$. This translation can be obtained by replacing x with $(x - h)$ and y with $(y - k)$.

$$(x - h)^2 = 4p(y - k). \qquad \text{Vertex } (h, k); \text{ vertical axis}$$
$$(y - k)^2 = 4p(x - h) \qquad \text{Vertex } (h, k); \text{ horizontal axis}$$

These two parabolas are shown in Figures 7.16 and 7.17, respectively.

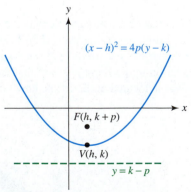

FIGURE 7.16 Vertex (h, k); Vertical Axis

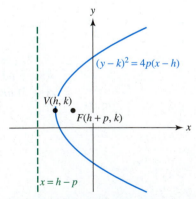

FIGURE 7.17 Vertex (h, k); Horizontal Axis

These results are summarized as follows.

> ### Equation of a parabola with vertex (h, k)
>
> $(x - h)^2 = 4p(y - k)$ Vertical axis; vertex: (h, k)
> $p > 0$: opens upward; $p < 0$: opens downward
> Focus: $(h, k + p)$; directrix: $y = k - p$
>
> $(y - k)^2 = 4p(x - h)$ Horizontal axis; vertex: (h, k)
> $p > 0$: opens to the right; $p < 0$: opens to the left
> Focus: $(h + p, k)$; directrix: $x = h - p$

In the next example we graph a parabola whose vertex is not $(0, 0)$.

EXAMPLE 4 *Graphing a parabola with vertex (h, k)*

Graph the parabola given by $x = -\frac{1}{8}(y + 3)^2 + 2$. Label the vertex, focus, and directrix.

SOLUTION Rewrite the equation in the form $(y - k)^2 = 4p(x - h)$.

$$x = -\frac{1}{8}(y + 3)^2 + 2 \qquad \text{Given equation}$$

$$x - 2 = -\frac{1}{8}(y + 3)^2 \qquad \text{Subtract 2.}$$

$$-8(x - 2) = (y + 3)^2 \qquad \text{Multiply by } -8.$$
$$(y + 3)^2 = -8(x - 2) \qquad \text{Rewrite equation.}$$

It follows that the vertex is $(2, -3)$, $4p = -8$ or $p = -2$, and the parabola opens to the left. The focus is located 2 units to the left of the vertex, and the directrix is located 2 units to the right of the vertex. Therefore the focus is $(0, -3)$, and the directrix is $x = 4$. See Figure 7.18.

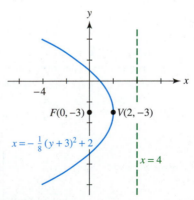

FIGURE 7.18

EXAMPLE 5 *Finding the equation of a parabola with vertex (h, k)*

Find the equation of the parabola with focus $(3, -4)$ and directrix $y = 2$. Sketch a graph of the parabola. Label the focus, directrix, and vertex.

SOLUTION   The focus and directrix are shown in Figure 7.19. The parabola opens downward ($p < 0$), and its equation has the form $(x - h)^2 = 4p(y - k)$. The vertex is located midway between the focus and directrix, so its coordinates are $(3, -1)$. The distance between the focus $(3, -4)$ and the vertex $(3, -1)$ is 3, so $p = -3 < 0$. The equation of the parabola is

$$(x - 3)^2 = -12(y + 1),$$

and its graph is shown in Figure 7.20.

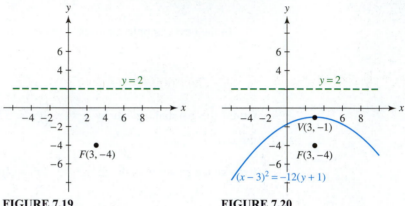

FIGURE 7.19   **FIGURE 7.20**

Graphing calculators can be used to graph parabolas with horizontal axes, as illustrated in the next example.

EXAMPLE 6   *Graphing a parabola with technology*

Graph the equation $(y - 1)^2 = -0.5(x - 2)$ with a graphing calculator.

SOLUTION   Begin by solving the equation for y.

$$(y - 1)^2 = -0.5(x - 2) \qquad \text{Given equation}$$
$$y - 1 = \pm\sqrt{-0.5(x - 2)} \qquad \text{Square root property}$$
$$y = 1 \pm \sqrt{-0.5(x - 2)} \qquad \text{Add 1.}$$

Let $Y_1 = 1 + \sqrt{(-0.5(X - 2))}$ and $Y_2 = 1 - \sqrt{(-0.5(X - 2))}$. The graph of Y_1 creates the upper portion of the parabola, and the graph of Y_2 creates the lower portion of the parabola, as shown in Figure 7.21.

$[-4.7, 4.7, 1]$ by $[-3.1, 3.1, 1]$

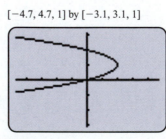

FIGURE 7.21

Putting it all Together 7.1

The following table summarizes some important concepts about parabolas.

Concept	Equation	Example
Parabola with vertex $(0, 0)$ and vertical axis	$x^2 = 4py$ $p > 0$: opens upward $p < 0$: opens downward Focus: $(0, p)$ Directrix: $y = -p$	$x^2 = -2y$ has $4p = -2$ or $p = -\dfrac{1}{2}$. The parabola opens downward with vertex $(0, 0)$, focus $\left(0, -\dfrac{1}{2}\right)$, and directrix $y = \dfrac{1}{2}$.
Parabola with vertex $(0, 0)$ and horizontal axis	$y^2 = 4px$ $p > 0$: opens to the right $p < 0$: opens to the left Focus: $(p, 0)$ Directrix: $x = -p$	$y^2 = 4x$ has $4p = 4$ or $p = 1$. The parabola opens to the right with vertex $(0, 0)$, focus $(1, 0)$, and directrix $x = -1$.
Parabola with vertex (h, k) and vertical axis	$(x - h)^2 = 4p(y - k)$ $p > 0$: opens upward $p < 0$: opens downward Focus: $(h, k + p)$ Directrix: $y = k - p$	$(x - 1)^2 = 8(y - 3)$ has $p = 2$. The parabola opens upward with vertex $(1, 3)$, focus $(1, 5)$, and directrix $y = 1$

Concept	Equation	Example
Parabola with vertex (h, k) and horizontal axis	$(y - k)^2 = 4p(x - h)$ $p > 0$: opens to the right $p < 0$: opens to the left Focus: $(h + p, k)$ Directrix: $x = h - p$	$(y + 1)^2 = -2(x + 2)$ has $p = -\dfrac{1}{2}$. The parabola opens to the left with vertex $(-2, -1)$, focus $\left(-\dfrac{5}{2}, -1\right)$, and directrix $x = -\dfrac{3}{2}$ 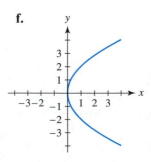

7.1 EXERCISES

Parabolas with Vertex (0, 0)

Exercises 1–6: Sketch a graph of the parabola.

1. $x^2 = y$

2. $x^2 = -y$

3. $y^2 = -x$

4. $y^2 = x$

5. $y^2 = -4x$

6. $x^2 = 4y$

Exercises 7–12: Match the equation of the parabola with its graph (a–f).

7. $x^2 = 2y$

8. $x^2 = -2y$

9. $y^2 = -8x$

10. $y^2 = 4x$

11. $x = \dfrac{1}{2}y^2$

12. $y = -2x^2$

a.
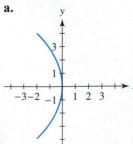

b.
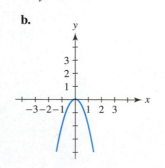

c.
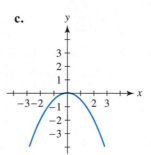

d.
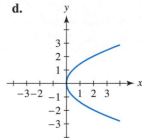

e.
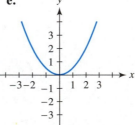

f.
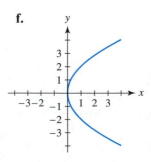

Exercises 13–22: Graph the parabola. Label the vertex, focus, and directrix.

13. $16y = x^2$

14. $y = -2x^2$

15. $x = \dfrac{1}{8}y^2$

16. $-y^2 = 6x$

17. $-4x = y^2$

18. $\dfrac{1}{2}y^2 = 3x$

19. $x^2 = -8y$

20. $-\dfrac{1}{8}x^2 = y$

21. $2y^2 = 6x$

22. $-\dfrac{1}{2}x^2 = 3y$

Exercises 23–26: Sketch a graph of a parabola with focus and directrix as shown in the figure. Find an equation of the parabola.

23.

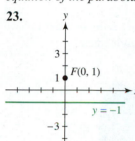

24.

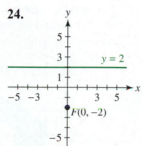

25.

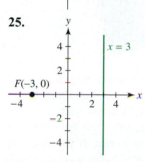

26.

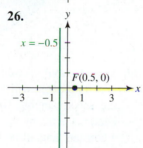

Exercises 27–34: Find an equation of the parabola with vertex (0, 0) that satisfies the given conditions. Sketch its graph.

27. Focus $\left(0, \dfrac{3}{4}\right)$

28. Directrix $y = 2$

29. Directrix $x = 2$

30. Focus $(-1, 0)$

31. Focus $(1, 0)$

32. Focus $\left(0, -\dfrac{1}{2}\right)$

33. Directrix $x = \dfrac{1}{4}$

34. Directrix $y = -1$

Exercises 35–38: Find an equation of a parabola that satisfies the given conditions.

35. Focus $(0, -3)$ and directrix $y = 3$

36. Focus $(0, 2)$ and directrix $y = -2$

37. Focus $(-1, 0)$ and directrix $x = 1$

38. Focus $(3, 0)$ and directrix $x = -3$

Parabolas with Vertex (*h*, *k*)

Exercises 39–42: Sketch a graph of the parabola.

39. $(x - 1)^2 = (y - 2)$

40. $(x - 2)^2 = -(y + 1)$

41. $(y - 1)^2 = -(x + 1)$ **42.** $(y + 2)^2 = 2x$

Exercises 43–46: Match the equation of the parabola with its graph (a–d).

43. $(x - 1)^2 = 4(y - 1)$

44. $(x + 1)^2 = -4(y - 2)$

45. $(y - 2)^2 = -8x$

46. $(y + 1)^2 = 8(x + 3)$

a.

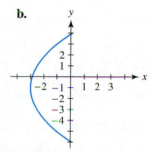

b.

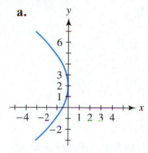

c.

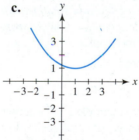

d.

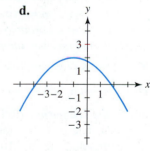

Exercises 47–52: Graph the parabola. Label the vertex, focus, and directrix.

47. $(x - 2)^2 = 8(y + 2)$

48. $\dfrac{1}{16}(x + 4)^2 = -(y - 4)$

49. $x = -\dfrac{1}{4}(y + 3)^2 + 2$

50. $x = 2(y - 2)^2 - 1$

51. $y = -\dfrac{1}{4}(x + 2)^2$

52. $-2(y + 1) = (x + 3)^2$

Exercises 53–58: Find an equation of a parabola that satisfies the given conditions.

53. Focus $(0, 2)$ and vertex $(0, 1)$

54. Focus $(-1, 2)$ and vertex $(3, 2)$

55. Focus $(0, 0)$ and directrix $x = -2$

56. Focus $(2, 1)$ and directrix $x = -1$

57. Focus $(-1, 3)$ and directrix $y = 7$

58. Focus $(1, 2)$ and directrix $y = 4$

Graphing Parabolas with Technology

Exercises 59–66: Graph the parabola with a graphing calculator.

59. $0.5y^2 = x$

60. $y^2 = -1.2x$

61. $(y + 0.75)^2 = -3x$ **62.** $(y - 3)^2 = \dfrac{1}{7}x$

63. $(y - 0.5)^2 = 3.1(x + 1.3)$

64. $1.4(y - 1.5)^2 = 0.5(x + 2.1)$

65. $x = 2.3(y + 1)^2$

66. $(y - 2.5)^2 = 4.1(x + 1)$

Applications

*Exercises 67 and 68: **Satellite Dishes**   (Refer to Example 3.) Use the dimensions of a television satellite dish in the shape of a paraboloid to calculate how far from the vertex the receiver should be located.*

67. Six-foot diameter, nine inches deep

68. Nine-inch radius, two inches deep

69. *Radio Telescope*   (Refer to Example 3.) The Parkes radio telescope has the shape of a parabolic dish with a diameter of 210 feet and a depth of 32 feet. The dish is shown in the figure.   (Source: J. Mar.)

Focus
210 ft
32 ft

(a) Determine an equation of the form $y = ax^2$ describing a cross section of the dish.

(b) The receiver is placed at the focus. How far from the vertex is the receiver located?

70. *Comets*   A comet sometimes travels along a parabolic path as it passes the sun. In this situation the sun is located at the focus of the parabola and the comet passes the sun once and does not orbit the sun. Suppose the path of a comet is given by $y^2 = 100x$, where units are in millions of miles.
(a) Find the coordinates of the sun.
(b) Find the minimum distance between the sun and the comet.

71. *Headlight*   A headlight is being constructed in the shape of a paraboloid with a depth of 4 inches and a diameter of 5 inches, as illustrated in the accompanying figure. Determine the distance d that the bulb should be from the vertex in order to have the beam of light shine straight ahead.

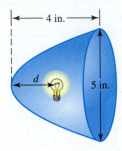

4 in.

d 5 in.

72. *Solar Heater*   A solar heater is being designed to heat a pipe that will contain water, as illustrated in the accompanying figure. A cross section of the heater is given by the equation $x^2 = ky$, where k is a constant and all units are in feet. If the pipe is to be placed 18 inches from the vertex of this cross section, find the value of k.

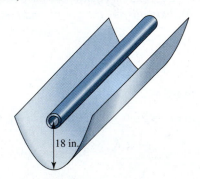

18 in.

Writing about Mathematics

73. Explain how the distance between the focus and the vertex of a parabola affects the shape of the parabola.

74. Explain how to determine the direction that a parabola opens, given the focus and the directrix.

7.2 ELLIPSES

Equations and Graphs of Ellipses • Translations of Ellipses • Reflective Property of Ellipses • Circles • Solving Systems of Equations and Inequalities

Introduction

When planets travel around the sun, they usually do not travel in circular orbits. Instead they travel in elliptical orbits, which look more like ovals than circles. This discovery by Johannes Kepler made it possible for astronomers to determine the precise position of all types of celestial objects, such as asteroids, comets, and moons. The predictions of both solar and lunar eclipses are made easier because of Kepler's discovery. Ellipses are also used in construction and medicine. In this section we learn about the basic properties of ellipses. (Source: *Historical Topics for the Mathematics Classroom, Thirty-first Yearbook,* NCTM.)

Equations and Graphs of Ellipses

One method for sketching an ellipse is to tie a string to two nails driven into a flat board. If a pencil is placed inside the loop formed by the string, the resulting curve shown in Figure 7.22 is an ellipse. The sum of the distances d_1 and d_2 between the pencil and each of the nails is always fixed by the string. The locations of the nails correspond to the *foci* of the ellipse. If the two nails coincide, the ellipse becomes a circle. As the nails spread farther apart, the ellipse becomes more elongated, or eccentric.

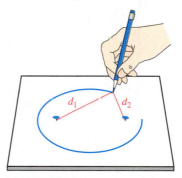

FIGURE 7.22

This method of sketching an ellipse suggests the following definition.

Ellipse

An **ellipse** is the set of points in a plane, the sum of whose distances from two fixed points is constant. Each fixed point is called a **focus** (plural **foci**) of the ellipse.

In Figures 7.23 and 7.24 on the following page the **major axis** and **minor axis** are labeled for each ellipse. The major axis is the line segment connecting V_1 and V_2, and the minor axis is the line segment connecting U_1 and U_2. Figure 7.23 shows an ellipse with a **horizontal major axis**, and Figure 7.24 illustrates an ellipse with a **vertical major axis**. The **vertices**, V_1 and V_2, of each ellipse are located at the endpoints of the major axis.

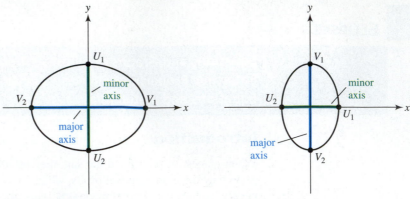

FIGURE 7.23   **FIGURE 7.24**

Since a vertical line can intersect the graph of an ellipse more than once, an ellipse cannot be modeled by a function. However, some ellipses can be represented by the following equations.

Standard equations for ellipses centered at (0, 0)

The ellipse with center at the origin, *horizontal* major axis, and equation

$$\frac{x^2}{a^2} + \frac{y^2}{b^2} = 1 \qquad (a > b > 0)$$

has vertices $(\pm a, 0)$, endpoints of the minor axis $(0, \pm b)$, and foci $(\pm c, 0)$, where $c^2 = a^2 - b^2$ and $c \geq 0$.

The ellipse with center at the origin, *vertical* major axis, and equation

$$\frac{x^2}{b^2} + \frac{y^2}{a^2} = 1 \qquad (a > b > 0)$$

has vertices $(0, \pm a)$, endpoints of the minor axis $(\pm b, 0)$, and foci $(0, \pm c)$, where $c^2 = a^2 - b^2$ and $c \geq 0$.

Note: If $a = b$, then the ellipse is a circle with radius $r = a$ and center $(0, 0)$.

Figures 7.25 and 7.26 show two ellipses. The first has a horizontal major axis and the second has a vertical major axis. The coordinates of the vertices V_1 and V_2, foci F_1 and F_2, and endpoints of the minor axis U_1 and U_2 are labeled.

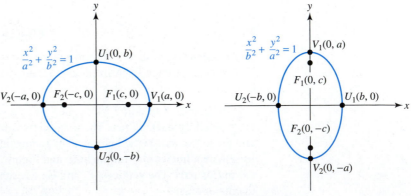

FIGURE 7.25   **FIGURE 7.26**

EXAMPLE 1   *Sketching graphs of ellipses*

Sketch a graph of each ellipse. Label the vertices, foci, and endpoints of the minor axes.

(a) $\dfrac{x^2}{9} + \dfrac{y^2}{4} = 1$ **(b)** $25x^2 + 16y^2 = 400$

SOLUTION   **(a)** The equation $\dfrac{x^2}{9} + \dfrac{y^2}{4} = 1$ describes an ellipse with $a = 3$ and $b = 2$. The ellipse has a horizontal major axis with vertices $(\pm 3, 0)$. The endpoints of the minor axis are $(0, \pm 2)$. To locate the foci, find c.

$$c^2 = a^2 - b^2 = 9 - 4 = 5 \qquad \text{or}$$
$$c = \sqrt{5} \approx 2.236.$$

The foci are located on the major axis with coordinates $(\pm\sqrt{5}, 0)$. See Figure 7.27.

(b) The equation $25x^2 + 16y^2 = 400$ can be put into standard form by dividing both sides by 400.

$$25x^2 + 16y^2 = 400 \qquad \text{Given equation}$$

$$\frac{25x^2}{400} + \frac{16y^2}{400} = \frac{400}{400} \qquad \text{Divide by 400.}$$

$$\frac{x^2}{16} + \frac{y^2}{25} = 1 \qquad \text{Reduce each fraction.}$$

This ellipse has a vertical major axis with $a = 5$ and $b = 4$. The value of c is given by

$$c^2 = 5^2 - 4^2 = 9 \qquad \text{or} \qquad c = 3.$$

The ellipse has foci $(0, \pm 3)$, vertices $(0, \pm 5)$, and endpoints of the minor axis located at $(\pm 4, 0)$. See Figure 7.28.

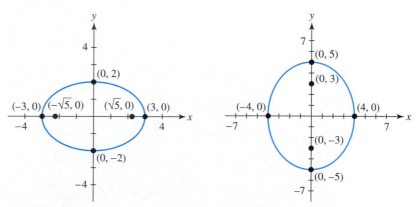

FIGURE 7.27   **FIGURE 7.28**

In the next two examples we find standard equations for ellipses.

EXAMPLE 2   *Finding the equation of an ellipse*

Find the standard equation of the ellipse shown in Figure 7.29 on the following page. Identify the coordinates of the vertices and the foci.

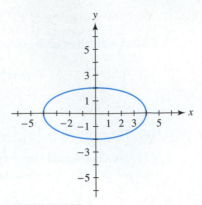

FIGURE 7.29

SOLUTION The ellipse is centered at $(0, 0)$ and has a horizontal major axis. Its standard equation has the form

$$\frac{x^2}{a^2} + \frac{y^2}{b^2} = 1.$$

The endpoints of the major axis are $(\pm 4, 0)$, and the endpoints of the minor axis are $(0, \pm 2)$. It follows that $a = 4$ and $b = 2$, and the standard equation is

$$\frac{x^2}{16} + \frac{y^2}{4} = 1.$$

The foci lie on the horizontal major axis and can be determined as follows.

$$c^2 = a^2 - b^2 = 16 - 4 = 12$$

Thus $c = \sqrt{12} \approx 3.46$, and the coordinates of the foci are $(\pm \sqrt{12}, 0)$. A graph of the ellipse with the vertices and foci plotted is shown in Figure 7.30.

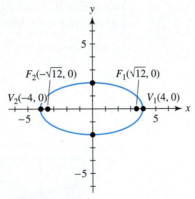

FIGURE 7.30

EXAMPLE 3   *Finding the equation of an ellipse*

Find the standard equation of the ellipse with foci $(0, \pm 1)$ and vertices $(0, \pm 3)$. Sketch its graph.

SOLUTION Since the foci and vertices lie on the y-axis, the ellipse has a vertical major axis. Its standard equation has the form

$$\frac{x^2}{b^2} + \frac{y^2}{a^2} = 1.$$

Because the foci are $(0, \pm 1)$ and the vertices are $(0, \pm 3)$, it follows that $c = 1$ and $a = 3$. The value of b^2 can be found by using the equation $c^2 = a^2 - b^2$.

$$b^2 = a^2 - c^2 = 9 - 1 = 8$$

Thus the equation of the ellipse is $\dfrac{x^2}{8} + \dfrac{y^2}{9} = 1$. Its graph is shown in Figure 7.31. To graph the ellipse, it is helpful to note that the endpoints of the minor axis are $(0, \pm b)$ or $(0, \pm\sqrt{8})$ where $\sqrt{8} \approx 2.83$.

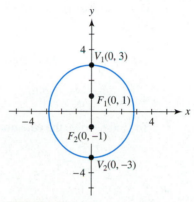

FIGURE 7.31

The planets travel around the sun in elliptical orbits. Although their orbits are nearly circular, many planets have a slight eccentricity to them. The **eccentricity** e of an ellipse is defined by

$$e = \frac{\sqrt{a^2 - b^2}}{a} = \frac{c}{a}.$$

Since the foci of an ellipse lie inside the ellipse, $0 \le c < a$ and $0 \le \dfrac{c}{a} < 1$. Therefore the eccentricity e of an ellipse satisfies $0 \le e < 1$. If $e = 0$, then $a = b$ and the ellipse is a circle. See Figure 7.32. As e increases, the foci spread apart and the ellipse becomes more elongated. See Figures 7.33 and 7.34.

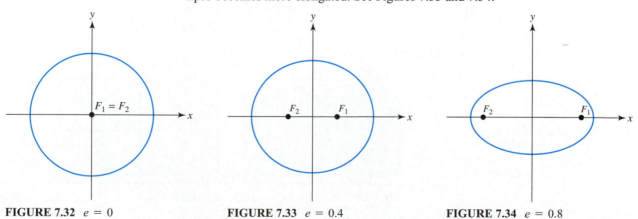

FIGURE 7.32 $e = 0$ **FIGURE 7.33** $e = 0.4$ **FIGURE 7.34** $e = 0.8$

■ MAKING CONNECTIONS

The number e and the variable e   Do not confuse the *variable e*, which is used to denote the eccentricity of an ellipse, with the irrational *number e* ≈ 2.72, which is the base of the natural exponential function $f(x) = e^x$ and of the natural logarithmic function $f(x) = \ln x$.

Astronomers have measured values of e and a for each planet. With this information and the fact that the sun is located at one focus of the ellipse, the equation of a planet's orbit can be found.

EXAMPLE 4 *Finding the orbital equation of the planet Pluto*

The planet Pluto has $e = 0.249$ and $a = 39.44$, the greatest eccentricity of any planet. (For Earth, $a = 1$.) Graph the orbit of Pluto and the position of the sun in $[-60, 60, 10]$ by $[-40, 40, 10]$. (Source: M. Zeilik, *Introductory Astronomy and Astrophysics.*)

SOLUTION Let the orbit of Pluto be given by $\dfrac{x^2}{a^2} + \dfrac{y^2}{b^2} = 1$. Then,

$$e = \frac{c}{a} = 0.249 \qquad \text{implies} \qquad c = 0.249a = 0.249(39.44) \approx 9.821.$$

To find b, solve the equation $c^2 = a^2 - b^2$ for b.

$$b = \sqrt{a^2 - c^2}$$
$$= \sqrt{39.44^2 - 9.821^2} \approx 38.20$$

Pluto's orbit is modeled by $\dfrac{x^2}{39.44^2} + \dfrac{y^2}{38.20^2} = 1$. Since $c \approx 9.821$, the foci are $(\pm 9.821, 0)$. The sun could be located at either focus. We locate the sun at $(9.821, 0)$.

To graph this ellipse on a graphing calculator, the equation must be solved for y. This results in two equations to graph. For example,

$$\frac{x^2}{39.44^2} + \frac{y^2}{38.20^2} = 1$$

$$\frac{y^2}{38.20^2} = 1 - \frac{x^2}{39.44^2}$$

$$\frac{y}{38.20} = \pm \sqrt{1 - \frac{x^2}{39.44^2}}$$

$$y = \pm 38.20\sqrt{1 - \frac{x^2}{39.44^2}}.$$

See Figures 7.35 and 7.36.

$[-60, 60, 10]$ by $[-40, 40, 10]$

Graphing Calculator Help

To learn how to access the variable Y_1, see Appendix B (page AP-15).

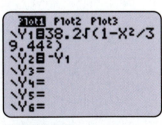

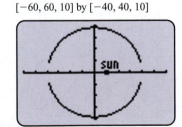

FIGURE 7.35   **FIGURE 7.36**

Translations of Ellipses

If the standard equation of an ellipse is given by either $\dfrac{x^2}{a^2} + \dfrac{y^2}{b^2} = 1$ or $\dfrac{x^2}{b^2} + \dfrac{y^2}{a^2} = 1$,

then the center of the ellipse is $(0, 0)$. We can use translations of graphs to find the equation of an ellipse centered at (h, k) by replacing x with $(x - h)$ and y with $(y - k)$.

Standard equations for ellipses centered at (*h*, *k*)

An ellipse with center (h, k), and either a horizontal or vertical major axis, satisfies one of the following equations, where $a > b > 0$ and $c^2 = a^2 - b^2$.

$$\frac{(x - h)^2}{a^2} + \frac{(y - k)^2}{b^2} = 1$$ Major axis: horizontal; foci: $(h \pm c, k)$
Vertices: $(h \pm a, k)$

$$\frac{(x - h)^2}{b^2} + \frac{(y - k)^2}{a^2} = 1$$ Major axis: vertical; foci: $(h, k \pm c)$
Vertices: $(h, k \pm a)$

EXAMPLE 5   *Translating an ellipse*

Translate the ellipse with equation $\dfrac{x^2}{9} + \dfrac{y^2}{4} = 1$ so that it is centered at $(-1, 2)$. Find this equation and sketch its graph.

SOLUTION   To translate the center from $(0, 0)$ to $(-1, 2)$, replace x with $(x + 1)$ and y with $(y - 2)$. This new equation is

$$\frac{(x + 1)^2}{9} + \frac{(y - 2)^2}{4} = 1.$$

This ellipse is congruent to the given ellipse, except that it is centered at $(-1, 2)$. The given ellipse is shown in Figure 7.37, and the translated ellipse is shown in Figure 7.38.

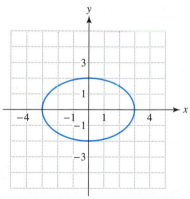

FIGURE 7.37

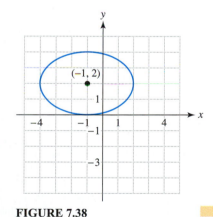

FIGURE 7.38

EXAMPLE 6   *Graphing an ellipse with center (h, k)*

Graph the ellipse whose equation is $\dfrac{(x + 2)^2}{16} + \dfrac{(y - 2)^2}{25} = 1$. Label the vertices and foci.

SOLUTION   The ellipse has a vertical major axis and its center is $(-2, 2)$. Since $a^2 = 25$ and $b^2 = 16$, it follows that $c^2 = a^2 - b^2 = 25 - 16 = 9$. Thus $a = 5$, $b = 4$, and $c = 3$. The vertices are located 5 units above and below the center of the ellipse and the foci are located 3 units above and below the center of the ellipse. That is, the vertices are $(-2, 2 \pm 5)$, or $(-2, 7)$ and $(-2, -3)$, and the foci are $(-2, 2 \pm 3)$, or $(-2, 5)$ and $(-2, -1)$. A graph of the ellipse is shown in Figure 7.39.

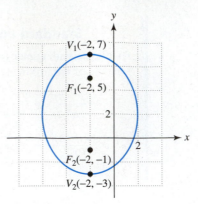

FIGURE 7.39

Reflective Property of Ellipses

Like parabolas, ellipses also have an important reflective property. If an ellipse is rotated about the x-axis, an **ellipsoid** is formed, which resembles the shell of an egg, as illustrated in Figure 7.40. If a light source is placed at focus F_1, then every beam of light emanating from the light source, regardless of its direction, is reflected at the surface of the ellipsoid toward focus F_2. See Figure 7.41.

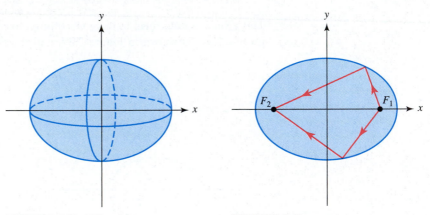

FIGURE 7.40 Ellipsoid   **FIGURE 7.41**

A *lithotripter* is a machine designed to break up kidney stones, without surgery, by producing powerful shock waves. To focus these shock waves accurately, a lithotripter uses the reflective property of ellipses. A patient is carefully positioned so that the kidney stone is at one focus, while the source of the shock waves is located at the other focus. When shock waves are emitted, they are reflected directly toward the kidney stone. As a result, it absorbs all of the energy from the shock wave and breaks up without harming the patient. This nonsurgical procedure reduces both risk and recovery time. In Exercises 71 and 72 this reflective property of ellipses is applied.

Circles

A circle is an ellipse where $a = b$. If a circle has radius r and center (h, k), then an

Geometry Review
To review circles, see Chapter R (page R-10).

equation for the circle is $\dfrac{(x - h)^2}{r^2} + \dfrac{(y - k)^2}{r^2} = 1$. Multiplying the equation by r^2 provides the following result.

> ### Standard equation of a circle
> The **standard equation of a circle** with center (h, k) and radius r is
> $$(x - h)^2 + (y - k)^2 = r^2.$$

EXAMPLE 7 *Finding the standard equation of a circle*

Find the standard equation of a circle with radius 4 and center $(5, -3)$.

SOLUTION Let $h = 5, k = -3$, and $r = 4$. The standard equation is
$$(x - 5)^2 + (y + 3)^2 = 16.$$

EXAMPLE 8 *Finding the center and radius of a circle*

Find the center and radius of the circle given by $x^2 + 6x + y^2 - 2y = -6$.

SOLUTION Use completing the square to write the standard equation of the circle.

Algebra Review
To review completing the square, see Section 3.2.

$$x^2 + 6x + y^2 - 2y = -6 \qquad \text{Given equation}$$
$$(x^2 + 6x + 9) + (y^2 - 2y + 1) = -6 + 9 + 1 \qquad \text{Complete the square.}$$
$$(x + 3)^2 + (y - 1)^2 = 4 \qquad \text{Factor.}$$

The center is $(-3, 1)$, and the radius is 2.

Solving Systems of Equations and Inequalities

In Sections 6.1 and 6.2 we discussed systems of equations and inequalities. In this subsection we revisit these topics.

EXAMPLE 9 *Solving a nonlinear system of equations*

Use substitution to solve the following system of equations. Give graphical support by making a sketch.
$$9x^2 + 4y^2 = 36$$
$$12x^2 + y^2 = 12$$

SOLUTION **STEP 1:** Begin by solving the second equation for y^2.
$$12x^2 + y^2 = 12 \quad \text{is equivalent to} \quad y^2 = 12 - 12x^2.$$

STEP 2: Next, substitute $(12 - 12x^2)$ for y^2 in the first equation and solve for x.
$$9x^2 + 4y^2 = 36 \qquad \text{First equation}$$
$$9x^2 + 4(12 - 12x^2) = 36 \qquad \text{Let } y^2 = 12 - 12x^2.$$
$$9x^2 + 48 - 48x^2 = 36 \qquad \text{Distributive property}$$
$$-39x^2 = -12 \qquad \text{Subtract 48; simplify.}$$
$$x^2 = \frac{4}{13} \qquad \text{Divide by } -39; \text{ reduce.}$$
$$x = \pm\sqrt{\frac{4}{13}} \qquad \text{Square root property}$$

STEP 3: To determine the corresponding y-values, substitute $x^2 = \dfrac{4}{13}$ into $y^2 = 12 - 12x^2$.

$$y^2 = 12 - 12\left(\frac{4}{13}\right) = \frac{108}{13} \quad \text{or} \quad y = \pm\sqrt{\frac{108}{13}}$$

There are four solutions: $\left(\sqrt{\dfrac{4}{13}},\ \sqrt{\dfrac{108}{13}}\right)$, $\left(\sqrt{\dfrac{4}{13}},\ -\sqrt{\dfrac{108}{13}}\right)$, $\left(-\sqrt{\dfrac{4}{13}},\ \sqrt{\dfrac{108}{13}}\right)$,

and $\left(-\sqrt{\dfrac{4}{13}},\ -\sqrt{\dfrac{108}{13}}\right)$.

To graph the system of equations by hand, begin by putting each equation in standard form by dividing the first equation by 36 and the second equation by 12 to obtain

$$\frac{x^2}{4} + \frac{y^2}{9} = 1 \quad \text{and} \quad x^2 + \frac{y^2}{12} = 1.$$

The graphs of these ellipses and the four solutions are shown in Figure 7.42.

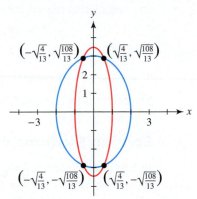

FIGURE 7.42

Note: The system of equations in Example 9 could also be solved by elimination. To do this, multiply the second equation by -4 and add.

The following formula can be used to calculate the area inside an ellipse.

Area inside an ellipse

Given the standard equation of an ellipse, the area A of the region contained inside is given by $A = \pi ab$.

This formula is applied in the next example.

EXAMPLE 10 *Finding the area of an ellipse*

Shade the region in the xy-plane that satisfies the inequality $x^2 + 4y^2 \leq 4$. Find the area of this region if units are in inches.

SOLUTION   Begin by dividing each term in the given inequality by 4.

$$x^2 + 4y^2 \leq 4 \qquad \text{Given inequality}$$

$$\frac{x^2}{4} + \frac{4y^2}{4} \leq \frac{4}{4} \qquad \text{Divide by 4.}$$

$$\frac{x^2}{4} + \frac{y^2}{1} \leq 1 \qquad \text{Reduce the fractions.}$$

The boundary of the region is the ellipse $\dfrac{x^2}{4} + \dfrac{y^2}{1} = 1$. The region *inside* the ellipse satisfies the inequality. To verify this fact, note that the test point $(0, 0)$, which is located inside the ellipse, satisfies the inequality. The solution set is shaded in Figure 7.43, and the area of this region is

$$A = \pi ab = \pi(2)(1) = 2\pi \approx 6.28 \text{ square inches.}$$

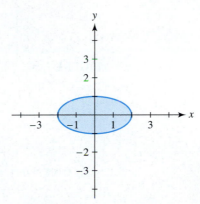

FIGURE 7.43

■ **CLASS DISCUSSION**
Explain why the area formula for an ellipse is a generalization of the area formula for a circle. ■

EXAMPLE 11   *Solving a nonlinear inequality*

Shade the region in the *xy*-plane that satisfies the system of inequalities.

$$36x^2 + 25y^2 \leq 900$$

$$x + (y + 2)^2 \leq 4$$

SOLUTION   Before sketching a graph, rewrite these two inequalities as follows.

First inequality: $\dfrac{36}{900}x^2 + \dfrac{25}{900}y^2 \leq \dfrac{900}{900} \qquad$ Divide each term by 900.

$\dfrac{x^2}{25} + \dfrac{y^2}{36} \leq 1 \qquad$ Simplify to standard form.

Second inequality: $(y + 2)^2 \leq -x + 4 \qquad$ Subtract x.

$(y + 2)^2 \leq -(x - 4) \qquad$ Rewrite in standard form.

The first inequality represents the region inside an ellipse, as shown in Figure 7.44. The second inequality represents the region left of a parabola that opens to the left with vertex $(4, -2)$, as shown in Figure 7.45. (Note that the test point $(-2, -2)$ satisfies the second inequality since $(-2 + 2)^2 \le -(-2 - 4)$ is a true statement. Thus we shade the region where $(-2, -2)$ is located.) The solution set for the system satisfies both inequalities and is shaded in Figure 7.46.

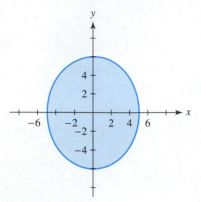

FIGURE 7.44

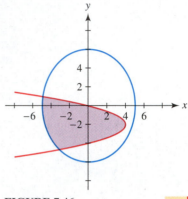

FIGURE 7.45   **FIGURE 7.46**

Putting it all Together

7.2

The following table summarizes some important concepts about ellipses.

Concept	Equation	Example
Ellipse with center $(0, 0)$	Standard equation with $a > b > 0$ Horizontal major axis: $$\frac{x^2}{a^2} + \frac{y^2}{b^2} = 1$$ Vertical major axis: $$\frac{x^2}{b^2} + \frac{y^2}{a^2} = 1$$	$\dfrac{x^2}{4} + \dfrac{y^2}{9} = 1; a = 3, b = 2$ Center: $(0, 0)$; major axis: vertical Vertices: $(0, \pm 3)$; foci: $(0, \pm \sqrt{5})$ $(c^2 = a^2 - b^2 = 9 - 4 = 5$, so $c = \sqrt{5}$.)

Concept	Equation	Example
Ellipse with center (h, k)	Standard equation with $a > b > 0$ Horizontal major axis: $$\frac{(x - h)^2}{a^2} + \frac{(y - k)^2}{b^2} = 1$$ Vertical major axis: $$\frac{(x - h)^2}{b^2} + \frac{(y - k)^2}{a^2} = 1$$	$\frac{(x - 1)^2}{4} + \frac{(y + 1)^2}{9} = 1$; $a = 3$, $b = 2$ Center: $(1, -1)$; major axis: vertical Vertices: $(1, -1 \pm 3)$; foci: $(1, -1 \pm \sqrt{5})$ $(c^2 = a^2 - b^2 = 9 - 4 = 5$, so $c = \sqrt{5}$.) 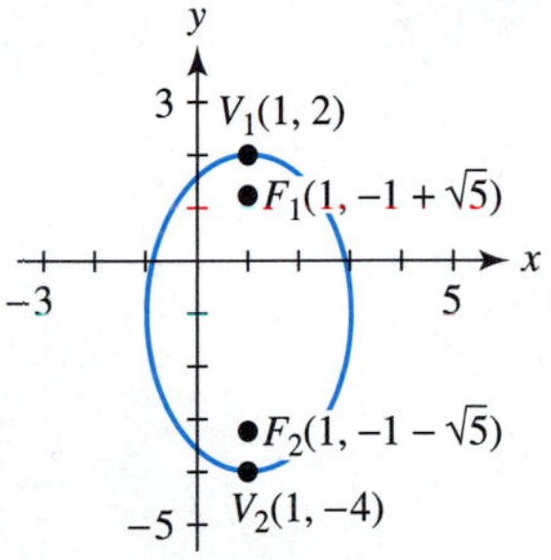
Circle with center (h, k) and radius r	Standard equation $$(x - h)^2 + (y - k)^2 = r^2$$ A circle is an ellipse with $a = b = r$.	$(x - 2)^2 + (y + 2)^2 = 9$ Center: $(2, -2)$; radius: $r = 3$ 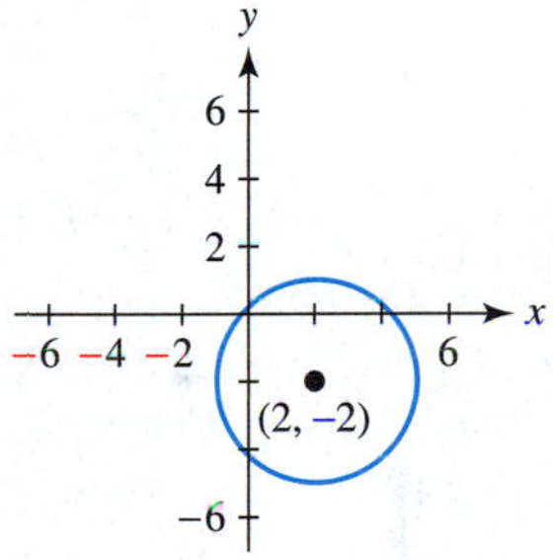
Area inside an ellipse	$A = \pi ab$	The area inside the ellipse given by $\frac{x^2}{49} + \frac{y^2}{9} = 1$ is $A = \pi(7)(3) = 21\pi$ square units.

7.2 EXERCISES

Ellipses with Center (0, 0)

Exercises 1–8: Graph the ellipse. Label the foci and the endpoints of each axis.

1. $\dfrac{x^2}{4} + \dfrac{y^2}{9} = 1$ **2.** $\dfrac{x^2}{9} + \dfrac{y^2}{4} = 1$

3. $\dfrac{x^2}{36} + \dfrac{y^2}{16} = 1$ **4.** $x^2 + \dfrac{y^2}{4} = 1$

5. $9x^2 + 5y^2 = 45$ **6.** $x^2 + 4y^2 = 400$

7. $25x^2 + 9y^2 = 225$

8. $5x^2 + 4y^2 = 20$

Exercises 9–12: Match the equation of the ellipse with its graph (a–d).

9. $\dfrac{x^2}{16} + \dfrac{y^2}{36} = 1$ **10.** $\dfrac{x^2}{4} + y^2 = 1$

11. $\dfrac{x^2}{16} + \dfrac{y^2}{4} = 1$

12. $\dfrac{x^2}{9} + \dfrac{y^2}{9} = 1$

a.

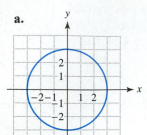

b.

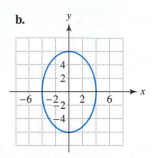

c.

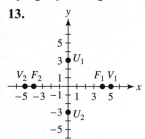

d.

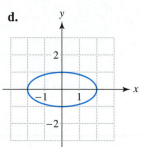

Exercises 13–16: The foci F_1 and F_2, vertices V_1 and V_2 and endpoints U_1 and U_2 of the minor axis of an ellipse are labeled in the figure. Graph the ellipse and find its standard equation. Note that the coordinates of $V_1, V_2, F_1,$ and F_2 are integers.

13.

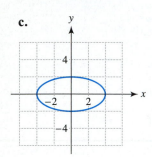

14.

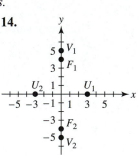

15.

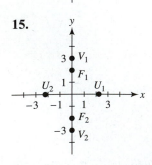

16.

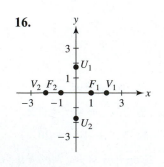

Exercises 17–24: Find an equation of the ellipse, centered at the origin, satisfying the conditions. Sketch its graph.

17. Foci $(0, \pm 2)$, vertices $(0, \pm 4)$

18. Foci $(0, \pm 3)$, vertices $(0, \pm 5)$

19. Foci $(\pm 5, 0)$, vertices $(\pm 6, 0)$

20. Foci $(\pm 4, 0)$, vertices $(\pm 6, 0)$

21. Horizontal major axis of length 8, minor axis of length 6

22. Vertical major axis of length 12, minor axis of length 8

23. Eccentricity $\dfrac{2}{3}$, horizontal major axis of length 6

24. Eccentricity $\dfrac{3}{4}$, vertices $(0, \pm 8)$

Ellipses with Center (*h, k*)

Exercises 25–30: Sketch a graph of the ellipse.

25. $\dfrac{(x-2)^2}{4} + \dfrac{(y-1)^2}{9} = 1$

26. $\dfrac{(x+1)^2}{16} + \dfrac{(y+3)^2}{9} = 1$

27. $\dfrac{(x+1)^2}{16} + \dfrac{(y+2)^2}{25} = 1$

28. $\dfrac{(x-4)^2}{9} + \dfrac{y^2}{4} = 1$

29. $\dfrac{(x+2)^2}{4} + y^2 = 1$

30. $x^2 + \dfrac{(y-3)^2}{4} = 1$

Exercises 31–34: Match the equation of the ellipse with its graph (a–d).

31. $\dfrac{(x-2)^2}{16} + \dfrac{(y+4)^2}{36} = 1$

32. $\dfrac{(x+1)^2}{4} + \dfrac{y^2}{9} = 1$

33. $\dfrac{(x+1)^2}{9} + \dfrac{(y-1)^2}{4} = 1$

34. $\dfrac{x^2}{25} + \dfrac{(y+1)^2}{10} = 1$

a.
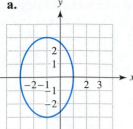

b.
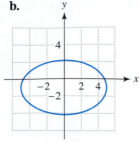

c.
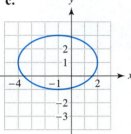

d.
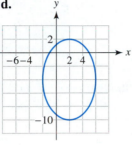

Circles

Exercises 45–48: Find the standard equation of the circle that satisfies the conditions. Graph the circle.

45. Center $(0, 0)$, radius of 4

46. Center $(1, -2)$, radius of 3

47. Center $(3, -4)$, radius of 1

48. Center $(-1, -3)$, passing through the point $(3, 0)$

Exercises 49–52: Find the center and radius of the circle.

49. $x^2 - 4x + y^2 - 2y = 11$

50. $x^2 + 6x + y^2 - 4y = 12$

51. $x^2 + y^2 + 10y = 0$

52. $x^2 - 2x + y^2 + 8y = 19$

Graphing Ellipses with Technology

Exercises 53–56: Graph the ellipse with a graphing calculator.

53. $\dfrac{x^2}{15} + \dfrac{y^2}{10} = 1$

54. $\dfrac{(x - 1.2)^2}{7.1} + \dfrac{y^2}{3.5} = 1$

55. $4.1x^2 + 6.3y^2 = 25$ **56.** $\dfrac{1}{2}x^2 + \dfrac{1}{3}y^2 = \dfrac{1}{6}$

Solving Equations and Inequalities

Exercises 57–62: Solve the system of equations. Give graphical support by making a sketch.

57. $\dfrac{x^2}{4} + \dfrac{y^2}{9} = 1$ **58.** $\dfrac{x^2}{16} + \dfrac{y^2}{25} = 1$
$\ x + y = 3$ $\ -2x + y = 5$

59. $4x^2 + 16y^2 = 64$ **60.** $4x^2 + y^2 = 4$
$\ x^2 + y^2 = 9$ $\ x^2 + y^2 = 2$

61. $x^2 + y^2 = 9$ **62.** $x^2 + y^2 = 4$
$\ 2x^2 + 3y^2 = 18$ $\ (x - 1)^2 + y^2 = 4$

Exercises 63–68: Shade the region in the xy-plane that satisfies the system of inequalities.

63. $\dfrac{x^2}{4} + \dfrac{y^2}{9} \le 1$ **64.** $\dfrac{x^2}{16} + \dfrac{y^2}{25} \le 1$
$\ x + y \ge 2$ $\ -x + y \le 4$

65. $x^2 + y^2 \le 4$ **66.** $x^2 + (y + 1)^2 \le 9$
$\ x^2 + (y - 2)^2 \le 4$ $(x + 1)^2 + y^2 \le 9$

67. $x^2 + y^2 \le 4$ **68.** $4x^2 + 9y^2 \le 36$
$(x + 1)^2 - y \le 0$ $\ x - (y - 2)^2 \ge 0$

Exercises 35–38: Sketch a graph of the ellipse. Identify the foci and vertices.

35. $\dfrac{(x - 1)^2}{9} + \dfrac{(y - 1)^2}{25} = 1$

36. $\dfrac{(x + 2)^2}{25} + \dfrac{(y + 1)^2}{16} = 1$

37. $\dfrac{(x + 4)^2}{16} + \dfrac{(y - 2)^2}{9} = 1$

38. $\dfrac{x^2}{4} + \dfrac{(y - 1)^2}{9} = 1$

Exercises 39–42: Find an equation of an ellipse that satisfies the given conditions.

39. Center $(2, 1)$, focus $(2, 3)$, and vertex $(2, 4)$

40. Center $(-3, -2)$, focus $(-1, -2)$, and vertex $(1, -2)$

41. Vertices $(\pm 3, 2)$ and foci $(\pm 2, 2)$

42. Vertices $(-1, \pm 3)$ and foci $(-1, \pm 1)$

Exercises 43–44: Find an (approximate) equation of the ellipse shown in the figure.

43.
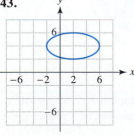

44.
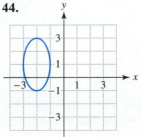

Applications

Exercises 69 and 70: (Refer to Example 4.) Find an equation of the orbit for the planet. Graph its orbit and the location of the sun at a focus on the positive x-axis.

69. Mercury: $e = 0.206$, $a = 0.387$

70. Mars: $e = 0.093$, $a = 1.524$

71. *Lithotripter* (Refer to the discussion in this section.) The source of a shock wave is placed at one focus of an ellipsoid with a major axis of 8 inches and a minor axis of 5 inches. Estimate, to the nearest thousandth of an inch, how far a kidney stone should be positioned from the source.

72. *Whispering Gallery* If a large room is constructed in the shape of the upper half of an ellipsoid, it will have a unique property. Two people, standing at each of the two foci, can hear each other whispering, even though there may be considerable distance between them. Any sound emanating from one focus is reflected directly toward the other focus. See the accompanying figure. If the foci are 100 feet apart, and the maximum height of the ceiling is 40 feet, estimate the area of the floor of the room.

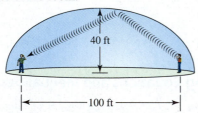

73. *Halley's Comet* One of the most famous comets is Halley's comet. It travels in an elliptical orbit with $a = 17.95$ and $b = 4.44$ and passes by Earth roughly every 76 years. Note that each unit represents 93 million miles. The most recent date that it passed by Earth was in February 1986. (Source: M. Zeilik, *Introductory Astronomy and Astrophysics.*)

(a) Write an equation for the orbit of Halley's comet, where the orbit is centered at $(0, 0)$, and the major axis lies on the x-axis.

(b) If the sun lies (at the focus) on the positive x-axis, finds its coordinates.

(c) Determine the maximum and minimum distances between Halley's comet and the sun.

74. *Orbit of Earth* (Refer to Example 4.) Earth has a nearly circular orbit with $e \approx 0.0167$ and $a = 93$ million miles. Find the minimum and maximum distances between Earth and the sun. (Source: M. Zeilik, *Introductory Astronomy and Astrophysics.*)

75. *Arch Bridge* An elliptical arch under a bridge is constructed so that it is 60 feet wide and has a maximum height of 25 feet, as illustrated in the accompanying figure. Find the height of the arch 15 feet from the center of the arch.

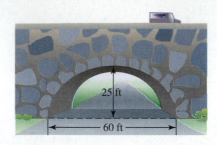

76. *Perimeter of an Ellipse* The perimeter P of an ellipse can be approximated by

$$P \approx 2\pi\sqrt{\frac{a^2 + b^2}{2}}.$$

(a) Approximate the distance in miles that Mercury travels in one orbit of the sun if $a = 36.0$, $b = 35.2$, and the units are in millions of miles.

(b) If a planet has a circular orbit, does this formula give the *exact* perimeter? Explain.

77. *Satellite Orbit* The orbit of Explorer VII and the outline of Earth's surface are shown in the figure. This orbit can be described by the equation $\frac{x^2}{a^2} + \frac{y^2}{b^2} = 1$, where $a = 4464$ and $b = 4462$. The surface of Earth can be described by the equation $(x - 164)^2 + y^2 = 3960^2$. Find the maximum and minimum heights of the satellite above Earth's surface if all units are in miles. (Sources: W. Loh; W. Thomson.)

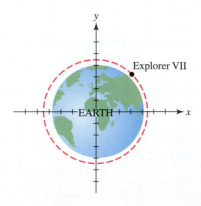

78. *Planet Velocity* The maximum and minimum velocities in kilometers per second of a planet moving in an elliptical orbit can be calculated by

$$v_{max} = \frac{2\pi a}{P}\sqrt{\frac{1+e}{1-e}} \quad \text{and} \quad v_{min} = \frac{2\pi a}{P}\sqrt{\frac{1-e}{1+e}}.$$

In these equations, a is half the length of the major axis of the orbit in kilometers, P is its orbital period in seconds, and e is the eccentricity of the orbit.   (Source: M. Zeilik.)

(a) Calculate v_{max} and v_{min} for Pluto if $a = 5.913 \times 10^9$ kilometers, $P = 2.86 \times 10^{12}$ seconds, and $e = 0.249$.

(b) If a planet has a circular orbit, what can be said about its orbital velocity?

Writing about Mathematics

79. Explain how the distance between the foci of an ellipse affects the shape of the ellipse.

80. Given the standard equation of an ellipse, explain how to determine the length of the major axis. How can you determine whether the major axis is vertical or horizontal?

CHECKING BASIC CONCEPTS FOR SECTIONS 7.1 AND 7.2

1. Graph the parabola defined by $x = \frac{1}{2}y^2$. Include the focus and directrix.

2. Find an equation of the parabola with focus $(-1, 0)$ and directrix $y = 3$.

3. Graph the ellipse defined by $\frac{x^2}{36} + \frac{y^2}{100} = 1$. Include the foci and label the major and minor axes.

4. Find an equation of the ellipse centered at $(3, -2)$ with a vertical major axis of length 6 and minor axis of length 4. What are the coordinates of the foci?

5. A parabolic reflector for a searchlight has a diameter of 4 feet and a depth of 1 foot. How far from the vertex of the reflector should the filament of the light bulb be located?

6. Solve the nonlinear system of equations.

$$x^2 + y^2 = 10$$
$$2x^2 + 3y^2 = 29$$

7.3 HYPERBOLAS

Equations and Graphs of Hyperbolas • Translations of Hyperbolas • Reflective Property of Hyperbolas

Introduction

Hyperbolas have several interesting properties. For example, if a comet passes by the sun with a high velocity, then the sun's gravity may not be strong enough to cause the comet to go into orbit. Instead, the comet will pass by the sun just once and follow a trajectory that can be described by a hyperbola. Hyperbolas also have a reflective property that is used in telescopes. In this section we learn some basic properties of hyperbolas.

Equations and Graphs of Hyperbolas

The third type of conic section is a hyperbola.

Hyperbola

A **hyperbola** is the set of points in a plane, the difference of whose distances from two fixed points is constant. Each fixed point is called a **focus** of the hyperbola.

In Figure 7.47 a point $P(x, y)$ is shown on a hyperbola with distance d_1 from focus $F_1(c, 0)$ and distance d_2 from focus $F_2(-c, 0)$. Regardless of the location of the point $P(x, y)$, $|d_2 - d_1| = 2a$. The **transverse axis** is the line segment connecting the **vertices** $V_1(a, 0)$ and $V_2(-a, 0)$, and its length equals $2a$.

Two hyperbolas are shown in Figures 7.48 and 7.49. The coordinates of the vertices, $(\pm a, 0)$ or $(0, \pm a)$, and foci, $(\pm c, 0)$ or $(0, \pm c)$, are labeled. The two parts of the hyperbola in Figure 7.48 are the **left branch** and the **right branch**, whereas in Figure 7.49 the hyperbola has an **upper branch** and a **lower branch**. A line segment connecting the points $(0, \pm b)$ in Figure 7.48 and $(\pm b, 0)$ in Figure 7.49 is the **conjugate axis**. The lines $y = \pm \frac{b}{a}x$ and $y = \pm \frac{a}{b}x$ are **asymptotes** for each respective hyperbola. They can be used as an aid in graphing. The dashed rectangle is sometimes called the **fundamental rectangle**.

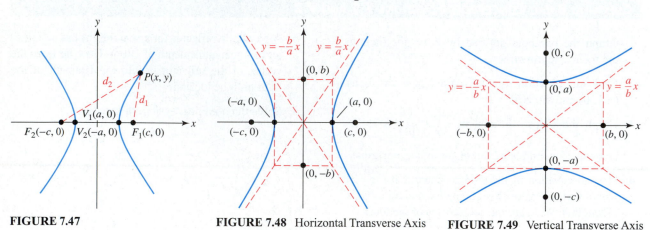

FIGURE 7.47

FIGURE 7.48 Horizontal Transverse Axis

FIGURE 7.49 Vertical Transverse Axis

By the vertical line test, a hyperbola cannot be represented by a function, but many can be described by the following equations. The constants a, b, and c are positive.

Standard equations for hyperbolas centered at (0, 0)

The hyperbola with center at the origin, *horizontal* transverse axis, and equation

$$\frac{x^2}{a^2} - \frac{y^2}{b^2} = 1$$

has asymptotes $y = \pm \frac{b}{a}x$, vertices $(\pm a, 0)$, and foci $(\pm c, 0)$, where $c^2 = a^2 + b^2$.

The hyperbola with center at the origin, *vertical* transverse axis, and equation

$$\frac{y^2}{a^2} - \frac{x^2}{b^2} = 1$$

has asymptotes $y = \pm \frac{a}{b}x$, vertices $(0, \pm a)$, and foci $(0, \pm c)$, where $c^2 = a^2 + b^2$.

Note: A hyperbola consists of two solid curves or branches. The asymptotes, foci, transverse axis, conjugate axis, and fundamental (dashed) rectangle are not part of the hyperbola, but are aids for sketching its graph.

One interpretation of an asymptote can be made using trajectories of comets as they approach the sun. Comets travel in parabolic, elliptic, or hyperbolic trajectories. If the speed of a comet is too slow, the gravitational pull of the sun captures the comet in an elliptical orbit. See Figure 7.50. If the speed of the comet is too fast, the sun's gravity is too weak and the comet passes by the sun in a hyperbolic trajectory. Near the sun the gravitational pull is stronger and the comet's trajectory is curved. Farther from the sun, gravity becomes weaker and the comet eventually returns to a straight-line trajectory that is determined by the *asymptote* of the hyperbola. See Figure 7.51. Finally, if the speed is neither too slow nor too fast, the comet will travel in a parabolic path. See Figure 7.52. In all three cases, the sun is located at a focus of the conic section.

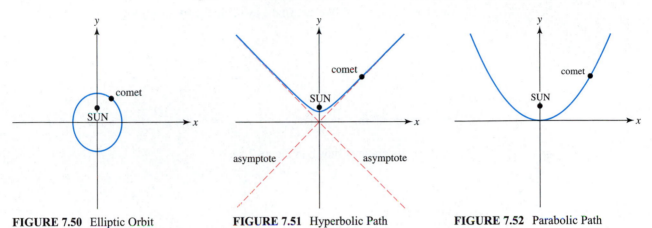

FIGURE 7.50 Elliptic Orbit   FIGURE 7.51 Hyperbolic Path   FIGURE 7.52 Parabolic Path

■ **CLASS DISCUSSION**
If a comet is seen at regular intervals, what type of path must it follow? ■

EXAMPLE 1   *Sketching the graph of a hyperbola*

Sketch a graph of $\dfrac{x^2}{4} - \dfrac{y^2}{9} = 1$. Label the vertices, foci, and asymptotes.

SOLUTION   The equation is in standard form with $a = 2$ and $b = 3$. It has a horizontal transverse axis with vertices $(\pm 2, 0)$. The endpoints of the conjugate axis are $(0, \pm 3)$. To locate the foci find c.

$$c^2 = a^2 + b^2 = 4 + 9 = 13 \qquad \text{or}$$
$$c = \sqrt{13} \approx 3.61.$$

The foci are $(\pm\sqrt{13}, 0)$. The asymptotes are $y = \pm\dfrac{b}{a}x$ or $y = \pm\dfrac{3}{2}x$. See Figure 7.53.

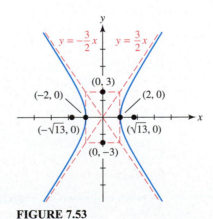

FIGURE 7.53

EXAMPLE 2   *Finding the equation of a hyperbola*

Find the equation of the hyperbola centered at the origin with a vertical transverse axis of length 6 and focus (0, 5). Sketch a graph of the hyperbola.

SOLUTION   Since the hyperbola is centered at the origin with a vertical axis, its equation is $\frac{y^2}{a^2} - \frac{x^2}{b^2} = 1$. The transverse axis has length $6 = 2a$, so $a = 3$. Since one focus is located at (0, 5), $c = 5$. We can find b by using the following equation.

$$b^2 = c^2 - a^2$$
$$b = \sqrt{c^2 - a^2}$$
$$b = \sqrt{5^2 - 3^2} = 4$$

The equation of this hyperbola is $\frac{y^2}{9} - \frac{x^2}{16} = 1$. Its asymptotes are $y = \pm\frac{a}{b}x$ or $y = \pm\frac{3}{4}x$. Its graph is shown in Figure 7.54.

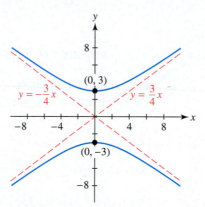

FIGURE 7.54

EXAMPLE 3   *Finding the equation of a hyperbola*

Find the standard equation of the hyperbola shown in Figure 7.55. Identify the vertices, foci, and asymptotes.

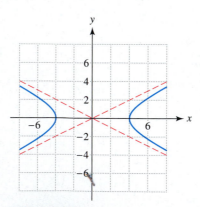

FIGURE 7.55

SOLUTION   Because the hyperbola is centered at $(0, 0)$ with a horizontal transverse axis, its equation has the form

$$\frac{x^2}{a^2} - \frac{y^2}{b^2} = 1.$$

The asymptotes are given by $y = \pm\frac{b}{a}$. In Figure 7.56 the fundamental rectangle is determined by the four points $(\pm4, 0)$ and $(0, \pm2)$, and its diagonals correspond to the asymptotes. It follows that $a = 4$ and $b = 2$. Thus the standard equation of the hyperbola is

$$\frac{x^2}{16} - \frac{y^2}{4} = 1.$$

The vertices are $(\pm4, 0)$, and the asymptotes are $y = \pm\frac{1}{2}x$. To find the coordinates of the foci, find c.

$$c^2 = a^2 + b^2 = 4^2 + 2^2 = 20$$

Thus $c = \sqrt{20} \approx 4.47$, and the coordinates of the foci are $(\pm\sqrt{20}, 0)$.

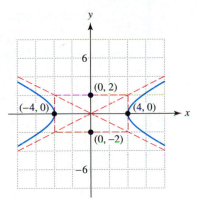

FIGURE 7.56

EXAMPLE 4   *Graphing a hyperbola with technology*

Use a graphing calculator to graph $\dfrac{y^2}{4.2} - \dfrac{x^2}{8.4} = 1$.

SOLUTION   Begin by solving the given equation for y.

$$\frac{y^2}{4.2} = 1 + \frac{x^2}{8.4} \qquad \text{Add } \frac{x^2}{8.4}.$$

$$y^2 = 4.2\left(1 + \frac{x^2}{8.4}\right) \qquad \text{Multiply by 4.2.}$$

$$y = \pm\sqrt{4.2\left(1 + \frac{x^2}{8.4}\right)} \qquad \text{Square root property}$$

Graph $Y_1 = \sqrt{(4.2(1 + X\text{^}2/8.4))}$ and $Y_2 = -\sqrt{(4.2(1 + X\text{^}2/8.4))}$. See Figures 7.57 and 7.58 on the following page.

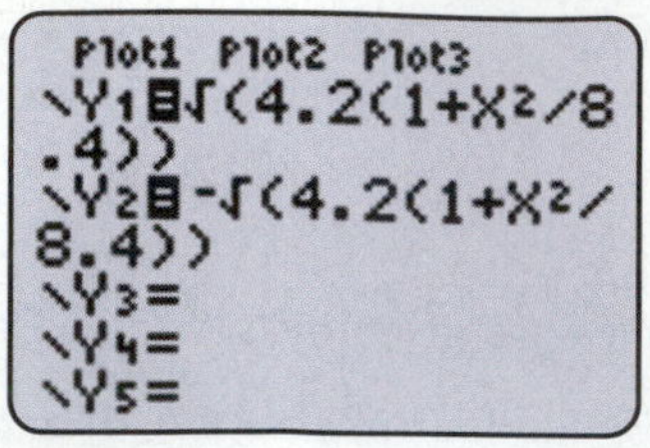

$[-10, 10, 1]$ by $[-10, 10, 1]$

FIGURE 7.57

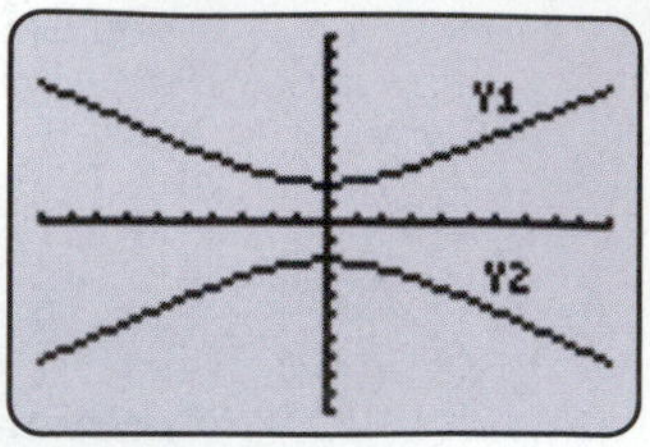

FIGURE 7.58

Translations of Hyperbolas

If the standard equation of a hyperbola is either $\dfrac{x^2}{a^2} - \dfrac{y^2}{b^2} = 1$ or $\dfrac{y^2}{a^2} - \dfrac{x^2}{b^2} = 1$, then the center of the hyperbola is $(0, 0)$. We can use translations of graphs to find the equation of a hyperbola centered at (h, k) by replacing x with $(x - h)$ and y with $(y - k)$.

Standard equations for hyperbolas centered at (h, k)

A hyperbola with center (h, k), and either a horizontal or vertical transverse axis, satisfies one of the following equations, where $c^2 = a^2 + b^2$.

$$\frac{(x - h)^2}{a^2} - \frac{(y - k)^2}{b^2} = 1$$

Transverse axes: horizontal

Vertices: $(h \pm a, k)$; foci: $(h \pm c, k)$
Asymptotes: $y = \pm\dfrac{b}{a}(x - h) + k$

$$\frac{(y - k)^2}{a^2} - \frac{(x - h)^2}{b^2} = 1$$

Transverse axes: vertical

Vertices: $(h, k \pm a)$; foci: $(h, k \pm c)$
Asymptotes: $y = \pm\dfrac{a}{b}(x - h) + k$

EXAMPLE 5   *Graphing a hyperbola with center (h, k)*

Graph the hyperbola whose equation is $\dfrac{(y + 2)^2}{9} - \dfrac{(x - 2)^2}{16} = 1$. Label the vertices, foci, and asymptotes.

SOLUTION   The hyperbola has a vertical transverse axis and its center is $(2, -2)$. Since $a^2 = 9$ and $b^2 = 16$, it follows that $c^2 = a^2 + b^2 = 9 + 16 = 25$. Thus $a = 3$, $b = 4$, and $c = 5$. The vertices are located 3 units above and below the center of the hyperbola and the foci are located 5 units above and below the center of the hyperbola. That is, the vertices are $(2, -2 \pm 3)$, or $(2, 1)$ and $(2, -5)$, and the foci are $(2, -2 \pm 5)$, or $(2, 3)$ and $(2, -7)$. The asymptotes are given by

$$y = \pm\frac{a}{b}(x - h) + k \quad \text{or} \quad y = \pm\frac{3}{4}(x - 2) - 2.$$

A graph of the hyperbola is shown in Figure 7.59.

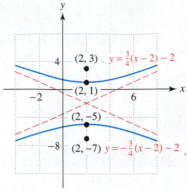

FIGURE 7.59

Reflective Property of Hyperbolas

Hyperbolas have an important reflective property. If a hyperbola is rotated about the x-axis, a **hyperboloid** is formed, as illustrated in Figure 7.60. Any beam of light that is directed toward focus F_1, will be reflected by the hyperboloid toward focus F_2. See Figure 7.61.

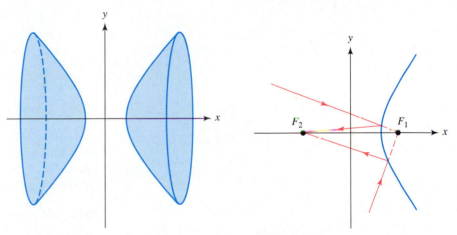

FIGURE 7.60 Hyperboloid   **FIGURE 7.61**

Telescopes sometimes make use of both parabolic and hyperbolic mirrors, as shown in Figure 7.62. When parallel rays of light from distant stars strike the large parabolic (primary) mirror, they are reflected toward its focus, F_1. A smaller hyperbolic (secondary) mirror is placed so that its focus is also located at F_1. Light rays striking the hyperbolic mirror are reflected toward its other focus, F_2, through a small hole in the parabolic mirror, and into an eye piece.

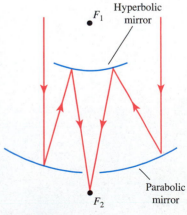

FIGURE 7.62

Putting it all Together

7.3

$\boldsymbol{T}$he following table summarizes some important concepts about hyperbolas.

Concept	Equation	Example
Hyperbola with center $(0, 0)$	Standard equation Transverse axis: horizontal $$\dfrac{x^2}{a^2} - \dfrac{y^2}{b^2} = 1$$ Transverse axis: vertical $$\dfrac{y^2}{a^2} - \dfrac{x^2}{b^2} = 1$$	$\dfrac{y^2}{4} - \dfrac{x^2}{9} = 1;\, a = 2,\, b = 3$ Transverse axis: vertical Vertices $(0, \pm 2)$; foci: $(0, \pm\sqrt{13})$ $(c^2 = a^2 + b^2 = 4 + 9 = 13$, so $c = \sqrt{13}$.) Asymptotes: $y = \pm\dfrac{2}{3}x$
Hyperbola with center (h, k)	Standard equation Transverse axis: horizontal $$\dfrac{(x - h)^2}{a^2} - \dfrac{(y - k)^2}{b^2} = 1$$ Transverse axis: vertical $$\dfrac{(y - k)^2}{a^2} - \dfrac{(x - h)^2}{b^2} = 1$$	$\dfrac{(x - 1)^2}{4} - \dfrac{(y + 1)^2}{9} = 1;\, a = 2,\, b = 3$ Transverse axis: horizontal; center $(1, -1)$ Vertices $(1 \pm 2, -1)$; foci: $(1 \pm \sqrt{13}, -1)$ $(c^2 = a^2 + b^2 = 4 + 9 = 13$, so $c = \sqrt{13}$.) Asymptotes: $y = \pm\dfrac{3}{2}(x - 1) - 1$

7.3 EXERCISES

Hyperbolas with Center (0, 0)

Exercises 1–8: Sketch a graph of the hyperbola, including the asymptotes. Give the coordinates of the foci.

1. $\dfrac{x^2}{9} - \dfrac{y^2}{49} = 1$

2. $\dfrac{x^2}{16} - \dfrac{y^2}{4} = 1$

3. $\dfrac{y^2}{36} - \dfrac{x^2}{16} = 1$

4. $\dfrac{y^2}{4} - \dfrac{x^2}{4} = 1$

5. $x^2 - y^2 = 9$

6. $49y^2 - 25x^2 = 1225$

7. $9y^2 - 16x^2 = 144$

8. $4x^2 - 4y^2 = 100$

Exercises 9–12: Match the equation of the hyperbola with its graph (a–d).

9. $\dfrac{x^2}{4} - \dfrac{y^2}{6} = 1$

10. $\dfrac{x^2}{9} - y^2 = 1$

11. $\dfrac{y^2}{9} - \dfrac{x^2}{16} = 1$

12. $\dfrac{y^2}{4} - \dfrac{x^2}{4} = 1$

a.
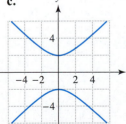

b.
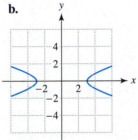

c.
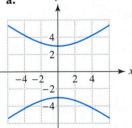

d.
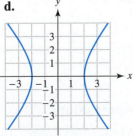

Exercises 13–16: Sketch a graph of a hyperbola, centered at the origin, with the given foci, vertices, and asymptotes shown in the figure. Find an equation of the hyperbola. Note that the coordinates of the foci and vertices are integers.

13.
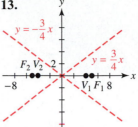

14.
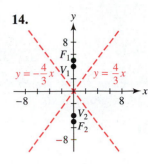

15.

16.
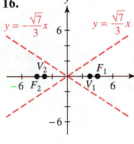

Exercises 17–24: Determine an equation of the hyperbola, centered at the origin, satisfying the conditions. Sketch its graph, including the asymptotes.

17. Foci $(0, \pm 13)$, vertices $(0, \pm 12)$

18. Foci $(\pm 13, 0)$, vertices $(\pm 5, 0)$

19. Vertical transverse axis of length 4, foci $(0, \pm 5)$

20. Horizontal transverse axis of length 12, foci $(\pm 10, 0)$

21. Vertices $(\pm 3, 0)$, asymptotes $y = \pm \dfrac{2}{3}x$

22. Vertices $(0, \pm 4)$, asymptotes $y = \pm \dfrac{1}{2}x$

23. Endpoints of conjugate axis $(0, \pm 3)$, vertices $(\pm 4, 0)$

24. Endpoints of conjugate axis $(\pm 4, 0)$, vertices $(0, \pm 2)$

Hyperbolas with Center (*h*, *k*)

Exercises 25–30: Sketch a graph of the hyperbola.

25. $\dfrac{(x - 1)^2}{16} - \dfrac{(y - 2)^2}{4} = 1$

26. $\dfrac{(y + 1)^2}{16} - \dfrac{(x + 3)^2}{9} = 1$

27. $\dfrac{(y - 2)^2}{36} - \dfrac{(x + 2)^2}{4} = 1$

28. $\dfrac{(x+1)^2}{4} - \dfrac{(y-1)^2}{4} = 1$

29. $\dfrac{x^2}{4} - (y-1)^2 = 1$

30. $(y+1)^2 - \dfrac{(x-3)^2}{4} = 1$

Exercises 31–34: Match the equation of the hyperbola with its graph (a–d).

31. $\dfrac{(x-2)^2}{4} - \dfrac{(y+4)^2}{4} = 1$

32. $\dfrac{(x+1)^2}{4} - \dfrac{y^2}{9} = 1$

33. $\dfrac{(y+1)^2}{16} - \dfrac{(x-2)^2}{16} = 1$

34. $\dfrac{y^2}{25} - \dfrac{(x+1)^2}{9} = 1$

a.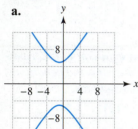

b.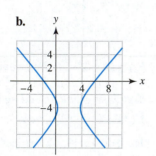

c.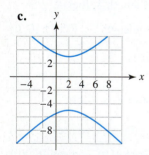

d.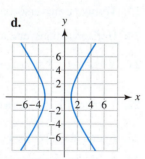

Exercises 35–36: Find an (approximate) equation of the hyperbola shown in the graph.

35.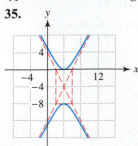

36.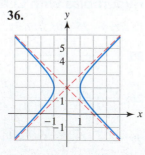

Exercises 37–40: Sketch a graph of the hyperbola including the asymptotes. Give the coordinates of the vertices and foci.

37. $\dfrac{(x-1)^2}{4} - \dfrac{(y-1)^2}{4} = 1$

38. $\dfrac{(x+2)^2}{4} - \dfrac{(y+1)^2}{16} = 1$

39. $\dfrac{(y+1)^2}{16} - \dfrac{(x-1)^2}{9} = 1$

40. $y^2 - \dfrac{(x-2)^2}{4} = 1$

Exercises 41–44: Find the standard equation of a hyperbola with center (h, k) that satisfies the given conditions.

41. Center $(2, -2)$, focus $(4, -2)$, and vertex $(3, -2)$

42. Center $(-1, 1)$, focus $(-1, 4)$, and vertex $(-1, 3)$

43. Vertices $(-1, \pm 1)$ and foci $(-1, \pm 3)$

44. Vertices $(2 \pm 1, 1)$ and foci $(2 \pm 3, 1)$

Graphing Hyperbolas with Technology

Exercises 45–48: Graph the hyperbola with a graphing calculator.

45. $\dfrac{(y-1)^2}{11} - \dfrac{x^2}{5.9} = 1$ **46.** $\dfrac{x^2}{5.3} - \dfrac{y^2}{6.7} = 1$

47. $3y^2 - 4x^2 = 15$ **48.** $2.1x^2 - 6y^2 = 12$

Solving Equations

Exercises 49–52: Solve the system of equations. Give graphical support by making a sketch.

49. $x^2 - y^2 = 4$ **50.** $x^2 - 4y^2 = 16$
$\ x^2 + y^2 = 9$ $\ x^2 + 4y^2 = 16$

51. $\dfrac{x^2}{4} - \dfrac{y^2}{9} = 1$ **52.** $x^2 - y^2 = 4$
$\ x + y = 2$ $\ x + y = 2$

Applications

53. *Satellite Orbits*   The trajectory of a satellite near Earth can trace a hyperbola, parabola, or ellipse. If the satellite follows either a hyperbolic or parabolic path, it escapes Earth's gravitational influence after a single pass. The path that a satellite travels near Earth depends both on its velocity V in meters per second and its distance D in meters from the center of Earth. Its path is hyperbolic if $V > \dfrac{k}{\sqrt{D}}$, parabolic if $V = \dfrac{k}{\sqrt{D}}$, and elliptic if $V < \dfrac{k}{\sqrt{D}}$, where

$k = 2.82 \times 10^7$ is a constant. (Sources: W. Loh, *Dynamics and Thermodynamics of Planetary Entry;* W. Thomson, *Introduction to Space Dynamics.*)

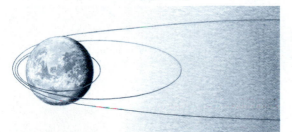

(a) When Explorer IV was at a maximum distance of 42.5×10^6 meters from Earth's center, it had a velocity of 2090 meters/second. Determine the shape of its trajectory.
(b) If an orbiting satellite is scheduled to escape Earth's gravity, so that it can travel to another planet, its velocity must be increased so its trajectory changes from elliptic to hyperbolic.

What range of velocities would allow Explorer IV to leave Earth's influence when it is at a maximum distance?

(c) Explain why it is easier to change a satellite's trajectory from an ellipse to a hyperbola when D is maximum rather than minimum.

54. *Telescopes* (Refer to Figure 7.62.) Suppose that the coordinates of F_1 are $(0, 5.2)$ and the coordinates of F_2 are $(0, -5.2)$. If the coordinates of the vertex of the hyperbolic mirror are $(0, 4.1)$, find the standard equation of a hyperbola, whose upper branch coincides with the hyperbolic mirror.

Writing about Mathematics

55. Explain how the center and the asymptotes of a hyperbola are related to the fundamental rectangle.

56. Given the standard equation of a hyperbola, explain how to determine the length of the transverse axis. How can you determine whether the transverse axis is vertical or horizontal?

CHECKING BASIC CONCEPTS FOR SECTION 7.3

1. Graph the hyperbola defined by $\dfrac{x^2}{9} - \dfrac{y^2}{16} = 1$. Include the foci and asymptotes.

2. Find an equation of the hyperbola centered at $(1, 3)$ with a horizontal transverse axis of length 6 and a conjugate axis of length 4. What are the coordinates of the foci?

3. Use the graph to write the equation of the hyperbola.

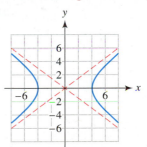

Chapter 7 Summary

CONCEPT	EXPLANATION AND EXAMPLES
Section 7.1	
CONIC SECTIONS	Three basic types: parabola, ellipse, and hyperbola (Circles are examples of ellipses.)
PARABOLAS WITH VERTEX (0, 0)	*Standard Forms* $x^2 = 4py$ (vertical axis) or $y^2 = 4px$ (horizontal axis) *Meaning of p* Both the vertex-focus distance and the vertex-directrix distance are p. The sign of p determines if the parabola opens upward or downward—left or right. The

CONCEPT	EXPLANATION AND EXAMPLES

focus is either $(0, p)$ or $(p, 0)$, and the directrix is either $y = -p$ or $x = -p$. Note that either $p > 0$ or $p < 0$ is possible.

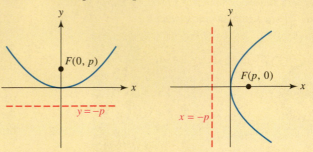

Vertical Axis: $x^2 = 4py$ Horizontal Axis: $y^2 = 4px$

PARABOLAS WITH VERTEX (h, k)

$(x - h)^2 = 4p(y - k)$ Vertical axis; focus $(h, k + p)$
Directrix: $y = k - p$

$(y - k)^2 = 4p(x - h)$ Horizontal axis; focus: $(h + p, k)$
Directrix: $x = h - p$

Section 7.2

ELLIPSES WITH CENTER $(0, 0)$

Standard Forms with $a > b > 0$

$\dfrac{x^2}{a^2} + \dfrac{y^2}{b^2} = 1$ (horizontal major axis)   or $\dfrac{y^2}{a^2} + \dfrac{x^2}{b^2} = 1$ (vertical major axis)

Meaning of a, b, and c

The distance from the center to a vertex is a, the distance from the center to an endpoint of the minor axis is b, and the distance from the center to a focus is c.

The ratio $\dfrac{c}{a}$ equals the eccentricity e. The values of a, b, and c are related by $c^2 = a^2 - b^2$. The foci are either $(\pm c, 0)$ or $(0, \pm c)$, and the vertices are either $(\pm a, 0)$ or $(0, \pm a)$. If $a = b$, then the ellipse is a circle.

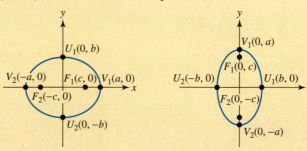

Horizontal Major Axis   Vertical Major Axis

STANDARD EQUATION OF A CIRCLE

$(x - h)^2 + (y - k)^2 = r^2$ Center: (h, k); radius: r

ELLIPSES WITH VERTEX (h, k)

$a > b > 0; c^2 = a^2 - b^2$

$\dfrac{(x - h)^2}{a^2} + \dfrac{(y - k)^2}{b^2} = 1$ Major axis: horizontal; foci: $(h \pm c, k)$
Vertices: $(h \pm a, k)$

CONCEPT	EXPLANATION AND EXAMPLES

$$\frac{(y-k)^2}{a^2} + \frac{(x-h)^2}{b^2} = 1$$

Major axis: vertical; foci: $(h, k \pm c)$
Vertices: $(h, k \pm a)$

AREA INSIDE AN ELLIPSE   Given the standard equation of an ellipse, the area A of the region contained inside is given by $A = \pi ab$.

Section 7.3

HYPERBOLAS WITH CENTER (0, 0)

Standard Forms with a, b > 0

$$\frac{x^2}{a^2} - \frac{y^2}{b^2} = 1 \text{ (horizontal transverse axis)} \quad \text{or}$$

$$\frac{y^2}{a^2} - \frac{x^2}{b^2} = 1 \text{ (vertical transverse axis)}$$

Meaning of a, b, and c

The distance from the center to a vertex is a, and the distance from the center to a focus is c. The asymptotes are $y = \pm\frac{b}{a}x$ if the transverse axis is horizontal, and $y = \pm\frac{a}{b}x$ if the transverse axis is vertical. The values of a, b, and c are related by $c^2 = a^2 + b^2$. The foci are either $(\pm c, 0)$ or $(0, \pm c)$ and the vertices are either $(\pm a, 0)$ or $(0, \pm a)$.

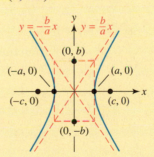

Horizontal Transverse Axis

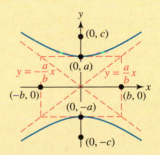

Vertical Transverse Axis

HYPERBOLAS WITH VERTEX (h, k)

$a, b > 0; c^2 = a^2 + b^2$

$$\frac{(x-h)^2}{a^2} - \frac{(y-k)^2}{b^2} = 1$$

Transverse axis: horizontal
Foci: $(h \pm c, k)$; vertices: $(h \pm a, k)$
Asymptotes: $y = \pm\frac{b}{a}(x - h) + k$

$$\frac{(y-k)^2}{a^2} - \frac{(x-h)^2}{b^2} = 1$$

Transverse axis: vertical
Foci: $(h, k \pm c)$; vertices: $(h, k \pm a)$
Asymptotes: $y = \pm\frac{a}{b}(x - h) + k$

Review Exercises

Exercises 1–6: Sketch a graph of the equation.

1. $-x^2 = y$

2. $y^2 = 2x$

3. $\dfrac{x^2}{25} + \dfrac{y^2}{49} = 1$

4. $\dfrac{y^2}{4} + \dfrac{x^2}{2} = 1$

5. $\dfrac{y^2}{4} - \dfrac{x^2}{9} = 1$

6. $x^2 - y^2 = 4$

Exercises 7–12: Match the equation with its graph (a–f).

7. $x^2 = 2y$

8. $y^2 = -3x$

9. $x^2 + y^2 = 4$

10. $\dfrac{x^2}{36} + \dfrac{y^2}{49} = 1$

11. $\dfrac{x^2}{4} - \dfrac{y^2}{9} = 1$

12. $\dfrac{y^2}{36} - \dfrac{x^2}{25} = 1$

a.
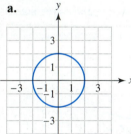

b.
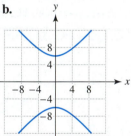

c.
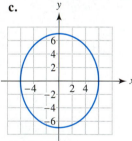

d.
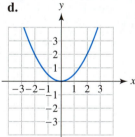

e.
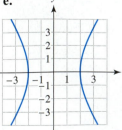

f.
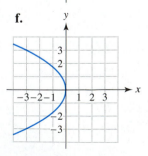

Exercises 13–18: Determine an equation of the conic section that satisfies the given conditions. Sketch its graph.

13. Parabola with focus $(2, 0)$ and vertex $(0, 0)$

14. Ellipse with foci $(\pm 4, 0)$ and vertices $(\pm 5, 0)$

15. Ellipse centered at the origin with vertical major axis of length 14 and minor axis of length 8

16. Hyperbola with foci $(0, \pm 10)$ and endpoints of the conjugate axis $(\pm 6, 0)$

17. Parabola with vertex $(5, 2)$ and focus $(5, 0)$

18. Hyperbola with vertices $(-2 \pm 3, 3)$ and foci $(-2 \pm 4, 3)$

Exercises 19–26: Sketch a graph of the conic section. Give the coordinates of any foci.

19. $-4y = x^2$

20. $y^2 = 8x$

21. $\dfrac{x^2}{25} + \dfrac{y^2}{4} = 1$

22. $49x^2 + 36y^2 = 1764$

23. $\dfrac{x^2}{16} - \dfrac{y^2}{9} = 1$

24. $\dfrac{y^2}{4} - x^2 = 1$

25. $x^2 + y^2 = 4$

26. $(x - 3)^2 + (y + 1)^2 = 9$

Exercises 27–30: Sketch a graph of the conic section. Identify the coordinates of its center when appropriate.

27. $\dfrac{(y - 2)^2}{4} + \dfrac{(x + 1)^2}{16} = 1$

28. $\dfrac{(x - 2)^2}{16} - \dfrac{y^2}{4} = 1$

29. $\dfrac{(x - 1)^2}{4} - \dfrac{(y + 1)^2}{4} = 1$

30. $(x + 2) = 4(y - 1)^2$

31. Sketch a graph of $(y - 4)^2 = -8(x - 8)$. Include the focus and the directrix.

32. Sketch a graph of $\dfrac{(x - 3)^2}{16} + \dfrac{(y + 2)^2}{9} = 1$. Include the foci.

Exercises 33–36: Use technology to graph the equation.

33. $y^2 = \dfrac{3}{4}x$

34. $7.1x^2 + 8.2y^2 = 60$

35. $9x^2 - 2y^2 = 17$

36. $\dfrac{(y - 1.4)^2}{7} - \dfrac{(x + 2.3)^2}{11} = 1$

37. Find the equation of a circle with center $(0, 0)$ and radius 7.

38. Find the equation of a circle with center $(-7, 6)$ and radius 20.

Exercises 39–40: Solve the system of equations.

39. $\dfrac{x^2}{4} + \dfrac{y^2}{4} = 1$

$\dfrac{x^2}{8} + \dfrac{y^2}{2} = 1$

40. $x^2 - y^2 = 1$

$x + y = 2$

Exercises 41–42. Shade the region in the xy-plane that satisfies the system of inequalities.

41. $\dfrac{x^2}{9} + \dfrac{y^2}{4} \le 1$

$x + y \le 3$

42. $y^2 - x^2 \le 9$

$y - x \le 0$

43. *Comets* A comet travels along an elliptical orbit around the sun. Its path can be described by the equation

$$\frac{x^2}{70^2} + \frac{y^2}{500^2} = 1,$$

where units are in millions of miles.
   (a) What is the comet's minimum and maximum distance from the sun?
   (b) Estimate the distance that the comet travels in one orbit around the sun. (Refer to Exercise 76, Section 7.2.)

44. *Searchlight* A searchlight is being constructed in the shape of a paraboloid with a depth of 7 inches and a diameter of 20 inches, as illustrated in the accompanying figure. Determine the distance d that the bulb should be from the vertex in order to have the beam of light shine straight ahead.

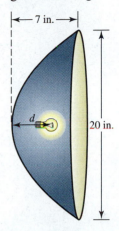

45. *Arch Bridge* An elliptical arch under a bridge is constructed so that it is 80 feet wide and has a maximum height of 30 feet, as illustrated in the accompanying figure. Find the height of the arch 10 feet from the center of the arch.

Extended and Discovery Exercises

Exercises 1–5: Farthest Planet from the Sun Both Neptune and Pluto travel around the sun in elliptical orbits. For Neptune's orbit, $a = 30.10$ and for Pluto's orbit, $a = 39.44$, where the variable a represents the planet's average distance from the sun in astronomical units. (One astronomical unit equals 93 million miles.) The value of the variable a also corresponds to half the length of the major axis. Pluto has a highly eccentric orbit with $e = 0.249$, and Neptune has a nearly circular orbit with $e = 0.009$.

1. Calculate the value of c for each planet's orbit.

2. Position the sun at the *origin* of the xy-plane. Find the coordinates of the center of Neptune's orbit and the coordinates of the center of Pluto's orbit. Assume that the centers lie on the positive x-axis.

3. Find equations for Neptune's orbit and for Pluto's orbit.

4. Graph both orbits in the same xy-plane.

5. Is Pluto always the farthest planet from the sun? Explain.

The art of asking the right questions in mathematics is more important than the art of solving them.

—

Georg Cantor

—

8

Further Topics in Algebra

Mathematics permeates the fabric of modern society. At times its presence is subtle and frequently missed by the average person, but its influence is nonetheless profound. Mathematics is the language of technology—it allows society to quantify its experiences.

In previous chapters we saw hundreds of examples where mathematics is used to describe physical phenomena and events. These included computers, CD players, cars, electricity, light, road construction, government data, telephones, medicine, ecology, business, sports, and psychology. Mathematics is di-

verse in its ability to adapt to new situations and solve complex problems. If any subject area is studied in enough detail, mathematics usually appears. Mathematics even describes aspects of seemingly intangible concepts such as color and photography.

This chapter introduces further topics in mathematics. It represents only a small fraction of the topics found in mathematics. Although it may be difficult to predict exactly what the twenty-first century will bring, one thing is certain—mathematics will continue to play a very important role.

8.1 SEQUENCES

Sequences and Functions • Representations of Sequences • Arithmetic Sequences • Geometric Sequences

Introduction

Sequences are a fundamental concept in mathematics and have many applications. A sequence is a function that computes an ordered list. For example, the average person in the United States uses 100 gallons of water each day. The function $f(n) = 100n$ generates the terms of the sequence

$$100, 200, 300, 400, 500, 600, 700, \ldots,$$

when $n = 1, 2, 3, 4, 5, 6, 7, \ldots$. This list represents the gallons of water used by the average person after n days.

Sequences and Functions

A second example of a sequence involves investing money. If \$100 is deposited into a savings account, paying 5% interest compounded annually, then the function defined by $g(n) = 100(1.05)^n$ calculates the account balance after n years, which is given by

$$g(1), g(2), g(3), g(4), g(5), g(6), g(7), \ldots.$$

These terms can be approximated as

$$105, 110.25, 115.76, 121.55, 127.63, 134.01, 140.71, \ldots.$$

We now define a sequence formally.

Sequence

An **infinite sequence** is a function that has the set of natural numbers as its domain. A **finite sequence** is a function with domain $D = \{1, 2, 3, \ldots, n\}$, for some fixed natural number n.

Since sequences are functions, many of the concepts discussed in previous chapters apply to sequences. Instead of letting y represent the output, it is common to write $a_n = f(n)$, where n is a natural number in the domain of the sequence. The **terms** of a sequence are

$$a_1, a_2, a_3, \ldots, a_n, \ldots.$$

The first term is $a_1 = f(1)$, the second term is $a_2 = f(2)$, and so on. The **nth term** or **general term** of a sequence is $a_n = f(n)$.

EXAMPLE 1 *Computing terms of a sequence*

The average distances of the planets from the sun display a pattern that was first described by Johann Bode in 1772. This relationship is called *Bode's law*, and was proposed even before Uranus, Neptune, and Pluto were discovered. It is a sequence defined by

$$f(1) = 0.4$$

$$f(n) = 0.3(2)^{n-2} + 0.4, \qquad \text{for } n = 2, 3, 4, \ldots, 10,$$

where $a_n = f(n)$. In this sequence, a distance of one unit corresponds to the average Earth–Sun distance. The number n represents the nth planet. The actual distances of the planets from the sun, including an average distance for the asteroids from the sun, are listed in Table 8.1. (Source: M. Zeilik, *Introductory Astronomy and Astrophysics.*)

(a) Find $a_4 = f(4)$, and interpret the result.

(b) Calculate the terms of the sequence defined by f. Compare them with the values in Table 8.1.

(c) If there is another planet beyond Pluto, use Bode's law to predict its distance from the sun.

TABLE 8.1

Planet	Distance
Mercury	0.39
Venus	0.72
Earth	1.00
Mars	1.52
Asteroids	2.8
Jupiter	5.20
Saturn	9.54
Uranus	19.2
Neptune	30.1
Pluto	39.5

SOLUTION   (a) $a_4 = f(4) = 0.3(2)^{4-2} + 0.4 = 1.6$. Bode's law predicts that the fourth planet, Mars, is located 1.6 times farther from the sun than Earth is.

(b) We will calculate $f(n)$ for $n = 1, 2, 3, \ldots, 10$. For example,

$$f(1) = 0.4 \qquad \text{and}$$

$$f(2) = 0.3(2)^{2-2} + 0.4 = 0.7.$$

Other values are found in a similar manner. The ten terms of the sequence are

$$0.4, \quad 0.7, \quad 1, \quad 1.6, \quad 2.8, \quad 5.2, \quad 10, \quad 19.6, \quad 38.8, \quad 77.2.$$

Bode's law works quite well for the seven inner planets, if we include the asteroids, but fails to predict the location of Neptune at 30.1. It appears to locate planet Pluto.

(c) Since Pluto's distance is 39.5, Bode's law predicts the next planet at 77.2. ▮

EXAMPLE 2   *Finding terms of a sequence*

Write the first four terms a_1, a_2, a_3, and a_4 of each sequence, where $a_n = f(n)$.

(a) $f(n) = 2n - 5$ (b) $f(n) = 4(2)^{n-1}$

(c) $f(n) = (-1)^n \left(\dfrac{n}{n+1} \right)$

SOLUTION   **(a)** Evaluate the following.

$$a_1 = f(1) = 2(1) - 5 = -3$$
$$a_2 = f(2) = 2(2) - 5 = -1$$

In a similar manner, $a_3 = f(3) = 1$ and $a_4 = f(4) = 3$.

(b) Since $f(n) = 4(2)^{n-1}$,

$$a_1 = f(1) = 4(2)^{1-1} = 4.$$

Similarly, $a_2 = 8$, $a_3 = 16$, and $a_4 = 32$.

(c) Let $f(n) = (-1)^n \left(\dfrac{n}{n+1} \right)$, and substitute $n = 1, 2, 3,$ and 4.

$$a_1 = f(1) = (-1)^1 \left(\frac{1}{1+1} \right) = -\frac{1}{2}$$

$$a_2 = f(2) = (-1)^2 \left(\frac{2}{2+1} \right) = \frac{2}{3}$$

$$a_3 = f(3) = (-1)^3 \left(\frac{3}{3+1} \right) = -\frac{3}{4}$$

$$a_4 = f(4) = (-1)^4 \left(\frac{4}{4+1} \right) = \frac{4}{5}$$

Notice that the factor $(-1)^n$ causes the terms of the sequence to alternate sign.

Graphing Calculator Help

To learn how to generate a sequence, see Appendix B (page AP-22).

Graphing calculators may be used to calculate the terms of a sequence. Figures 8.1–8.3 show the first four terms of each sequence in Example 2. The graphing calculator is in sequence mode.

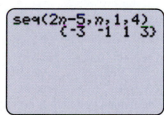

FIGURE 8.1

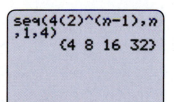

FIGURE 8.2

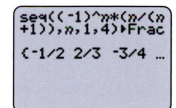

FIGURE 8.3

Some sequences are not defined using a general term. Instead they are defined *recursively*. With a recursive formula, we must find terms a_1 through a_{n-1} before we can find a_n.

A population model for an insect with a life span of 1 year can be described using a **recursive sequence**. Suppose each adult female insect produces r female offspring that survive to reproduce the following year. Let $f(n)$ calculate the insect population during year n. Then, the number of female insects is given recursively by

$$f(n) = rf(n-1) \qquad \text{for } n > 1.$$

The number of female insects in the year n is equal to r times the number of female insects in the previous year $n - 1$. This symbolic representation of a function is fundamentally different from any other equation we have encountered thus far. The reason is that the function f is defined symbolically in terms of itself. To evaluate $f(n)$, we evaluate $f(n-1)$. To evaluate $f(n-1)$, we evaluate $f(n-2)$, and so on. If we know the number of adult female insects during the first year, then we can determine the sequence. That is, if $f(1)$ is given, we can determine $f(n)$ by first computing

$$f(1), f(2), f(3), \ldots, f(n-1).$$

The next example illustrates this recursively defined sequence. (Source: D. Brown and P. Rothery, *Models in Biology: Mathematics, Statistics and Computing.*)

EXAMPLE 3 *Modeling insect population*

Suppose that the initial density of adult female insects is 1000 per acre and $r = 1.1$. Then, the density of female insects is described by

$$f(1) = 1000$$
$$f(n) = rf(n-1), \qquad n > 1.$$

(a) Rewrite this symbolic representation in terms of a_n.
(b) Find a_4 and interpret the result. Is the density of female insects increasing or decreasing?
(c) A general term for this sequence is given by $f(n) = 1000(1.1)^{n-1}$. Use this representation to find a_4.

SOLUTION (a) Since $a_n = f(n)$ for all n, $a_{n-1} = f(n-1)$. The sequence can be expressed as

$$a_1 = 1000$$
$$a_n = ra_{n-1}, \qquad n > 1.$$

(b) In order to calculate the fourth term, a_4, we must first determine a_1, a_2, and a_3. Since $r = 1.1$, $a_n = 1.1a_{n-1}$.

$$a_1 = 1000$$
$$a_2 = 1.1a_1 = 1.1(1000) = 1100$$
$$a_3 = 1.1a_2 = 1.1(1100) = 1210$$
$$a_4 = 1.1a_3 = 1.1(1210) = 1331$$

The fourth term is $a_4 = 1331$. The female population density is increasing and reaches 1331 per acre during the fourth year.

(c) Since $a_4 = f(4)$,

$$a_4 = 1000(1.1)^{4-1} = 1331.$$

It is less work to find a_n using a formula for a general term than a recursive formula—particularly if n is large.

■ **CLASS DISCUSSION**

How does the value of r in Example 3 affect the population density in future years? ■

Representations of Sequences

Sequences are functions. Therefore they have graphical, numerical, and symbolic representations.

EXAMPLE 4   *Representing a sequence numerically and graphically*

Let a recursive sequence be defined as follows.

$$a_1 = 3$$

$$a_n = 2a_{n-1} - 2, \quad n > 1$$

(a) Give a numerical representation of this sequence for $n = 1, 2, 3, 4, 5$.

(b) Graph the first five terms of this sequence.

SOLUTION   **(a)** *Numerical Representation*   Start by calculating the first five terms of the sequence.

$$a_1 = 3$$

$$a_2 = 2a_1 - 2 = 2(3) - 2 = 4$$

$$a_3 = 2a_2 - 2 = 2(4) - 2 = 6$$

$$a_4 = 2a_3 - 2 = 2(6) - 2 = 10$$

$$a_5 = 2a_4 - 2 = 2(10) - 2 = 18$$

TABLE 8.2

n	1	2	3	4	5
a_n	3	4	6	10	18

The first five terms are 3, 4, 6, 10, and 18. A numerical representation of the sequence is shown in Table 8.2.

(b) *Graphical Representation*   To represent these terms graphically, plot the points $(1, 3)$, $(2, 4)$, $(3, 6)$, $(4, 10)$, and $(5, 18)$, as shown in Figure 8.4. Because the domain of a sequence contains only natural numbers, the graph of a sequence is a scatterplot.

FIGURE 8.4

A graphing calculator set in sequence mode may be used to calculate the terms of the recursive sequence in Example 4. See Figures 8.5 and 8.6.

Graphing Calculator Help

To make a table or graph of a sequence, see Appendix B (page AP-22).

[0, 6, 1] by [0, 20, 4]

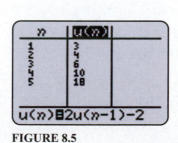

FIGURE 8.5

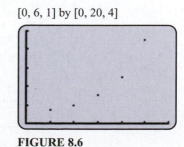

FIGURE 8.6

The next example illustrates numerical and graphical representations for a sequence involving population growth.

EXAMPLE 5 *Representing sequences numerically and graphically*

Frequently the population of a particular insect does not continue to grow indefinitely, as it does in Example 3. Instead, its population grows rapidly at first, and then levels off because of competition for limited resources. In one study, the behavior of the winter moth was modeled with a sequence similar to the following, where a_n represents the population density in thousands per acre during year n. (Source: G. Varley and G. Gradwell, "Population models for the winter moth.")

$$a_1 = 1$$
$$a_n = 2.85a_{n-1} - 0.19a_{n-1}^2, \qquad n \geq 2$$

(a) Give a numerical representation for $n = 1, 2, 3, \ldots, 10$. Describe what happens to the population density of the winter moth.

(b) Use the numerical representation to graph the sequence.

SOLUTION (a) Evaluate $a_1, a_2, a_3, \ldots, a_{10}$ recursively. Since $a_1 = 1$,

$$a_2 = 2.85a_1 - 0.19a_1^2 = 2.85(1) - 0.19(1)^2 = 2.66, \qquad \text{and}$$
$$a_3 = 2.85a_2 - 0.19a_2^2 = 2.85(2.66) - 0.19(2.66)^2 \approx 6.24.$$

Approximate values for other terms are shown in Table 8.3. Figure 8.7 shows the computation of the sequence using a calculator. The sequence is denoted by $u(n)$ rather than a_n.

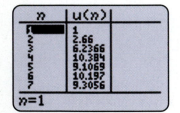

FIGURE 8.7

TABLE 8.3

n	1	2	3	4	5	6	7	8	9	10
a_n	1	2.66	6.24	10.4	9.11	10.2	9.31	10.1	9.43	9.98

(b) The graph of a sequence is a set of discrete points. Plot the points

$$(1, 1), (2, 2.66), (3, 6.24), \ldots, (10, 9.98),$$

as shown in Figure 8.8. At first, the insect population increases rapidly, and then oscillates about the line $y = 9.7$. (See the following Class Discussion.) The oscillations become smaller as n increases, indicating that the population density may stabilize near 9.7 thousand per acre. Some calculators can plot sequences, as shown in Figure 8.9. In this figure, the first 20 terms have been plotted.

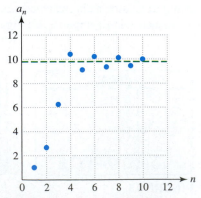

FIGURE 8.8 Insect Population

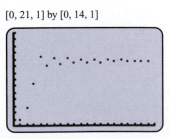

[0, 21, 1] by [0, 14, 1]

FIGURE 8.9

■ **CLASS DISCUSSION**
In Example 5, the insect population stabilizes near the value $k = 9.74$ thousand. This value of k can be found by solving the quadratic equation $k = 2.85k - 0.19k^2$. Try to explain why this is true. ■

Arithmetic Sequences

Suppose that a person's starting salary was $20,000 per year. Thereafter this person receives a $1000 raise each year. The salary *after n* years of experience is represented by

$$f(n) = 1000n + 20,000$$

where f is a linear function. After 10 years of experience, the annual salary would be

$$f(10) = 1000(10) + 20,000 = \$30,000.$$

If a sequence can be defined by a linear function, it is an *arithmetic sequence*.

Algebra Review
To review linear functions, see Section 2.1.

Infinite arithmetic sequence

An **infinite arithmetic sequence** is a linear function whose domain is the set of natural numbers.

An arithmetic sequence can be defined recursively by $a_n = a_{n-1} + d$, where d is a constant. Since $d = a_n - a_{n-1}$ for each valid n, d is called the **common difference**. If $d = 0$, then the sequence is a **constant sequence**. A **finite arithmetic sequence** is similar to an infinite arithmetic sequence except its domain is $D = \{1, 2, 3, \ldots, n\}$, where n is a fixed natural number.

EXAMPLE 6 *Determining arithmetic sequences*

Determine if f is an arithmetic sequence for each situation.
(a) $f(n) = n^2 + 3n$
(b) A graph of f is shown in Figure 8.10.

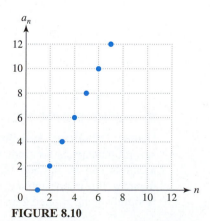

FIGURE 8.10

(c) The function f is represented numerically in Table 8.4.

TABLE 8.4

n	1	2	3	4	5	6	7
$f(n)$	-1.5	0	1.5	3	4.5	6	7.5

SOLUTION
(a) This sequence is not arithmetic, because $f(n) = n^2 + 3n$ is nonlinear.
(b) The sequence in Figure 8.10 represents an arithmetic sequence because the points lie on a line. A linear function could generate these points. Notice that the slope between points is always equal to 2. This slope represents the common difference d of the sequence.
(c) The successive terms

$$-1.5, 0, 1.5, 3, 4.5, 6, 7.5$$

increase by precisely 1.5. Therefore the common difference is $d = 1.5$. Since $a_n = a_{n-1} + 1.5$ for each valid n, the sequence is arithmetic. ∎

Since an arithmetic sequence is a linear function, it can always be represented by $f(n) = dn + c$, where d is the common difference and c is a constant.

EXAMPLE 7 *Finding a symbolic representation for an arithmetic sequence*

Find a general term $a_n = f(n)$ for each arithmetic sequence.
(a) $a_1 = 3$ and $d = -2$
(b) $a_3 = 4$ and $a_9 = 17$

SOLUTION
(a) Let $f(n) = dn + c$. Since $d = -2$, $f(n) = -2n + c$.

$$a_1 = f(1) = -2(1) + c = 3 \qquad \text{or} \qquad c = 5$$

Thus $a_n = -2n + 5$.
(b) Since $a_3 = 4$ and $a_9 = 17$, we find a linear function $f(n) = dn + c$ that satisfies the equations $f(3) = 4$ and $f(9) = 17$. The common difference is equal to the slope between the points $(3, 4)$ and $(9, 17)$.

$$d = \frac{17 - 4}{9 - 3} = \frac{13}{6}$$

It follows that $f(n) = \frac{13}{6}n + c$.

$$a_3 = f(3) = \frac{13}{6}(3) + c = 4 \qquad \text{or} \qquad c = -\frac{5}{2}$$

Thus $a_n = \frac{13}{6}n - \frac{5}{2}$. ∎

■ **MAKING CONNECTIONS**

Linear Functions and Arithmetic Sequences   In Chapter 2 we discussed several techniques for finding symbolic representations of linear functions. These methods can be applied to finding symbolic representations of arithmetic sequences. It is important to realize that the mathematical concept of a linear function is simply being applied to the new topic of sequences.

■ **CLASS DISCUSSION**
Explain why the nth term of an arithmetic sequence is given by

$$a_n = a_1 + (n - 1)d. \quad ■$$

Geometric Sequences

Suppose that a person with a starting salary of $20,000 per year receives a 5% raise each year. If $a_n = f(n)$ computes this salary at the *beginning* of the nth year, then

$$f(1) = 20,000$$
$$f(2) = 20,000(1.05) = 21,000$$
$$f(3) = 21,000(1.05) = 22,050$$
$$f(4) = 22,050(1.05) = 23,152.50$$

are salaries for the first 4 years. Each salary results from multiplying the previous salary by 1.05. A general term in the sequence can be written as

$$f(n) = 20,000(1.05)^{n-1}.$$

During the 10th year, the annual salary is

$$f(10) = 20,000(1.05)^{10-1} \approx \$31,027.$$

This type of sequence is a *geometric sequence* given by $f(n) = cr^{n-1}$, where c and r are constants. Geometric sequences are capable of either rapid growth or decay. The first five terms from some geometric sequences are shown in Table 8.5. The corresponding values of c and r have been included.

TABLE 8.5   Terms of Geometric Sequences

c	r	a_1, a_2, a_3, a_4, a_5
1	2	1, 2, 4, 8, 16
1	$\dfrac{1}{2}$	$1, \dfrac{1}{2}, \dfrac{1}{4}, \dfrac{1}{8}, \dfrac{1}{16}$
2	-4	2, -8, 32, -128, 512
3	$\dfrac{1}{10}$	3, 0.3, 0.03, 0.003, 0.0003

The terms of a geometric sequence can be found by multiplying the previous term by r. We now define a geometric sequence formally.

Infinite geometric sequence

An **infinite geometric sequence** is a function defined by $f(n) = cr^{n-1}$, where c and r are nonzero constants. The domain of f is the set of natural numbers.

A geometric sequence can be defined recursively by $a_n = ra_{n-1}$, where $a_n = f(n)$ and the first term is $a_1 = c$. Since $r = \dfrac{a_n}{a_{n-1}}$ for each valid n, r is referred to as the **common ratio**.

■ MAKING CONNECTIONS ─────────

Exponential Function and Geometric Sequences   If the domain of an exponential function, $f(x) = Ca^x$, is restricted to the natural numbers, then f represents a geometric sequence. For example, $f(x) = 3(2)^x$ generates the sequence 3, 6, 12, 24, 48, ... when $x = 1, 2, 3, 4, 5, \ldots$. ■

The next example illustrates how to recognize symbolic, graphical, and numerical representations of geometric sequences.

EXAMPLE 8 *Representing geometric sequences*

Decide which of the following represents a geometric sequence, where $a_n = f(n)$.
(a) $f(n) = 4(0.5)^n$
(b) The function f is represented graphically in Figure 8.11.

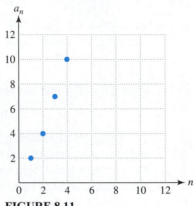

FIGURE 8.11

(c) Table 8.6 is a numerical representation of f.

TABLE 8.6

n	1	2	3	4	5	6
$f(n)$	1	-3	9	-27	81	-243

SOLUTION **(a)** The symbolic representation of f can be written as

$$f(n) = 4(0.5)^n = 4(0.5)(0.5)^{n-1} = 2(0.5)^{n-1}.$$

Thus $f(n)$ represents a geometric sequence with $c = 2$ and $r = 0.5$.
(b) The points on the graph are $(1, 2)$, $(2, 4)$, $(3, 7)$, and $(4, 10)$. Thus $a_1 = 2$, $a_2 = 4$, $a_3 = 7$, and $a_4 = 10$. Taking ratios of successive terms results in

$$\frac{a_2}{a_1} = 2, \qquad \frac{a_3}{a_2} = \frac{7}{4}, \qquad \text{and} \qquad \frac{a_4}{a_3} = \frac{10}{7}.$$

Since these ratios are not equal, there is no *common* ratio. The sequence is not geometric.
(c) Notice that

$$1, -3, 9, -27, 81, -243$$

result from multiplying the previous term by -3. This sequence can be expressed either as

$$a_n = -3a_{n-1} \qquad \text{with} \qquad a_1 = 1.$$

or as

$$a_n = (-3)^{n-1}$$

Therefore the sequence is geometric.

EXAMPLE 9   *Finding a symbolic representation for a geometric sequence*

Find a general term a_n for each geometric sequence.
(a) $a_1 = 5$ and $r = 1.12$ **(b)** $a_2 = 8$ and $a_5 = 512$

SOLUTION   **(a)** Since the first term is 5 and the common ratio is 1.12, $a_n = 5(1.12)^{n-1}$.
(b) We must find $a_n = cr^{n-1}$ so that $a_2 = 8$ and $a_5 = 512$. Start by determining the common ratio r. Since

$$\frac{a_5}{a_2} = \frac{cr^{5-1}}{cr^{2-1}} = \frac{r^4}{r^1} = r^3 \quad \text{and} \quad \frac{a_5}{a_2} = \frac{512}{8} = 64,$$

$r^3 = 64$ or $r = 4$. So, $a_n = c(4)^{n-1}$. Now,

$$a_2 = c(4)^{2-1} = 8 \quad \text{or} \quad c = 2.$$

Thus $a_n = 2(4)^{n-1}$.

Putting it all Together
8.1

*T*he following table summarizes some fundamental concepts about sequences.

Concept	Explanation	Example
Infinite sequence	A function f whose domain is the set of natural numbers; denoted $a_n = f(n)$; the terms are $a_1, a_2, a_3, \ldots$.	$f(n) = n^2 - 2n$, where $a_n = f(n)$. The first three terms are $a_1 = 1^2 - 2(1) = -1$ $a_2 = 2^2 - 2(2) = 0$ $a_3 = 3^2 - 2(3) = 3$. Graphs of sequences are scatterplots.
Recursive sequence	Defined in terms of previous terms; a_1 through a_{n-1} must be calculated before a_n can be found.	$a_n = 2a_{n-1}, a_1 = 1$ $a_1 = 1, a_2 = 2, a_3 = 4,$ and $a_4 = 8$ A new term is found by multiplying the previous term by 2.
Arithmetic sequence	A *linear* function f whose domain is the natural numbers; $a_n = dn + c$ or $a_n = a_{n-1} + d$, with common difference d	$f(n) = 2n - 1$, where $a_n = f(n)$. $a_1 = 1, a_2 = 3, a_3 = 5,$ and $a_4 = 7$ Consecutive terms increase by the common difference $d = 2$. The points on the graph of this sequence lie on a line with slope 2.
Geometric sequence	$f(n) = cr^{n-1}$, where c is constant and r is the common ratio; may also be written as $a_n = ra_{n-1}$	$f(n) = 2(3)^{n-1}$, where $a_n = f(n)$. $a_1 = 2, a_2 = 6, a_3 = 18,$ and $a_4 = 54$ Consecutive terms are found by multiplying the previous term by the common ratio r. Points on the graph of a geometric sequence do not lie on a line.

8.1 EXERCISES

Finding Terms of Sequences

Exercises 1–12: Find the first four terms of the sequence.

1. $a_n = 2n + 1$

2. $a_n = 3(n - 1) + 5$

3. $a_n = 4(-2)^{n-1}$

4. $a_n = 2(3)^n$

5. $a_n = \dfrac{n}{n^2 + 1}$

6. $a_n = 5 - \dfrac{1}{n^2}$

7. $a_n = (-1)^n \left(\dfrac{1}{2}\right)^n$

8. $a_n = (-1)^n \left(\dfrac{1}{n}\right)$

9. $a_n = (-1)^{n-1}\left(\dfrac{2^n}{1 + 2^n}\right)$

10. $a_n = (-1)^{n-1}\left(\dfrac{1}{3^n}\right)$

11. $a_n = 2^n + n^2$

12. $a_n = \dfrac{1}{n} + \dfrac{1}{3n}$

Exercises 13 and 14: Use the graphical representation to write the terms of the sequence.

13.

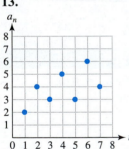

14.

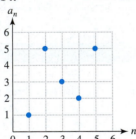

Exercises 15–24: Complete the following for the recursively defined sequence.

   (a) Find the first four terms.
   (b) Graph these terms.

15. $a_n = 2a_{n-1};\ a_1 = 1$

16. $a_n = a_{n-1} + 5;\ a_1 = -4$

17. $a_n = a_{n-1} - a_{n-2};\ a_1 = 2,\ a_2 = 5$

18. $a_n = 2a_{n-1} + a_{n-2};\ a_1 = 0,\ a_2 = 1$

19. $a_n = a_{n-1}^2;\ a_1 = 2$

20. $a_n = \dfrac{1}{2}a_{n-1}^3 + 1;\ a_1 = 0$

21. $a_n = a_{n-1} + n;\ a_1 = 1$

22. $a_n = 3a_{n-1}^2;\ a_1 = 2$

23. $a_n = a_{n-1}a_{n-2};\ a_1 = 2,\ a_2 = 3$

24. $a_n = 2a_{n-1}^2 + a_{n-2};\ a_1 = 2,\ a_2 = 1$

Modeling Insect and Bacteria Populations

*Exercises 25 and 26: **Insect Population**   The annual population density of a species of insect after n years is modeled by a sequence. Use the graph to discuss trends in the insect population.*

25.

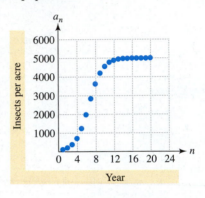

26.

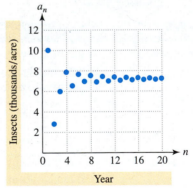

27. **Insect Population**   (Refer to Example 3.) Suppose the initial density of female insects is 500 per acre with $r = 0.8$.

   (a) Write a recursive sequence that describes this data, where a_n denotes the female insect density during year n.

   (b) Find the terms $a_1, a_2, a_3, \ldots, a_6$. Interpret the results.

   (c) Find a formula for a_n.

28. **Insect Population**   (Refer to Example 5.) Suppose an insect population density during year n can be modeled by the recursively defined sequence

$$a_1 = 8$$
$$a_n = 2.9a_{n-1} - 0.2a_{n-1}^2, \qquad n > 1.$$

(a) Find the population for $n = 1, 2, 3$.

(b) Use technology to graph the sequence for $n = 1, 2, 3, \ldots, 20$. Interpret the graph.

29. *Bacteria Growth* Some strains of bacteria are incapable of producing an amino acid necessary for cell division, called *histidine*. If such bacteria are cultured in a medium with sufficient histidine, they can double their size and divide every 40 minutes. (Source: F. Hoppensteadt and C. Peskin, *Mathematics in Medicine and the Life Sciences.*)

(a) Write a recursive sequence that describes this growth where each value of n represents a 40-minute interval. Let $a_1 = 300$ represent the initial number of bacteria per milliliter. Find the first five terms.

(b) Determine the number of bacteria after 10 hours have elapsed.

(c) Is this sequence arithmetic or geometric? Explain.

30. *Bacteria Growth* (Refer to the previous exercise.) If bacteria are cultured in a medium with limited nutrients, competition ensues and growth slows. According to *Verhulst's model*, the number of bacteria at 40-minute intervals is given by

$$a_n = \left(\frac{2}{1 + a_{n-1}/K}\right)a_{n-1},$$ where K is a constant.

(a) Let $a_1 = 200$ and $K = 10,000$. Use technology to graph the sequence for $n = 1, 2, 3, \ldots, 20$.

(b) Describe the growth of this bacteria.

(c) Trace the graph of the sequence. Make a conjecture as to why K is called the **saturation constant**. Test your conjecture by changing the value of K.

Representations of Sequences

Exercises 31–34: The first five terms of an arithmetic sequence are given. Find

    (a) *numerical,*
    (b) *graphical, and*
    (c) *symbolic*

representations of the sequence. Include at least eight terms of the sequence.

31. $1, 3, 5, 7, 9$ **32.** $4, 1, -2, -5, -8$

33. $7.5, 6, 4.5, 3, 1.5$ **34.** $5.1, 5.5, 5.9, 6.3, 6.7$

Exercises 35–38: The first five terms of a geometric sequence are given. Find

    (a) *numerical,*
    (b) *graphical, and*
    (c) *symbolic*

representations of the sequence. Include at least eight terms of the sequence.

35. $8, 4, 2, 1, \dfrac{1}{2}$ **36.** $32, -8, 2, -\dfrac{1}{2}, \dfrac{1}{8}$

37. $\dfrac{3}{4}, \dfrac{3}{2}, 3, 6, 12$ **38.** $\dfrac{1}{27}, \dfrac{1}{9}, \dfrac{1}{3}, 1, 3$

Exercises 39–44: Find a general term a_n for each arithmetic sequence.

39. $a_1 = 5,$ $d = -2$

40. $a_1 = -3,$ $d = 5$

41. $a_3 = 1,$ $d = 3$

42. $a_4 = 12,$ $d = -10$

43. $a_2 = 5,$ $a_6 = 13$

44. $a_3 = 22,$ $a_{17} = -20$

Exercises 45–50: Find a general term a_n for each geometric sequence.

45. $a_1 = 2,$ $r = \dfrac{1}{2}$

46. $a_1 = 0.8,$ $r = -3$

47. $a_3 = \dfrac{1}{32},$ $r = -\dfrac{1}{4}$

48. $a_4 = 3,$ $r = 3$

49. $a_3 = 2,$ $a_6 = \dfrac{1}{4}$

50. $a_2 = 6,$ $a_4 = 24$

Identifying Types of Sequences

Exercises 51–56: Given the terms of a finite sequence, classify it as arithmetic, geometric, or neither.

51. $-5, 2, 9, 16, 23, 30$ **52.** $5, 2, -2, -6, -11$

53. $2, 8, 32, 128, 512$

54. $5.75, 5.5, 5.25, 5, 4.75, 4.5$

55. $100, 110, 130, 160, 200$

56. $0.7, 0.21, 0.063, 0.0189, 0.00567$

Exercises 57–60: Use the graph to determine if the sequence is arithmetic or geometric. If the sequence is arithmetic, state the sign of the common difference d and estimate its value. If the sequence is geometric, give the sign of the common ratio r and state if $|r| < 1$.

57. **58.**

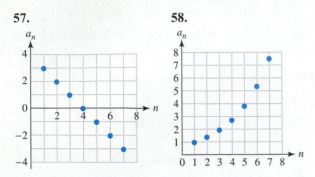

59. **60.**

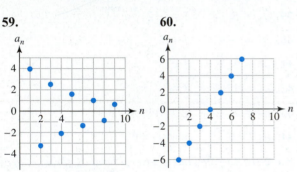

61. Fibonacci Sequence The *Fibonacci sequence* dates back to 1202. It is one of the most famous sequences in mathematics and can be defined recursively by

$$a_1 = 1, a_2 = 1$$
$$a_n = a_{n-1} + a_{n-2} \quad \text{for } n > 2.$$

(a) Find the first 12 terms of this sequence.

(b) Compute $\dfrac{a_n}{a_{n-1}}$ when $n = 2, 3, 4, \ldots, 12$. What happens to this ratio?

(c) Show that for $n = 2, 3$, and 4 the terms of the Fibonacci sequence satisfy the equation

$$a_{n-1} \cdot a_{n+1} - a_n^2 = (-1)^n.$$

62. Bouncing Ball If a tennis ball is dropped, it bounces or rebounds to 80% of its initial height.

(a) Write the first five terms of a sequence that gives the maximum height attained by the tennis ball on each rebound when it is dropped from an initial height of 5 feet. Let $a_1 = 5$. What type of sequence is this?

(b) Give a graphical representation of these terms.

(c) Find a general term a_n.

63. Salary Increases Suppose an employee's initial salary is $30,000.

(a) If this person receives a $2000 raise for each year of experience, determine a sequence that gives the salary at the beginning of the nth year. What type of sequence is this?

(b) Suppose another employee has the same starting salary and receives a 5% raise after each year. Find a sequence that computes the salary at the beginning of the nth year. What type of sequence is this?

(c) Which salary is larger at the beginning of the 10th year, and the 20th year?

(d) Graph both sequences in the same viewing rectangle. Compare the two salaries.

64. Area A sequence of smaller squares is formed by connecting the midpoints of the sides of a larger square, as shown in the figure. If the area of the largest square is one square unit, give the first five terms of a sequence that describes the area of each successive square. What type of sequence is this? Write an expression for the area of the nth square.

Exercises 65 and 66: **Computing Square Roots** *The following recursively defined sequence can be used to compute $\sqrt{k}$ for any positive number k.*

$$a_1 = k$$
$$a_n = \frac{1}{2}\left(a_{n-1} + \frac{k}{a_{n-1}}\right)$$

This sequence was known to Sumerian mathematicians 4000 years ago, but it is still used today. Use this sequence to approximate the given square root by finding a_6. Compare your result with the actual value. (Source: P. Heinz-Otto, *Chaos and Fractals.*)

65. $\sqrt{2}$

66. $\sqrt{11}$

Writing about Mathematics

67. Explain how we can distinguish between an arithmetic and a geometric sequence. Give examples.

68. Compare and contrast a sequence whose nth term is given by $a_n = f(n)$, and a sequence that is defined recursively. Give examples. Which symbolic representation for defining a sequence is usually more convenient to use? Explain why.

8.2 SERIES

Basic Concepts • Arithmetic Series • Geometric Series • Summation Notation

Introduction

Although the terms *sequence* and *series* are sometimes used interchangeably in everyday English, they represent different mathematical concepts. In mathematics, a sequence is a function whose domain is the set of natural numbers, whereas a series is a summation of the terms in a sequence. Series have played a central role in the development of modern mathematics. Today series are often used to approximate functions that are too complicated to have simple formulas. Series are also instrumental in calculating approximations of numbers like π and e.

Basic Concepts

Suppose a person has a starting salary of $30,000 per year and receives a $2000 raise each year. Then,

$$30{,}000, 32{,}000, 34{,}000, 36{,}000, 38{,}000$$

are terms of the sequence that describe this person's salaries over a 5-year period. The total earned is given by the finite *series*

$$30{,}000 + 32{,}000 + 34{,}000 + 36{,}000 + 38{,}000,$$

whose sum is $170,000. Any sequence can be used to define a series. For example, the infinite sequence

$$1, \frac{1}{3}, \frac{1}{9}, \frac{1}{27}, \frac{1}{81}, \frac{1}{243}, \cdots$$

defines the terms of the infinite series

$$1 + \frac{1}{3} + \frac{1}{9} + \frac{1}{27} + \frac{1}{81} + \frac{1}{243} + \cdots .$$

We now define the concept of a series, where $a_1, a_2, a_3, \ldots, a_n, \ldots$ represent terms of a sequence.

> ## Series
>
> A **finite series** is an expression of the form
>
> $$a_1 + a_2 + a_3 + \cdots + a_n,$$
>
> and an **infinite series** is an expression of the form
>
> $$a_1 + a_2 + a_3 + \cdots + a_n + \cdots .$$

An infinite series contains infinitely many terms. We must define what is meant by the sum of an *infinite* series. Let the following be a **sequence of partial sums**.

$$S_1 = a_1$$
$$S_2 = a_1 + a_2$$
$$S_3 = a_1 + a_2 + a_3$$
$$\vdots$$
$$S_n = a_1 + a_2 + a_3 + \cdots + a_n$$

If S_n approaches a real number S as $n \to \infty$, then the sum of the infinite series is S. For example, let $S_1 = 0.3$, $S_2 = 0.3 + 0.03$, $S_3 = 0.3 + 0.03 + 0.003$, and so on. Then, as $n \to \infty$, $S_n \to \frac{1}{3}$. We say that the infinite series

$$0.3 + 0.03 + 0.003 + 0.0003 + \cdots = 0.3333\ldots = 0.\overline{3}$$

has sum $\frac{1}{3}$. Some infinite series do not have a sum S. For example, the series $1 + 2 + 3 + 4 + 5 + \cdots$ would have an unbounded or "infinite" sum.

EXAMPLE 1 *Interpreting a sequence of partial sums*

Suppose that a_1 represents the number of U.S. AIDS deaths reported in 1991, a_2 the number in 1992, a_3 in 1993, and so on.
(a) Write a series that represents the total number of AIDS deaths from 1995 to 2002.
(b) Interpret S_{10}.

SOLUTION **(a)** The series $a_5 + a_6 + a_7 + a_8 + a_9 + a_{10} + a_{11} + a_{12}$ represents the total number of AIDS deaths reported from 1995 to 2002.
(b) Since $S_{10} = a_1 + a_2 + a_3 + \cdots + a_{10}$, it follows that S_{10} represents the total number of AIDS deaths reported from 1991 to 2000.

EXAMPLE 2 *Finding partial sums*

For each a_n, calculate S_4
(a) $a_n = 2n + 1$ **(b)** $a_n = n^2$

SOLUTION **(a)** Since $S_4 = a_1 + a_2 + a_3 + a_4$, start by calculating the first four terms of the sequence $a_n = 2n + 1$.

$$a_1 = 2(1) + 1 = 3; \quad a_2 = 2(2) + 1 = 5;$$
$$a_3 = 2(3) + 1 = 7; \quad a_4 = 2(4) + 1 = 9$$

Thus $S_4 = 3 + 5 + 7 + 9 = 24$.

(b) $a_1 = 1^2 = 1;$ $a_2 = 2^2 = 4;$ $a_3 = 3^2 = 9;$ $a_4 = 4^2 = 16$
Thus $S_4 = 1 + 4 + 9 + 16 = 30.$

The techniques used to calculate π throughout history is a fascinating story. Since π is an irrational number, it cannot be represented exactly by a fraction. Its decimal expansion neither repeats nor does it have a discernible pattern. The ability to compute π was essential to the development of a society, because π appears in formulas used in construction, surveying, and geometry. In early historical records, π was given the value of 3. Later the Egyptians used a value of

$$\frac{256}{81} \approx 3.1605.$$

It was not until the discovery of series that exceedingly accurate decimal approximations of π were possible. In 1989, after 100 hours of supercomputer time, π was computed to 1,073,740,000 digits. Why would anyone want to compute π to so many decimal places? One practical reason is to test electrical circuits in new computers. If a computer has a small defect in its hardware, there is a good chance that an error will appear after performing trillions of arithmetic calculations during the computation of π.   (Sources: P. Beckmann, *A History of PI;* P. Heinz-Otto, *Chaos and Fractals.*)

EXAMPLE 3   *Computing π with a series*

The infinite series given by

$$\frac{\pi^4}{90} = \frac{1}{1^4} + \frac{1}{2^4} + \frac{1}{3^4} + \frac{1}{4^4} + \frac{1}{5^4} + \cdots + \frac{1}{n^4} + \cdots$$

can be used to estimate π.
(a) Approximate π by finding the sum of the first four terms.
(b) Use technology to approximate π by summing the first 50 terms. Compare the result to the actual value of π.

SOLUTION   **(a)** Summing the first four terms results in the following.

Graphing Calculator Help

To find the sum of the terms of a series, see Appendix B (page AP-23).

```
sum(seq(1/n^4,n,
1,50)
         1.082320646
(90*Ans)^(1/4)
         3.141590776
π
         3.141592654
```

FIGURE 8.12

$$\frac{\pi^4}{90} \approx \frac{1}{1^4} + \frac{1}{2^4} + \frac{1}{3^4} + \frac{1}{4^4} \approx 1.078751929$$

This approximation can be solved for π by multiplying by 90 and then taking the fourth root. Thus

$$\pi \approx \sqrt[4]{90(1.078751929)} \approx 3.139.$$

(b) Some calculators are capable of summing the terms of a sequence, as shown in Figure 8.12. (Summing the terms of a sequence is equivalent to finding the sum of a series.) The first 50 terms of the series provides an approximation of $\pi \approx 3.141590776$. This computation matches the actual value of π for the first five decimal places.

Arithmetic Series

Summing the terms of an arithmetic sequence results in an **arithmetic series**. For example, the sequence defined by $a_n = 2n - 1$ for $n = 1, 2, 3, \ldots, 7$ is the arithmetic sequence

$$1, 3, 5, 7, 9, 11, 13.$$

The corresponding arithmetic series is

$$1 + 3 + 5 + 7 + 9 + 11 + 13.$$

The following formula can be used to sum the first n terms of an arithmetic sequence. (For a proof see Exercise 3 in the Extended and Discovery Exercises.)

> ### Sum of the first n terms of an arithmetic sequence
>
> The **sum of the first n terms of an arithmetic sequence**, denoted S_n, is found by averaging the first and nth terms and then multiplying by n. That is,
>
> $$S_n = a_1 + a_2 + a_3 + \cdots + a_n = n\left(\frac{a_1 + a_n}{2}\right).$$

Since $a_n = a_1 + (n - 1)d$, S_n can also be written as follows.

$$S_n = n\left(\frac{a_1 + a_n}{2}\right)$$

$$= \frac{n}{2}(a_1 + a_1 + (n - 1)d)$$

$$= \frac{n}{2}(2a_1 + (n - 1)d)$$

EXAMPLE 4 *Finding the sum of a finite arithmetic series*

A person has a starting annual salary of $30,000 and receives a $1500 raise each year.
(a) Calculate the total amount earned over 10 years.
(b) Verify this value using a calculator.

SOLUTION   **(a)** The arithmetic sequence describing the salary during year n is computed by

$$a_n = 30,000 + 1500(n - 1).$$

The first and tenth year's salaries are:

$$a_1 = 30,000 + 1500(1 - 1) = 30,000$$
$$a_{10} = 30,000 + 1500(10 - 1) = 43,500.$$

Thus the total amount earned during this 10-year period is

$$S_{10} = 10\left(\frac{30,000 + 43,500}{2}\right) = \$367,500.$$

This sum can also be found using $S_n = \frac{n}{2}(2a_1 + (n - 1)d)$.

$$S_{10} = \frac{10}{2}(2 \cdot 30,000 + (10 - 1)1500) = \$367,500$$

(b) To verify this with a calculator, compute the sum

$$a_1 + a_2 + a_3 + \cdots + a_{10},$$

where $a_n = 30,000 + 1500(n - 1)$. This calculation is shown in Figure 8.13. The result of 367,500 agrees with part (a).

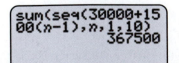

```
sum(seq(30000+15
00(n-1),n,1,10)
            367500
```

FIGURE 8.13

For tax purposes, businesses frequently depreciate equipment. Two methods of depreciation are *straight-line depreciation* and *sum-of-the-years'-digits*.

Suppose a college student buys a $2000 computer to start a business that develops Web sites. This student estimates the life of the computer at 5 years, after which its value will be $200. The difference between $2000 and $200 is $1800, which may be deducted from his or her taxable income over a 5-year period.

In straight-line depreciation, equal portions of $1800 are deducted over 5 years. The *sum-of-the-years'-digits* method calculates depreciation differently. A computer with a useful life of 5 years has a sum-of-the-years computed by

$$1 + 2 + 3 + 4 + 5 = 15.$$

With this method, $\frac{5}{15}$ of $1800 is deduced the first year, $\frac{4}{15}$ the second year, and so on, until $\frac{1}{15}$ is deducted the fifth year. Both depreciation methods deduct a total of $1800 over 5 years.   (Source: Sharp Electronics Corporation, *Conquering the Sciences.*)

EXAMPLE 5   *Calculating depreciation*

For each of the previous depreciation methods, complete the following.
(a) Find an arithmetic sequence that gives the amount depreciated each year.
(b) Write a series whose sum is the amount depreciated over 5 years.

SOLUTION   **(a)** With straight-line depreciation, $\frac{1}{5}(1800) = \$360$ is depreciated each year. The sequence describing this depreciation is

$$360, 360, 360, 360, 360.$$

For sum-of-the-years'-digits, the following depreciation schedule is computed.

$$\frac{5}{15} \text{ of } \$1800 = \$600$$

$$\frac{4}{15} \text{ of } \$1800 = \$480$$

$$\frac{3}{15} \text{ of } \$1800 = \$360$$

$$\frac{2}{15} \text{ of } \$1800 = \$240$$

$$\frac{1}{15} \text{ of } \$1800 = \$120$$

The arithmetic sequence describing these amounts is

$$600, 480, 360, 240, 120.$$

(b) The arithmetic series that describes the total amount deducted for straight-line depreciation is

$$360 + 360 + 360 + 360 + 360.$$

Although we know that the amount depreciated is equal to $1800, it can also be verified by using the following formula.

$$S_5 = 5\left(\frac{360 + 360}{2}\right) = \$1800$$

For the sum-of-the-years'-digits, the total depreciation is calculated by the series

$$600 + 480 + 360 + 240 + 120.$$

Notice that this series also sums to $1800.

$$S_5 = 5\left(\frac{600 + 120}{2}\right) = \$1800$$

EXAMPLE 6   *Finding a term of an arithmetic series*

The sum of an arithmetic series with 15 terms is 285. If $a_{15} = 40$, find a_1.

SOLUTION   To find a_1, we apply the sum formula

$$S_n = n\left(\frac{a_1 + a_n}{2}\right) = a_1 + a_2 + a_3 + \cdots + a_n,$$

with $n = 15$ and $a_{15} = 40$.

$$15\left(\frac{a_1 + 40}{2}\right) = 285$$

$$15(a_1 + 40) = 570 \qquad \text{Multiply by 2.}$$

$$a_1 + 40 = 38 \qquad \text{Divide by 15.}$$

$$a_1 = -2 \qquad \text{Subtract 40.}$$

■ **CLASS DISCUSSION**

Explain why a formula for the sum of an infinite arithmetic series is not given. ■

Geometric Series

What will happen if we attempt to find the sum of the terms of an infinite geometric sequence? The mathematical concept of infinity dates back to at least the paradoxes of Zeno (450 B.C.). The following illustrates a difficulty that occurred in attempting to comprehend infinity.

Suppose that a person walks 1 mile on the first day, $\frac{1}{2}$ mile the second day, $\frac{1}{4}$ mile the third day, and so on. How far down the road would this person travel? This distance is described by the infinite series

$$1 + \frac{1}{2} + \frac{1}{4} + \frac{1}{8} + \frac{1}{16} + \frac{1}{32} + \frac{1}{64} + \cdots.$$

Does the sum of an infinite number of positive values always become infinitely large? Problems like this took centuries for mathematicians to solve.   (Source: H. Eves, *An Introduction to the History of Mathematics.*)

Finite Geometric Series.   In a manner analogous to the way an arithmetic series was defined, a **geometric series** is the sum of the terms of a geometric sequence. In order to calculate sums of infinite geometric series, we begin by finding sums of finite geometric series.

Any finite geometric sequence can be written as

$$a_1, a_1r, a_1r^2, a_1r^3, \ldots, a_1r^{n-1}.$$

The summation of these n terms is a finite geometric series. Its sum S_n is expressed by

$$S_n = a_1 + a_1r + a_1r^2 + a_1r^3 + \cdots + a_1r^{n-1}.$$

To find the value of S_n, multiply this equation by r:

$$rS_n = a_1r + a_1r^2 + a_1r^3 + \cdots + a_1r^{n-1} + a_1r^n$$

Subtracting this equation from the previous equation results in

$$S_n - rS_n = a_1 - a_1r^n$$
$$S_n(1 - r) = a_1(1 - r^n)$$
$$S_n = a_1\left(\frac{1 - r^n}{1 - r}\right), \qquad \text{provided } r \neq 1.$$

This formula can be used to find the sum of the first n terms of a geometric sequence.

Sum of the first n terms of a geometric sequence

If a geometric sequence has first term a_1 and common ratio r, then the sum of the first n terms is given by

$$S_n = a_1\left(\frac{1 - r^n}{1 - r}\right),$$

provided $r \neq 1$.

EXAMPLE 7   *Finding the sum of a finite geometric series*

Approximate the sum for the given values of n.

(a) $1 + \frac{1}{2} + \frac{1}{4} + \cdots + \left(\frac{1}{2}\right)^{n-1}$; $n = 5$, 10, and 20

(b) $3 - 6 + 12 - 24 + 48 - \cdots + 3(-2)^{n-1}$; $n = 3$, 8, and 13

SOLUTION   (a) This geometric series has $a_1 = 1$ and $r = \frac{1}{2} = 0.5$.

$$S_5 = 1\left(\frac{1 - 0.5^5}{1 - 0.5}\right) = 1.9375$$

$$S_{10} = 1\left(\frac{1 - 0.5^{10}}{1 - 0.5}\right) \approx 1.998047$$

$$S_{20} = 1\left(\frac{1 - 0.5^{20}}{1 - 0.5}\right) \approx 1.999998$$

(b) This geometric series has $a_1 = 3$ and $r = -2$.

$$S_3 = 3\left(\frac{1 - (-2)^3}{1 - (-2)}\right) = 9$$

$$S_8 = 3\left(\frac{1 - (-2)^8}{1 - (-2)}\right) = -255$$

$$S_{13} = 3\left(\frac{1 - (-2)^{13}}{1 - (-2)}\right) = 8193$$

Annuities.   A sequence of deposits made at equal periods of time is called an **annuity**. Suppose A_0 dollars is deposited at the end of each year into an account that pays an annual interest rate i compounded annually. At the end of the first year the account contains A_0 dollars. At the end of the second year A_0 dollars would be deposited again. In addition, the first deposit of A_0 dollars would have received interest during the second year. Therefore the value of the annuity after 2 years is

$$A_0 + A_0(1 + i).$$

After 3 years the balance is

$$A_0 + A_0(1 + i) + A_0(1 + i)^2,$$

and after n years this amount is given by

$$A_0 + A_0(1 + i) + A_0(1 + i)^2 + \cdots + A_0(1 + i)^{n-1}.$$

This is a geometric series with first term $a_1 = A_0$ and common ratio $r = (1 + i)$. The sum of the first n terms is given by

$$S_n = a_1\left(\frac{1 - (1 + i)^n}{1 - (1 + i)}\right) = a_1\left(\frac{(1 + i)^n - 1}{i}\right).$$

EXAMPLE 8   *Finding the future value of an annuity*

Suppose that a 20-year-old worker deposits $1000 into an account at the end of each year until age 65. If the interest rate is 12%, find the future value of the annuity.

SOLUTION   Let $a_1 = 1000$, $i = 0.12$, and $n = 45$. The future value of the annuity is given by

$$S_n = a_1\left(\frac{(1 + i)^n - 1}{i}\right)$$

$$= 1000\left(\frac{(1 + 0.12)^{45} - 1}{0.12}\right)$$

$$\approx \$1{,}358{,}230.$$

Infinite Geometric Series.   In Example 7 the value of r affects the sum of a finite geometric series. If $|r| > 1$, as in part (b), then r^n becomes large in absolute value for increasing n. As a result, the sum of the series also becomes large in absolute value. On the other hand, in part (a) the common ratio satisfies $|r| < 1$. The values of r^n become closer to 0 as n increases. For large values of n,

$$S_n = a_1\left(\frac{1 - r^n}{1 - r}\right) \approx a_1\left(\frac{1 - 0}{1 - r}\right) = \frac{a_1}{1 - r}.$$

This result is summarized in the following.

Sum of an infinite geometric sequence

The sum of the infinite geometric sequence with first term a_1 and common ratio r is given by

$$S = \frac{a_1}{1 - r},$$

provided $|r| < 1$. If $|r| \geq 1$, then this sum does not exist.

Infinite series can be used to describe repeating decimals. For example, the fraction $\frac{5}{9}$ can be written as the repeating decimal $0.555555\ldots$. This decimal can be expressed as an infinite series.

$$\frac{5}{9} = 0.5 + 0.05 + 0.005 + 0.0005 + \cdots + 0.5(0.1)^{n-1} + \cdots$$

In this series $a_1 = 0.5$ and $r = 0.1$. Since $|r| < 1$, the sum exists and is given by

$$S = \frac{0.5}{1 - 0.1} = \frac{5}{9},$$

as expected.

We are now able to answer the question concerning how far a person will walk if he or she travels 1 mile on the first day, $\frac{1}{2}$ mile on the second day, $\frac{1}{4}$ mile the third day, and so on.

EXAMPLE 9   *Finding the sum of an infinite geometric series*

Find the sum of the infinite geometric series

$$1 + \frac{1}{2} + \frac{1}{4} + \cdots + \left(\frac{1}{2}\right)^{n-1} + \cdots.$$

SOLUTION   In this series, the first term is $a_1 = 1$ and the common ratio is $\frac{1}{2} = 0.5$. Its sum is

$$S = \frac{a_1}{1 - r} = \frac{1}{1 - 0.5} = 2.$$

If it were possible to walk in the prescribed manner, the total distance traveled after each day would always be less than 2 miles.

Summation Notation

Summation notation is used to write series efficiently. The symbol Σ, the uppercase Greek letter *sigma*, indicates a sum.

> ### Summation notation
>
> $$\sum_{k=1}^{n} a_k = a_1 + a_2 + a_3 + \cdots + a_n.$$

The letter k is called the **index of summation**. The numbers 1 and n represent the subscripts of the first and last terms in the series. They are called the **lower limit** and **upper limit** of the summation, respectively.

EXAMPLE 10   *Using summation notation*

Evaluate each series.

(a) $\displaystyle\sum_{k=1}^{5} k^2$ (b) $\displaystyle\sum_{k=1}^{4} 5$ (c) $\displaystyle\sum_{k=3}^{6} (2k - 5)$

SOLUTION **(a)** $\displaystyle\sum_{k=1}^{5} k^2 = 1^2 + 2^2 + 3^2 + 4^2 + 5^2 = 55$

(b) $\displaystyle\sum_{k=1}^{4} 5 = 5 + 5 + 5 + 5 = 20$

(c) $\displaystyle\sum_{k=3}^{6} (2k - 5) = \underset{k=3}{(2(3) - 5)} + \underset{k=4}{(2(4) - 5)} + \underset{k=5}{(2(5) - 5)} + \underset{k=6}{(2(6) - 5)}$

$$= 1 + 3 + 5 + 7 = 16$$

EXAMPLE 11 *Writing a series in summation notation*

Write the series using summation notation.

(a) $\dfrac{1}{2^3} + \dfrac{1}{3^3} + \dfrac{1}{4^3} + \dfrac{1}{5^3} + \dfrac{1}{6^3} + \dfrac{1}{7^3} + \dfrac{1}{8^3}$

(b) $\dfrac{1}{2} + \dfrac{2}{3} + \dfrac{3}{4} + \dfrac{4}{5} + \dfrac{5}{6} + \dfrac{6}{7} + \dfrac{7}{8}$

SOLUTION **(a)** The terms of the series can be written as $\dfrac{1}{k^3}$ for $k = 2, 3, 4, \ldots, 8$. Thus

$$\frac{1}{2^3} + \frac{1}{3^3} + \frac{1}{4^3} + \frac{1}{5^3} + \frac{1}{6^3} + \frac{1}{7^3} + \frac{1}{8^3} = \sum_{k=2}^{8} \frac{1}{k^3}.$$

(b) The terms of the series can be written as $\dfrac{k}{k+1}$ for $k = 1, 2, 3, \ldots, 7$. Thus

$$\frac{1}{2} + \frac{2}{3} + \frac{3}{4} + \frac{4}{5} + \frac{5}{6} + \frac{6}{7} + \frac{7}{8} = \sum_{k=1}^{7} \frac{k}{k+1}.$$

Series play an essential role in mathematics and its applications, as illustrated by the next example.

EXAMPLE 12 *Expressing a series in summation notation*

Suppose that an air filter removes 90% of the impurities that enter it.
(a) Find a series that represents the amount of impurities removed by a sequence of n air filters. Express this answer in summation notation.
(b) How many air filters would be necessary to remove 99.99% of the impurities? Could 100% of the impurities be removed from the air?

SOLUTION **(a)** The first filter removes 90% of the impurities so 10%, or 0.1, passes through it. Of the 0.1 that passes through the first filter, 90% is removed by the second filter, while 10% of 10%, or 0.01, passes through. Then, 10% of 0.01, or 0.001, passes through the third filter. See Figure 8.14. From this we can establish a pattern. Let 100% or 1 represent the amount of impurities entering the first air filter. The amount removed by n filters would equal

$$(0.9)(1) + (0.9)(0.1) + (0.9)(0.01) + (0.9)(0.001) + \cdots + (0.9)(0.1)^{n-1}.$$

In summation notation, this series can be written as $\displaystyle\sum_{k=1}^{n} 0.9(0.1)^{k-1}$.

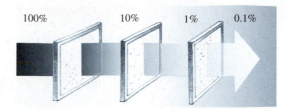

FIGURE 8.14   Percentages of Impurities Passing through Air Filters

(b) To remove 99.99% or 0.9999 of the impurities requires four air filters, since

$$\sum_{k=1}^{4} 0.9(0.1)^{k-1} = (0.9)(1) + (0.9)(0.1) + (0.9)(0.01) + (0.9)(0.001)$$

$$= 0.9 + 0.09 + 0.009 + 0.0009$$

$$= 0.9999.$$

Intuition tells us that it is impossible to remove 100% of the impurities. If k increases without a maximum value, then the fraction of impurities removed is

$$\sum_{k=1}^{\infty} (0.9)(0.1)^{k-1} = (0.9)(1) + (0.9)(0.1) + (0.9)(0.1)^2 + \cdots$$

$$= 0.99999 \ldots .$$

Notice that since the upper limit has no maximum, the symbol ∞ is used. If k is allowed to increase without bound, then the series equals the repeating decimal $0.\overline{9}$. This is an infinite geometric series with $a_1 = 0.9$ and $r = 0.1$. Its sum is

$$S = \frac{a_1}{1-r} = \frac{0.9}{1-0.1} = \frac{0.9}{0.9} = 1.$$

The mathematics of this problem is telling us that it would require an infinite number of air filters to remove 100% of the impurities.

The following is a list of properties for summation notation.

Properties for summation notation

Let $a_1, a_2, a_3, \ldots , a_n$ and $b_1, b_2, b_3, \ldots , b_n$ be sequences, and c be a constant.

1. $\displaystyle\sum_{k=1}^{n} ca_k = c \sum_{k=1}^{n} a_k$ **4.** $\displaystyle\sum_{k=1}^{n} c = nc$

2. $\displaystyle\sum_{k=1}^{n} (a_k + b_k) = \sum_{k=1}^{n} a_k + \sum_{k=1}^{n} b_k$ **5.** $\displaystyle\sum_{k=1}^{n} k = \frac{n(n+1)}{2}$

3. $\displaystyle\sum_{k=1}^{n} (a_k - b_k) = \sum_{k=1}^{n} a_k - \sum_{k=1}^{n} b_k$ **6.** $\displaystyle\sum_{k=1}^{n} k^2 = \frac{n(n+1)(2n+1)}{6}$

These properties can be used to find sums, as illustrated in the next example.

EXAMPLE 13 *Applying summation notation*

Use properties for summation notation to find each sum. Support your results using a graphing calculator.

(a) $\displaystyle\sum_{k=1}^{40} 5$ (b) $\displaystyle\sum_{k=1}^{22} 2k$ (c) $\displaystyle\sum_{k=1}^{14} (2k^2 - 3)$

SOLUTION (a) $\displaystyle\sum_{k=1}^{40} 5 = 40(5) = 200$ Property 4 with $n = 40$ and $c = 5$

(b) $\displaystyle\sum_{k=1}^{22} 2k = 2\sum_{k=1}^{22} k$ Property 1 with $c = 2$ and $a_k = k$

$\quad = 2 \cdot \dfrac{22(22 + 1)}{2}$ Property 5 with $n = 22$

$\quad = 506$ Simplify.

(c) $\displaystyle\sum_{k=1}^{14} (2k^2 - 3) = \sum_{k=1}^{14} 2k^2 - \sum_{k=1}^{14} 3$ Property 3 with $a_k = 2k^2$ and $b_k = 3$

$\quad = 2\displaystyle\sum_{k=1}^{14} k^2 - \sum_{k=1}^{14} 3$ Property 1 with $c = 2$ and $a_k = k^2$

$\quad = 2 \cdot \dfrac{14(14 + 1)(2 \cdot 14 + 1)}{6} - 14(3)$ Properties 6 and 4

$\quad = 1988$ Simplify.

Graphing Calculator Help

To find the sum of the terms of a sequence, see Appendix B (page AP-23).

Support for each of these calculations is shown in Figures 8.15–8.17.

```
sum(seq(5,n,1,40
))
             200
```
FIGURE 8.15

```
sum(seq(2n,n,1,2
2))
             506
```
FIGURE 8.16

```
sum(seq(2n^2-3,n
,1,14))
            1988
```
FIGURE 8.17

Putting it all Together **8.2**

*T*he following table summarizes concepts related to series.

Concept	Explanation	Example
Series	A series is the summation of the terms of a sequence. A finite series always has a sum, but an infinite series may not have a sum.	$2 + 4 + 6 + 8 + \cdots + 20$ $S_4 = a_1 + a_2 + a_3 + a_4$ (partial sum) $\quad = 2 + 4 + 6 + 8$ $\quad = 20$

Concept	Explanation	Example
Arithmetic series	An arithmetic series is the summation of the terms of an arithmetic sequence. The sum of the first n terms is given by $$S_n = n\left(\frac{a_1 + a_n}{2}\right) \quad \text{or}$$ $$S_n = \frac{n}{2}(2a_1 + (n-1)d).$$	$1 + 4 + 7 + 10 + 13 + 16 + 19$ (7 terms) $$S_7 = 7\left(\frac{a_1 + a_7}{2}\right) = 7\left(\frac{1 + 19}{2}\right) = 70 \quad \text{or}$$ $$S_7 = \frac{7}{2}(2a_1 + 6d) = \frac{7}{2}(2(1) + 6(3)) = 70$$
Geometric series	A geometric series is the summation of the terms of a geometric sequence. The sum of the first n terms is given by $$S_n = a_1\left(\frac{1 - r^n}{1 - r}\right).$$ If $|r| < 1$ then an infinite geometric series has the sum, $S = \dfrac{a_1}{1 - r}.$	$3 + 6 + 12 + 24 + 48 + 96$ (6 terms) $$S_6 = 3\left(\frac{1 - 2^6}{1 - 2}\right) = 189$$ The infinite geometric series $$1 + \frac{1}{4} + \frac{1}{16} + \frac{1}{64} + \cdots$$ has a sum $$S = \frac{1}{1 - \dfrac{1}{4}} = \frac{4}{3}.$$
Summation notation	The series $a_1 + a_2 + \cdots + a_n$ can be written as $\displaystyle\sum_{k=1}^{n} a_k.$	$1^2 + 2^2 + 3^2 + 4^2 + \cdots + 10^2$ can be written as $\displaystyle\sum_{k=1}^{10} k^2.$

8.2 EXERCISES

Finding Sums of Series

1. **Prison Escapees**   The table lists the number of escapees from state prisons each year.   (Source: Bureau of Justice Statistics.)

Year	1990	1991	1992
Escapees	8518	9921	10,706

Year	1993	1994	1995
Escapees	14,035	14,307	12,249

(a) Write a series whose sum is the total number of escapees from 1990 to 1995. Find its sum.

(b) Is the series finite or infinite?

2. **Captured Prison Escapees**   (Refer to the previous exercise.) The table lists the number of es-capees from state prisons who were captured, in-cluding inmates who may have escaped during a previous year.   (Source: Bureau of Justice Statistics.)

Year	1990	1991	1992
Captured	9324	9586	10,031

Year	1993	1994	1995
Captured	12,872	13,346	12,166

(a) Write a series whose sum is the total number of escapees captured from 1990 to 1995. Find its sum.

(b) Compare the number of escapees to the num-ber captured during this time.

Exercises 3–10: Use a formula to find the sum of the arithmetic series.

3. $3 + 5 + 7 + 9 + 11 + 13 + 15 + 17$

4. $7.5 + 6 + 4.5 + 3 + 1.5 + 0 + (-1.5)$

5. $1 + 2 + 3 + 4 + \cdots + 50$

6. $1 + 3 + 5 + 7 + \cdots + 97$

7. $-7 + (-4) + (-1) + 2 + 5 + \cdots + 98 + 101$

8. $89 + 84 + 79 + 74 + \cdots + 9 + 4$

9. The first 40 terms of the series whose terms are $a_n = 5n$

10. The first 50 terms of the series whose terms are $a_n = 1 - 3n$

Exercises 11–16: Use $S_n = a_1\left(\dfrac{1 - r^n}{1 - r}\right)$ to find the sum of the finite geometric series.

11. $1 + 2 + 4 + 8 + 16 + 32 + 64 + 128$

12. $2 + \dfrac{1}{2} + \dfrac{1}{8} + \dfrac{1}{32} + \dfrac{1}{128} + \dfrac{1}{512}$

13. $0.5 + 1.5 + 4.5 + 13.5 + 40.5 + 121.5 + 364.5$

14. $0.6 + 0.3 + 0.15 + 0.075 + 0.0375$

15. The first 20 terms of the series whose terms are $a_n = 3(2)^{n-1}$

16. The first 15 terms of the series whose terms are $a_n = 2\left(\dfrac{1}{3}\right)^n$

Exercises 17–20: Find the sum of the infinite geometric series.

17. $1 + \dfrac{1}{3} + \dfrac{1}{9} + \dfrac{1}{27} + \dfrac{1}{81} + \cdots$

18. $5 + \dfrac{5}{2} + \dfrac{5}{4} + \dfrac{5}{8} + \dfrac{5}{16} + \cdots$

19. $6 - 4 + \dfrac{8}{3} - \dfrac{16}{9} + \dfrac{32}{27} - \dfrac{64}{81} + \cdots$

20. $-2 + \dfrac{1}{2} - \dfrac{1}{8} + \dfrac{1}{32} - \dfrac{1}{128} + \cdots$

21. *Area* (Refer to Exercise 64, Section 8.1.) Use a geometric series to find the sum of the areas of the squares if they continue indefinitely.

22. *Perimeter* (Refer to Exercise 64, Section 8.1.) Use a geometric series to find the sum of the perimeters of the squares if they continue indefinitely.

Exercises 23–26: Annuities (Refer to Example 8.) Find the future value of each annuity. Interpret the result.

23. $A_0 = \$2000,$ $\quad i = 0.08, n = 20$

24. $A_0 = \$500,$ $\quad i = 0.15, n = 10$

25. $A_0 = \$10{,}000,$ $\quad i = 0.11, n = 5$

26. $A_0 = \$3000,$ $\quad i = 0.19, n = 45$

Decimal Numbers and Geometric Series

Exercises 27–30: Write each rational number in the form of an infinite geometric series.

27. $\dfrac{2}{3}$ **28.** $\dfrac{1}{9}$

29. $\dfrac{9}{11}$ **30.** $\dfrac{14}{33}$

Exercises 31–34: Write the sum of each geometric series as a rational number.

31. $0.8 + 0.08 + 0.008 + 0.0008 + \cdots$

32. $0.9 + 0.09 + 0.009 + 0.0009 + \cdots$

33. $0.45 + 0.0045 + 0.000045 + \cdots$

34. $0.36 + 0.0036 + 0.000036 + \cdots$

Summation Notation

Exercises 35–42: Write out the terms of the series.

35. $\displaystyle\sum_{k=1}^{4} (k + 1)$ **36.** $\displaystyle\sum_{k=1}^{6} (3k - 1)$

37. $\displaystyle\sum_{k=1}^{8} 4$ **38.** $\displaystyle\sum_{k=2}^{6} (5 - 2k)$

39. $\displaystyle\sum_{k=1}^{7} k^3$ **40.** $\displaystyle\sum_{k=1}^{4} 5(2)^{k-1}$

41. $\displaystyle\sum_{k=4}^{5} (k^2 - k)$ **42.** $\displaystyle\sum_{k=1}^{5} \log k$

Exercises 43–46: Write each series with summation notation.

43. $1^4 + 2^4 + 3^4 + 4^4 + 5^4 + 6^4$

44. $1 + \dfrac{1}{5} + \dfrac{1}{25} + \dfrac{1}{125} + \dfrac{1}{625}$

45. $1 + \dfrac{1}{2^2} + \dfrac{1}{3^2} + \dfrac{1}{4^2} + \dfrac{1}{5^2} + \cdots$

46. $1 + \dfrac{1}{10} + \dfrac{1}{100} + \dfrac{1}{1000} + \dfrac{1}{10{,}000} + \cdots$

Exercises 47–58: (Refer to Example 13.) Use properties of summation notation to find the sum. Support your result with a graphing calculator.

47. $\displaystyle\sum_{k=1}^{60} 9$ **48.** $\displaystyle\sum_{k=1}^{43} -4$

49. $\displaystyle\sum_{k=1}^{15} 5k$ **50.** $\displaystyle\sum_{k=1}^{22} -2k$

51. $\displaystyle\sum_{k=1}^{31} (3k - 3)$

52. $\displaystyle\sum_{k=1}^{17} (1 - 4k)$

53. $\displaystyle\sum_{k=1}^{25} k^2$

54. $\displaystyle\sum_{k=1}^{12} 3k^2$

55. $\displaystyle\sum_{k=1}^{16} (k^2 - k)$

56. $\displaystyle\sum_{k=1}^{18} (k^2 - 4k + 3)$

57. $\displaystyle\sum_{k=5}^{24} k$

58. $\displaystyle\sum_{k=7}^{19} (k^2 + 1)$

59. Verify the formula $\displaystyle\sum_{k=1}^{n} k = \frac{n(n + 1)}{2}$ by using the formula for the sum of the first n terms of a finite arithmetic sequence.

60. Use the previous exercise to find the sum of the series $\displaystyle\sum_{k=1}^{200} k$.

61. *Stacking Logs*   A stack of logs is made in layers, with one less log in each layer. See the accompanying figure. If the top layer has 7 logs and the bottom layer has 15 logs, what is the total number of logs in the pile? Use a formula to find the sum.

62. (Refer to the previous exercise.) Suppose a stack of logs has 13 logs in the top layer and a total of 7 layers. How many logs are in the stack?

Exercises 63 and 64: **Depreciation**   *(Refer to Example 5.) Let the total depreciation be T over n years.*

(a) Find an arithmetic sequence that gives the amount depreciated each year by using straight-line and sum-of-the-years'-digits depreciation.

(b) Write a series for each method whose sum is the total amount depreciated over n years.

63. $T = \$10,000$, $n = 4$

64. $T = \$42,000$, $n = 7$

Exercises 65 and 66: **The Natural Exponential Function**   *The following series can be used to estimate the value of e^a for any real number a.*

$$e^a \approx 1 + a + \frac{a^2}{2!} + \frac{a^3}{3!} + \cdots + \frac{a^n}{n!},$$

where $n! = 1 \cdot 2 \cdot 3 \cdot 4 \cdots \cdots n$. Use the first eight terms of this series to approximate the given expression. Compare this estimate with the actual value.

65. e **66.** e^{-1}

Computing Partial Sums

Exercises 67–70: Use a_k and n to find $S_n = \displaystyle\sum_{k=1}^{n} a_k$. Then, evaluate the infinite geometric series $S = \displaystyle\sum_{k=1}^{\infty} a_k$. Compare S to the values for S_n.

67. $a_k = \left(\dfrac{1}{3}\right)^{k-1}$; $n = 2, 4, 8, 16$

68. $a_k = 3\left(\dfrac{1}{2}\right)^{k-1}$; $n = 5, 10, 15, 20$

69. $a_k = 4\left(-\dfrac{1}{10}\right)^{k-1}$; $n = 1, 2, 3, 4, 5, 6$

70. $a_k = 2(-0.02)^{k-1}$; $n = 1, 2, 3, 4, 5, 6$

Writing about Mathematics

71. Discuss the difference between a sequence and a series. Give examples of each.

72. Under what circumstances can we find the sum of a geometric series? Give examples to illustrate your answer.

CHECKING BASIC CONCEPTS FOR SECTIONS 8.1 AND 8.2

1. Give graphical and numerical representations of the sequence defined by $a_n = -2n + 3$, where $a_n = f(n)$. Include the first six terms.

2. Determine if the sequence is arithmetic or geometric. If it is arithmetic, state the common difference; if it is geometric, give the common ratio.
   (a) $2, -4, 8, -16, 32, -64, 128, \ldots$
   (b) $-3, 0, 3, 6, 9, 12, \ldots$
   (c) $4, 2, 1, \dfrac{1}{2}, \dfrac{1}{4}, \dfrac{1}{8}, \dfrac{1}{16}, \ldots$

3. Determine if the series is arithmetic or geometric. Use a formula to find its sum.

   (a) $1 + 5 + 9 + 13 + \cdots + 37$
   (b) $3 + 1 + \dfrac{1}{3} + \dfrac{1}{9} + \dfrac{1}{27} + \dfrac{1}{81}$
   (c) $2 + \dfrac{1}{2} + \dfrac{1}{8} + \dfrac{1}{32} + \cdots$
   (d) $0.9 + 0.09 + 0.009 + 0.0009 + \cdots$

4. Write each series in the previous exercise in summation notation.

5. Use properties of summation notation to find the sum. Support your results with a graphing calculator.
   (a) $\displaystyle\sum_{k=1}^{15} (k + 2)$ (b) $\displaystyle\sum_{k=1}^{21} 2k^2$

8.3 COUNTING

Fundamental Counting Principle • Permutations • Combinations

Introduction

The notion of counting in mathematics includes much more than simply counting from 1 to 100. It also includes determining the number of ways that an event can occur. For example, how many ways are there to answer a true-false quiz with ten questions? The answer involves counting the different ways that a student could answer such a quiz. Counting is an important concept that is used to calculate probabilities. Probability is discussed in Section 8.5.

Fundamental Counting Principle

Suppose a quiz has only two questions. The first is a multiple-choice question with four choices: A, B, C, or D, and the second is a true-false (T-F) question. The tree diagram in Figure 8.18 can be used to count the ways that this quiz can be answered.

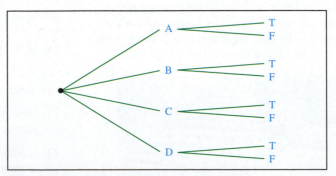

FIGURE 8.18   Different Ways to Answer a Quiz

A tree diagram is a systematic way of listing every possibility. From Figure 8.18, we can see that there are eight ways to answer the test. They are

AT, AF, BT, BF, CT, CF, DT, and DF.

For instance, CF indicates a quiz with answers of C on the first question and F on the second question.

A tree diagram is not always practical, because it can quickly become very large. For this reason mathematicians have developed more efficient ways of counting. Since the multiple-choice question has four possible answers, after which the true-false question has two possible answers, there are $4 \cdot 2 = 8$ possible ways of answering the test. This is an application of the *fundamental counting principle*, which applies to independent events. Two events are **independent** if neither event influences the outcome of the other.

Fundamental counting principle

Let $E_1, E_2, E_3, \ldots, E_n$ be a sequence of n independent events. If event E_k can occur m_k ways for $k = 1, 2, 3, \ldots, n$, then there are

$$m_1 \cdot m_2 \cdot m_3 \cdot \cdots \cdot m_n$$

ways for all n events to occur.

EXAMPLE 1   *Counting ways to answer an exam*

An exam contains four true-false questions and six multiple-choice questions. Each multiple-choice question has five possible answers. Count the number of ways that the exam can be answered.

SOLUTION   Answering these ten questions can be thought of as a sequence of ten independent events. There are two ways to answer each of the first four questions, and five ways of answering the next six questions. The number of ways to answer the exam is

$$2 \cdot 2 \cdot 2 \cdot 2 \cdot 5 \cdot 5 \cdot 5 \cdot 5 \cdot 5 \cdot 5 = 2^4 5^6 = 250,000.$$

In the next example we count the number of different license plates possible with a given format.

EXAMPLE 2   *Counting license plates*

Sometimes a license plate is limited to 3 uppercase letters (A through Z) followed by 3 digits (0 through 9). For example, ABB 112 would be a valid license plate. Would this format provide enough license plates for a state with 8 million vehicles?

SOLUTION   Since there are 26 letters of the alphabet, it follows that there are 26 ways to choose each of the first three letters of the license plate. Similarly, there are 10 digits, so there are 10 ways to choose each of the three digits in the license plate. By the fundamental counting principle, there are

$$26 \cdot 26 \cdot 26 \cdot 10 \cdot 10 \cdot 10 = 17,576,000$$

unique license plates that could be issued. This format for license plates could accommodate 8 million vehicles.

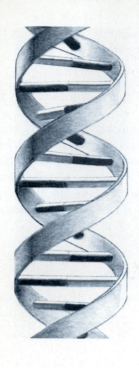

One of the most important discoveries of our time was made in 1953 when the double helical structure of DNA (deoxyribonucleic acid) was identified. DNA is the material of heredity. It is composed of four bases called adenine (A), cytosine (C), guanine (G), and thymine (T). The genetic code of a person is composed of approximately six billion of these bases stored in a long list. The nucleus of every cell contains a copy of this genetic code, tightly coiled in the shape of a double helix. The DNA of a single cell has the potential to store 30 encyclopedic volumes three times over. Any one of the four bases can occur anytime in the genetic code.

Small changes in this code can have dramatic effects. In a normal red blood cell, hemoglobin contains the genetic code

ACTCCTGAGGAGGAGT,

whereas a person with sickle cell anemia has the same sequence except that thymine (T) has been substituted for adenine (A):

ACTCCTGTGGAGGAGT.

Thus a single error in a list of six billion letters causes this serious disease. This unique list represented by the four letters A, C, G, and T defines the characteristics of an individual. Approximately 99–99.9% of this genetic code is identical for all humans. Only 0.1–1% is unique to a particular individual. (Source: S. Easteal, *DNA Profiling: Principles, Pitfalls and Potential.*)

EXAMPLE 3 *Counting genetic codes in DNA*

Using the four letters A, C, G, and T, write a numeric expression that represents the total number of genetic codes that are possible.

SOLUTION A human genetic code is composed of a list of six billion letters. For each letter there are four choices. By the fundamental counting principle, there are $4^{6,000,000,000}$ genetic codes of this type. (Many of these would not be valid for human beings.) This number is too large to approximate with a calculator.

EXAMPLE 4 *Counting toll-free 800 telephone numbers*

The number of 800 telephone numbers available for new businesses and individuals is decreasing rapidly. Count the total number of 800 numbers, if the local portion of a telephone number does not start with a 0 or 1. (Source: Database Services Management.)

SOLUTION A toll-free 800 number assumes the following form.

Local Number

We can think of choosing the remaining digits for the local number as seven independent events. Since the local number cannot begin with a 0 or 1, there are eight possibilities (2 to 9) for the first digit. The remaining six digits can be any number from 0 to 9, so there are ten possibilities for each of these digits. The total is given by

$$8 \cdot 10 \cdot 10 \cdot 10 \cdot 10 \cdot 10 \cdot 10 = 8 \cdot 10^6 = 8,000,000.$$

Note: Toll-free numbers also begin with 888 and 877.

Permutations

A **permutation** is an *ordering* or *arrangement*. For example, suppose that three students are scheduled to give a speech in a class. The different arrangements of how these speeches can be ordered are called permutations. Initially, any one of the three students could give the first speech. After the first speech, there are two students remaining for the second speech. For the third speech there is only one possibility. By the fundamental counting principle, the total number of permutations is equal to

$$3 \cdot 2 \cdot 1 = 6.$$

If the students are denoted as A, B, and C, then these six permutations are ABC, ACB, BAC, BCA, CAB, and CBA. In a similar manner, if there were ten students scheduled to give speeches, the number of permutations would increase to

$$10 \cdot 9 \cdot 8 \cdot 7 \cdot 6 \cdot 5 \cdot 4 \cdot 3 \cdot 2 \cdot 1 = 3,628,800.$$

A more efficient way of writing the previous two products is to use *factorial notation*. The number $n!$ (read "n-factorial") is defined as follows.

---

n-factorial

For any natural number n,

$$n! = n(n - 1)(n - 2) \cdots (3)(2)(1)$$

and

$$0! = 1.$$

The reason for the definition of $0!$ will become apparent later. Factorials grow rapidly and can be computed on most calculators.

EXAMPLE 5   *Calculating factorials*

Compute $n!$ for $n = 0, 1, 2, 3, 4,$ and 5 by hand. Use a calculator to find $8!$, $13!$, and $25!$.

SOLUTION   The values for $n!$ can be calculated as

$$0! = 1, \quad 1! = 1, \quad 2! = 2 \cdot 1 = 2, \quad 3! = 3 \cdot 2 \cdot 1 = 6,$$
$$4! = 4 \cdot 3 \cdot 2 \cdot 1 = 24, \quad \text{and} \quad 5! = 5 \cdot 4 \cdot 3 \cdot 2 \cdot 1 = 120.$$

Graphing Calculator Help

To calculate $n!$, see Appendix B (page AP-24).

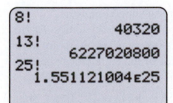

FIGURE 8.19

Figure 8.19 shows the values of $8!$, $13!$, and $25!$. The value for $25!$ is an approximation. Notice how dramatically $n!$ increases.

One of the most famous unanswered questions in computing today is called the *traveling salesperson problem*. It is a relatively simple problem to state, but if someone could design a procedure to solve this problem *efficiently*, he or she would not only become famous, but would also provide a valuable method for businesses to save millions of dollars on scheduling problems, such as bus routes and truck deliveries.

One instance of the traveling salesperson problem can be stated as follows. A salesperson must begin and end at home and travel to three cities. Assuming that the salesperson can travel between any pair of cities, what route would minimize the salesperson's mileage? In Figure 8.20, the four cities are labeled A, B, C, and D. Let the salesperson live in city A. There are six routes that could be tried. They are listed in Table 8.7 with the appropriate mileage for each.

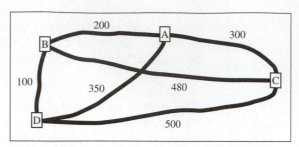

FIGURE 8.20   Traveling Salesperson Problem

TABLE 8.7

Route	Mileage
A B C D A	200 + 480 + 500 + 350 = 1530
A B D C A	200 + 100 + 500 + 300 = 1100
A C B D A	300 + 480 + 100 + 350 = 1230
A C D B A	300 + 500 + 100 + 200 = 1100
A D B C A	350 + 100 + 480 + 300 = 1230
A D C B A	350 + 500 + 480 + 200 = 1530

The shortest route of 1100 miles occurs when the salesperson either starts at A and travels through B, D, and C, and back to A, or reverses this route. Currently, this method of listing all possible routes to find the minimum distance is the only known way to consistently find the optimal solution for any general map containing *n* cities. In fact, people have not been able to determine whether a *significantly* faster method even exists.   (Source: J. Smith, *Design and Analysis of Algorithms.*)

Counting the number of routes involves the fundamental counting principle. At the first step, the salesperson can travel to any one of three cities. Once this city is selected, there are then two possible cities to choose, and then one. Finally the salesperson returns home. The total number of routes is given by

$$3! = 3 \cdot 2 \cdot 1 = 6$$

If the salesperson must travel to 30 cities, then there are

$$30! \approx 2.7 \times 10^{32}$$

routes to check, far too many to check even with the largest supercomputers.

Next suppose that a salesperson must visit three of eight possible cities. At first there are eight cities to choose. After the first city has been visited, there are seven cities to select. Since the salesperson only travels to three of the eight cities, there are

$$8 \cdot 7 \cdot 6 = 336$$

possible routes. This number of permutations is denoted by $P(8, 3)$. It represents the number of arrangements that can be made using three elements taken from a sample of eight.

> ### Permutations of n elements taken r at a time
>
> If $P(n, r)$ denotes the number of permutations of n elements taken r at a time, with $r \leq n$, then
>
> $$P(n, r) = \frac{n!}{(n - r)!} = n(n - 1)(n - 2) \cdots (n - r + 1).$$

EXAMPLE 6   *Calculating P(n, r)*

Calculate each of the following by hand. Then, support your answers by using a calculator.
(a) $P(7, 3)$
(b) $P(100, 2)$

SOLUTION   **(a)** $P(7, 3) = 7 \cdot 6 \cdot 5 = 210$
(b) $P(100, 2) = 100 \cdot 99 = 9900$. One also can compute this number as follows.

$$P(100, 2) = \frac{100!}{(100 - 2)!} = \frac{100 \cdot 99 \cdot 98!}{98!} = 100 \cdot 99 = 9900$$

In this case, it is helpful to cancel 98! before performing the arithmetic. Both of these computations are performed by a calculator in Figure 8.21.

Graphing Calculator Help
To calculate $P(n, r)$, see Appendix B (page AP-24).

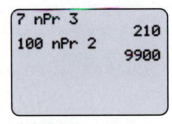

FIGURE 8.21

EXAMPLE 7   *Calculating permutations*

If there is a class of 30 students, how many arrangements are there for 4 students to give a speech?

SOLUTION   The number of permutations of 30 elements taken 4 at a time is given by

$$P(30, 4) = 30 \cdot 29 \cdot 28 \cdot 27 = 657,720$$

Thus there are 657,720 ways to arrange the four speeches.

In the next example we determine the number of ways that 4 people can have different birthdays. For example, if the birthdays of 4 people are February 29, May 2, June 30, and July 11, then these dates would be one way that 4 people could have different birthdays.

EXAMPLE 8 *Counting birthdays*

Count the possible ways that 4 people can have different birthdays.

SOLUTION Counting February 29, there are 366 possible birthdays. The first person could have any of the 366 possible birthdays. The second person could have any of 365 birthdays, because the first person's birthday cannot be duplicated. Similarly, the third person could have any of 364 birthdays, and the fourth person could have any of 363 birthdays. The total number of ways that 4 people could have different birthdays equals

$$P(366, 4) = 366 \cdot 365 \cdot 364 \cdot 363 \approx 1.77 \times 10^{10}.$$

■ CLASS DISCUSSION

Count the number of arrangements of 52 cards in a standard deck. Is it likely that there are arrangements that no one has ever shuffled at any time in the history of the world? Explain. ■

Combinations

Unlike a permutation, a **combination** is not an ordering or arrangement, but rather a subset of a set of elements. Order is unimportant when finding combinations. For example, suppose we want to select a tennis team of two players from four people. The order in which the selection is made does not affect the final team of two players. From a set of four people, we select a subset of two players. This number of possible subsets or combinations is denoted either $C(4, 2)$ or $\binom{4}{2}$.

To calculate $C(4, 2)$, we first consider $P(4, 2)$. Denote the four players by the letters A, B, C, and D. There are $P(4, 2) = 4 \cdot 3 = 12$ permutations given by

AB, BA, AC, CA, AD, DA, BC, CB, BD, DB, CD, DC.

However, the team that comprises person A and person B is equivalent to the team with person B and person A. The sets {AB} and {BA} are equal. The valid combinations are the following two-element subsets of {A, B, C, D}.

{AB}   {AC}   {AD}   {BC}   {BD}   {CD}

That is, $C(4, 2) = \dfrac{P(4, 2)}{2!} = 6$. The relationship between $P(n, r)$ and $C(n, r)$ is now given.

> ### Combinations of *n* elements taken *r* at a time
>
> If $C(n, r)$ denotes the number of combinations of *n* elements taken *r* at a time, with $r \leq n$, then
>
> $$C(n, r) = \frac{P(n, r)}{r!} = \frac{n!}{(n - r)!r!}.$$

EXAMPLE 9   *Calculating C(n, r)*

Calculate each of the following. Support your answers by using a calculator.

(a) $C(7, 3)$ **(b)** $\binom{50}{47}$

SOLUTION   **(a)** $C(7, 3) = \dfrac{7!}{(7 - 3)!\,3!} = \dfrac{7!}{4!\,3!} = \dfrac{7 \cdot 6 \cdot 5 \cdot 4!}{4!\,3!} = \dfrac{7 \cdot 6 \cdot 5}{3!} = \dfrac{210}{6} = 35$

(b) The notation $\binom{50}{47}$ is equivalent to $C(50, 47)$.

$$\binom{50}{47} = \frac{50!}{(50 - 47)!\,47!} = \frac{50!}{3!\,47!} = \frac{50 \cdot 49 \cdot 48 \cdot 47!}{3!\,47!}$$

$$= \frac{50 \cdot 49 \cdot 48}{3!} = \frac{117{,}600}{6} = 19{,}600$$

These computations are performed by a calculator in Figure 8.22.

Graphing Calculator Help

To calculate $C(n, r)$, see Appendix B (page AP-24).

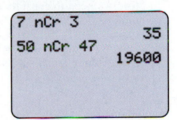

FIGURE 8.22

EXAMPLE 10   *Counting combinations*

A college student has 5 courses left in her major and plans to take 2 courses this semester. Assuming that this student has the prerequisites for all 5 courses, determine how many ways these 2 courses can be selected.

SOLUTION   The order in which the courses are selected is unimportant. From a set of 5 courses, the student selects a subset of 2 courses. This number of subsets is given by $C(5, 2)$.

$$C(5, 2) = \frac{5!}{(5 - 2)!\,2!} = \frac{5!}{3!\,2!} = 10$$

There are 10 ways to select 2 courses from a set of 5.

EXAMPLE 11   *Calculating the number of ways to play the lottery*

To win the jackpot in a lottery, a person must select 5 different numbers from 1 to 49 and then pick the powerball, which is numbered from 1 to 42. Count the ways to play the game.   (Source: Minnesota State Lottery.)

SOLUTION   From 49 numbers a player picks 5 numbers. There are $C(49, 5)$ ways of doing this. There are 42 ways to choose the powerball. Using the fundamental counting principle, the number of ways to play the game equals $C(49, 5) \cdot 42 = 80{,}089{,}128$.

■ MAKING CONNECTIONS

Permutations and Combinations A *permutation* is an arrangement (or list) of objects, where the ordering of the objects is important. For example, the expression $P(20, 9)$ would give the number of batting orders possible for 9 players selected from a team of 20 players. A *combination* is a subset of a set of objects, where the ordering of the objects is *not* important. For example, the expression $C(20, 9)$ would give the number of committees possible when 9 people are selected from a group of 20 people. The order in which committee members are selected does not affect the makeup of the committee.

EXAMPLE 12 *Counting committees*

How many committees of 6 people can be selected from 6 women and 3 men, if a committee must consist of at least 2 men?

SOLUTION Because there are 3 men and each committee must consist of at least 2 men, a committee can include either 2 or 3 men. The order of selection is not important. Therefore we need to consider combinations of committee members rather than permutations.

Two Men: This committee would consist of 4 women and 2 men. We can select 2 men from a group of 3 in $C(3, 2) = 3$ ways. Four women can be selected from a group of 6 in $C(6, 4) = 15$ ways. By the fundamental counting principle, there is a total of

$$C(3, 2) \cdot C(6, 4) = 3 \cdot 15 = 45$$

committees that have 2 men.

Three Men: This committee would have 3 women and 3 men. We can select 3 men from a group of 3 in $C(3, 3) = 1$ way. We can select 3 women from a group of 6 in $C(6, 3) = 20$ ways. By the fundamental counting principle, there is a total of

$$C(3, 3) \cdot C(6, 3) = 1 \cdot 20 = 20$$

committees that have 3 men.

The total number of possible committees would be $45 + 20 = 65$.

Putting it all Together 8.3

*T*he fundamental counting principle may be used to determine the number of ways a sequence of independent events can occur. Permutations are arrangements or listings, whereas combinations are subsets of a set of events.

The following table summarizes some concepts related to counting in mathematics.

Notation	Meaning	Examples
n-factorial $n!$	$n!$ represents the product $n(n - 1) \cdots (3)(2)(1)$.	$6! = 6 \cdot 5 \cdot 4 \cdot 3 \cdot 2 \cdot 1 = 720$ $0! = 1$

Notation	Meaning	Examples
$P(n, r) = \dfrac{n!}{(n-r)!}$	$P(n, r)$ represents the number of permutations of n elements taken r at a time.	The number of two-letter strings that can be formed using the four letters A, B, C, and D exactly once is given by $$P(4, 2) = \frac{4!}{(4-2)!} = \frac{4!}{2!} = 12$$ or $P(4, 2) = 4 \cdot 3 = 12.$
$C(n, r) = \dfrac{n!}{(n-r)!\,r!}$	$C(n, r)$ represents the number of combinations of n elements taken r at a time.	The number of committees of 3 people that can be formed from 5 people is given by $$C(5, 3) = \frac{5!}{(5-3)!\,3!} = \frac{5!}{2!\,3!} = 10.$$

8.3 EXERCISES

Counting

Exercises 1–4: **Exam Questions** *Count the number of ways that the questions on an exam could be answered.*

1. Ten true-false questions

2. Ten multiple-choice questions with five choices each

3. Five true-false questions and ten multiple-choice questions with four choices each

4. One question involving matching ten items in one column with ten items in another column, using a one-to-one correspondence

Exercises 5–8: **License Plates** *Count the number of possible license plates with the given constraints.*

5. Three digits followed by three letters

6. Two letters followed by four digits

7. Three letters followed by three digits or letters

8. Two letters followed by either three or four digits

Exercises 9 and 10: **Counting Strings** *Count the number of five-letter strings that can be formed with the given letters, assuming a letter can be used more than once.*

9. A, B, C

10. W, X, Y, Z

Exercises 11–14: **Counting Strings** *Count the number of strings that can be formed with the given letters, assuming each letter is used exactly once.*

11. A, B

12. A, B, C

13. W, X, Y, Z

14. V, W, X, Y, Z

15. **Combination Lock** A briefcase has two locks. The combination to each lock consists of a three-digit number, where digits may be repeated. See the accompanying figure. How many combinations are possible? (*Hint:* The word *combination* is a misnomer. Lock combinations are permutations where the arrangement of the numbers is important.)

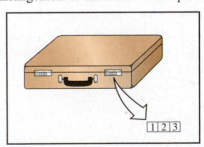

16. **Combination Lock** A typical combination for a padlock consists of three numbers from 0 to 39. Count the number of combinations that are possible with this type of lock, if a number may be repeated.

17. *Garage Door Openers* The code for some garage door openers consists of 12 electrical switches that can be set to either 0 or 1 by the owner. With this type of opener, how many codes are possible? (Source: Promax.)

18. *Lottery* To win the jackpot in a lottery game, a person must pick three numbers from 0 to 9 in the correct order. If a number can be repeated, how many ways are there to play the game?

19. *Radio Stations* Call letters for a radio station usually begin with either a K or W, followed by three letters. In 1995 there were 11,834 radio stations on the air. Is there any shortage of call letters for new radio stations? (Source: M. Street Corporation.)

20. *Access Codes* ATM access codes often consist of a four-digit number. How many codes are possible without giving two accounts the same access code?

21. *Computer Sale* A computer sale offers a choice of 2 monitors, 3 printers, and 4 types of software. How many different packages can be purchased?

22. *Dice* A red die and a blue die are thrown. How many ways are there for both dice to show an even number?

23. *Dinner Choices* A menu offers 5 different salads, 10 different entrées (main courses), and 4 different desserts. How many ways are there to order a salad, an entrée, and a dessert?

24. *Telephone Numbers* How many different 7-digit telephone numbers are possible if the first digit cannot be a 0 or a 1?

Permutations

Exercises 25–30: Evaluate the expression.

25. $P(5, 3)$ **26.** $P(10, 2)$

27. $P(8, 1)$ **28.** $P(6, 6)$

29. $P(7, 3)$ **30.** $P(12, 3)$

31. How many ways can 4 people stand in a line?

32. How many arrangements are there of 6 different books on a shelf?

33. How many ways could 5 basketball players be introduced at a game?

34. In how many arrangements can 3 students from a class of 15 give a speech?

35. *Traveling Salesperson* (Refer to the discussion after Example 5.) A salesperson must travel to 3 of 7 cities. Direct travel is possible between every pair of cities. How many arrangements are there for the salesperson to visit these 3 cities? Assume that traveling a route in reverse order constitutes a different route.

36. *Keys* How many distinguishable ways can 4 keys be put on a key ring?

37. *Sitting at a Round Table* How many ways can 7 people sit at a round table? Assume that a different way means that at least one person is sitting next to someone different.

38. *Batting Orders* A softball team has 10 players. How many batting orders are possible?

Combinations

Exercises 39–44: Evaluate the expression.

39. $C(3, 1)$

40. $C(4, 3)$

41. $C(6, 3)$

42. $C(7, 5)$

43. $C(5, 0)$

44. $C(10, 2)$

45. *Lottery* To win the jackpot in a lottery, one must select 5 different numbers from 1 to 39. How many ways are there to play this game?

46. *Selecting a Committee* How many ways can a committee of 5 be selected from 8 people?

47. *Selecting a Committee* How many committees of 4 people can be selected from 5 women and 3 men, if a committee must have 2 people of each sex on it?

48. *Essay Questions* On a test involving 6 essay questions, students are asked to answer 4 questions. How many ways can the essay questions be selected?

49. *Test Questions*   A test consists of two parts. In the first part a student must choose 3 of 5 essay questions, and in the second part a student must choose 4 of 5 essay questions. How many ways can the essay questions be selected?

50. *Cards*   How many ways are there to draw a 5-card hand from a 52-card deck?

51. *Selecting Marbles*   How many ways are there to draw 3 red marbles and 2 blue marbles from a jar that contains 10 red marbles and 12 blue marbles?

52. *Book Arrangements*   A professor has 3 copies of an algebra book and 4 copies of a calculus text. How many distinguishable ways can the books be placed on a shelf?

Writing about Mathematics

53. Explain the difference between a permutation and a combination. Give examples.

54. Explain what counting is, as presented in this section.

8.4 THE BINOMIAL THEOREM

Derivation of the Binomial Theorem • Pascal's Triangle

Introduction

In this section we discuss how to expand expressions in the form $(a + b)^n$, where n is a natural number. These expressions occur in statistics, finite mathematics, computer science, and calculus.

Derivation of the Binomial Theorem

Combinations play a central role in the development of the binomial theorem. The binomial theorem can be used to expand expressions of the form $(a + b)^n$. Before stating the binomial theorem, we begin by counting the number of strings of a given length that can be formed with only the letters a and b.

EXAMPLE 1   *Calculating distinguishable strings*

Count the number of distinguishable strings that can be formed with the given number of a's and b's. List these strings.
(a) Two a's, one b
(b) Two a's, three b's
(c) Four a's, no b's

SOLUTION   **(a)** Using two a's and one b, we can form strings of length three. Once the b has been positioned, the string is determined. For example, if the b is placed in the middle position,

$$\boxed{}\;\boxed{b}\;\boxed{},$$

then the string must be *aba*. From a set of three slots, we choose one slot to place the b. This is computed by $C(3, 1) = 3$. The strings are *aab*, *aba*, and *baa*.

(b) With two a's and three b's we can form strings of length five. Once the locations of the three b's have been selected, the string is determined. For instance, if the b's are placed in the first, third, and fifth positions,

$$\boxed{b}\;\boxed{}\;\boxed{b}\;\boxed{}\;\boxed{b},$$

then the string becomes *babab*. From a set of five slots, we select three to place the *b*'s. This is computed by $C(5, 3) = 10$. The ten strings are

bbbaa, bbaba, bbaab, babba, babab, baabb, abbba, abbab, ababb, and *aabbb.*

(c) There is only one string of length four that contains no *b*'s. This is *aaaa* and is computed by

$$C(4, 0) = \frac{4!}{(4-0)!\, 0!} = \frac{24}{24(1)} = 1.$$

Next we expand $(a + b)^n$ for a few values of *n*, without simplifying.

$$(a + b)^1 = a + b$$
$$(a + b)^2 = (a + b)(a + b)$$
$$= aa + ab + ba + bb$$
$$(a + b)^3 = (a + b)(a + b)^2$$
$$= (a + b)(aa + ab + ba + bb)$$
$$= aaa + aab + aba + abb + baa + bab + bba + bbb$$

Notice that $(a + b)^1$ is the sum of all possible strings of length one that can be formed using *a* and *b*. The only possibilities are *a* and *b*. The expression $(a + b)^2$ is the sum of all possible strings of length two using *a* and *b*. The strings are *aa, ab, ba,* and *bb*. In a similar manner, $(a + b)^3$ is the sum of all possible strings of length three using *a* and *b*. This pattern continues for higher powers of $(a + b)$.

Strings with equal numbers of *a*'s and equal numbers of *b*'s can be combined into one term. For example, in $(a + b)^2$ the terms *ab* and *ba* can be combined as $2ab$. Notice that there are $C(2, 1) = 2$ distinguishable strings of length two containing one *b*. Similarly, in the expansion of $(a + b)^3$, the terms containing one *a* and two *b*'s can be combined as

$$abb + bab + bba = 3ab^2.$$

There are $C(3, 2) = 3$ strings of length three that contain two *b*'s.

We can use these concepts to expand $(a + b)^4$. The expression $(a + b)^4$ consists of the sum of all strings of length four using only the letters *a* and *b*. There is $C(4, 0) = 1$ string containing no *b*'s, $C(4, 1) = 4$ strings containing one *b*, and so on, until there is $C(4, 4) = 1$ string containing four *b*'s. Thus

$$(a + b)^4 = \binom{4}{0}a^4b^0 + \binom{4}{1}a^3b^1 + \binom{4}{2}a^2b^2 + \binom{4}{3}a^1b^3 + \binom{4}{4}a^0b^4$$

$$= a^4 + 4a^3b + 6a^2b^2 + 4ab^3 + b^4.$$

These results are summarized by the binomial theorem.

Binomial theorem

For any positive integer *n* and numbers *a* and *b*,

$$(a + b)^n = \binom{n}{0}a^n + \binom{n}{1}a^{n-1}b^1 + \cdots + \binom{n}{n-1}a^1b^{n-1} + \binom{n}{n}b^n.$$

We can use the binomial theorem to expand $(a + b)^n$. For example,

$$(a + b)^3 = \binom{3}{0}a^3 + \binom{3}{1}a^2b^1 + \binom{3}{2}a^1b^2 + \binom{3}{3}b^3$$
$$= 1a^3 + 3a^2b + 3ab^2 + 1b^3$$
$$= a^3 + 3a^2b + 3ab^2 + b^3.$$

Since $\binom{n}{r} = C(n, r)$, we can use the combination formula $C(n, r) = \dfrac{n!}{(n - r)! \, r!}$ to evaluate the binomial coefficients.

EXAMPLE 2   *Applying the binomial theorem*

Use the binomial theorem to expand the expression $(2a + 1)^5$.

SOLUTION   Using the binomial theorem, we arrive at the following result.

$$(2a + 1)^5 = \binom{5}{0}(2a)^5 + \binom{5}{1}(2a)^4 1^1 + \binom{5}{2}(2a)^3 1^2 + \binom{5}{3}(2a)^2 1^3 + \binom{5}{4}(2a)^1 1^4 + \binom{5}{5}1^5$$

$$= \frac{5!}{5!0!}(32a^5) + \frac{5!}{4!1!}(16a^4) + \frac{5!}{3!2!}(8a^3) + \frac{5!}{2!3!}(4a^2) + \frac{5!}{1!4!}(2a) + \frac{5!}{0!5!}$$

$$= 32a^5 + 80a^4 + 80a^3 + 40a^2 + 10a + 1$$

Pascal's Triangle

Expanding $(a + b)^n$ for increasing values of n gives the following results.

$$(a + b)^0 = \qquad\qquad\qquad 1$$
$$(a + b)^1 = \qquad\qquad\qquad 1a + 1b$$
$$(a + b)^2 = \qquad\qquad\quad 1a^2 + 2ab + 1b^2$$
$$(a + b)^3 = \qquad\qquad 1a^3 + 3a^2b + 3ab^2 + 1b^3$$
$$(a + b)^4 = \qquad\quad 1a^4 + 4a^3b + 6a^2b^2 + 4ab^3 + 1b^4$$
$$(a + b)^5 = \quad 1a^5 + 5a^4b + 10a^3b^2 + 10a^2b^3 + 5ab^4 + 1b^5$$

Notice that $(a + b)^1$ has two terms starting with a and ending with b, $(a + b)^2$ has three terms starting with a^2 and ending with b^2, and in general $(a + b)^n$ has $n + 1$ terms starting with a^n and ending with b^n. The exponent on a decreases by 1 each successive term, and the exponent on b increases by 1 each successive term.

The triangle formed by the highlighted numbers is called **Pascal's triangle**. It can be used to efficiently compute the binomial coefficients, $C(n, r)$. The triangle consists of 1's along the sides. Each element inside the triangle is the sum of the two numbers above it. Pascal's triangle is usually written without variables as follows. It can be extended to include as many rows as needed.

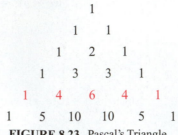

```
                1
              1   1
            1   2   1
          1   3   3   1
        1   4   6   4   1
      1   5   10  10  5   1
```

FIGURE 8.23 Pascal's Triangle

We can use this triangle to expand powers of binomials in the form $(a + b)^n$, where n is a natural number. For example, the expression $(m + n)^4$ consists of five terms written as follows.

$$(m + n)^4 = \underline{}\, m^4 + \underline{}\, m^3n^1 + \underline{}\, m^2n^2 + \underline{}\, m^1n^3 + \underline{}\, n^4$$

Since there are five terms, the coefficients can be found in the fifth row of Pascal's triangle, which is

$$1 \quad 4 \quad 6 \quad 4 \quad 1.$$

Thus,

$$(m + n)^4 = \underline{1}\, m^4 + \underline{4}\, m^3n^1 + \underline{6}\, m^2n^2 + \underline{4}\, m^1n^3 + \underline{1}\, n^4.$$
$$= m^4 + 4m^3n + 6m^2n^2 + 4mn^3 + n^4.$$

■ **EXAMPLE 3** *Expanding expressions with the binomial theorem*

Expand each of the following.
(a) $(2x + 1)^5$ **(b)** $(3x - y)^3$

SOLUTION **(a)** To expand $(2x + 1)^5$, let $a = 2x$ and $b = 1$ in the binomial theorem. We can use the sixth row of Pascal's triangle to obtain the coefficients 1, 5, 10, 10, 5, and 1. Compare this solution with the solution for Example 2.

$$(2x + 1)^5 = 1(2x)^5 + 5(2x)^4(1)^1 + 10(2x)^3(1)^2 + 10(2x)^2(1)^3 + 5(2x)^1(1)^4 + 1(1)^5$$
$$= 32x^5 + 80x^4 + 80x^3 + 40x^2 + 10x + 1$$

(b) Let $a = 3x$ and $b = -y$ in the binomial theorem. Use the coefficients 1, 3, 3, and 1 from the fourth row of Pascal's triangle.

$$(3x - y)^3 = 1(3x)^3 + 3(3x)^2(-y)^1 + 3(3x)^1(-y)^2 + 1(-y)^3$$
$$= 27x^3 - 27x^2y + 9xy^2 - y^3$$

Putting it all Together 8.4

*T*he following table summarizes topics related to the binomial theorem.

Concept	Explanation	Example
Binomial coefficient	$\dbinom{n}{r} = C(n, r) = \dfrac{n!}{(n - r)!\, r!}$	$\dbinom{5}{3} = \dfrac{5!}{(5 - 3)!\, 3!} = \dfrac{120}{2 \cdot 6} = 10$ $\dbinom{4}{0} = \dfrac{4!}{(4 - 0)!\, 0!} = \dfrac{24}{24 \cdot 1} = 1$
Binomial theorem	$(a + b)^n =$ $\dbinom{n}{0}a^n + \dbinom{n}{1}a^{n-1}b + \cdots + \dbinom{n}{n}b^n,$ for any positive integer n and real numbers a and b.	$(a + b)^3 = \dbinom{3}{0}a^3 + \dbinom{3}{1}a^2b + \dbinom{3}{2}ab^2 + \dbinom{3}{3}b^3$ $= a^3 + 3a^2b + 3ab^2 + b^3$ The binomial coefficients can also be found using row 4 of Pascal's triangle, which is shown below.
Pascal's triangle	A triangle of numbers that can be used to find the binomial coefficients needed to expand an expression of the form $(a + b)^n$. To expand $(a + b)^n$, use row $n + 1$ of Pascal's triangle.	$\begin{array}{ccccccccccc} & & & & & 1 & & & & & \\ & & & & 1 & & 1 & & & & \\ & & & 1 & & 2 & & 1 & & & \\ & & 1 & & 3 & & 3 & & 1 & & \\ & 1 & & 4 & & 6 & & 4 & & 1 & \\ 1 & & 5 & & 10 & & 10 & & 5 & & 1 \end{array}$

8.4 EXERCISES

Binomial Coefficients

Exercises 1–8: Evaluate the expression.

1. $\dbinom{5}{4}$ 2. $\dbinom{6}{2}$

3. $\dbinom{4}{0}$ 4. $\dbinom{4}{2}$

5. $\dbinom{6}{5}$ 6. $\dbinom{6}{3}$

7. $\dbinom{3}{3}$ 8. $\dbinom{5}{2}$

Binomial Theorem

Exercises 9–12: (Refer to Example 1.) Calculate the number of distinguishable strings that can be formed with the given number of a's and b's.

9. Three *a*'s, two *b*'s   10. Five *a*'s, three *b*'s

11. Four *a*'s, four *b*'s   12. One *a*, five *b*'s

Exercises 13–22: Use the binomial theorem to expand each expression.

13. $(x + y)^2$ 14. $(x + y)^4$

15. $(m + 2)^3$ 16. $(m + 2n)^5$

17. $(2x - 3)^3$

18. $(x + y^2)^3$

19. $(p - q)^6$

20. $(p^2 - 3)^4$

21. $(2m + 3n)^3$

22. $(3a - 2b)^5$

Pascal's Triangle

Exercises 23–32: Use Pascal's triangle to help expand the expression.

23. $(x + y)^2$

24. $(m + n)^3$

25. $(3x + 1)^4$

26. $(2x - 1)^4$

27. $(2 - x)^5$

28. $(2a + 3b)^3$

29. $(x^2 + 2)^4$

30. $(5 - x^2)^3$

31. $(4x - 3y)^4$

32. $(3 - 2x)^5$

Exercises 33–38: The $(r + 1)$st term of $(a + b)^n$, where $0 \leq r \leq n$, is given by $\binom{n}{n-r}a^{n-r}b^r$. Find the specified term for the following.

33. The fourth term of $(a + b)^9$
(*Hint:* Let $n = 9$ and $r = 3$.)

34. The second term of $(m - n)^9$

35. The fifth term of $(x + y)^8$

36. The third term of $(a + b)^7$

37. The fourth term of $(2x + y)^5$

38. The eighth term of $(2a - b)^9$

Writing about Mathematics

39. Explain how to find the numbers in Pascal's triangle.

40. Compare the expansion of $(a + b)^n$ with the expansion of $(a - b)^n$. Give an example.

CHECKING BASIC CONCEPTS FOR SECTIONS 8.3 AND 8.4

1. Count the ways to answer a quiz that consists of 8 true-false questions.

2. Count the number of 5-card poker hands that are possible using a standard deck of 52 cards.

3. How many distinct license plates could be made using a letter followed by 5 digits or letters?

4. Expand each expression.
(a) $(2x + 1)^4$ **(b)** $(4 - 3x)^3$

8.5 PROBABILITY

Definition of Probability • Compound Events • Independent and Dependent Events

Introduction

Questions of chance have no doubt engaged the minds of people since antiquity. However, the mathematical treatment of probability did not occur until the fifteenth century. The birth of probability theory as a mathematical discipline began in the seventeenth century with the work of Blaise Pascal and Pierre Fermat. Today probability pervades modern society. It is used not only to determine outcomes in gambling, but also to predict weather, genetic outcomes, and the risk involved with various types of substances and behaviors.

Risk is the chance or probability that a harmful event will occur. The following activities carry an annual increased risk of death by one chance in a million: flying

1000 miles in a jet, traveling 300 miles in a car, riding 10 miles on a bicycle, smoking 1.4 cigarettes, living 2 days in New York, having one chest x-ray, or living 2 months with a cigarette smoker. In Section 8.3 we saw that the likelihood of winning the jackpot in a lottery was one chance in 80,089,128.

Ideally, we would live in a risk-free world. However, almost every action or substance exposes people to some risk. Recognizing relative risks is important for a long and healthy life. Probability provides us with a measure of the likelihood that an event will occur. Knowledge about probability allows individuals to make informed decisions about their lives.   (Sources: *Historical Topics for the Mathematics Classroom, Thirty-first Yearbook, NCTM; J. Rodricks, Calculated Risk.*)

Definition of Probability

In the study of probability, experiments often are performed. An experiment might involve tossing a coin or measuring the cholesterol level of a patient who has had a heart attack. A result from an experiment is called an **outcome**. The set of all possible outcomes is the **sample space**. Any subset of a sample space is an **event**.

For example, if an experiment involves rolling a die, then the sample space consists of $S = \{1, 2, 3, 4, 5, 6\}$. The event $E = \{1, 6\}$ contains the outcomes of 1 or 6 showing on the die. If $n(E)$ and $n(S)$ denote the number of outcomes in E and S, then $n(E) = 2$ and $n(S) = 6$. The probability of rolling a 1 or 6 is given by $P(E) = \dfrac{2}{6}$. That is, the likelihood of event E occurring is two chances in six. These concepts are summarized in the following.

Probability of an event

If the outcomes of a finite sample space S are equally likely, and if E is an event in S, then the **probability of E** is given by

$$P(E) = \frac{n(E)}{n(S)},$$

where $n(E)$ and $n(S)$ represent the number of outcomes in E and S, respectively.

Since $n(E) \leq n(S)$, the probability of an event E satisfies $0 \leq P(E) \leq 1$. If $P(E) = 1$, then event E is certain to occur. If $P(E) = 0$, then event E is impossible.

EXAMPLE 1   *Drawing a card*

One card is drawn from a standard deck of 52 cards. Find the probability that the card is an ace.

SOLUTION   The sample space S consists of 52 outcomes that correspond to drawing any one of 52 cards. Each outcome is equally likely. Let E represent the event of drawing an ace. There are four aces in the deck, so event E contains four outcomes. Therefore $n(S) = 52$ and $n(E) = 4$. The probability of drawing an ace is given by

$$P(E) = \frac{n(E)}{n(S)} = \frac{4}{52} = \frac{1}{13}.$$

■ **EXAMPLE 2** *Estimating probability of organ transplants*

In 1997 there were 51,277 people waiting for an organ transplant. Table 8.8 lists the number of patients waiting for the most common types of transplants. Assuming none of these people need two or more transplants, approximate the probability that a transplant patient chosen at random will need

(a) a kidney or a heart;   (b) neither a kidney nor a heart.

TABLE 8.8

Heart	3774
Kidney	35,025
Liver	7920
Lung	2340

Source: Coalition on Organ
and Tissue Donation.

SOLUTION

(a) Let each patient represent an outcome in a sample space S. The event E of a transplant patient needing either a kidney or a heart contains $35,025 + 3774 = 38,779$ outcomes. The desired probability is

$$P(E) = \frac{n(E)}{n(S)} = \frac{38,779}{51,277} \approx 0.76.$$

In 1997, about 76% of transplant patients needed either a kidney or a heart.

(b) Let F be the event of a patient waiting for an organ other than a kidney or a heart. Then,

$$n(F) = n(S) - n(E) = 51,277 - 38,779 = 12,498.$$

The probability of F is

$$P(F) = \frac{n(F)}{n(S)} = \frac{12,498}{51,277} \approx 0.24 \quad \text{or} \quad 24\%. \qquad ■$$

Notice that $P(E) + P(F) = 1$ in Example 2. The events E and F are **complements** because $E \cap F = \varnothing$ and $E \cup F = S$, where $\varnothing$ denotes the *empty set*. That is, a transplant patient is either waiting for a kidney or a heart (event E), or not waiting for a kidney or a heart (event F). The complement of E may be denoted by E'.

Probability concepts can be illustrated using **Venn diagrams**. In Figure 8.24 the sample space S of an experiment is the disjoint union of event E and its complement E'. That is, $E \cup E' = S$ and $E \cap E' = \varnothing$.

If $P(E)$ is known, then $P(E')$ can be calculated as follows.

$$P(E') = \frac{n(E')}{n(S)} = \frac{n(S) - n(E)}{n(S)} = 1 - \frac{n(E)}{n(S)} = 1 - P(E)$$

In part (b) of Example 2, the probability of $F = E'$ could have also been calculated by using

$$P(F) = 1 - P(E) \approx 1 - 0.76 = 0.24.$$

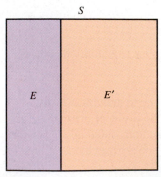

FIGURE 8.24

> ## Probability of a complement
>
> Let E be an event and E' be its complement. If the probability of E is $P(E)$, then the probability of its complement is given by
>
> $$P(E') = 1 - P(E).$$

EXAMPLE 3   *Finding probabilities of human eye color*

In 1865 Gregor Mendel performed important research in genetics. His work led to a better understanding of dominant and recessive genetic traits. One example is human eye color, which is determined by a pair of genes called a *genotype*. Brown eye color B is dominant over blue eye color b. If a person has the genotype of BB, Bb, or bB, he or she will have brown eyes. A genotype of bb will result in blue eyes. A person receives one gene (B or b) from each parent. Table 8.9 shows how these two genes can be paired.   (Source: H. Lancaster, *Quantitative Methods in Biology and Medical Sciences.*)

(a) Assuming that each genotype is equally likely, find the probability of blue eyes.

(b) What is the probability that a person has brown eyes?

TABLE 8.9

	B	b
B	BB	Bb
b	bB	bb

SOLUTION   **(a)** The sample space S consists of four equally likely outcomes denoted BB, Bb, bB, and bb. The event E of blue eye color (bb) occurs once. Therefore

$$P(E) = \frac{n(E)}{n(S)} = \frac{1}{4} = 0.25.$$

(b) In this chart, brown eyes are the complement of blue eyes. The probability of brown eyes is

$$P(E') = 1 - P(E) = 1 - 0.25 = 0.75.$$

This probability also could be computed as

$$P(E') = \frac{n(E')}{n(S)} = \frac{3}{4} = 0.75,$$

since there are three genotypes that result in brown eye color.

Compound Events

Frequently the probability of more than one event is needed. For example, suppose a college has a total of 2500 students with 225 students enrolled in college algebra, 75 in trigonometry, and 30 in both. Let E_1 denote the event that a student is enrolled in college algebra, and E_2 the event that a student is enrolled in trigonometry. Then the Venn diagram in Figure 8.25 visually describes this situation.

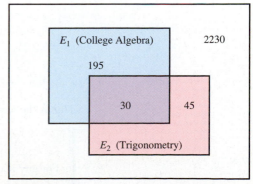

FIGURE 8.25

In this Venn diagram, it is important not to count the 30 students taking both courses twice. The set E_1 has a total of $195 + 30 = 225$ students, and set E_2 contains $45 + 30 = 75$ students.

EXAMPLE 4   *Calculating the probability of a union*

In the preceding scenario, suppose a student is selected at random. What is the probability that this student is enrolled in college algebra, trigonometry, or both?

SOLUTION   We would like to find the probability $P(E_1 \text{ or } E_2)$, which is denoted $P(E_1 \cup E_2)$. Since $n(E_1 \cup E_2) = 195 + 30 + 45 = 270$ and $n(S) = 2500$,

$$P(E_1 \cup E_2) = \frac{n(E_1 \cup E_2)}{n(S)} = \frac{270}{2500} = 0.108.$$

There is a 10.8% chance that a student selected at random will be taking algebra, trigonometry, or both.

In the previous example, it would have been incorrect to simply add the probability of a student taking algebra and the probability of a student taking trigonometry. Their sum would be

$$P(E_1) + P(E_2) = \frac{n(E_1)}{n(S)} + \frac{n(E_2)}{n(S)}$$

$$= \frac{225}{2500} + \frac{75}{2500}$$

$$= \frac{300}{2500}$$

$$= 0.12 \quad \text{or} \quad 12\%.$$

This sum is greater than $P(E_1 \cup E_2)$ because this calculation counts the 30 students taking both courses twice. In order to find the correct probability for $P(E_1 \cup E_2)$, we must subtract the probability of the intersection $E_1 \cap E_2$.

$$P(E_1 \cup E_2) = P(E_1) + P(E_2) - P(E_1 \cap E_2)$$

$$= \frac{n(E_1)}{n(S)} + \frac{n(E_2)}{n(S)} - \frac{n(E_1 \cap E_2)}{n(S)}$$

$$= \frac{225}{2500} + \frac{75}{2500} - \frac{30}{2500}$$

$$= \frac{270}{2500}$$

$$= 0.108 \quad \text{or} \quad 10.8\%$$

This is the same result obtained in Example 4 and suggests the following property.

Probability of the union of two events

For any two events E_1 and E_2,

$$P(E_1 \cup E_2) = P(E_1) + P(E_2) - P(E_1 \cap E_2).$$

EXAMPLE 5 *Rolling dice*

Suppose two dice are rolled. Find the probability that the dice show either a sum of eight or a pair.

SOLUTION   In Table 8.10 the roll of the dice is represented by an ordered pair. For example, the ordered pair (3, 6) represents the first die showing 3 and the second die 6.

TABLE 8.10   Rolling Two Dice

(1, 1)	(1, 2)	(1, 3)	(1, 4)	(1, 5)	(1, 6)
(2, 1)	**(2, 2)**	(2, 3)	(2, 4)	(2, 5)	**(2, 6)**
(3, 1)	(3, 2)	**(3, 3)**	(3, 4)	**(3, 5)**	(3, 6)
(4, 1)	(4, 2)	(4, 3)	**(4, 4)**	(4, 5)	(4, 6)
(5, 1)	(5, 2)	**(5, 3)**	(5, 4)	**(5, 5)**	(5, 6)
(6, 1)	**(6, 2)**	(6, 3)	(6, 4)	(6, 5)	**(6, 6)**

Since each die can show six different outcomes, there is a total of $6 \cdot 6 = 36$ outcomes in the sample space S. Let E_1 denote the event of rolling a sum of eight, and E_2 the event of rolling a pair. Then

$$E_1 = \{(6, 2), (5, 3), (4, 4), (3, 5), (2, 6)\} \quad \text{and}$$
$$E_2 = \{(1, 1), (2, 2), (3, 3), (4, 4), (5, 5), (6, 6)\}.$$

The intersection of E_1 and E_2 is

$$E_1 \cap E_2 = \{(4, 4)\}.$$

Since $n(S) = 36$, $n(E_1) = 5$, $n(E_2) = 6$, and $n(E_1 \cap E_2) = 1$, the following can be computed.

$$P(E_1 \cup E_2) = P(E_1) + P(E_2) - P(E_1 \cap E_2)$$

$$= \frac{n(E_1)}{n(S)} + \frac{n(E_2)}{n(S)} - \frac{n(E_1 \cap E_2)}{n(S)}$$

$$= \frac{5}{36} + \frac{6}{36} - \frac{1}{36}$$

$$= \frac{10}{36} \quad \text{or} \quad \frac{5}{18}$$

This result can be verified by counting the number of boldfaced outcomes in Table 8.10. Of a total of 36, 10 outcomes satisfy the conditions so the probability is $\frac{10}{36}$ or $\frac{5}{18}$.

If $E_1 \cap E_2 = \varnothing$, then the events E_1 and E_2 are **mutually exclusive**. Mutually exclusive events have no outcomes in common so $P(E_1 \cap E_2) = 0$. Thus $P(E_1 \cup E_2) = P(E_1) + P(E_2)$.

EXAMPLE 6 *Drawing cards*

Find the probability of drawing either an ace or a king from a standard deck of 52 playing cards.

SOLUTION   The event E_1 of drawing an ace and the event E_2 of drawing a king are mutually exclusive. No card can be both an ace and a king. Therefore $P(E_1 \cap E_2) = 0$. The probability of drawing an ace is $P(E_1) = \dfrac{4}{52}$, since there are 4 aces in 52 cards. Similarly, the probability of drawing a king is $P(E_2) = \dfrac{4}{52}$.

$$P(E_1 \cup E_2) = P(E_1) + P(E_2)$$
$$= \frac{4}{52} + \frac{4}{52}$$
$$= \frac{8}{52} \quad \text{or} \quad \frac{2}{13}$$

Thus the probability of drawing either an ace or a king is $\dfrac{2}{13}$.

The next example uses concepts from both counting and probability.

EXAMPLE 7   *Drawing a poker hand*

A standard deck of cards contains 52 cards, consisting of four suits: hearts, diamonds, spades, and clubs. Each suit contains 13 different cards. A poker hand consists of 5 cards drawn from a standard deck of cards, and a flush occurs when the 5 cards are all the same suit. Find an expression for the probability of drawing 5 cards with the same suit in one try. Assume that the cards are not replaced.

SOLUTION   Let E be the event of drawing 5 cards with the same suit. To determine $n(E)$, start by calculating the number of ways to draw a flush in a particular suit, such as hearts. From a set of 13 hearts, 5 hearts need to be drawn. There are $\dbinom{13}{5}$ ways to draw this hand. Because there are 4 suits, there are $4\dbinom{13}{5}$ ways to draw 5 cards with the same suit. Thus $n(E) = 4\dbinom{13}{5}$. The sample space S consists of all 5-card poker hands that can be drawn from a deck of 52 cards. There are $\dbinom{52}{5}$ different poker hands.

Graphing Calculator Help

To calculate $C(n, r)$, see Appendix B (page AP-24).

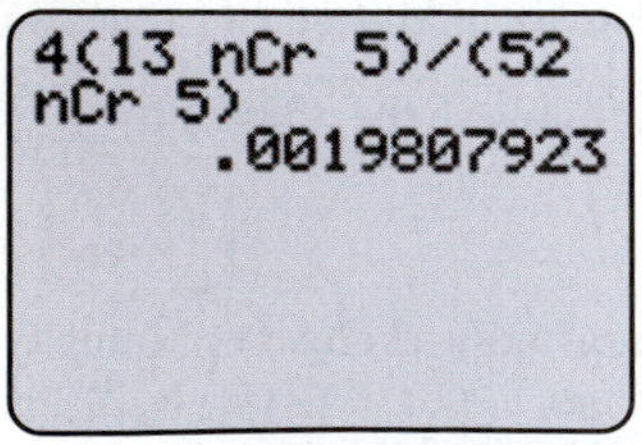

```
4(13 nCr 5)/(52
nCr 5)
        .0019807923
```

Thus $n(S) = \dbinom{52}{5}$. The probability of a flush can now be calculated as follows.

$$P(E) = \frac{n(E)}{n(S)} = \frac{4\dbinom{13}{5}}{\dbinom{52}{5}}$$

In Figure 8.26, we see that $P(E) \approx 0.00198$. Thus there is about a 0.2% chance of drawing 5 cards of the same suit (in one try) from a standard deck of 52 cards.

FIGURE 8.26

Independent and Dependent Events

Two events are **independent** if they do not influence one another. Otherwise they are **dependent**. An example of independent events would be one coin being tossed twice. The result of the first toss does not affect the second toss.

> **Probability of independent events**
>
> If E_1 and E_2 are independent events, then
> $$P(E_1 \cap E_2) = P(E_1) \cdot P(E_2).$$

EXAMPLE 8   *Tossing a coin*

Suppose a coin is tossed twice. Determine the probability that the result is two heads.

SOLUTION   Let E_1 be the event of a head on the first toss, and let E_2 be the event of a head on the second toss. Then $P(E_1) = P(E_2) = \frac{1}{2}$. The two events are independent. The probability of two heads occurring is

$$P(E_1 \cap E_2) = \frac{1}{2} \cdot \frac{1}{2} = \frac{1}{4}.$$

This probability of $\frac{1}{4}$ also can be found using a tree diagram, as shown in Figure 8.27. There are four equally likely outcomes. Tosses resulting in two heads occur once.

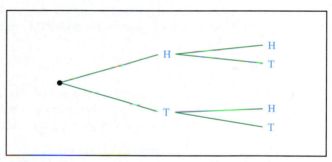

FIGURE 8.27

EXAMPLE 9   *Rolling dice*

What is the probability of rolling a sum of 12 with two dice?

SOLUTION   The roll of one die does not influence the roll of the other. They are independent events. To obtain a sum of 12, both dice must show a six. Let E_1 be the event of rolling a 6 with the first die, and E_2 the event of rolling a 6 with the second die. Then $P(E_1) = P(E_2) = \frac{1}{6}$. The probability of rolling a 12 is

$$P(E_1 \cap E_2) = P(E_1) \cdot P(E_2) = \frac{1}{6} \cdot \frac{1}{6} = \frac{1}{36}.$$

This can be verified by using Table 8.10. There is only 1 outcome out of 36 that results in a sum of 12.

■ **CLASS DISCUSSION**

When rolling a pair of dice, what sum is most likely to appear? ■

If events E_1 and E_2 influence each other, they are *dependent*. The probability of dependent events is given as follows.

> ## Probability of dependent events
>
> If E_1 and E_2 are dependent events, then
>
> $$P(E_1 \cap E_2) = P(E_1) \cdot P(E_2, \text{given that } E_1 \text{ occurred}).$$

Note: $P(E_2$, given that E_1 occurred) is called the **conditional probability of E_2, given E_1**.

EXAMPLE 10 *Drawing cards*

Find the probability of drawing two hearts from a standard deck of 52 cards, when the first card is
(a) replaced before drawing the second card;   **(b)** not replaced.

SOLUTION   **(a)** Let E_1 denote the event of the first card being a heart, and E_2 the event of the second card being a heart. If the first card is replaced before the second card is drawn, the two events are independent. Since there are 13 hearts in a standard deck of 52 cards, the probability of two hearts being drawn is

$$P(E_1 \cap E_2) = P(E_1) \cdot P(E_2) = \frac{13}{52} \cdot \frac{13}{52} = \frac{1}{16}.$$

(b) If the first card is not replaced, then the outcome for the second card is influenced by the first card. Therefore the events E_1 and E_2 are dependent. The probability of drawing a second heart, given that the first card is a heart, is represented by $P(E_2$, given E_1 has occurred$) = \dfrac{12}{51}$. That is, if the first card drawn were a heart and removed from the deck, then there would be 12 hearts in a sample space of 51 cards.

$$P(E_1 \cap E_2) = P(E_1) \cdot P(E_2, \text{given that } E_1 \text{ occurred}) = \frac{13}{52} \cdot \frac{12}{51} = \frac{1}{17}$$

Thus the probability of drawing two hearts is slightly less if the first card is not replaced.

EXAMPLE 11 *Calculating the probability of dependent events*

Table 8.11 shows the number of students (by gender) registered for either a Spanish class or a French class. No student is taking both languages.

TABLE 8.11

	Spanish	French	Totals
Females	20	25	45
Males	40	25	65
Totals	60	50	110

If one student is selected at random, calculate each of the following.

(a) The probability that the student is female

(b) The probability that the student is taking Spanish, given that the student is female

(c) The probability that the student is female and taking Spanish

SOLUTION **(a)** Let F represent the event that the student is female. Since 45 of the 110 students are female, $P(F) = \dfrac{45}{110} = \dfrac{9}{22}$.

(b) Let S be the event that the student is taking Spanish. The probability that the student is taking Spanish, given that the student is female, is $P(S, \text{given } F) = \dfrac{20}{45} = \dfrac{4}{9}$, because 20 of the 45 female students are taking Spanish.

(c) The probability that the student is female and taking Spanish is calculated by

$$P(F \cap S) = P(F) \cdot P(S, \text{given } F) = \frac{45}{110} \cdot \frac{20}{45} = \frac{2}{11}.$$

Table 8.11 shows that 20 of the 110 students are female and taking Spanish. Thus the required probability is $\dfrac{20}{110} = \dfrac{2}{11}$, which is in agreement with our previous calculation.

EXAMPLE 12   *Analyzing a polygraph test*

Suppose there is a 6% chance that a polygraph test will incorrectly say a person is lying when he or she is actually telling the truth. If a person tells the truth 95% of the time, what percentage of the time will the polygraph test incorrectly indicate is a lie?

SOLUTION Let E_1 be the event that the person is telling the truth and E_2 be the event that the polygraph test is incorrect. Then,

$$P(E_1 \cap E_2) = P(E_1) \cdot P(E_2, \text{given the person is telling the truth})$$
$$= (0.95)(0.06)$$
$$= 0.057 \text{ or } 5.7\%.$$

Putting it all Together 8.5

*T*he following table summarizes some concepts about probability. In this table, S denotes a sample space of equally likely outcomes, and E, E_1, and E_2 denote events in a sample space S.

Concept	Comments	Example
Probability P	A number P satisfying $0 \leq P \leq 1$.	$P(E) = 1$ indicates that an event E is certain to occur, $P(E) = 0$ indicates an event E is impossible, and $P(E) = 0.3$ indicates that an event E has a 30% chance of occurring.

Concept	Comments	Example
Probability of an event	The probability of event E is $$P(E) = \frac{n(E)}{n(S)},$$ where $n(E)$ and $n(S)$ denote the number of equally likely outcomes in E and S, respectively.	The probability of rolling two dice and their sum being 3 is $\frac{2}{36}$. This is because there are $6 \cdot 6 = 36$ ways to roll two dice, and only 2 ways to roll a sum of 3. (They are a 1 and a 2, or a 2 and a 1.)
Probability of a complement	If E and E' are complementary events, then $E \cap E' = \varnothing$, $E \cup E' = S$, and $P(E') = 1 - P(E)$.	The probability of rolling a 6 with one die is $\frac{1}{6}$. Therefore the probability of *not* rolling a 6 is $1 - \frac{1}{6} = \frac{5}{6}$.
Probability of the union of two events	The probability of both E_1 and E_2 occurring is $P(E_1 \cup E_2) = P(E_1) + P(E_2) - P(E_1 \cap E_2)$.	If $P(E_1) = 0.5$, $P(E_2) = 0.2$, and $P(E_1 \cap E_2) = 0.1$, then $P(E_1 \cup E_2) = 0.5 + 0.2 - 0.1 = 0.6$.
Probability of independent events	Two events E_1 and E_2 are independent if they do not influence one another. $$P(E_1 \cap E_2) = P(E_1) \cdot P(E_2)$$	If E_1 represents a head on the first toss of a coin and E_2 represents a tail on the second toss, then E_1 and E_2 are independent events and $$P(E_1 \cap E_2) = \frac{1}{2} \cdot \frac{1}{2} = \frac{1}{4}.$$
Probability of dependent events	Two events E_1 and E_2 are dependent if they are not independent. $P(E_1 \cap E_2) = $ $P(E_1) \cdot P(E_2,$ given that E_1 occurred.)	Drawing two hearts without replacement from a standard deck of cards is an example of dependent events. See Example 10.

8.5 EXERCISES

Probability of an Event

Exercises 1–8: Determine if the number could represent a probability.

1. $\dfrac{11}{13}$

2. 0.995

3. 2.5

4. 1

5. 0

6. 110%

7. -0.375 8. $\dfrac{9}{8}$

Exercises 9–18: Find the probability of each event.

9. Tossing a head with a fair coin

10. Tossing a tail with a fair coin

11. Rolling a 2 with a fair die

12. Rolling a 5 or 6 with a fair die

13. Guessing the correct answer for a true-false question

14. Guessing the correct answer for a multiple-choice question with five choices

15. Drawing an ace from a standard deck of 52 cards

16. Drawing a club from a standard deck of 52 cards

17. Randomly guessing a four-digit ATM access code

18. Randomly picking the winning team at a basketball game

19. The table shows some of the favorite pizza toppings. (Source: *USA Today.*)

Pepperoni	43%
Sausage	19%
Mushrooms	14%
Vegetables	13%

(a) If a person is selected at random, what is the probability that pepperoni is not his or her favorite topping?

(b) Find the probability that a person's favorite topping is either mushrooms or sausage.

20. (Refer to Example 2.) Find the probability that a transplant patient in 1997 was waiting for the following.
(a) A lung
(b) A lung or a liver

Probability of Compound Events

Exercises 21–30: Find the probability of each compound event.

21. Tossing a coin twice with the outcomes of two tails

22. Tossing a coin three times with the outcomes of three heads

23. Rolling a die three times and obtaining a 5 or 6 on each roll

24. Rolling a sum of 2 with two dice

25. Rolling a sum of 7 with two dice

26. Rolling a sum other than 7 with two dice

27. Rolling a die four times without obtaining a 6

28. Rolling a die four times and obtaining at least one 6

29. Drawing four consecutive aces from a standard deck of 52 cards without replacement

30. Drawing a pair (two cards with the same value) from a standard deck of 52 cards without replacement

31. *Poker Hands* (Refer to Example 7.) Calculate the probability of drawing 3 hearts and 2 diamonds in a 5-card poker hand. Assume that drawn cards are not replaced and that the 5 cards are drawn only once.

32. *Poker Hands* (Refer to Example 7.) Calculate the probability of drawing three kings and two queens in a 5-card poker hand. Assume that drawn cards are not replaced and that the 5 cards are drawn only once.

33. *Quality Control* A quality control experiment involves selecting one string of decorative lights from a box of 20. If the string is defective, the entire box of 20 is rejected. Suppose the box contains four defective strings of lights. What is the probability of rejecting the box?

34. *Quality Control* (Refer to the previous exercise.) Suppose three strings of lights are tested. If any of the strings are defective, the entire box of 20 is rejected. What is the probability of rejecting the box if there are four defective strings of lights in a box? (*Hint:* Start by finding the probability that the box is not rejected.)

35. *Entrance Exams* A group of students are preparing for college entrance exams. It is estimated that 50% need help with mathematics, 45% with English, and 25% with both.
(a) Draw a Venn diagram representing this data.
(b) Use this diagram to find the probability that a student needs help with mathematics, English, or both.
(c) Solve part (b) symbolically by applying a probability formula.

36. *College Classes* In a college of 5500 students, 950 students are enrolled in English classes, 1220 in business classes, and 350 in both. If a student is chosen at random, find the probability that he or she is enrolled in an English class, a business class, or both.

37. *New Books Published* In 1999 a total of 100,405 new books and editions were published. The table lists the number of books published in specific areas. If a new book or edition is selected at random, find the probability that its subject area satisfies the following. (Source: R. R. Bowker Co.)

Art	4272
Business	3154
History	6018
Music	1155
Religion	4806
Science	6410

(a) Art or music **(b)** Neither science nor religion

38. *Death Rates*   In 1998, the U.S. death rate per 100,000 people was 864.7. What is the probability that a person selected at random in 1998 died? (Source: Department of Health and Human Services.)

39. *Death Rates*   In 1998, the death rate per 100,000 people between the ages of 15 and 24 was 95.3. What is the probability that a person selected at random from this age group died during 1998? (Source: Department of Health and Human Services.)

40. *AIDS*   By 1999, a total of 733,374 cases of AIDS had been diagnosed. The table lists AIDS cases diagnosed in certain cities. Estimate the probability that a person diagnosed with AIDS satisfied the following conditions.   (Source: Department of Health and Human Services.)
(a) Resided in New York
(b) Did not reside in New York
(c) Resided in Los Angeles or Miami

New York	115,059
Los Angeles	40,709
San Francisco	27,151
Miami	22,872

41. *Tossing a Coin*   Find the probability of tossing a coin n times and obtaining n heads. What happens to this probability as n increases? Does this agree with your intuition? Explain.

42. *Rolling Dice*   Find the probability of rolling a die five times and obtaining a 6 on the first two rolls, a 5 on the third roll, and a 1, 2, 3, or 4 on the last two rolls.

43. *Unfair Coin*   Suppose a coin is not fair, but instead the probability of obtaining a head (H) is $\frac{3}{4}$ and a tail (T) is $\frac{1}{4}$. What is the probability of the following events?
(a) HT
(b) HH
(c) HHT
(d) THT

44. *Unfair Die*   Suppose a die is not fair, but instead the probability P of each number n is listed in the table. Find the probability of each event.

n	1	2	3	4	5	6
P	0.1	0.1	0.1	0.2	0.2	0.3

(a) Rolling a number that is 4 or higher
(b) Rolling a 6 twice on consecutive rolls

45. *Dice*   Suppose there are two dice, one red and one blue, having the probabilities shown in the table in the previous exercise. If both dice are rolled, find the probability of the given sum.
(a) 12   **(b)** 11

46. *Garage Door Code*   The code for some garage door openers consists of 12 electrical switches that can be set to either 0 or 1 by the owner. Each setting represents a different code. What is the probability of guessing someone's code at random? (Source: Promax.)

47. *Lottery*   To win a lottery, a person must pick three numbers from 0 to 9 in the correct order. If a number may be repeated, what is the probability of winning this game with one play?

48. *Lottery*   To win the jackpot in a lottery, a person must pick five numbers from 1 to 49, and then pick the powerball number from 1 to 42. If the numbers are picked at random, what is the probability of winning this game with one play?

49. *Marbles*   A jar contains 22 red marbles, 18 blue marbles, and 10 green marbles. If a marble is drawn from the jar at random, find the probability that the ball is the following.
(a) Red   **(b)** Not red   **(c)** Blue or green

50. *Marbles*   A jar contains 55 red marbles and 45 blue marbles. If 2 marbles are drawn from the jar at random without replacement, find the probability that the marbles satisfy the following.
(a) Both are blue.   **(b)** Neither is blue.
(c) The first marble is red and the second marble is blue.

51. *Cancer and Saccharin*   Saccharin is the *least* potent carcinogen ever detected in an animal study. The dose-risk curve is shown in the figure for saccharin-induced bladder tumors in rats. Doses on the x-axis represent the percentage of the animals' diets consisting of saccharin. The associated lifetime risk R or probability of the animal developing bladder cancer is shown on the y-axis. (Source: J. Rodricks.)

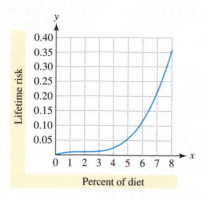

Percent of diet

(a) If the diet of a rat consists of 6% saccharin, estimate the risk of developing bladder cancer.
(b) Discuss the information shown in this graph.

52. *Horse Racing* The favorite to win the Kentucky Derby has won the race 48 out of 122 times. If a past favorite is selected at random, what is the probability that the horse won the Kentucky Derby? (Source: Churchill Downs.)

Conditional Probability and Dependent Events

53. *Drawing a Card* Find the probability of drawing a queen from a standard deck of cards given that one card, a queen, has already been drawn and not replaced.

54. *Drawing a Card* Find the probability of drawing a king from a standard deck of cards given that two cards, both kings, have already been drawn and not replaced.

55. *Drawing a Card* A card is drawn from a standard deck of 52 cards. Given that the card is a face card, what is the probability that the card is a king? (*Hint:* A face card is a jack, queen, or king.)

56. *Drawing a Card* Three cards are drawn from a deck without replacement. Find the probability that the three cards are an ace, king, and queen.

57. *Drawing Marbles* A jar initially contains 10 red marbles and 23 blue marbles. What is the probability of drawing a blue marble, given that 2 red marbles and 4 blue marbles have already been drawn?

58. *Volleyball Serve* The probability that the first serve of a volleyball is out of bounds is 0.3, and the probability that the second serve of a volleyball is

in bounds, given that the first serve was out of bounds, is 0.8. Find the probability that the first serve is out of bounds and the second serve is in bounds.

59. *Cloudy and Windy* The probability of it being cloudy is 30% and the probability of it being cloudy and windy is 12%. Given that the day is cloudy, what is the probability that it will be windy?

60. *Rainy and Windy* The probability of it being rainy is 80% and the probability of it being windy and rainy is 72%. Given that the day is rainy, what is the probability that it will be windy?

61. *Rolling Dice* Two dice are rolled. If one die is a 2, find the probability that the sum of the dice is 7 or more.

62. *Rolling Dice* Three dice are rolled. If one die is a 4, find the probability that the sum of the three dice is less than 12.

63. *Defective Parts* The accompanying table shows numbers of automobile parts that are either defective or not defective.

	Type A	Type B	Totals
Defective	7	11	18
Not Defective	123	94	217
Totals	130	105	235

If one part is selected at random, calculate each of the following.
(a) The probability that the part is defective
(b) The probability that the part is type A, given that it is defective
(c) The probability that the part is type A and defective

64. *Health* The accompanying table shows numbers of patients with two different diseases by gender. (Assume that a person cannot have both diseases.)

	Disease A	Disease B	Totals
Females	145	851	996
Males	256	355	611
Totals	401	1206	1607

If one patient is selected at random, find the probability that the patient is female and has disease B.

65. *Prime Numbers* Suppose a number between 1 and 15 is selected at random. Find the probability of each event.

(a) The number is odd.

(b) The number is even.

(c) The number is prime. (*Hint:* A natural number greater than 1 that has only itself and 1 as factors is called a **prime number**.)

(d) The number is prime and odd.

(e) The number is prime and even.

66. *Students and Classes* (Refer to Example 11.) If one student is selected at random, use Table 8.11 to calculate each of the following.

(a) The probability that the student is male

(b) The probability that the student is taking French, given that the student is male

(c) The probability that the student is male and taking French

Writing about Mathematics

67. What values are possible for a probability? Interpret different probabilities and give examples.

68. Discuss the difference between dependent and independent events. How are their probabilities calculated?

CHECKING BASIC CONCEPTS FOR SECTION 8.5

1. Find the probability of tossing a coin four times and obtaining a head every time.

2. Find the probability of rolling a sum of 11 with two dice.

3. Find the probability of drawing four aces and a queen from a standard deck of 52 cards.

4. In 1995 there were 2.6 million high school graduates, of which 1.2 million were male. If a 1995 high school graduate is selected at random, estimate the probability that this graduate is female. (Source: The American College Testing Program.)

Chapter 8 Summary

CONCEPT	EXPLANATION AND EXAMPLES

Section 8.1

SEQUENCES

An infinite sequence is a function whose domain is the natural numbers. Its graph is a scatterplot.

Example: $a_n = \frac{1}{2}n^2 - 2$; the first 4 terms are as follows.

$$a_1 = \frac{1}{2}(1)^2 - 2 = -\frac{3}{2}, \qquad a_2 = \frac{1}{2}(2)^2 - 2 = 0$$

$$a_3 = \frac{1}{2}(3)^2 - 2 = \frac{5}{2}, \qquad a_4 = \frac{1}{2}(4)^2 - 2 = 6$$

A graph of the first 4 terms is shown here.

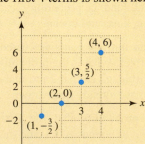

CONCEPT	EXPLANATION AND EXAMPLES

ARITHMETIC SEQUENCE

Recursive Definition:
$a_n = a_{n-1} + d$, where d is the common difference.

Function Definition:
$f(n) = dn + c$, or equivalently, $f(n) = a_1 + d(n - 1)$, where $a_n = f(n)$ and d is the common difference.

Example: $a_n = a_{n-1} + 3$, $a_1 = 4$, and $f(n) = 3n + 1$ describe the same sequence. The common difference is $d = 3$. The terms of the sequence are

$$4, 7, 10, 13, 16, 19, 22, \ldots.$$

GEOMETRIC SEQUENCE

Recursive Definition:
$a_n = ra_{n-1}$, where r is the common ratio.

Function Definition:
$f(n) = cr^{n-1}$, where $c = a_1$ and r is the common ratio.

Example: $a_n = -2a_{n-1}$, $a_1 = 3$, and $f(n) = 3(-2)^{n-1}$ describe the same sequence. The common ratio is $r = -2$. The terms of the sequence are

$$3, -6, 12, -24, 48, -96, 192, \ldots.$$

Section 8.2

SERIES

A series is the summation of the terms of a sequence.

Examples: $1 + \dfrac{1}{2} + \dfrac{1}{4} + \dfrac{1}{8} + \cdots + \dfrac{1}{2^{n-1}} + \cdots$ (Infinite series)

$$\sum_{k=1}^{5} 2k = 2 + 4 + 6 + 8 + 10$$ (Finite series)

ARITHMETIC SERIES

Finite Arithmetic Series:

$$\sum_{k=1}^{n} a_k = a_1 + a_2 + a_3 + \cdots + a_n, \text{ where } a_k = dk + c \text{ for some}$$
constants c and d.

Sum of the First n Terms:
$$S_n = n\left(\frac{a_1 + a_n}{2}\right) \quad \text{or} \quad S_n = \frac{n}{2}(2a_1 + (n - 1)d).$$

Example: The series $4 + 7 + 10 + 13 + 16 + 19 + 22$ is defined by
$$a_k = 3k + 1. \text{ Its sum is } S_7 = 7\left(\frac{4 + 22}{2}\right) = 91.$$

GEOMETRIC SERIES

Infinite Geometric Series:
$$\sum_{k=1}^{n} a_k = a_1 + a_2 + a_3 + \cdots + a_n + \cdots$$
where $a_k = a_1 r^{k-1}$ for some nonzero constants a_1 and r.

CONCEPT	EXPLANATION AND EXAMPLES

Section 8.2 (continued)

GEOMETRIC SERIES
(continued)

Sum of First n Terms:

$S_n = a_1 \left(\dfrac{1 - r^n}{1 - r} \right)$, where a_1 is the first term and r is the common ratio.

Example: The series $3 + 6 + 12 + 24 + 48 + 96$ has $a_1 = 3$ and $r = 2$. Its sum is $S_6 = 3 \left(\dfrac{1 - 2^6}{1 - 2} \right) = 189$.

Sum of Infinite Geometric Series:

$S = \dfrac{a_1}{1 - r}$, if $|r| < 1$. S does not exist if $|r| \geq 1$.

Example: The series $4 + 1 + \dfrac{1}{4} + \dfrac{1}{16} + \dfrac{1}{64} + \cdots$ has sum

$$S = \frac{4}{1 - \frac{1}{4}} = \frac{16}{3}.$$

Section 8.3

FUNDAMENTAL COUNTING PRINCIPLE

Let E_1 and E_2 be independent events. If event E_1 can occur m_1 ways and if event E_2 can occur m_2 ways, then there are $m_1 \cdot m_2$ ways for both events to occur.

Example: If two multiple-choice questions have 5 choices each, then there are $5 \cdot 5 = 25$ ways to answer the two questions.

FACTORIAL NOTATION

$n! = n(n - 1)(n - 2) \cdots (3)(2)(1)$

Examples: $0! = 1$; $1! = 1$; $5! = 5 \cdot 4 \cdot 3 \cdot 2 \cdot 1 = 120$

PERMUTATIONS

$P(n, r) = \dfrac{n!}{(n - r)!}$ represents the number of permutations of n elements taken r at a time. Order is important when calculating a permutation.

Example: $P(5, 3) = \dfrac{5!}{(5 - 3)!} = \dfrac{120}{2} = 60$

COMBINATIONS

$C(n, r) = \dbinom{n}{r} = \dfrac{n!}{(n - r)! \, r!}$ represents the number of combinations of n elements taken r at a time. Order is unimportant when calculating a combination.

Example: $C(5, 3) = \dbinom{5}{3} = \dfrac{5!}{(5 - 3)! \, 3!} = \dfrac{120}{2 \cdot 6} = 10$

CONCEPT	EXPLANATION AND EXAMPLES

Section 8.4

BINOMIAL THEOREM

$$(a + b)^n = \binom{n}{0}a^n + \binom{n}{1}a^{n-1}b + \cdots + \binom{n}{n-1}ab^{n-1} + \binom{n}{n}b^n$$

Example: $(x + y)^4 = 1x^4 + 4x^3y + 6x^2y^2 + 4xy^3 + 1y^4$

The coefficients can be found in the fifth row of Pascal's triangle.

PASCAL'S TRIANGLE

Pascal's triangle can be used to calculate the binomial coefficients when expanding $(a + b)^n$.

$$
\begin{array}{ccccccccccc}
 & & & & & 1 & & & & & \\
 & & & & 1 & & 1 & & & & \\
 & & & 1 & & 2 & & 1 & & & \\
 & & 1 & & 3 & & 3 & & 1 & & \\
 & 1 & & 4 & & 6 & & 4 & & 1 & \\
1 & & 5 & & 10 & & 10 & & 5 & & 1
\end{array}
$$

Section 8.5

PROBABILITY

$P(E) = \dfrac{n(E)}{n(S)}$, where $n(E)$ is the number of outcomes in event E and $n(S)$ is the number of equally likely outcomes in the sample space S. Note that $0 \le P(E) \le 1$.

COMPOUND EVENTS

Probability of either E_1 or E_2 (or both) occurring:

$$P(E_1 \cup E_2) = P(E_1) + P(E_2) - P(E_1 \cap E_2).$$

If E_1 and E_2 are *mutually exclusive,* then $E_1 \cap E_2 = \varnothing$, and

$$P(E_1 \cup E_2) = P(E_1) + P(E_2).$$

Probability of *both* E_1 and E_2 occurring:
If E_1 and E_2 are *independent,* then

$$P(E_1 \cap E_2) = P(E_1) \cdot P(E_2).$$

If E_1 and E_2 are *dependent,* then

$$P(E_1 \cap E_2) = P(E_1) \cdot P(E_2, \text{given that } E_1 \text{ has occurred}).$$

Review Exercises

Exercises 1–4: Find the first four terms of the sequence.

1. $a_n = -3n + 2$

2. $a_n = n^2 + n$

3. $a_n = 2a_{n-1} + 1$; $a_1 = 0$

4. $a_n = a_{n-1} + 2a_{n-2}$; $a_1 = 1$, $a_2 = 4$

Exercises 5 and 6: Use the graphical representation to identify the terms of the finite sequence.

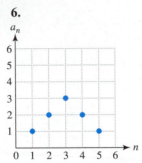

5.

6.

Exercises 7–10: The first five terms of an infinite arithmetic or geometric sequence are given. Find

    (a) numerical,
    (b) graphical, and
    (c) symbolic

representations of the sequence. Include at least the first eight terms of the sequence.

7. 2, 4, 6, 8, 10

8. 3, 1, −1, −3, −5

9. 81, 27, 9, 3, 1

10. 1.5, −3, 6, −12, 24

11. Find a general term a_n for the arithmetic sequence with $a_3 = -3$ and $d = 4$.

12. Find a general term a_n for the geometric sequence with $a_1 = 2.5$ and $a_6 = -80$.

13. *Height of a Ball* When a Ping-Pong ball is dropped, it rebounds to 90% of its initial height.

    (a) Write the first five terms of a sequence that gives the maximum height attained by the ball on each rebound when it is dropped from an initial height of 4 feet. Let $a_1 = 4$. What type of sequence describes these maximums?

    (b) Give a graphical representation of this sequence for the first five terms.

    (c) Find a formula for a_n.

14. *Falling Object* If air resistance is ignored, an object falls 16, 48, 80, and 112 feet during each successive 1-second interval.

    (a) What type of sequence describes these distances?

    (b) Determine how far an object falls during the sixth second.

    (c) Find a formula for the nth term of this sequence.

Exercises 15–18: Use a formula to find the sum of the series.

15. $-2 + 1 + 4 + 7 + 10 + 13 + 16 + 19 + 22$

16. $2 + 4 + 6 + 8 + \cdots + 98 + 100$

17. $1 + 3 + 9 + 27 + 81 + 243 + 729 + 2187$

18. $64 + 16 + 4 + 1 + \dfrac{1}{4} + \dfrac{1}{16}$

Exercises 19–22: Find the sum of the infinite geometric series.

19. $2 + 1 + \dfrac{1}{2} + \dfrac{1}{4} + \dfrac{1}{8} + \dfrac{1}{16} + \cdots$

20. $4 - \dfrac{4}{3} + \dfrac{4}{9} - \dfrac{4}{27} + \dfrac{4}{81} - \dfrac{4}{243} + \cdots$

21. $0.2 + 0.02 + 0.002 + 0.0002 + 0.00002 + \cdots$

22. $0.25 + 0.0025 + 0.000025 + 0.00000025 + \cdots$

Exercises 23 and 24: Write out the terms of the series.

23. $\displaystyle\sum_{k=1}^{5} (5k + 1)$

24. $\displaystyle\sum_{k=1}^{4} (2 - k^2)$

Exercises 25 and 26: Write each series using summation notation.

25. $1^3 + 2^3 + 3^3 + 4^3 + 5^3 + 6^3$

26. $1 + \dfrac{1}{10} + \dfrac{1}{100} + \dfrac{1}{1000} + \dfrac{1}{10,000}$

27. *Exam Questions* Count the ways that an exam, consisting of 20 multiple-choice questions with 4 choices each, could be answered.

28. *License Plates* Count the different license plates having 4 numeric digits followed by 2 letters.

29. *Combination Lock* A combination lock consists of 4 numbers from 0 to 49. If a number may be repeated, find the number of possible combinations.

30. *Dice*   A red die and a blue die are rolled. How many ways are there for the sum to equal 4?

Exercises 31 and 32: Evaluate the expression.

31. $P(6, 3)$ **32.** $C(7, 4)$

33. *Standing in Line*   In how many arrangements can 5 people stand in a line?

34. *Giving a Speech*   How many arrangements are possible for 4 students to give a speech out of a class of 15?

35. *Committees*   In how many ways can a committee of 3 be selected from 6 people?

36. *Committees*   Find the number of committees with 3 women and 3 men that can be selected from a group of 7 women and 5 men.

37. *Test Questions*   On a test involving 10 essay questions, students are asked to answer 6 questions. How many ways can the essay questions be selected?

38. *Binomial Theorem*   Use the binomial theorem to expand the expression $(2x - y)^4$.

Exercises 39 and 40: Find the probability of each event.

39. A 1, 2, or 3 appears when rolling a die

40. Tossing three heads in a row using a fair coin

41. *Quality Control*   A quality control experiment involves selecting 2 batteries from a pack of 16. If either battery is defective, the entire pack of 16 is rejected. Suppose a pack contains 2 defective batteries. What is the probability that this pack will not be rejected?

42. *Venn Diagram*   From a group of 82 students, 19 are enrolled in music, 22 in art, and 10 in both.
  (a) Draw a Venn diagram representing these data.
  (b) Use this diagram to determine the probability that a student selected at random is enrolled in music, art, or both.
  (c) Solve part (b) symbolically by applying a probability formula.

43. *Marbles*   A jar contains 13 red, 27 blue, and 20 green marbles. If a marble is drawn from the jar at random, find the probability that the marble is the following.
  (a) Blue
  (b) Not blue
  (c) Red

44. *Cards*   Find the probability of drawing 2 diamonds from a standard deck of 52 cards without replacing the first card.

45. *Insect Populations*   The monthly density of an insect population, measured in thousands per acre, is described by the recursive sequence

$$a_1 = 100$$

$$a_n = \frac{2a_{n-1}}{1 + (a_{n-1}/4000)}, \qquad n > 1.$$

Use your calculator to graph the sequence in $[0, 16, 1]$ by $[0, 5000, 1000]$. Include the first 15 terms and discuss any trends illustrated by the graph.

Extended and Discovery Exercises

Exercises 1 and 2: Antibiotic Resistance   Due to the frequent use of antibiotics in society, many strains of bacteria are becoming resistant. Some types of haploid bacteria contain genetic material called plasmids. Plasmids are capable of making a strain of bacterium resistant to antibiotic drugs. Genetic engineers want to predict the resistance of various bacteria after many generations.

1. Suppose a strain of bacterium contains two plasmids R_1 and R_2. Plasmid R_1 is resistant to the antibiotic ampicillin, whereas plasmid R_2 is resistant to the antibiotic tetracycline. When bacteria reproduce through cell division, the type of plasmids passed on to each new cell is random. For example, a daughter cell could have two plasmids of type R_1 and no plasmid of type R_2, one of each type, or no plasmid of type R_1 and two plasmids of type R_2. The probability $P_{k,j}$ that a mother cell with k plasmids of type R_1 produces a daughter cell with j plasmids of type R_1 can be calculated by the formula

$$P_{k,j} = \frac{\binom{2k}{j}\binom{4-2k}{2-j}}{\binom{4}{2}}.$$

(Source: F. Hoppensteadt and C. Peskin, Mathematics in Medicine and the Life Sciences.)

(a) Compute $P_{k,j}$ for $0 \le k, j \le 2$. Assume that $\binom{0}{0} = 1$ and $\binom{k}{j} = 0$ whenever $k < j$.

Record your results in the matrix

$$P = \begin{bmatrix} P_{00} & P_{01} & P_{02} \\ P_{10} & P_{11} & P_{12} \\ P_{20} & P_{21} & P_{22} \end{bmatrix}.$$

(b) Which elements in P are the greatest? Interpret the result.

2. (Continuation of the previous exercise.) The genetic makeup of future generations of the haploid bacterium can be modeled using matrices. Let $A = [a_1, a_2, a_3]$ be a 1×3 matrix containing three probabilities. The value of a_1 is the probability that a cell has two R_1 plasmids and no R_2 plas-

mid; a_2 is the probability that it has one R_1 plasmid and one R_2 plasmid; a_3 is the probability that a cell has no R_1 plasmid and two R_2 plasmids. If an entire generation of bacterium has one plasmid of each type then $A_1 = [0, 1, 0]$. In this case the bacterium is resistant to both antibiotics. The probabilities A_n for plasmids R_1 and R_2 in nth generation of bacterium can be calculated with the matrix recurrence equation $A_n = A_{n-1}P$, where $n > 1$ and P is the 3×3 matrix determined in the previous exercise. The resulting phenomenon was not well understood until quite recently. It is now used in the genetic engineering of plasmids.

(a) If an entire strain of bacterium is resistant to both the antibiotics ampicillin and tetracycline, make a conjecture as to the drug resistance of future generations of this bacterium.

(b) Test your conjecture by repeatedly computing the matrix product $A_n = A_{n-1}P$ with your graphing calculator. Let $A_1 = [0, 1, 0]$ and $n = 2, 3, \ldots, 12$. Interpret the result. (It may surprise you.)

3. The sum of the first n terms of an arithmetic series is give by $S_n = n\left(\dfrac{a_1 + a_n}{2}\right)$. Justify this formula using a geometric discussion. (*Hint:* Start by graphing an arithmetic sequence where n is an odd number.)

R

Reference:
Basic Concepts from
Algebra and Geometry

Throughout the text there are algebra and geometry review notes that direct students to "see Chapter R." This reference chapter contains six sections, which provide a review of important topics from algebra and geometry. Students can refer to these sections for more explanation or extra practice. Instructors can use these sections to emphasize a variety of mathematical skills.

R.1 FORMULAS FROM GEOMETRY

> Geometric Shapes in a Plane • The Pythagorean Theorem • Three-Dimensional Objects • Similar Triangles • A Summary of Geometric Formulas

Geometric Shapes in a Plane

In this subsection we discuss formulas related to rectangles, triangles, and circles.

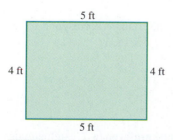

FIGURE R.1 Rectangle

Rectangles. The distance around the boundary of a geometric shape in a plane is called its **perimeter**. The perimeter of a rectangle equals the sum of the lengths of its four sides. For example, the perimeter of the rectangle shown in Figure R.1 is $5 + 4 + 5 + 4 = 18$ feet. In general, the perimeter P of a rectangle with length L and width W is given by $P = 2L + 2W$.

The area A of a rectangle equals the product of its length and width: $A = LW$, so the rectangle in Figure R.1 has an area of $4 \cdot 5 = 20$ square feet.

Many times the perimeter or area of a rectangle is written in terms of variables. This is demonstrated in the next example.

EXAMPLE 1   *Finding the perimeter and area of a rectangle*

The length of a rectangle is three times greater than its width. If the width is x inches, write expressions that give the perimeter and area.

SOLUTION The width of the rectangle is x inches, so its length is $3x$ inches. A sketch is shown in Figure R.2. The perimeter is

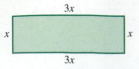

$$P = 2W + 2L$$
$$= 2(x) + 2(3x)$$
$$= 8x \text{ inches.}$$

FIGURE R.2

The area is

$$A = LW$$
$$= 3x \cdot x$$
$$= 3x^2 \text{ square inches.}$$

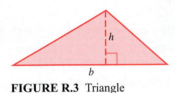

FIGURE R.3 Triangle

Triangles. If the base of a triangle is b and its height is h, as illustrated in Figure R.3, then the area A of the triangle is given by

$$A = \frac{1}{2}bh.$$

EXAMPLE 2 *Finding areas of triangles*

Calculate the area of each triangle.

(a)

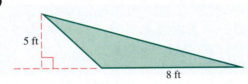

(b)

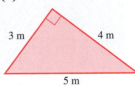

SOLUTION **(a)** The triangle has a base of 8 feet and a height of 5 feet. Therefore its area is

$$A = \frac{1}{2} \cdot 8 \cdot 5 = 20 \text{ square feet.}$$

(b) The triangle shown is a right triangle. The legs of the triangle correspond to its height and base. If we let $b = 4$ meters and $h = 3$ meters, then

$$A = \frac{1}{2}(3)(4) = 6 \text{ square meters.}$$

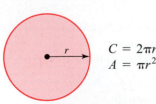

$C = 2\pi r$
$A = \pi r^2$

Circles. The perimeter of a circle is called its **circumference** C and equals $C = 2\pi r$, where r is the radius of the circle. The area A of a circle is given by $A = \pi r^2$. See Figure R.4.

FIGURE R.4 Circle

EXAMPLE 3 *Finding the circumference and area of a circle*

A circle has a radius of 12.5 inches. Approximate its circumference and area.

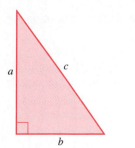

FIGURE R.5 $a^2 + b^2 = c^2$

SOLUTION   *Circumference:* $C = 2\pi r = 2\pi(12.5) = 25\pi \approx 78.5$ inches.

Area: $A = \pi r^2 = \pi(12.5)^2 = 156.25\pi \approx 491$ square inches.

The Pythagorean Theorem

One of the most famous theorems in mathematics is the Pythagorean theorem. It states that a triangle with legs a and b and hypotenuse c is a right triangle if and only if

$$a^2 + b^2 = c^2,$$

as illustrated in Figure R.5.

EXAMPLE 4   *Applying the Pythagorean theorem*

A rectangle has sides of 5 feet and 12 feet, as shown in Figure R.6. Find the diagonal c of the rectangle.

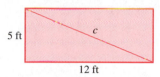

5 ft

c

12 ft

FIGURE R.6

SOLUTION   The diagonal of the rectangle corresponds to the hypotenuse of a right triangle with legs of 5 feet and 12 feet. Let $a = 5$ and $b = 12$.

$$c^2 = a^2 + b^2 \qquad \text{Pythagorean theorem}$$
$$c^2 = 5^2 + 12^2 \qquad \text{Substitute } a = 5 \text{ and } b = 12.$$
$$c^2 = 169 \qquad \text{Simplify.}$$
$$c = 13 \qquad \text{Solve for } c > 0.$$

The diagonal of the rectangle is 13 feet.

EXAMPLE 5   *Finding the perimeter of a right triangle*

Find the perimeter of the triangle shown in Figure R7.

SOLUTION   Given one leg and the hypotenuse of a right triangle, we can use the Pythagorean theorem to find the other leg. Let $a = 7$ and $c = 25$.

$$a^2 + b^2 = c^2 \qquad \text{Pythagorean theorem}$$
$$b^2 = c^2 - a^2 \qquad \text{Subtract } a^2.$$
$$b^2 = 25^2 - 7^2 \qquad \text{Let } a = 7 \text{ and } c = 25.$$
$$b^2 = 576 \qquad \text{Simplify.}$$
$$b = 24 \qquad \text{Solve for } b > 0.$$

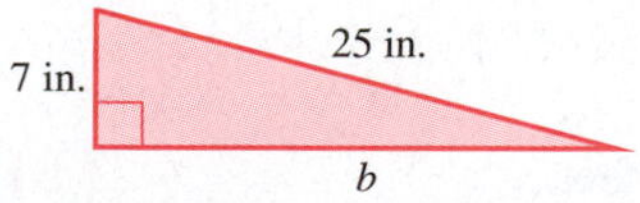

7 in.   25 in. b

FIGURE R.7

The perimeter of the triangle is

$$a + b + c = 7 + 24 + 25 = 56 \text{ inches.}$$

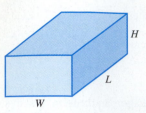

FIGURE R.8 Rectangular Box

Three-Dimensional Objects

Objects that occupy space have both volume and surface area. In this subsection we discuss rectangular boxes, spheres, cylinders, and cones.

Rectangular Boxes. The volume V of a rectangular box with length L, width W, and height H equals $V = LWH$. See Figure R.8. The surface area S of the box equals the sum of the areas of the six sides:

$$S = 2LW + 2WH + 2LH.$$

EXAMPLE 6 *Finding the volume and surface area of a box*

The box shown in Figure R.9 has dimensions x by $2x$ by y. Find its volume and surface area.

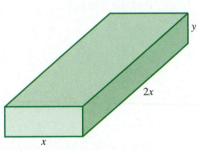

FIGURE R.9

SOLUTION **Volume:** $LWH = 2x \cdot x \cdot y = 2x^2y$ cubic units

Surface Area: Base and top: $2x \cdot x + 2x \cdot x = 4x^2$
Front and back: $xy + xy = 2xy$
Left and right sides: $2xy + 2xy = 4xy$
Total surface area: $4x^2 + 2xy + 4xy = 4x^2 + 6xy$ square units

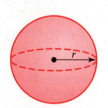

FIGURE R.10 Sphere

Spheres. The volume V of a sphere with radius r is $V = \frac{4}{3}\pi r^3$, and its surface area S is $S = 4\pi r^2$. See Figure R.10.

EXAMPLE 7 *Finding the volume and surface area of a sphere*

Estimate, to the nearest tenth, the volume and surface area of a sphere with a radius of 5.1 feet.

SOLUTION **Volume:** $V = \frac{4}{3}\pi r^3 = \frac{4}{3}\pi(5.1)^3 \approx 555.6$ cubic feet

Surface Area: $S = 4\pi r^2 = 4\pi(5.1)^2 \approx 326.9$ square feet.

Cylinders. A soup can is usually made in the shape of a cylinder. The volume of a cylinder with radius r and height h is $V = \pi r^2h$. See Figure R.11. To find the total surface area of a cylinder, we add the area of the top and bottom to the area of the side. Figure R.12 illustrates a can cut open to determine its surface area. The top and bottom are circular with areas of πr^2 each, and the side has a surface area of $2\pi rh$. The total surface area is $S = 2\pi r^2 + 2\pi rh$.

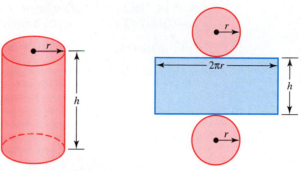

FIGURE R.11 Cylinder   **FIGURE R.12** A Cylinder Cut Open

EXAMPLE 8   *Finding the volume and surface area of a cylinder*

A cylinder has radius $r = 3$ inches and a height $h = 2.5$ feet. Find its volume and total surface area.

SOLUTION   Begin by changing 2.5 feet to 30 inches so that all units are in inches.

Volume: $V = \pi r^2 h = \pi(3)^2(30) = 270\pi \approx 848.2$ cubic inches

Total Surface Area: $S = 2\pi r^2 + 2\pi rh = 2\pi(3)^2 + 2\pi(3)(30) = 198\pi \approx 622.0$ square inches

Cones.   The volume of a cone with radius r and height h is $V = \dfrac{1}{3}\pi r^2 h$, as shown in Figure R.13. (Compare this formula with the formula for the volume of a cylinder.) Excluding the bottom of the cone, the side (or lateral) surface area is $S = \pi r\sqrt{r^2 + h^2}$. The bottom of the cone is circular and has a surface area of πr^2.

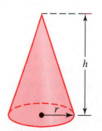

FIGURE R.13 Cone

EXAMPLE 9   *Finding the volume and surface area of a cone*

Approximate the volume and surface area (side only) of a cone with a radius of 1.45 inches and a height of 5.12 inches.

SOLUTION   *Volume:* $V = \dfrac{1}{3}\pi r^2 h = \dfrac{1}{3}\pi(1.45)^2(5.12) \approx 11.3$ cubic inches

Surface Area (side only): $S = \pi r\sqrt{r^2 + h^2} = \pi(1.45)\sqrt{(1.45)^2 + (5.12)^2} \approx$ 24.2 square inches

Similar Triangles

The corresponding angles of **similar triangles** have equal measure, but similar triangles are not necessarily the same size. An example of two similar triangles is shown in Figure R.14. Notice that both triangles have angles of 30°, 60°, and 90°. Corresponding sides are not equal in length; however, corresponding sides are proportional. For example, in triangle ABC the ratio of the shortest leg to the hypotenuse equals $\frac{2}{4} = \frac{1}{2}$, and in triangle DEF this ratio is $\frac{3}{6} = \frac{1}{2}$.

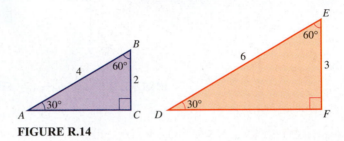

FIGURE R.14

EXAMPLE 10   *Using similar triangles*

Find the length of BC in Figure R.15.

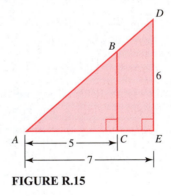

FIGURE R.15

SOLUTION Notice that triangle ABC and triangle ADE are both right triangles. These triangles share an angle at vertex A. Therefore triangles ABC and ADE have two corresponding angles that are congruent. Because the sum of the angles in a triangle equals 180°, all three corresponding angles in these two triangles are congruent. Thus triangles ABC and ADE are similar.

Since corresponding sides are proportional, we can find BC as follows.

$$\frac{BC}{AC} = \frac{DE}{AE}$$

$$\frac{BC}{5} = \frac{6}{7}$$

Solving this equation for BC gives $BC = \frac{30}{7} \approx 4.3$.

A Summary of Geometric Formulas

The following table provides a summary of some important formulas from geometry.

Geometric Shape	Related Formulas
Rectangle	Let W be the width and L be the length. **Perimeter:** $P = 2W + 2L$ **Area:** $A = LW$
Triangle	Let s_1, s_2, and s_3 be the sides of a triangle. **Perimeter:** $P = s_1 + s_2 + s_3$ Let b be the base and h be the height. **Area:** $A = \dfrac{1}{2}bh$
Circle	Let r be the radius. **Circumference:** $C = 2\pi r$ **Area:** $A = \pi r^2$
Rectangular box	Let L, W, and H be the length, width, and height. **Area:** $A = 2LW + 2WH + 2LH$ **Volume:** $V = LWH$
Sphere	Let r be the radius. **Surface Area:** $S = 4\pi r^2$ **Volume:** $V = \dfrac{4}{3}\pi r^3$
Cylinder	Let h be the height and r be the radius. **Surface Area (side only):** $S = 2\pi rh$ **Total Surface Area:** $S = 2\pi rh + 2\pi r^2$ **Volume:** $V = \pi r^2 h$
Cone	Let h be the height and r be the radius. **Surface Area (side only):** $S = \pi r \sqrt{r^2 + h^2}$ **Total Surface Area:** $S = \pi r \sqrt{r^2 + h^2} + \pi r^2$ **Volume:** $V = \dfrac{1}{3}\pi r^2 h$
Units	**Perimeter (length):** meters, inches, feet **Surface Area (square units):** square yards, square miles **Volume (cubic units):** cubic inches, cubic centimeters

R.1 EXERCISES

Rectangles

Exercises 1–6: Find the area and perimeter of the rectangle with length L and width W.

1. $L = 15$ feet, $W = 7$ feet

2. $L = 16$ inches, $W = 10$ inches

3. $L = 100$ meters, $W = 35$ meters

4. $L = 80$ yards, $W = 13$ yards

5. $L = 3x, W = y$

6. $L = a + 5, W = a$

Exercises 7–10: Find the area and perimeter of the rectangle in terms of the width W.

7. The width W is half the length.

8. Triple the width W minus 3 equals the length.

9. The length equals the width W plus 5.

10. The length is 2 less than twice the width W.

Triangles

Exercises 11–12: Find the area of the triangle shown in the figure.

11.
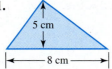

12.
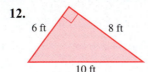

Exercises 13–20: Find the area of the triangle with base b and height h.

13. $b = 5$ inches, $h = 8$ inches

14. $b = 24$ inches, $h = 9$ feet

15. $b = 10.1$ meters, $h = 730$ meters

16. $b = 52$ yards, $h = 102$ feet

17. $b = 2x, h = 6x$

18. $b = x, h = x + 4$

19. $b = z, h = 5z$

20. $b = y + 1, h = 2y$

Circles

Exercises 21–26: Find the circumference and area of the circle. Approximate each value to the nearest tenth when appropriate.

21. $r = 4$ meters

22. $r = 1.5$ feet

23. $r = 19$ inches

24. $r = 22$ miles

25. $r = 2x$

26. $r = 5z$

Pythagorean Theorem

Exercises 27–34: Use the Pythagorean theorem to find the missing side of the right triangle with legs a and b, and hypotenuse c. Then calculate the perimeter. Approximate values to the nearest tenth when appropriate.

27. $a = 60$ feet, $b = 11$ feet

28. $a = 16$ inches, $b = 9$ inches

29. $a = 21$ feet, $b = 11$ yards

30. $a = 5$ centimeters, $c = 13$ centimeters

31. $a = 6$ meters, $c = 15$ meters

32. $b = 7$ millimeters, $c = 10$ millimeters

33. $b = 1.2$ miles, $c = 2$ miles

34. $a = 0.5$ kilometers, $b = 1.2$ kilometers

Exercises 35–38: Find the area of the right triangle that satisfies the conditions. Approximate values to the nearest tenth when appropriate.

35. Legs with lengths 3 feet and 6 feet

36. Hypotenuse 10 inches and leg 6 inches

37. Hypotenuse 15 inches and leg 11 inches

38. Shorter leg 40 centimeters and hypotenuse twice the shorter leg

Rectangular Boxes

Exercises 39–46: Find the volume and surface area of a rectangular box with length L, width W, and height H.

39. $L = 4$ feet, $W = 3$ feet, $H = 2$ feet

40. $L = 6$ meters, $W = 4$ meters, $H = 1.5$ meters

41. $L = 4.5$ inches, $W = 4$ inches, $H = 1$ foot

42. $L = 9.1$ yards, $W = 8$ yards, $H = 6$ feet

43. $L = 3x, W = 2x, H = x$

44. $L = 6z, W = 5z, H = 7z$

45. $L = x, W = 2y, H = 3z$

46. $L = 8x, W = y, H = z$

Exercises 47–48: Find the volume of the rectangular box in terms of the width W.

47. The length is twice the width W, and the height is half the width.

48. The width W is three times the height and one-third of the length.

Spheres

Exercises 49–54: Find the volume and surface area of the sphere satisfying the given condition, where r is the radius and d is the diameter. Approximate values to the nearest tenth.

49. $r = 3$ feet

50. $r = 4.1$ inches

51. $r = 5$ centimeters

52. $d = 6.4$ meters

53. $d = 10$ millimeters

54. $d = 16$ feet

Cylinders

Exercises 55–60: Find the volume, the surface area of the side, and the total surface area of the cylinder that satisfies the given conditions, where r is the radius and h is the height. Approximate values to the nearest tenth.

55. $r = 0.5$ feet, $h = 2$ feet

56. $r = 4.1$ centimeters, $h = 30.5$ centimeters

57. $r = 5$ inches, $h = 1.5$ feet

58. $r = \dfrac{1}{3}$ yard, $h = 2$ feet

59. $r = 12$ millimeters, and h is twice r

60. r is one-fourth of h, and $h = 2.1$ feet

Exercises 61–62: Use r to write an expression that calculates the volume of the cylinder.

61. The radius r is half the height.

62. The height is 2 units longer than the radius r.

Cones

Exercises 63–68: Approximate, to the nearest tenth, the volume and surface area (side only) of the cone satisfying the given conditions, where r is the radius and h is the height.

63. $r = 5$ centimeters, $h = 6$ centimeters

64. $r = 8$ inches, $h = 30$ inches

65. $r = 24$ inches, $h = 3$ feet

66. $r = 100$ centimeters, $h = 1.3$ meters

67. Three times r equals h, and $r = 2.4$ feet

68. Twice h equals r, and $h = 3$ centimeters

Similar Triangles

Exercises 69–74: Use the fact that triangles ABC and DEF are similar to find the value of x.

69.

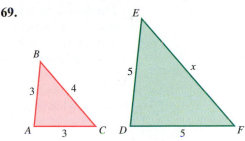

70.

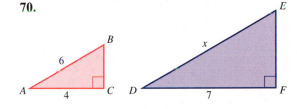

71.

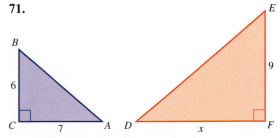

72.

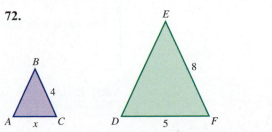

73.

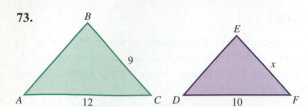

74.

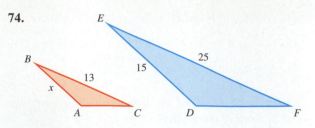

$R.2$ CIRCLES

Equations and Graphs of Circles • Finding the Center and Radius of a Circle

Equations and Graphs of Circles

A **circle** consists of the set of points in a plane that are equidistant from a fixed point. The distance is called the **radius** of the circle, and the fixed point is called the **center**. If we let the center of the circle be (h, k), the radius be r, and (x, y) be any point on the circle, then the distance between (x, y) and (h, k) must equal r. See Figure R.16. By the distance formula we have

$$\sqrt{(x - h)^2 + (y - k)^2} = r.$$

Squaring both sides gives

$$(x - h)^2 + (y - k)^2 = r^2.$$

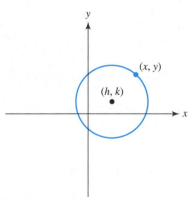

FIGURE R.16

Standard equation of a circle

The circle with center (h, k) and radius r has equation

$$(x - h)^2 + (y - k)^2 = r^2.$$

Note: If the center of a circle is the origin $(0, 0)$, then the equation reduces to $x^2 + y^2 = r^2$.

EXAMPLE 1   *Finding the center and radius of a circle*

Find the center and radius of the circle with the given equation. Graph each circle.
(a) $x^2 + y^2 = 9$ **(b)** $(x - 1)^2 + (y + 2)^2 = 4$

SOLUTION   **(a)** Because the equation can be written as $(x - 0)^2 + (y - 0)^2 = 3^2$, the center is $(0, 0)$ and its radius is 3. The graph of this circle is shown in Figure R.17.
(b) The center is $(1, -2)$ and its radius is 2. Its graph is shown in Figure R.18.

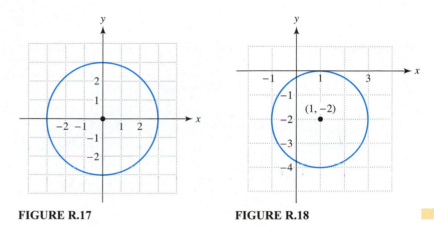

FIGURE R.17   **FIGURE R.18**

EXAMPLE 2   *Finding the equation of a circle*

Find the equation of the circle that satisfies the conditions. Graph each circle.
(a) Radius 4, center $(-3, 5)$
(b) Center $(6, -3)$ with the point $(1, 2)$ on the circle.

SOLUTION   **(a)** Let $r = 4$ and $(h, k) = (-3, 5)$. The equation of this circle is

$$(x - (-3))^2 + (y - 5)^2 = 4^2 \quad \text{or} \quad (x + 3)^2 + (y - 5)^2 = 16$$

A graph of the circle is shown in Figure R.19.

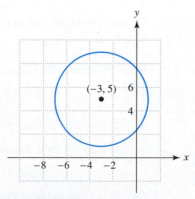

FIGURE R.19

(b) First we must find the distance between the points $(6, -3)$ and $(1, 2)$ to determine r.

$$r = \sqrt{(6 - 1)^2 + (-3 - 2)^2} = \sqrt{50} \approx 7.1$$

Since $r^2 = 50$, the equation of the circle is

$$(x - 6)^2 + (y + 3)^2 = 50.$$

Its graph is shown in Figure R.20.

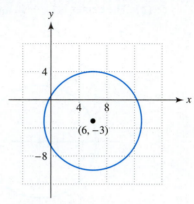

FIGURE R.20

Finding the Center and Radius of a Circle

When a circle's equation is not written in standard form, we can complete the square to find its center and radius. To complete the square, we use the fact that

$$x^2 + kx + \left(\frac{k}{2}\right)^2 = \left(x + \frac{k}{2}\right)^2.$$

For example, to complete the square on the expression $x^2 + 8x$, add $\left(\frac{8}{2}\right)^2 = 16$. Then,

$$x^2 + 8x + 16 = (x + 4)^2.$$

EXAMPLE 3   *Finding the center and radius of a circle*

Find the center and radius of a circle whose equation is $x^2 + 6x + y^2 - 4y - 12 = 0$. Graph the circle.

SOLUTION   Begin by writing the equation as

$$(x^2 + 6x + \underline{}) + (y^2 - 4y + \underline{}) = 12.$$

To complete the square, add $\left(\frac{6}{2}\right)^2 = 9$ to the first quantity and $\left(\frac{-4}{2}\right)^2 = 4$ to the second quantity. Add the same values to the right side of the equation.

$$(x^2 + 6x + \underline{\;9\;}) + (y^2 - 4y + \underline{\;4\;}) = 12 + 9 + 4$$

Factoring gives the following result.

$$(x + 3)^2 + (y - 2)^2 = 5^2$$

The center is $(-3, 2)$ and the radius is $r = 5$. Its graph is shown in Figure R.21.

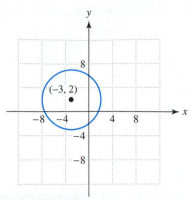

FIGURE R.21

$R.2$ EXERCISES

Equations and Graphs of Circles

Exercises 1–8: Find the center and radius of the circle.

1. $x^2 + y^2 = 25$

2. $x^2 + y^2 = 100$

3. $x^2 + y^2 = 7$

4. $x^2 + y^2 = 20$

5. $(x - 2)^2 + (y + 3)^2 = 9$

6. $(x + 1)^2 + (y - 1)^2 = 16$

7. $x^2 + (y + 1)^2 = 100$

8. $(x - 5)^2 + y^2 = 19$

Exercises 9–12: Use the graph to find the standard equation of a circle.

9.

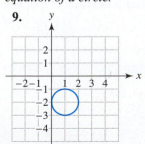

10.

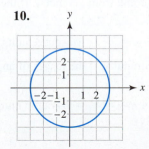

11.

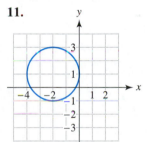

12.

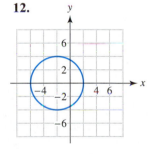

Exercises 13–20: Find the standard equation of a circle that satisfies the conditions.

13. Radius 8, center $(3, -5)$

14. Radius 5, center $(-1, 4)$

15. Radius 7, center $(3, 0)$

16. Radius 1, center $(0, 0)$

17. Center $(0, 0)$ with the point $(-3, -1)$ on the circle

18. Center $(3, -5)$ with the point $(4, 2)$ on the circle

19. Endpoints of a diameter $(-5, -7)$ and $(1, 1)$.

20. Endpoints of a diameter $(-3, -2)$ and $(1, -4)$.

Finding the Center and Radius of a Circle

Exercises 21–28: Find the center and radius of the circle. Graph the circle.

21. $x^2 + y^2 = 9$

22. $x^2 + y^2 = 1$

23. $x^2 + 2x + y^2 - 2y = 2$

24. $x^2 - 6x + y^2 - 8y = 0$

25. $x^2 + y^2 - 2y = 0$

26. $x^2 - 20x + y^2 + 64 = 0$

27. $x^2 + 18x + y^2 + 10y = -97$

28. $x^2 + 16x + y^2 - 18y + 141 = 0$

$R.3$ INTEGER EXPONENTS

Bases and Positive Exponents • Zero and Negative Exponents • Product, Quotient, and Power Rules

Bases and Positive Exponents

The area of a square that is 8 inches on a side is given by the expression

$$8 \cdot 8 = 8^2 = 64 \text{ square inches.}$$

The expression 8^2 is an exponential expression with base 8 and exponent 2. Exponential expressions occur frequently in a variety of applications. For example, suppose that an investment doubles its initial value 3 times. Then its final value is

$$2 \cdot 2 \cdot 2 = 2^3 = 8$$

times larger than its original value. Table R.1 contains examples of exponential expressions.

TABLE R.1

Expression	Base	Exponent
$2 \cdot 2 \cdot 2 = 2^3$	2	3
$6 \cdot 6 \cdot 6 \cdot 6 = 6^4$	6	4
$7 = 7^1$	7	1
$0.5 \cdot 0.5 = 0.5^2$	0.5	2
$x \cdot x \cdot x = x^3$	x	3

4

4

FIGURE R.22 4 Squared

We read 0.5^2 as "0.5 squared," 2^3 as "2 cubed," and 6^4 as "6 to the fourth power." The terms *squared* and *cubed* come from geometry. If the length of a side of a square is 4, then its area is

$$4 \cdot 4 = 4^2 = 16 \text{ square units,}$$

as illustrated in Figure R.22. Similarly, if the length of an edge of a cube is 4, then its volume is

$$4 \cdot 4 \cdot 4 = 4^3 = 64 \text{ cubic units,}$$

as shown in Figure R.23.

4

4

4

FIGURE R.23 4 Cubed

EXAMPLE 1   *Writing numbers in exponential notation*

Use the given base to write each number as an exponential expression. Check your results with a calculator.

(a) 10,000   (base 10)
(b) 27   (base 3)
(c) 32   (base 2)

SOLUTION   **(a)** $10{,}000 = 10 \cdot 10 \cdot 10 \cdot 10 = 10^4$ **(b)** $27 = 3 \cdot 3 \cdot 3 = 3^3$
(c) $32 = 2 \cdot 2 \cdot 2 \cdot 2 \cdot 2 = 2^5$

These values are supported in Figure R.24, where exponential expressions are evaluated with a graphing calculator, using the $\boxed{\wedge}$ key.

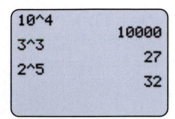

FIGURE R.24

Zero and Negative Exponents

Exponents can be defined for any integer. If a is any nonzero real number, we define

$$a^0 = 1.$$

For example, $3^0 = 1$ and $\left(\dfrac{1}{7}\right)^0 = 1$. We can also define negative integer exponents as follows.

$$5^{-4} = \frac{1}{5^4}, \qquad y^{-2} = \frac{1}{y^2}, \qquad \left(\frac{1}{2}\right)^{-5} = \left(\frac{1}{1/2}\right)^5 = 2^5$$

Powers of 10 are important because they are used frequently in mathematics to express numbers that are either small or large in absolute value. Table R.2 may be used to simplify powers of 10. Note that if the power decreases by 1, the result is decreased by $\dfrac{1}{10}$.

TABLE R.2   Powers of Ten

Power of 10	Value
10^3	1000
10^2	100
10^1	10
10^0	1
10^{-1}	$\dfrac{1}{10} = 0.1$
10^{-2}	$\dfrac{1}{100} = 0.01$
10^{-3}	$\dfrac{1}{1000} = 0.001$

The following summarizes integer exponents with a nonzero base.

> ### Integer exponents
>
> Let a be a nonzero real number and n be a positive integer. Then
>
> $$a^n = a \cdot a \cdot a \cdot \cdots \cdot a \quad (n \text{ factors of } a),$$
>
> $$a^0 = 1, \quad \text{and}$$
>
> $$a^{-n} = \frac{1}{a^n}.$$

EXAMPLE 2 *Evaluating numbers in exponential notation*

Evaluate each expression by hand and then check your result with a calculator when possible.

(a) 3^{-4} **(b)** $\left(\dfrac{5}{7}\right)^{-2}$ **(c)** $\dfrac{1}{2^{-3}}$ **(d)** $(xy)^{-1}$

SOLUTION

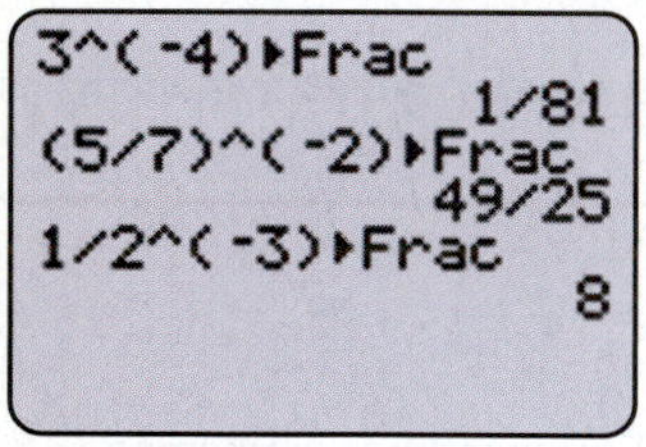

```
3^(-4)▶Frac
            1/81
(5/7)^(-2)▶Frac
           49/25
1/2^(-3)▶Frac
               8
```

FIGURE R.25

(a) $3^{-4} = \dfrac{1}{3^4} = \dfrac{1}{3 \cdot 3 \cdot 3 \cdot 3} = \dfrac{1}{81}$

(b) $\left(\dfrac{5}{7}\right)^{-2} = \dfrac{1}{(5/7)^2} = \dfrac{1}{25/49} = \dfrac{49}{25}$

(c) $\dfrac{1}{2^{-3}} = \dfrac{1}{1/2^3} = 2^3 = 8$

(d) $(xy)^{-1} = \dfrac{1}{(xy)^1} = \dfrac{1}{xy}$

The results for parts (a) – (c) are supported by Figure R.25.

Product, Quotient, and Power Rules

We can calculate products and quotients of exponential expressions *provided their bases are the same.* For example,

$$3^2 \cdot 3^3 = (3 \cdot 3) \cdot (3 \cdot 3 \cdot 3) = 3^5.$$

This expression has a total of $2 + 3 = 5$ factors of 3, so the result is 3^5. To multiply exponential expressions with like bases, add exponents.

> ### The product rule
>
> For any nonzero number a and integers m and n,
>
> $$a^m \cdot a^n = a^{m+n}.$$

The product rule holds for negative exponents. For example,

$$10^5 \cdot 10^{-2} = 10^{5+(-2)} = 10^3.$$

EXAMPLE 3 *Using the product rule*

Multiply and simplify.

(a) $10^2 \cdot 10^4$ **(b)** $7^3 \cdot 7^{-4}$ **(c)** $x^3 \cdot x^{-2} \cdot x^4$ **(d)** $3y^2 \cdot 2y^{-4}$

SOLUTION **(a)** $10^2 \cdot 10^4 = 10^{2+4} = 10^6 = 1,000,000$ **(b)** $7^3 \cdot 7^{-4} = 7^{3+(-4)} = 7^{-1} = \dfrac{1}{7}$

(c) $x^3 \cdot x^{-2} \cdot x^4 = x^{3+(-2)+4} = x^5$

(d) $3y^2 \cdot 2y^{-4} = 3 \cdot 2 \cdot y^2 \cdot y^{-4} = 6y^{2+(-4)} = 6y^{-2} = \dfrac{6}{y^2}$

Note that 6 is not raised to the power of -2 in the expression $6y^{-2}$.

Consider division of exponential expressions using the following example.

$$\frac{6^5}{6^3} = \frac{6 \cdot 6 \cdot \cancel{6}^1 \cdot \cancel{6}^1 \cdot \cancel{6}^1}{\cancel{6} \cdot \cancel{6} \cdot \cancel{6}} = 6 \cdot 6 = 6^2$$

After canceling three 6s, there are two 6s left in the numerator. The result is $6^{5-3} = 6^2 = 36$. To divide exponential expressions with like bases, subtract exponents.

The quotient rule

For any nonzero number a and integers m and n,

$$\frac{a^m}{a^n} = a^{m-n}.$$

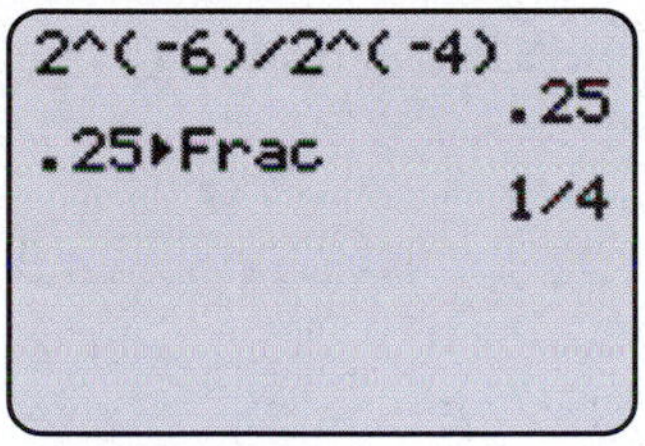

```
2^(-6)/2^(-4)
                  .25
.25▸Frac
                  1/4
```

FIGURE R.26

The quotient rule holds true for negative exponents. For example,

$$\frac{2^{-6}}{2^{-4}} = 2^{-6-(-4)} = 2^{-2} = \frac{1}{2^2} = \frac{1}{4}.$$

This result is supported by Figure R.26.

EXAMPLE 4   *Using the quotient rule*

Simplify the expression. Use positive exponents.

(a) $\dfrac{10^4}{10^6}$ **(b)** $\dfrac{x^5}{x^2}$ **(c)** $\dfrac{15x^2y^3}{5x^4y}$

SOLUTION **(a)** $\dfrac{10^4}{10^6} = 10^{4-6} = 10^{-2} = \dfrac{1}{10^2}$ **(b)** $\dfrac{x^5}{x^2} = x^{5-2} = x^3$

(c) $\dfrac{15x^2y^3}{5x^4y} = \dfrac{15}{5} \cdot \dfrac{x^2}{x^4} \cdot \dfrac{y^3}{y^1} = 3 \cdot x^{(2-4)}y^{(3-1)} = 3x^{-2}y^2 = \dfrac{3y^2}{x^2}$

How should we evaluate $(4^3)^2$? To answer this question consider

$$(4^3)^2 = 4^3 \cdot 4^3 = 4^{3+3} = 4^6.$$

Similarly,

$$(x^4)^3 = x^4 \cdot x^4 \cdot x^4 = x^{4+4+4} = x^{12}.$$

These results suggest that to raise a power to a power, we must multiply the exponents.

> ## Raising powers to powers
>
> For any real number a and integers m and n,
>
> $$(a^m)^n = a^{mn}.$$

EXAMPLE 5 *Raising powers to powers*

Simplify each expression. Use positive exponents.

(a) $(5^2)^3$ (b) $(2^4)^{-2}$ (c) $(b^{-7})^5$

SOLUTION   (a) $(5^2)^3 = 5^{2 \cdot 3} = 5^6$ (b) $(2^4)^{-2} = 2^{4 \cdot (-2)} = 2^{-8} = \dfrac{1}{2^8}$

(c) $(b^{-7})^5 = b^{-7 \cdot 5} = b^{-35} = \dfrac{1}{b^{35}}$

How can we simplify the expression $(2x)^3$? Consider the following.

$$(2x)^3 = 2x \cdot 2x \cdot 2x = (2 \cdot 2 \cdot 2) \cdot (x \cdot x \cdot x) = 2^3 x^3$$

This result suggests that to cube a product, we can cube each factor.

> ## Raising products to powers
>
> For any real numbers a and b and integer n,
>
> $$(ab)^n = a^n b^n.$$

EXAMPLE 6 *Raising products to powers*

Simplify each expression. Use positive exponents.

(a) $(6y)^2$ (b) $(x^2 y)^{-2}$ (c) $(2xy^3)^4$

SOLUTION   (a) $(6y)^2 = 6^2 y^2 = 36y^2$ (b) $(x^2 y)^{-2} = (x^2)^{-2} y^{-2} = x^{-4} y^{-2} = \dfrac{1}{x^4 y^2}$

(c) $(2xy^3)^4 = 2^4 x^4 (y^3)^4 = 16 x^4 y^{12}$

To simplify a power of a quotient use the following rule.

> ## Raising quotients to powers
>
> For nonzero numbers a and b and any integer n,
>
> $$\left(\frac{a}{b}\right)^n = \frac{a^n}{b^n}.$$

EXAMPLE 7 *Raising quotients to powers*

Simplify each expression. Use positive exponents.

(a) $\left(\dfrac{3}{x}\right)^3$ (b) $\left(\dfrac{1}{2^3}\right)^{-2}$ (c) $\left(\dfrac{3x^{-3}}{y^2}\right)^4$

SOLUTION

(a) $\left(\dfrac{3}{x}\right)^3 = \dfrac{3^3}{x^3} = \dfrac{27}{x^3}$ (b) $\left(\dfrac{1}{2^3}\right)^{-2} = \dfrac{1^{-2}}{(2^3)^{-2}} = \dfrac{1}{2^{-6}} = 2^6 = 64$

(c) $\left(\dfrac{3x^{-3}}{y^2}\right)^4 = \dfrac{3^4(x^{-3})^4}{(y^2)^4} = \dfrac{81x^{-12}}{y^8} = \dfrac{81}{x^{12}y^8}$

$R.3$ EXERCISES

Concepts

1. Identify the base and the exponent in the expression 8^3.

2. Evaluate 97^0 and 2^{-1}.

3. Write 7 cubed, using symbols.

4. Write 5 squared, using symbols.

5. Are the expressions 2^3 and 3^2 equal? Explain your answer.

6. Are the expressions -4^2 and $(-4)^2$ equal? Explain your answer.

7. $7^{-n} = $ ____.

8. $6^m \cdot 6^n = $ ____.

9. $\dfrac{5^m}{5^n} = $ ____.

10. $(3x)^k = $ ____.

11. $(2^m)^k = $ ____.

12. $\left(\dfrac{x}{y}\right)^m = $ ____.

13. $5 \times 10^3 = $ ____.

14. $5 \times 10^{-3} = $ ____.

Properties of Exponents

Exercises 15–20: (Refer to Example 1.) Write the number as an exponential expression, using the base shown. Check your result with a calculator.

15. 8 (base 2)

16. 1000 (base 10)

17. 256 (base 4)

18. $\dfrac{1}{64}$ (base 4)

19. 1 (base 3)

20. $\dfrac{1}{49}$ (base 7)

Exercises 21–30: Evaluate the expression by hand. Check your result with a calculator.

21. 5^3

22. 5^{-3}

23. -2^4

24. $(-2)^4$

25. 5^0

26. $\left(-\dfrac{2}{3}\right)^{-3}$

27. $\left(\dfrac{2}{3}\right)^3$

28. $\dfrac{1}{4^{-2}}$

29. $\left(-\dfrac{1}{2}\right)^4$

30. $\left(-\dfrac{3}{4}\right)^3$

Exercises 31–42: Use the product rule to simplify the expression.

31. $6^3 \cdot 6^{-4}$

32. $10^2 \cdot 10^5 \cdot 10^{-3}$

33. $5^{-5} \cdot 5^2$

34. $z^2 \cdot z^3$

35. $2x^2 \cdot 3x^{-3} \cdot x^4$

36. $3y^4 \cdot 6y^{-4} \cdot y$

37. $10^0 \cdot 10^6 \cdot 10^2$

38. $y^3 \cdot y^{-5} \cdot y^4$

39. $5^{-2} \cdot 5^3 \cdot 2^{-4} \cdot 2^3$

40. $2^{-3} \cdot 3^4 \cdot 3^{-2} \cdot 2^5$

41. $2a^3 \cdot b^2 \cdot a^{-4} \cdot 4b^{-5}$

42. $3x^{-4} \cdot 2x^2 \cdot 5y^4 \cdot y^{-3}$

Exercises 43–54: Use the quotient rule to simplify the expression. Use positive exponents to write your answer.

43. $\dfrac{5^4}{5^2}$

44. $\dfrac{6^2}{6^{-7}}$

45. $\dfrac{10^{-2} \cdot 10^3}{10^{-5}}$

46. $\dfrac{6^{-5} \cdot 6^2}{6}$

47. $\dfrac{a^{-3}}{a^2 \cdot a}$

48. $\dfrac{y^0 \cdot y \cdot y^5}{y^{-2} \cdot y^{-3}}$

49. $\dfrac{24x^3}{6x}$

50. $\dfrac{10x^5}{5x^{-3}}$

51. $\dfrac{12a^2b^3}{18a^4b^2}$

52. $\dfrac{-6x^7y^3}{3x^2y^{-5}}$

53. $\dfrac{21x^{-3}y^4}{7x^4y^{-2}}$

54. $\dfrac{32x^3y}{-24x^5y^{-3}}$

Exercises 55–64: Use the power rules to simplify the expression. Use positive exponents to write your answer.

55. $(5^3)^{-1}$

56. $(-4^2)^3$

57. $(3y^4)^{-2}$

58. $(2x^2)^4$

59. $(4y^2)^3$

60. $(-2xy^3)^{-4}$

61. $\left(\dfrac{4}{x}\right)^3$

62. $\left(\dfrac{-3}{x^3}\right)^2$

63. $\left(\dfrac{2x}{z^4}\right)^{-5}$

64. $\left(\dfrac{2xy}{3z^5}\right)^{-1}$

$R.4$ POLYNOMIAL EXPRESSIONS

Addition and Subtraction of Monomials • Addition and Subtraction of Polynomials • Distributive Properties • Multiplying Polynomials • Some Special Products

Addition and Subtraction of Monomials

A term is a number, a variable, or a *product* of numbers and variables raised to powers. Examples of terms include

$$-15, \qquad y, \qquad x^4, \qquad 3x^3z, \qquad x^{-1/2}y^{-2}, \qquad \text{and} \qquad 6x^{-1}y^3.$$

If the variables in a term have only *nonnegative integer* exponents, the term is called a *monomial*. Examples of monomials include

$$-4, \qquad 5y, \qquad x^2, \qquad 5x^2z^6, \qquad -xy^7, \qquad \text{and} \qquad 6xy^3.$$

To learn how to add monomials, consider the following example. Suppose that we have 3 rectangles of the same dimension having length x and width y, as shown in Figure R.27. The total area is given by

$$xy + xy + xy.$$

This sum is equivalent to 3 times xy, which can be expressed as $3xy$. In symbols we write

$$xy + xy + xy = 3xy.$$

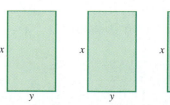

FIGURE R.27 Total Area: $3xy$

We can add these three terms because they are like terms. If two terms contain the same variables raised to the same powers, we call them like terms. We can add or subtract *like* terms, but not *unlike* terms. For example, if one cube has sides of length x, and another cube has sides of length y, their respective volumes are x^3 and y^3. The total volume of the two cubes equals

$$x^3 + y^3,$$

but we cannot combine these terms into one term because they are unlike terms. However,

$$x^3 + 3x^3 = (1 + 3)x^3 = 4x^3$$

because x^3 and $3x^3$ are like terms. To add or subtract monomials we simply combine like terms as illustrated in the next example.

EXAMPLE 1   *Adding and subtracting monomials*

Simplify each of the following expressions by combining like terms.
(a) $8x^2 - 4x^2 + x^3$ **(b)** $9x - 6xy^2 + 2xy^2 + 4x$

SOLUTION   **(a)** The terms $8x^2$ and $-4x^2$ are like terms, so they may be combined.

$$8x^2 - 4x^2 + x^3 = (8 - 4)x^2 + x^3 \qquad \text{Combine like terms.}$$
$$= 4x^2 + x^3 \qquad \text{Subtract.}$$

However, $4x^2$ and x^3 are unlike terms and cannot be combined.
(b) The terms $9x$ and $4x$ may be combined, as can $-6xy^2$ and $2xy^2$.

$$9x - 6xy^2 + 2xy^2 + 4x = 9x + 4x - 6xy^2 + 2xy^2 \qquad \text{Commutative property}$$
$$= (9 + 4)x + (-6 + 2)xy^2 \qquad \text{Combine like terms.}$$
$$= 13x - 4xy^2 \qquad \text{Add.}$$

Addition and Subtraction of Polynomials

A polynomial is either a monomial or a sum of monomials. Examples of polynomials include

$$5x^4z^2, \qquad 9x^4 - 5, \qquad 4x^2 + 5xy - y^2, \qquad \text{and} \qquad 4 - y^2 + 5y^4 + y^5$$

1 term   2 terms   3 terms   4 terms

Polynomials containing one variable are called polynomials of one variable. The second and fourth polynomials shown above are examples of polynomials of one variable. The leading coefficient of a polynomial of one variable is the coefficient of the monomial with highest degree. The degree of a polynomial with one variable equals the exponent of the monomial with the highest power. Table R.3 shows several polynomials of one variable along with their degrees and leading coefficients. A polynomial of degree 1 is a linear polynomial, a polynomial of degree 2 is a quadratic polynomial, and a polynomial of degree 3 is a cubic polynomial.

TABLE R.3

Polynomial	Degree	Leading Coefficient
-98	0	-98
$2x - 7$	1	2
$-5z + 9z^2 + 7$	2	9
$-2x^3 + 4x^2 + x - 1$	3	-2
$7 - x + 4x^2 + x^5$	5	1

To add two polynomials, we combine like terms.

EXAMPLE 2   *Adding polynomials*

Simplify each expression.
(a) $(2x^2 - 3x + 7) + (3x^2 + 4x - 2)$ (b) $(z^3 + 4z + 8) + (4z^2 - z + 6)$

SOLUTION   (a) $(2x^2 - 3x + 7) + (3x^2 + 4x - 2) = 2x^2 + 3x^2 - 3x + 4x + 7 - 2$
$$= (2 + 3)x^2 + (-3 + 4)x + (7 - 2)$$
$$= 5x^2 + x + 5$$

(b) $(z^3 + 4z + 8) + (4z^2 - z + 6) = z^3 + 4z^2 + 4z - z + 8 + 6$
$$= z^3 + 4z^2 + (4 - 1)z + (8 + 6)$$
$$= z^3 + 4z^2 + 3z + 14$$

To subtract integers we add the first integer with the *additive inverse* or *opposite* of the second integer. For example, to evaluate $3 - 5$ we perform the following operations.

$$3 - 5 = 3 + (-5) \qquad \text{Add the opposite.}$$
$$= -2 \qquad \text{Simplify.}$$

Similarly, to subtract two polynomials we add the first polynomial and the opposite of the second polynomial. To find the opposite of a polynomial, we simply negate each term. Table R.4 shows three polynomials and their opposites.

TABLE R.4

Polynomial	Opposite
$9 - x$	$-9 + x$
$5x^2 + 4x - 1$	$-5x^2 - 4x + 1$
$-x^4 + 5x^3 - x^2 + 5x - 1$	$x^4 - 5x^3 + x^2 - 5x + 1$

EXAMPLE 3   *Subtracting polynomials*

Simplify.
(a) $(y^5 + 3y^3) - (-y^4 + 2y^3)$ (b) $(5x^3 + 9x^2 - 6) - (5x^3 - 4x^2 - 7)$

SOLUTION   (a) The opposite of $(-y^4 + 2y^3)$ is $(y^4 - 2y^3)$.
$$(y^5 + 3y^3) - (-y^4 + 2y^3) = (y^5 + 3y^3) + (y^4 - 2y^3)$$
$$= y^5 + y^4 + (3 - 2)y^3$$
$$= y^5 + y^4 + y^3$$

(b) The opposite of $(5x^3 - 4x^2 - 7)$ is $(-5x^3 + 4x^2 + 7)$
$$(5x^3 + 9x^2 - 6) - (5x^3 - 4x^2 - 7) = (5x^3 + 9x^2 - 6) + (-5x^3 + 4x^2 + 7)$$
$$= (5 - 5)x^3 + (9 + 4)x^2 + (-6 + 7)$$
$$= 0x^3 + 13x^2 + 1$$
$$= 13x^2 + 1$$

Distributive Properties

Distributive properties are used frequently in the multiplication of polynomials. For all real numbers a, b, and c

$$a(b + c) = ab + ac \qquad \text{and}$$

$$a(b - c) = ab - ac.$$

In the next example we use these distributive properties to multiply expressions.

EXAMPLE 4   *Using distributive properties*

Multiply.
(a) $4(5 + x)$ **(b)** $-3(x - 4y)$ **(c)** $(2x - 5)(6)$

SOLUTION

(a) $4(5 + x) = 4 \cdot 5 + 4 \cdot x = 20 + 4x$

(b) $-3(x - 4y) = -3 \cdot x - (-3) \cdot (4y) = -3x + 12y$

(c) $(2x - 5)(6) = 2x \cdot 6 - 5 \cdot 6 = 12x - 30$

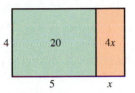

FIGURE R.28   Area: $20 + 4x$

You can visualize the solution in part (a) of the preceding example by using areas of rectangles. If a rectangle has width 4 and length $5 + x$, its area is $20 + 4x$, as shown in Figure R.28.

Multiplying Polynomials

A polynomial with two terms is a binomial and a polynomial with three terms is a trinomial. Examples are shown in Table R.5.

TABLE R.5

Monomials	$2x^2$	$-3x^4$	9
Binomials	$3x - 1$	$2x^3 - x$	$x^2 + 5$
Trinomials	$x^2 - 3x + 5$	$5x^2 - 2x + 10$	$2x^3 - x^2 - 2$

In the next example we multiply two binomials.

EXAMPLE 5   *Multiplying binomials*

Multiply $(x + 1)(x + 3)$.

SOLUTION   To multiply $(x + 1)(x + 3)$ symbolically, we apply the distributive property.

$$\begin{aligned}
(x + 1)(x + 3) &= (x + 1)(x) + (x + 1)(3) \\
&= x \cdot x + 1 \cdot x + x \cdot 3 + 1 \cdot 3 \\
&= x^2 + x + 3x + 3 \\
&= x^2 + 4x + 3
\end{aligned}$$

To multiply $(x + 1)$ by $(x + 3)$, we multiplied every term in $x + 1$ by every term in $x + 3$. That is,

$$(x + 1)(x + 3) = x^2 + 3x + x + 3$$
$$= x^2 + 4x + 3.$$

Note: This process of multiplying binomials is called *FOIL*. You may use it to remind yourself to multiply the first terms (F), outside terms (O), inside terms (I), and last terms (L).

Multiply the *First terms* to obtain x^2. $(x + 1)(x + 3)$

Multiply the *Outside terms* to obtain $3x$. $(x + 1)(x + 3)$

Multiply the *Inside terms* to obtain x. $(x + 1)(x + 3)$

Multiply the *Last terms* to obtain 3. $(x + 1)(x + 3)$

The following method summarizes how to multiply two polynomials in general.

Multiplying polynomials

The product of two polynomials may be found by multiplying every term in the first polynomial by every term in the second polynomial.

EXAMPLE 6   *Multiplying polynomials*

Multiply each binomial.
(a) $(2x - 1)(x + 2)$ **(b)** $(1 - 3x)(2 - 4x)$ **(c)** $(x^2 + 1)(5x - 3)$

SOLUTION   **(a)** $(2x - 1)(x + 2) = 2x \cdot x + 2x \cdot 2 - 1 \cdot x - 1 \cdot 2$
$$= 2x^2 + 4x - x - 2$$
$$= 2x^2 + 3x - 2$$

(b) $(1 - 3x)(2 - 4x) = 1 \cdot 2 - 1 \cdot 4x - 3x \cdot 2 + 3x \cdot 4x$
$$= 2 - 4x - 6x + 12x^2$$
$$= 2 - 10x + 12x^2$$

(c) $(x^2 + 1)(5x - 3) = x^2 \cdot 5x - x^2 \cdot 3 + 1 \cdot 5x - 1 \cdot 3$
$$= 5x^3 - 3x^2 + 5x - 3$$

EXAMPLE 7   *Multiplying polynomials*

Multiply each expression.
(a) $3x(x^2 + 5x - 4)$ **(b)** $-x^2(x^4 - 2x + 5)$ **(c)** $(x + 2)(x^2 + 4x - 3)$

SOLUTION   **(a)** $3x(x^2 + 5x - 4) = 3x \cdot x^2 + 3x \cdot 5x - 3x \cdot 4$
$$= 3x^3 + 15x^2 - 12x$$

(b) $-x^2(x^4 - 2x + 5) = -x^2 \cdot x^4 + x^2 \cdot 2x - x^2 \cdot 5$
$$= -x^6 + 2x^3 - 5x^2$$

(c) $(x + 2)(x^2 + 4x - 3) = x \cdot x^2 + x \cdot 4x - x \cdot 3 + 2 \cdot x^2 + 2 \cdot 4x - 2 \cdot 3$
$$= x^3 + 4x^2 - 3x + 2x^2 + 8x - 6$$
$$= x^3 + 6x^2 + 5x - 6$$

Some Special Products

The following special product often occurs in mathematics.

$$(a - b)(a + b) = a \cdot a + a \cdot b - b \cdot a - b \cdot b$$
$$= a^2 + ab - ba - b^2$$
$$= a^2 - b^2$$

That is, the product of a sum and difference equals the difference of their squares.

EXAMPLE 8   *Finding the product of a sum and difference*

Multiply.
(a) $(x - 3)(x + 3)$ **(b)** $(5 - 4x^2)(5 + 4x^2)$

SOLUTION   **(a)** If we let $a = x$ and $b = 3$, we can apply the rule

$$(a - b)(a + b) = a^2 - b^2.$$

Thus

$$(x - 3)(x + 3) = (x)^2 - (3)^2$$
$$= x^2 - 9.$$

(b) Similarly, we can multiply as follows.

$$(5 - 4x^2)(5 + 4x^2) = (5)^2 - (4x^2)^2$$
$$= 25 - 16x^4$$

Two other special products involve *squaring a binomial:*

$$(a + b)^2 = (a + b)(a + b)$$
$$= a^2 + ab + ba + b^2$$
$$= a^2 + 2ab + b^2$$

and

$$(a - b)^2 = (a - b)(a - b)$$
$$= a^2 - ab - ba + b^2$$
$$= a^2 - 2ab + b^2.$$

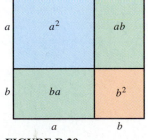

FIGURE R.29
$(a + b)^2 = a^2 + 2ab + b^2$

The first product is illustrated geometrically in Figure R.29, where each side of a square has length $(a + b)$. The area of the square is

$$(a + b)(a + b) = (a + b)^2.$$

This area can also be computed by adding the area of the four small rectangles.

$$a^2 + ab + ba + b^2 = a^2 + 2ab + b^2$$

Thus $(a + b)^2 = a^2 + 2ab + b^2$. Note that to obtain the middle term, we multiply the two terms in the binomial and double the result.

EXAMPLE 9   *Squaring a binomial*

Multiply.
(a) $(x + 5)^2$ **(b)** $(3 - 2x)^2$

SOLUTION   **(a)** If we let $a = x$ and $b = 5$, we can apply the formula

$$(a + b)^2 = a^2 + 2ab + b^2.$$

Thus

$$(x + 5)^2 = (x)^2 + 2(x)(5) + (5)^2 \qquad \text{To find the middle term, multiply}$$
$$= x^2 + 10x + 25. \qquad \text{a and b and double the result.}$$

(b) Applying the formula $(a - b)^2 = a^2 - 2ab + b^2$, we find

$$(3 - 2x)^2 = (3)^2 - 2(3)(2x) + (2x)^2$$
$$= 9 - 12x + 4x^2.$$

Note: If you forget these special products, you can still use techniques learned earlier to multiply the polynomials in Examples 8 and 9. For example,

$$(3 - 2x)^2 = (3 - 2x)(3 - 2x)$$
$$= 3 \cdot 3 - 3 \cdot 2x - 2x \cdot 3 + 2x \cdot 2x$$
$$= 9 - 6x - 6x + 4x^2$$
$$= 9 - 12x + 4x^2.$$

$R.4$ EXERCISES

Concepts

1. Give an example of a term that is a monomial.

2. What are the degree and leading coefficient of $3x^2 - x^3 + 1$?

3. Are $-5x^3y$ and $6xy^3$ like terms? Explain.

4. Give an example of a polynomial that has 3 terms and is degree 4.

5. Does the opposite of $x^2 + 1$ equal $-x^2 + 1$? Explain.

6. The equation $5(x - 4) = 5x - 20$ illustrates what property?

7. $(a - b)(a + b) = $ ___.   8. $(a + b)^2 = $ ___.

Monomials and Polynomials

Exercises 9–18: Combine like terms whenever possible.

9. $3x^3 + 5x^3$

10. $-9z + 6z$

11. $5y^7 - 8y^7$

12. $9x - 7x$

13. $5x^2 + 8x + x^2$

14. $5x + 2x + 10x$

15. $9x^2 - x + 4x - 6x^2$

16. $-y^2 - \dfrac{1}{2}y^2$

17. $x^2 + 9x - 2 + 4x^2 + 4x$

18. $6y + 4y^2 - 6y + y^2$

Exercises 19–24: Identify the degree and leading coefficient of the polynomial.

19. $5x^2 - 4x + \dfrac{3}{4}$

20. $-9y^4 + y^2 + 5$

21. $5 - x + 3x^2 - \dfrac{2}{5}x^3$

22. $7x + 4x^4 - \dfrac{4}{3}x^3$

23. $8x^4 + 3x^3 - 4x + x^5$

24. $5x^2 - x^3 + 7x^4 + 10$

Exercises 25–32: Add the polynomials.

25. $(5x + 6) + (-2x + 6)$

26. $(5y^2 + y^3) + (12y^2 - 5y^3)$

27. $(2x^2 - x + 7) + (-2x^2 + 4x - 9)$

28. $(x^3 - 5x^2 + 6) + (5x^2 + 3x + 1)$

29. $(4x) + (1 - 4.5x)$

30. $(y^5 + y) + \left(5 - y + \dfrac{1}{3}y^2\right)$

31. $(x^4 - 3x^2 - 4) + \left(-8x^4 + x^2 - \dfrac{1}{2}\right)$

32. $(3z + z^4 + 2) + (-3z^4 - 5 + z^2)$

Exercises 33–38: Find the opposite of the polynomial.

33. $7x^3$ **34.** $-3z^8$

35. $19z^5 - 5z^2 + 3z$ **36.** $-x^2 - x + 6$

37. $z^4 - z^2 - 9$

38. $1 - 8x + 6x^2 - \dfrac{1}{6}x^3$

Exercises 39–46: Subtract the polynomials.

39. $(5x - 3) - (2x + 4)$

40. $(10x + 5) - (-6x - 4)$

41. $(x^2 - 3x + 1) - (-5x^2 + 2x - 4)$

42. $(-x^2 + x - 5) - (x^2 - x + 5)$

43. $(4x^4 + 2x^2 - 9) - (x^4 - 2x^2 - 5)$

44. $(8x^3 + 5x^2 - 3x + 1) - (-5x^3 + 6x - 11)$

45. $(x^4 - 1) - (4x^4 + 3x + 7)$

46. $(5x^4 - 6x^3 + x^2 + 5) - (x^3 + 11x^2 + 9x - 3)$

Exercises 47–56: Apply the distributive property.

47. $5x(x - 5)$ **48.** $3x^2(-2x + 2)$

49. $-5(3x + 1)$ **50.** $-(-3x + 1)$

51. $5(y + 2)$ **52.** $4(x - 7)$

53. $-2(5x + 9)$ **54.** $-3x(5 + x)$

55. $(y - 3)6y$ **56.** $(2x - 5)8x^3$

Exercises 57–68: Multiply the binomials.

57. $(y + 5)(y - 7)$ **58.** $(3x + 1)(2x + 1)$

59. $(7x - 3)(4 - 7x)$ **60.** $(3 - 2x)(3 + x)$

61. $(-2x + 3)(x - 2)$ **62.** $(z - 2)(4z + 3)$

63. $\left(x - \dfrac{1}{2}\right)\left(x + \dfrac{1}{4}\right)$ **64.** $\left(z - \dfrac{1}{3}\right)\left(z - \dfrac{1}{6}\right)$

65. $(x^2 + 1)(2x^2 - 1)$ **66.** $(x^2 - 2)(x^2 + 4)$

67. $(x + y)(x - 2y)$ **68.** $(x^2 + y^2)(x - y)$

Exercises 69–76: Multiply the polynomials.

69. $3x(2x^2 - x - 1)$ **70.** $-2x(3 - 2x + 5x^2)$

71. $-x(2x^4 - x^2 + 10)$ **72.** $-2x^2(5x^3 + x^2 - 2)$

73. $(2x^2 - 4x + 1)(3x^2)$ **74.** $(x - y + 5)(xy)$

75. $(x + 1)(x^2 + 2x - 3)$

76. $(2x - 1)(3x^2 - x + 6)$

Exercises 77–92: Multiply the expressions.

77. $(x - 7)(x + 7)$ **78.** $(x + 9)(x - 9)$

79. $(3x - 4)(3x + 4)$ **80.** $(9x - 4)(9x + 4)$

81. $(2x - 3y)(2x + 3y)$ **82.** $(x + 2y)(x - 2y)$

83. $(x + 4)^2$ **84.** $(z + 9)^2$

85. $(2x + 1)^2$ **86.** $(3x + 5)^2$

87. $(x - 1)^2$ **88.** $(x - 7)^2$

89. $(3x - 2)^2$ **90.** $(6x - 5)^2$

91. $3x(x + 1)(x - 1)$ **92.** $-4x(3x - 5)^2$

R.5 FACTORING POLYNOMIALS

Common Factors • Factoring by Grouping • Factoring Trinomials • Difference of Two Squares • Perfect Square Trinomials • Sum and Difference of Two Cubes

Common Factors

When factoring a polynomial, we first look for factors that are common to all terms in an expression. By applying the distributive property, we can write a polynomial as two factors. For example, each term in the polynomial $2x^2 + 4x$ contains a factor of $2x$.

$$2x^2 + 4x = 2x(x + 2) \qquad \text{Distributive property}$$

Thus the product $2x(x + 2)$ equals $2x^2 + 4x$. We can check this result by multi-plying the two factors.

$$2x(x + 2) = 2x \cdot x + 2x \cdot 2$$
$$= 2x^2 + 4x \qquad \text{It checks.}$$

EXAMPLE 1   *Finding common factors*

Factor.
(a) $4x^2 + 5x$ **(b)** $12x^3 - 4x^2$ **(c)** $6z^3 - 2z^2 + 4z$

SOLUTION   **(a)** Both $4x^2$ and $5x$ contain a common factor of x.

$$4x^2 + 5x = x(4x + 5)$$

(b) Both $12x^3$ and $4x^2$ contain a common factor of $4x^2$.

$$12x^3 - 4x^2 = 4x^2(3x - 1)$$

(c) The terms $6z^3$, $2z^2$, and $4z$ contain a common factor of $2z$.

$$6z^3 - 2z^2 + 4z = 2z(3z^2 - z + 2)$$

In most situations we factor out the *greatest common factor* (GCF). For example, the polynomial $15x^4 - 5x^2$ has a common factor of $5x$. We could write this polynomial as

$$15x^4 - 5x^2 = 5x(3x^3 - x).$$

However, we can also factor out $5x^2$ to obtain

$$15x^4 - 5x^2 = 5x^2(3x^2 - 1).$$

Because no obvious common factors are left in the expression $(3x^2 - 1)$, we say that $5x^2$ is the greatest common factor of $15x^4 - 5x^2$.

Factoring by Grouping

Factoring by grouping is a technique that makes use of the distributive property. Consider the polynomial

$$3t^3 + 6t^2 + 2t + 4.$$

We can factor it as follows.

$$(3t^3 + 6t^2) + (2t + 4) \qquad \text{Associative property}$$
$$3t^2(t + 2) + 2(t + 2) \qquad \text{Factor out common factors.}$$
$$(3t^2 + 2)(t + 2) \qquad \text{Distributive property}$$

EXAMPLE 2   *Factoring by grouping*

Factor the polynomial

$$x^3 - 2x^2 + 3x - 6$$

SOLUTION

$$x^3 - 2x^2 + 3x - 6 \qquad \text{Given expression}$$
$$(x^3 - 2x^2) + (3x - 6) \qquad \text{Associative property}$$
$$x^2(x - 2) + 3(x - 2) \qquad \text{Factor out common factors.}$$
$$(x^2 + 3)(x - 2) \qquad \text{Distributive property}$$

Factoring Trinomials

A *trinomial* is a polynomial that has three terms. For now we focus on factoring trinomials of degree 2. We begin by reviewing multiplication of binomials, which was introduced in Section R.4.

$$(x + 1)(x + 2) = x \cdot x + x \cdot 2 + 1 \cdot x + 1 \cdot 2$$
$$= x^2 + 3x + 2$$

Note that x^2 results from multiplying the *first* terms of each binomial. The middle term of $3x$ results from adding the product of the *outside* terms to the product of the *inside* terms. Finally, the last term of 2 results from multiplying the *last* terms of each binomial. Recall that this method of multiplying binomials is abbreviated *FOIL* and is applied as follows.

$$(x + 1)(x + 2) = x^2 + 3x + 2$$
$$1x$$
$$+2x$$
$$3x$$

Now suppose that we want to factor $x^2 + 3x + 2$. We can apply this process in reverse. The factors of x^2 are x and x, so we do the following.

$$x^2 + 3x + 2 \overset{?}{=} (x + \underline{\quad})(x + \underline{\quad})$$

Next we consider the factors of 2. The integer factors are 1 and 2, so we write them in the blanks.

$$x^2 + 3x + 2 \overset{?}{=} (x + 1)(x + 2)$$

For the factorization to be valid, *we must add the product of the outside terms to the product of the inside terms.* The outside terms give $2x$, and the inside terms give $1x$. Their sum is $2x + 1x = 3x$, which is the correct middle term.

EXAMPLE 3   *Factoring the form $x^2 + bx + c$*

Factor each trinomial.
(a) $x^2 + 8x + 12$ **(b)** $x^2 - 7x + 12$ **(c)** $x^2 - 4x - 12$

SOLUTION   **(a)** The factors of x^2 are x and x, so we begin by writing

$$x^2 + 8x + 12 \overset{?}{=} (x + \underline{\quad})(x + \underline{\quad}).$$

There are several factors for 12, such as 1 and 12, 2 and 6, and 3 and 4. If we try 3 and 4, then

$$(x + 3)(x + 4) = x^2 + 4x + 3x + 12$$
$$= x^2 + 7x + 12.$$

The middle term, $7x$, is incorrect because the sum of these factors is $3 + 4 = 7$ and a middle term of $8x$ is needed. We try the factors 2 and 6 because their sum is 8.

$$x^2 + 8x + 12 = (x + 6)(x + 2)$$
$$6x$$
$$+2x$$
$$8x \qquad \text{It checks.}$$

Thus $x^2 + 8x + 12 = (x + 2)(x + 6)$.

(b) To obtain a middle term of $-7x$ and a last term of *positive* 12, we need both factors of 12 to be negative and sum to -7. We try the factors -3 and -4.

$$(x - 3)(x - 4) = x^2 - 4x - 3x + 12$$
$$= x^2 - 7x + 12 \qquad \text{It checks.}$$

Thus

$$x^2 - 7x + 12 = (x - 3)(x - 4).$$

(c) For the last term to be -12, we need one factor of 12 to be positive and one to be negative. The middle term is $-4x$, so we need the sum of the two factors to be -4. We try the factors 2 and -6.

$$(x + 2)(x - 6) = x^2 - 6x + 2x - 12$$
$$= x^2 - 4x - 12 \qquad \text{It checks.}$$

Thus

$$x^2 - 4x - 12 = (x + 2)(x - 6).$$

In the next example we factor expressions of the form $ax^2 + bx + c$, where $a \neq 1$. In this situation, we may need to *guess and check* or use *trial and error* a few times before finding the correct factors.

EXAMPLE 4 *Factoring the form $ax^2 + bx + c$*

Factor each trinomial.
(a) $2x^2 + 9x + 4$ **(b)** $6x^2 - x - 2$ **(c)** $4x^3 - 14x^2 + 6x$

SOLUTION **(a)** The factors of $2x^2$ are $2x$ and x, so we begin by writing

$$2x^2 + 9x + 4 \overset{?}{=} (2x + \underline{\quad})(x + \underline{\quad}).$$

The factors of the last term, 4, are either 1 and 4 or 2 and 2. Selecting the factors 2 and 2 results in a middle term of $6x$ rather than $9x$.

$$(2x + 2)\,(x + 2) = 2x^2 + 6x + 4$$

Next we try the factors 1 and 4.

$$(2x + 4)\,(x + 1) = 2x^2 + 6x + 4$$

Again we obtain the wrong middle term. By interchanging the 1 and 4, we find the correct factorization.

$$(2x + 1)\,(x + 4) = 2x^2 + 9x + 4$$

(b) The factors of $6x^2$ are either $2x$ and $3x$ or $6x$ and x. The factors of -2 are either -1 and 2 or 1 and -2. To obtain a middle term of $-x$, we use the following factors.

$$(3x - 2)(2x + 1) = 6x^2 + 3x - 4x - 2$$
$$= 6x^2 - x - 2 \qquad \text{It checks.}$$

To find the correct factorization, we may need to guess and check a few times.

(c) The terms contain a common factor of $2x$, so we do the following step first.

$$4x^3 - 14x^2 + 6x = 2x(2x^2 - 7x + 3)$$

Next we factor $2x^2 - 7x + 3$. The factors of $2x^2$ are $2x$ and x. Because the middle term is negative and the last term is positive, we use -1 and -3 for factors of 3.

$$4x^3 - 14x^2 + 6x = 2x(2x^2 - 7x + 3)$$
$$= 2x(2x - 1)(x - 3)$$

Again we may need to guess and check more than once before obtaining the correct factors.

Difference of Two Squares

In Section R.4 we learned that

$$(a - b)(a + b) = a^2 - b^2.$$

We can use this equation to factor a difference of two squares. For example, if we want to factor $x^2 - 25$, we can think of it being in the form $a^2 - b^2$, where $a = x$ and $b = 5$. Thus substituting into

$$a^2 - b^2 = (a - b)(a + b)$$

gives

$$x^2 - 25 = (x - 5)(x + 5).$$

EXAMPLE 5   *Factoring the difference of two squares*

Factor each polynomial.
(a) $9x^2 - 64$ **(b)** $4x^2 - 9y^2$

SOLUTION   **(a)** Note that $9x^2 = (3x)^2$ and $64 = 8^2$.

$$9x^2 - 64 = (3x)^2 - (8)^2$$
$$= (3x - 8)(3x + 8)$$

(b) $4x^2 - 9y^2 = (2x)^2 - (3y)^2$
$$= (2x - 3y)(2x + 3y)$$

Perfect Square Trinomials

In Section R.4 we also learned how to expand $(a + b)^2$ and $(a - b)^2$.

$$(a + b)^2 = a^2 + 2ab + b^2$$
$$(a - b)^2 = a^2 - 2ab + b^2$$

The expressions $a^2 + 2ab + b^2$ and $a^2 - 2ab + b^2$ are called perfect square trinomials. If we can recognize a perfect square trinomial, we can use these formulas to factor it, as demonstrated in the next example.

EXAMPLE 6 *Factoring perfect square trinomials*

Factor.
(a) $x^2 + 6x + 9$ (b) $81x^2 - 72x + 16$

SOLUTION (a) Because $x^2 = (x)^2$ and $9 = (3)^2$, we let $a = x$, $b = 3$. Thus

$$a^2 + 2ab + b^2 = (a + b)^2$$

$$x^2 + 6x + 9 = (x)^2 + 2(x)(3) + (3)^2 = (x + 3)^2.$$

(b) Because $81x^2 = (9x)^2$ and $16 = 4^2$, we let $a = 9x$, $b = 4$. Thus

$$a^2 - 2ab + b^2 = (a - b)^2$$

$$81x^2 - 72x + 16 = (9x)^2 - 2(9x)(4) + (4)^2 = (9x - 4)^2.$$

Sum and Difference of Two Cubes

The sum or difference of two cubes may be factored, which is a result of the following two equations.

$$(a + b)(a^2 - ab + b^2) = a^3 + b^3$$
$$(a - b)(a^2 + ab + b^2) = a^3 - b^3$$

These equations can be verified by multiplying the left side to obtain the right side. For example,

$$(a + b)(a^2 - ab + b^2) = a \cdot a^2 - a \cdot ab + a \cdot b^2 + b \cdot a^2 - b \cdot ab + b \cdot b^2$$
$$= a^3 - a^2b + ab^2 + a^2b - ab^2 + b^3$$
$$= a^3 + b^3.$$

EXAMPLE 7 *Factoring the sum or difference of two cubes*

Factor each polynomial.
(a) $x^3 + 8$ (b) $27x^3 - 64y^3$

SOLUTION (a) Because $x^3 = (x)^3$ and $8 = 2^3$, we let $a = x$, $b = 2$, and factor. Substituting into

$$a^3 + b^3 = (a + b)(a^2 - ab + b^2)$$

gives

$$x^3 + 2^3 = (x + 2)(x^2 - x \cdot 2 + 2^2)$$
$$= (x + 2)(x^2 - 2x + 4).$$

(b) Here, $27x^3 = (3x)^3$ and $64y^3 = (4y)^3$, so

$$27x^3 - 64y^3 = (3x)^3 - (4y)^3.$$

Substituting $a = 3x$ and $b = 4y$ into

$$a^3 - b^3 = (a - b)(a^2 + ab + b^2)$$

gives

$$(3x)^3 - (4y)^3 = (3x - 4y)((3x)^2 + 3x \cdot 4y + (4y)^2)$$
$$= (3x - 4y)(9x^2 + 12xy + 16y^2).$$

$R.5$ EXERCISES

Greatest Common Factor

Exercises 1–10: Factor out the greatest common factor.

1. $10x - 15$
2. $32 - 16x$
3. $2x^3 - 5x$
4. $3y - 9y^2$
5. $8x^3 - 4x^2 + 16x$
6. $-5x^3 + x^2 - 4x$
7. $5x^4 - 15x^3 + 15x^2$
8. $21y + 14y^3 - 7y^5$
9. $15x^3 + 10x^2 - 25x$
10. $14a^4 - 21a^2 + 35a$

Factoring by Grouping

Exercises 11–18: Use grouping to factor the polynomial.

11. $x^3 + 3x^2 + 2x + 6$
12. $4x^3 + 3x^2 + 8x + 6$
13. $6x^3 - 4x^2 + 9x - 6$
14. $x^3 - 3x^2 - 5x + 15$
15. $z^3 - 5z^2 + z - 5$
16. $y^3 - 7y^2 + 8y - 56$
17. $y^4 + 2y^3 - 5y^2 - 10y$
18. $4z^4 + 4z^3 + z^2 + z$

Factoring Trinomials

Exercises 19–38: Factor each expression.

19. $x^2 + 7x + 10$
20. $x^2 + 3x - 10$
21. $x^2 + 8x + 12$
22. $x^2 - 8x + 12$
23. $z^2 + z - 42$
24. $z^2 - 9z + 20$
25. $z^2 + 11z + 24$
26. $z^2 + 15z + 54$
27. $24x^2 + 14x - 3$
28. $25x^2 - 5x - 6$
29. $6x^2 - x - 2$
30. $10x^2 + 3x - 1$
31. $1 + x - 2x^2$
32. $3 - 5x - 2x^2$
33. $20 + 7x - 6x^2$
34. $4 + 13x - 12x^2$
35. $5x^3 + x^2 - 6x$
36. $2x^3 + 8x^2 - 24x$
37. $6x^3 + 21x^2 + 9x$
38. $12x^3 - 8x^2 - 20x$

Difference of Two Squares

Exercises 39–48: Factor the expression completely.

39. $x^2 - 25$
40. $z^2 - 169$
41. $4x^2 - 25$
42. $36 - y^2$
43. $36x^2 - 100$
44. $9x^2 - 4y^2$
45. $64z^2 - 25z^4$
46. $100x^3 - x$
47. $16x^4 - y^4$
48. $x^4 - 9y^2$

Perfect Square Trinomials

Exercises 49–56: Factor the expression.

49. $x^2 + 2x + 1$
50. $x^2 - 6x + 9$
51. $4x^2 + 20x + 25$
52. $x^2 + 10x + 25$
53. $x^2 - 12x + 36$
54. $16z^4 - 24z^3 + 9z^2$
55. $9z^3 - 6z^2 + z$
56. $49y^2 + 42y + 9$

Sum and Difference of Two Cubes

Exercises 57–64: Factor the expression.

57. $x^3 - 1$
58. $x^3 + 1$
59. $y^3 + z^3$
60. $y^3 - z^3$
61. $8x^3 - 27$
62. $8 - z^3$
63. $x^4 + 125x$
64. $3x^4 - 81x$

General Factoring

Exercises 65–76: Factor the expression completely.

65. $16x^2 - 25$
66. $25x^2 - 30x + 9$
67. $x^3 - 64$
68. $1 + 8y^3$
69. $x^2 + 16x + 64$
70. $12x^2 + x - 6$
71. $5x^2 - 38x - 16$
72. $125x^3 - 1$
73. $x^4 + 8x$
74. $2x^3 - 12x^2 + 18x$
75. $64x^3 + 8y^3$
76. $54 - 16x^3$

R.6 RATIONAL EXPRESSIONS

Multiplication and Division of Rational Expressions • Common Denominators • Addition and Subtraction of Rational Expressions • Clearing Fractions • Complex Fractions

Multiplication and Division of Rational Expressions

Multiplying and dividing rational expressions is similar to multiplying and dividing fractions.

Products and quotients of rational expressions

To multiply two rational expressions, multiply numerators and multiply denominators.

$$\frac{A}{B} \cdot \frac{C}{D} = \frac{AC}{BD} \qquad B \text{ and } D \text{ are nonzero.}$$

To divide two rational expressions, multiply by the reciprocal of the divisor.

$$\frac{A}{B} \div \frac{C}{D} = \frac{A}{B} \cdot \frac{D}{C} \qquad B, C, \text{ and } D \text{ are nonzero.}$$

EXAMPLE 1   *Multiplying rational expressions*

Multiply.

(a) $\dfrac{1}{x} \cdot \dfrac{x+1}{2x}$ **(b)** $\dfrac{x-1}{x} \cdot \dfrac{x-1}{x+2}$

SOLUTION **(a)** $\dfrac{1}{x} \cdot \dfrac{x+1}{2x} = \dfrac{1 \cdot (x+1)}{x \cdot 2x} = \dfrac{x+1}{2x^2}$

(b) $\dfrac{x-1}{x} \cdot \dfrac{x-1}{x+2} = \dfrac{(x-1)(x-1)}{x(x+2)}$

We often simplify, or reduce, fractions. For example,

$$\frac{10}{15} = \frac{2 \cdot 5}{3 \cdot 5} = \frac{2}{3}.$$

To reduce rational expressions, we often factor the numerator and denominator. For example,

$$\frac{x^2}{x^2+5x} = \frac{x \cdot x}{x(x+5)} = \frac{x}{x+5}.$$

Note: We factor *first* and then reduce.

EXAMPLE 2   *Simplifying rational expressions*

Simplify.

(a) $\dfrac{(x-1)(x+2)}{(x-3)(x-1)}$ (b) $\dfrac{x^2-4}{2x^2-3x-2}$

SOLUTION   (a) $\dfrac{(x-1)(x+2)}{(x-3)(x-1)} = \dfrac{(x+2)(x-1)}{(x-3)(x-1)}$ Commutative property

$= \dfrac{x+2}{x-3}$ Reduce.

(b) $\dfrac{x^2-4}{2x^2-3x-2} = \dfrac{(x+2)(x-2)}{(2x+1)(x-2)}$ Factor.

$= \dfrac{x+2}{2x+1}$ Reduce.

In the next example we divide and simplify rational expressions.

EXAMPLE 3   *Dividing two rational expressions*

Divide and simplify.

(a) $\dfrac{2}{x} \div \dfrac{2x-1}{4x}$

(b) $\dfrac{x^2-1}{x^2+x-6} \div \dfrac{x-1}{x+3}$

SOLUTION   (a) $\dfrac{2}{x} \div \dfrac{2x-1}{4x} = \dfrac{2}{x} \cdot \dfrac{4x}{2x-1}$ "Invert and multiply."

$= \dfrac{8x}{x(2x-1)}$ Multiply.

$= \dfrac{8}{2x-1}$ Reduce.

(b) $\dfrac{x^2-1}{x^2+x-6} \div \dfrac{x-1}{x+3} = \dfrac{x^2-1}{x^2+x-6} \cdot \dfrac{x+3}{x-1}$ "Invert and multiply."

$= \dfrac{(x+1)(x-1)}{(x-2)(x+3)} \cdot \dfrac{x+3}{x-1}$ Factor.

$= \dfrac{(x+1)(x-1)(x+3)}{(x-2)(x-1)(x+3)}$ Commutative property

$= \dfrac{(x+1)}{(x-2)}$ Reduce.

Common Denominators

If each of the denominators is a factor of an expression, that expression is a *common denominator*. For example, $4x^2$ is a common denominator of x, x^2, and 4 because all are factors of $4x^2$.

$4x^2 = x \cdot 4x$ x is a factor.

$4x^2 = x^2 \cdot 4$ x^2 is a factor.

$4x^2 = 4 \cdot x^2$ 4 is a factor.

The expression $4x^3$ is also a common denominator of x, x^2, and 4. However, we say that $4x^2$ is the *least common denominator* (LCD) because it is the common denominator with fewest factors.

Note: A common denominator can always be found by taking the product of the denominators. However, it may not be the *least* common denominator.

EXAMPLE 4   *Finding a least common denominator*

Find the LCD for the given expressions.

(a) $\dfrac{1}{x}, \dfrac{1}{x-1}$ **(b)** $\dfrac{1}{x+2}, \dfrac{2x}{x^2-4}, \dfrac{1}{3}$

SOLUTION   **(a)** The LCD of x and $x-1$ is their product, $x(x-1)$.
(b) A common denominator of $x+2$, x^2-4, and 3 would be their product. However, because

$$x^2 - 4 = (x+2)(x-2),$$

the LCD is

$$(x+2)(x-2)(3).$$

Addition and Subtraction of Rational Expressions

Addition and subtraction of rational expressions with like denominators are performed in the following manner.

> ### Sums and differences of rational expressions
>
> To add (or subtract) two rational expressions with like denominators, add (or subtract) their numerators. The denominator does not change.
>
> $$\frac{A}{C} + \frac{B}{C} = \frac{A+B}{C}$$
>
> $$\frac{A}{C} - \frac{B}{C} = \frac{A-B}{C}, \qquad C \neq 0$$

Note: If the denominators are not alike, begin by writing each rational expression, using a common denominator. Then add or subtract the numerators.

EXAMPLE 5   *Adding rational expressions*

Add and simplify.

(a) $\dfrac{x}{x+2} + \dfrac{3x+1}{x+2}$ **(b)** $\dfrac{1}{x} + \dfrac{2}{x^2}$ **(c)** $\dfrac{1}{x-1} + \dfrac{2x}{x+1}$

SOLUTION **(a)** The denominators are alike, so we add the numerators and keep the same denominator.

$$\frac{x}{x+2}+\frac{3x+1}{x+2}=\frac{x+3x+1}{x+2}$$ Add numerators.

$$=\frac{4x+1}{x+2}$$ Combine like terms.

(b) The LCD for x and x^2 is x^2.

$$\frac{1}{x}+\frac{2}{x^2}=\frac{1}{x}\cdot\frac{x}{x}+\frac{2}{x^2}$$ Change to a common denominator.

$$=\frac{x}{x^2}+\frac{2}{x^2}$$ Multiply.

$$=\frac{x+2}{x^2}$$ Add numerators.

(c) The LCD for $x-1$ and $x+1$ is their product, $(x-1)(x+1)$.

$$\frac{1}{x-1}+\frac{2x}{x+1}=\frac{1}{x-1}\cdot\frac{x+1}{x+1}+\frac{2x}{x+1}\cdot\frac{x-1}{x-1}$$ Change to a common denominator.

$$=\frac{x+1}{(x-1)(x+1)}+\frac{2x(x-1)}{(x+1)(x-1)}$$ Multiply.

$$=\frac{x+1+2x^2-2x}{(x-1)(x+1)}$$ Add numerators; distributive property

$$=\frac{2x^2-x+1}{(x-1)(x+1)}$$ Combine like terms.

Subtraction of rational expressions is similar.

EXAMPLE 6   *Subtracting rational expressions*

Subtract and simplify.

(a) $\dfrac{3}{x^2}-\dfrac{x+3}{x^2}$ **(b)** $\dfrac{x-1}{x}-\dfrac{5}{x+5}$ **(c)** $\dfrac{1}{x^2-3x+2}-\dfrac{1}{x^2-x-2}$

SOLUTION **(a)** The denominators are alike, so we subtract the numerators and keep the same denominator.

$$\frac{3}{x^2}-\frac{x+3}{x^2}=\frac{3-(x+3)}{x^2}$$ Subtract numerators.

$$=\frac{3-x-3}{x^2}$$ Distributive property

$$=\frac{-x}{x^2}$$ Simplify numerator.

$$=-\frac{1}{x}$$ Reduce.

(b) The LCD is $x(x + 5)$.

$$\frac{x - 1}{x} - \frac{5}{x + 5} = \frac{x - 1}{x} \cdot \frac{x + 5}{x + 5} - \frac{5}{x + 5} \cdot \frac{x}{x}$$ Change to a common denominator.

$$= \frac{(x - 1)(x + 5)}{x(x + 5)} - \frac{5x}{x(x + 5)}$$ Multiply.

$$= \frac{(x - 1)(x + 5) - 5x}{x(x + 5)}$$ Subtract numerators.

$$= \frac{x^2 + 4x - 5 - 5x}{x(x + 5)}$$ Multiply binomials.

$$= \frac{x^2 - x - 5}{x(x + 5)}$$ Combine like terms.

(c) Because $x^2 - 3x + 2 = (x - 2)(x - 1)$ and $x^2 - x - 2 = (x - 2)(x + 1)$, the LCD is $(x - 2)(x - 1)(x + 1)$.

$$\frac{1}{x^2 - 3x + 2} - \frac{1}{x^2 - x - 2}$$

$$= \frac{1}{(x - 2)(x - 1)} \cdot \frac{(x + 1)}{(x + 1)} - \frac{1}{(x - 2)(x + 1)} \cdot \frac{(x - 1)}{(x - 1)}$$ Change to a common denominator.

$$= \frac{(x + 1)}{(x - 2)(x - 1)(x + 1)} - \frac{(x - 1)}{(x - 2)(x + 1)(x - 1)}$$ Multiply.

$$= \frac{(x + 1) - (x - 1)}{(x - 2)(x - 1)(x + 1)}$$ Subtract numerators.

$$= \frac{x + 1 - x + 1}{(x - 2)(x - 1)(x + 1)}$$ Distributive property

$$= \frac{2}{(x - 2)(x - 1)(x + 1)}$$ Simplify numerator.

Clearing Fractions

To solve rational equations or simplify complex fractions, it is sometimes advantageous to multiply rational expressions by the least common denominator to clear fractions from an expression. For example, the least common denominator of $\dfrac{1}{x + 2} + \dfrac{1}{x - 2}$ is $(x + 2)(x - 2)$. Multiplying this expression by its least common denominator results in the following.

$$(x + 2)(x - 2)\left(\frac{1}{x + 2} + \frac{1}{x - 2}\right)$$

$$= \left(\frac{(x + 2)(x - 2)}{x + 2} + \frac{(x + 2)(x - 2)}{x - 2}\right)$$ Distributive property

$$= (x - 2) + (x + 2)$$ Reduce.

$$= 2x$$ Combine like terms.

This technique is applied in the next example.

Note: The results are *not* equal to the given expression.

EXAMPLE 7   *Clearing fractions from rational expressions*

Multiply each expression by its least common denominator.

(a) $\dfrac{1}{x} - \dfrac{2}{x^2}$ **(b)** $\dfrac{1}{x+3} + \dfrac{3x}{x^2-9}$ **(c)** $\dfrac{1}{x} + \dfrac{x}{x^2-1} - \dfrac{4}{x+1}$

SOLUTION   **(a)** The LCD of x and x^2 is x^2.

$$x^2\left(\frac{1}{x} - \frac{2}{x^2}\right) = \left(\frac{x^2}{x} - \frac{2x^2}{x^2}\right)$$

$$= x - 2$$

(b) The LCD of $x+3$ and x^2-9 is $x^2-9 = (x-3)(x+3)$.

$$(x^2-9)\left(\frac{1}{x+3} - \frac{3x}{x^2-9}\right) = \left(\frac{x^2-9}{x+3} - \frac{3x(x^2-9)}{x^2-9}\right)$$

$$= \left(\frac{(x-3)(x+3)}{x+3} - \frac{3x(x^2-9)}{x^2-9}\right)$$

$$= x - 3 - 3x$$

$$= -3 - 2x$$

(c) The LCD of x, x^2-1, and $x+1$ is $x(x^2-1) = x(x-1)(x+1)$.

$$x(x^2-1)\left(\frac{1}{x} + \frac{x}{x^2-1} - \frac{4}{x+1}\right)$$

$$= \left(\frac{x(x^2-1)}{x} + \frac{x(x)(x^2-1)}{x^2-1} - \frac{4x(x^2-1)}{x+1}\right)$$

$$= \left(\frac{x(x^2-1)}{x} + \frac{x^2(x^2-1)}{x^2-1} - \frac{4x(x-1)(x+1)}{x+1}\right)$$

$$= (x^2-1) + x^2 - 4x(x-1)$$

$$= x^2 - 1 + x^2 - 4x^2 + 4x$$

$$= -2x^2 + 4x - 1$$

Complex Fractions

A complex fraction is a rational expression that contains fractions in its numerator, denominator, or both. Examples of complex fractions include

$$\frac{1 + \dfrac{1}{x}}{1 - \dfrac{1}{x}}, \qquad \frac{2x}{\dfrac{4}{x} + \dfrac{3}{x}}, \qquad \text{and} \qquad \frac{\dfrac{a}{3} + \dfrac{a}{4}}{a - \dfrac{1}{a-1}}.$$

One strategy for simplifying a complex fraction is to multiply the numerator and denominator by the LCD of the fractions in the numerator and denominator. For example, the LCD for the complex fraction

$$\frac{1 - \dfrac{1}{x}}{1 + \dfrac{1}{2x}}$$

is $2x$. To simplify, multiply the complex fraction by 1, expressed in the form $\dfrac{2x}{2x}$.

$$\frac{\left(1 - \dfrac{1}{x}\right) \cdot 2x}{\left(1 + \dfrac{1}{2x}\right) \cdot 2x} = \frac{2x - \dfrac{2x}{x}}{2x + \dfrac{2x}{2x}} \qquad \text{Distributive property}$$

$$= \frac{2x - 2}{2x + 1} \qquad \text{Reduce.}$$

In the next example we simplify other complex fractions.

EXAMPLE 8 *Simplifying complex fractions*

Simplify.

(a) $\dfrac{\dfrac{1}{x} - \dfrac{1}{y}}{x - y}$ (b) $\dfrac{\dfrac{3}{x - 1} - \dfrac{2}{x}}{\dfrac{1}{x - 1} + \dfrac{3}{x}}$

SOLUTION (a) The LCD of x and y is their product, xy. Multiply the expression by $\dfrac{xy}{xy}$.

$$\frac{\left(\dfrac{1}{x} - \dfrac{1}{y}\right) \cdot xy}{(x - y) \cdot xy} = \frac{\dfrac{xy}{x} - \dfrac{xy}{y}}{xy(x - y)} \qquad \begin{array}{l}\text{Distributive and commutative} \\ \text{properties}\end{array}$$

$$= \frac{y - x}{xy(x - y)} \qquad \text{Reduce.}$$

$$= \frac{-1(x - y)}{xy(x - y)} \qquad \text{Factor out } -1.$$

$$= -\frac{1}{xy} \qquad \text{Reduce.}$$

(b) The LCD of x and $x - 1$ is their product, $x(x - 1)$. Multiply the expression by $\dfrac{x(x - 1)}{x(x - 1)}$.

$$\frac{\left(\dfrac{3}{x - 1} - \dfrac{2}{x}\right)}{\left(\dfrac{1}{x - 1} + \dfrac{3}{x}\right)} \cdot \frac{x(x - 1)}{x(x - 1)} = \frac{\dfrac{3x(x - 1)}{x - 1} - \dfrac{2x(x - 1)}{x}}{\dfrac{x(x - 1)}{x - 1} + \dfrac{3x(x - 1)}{x}} \qquad \text{Distributive property}$$

$$= \frac{3x - 2(x - 1)}{x + 3(x - 1)} \qquad \text{Reduce.}$$

$$= \frac{3x - 2x + 2}{x + 3x - 3} \qquad \text{Distributive property}$$

$$= \frac{x + 2}{4x - 3} \qquad \text{Combine like terms.}$$

R.6 EXERCISES

Simplifying Rational Expressions

Exercises 1–8: Simplify the expression.

1. $\dfrac{10x^3}{5x^2}$

2. $\dfrac{24t^3}{6t^2}$

3. $\dfrac{(x-5)(x+5)}{x-5}$

4. $(3x+2) \cdot \dfrac{2x-6}{3x+2}$

5. $\dfrac{x^2-16}{x-4}$

6. $\dfrac{(x+5)(x-4)}{(x+7)(x+5)}$

7. $\dfrac{x+3}{2x^2+5x-3}$

8. $\dfrac{x^2-9}{x^2+6x+9} \cdot (x+3)$

Multiplication and Division of Rational Expressions

Exercises 9–24: Simplify the expression.

9. $\dfrac{1}{x^2} \cdot \dfrac{3x}{2}$

10. $\dfrac{6a}{5} \cdot \dfrac{5}{12a^2}$

11. $\dfrac{5x}{3} \div \dfrac{10x}{6}$

12. $\dfrac{2x^2+x}{3x+9} \div \dfrac{x}{x+3}$

13. $\dfrac{x+1}{2x-5} \cdot \dfrac{x}{x+1}$

14. $\dfrac{4x+8}{2x} \cdot \dfrac{x^2}{x+2}$

15. $\dfrac{(x-5)(x+3)}{3x-1} \cdot \dfrac{x(3x-1)}{(x-5)}$

16. $\dfrac{b^2+1}{b^2-1} \cdot \dfrac{b-1}{b+1}$

17. $\dfrac{x^2-2x-35}{2x^3-3x^2} \cdot \dfrac{x^3-x^2}{2x-14}$

18. $\dfrac{2x+4}{x+1} \cdot \dfrac{x^2+3x+2}{4x+2}$

19. $\dfrac{6b}{b+2} \div \dfrac{3b^4}{2b+4}$

20. $\dfrac{5x^5}{x-2} \div \dfrac{10x^3}{5x-10}$

21. $\dfrac{3a+1}{a^7} \div \dfrac{a+1}{3a^8}$

22. $\dfrac{x^2-16}{x+3} \div \dfrac{x+4}{x^2-9}$

23. $\dfrac{x+5}{x^3-x} \div \dfrac{x^2-25}{x^3}$

24. $\dfrac{x^2+x-12}{2x^2-9x-5} \div \dfrac{x^2+7x+12}{2x^2-7x-4}$

Common Denominators

Exercises 25–30: Find the LCD for the rational expression.

25. $\dfrac{1}{x+1}, \dfrac{1}{7}$

26. $\dfrac{1}{2x-1}, \dfrac{1}{x+1}$

27. $\dfrac{1}{x+4}, \dfrac{1}{x^2-16}$

28. $\dfrac{x}{2x^2}, \dfrac{1}{2x+2}$

29. $\dfrac{3}{2}, \dfrac{x}{2x+1}, \dfrac{x}{2x-4}$

30. $\dfrac{1}{x}, \dfrac{1}{x^2-4x}, \dfrac{1}{2x}$

Addition and Subtraction of Rational Expressions

Exercises 31–48: Simplify.

31. $\dfrac{4}{x+1} + \dfrac{3}{x+1}$

32. $\dfrac{2}{x^2} + \dfrac{5}{x^2}$

33. $\dfrac{2}{x^2-1} - \dfrac{x+1}{x^2-1}$

34. $\dfrac{2x}{x^2+x} - \dfrac{2x}{x+1}$

35. $\dfrac{x}{x+4} - \dfrac{x+1}{x(x+4)}$

36. $\dfrac{4x}{x+2} + \dfrac{x-5}{x-2}$

37. $\dfrac{2}{x^2} - \dfrac{4x-1}{x}$

38. $\dfrac{2x}{x-5} - \dfrac{x}{x+5}$

39. $\dfrac{x+3}{x-5} + \dfrac{5}{x-3}$

40. $\dfrac{x}{2x-1} + \dfrac{1-x}{3x}$

41. $\dfrac{3}{x-5} - \dfrac{1}{x-3} - \dfrac{2x}{x-5}$

42. $\dfrac{2x+1}{x-1} - \dfrac{3}{x+1} + \dfrac{x}{x-1}$

43. $\dfrac{x}{x^2-9} + \dfrac{5x}{x-3}$

44. $\dfrac{a^2+1}{a^2-1} + \dfrac{a}{1-a^2}$

45. $\dfrac{b}{2b-4} - \dfrac{b-1}{b-2}$

46. $\dfrac{y^2}{2-y} - \dfrac{y}{y^2-4}$

47. $\dfrac{2x}{x-5} + \dfrac{2x-1}{3x^2-16x+5}$

48. $\dfrac{x+3}{2x-1} + \dfrac{3}{10x^2-5x}$

Clearing Fractions

Exercises 49–58: (Refer to Example 7.) Multiply the expression by its least common denominator and simplify. Note that the final result does not equal the given expression.

49. $\dfrac{1}{x} + \dfrac{3}{x^2}$

50. $\dfrac{1}{x - 2} + \dfrac{3}{x + 1}$

51. $\dfrac{1}{x} + \dfrac{5x}{2x - 1}$

52. $\dfrac{x}{2x - 5} + \dfrac{3}{x}$

53. $\dfrac{2x}{9 - x^2} + \dfrac{1}{3 - x}$

54. $\dfrac{1}{1 - x^2} + \dfrac{1}{1 + x}$

55. $\dfrac{1}{2x} + \dfrac{1}{3x^2} - \dfrac{1}{x^3}$

56. $\dfrac{1}{x^2 - 16} + \dfrac{4}{x + 4} - \dfrac{5}{x - 4}$

57. $\dfrac{1}{x} + \dfrac{3x}{x + 5} - \dfrac{1}{x - 5}$

58. $\dfrac{1}{x - 2} + \dfrac{3}{x - 3} - \dfrac{1}{x}$

Complex Fractions

Exercises 59–68: Simplify the expression.

59. $\dfrac{1 + \dfrac{1}{x}}{1 - \dfrac{1}{x}}$

60. $\dfrac{\dfrac{1}{2} - x}{\dfrac{1}{x} - 2}$

61. $\dfrac{\dfrac{1}{x - 5}}{\dfrac{4}{x} - \dfrac{1}{x - 5}}$

62. $\dfrac{1 + \dfrac{1}{x - 3}}{\dfrac{1}{x - 3} - 1}$

63. $\dfrac{\dfrac{1}{x} + \dfrac{2 - x}{x^2}}{\dfrac{3}{x^2} - \dfrac{1}{x}}$

64. $\dfrac{\dfrac{1}{x - 1} + \dfrac{2}{x}}{2 - \dfrac{1}{x}}$

65. $\dfrac{\dfrac{1}{x + 3} + \dfrac{2}{x - 3}}{2 - \dfrac{1}{x - 3}}$

66. $\dfrac{\dfrac{1}{x} + \dfrac{2}{x}}{\dfrac{1}{x - 1} + \dfrac{x}{2}}$

67. $\dfrac{\dfrac{4}{x - 5}}{\dfrac{1}{x + 5} + \dfrac{1}{x}}$

68. $\dfrac{\dfrac{2}{x - 4}}{1 - \dfrac{1}{x + 4}}$

Appendix

A Library of Functions

Basic Functions

Several important functions are used in algebra. The following provides symbolic, numerical, and graphical representations for some of these basic functions.

Identity Function: $f(x) = x$

x	−2	−1	0	1	2
$y = x$	−2	−1	0	1	2

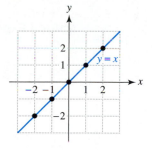

Absolute Value Function: $f(x) = |x|$

x	−2	−1	0	1	2
$y = |x|$	2	1	0	1	2

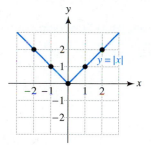

Square Function: $f(x) = x^2$

x	−2	−1	0	1	2
$y = x^2$	4	1	0	1	4

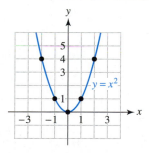

Cube Function: $f(x) = x^3$

x	−2	−1	0	1	2
$y = x^3$	−8	−1	0	1	8

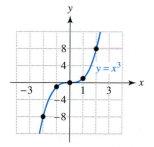

Square Root Function: $f(x) = \sqrt{x}$

x	0	1	4	9
$y = \sqrt{x}$	0	1	2	3

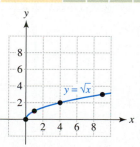

Cube Root Function: $f(x) = \sqrt[3]{x}$

x	−8	−1	0	1	8
$y = \sqrt[3]{x}$	−2	−1	0	1	2

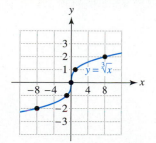

Greatest Integer Function: $f(x) = [x]$

x	-2.5	-1.5	0	1.5	2.5
$y = [x]$	-3	-2	0	1	2

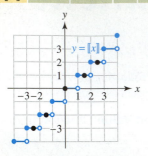

Reciprocal Function: $f(x) = \dfrac{1}{x}$

x	-2	-1	0	1	2
$y = \dfrac{1}{x}$	$-\dfrac{1}{2}$	-1	—	1	$\dfrac{1}{2}$

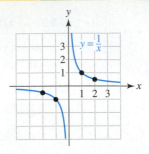

Base-2 Exponential Function: $f(x) = 2^x$

x	-2	-1	0	1	2
$y = 2^x$	$\dfrac{1}{4}$	$\dfrac{1}{2}$	1	2	4

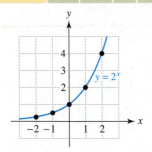

Natural Exponential Function: $f(x) = e^x$

x	-2	-1	0	1	2
$y = e^x$	e^{-2}	e^{-1}	1	e^1	e^2

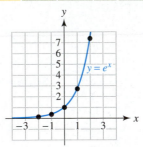

Common Logarithmic Function: $f(x) = \log x$

x	0.1	1	4	7	10
$y = \log x$	-1	0	$\log 4$	$\log 7$	1

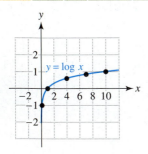

Natural Logarithmic Function: $f(x) = \ln x$

x	$\dfrac{1}{2}$	1	2	e	e^2
$y = \ln x$	$\ln \dfrac{1}{2}$	0	$\ln 2$	1	2

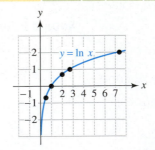

Families of Functions

Some examples of different families of functions are linear, quadratic, and exponential functions. This subsection shows the formulas and graphs of some families of functions. Notice that the graphs of these functions change when their formulas are changed.

Constant Functions: $f(x) = k$

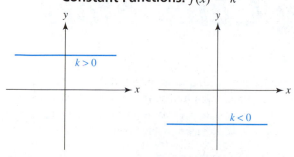

Linear Functions: $f(x) = mx + b$

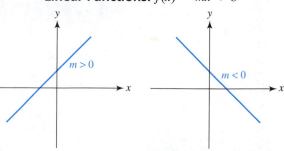

Quadratic Functions: $f(x) = ax^2 + bx + c$

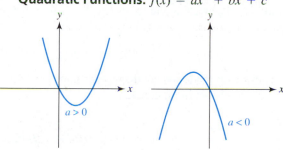

Cubic Functions: $f(x) = ax^3 + bx^2 + cx + d$

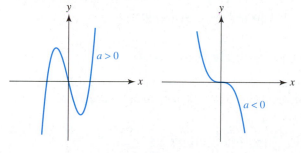

Power Functions: $f(x) = x^a, x > 0$

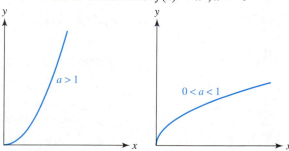

Exponential Functions: $f(x) = Ca^x$

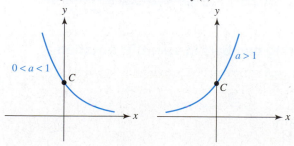

Logarithmic Functions: $f(x) = \log_a x$

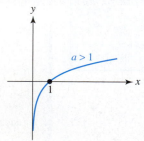

The TI-83/83 Plus Graphing Calculators

Overview of the Appendix

The intent of this appendix is to provide instruction for the TI-83 and TI-83 Plus graphing calculators that may be used in conjunction with this textbook. It includes specific keystrokes needed to work several examples from the text. Students are also advised to consult the *Graphing Calculator Guidebook* provided by the manufacturer.

FIGURE B.1

Displaying Numbers in Scientific Notation

To display numbers in scientific notation, set the graphing calculator in scientific mode (Sci), by using the following keystrokes. See Figure B.1. (These keystrokes assume that the calculator is in normal mode.)

$$\boxed{\text{MODE}}\; \boxed{\triangleright}\; \boxed{\text{ENTER}}\; \boxed{\text{2nd}}\; \boxed{\text{MODE [QUIT]}}$$

In scientific mode we can display the numbers 5432 and 0.00001234 in scientific notation, as shown in Figure B.2.

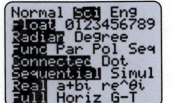

FIGURE B.2

> ### Summary: Setting scientific mode
>
> If your calculator is in normal mode, it can be set in scientific mode by pressing
>
> $$\boxed{\text{MODE}}\; \boxed{\triangleright}\; \boxed{\text{ENTER}}\; \boxed{\text{2nd}}\; \boxed{\text{MODE [QUIT]}}.$$
>
> These keystrokes return the graphing calculator to the home screen.

Entering Numbers in Scientific Notation

Numbers can be entered in scientific notation. For example, to enter 4.2×10^{-3} in scientific notation, use the following keystrokes. (Be sure to use the negation key $(-)$ rather than the subtraction key.)

$$\boxed{4}\; \boxed{.}\; \boxed{2}\; \boxed{\text{2nd}}\; \boxed{,}\; \boxed{(-)}\; \boxed{3}$$

This number can also be entered using the following keystrokes. See Figure B.3.

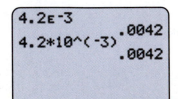

FIGURE B.3

> ### Summary: Entering numbers in scientific notation
>
> One way to enter a number in scientific notation is to use the keystrokes
>
> $$\boxed{\text{2nd}}\; \boxed{,}$$
>
> to access an exponent (EE) of 10.

Entering Mathematical Expressions

Several expressions are evaluated in Example 7, Section 1.1. To evaluate $\sqrt[3]{131}$, use the following keystrokes from the home screen.

To calculate $\pi^3 + 1.2^2$, use the following keystrokes. (Do *not* use 3.14 for π.)

(2nd) (^[π]) (^) (3) (+) (1) (.) (2) (x^2) (ENTER)

To calculate $|\sqrt{3} - 6|$, use the following keystrokes.

(MATH) (▷) (1) (2nd) (x^2[√]) (3) ()) (−) (6) ()) (ENTER)

Calculating One-Variable Statistics

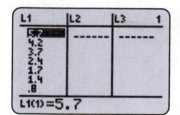

FIGURE B.4

In Example 2, Section 1.2, the mean, median, minimum, and maximum for a data set are found with a graphing calculator. Begin by entering the data into list L1 with the "STAT EDIT" menus, as shown in Figure B.4. Use the following keystrokes.

(STAT) (1) (5) (.) (7) (ENTER) (4) (.) (2) (ENTER) (3) (.) (7) (ENTER)

(2) (.) (4) (ENTER) . . . .

Continue until all 12 numbers have been entered into list L1. To calculate one-variable statistics press the following keys.

(STAT) (▷) (1) (ENTER)

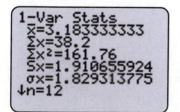

FIGURE B.5

The mean (or average) of the data is given by $\bar{x} = 3.183333333$, as shown in Figure B.5. (If you enroll in a statistics class you will learn more about some of the numbers that appear.) In Figure B.6, the minimum is given by $\text{minX} = 0.8$, the maximum by $\text{maxX} = 5.9$, and the median by $\text{Med} = 2.95$.

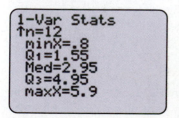

FIGURE B.6

Setting the Viewing Rectangle

In Example 7, Section 1.2, there are at least two ways to set the standard viewing rectangle to $[-10, 10, 1]$ by $[-10, 10, 1]$. The first involves pressing (ZOOM) followed by (6). See Figure B.7. The second method for setting the standard viewing rectangle is to press (WINDOW), and enter the following keystrokes. See Figure B.8.

((−))(1)(0)(ENTER)(1)(0)(ENTER)(1)(ENTER)

((−))(1)(0)(ENTER)(1)(0)(ENTER)(1)(ENTER)

(Be sure to use the negation key $(-)$ rather than the subtraction key.) The viewing rectangle $[-30, 40, 10]$ by $[-400, 800, 100]$ can be set in a similar manner by pressing (WINDOW), as shown in Figure B.9. To see the viewing rectangle, press (GRAPH).

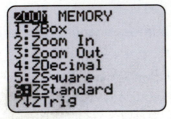

FIGURE B.7

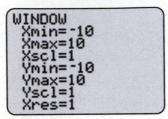

FIGURE B.8

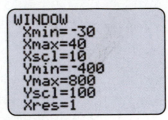

FIGURE B.9

Summary: Setting the viewing rectangle

To set the standard viewing rectangle, press (ZOOM)(6). To set any viewing rectangle, press (WINDOW) and enter the necessary values. To see the viewing rectangle, press (GRAPH).

Note: You do not need to change "Xres".

Making a Scatterplot or a Line Graph

Example 8, Section 1.2 asks us to make a scatterplot with $(-5, -5)$, $(-2, 3)$, $(1, -7)$, and $(4, 8)$. To accomplish this task, begin by following these steps.

1. Press (STAT) followed by (1).
2. If list L1 is not empty, use the arrow keys to place the cursor on L1, as shown in Figure B.10. Then press (CLEAR) followed by (ENTER). This deletes all elements in the list. Similarly, if L2 is not empty, clear the list.
3. Input each x-value into list L1 followed by (ENTER). Input each y-value into list L2 followed by (ENTER). See Figure B.11.

It is essential that both lists have the same number of values—otherwise an error message appears when a scatterplot is attempted. Before these four points can be plotted, "STATPLOT" must be turned on. It is accessed by pressing

(2nd)(Y = [STAT PLOT]),

as shown in Figure B.12.

There are three possible "STATPLOTS", numbered 1, 2, and 3. Any one of the three can be selected. The first plot can be selected by pressing (1). Next, place the cursor over "On" and press (ENTER) to turn "Plot1" on. There are six types of plots that can be selected. The first type is a *scatterplot* and the second type is a *line graph*, so place the cursor over the first type of plot and press (ENTER) to select a scatterplot. (To make the line graph in Example 9, Section 1.2, be sure to select the line graph.) The *x*-values are stored in list L1, so select L1 for "Xlist" by pressing (2nd)(1). Similarly press (2nd)(2) for the "Ylist," since the *y*-values are stored in list L2. Finally, there are three styles of marks that can be used to show data points in the graph. We will usually use the first, because it is largest and shows up the best. Make the screen appear as in Figure B.13. Before plotting the four data points, be sure to set an appropriate viewing rectangle. Then press (GRAPH). The data points appear as in Figure B.14.

Remark 1: A fast way to set the viewing rectangle for any scatterplot is to select the "ZOOMSTAT" feature by pressing (ZOOM)(9). This feature automatically scales the viewing rectangle so that all data points are shown.

Remark 2: If an equation has been entered into the (Y =) menu and selected, it will be graphed with the data. This feature is used frequently to model data throughout the textbook.

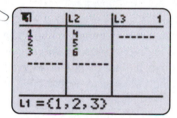

FIGURE B.10

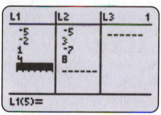

FIGURE B.11

FIGURE B.12

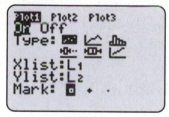

FIGURE B.13

$[-10, 10, 1]$ by $[-10, 10, 1]$

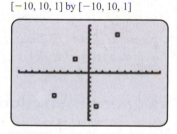

FIGURE B.14

Summary: Making a scatterplot or a line graph

The following are basic steps necessary to make either a scatterplot or line graph.

1. Use (STAT)(1) to access lists L1 and L2.
2. If list L1 is not empty, place the cursor on L1 and press (CLEAR)(ENTER). Repeat for list L2, if it is not empty.
3. Enter the *x*-values into list L1 and the *y*-values into list L2.
4. Use (2nd)(Y = [STAT PLOT]) to set the appropriate parameters for the scatterplot or line graph.
5. Set an appropriate viewing rectangle. Otherwise, press (ZOOM)(9). This feature automatically sets the viewing rectangle and plots the data.

Note: (ZOOM)(9) *cannot* be used to set a viewing rectangle for the graph of a function.

FIGURE B.15

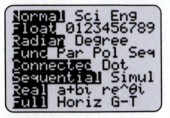

FIGURE B.16

Entering a Formula for a Function

To enter the formula for a function f, press $\boxed{Y=}$. For example, use the following keystrokes after "$Y_1 =$" to enter $f(x) = 2x^2 - 3x + 7$. See Figure B.15.

$$\boxed{Y=}\ \boxed{\text{CLEAR}}\ \boxed{2}\ \boxed{X, T, \theta, n}\ \boxed{\wedge}\ \boxed{2}\ \boxed{-}\ \boxed{3}\ \boxed{X, T, \theta, n}\ \boxed{+}\ \boxed{7}$$

Note that there is a built-in key to enter the variable X. If "$Y_1 =$" does not appear after pressing $\boxed{Y=}$, press $\boxed{\text{MODE}}$ and make sure the calculator is set in function mode denoted "Func". See Figure B.16.

> ## Summary: Entering a formula for a function
>
> To enter the formula for a function, press $\boxed{Y=}$. To delete an existing formula, press $\boxed{\text{CLEAR}}$. Then enter the symbolic representation for the function.

FIGURE B.17

[−10, 10, 1] by [−10, 10, 1]

FIGURE B.18

Graphing a Function

To graph a function, such as $f(x) = x^2 - 4$, start by pressing $\boxed{Y=}$ and enter $Y_1 = X^2 - 4$. If there is an equation already entered, remove it by pressing $\boxed{\text{CLEAR}}$. The equals signs in "$Y_1 =$" should be in reverse video (a dark rectangle surrounding a white equals sign), which indicates that the equation will be graphed. If the equals sign is not in reverse video, place the cursor over it and press $\boxed{\text{ENTER}}$. Set an appropriate viewing rectangle, and then press $\boxed{\text{GRAPH}}$. The graph of f will appear in the specified viewing rectangle. See Figures B.17 and B.18.

> ## Summary: Graphing a function
>
> Use the $\boxed{Y=}$ menu to enter the formula for the function and the $\boxed{\text{WINDOW}}$ menu to set an appropriate viewing rectangle. Then press $\boxed{\text{GRAPH}}$.

Evaluating a Function Graphically

In Example 7, Section 1.3, we evaluate the function $f(x) = 0.72x + 2$ graphically at $x = 65$. Begin by entering the formula $Y_1 = .72X + 2$ into the $\boxed{Y=}$ menu. Then graph f in the appropriate viewing rectangle, as shown in Figure 1.40. To evaluate $f(65)$, use the following keystrokes, which access the CALCULATE menu.

$$\boxed{\text{2nd}}\ \boxed{\text{TRACE [CALC]}}\ \boxed{1}\ \boxed{6}\ \boxed{5}\ \boxed{\text{ENTER}}$$

See Figures 1.41 and 1.42.

> ## Summary: Evaluating a function graphically
>
> To evaluate a function graphically, begin by graphing the function. Then use the following keystrokes to use the "value" routine in the CALCULATE menu.
>
> $$\boxed{\text{2nd}}\ \boxed{\text{TRACE [CALC]}}\ \boxed{1}$$
>
> Then enter the x-value where the function should be evaluated, and press $\boxed{\text{ENTER}}$.

Making a Table

In Example 7, Section 1.3, a table of values is requested. Start by pressing $\boxed{Y=}$ and then entering the formula $Y_1 = .72X + 2$, as shown in Figure B.19. To set the table parameters, press the following keystrokes. See Figure B.20.

$$\boxed{2nd}\ \boxed{WINDOW\ [TBLSET]}\ \boxed{6}\ \boxed{0}\ \boxed{ENTER}\ \boxed{1}$$

These keystrokes specify a table that starts at $x = 60$ and increments the x-values by 1. Therefore, the values of Y_1 at $x = 60, 61, 62, \ldots$ appear in the table. To create this table, press the following keys.

$$\boxed{2nd}\ \boxed{GRAPH\ [TABLE]}$$

One can stroll through x- and y-values by using the arrow keys. See Figure B.21. Note that there is no first or last x-value in the table.

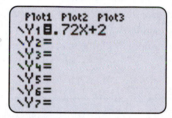

FIGURE B.19

FIGURE B.20

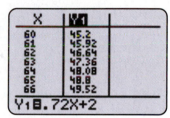

FIGURE B.21

Summary: Making a table of a function

Enter the formula for the function using $\boxed{Y=}$. Then press

$$\boxed{2nd}\ \boxed{WINDOW\ [TBLSET]}$$

to set the starting x-value and the increment between x-values appearing in the table. Create the table by pressing

$$\boxed{2nd}\ \boxed{GRAPH\ [TABLE]}.$$

Squaring a Viewing Rectangle

In a square viewing rectangle the graph of $y = x$ is a line that makes a 45° angle with the positive x-axis, a circle appears circular, and all sides of a square have the same length. An approximate square viewing rectangle can be set if the distance along the x-axis is 1.5 times the distance along the y-axis. Examples of viewing rectangles that are (approximately) square include

$$[-6, 6, 1] \text{ by } [-4, 4, 1] \quad \text{and} \quad [-9, 9, 1] \text{ by } [-6, 6, 1]$$

Square viewing rectangles can be set automatically by pressing either

$$\boxed{ZOOM}\ \boxed{4} \quad \text{or} \quad \boxed{ZOOM}\ \boxed{5}$$

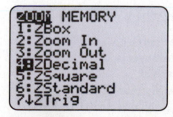

FIGURE B.22

ZOOM 4 provides a decimal window, which is discussed later. See Figure B.22.

Summary: Squaring a viewing rectangle

Either ⟨ZOOM⟩⟨4⟩ or ⟨ZOOM⟩⟨5⟩ may be used to produce a square viewing rectangle. An (approximately) square viewing rectangle has the form

$$[-1.5k, 1.5k, 1] \text{ by } [-k, k, 1],$$

where k is a positive number.

Plotting Data and an Equation

In Example 1, Section 2.1, we plot data and graph a modeling function in the same viewing rectangle. (You may want to refer to the subsection on making a scatterplot and line graph in this appendix.) Start by entering the x-values into list L1 and the y-values into list L2, as shown in Figure B.23. Then press ⟨Y =⟩ and enter the formula $Y_1 = .062X - 122.57$ for $f(x)$. Make sure that "STATPLOT" is on and set an appropriate viewing rectangle. See Figures B.24 and B.25, and note that Figure B.24 shows "Plot1" in reverse video, which indicates that the scatterplot is on. Now press ⟨GRAPH⟩ to have both the scatterplot and the graph of Y_1 appear in the same viewing rectangle, as shown in Figure B.26.

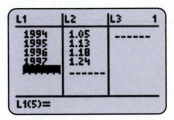

FIGURE B.23

FIGURE B.24

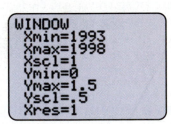

FIGURE B.25

[1993, 1998, 1] by [0, 1.5, 0.5]

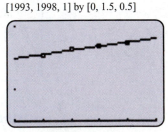

FIGURE B.26

Summary: Plotting data and an equation

1. Enter the x-values into the list L1 and the y-values into list L2 using the STAT EDIT menu. Turn on "Plot1" so that the scatterplot appears.
2. Use the ⟨Y =⟩ menu to enter the equation to be graphed.
3. Use ⟨WINDOW⟩ or ⟨ZOOM⟩ to set an appropriate viewing rectangle.
4. Press ⟨GRAPH⟩ to graph both the scatterplot and the equation in the same viewing rectangle.

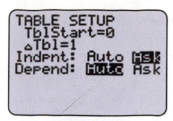

FIGURE B.27

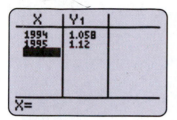

FIGURE B.28

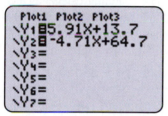

FIGURE B.29

Using the Ask Table Feature

In Example 1, Section 2.1, a table with x-values of 1994, 1995, 1996, 1997 is created. Start by entering $Y_1 = .062X - 122.57$. To obtain the table shown in Figure 2.4, use the "Ask" feature rather than the "Auto" feature for the independent variable (Indpnt:). Press ⟨2nd⟩⟨GRAPH [TABLE]⟩. Whenever an x-value is entered, the corresponding y-value is calculated automatically. See Figures B.27 and B.28.

Summary: Using the ask feature for a table

1. Enter the formula for $f(x)$ into Y_1 by using the ⟨Y =⟩ menu.
2. Press ⟨2nd⟩⟨WINDOW [TBLSET]⟩ to access "TABLE SETUP" and then select "Ask" for the independent variable (Indpnt:). "TblStart" and "ΔTbl" do not need to be set.
3. Enter x-values of your choice. The corresponding y-values will be calculated automatically.

Locating a Point of Intersection

In Example 1, Section 2.3, we find the point of intersection for two lines. To find the point of interesection for the graphs of

$$f(x) = 5.91x + 13.7 \quad \text{and} \quad g(x) = -4.71x + 64.7,$$

start by entering Y_1 and Y_2, as shown in Figure B.29. Set the viewing rectangle to [0, 12, 1] by [0, 100, 10], and graph both equations in the same viewing rectangle, as shown in Figure 2.25. Then press the following keys to find the intersection point.

⟨2nd⟩⟨TRACE [CALC]⟩⟨5⟩

See Figure B.30, where the "intersect" utility is being selected. The calculator prompts for the first curve, as shown in Figure B.31. Use the arrow keys to locate the cursor near the point of intersection and press ⟨ENTER⟩. Repeat these steps for the second curve, as shown in Figure B.32. Finally we are prompted for a guess. For each of the three prompts, place the free-moving cursor near the point of intersection and press ⟨ENTER⟩. The approximate coordinates of the point of intersection are shown in Figure 2.26.

FIGURE B.30

[0, 12, 1] by [0, 100, 10]

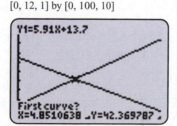

FIGURE B.31

[0, 12, 1] by [0, 100, 10]

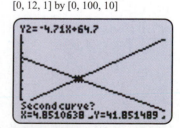

FIGURE B.32

Summary: Finding a point of intersection

1. Graph the two functions in an appropriate viewing rectangle.
2. Press ⟨2nd⟩⟨TRACE [CALC]⟩⟨5⟩.

---

3. Use the arrow keys to select an approximate location for the point of intersection. Press $\boxed{\text{ENTER}}$ to make the three selections for "First curve?", "Second curve?", and "Guess?". (Note that if the cursor is near the point of intersection, you usually do not need to move the cursor for each selection.)

Locating a Zero of a Function

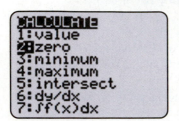

FIGURE B.33

In Example 3, Section 2.4, we locate an x-intercept or *zero* of $f(x) = 1 - x - \left(\frac{1}{2}x - 2\right)$. Start by entering $Y_1 = 1 - X - (.5X - 2)$ into the $\boxed{Y =}$ menu. Set the viewing rectangle to $[-6, 6, 1]$ by $[-4, 4, 1]$ and graph Y_1. Afterwards, press the following keys to invoke the zero finder. See Figure B.33.

$$\boxed{\text{2nd}}\ \boxed{\text{TRACE [CALC]}}\ \boxed{2}$$

The graphing calculator prompts for a left bound. Use the arrow keys to set the cursor to the left of the x-intercept and press $\boxed{\text{ENTER}}$. The graphing calculator then prompts for a right bound. Set the cursor to the right of the x-intercept and press $\boxed{\text{ENTER}}$. Finally the graphing calculator prompts for a guess. Set the cursor roughly at the x-intercept and press $\boxed{\text{ENTER}}$. See Figures B.34–B.36. The calculator then approximates the x-intercept or zero automatically, as shown in Figure 2.39.

$[-6, 6, 1]$ by $[-4, 4, 1]$

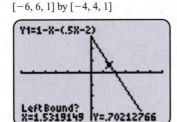

FIGURE B.34

$[-6, 6, 1]$ by $[-4, 4, 1]$

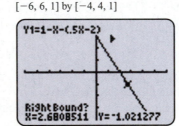

FIGURE B.35

$[-6, 6, 1]$ by $[-4, 4, 1]$

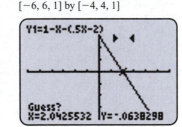

FIGURE B.36

Summary: Locating a zero of a function

1. Graph the function in an appropriate viewing rectangle.
2. Press $\boxed{\text{2nd}}\ \boxed{\text{TRACE [CALC]}}\ \boxed{2}$.
3. Select the left and right bounds, followed by a guess. Press $\boxed{\text{ENTER}}$ after each selection. The calculator then approximates the zero.

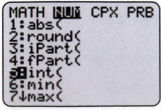

FIGURE B.37

Accessing the Greatest Integer Function

To access the greatest integer function, enter the following keystrokes from the home screen.

$$\boxed{\text{MATH}}\ \boxed{\triangleright}\ \boxed{5}.$$

See Figure B.37.

Setting Connected and Dot Mode

FIGURE B.38

In Figure 2.56 of Section 2.5 the greatest integer function is graphed in dot mode, and in Figure 2.57 it is graphed in connected mode. To set your graphing calculator in dot mode, press MODE, position the cursor over "Dot", and press ENTER. See Figure B.38. Graphs will now appear in dot mode rather than connected mode.

Accessing the Absolute Value

In Example 4, Section 2.5, the absolute value is used to graph $f(x) = |x + 2|$. To graph f, begin by entering $Y_1 = $ abs(X + 2). The absolute value (abs) is accessed by pressing

See Figure B.39.

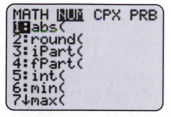

FIGURE B.39

Finding the Line of Least-Squares Fit

In Example 3, Section 2.6, the line of least-squares fit for the points (1, 1), (2, 3), and (3, 4) is found. Begin by entering the points in the same way a scatterplot is entered. See Figure 2.78, where the x-values are in list L1 and the y-values are in list L2.

After the data has been entered, perform the following keystrokes from the home screen.

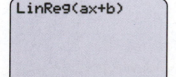

FIGURE B.40

See Figure 2.79. This causes "LinReg(ax+b)" to appear on the home screen, as shown in Figure B.40. The graphing calculator assumes that the x-values are in list L1 and the y-values are in list L2. Now press ENTER. The result is shown in Figure 2.80.

If the correlation coefficient r does not appear, then enter the following key-strokes.

$$\boxed{\text{2nd}}\ \boxed{0\ [\text{CATALOG}]}$$

and scroll down until "DiagnosticsOn" is found. Press $\boxed{\text{ENTER}}$ twice. See Figures B.41 and B.42. The graph of the data and the least-squares regression line are shown in Figure 2.81.

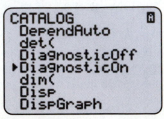

FIGURE B.41

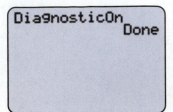

FIGURE B.42

Summary: Linear least-squares fit

1. Enter the data using $\boxed{\text{STAT}}\ \boxed{1}$, as is done for a scatterplot. Input the x-values into list L1 and the y-values into list L2.
2. Press $\boxed{\text{STAT}}\ \boxed{\triangleright}\ \boxed{4}$ from the home screen to access the least-squares regression line. Press $\boxed{\text{ENTER}}$ to start the computation.

Finding Extrema (Minima and Maxima)

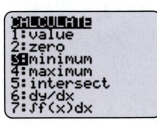

FIGURE B.43

In Example 5, Section 3.1, we are asked to find a minimum point (or vertex) on the graph of $f(x) = 1.5x^2 - 6x + 4$. Start by entering $Y_1 = 1.5X^2 - 6X + 4$ from the $\boxed{\text{Y}=}$ menu. Set the viewing rectangle to $[-4.7, 4.7, 1]$ by $[-3.1, 3.1, 1]$. Then perform the following keystrokes to find the minimum y-value.

$$\boxed{\text{2nd}}\ \boxed{\text{TRACE [CALC]}}\ \boxed{3}$$

See Figure B.43.

    The calculator prompts for a left bound. Use the arrow keys to position the cursor left of the vertex and press $\boxed{\text{ENTER}}$. Similarly, position the cursor to the right of the vertex for the right bound and press $\boxed{\text{ENTER}}$. Finally, the graphing calculator asks for a guess between the left and right bounds. Place the cursor near the minimum point and press $\boxed{\text{ENTER}}$. See Figures B.44–B.46. The minimum point (or vertex) is shown in Figure 3.20.

$[-4.7, 4.7, 1]$ by $[-3.1, 3.1, 1]$

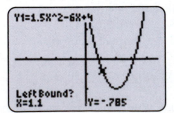

FIGURE B.44

$[-4.7, 4.7, 1]$ by $[-3.1, 3.1, 1]$

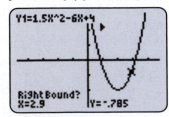

FIGURE B.45

$[-4.7, 4.7, 1]$ by $[-3.1, 3.1, 1]$

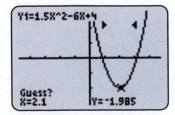

FIGURE B.46

A maximum of the function f on an interval can be found in a similar manner, except enter

$$\boxed{\text{2nd}}\,\boxed{\text{TRACE [CALC]}}\,\boxed{4}\,.$$

The calculator prompts for left and right bounds, followed by a guess. Press $\boxed{\text{ENTER}}$ after the cursor has been located appropriately for each prompt. The maximum point is displayed as it is in Figure 3.27.

Summary: Finding extrema (maxima and minima)

1. Graph the function in an appropriate viewing rectangle.
2. Press $\boxed{\text{2nd}}\,\boxed{\text{TRACE [CALC]}}\,\boxed{3}$ to find a minimum point.
3. Press $\boxed{\text{2nd}}\,\boxed{\text{TRACE [CALC]}}\,\boxed{4}$ to find a maximum point.
4. Use the arrow keys to locate the left and right x-bounds, followed by a guess. Press $\boxed{\text{ENTER}}$ to select each position of the cursor.

Finding a Nonlinear Function of Least-Squares Fit

In Example 10, Section 3.1, a quadratic function of least-squares fit is found in a manner similar to the way a linear function of least-squares fit is found. To solve Example 10, start by pressing $\boxed{\text{STAT}}\,\boxed{1}$ and then enter the data points from Table 3.4, as shown in Figure 3.31. Input the x-values into list L1 and the y-values into list L2. To find the equation for a quadratic polynomial of least-squares fit, perform the following keystrokes from the home screen.

$$\boxed{\text{CLEAR}}\,\boxed{\text{STAT}}\,\boxed{\triangleright}\,\boxed{5}$$

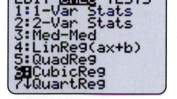

FIGURE B.47

This causes the "Quadreg" to appear on the home screen. The graphing calculator assumes that the x-values are in list L1 and the y-values are in list L2, unless otherwise designated. Press $\boxed{\text{ENTER}}$ to obtain the quadratic regression equation, as shown in Figure 3.33. A graph of the data and the regression equation are shown in Figure 3.34.

Other types of regression equations, such as cubic, quartic, power, and exponential, can be selected from the STAT CALC menu. See Figure B.47.

Summary: Nonlinear least-squares fit

1. Enter the data using $\boxed{\text{STAT}}\,\boxed{1}$. Input the x-values into list L1 and the y-values into list L2, as is done for a scatterplot.
2. From the home screen press $\boxed{\text{STAT}}\,\boxed{\triangleright}$ and select a type of least-squares modeling function from the menu. Press $\boxed{\text{ENTER}}$ to initiate the computation.

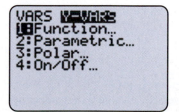

FIGURE B.48

Accessing the Variable Y_1

In Figure 3.94, Section 3.4, the expressions $-Y_1$ and $Y_1(-X)$ in the $\boxed{Y=}$ menu are used to graph reflections. The Y_1 variable can be found by pressing the following keys. See Figures B.48 and B.49.

$$\boxed{\text{VARS}}\,\boxed{\triangleright}\,\boxed{1}\,\boxed{1}$$

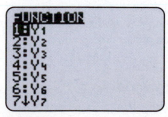

FIGURE B.49

Summary: Accessing the variable Y₁

1. Press (VARS).
2. Position the cursor over "Y-VARS".
3. Press (1) twice.
These keystrokes will make Y_1 appear on the screen.

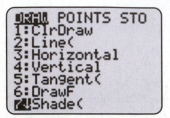

FIGURE B.50

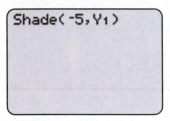

FIGURE B.51

Shading between Two Graphs

In Example 4, Section 3.4, the region below the graph of $f(x) = -0.4x^2 + 4$ is shaded to make it appear like a mountain, as illustrated in Figure 3.104. One way to shade below the graph of f is to begin by entering $Y_1 = -.4X^2 + 4$ after pressing (Y=). Then use the following keystrokes from the home screen.

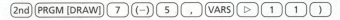

The expression Shade(-5, Y_1) should appear on your home screen. See Figures B.50 and B.51. The shading utility is accessed from the DRAW menu, where this shading option requires a lower function and an upper function, respectively, separated by a comma. When (ENTER) is pressed the graphing calculator shades between the graph of the lower function and the graph of the upper function. For the lower function we have arbitrarily selected $y = -5$ because its graph lies below the graph of f and does not appear in the viewing rectangle in the Figure 3.104. Instead of entering the variable Y_1, we could enter the formula $-.4X^2 + 4$ for the upper function.

Summary: Shading a graph

1. Press (2nd) (PRGM [DRAW]) (7) from the home screen.
2. Enter a formula or a variable such as Y_1 for the lower function followed by a comma.
3. Enter a formula or a variable such as Y_2 for the upper function followed by a right parenthesis.
4. Set an appropriate viewing rectangle.
5. Press (ENTER). The region between the two graphs will be shaded.

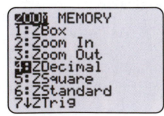

FIGURE B.52

Setting a Decimal Window

In Example 3, Section 4.1, a decimal (or friendly) window is used to trace the graph of f. With a decimal window, the cursor stops on convenient x-values. In the decimal window $[-9.4, 9.4, 1]$ by $[-6.2, 6.2, 1]$ the cursor stops on x-values that are multiples of 0.2. If we reduce the viewing rectangle to $[-4.7, 4.7, 1]$ by $[-3.1, 3.1, 1]$ the cursor stops on x-values that are multiples of 0.1. To set this smaller window automatically, press (ZOOM)(4). See Figure B.52. Decimal windows are also useful when graphing rational functions with asymptotes in connected mode.

Summary: Setting a decimal window

1. Press (ZOOM)(4) to set the viewing rectangle $[-4.7, 4.7, 1]$ by $[-3.1, 3.1, 1]$.
2. A larger decimal window is $[-9.4, 9.4, 1]$ by $[-6.2, 6.2, 1]$.

FIGURE B.53

FIGURE B.54

Copying a Regression Equation into Y_1

In Example 3, Section 4.2, we use the cubic regression to model real data. The resulting formula for the cubic function shown in Figure 4.55 is quite complicated, and tedious to enter into "$Y_1 =$" by hand. A graphing calculator has the capability of copying this equation into Y_1 automatically. To do this, clear the equation for "$Y_1 =$". Then enter Y_1 after "CubicReg", as shown in Figure B.53. When (ENTER) is pressed, the regression equation is calculated, and then it is copied into "$Y_1 =$", as shown in Figure B.54. The following keystrokes may be used from the home screen. Be sure to enter the data into lists L1 and L2.

(STAT) (▷) (6) (VARS) (▷) (1) (1) (ENTER)

> **Summary: Copying a regression equation into Y_1**
>
> 1. Clear Y_1 in the (Y =) menu, if an equation is present. Return to the home screen.
> 2. Select a type of regression from the STAT CALC menu.
> 3. Press (VARS) (▷) (1) (1) (ENTER).

Evaluating Complex Arithmetic

Complex arithmetic can be performed much like other arithmetic expressions. This is done by entering

(2nd) (. [i])

to obtain the imaginary unit i from the home screen. For example, to add $(-2 + 3i) + (4 - 6i)$, perform the following keystrokes on the home screen.

(((-) (2) (+) (3) (2nd) (. [i])) (+) ((4) (−) (6) (2nd) (. [i])) (ENTER)

The result is shown in the first two lines of Figure 4.80. Other complex arithmetic operations are done similarly.

> **Summary: Evaluating complex arithmetic**
>
> Enter a complex expression in the same way as any arithmetic expression. To obtain the complex number i, use (2nd) (. [i]).

Graphing an Inverse Function

In Example 7, Section 5.2, the inverse function of $f(x) = x^3 + 2$ is graphed. A graphing calculator can graph the inverse of a function without a formula for $f^{-1}(x)$. Begin by entering $Y_1 = X^3 + 2$ into the (Y =) menu. Then return to the home screen by pressing

(2nd) (MODE [QUIT]).

The "DrawInv" utility may be accessed by pressing

(2nd) (PRGM [DRAW]) (8),

followed by

(VARS) (▷) (1) (1)

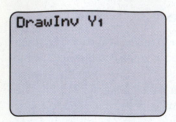

FIGURE B.55

[−6, 6, 1] by [−4, 4, 1]

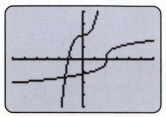

FIGURE B.56

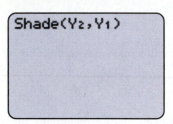

FIGURE B.57

[−6, 6, 1] by [−6, 6, 1]

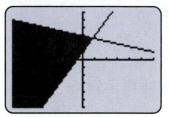

FIGURE B.58

to obtain the variable Y_1. See Figure B.55. Pressing (ENTER) causes both Y_1 and its inverse to be graphed, as shown in Figure B.56.

> ### Summary: Graphing an inverse function
> 1. Enter the formula for $f(x)$ into Y_1 using the (Y =) menu.
> 2. Set an appropriate viewing rectangle by pressing (WINDOW).
> 3. Return to the home screen by pressing (2nd)(MODE [QUIT]).
> 4. Press (2nd)(PRGM [DRAW])(8)(VARS)(▷)(1)(1)(ENTER) to create the graphs of f and f^{-1}.

Shading a System of Inequalities

In Example 5(b), Section 6.2, we are asked to shade the solution set for the system of linear inequalities $x + 3y \le 9, 2x - y \le -1$. Begin by solving each system for y to obtain $y \le (9 - x)/3$ and $y \ge 2x + 1$. Then let $Y_1 = (9 - X)/3$ and $Y_2 = 2X + 1$, as shown in Figure 6.30. Position the cursor to left of Y_1, and press (ENTER) three times. The triangle that appears indicates that the calculator will shade the region below the graph of Y_1. Next locate the cursor to the left of Y_2 and press (ENTER) twice. This triangle indicates that the calculator will shade the region above the graph of Y_2. After setting the viewing rectangle to [−6, 6, 1] by [−6, 6, 1] press (GRAPH). The result is shown in Figure 6.31. The solution set could also be shaded using Shade(Y_2, Y_1) from the home screen. See Figures B.57 and B.58.

> ### Summary: Shading a system of equations
> 1. Solve each inequality for y.
> 2. Enter each formula as Y_1 and Y_2 in the (Y =) menu.
> 3. Locate the cursor to the left of Y_1 and press (ENTER) two or three times, to shade either above or below the graph of Y_1. Repeat for Y_2.
> 4. Set an appropriate viewing rectangle.
> 5. Press (GRAPH).
>
> *Note:* The "Shade" utility under the DRAW menu can also be used to shade the region *between* two graphs.

Entering the Elements of a Matrix

In Example 8, Section 6.3, the elements of a matrix are entered. The augmented matrix A is given by

$$A = \begin{bmatrix} 2 & 1 & 2 & | & 10 \\ 1 & 0 & 2 & | & 5 \\ 1 & -2 & 2 & | & 1 \end{bmatrix}.$$

Use the following keystrokes on the TI-83 Plus to define a matrix A with dimension 3×4. (*Note:* On the TI-83 graphing calculator the matrix menu is found by pressing (MATRIX).)

(2nd)(x^{-1} [MATRIX])(▷)(▷)(1)(3)(ENTER)(4)(ENTER)

See Figures 6.36 and 6.37.

Then input the 12 elements of the matrix A, row by row, as shown in Figure 6.37. Finish each entry by pressing (ENTER). After these elements have been entered, press

$$\text{(2nd) (MODE [QUIT])}$$

to return to the home screen. To display the matrix A, press

$$\text{(2nd) } (x^{-1} \text{ [MATRIX]}) \text{ (1) (ENTER)}.$$

See Figure 6.38.

Summary: Entering the elements of a matrix A

1. Begin by accessing the matrix A by pressing (2nd) $(x^{-1}$ [MATRIX]) $(\triangleright)$ $(\triangleright)$ (1).
2. Enter the dimension of A by pressing (m) (ENTER) (n) (ENTER), where the dimension of the matrix is $m \times n$.
3. Input each element of the matrix, row by row. Finish each entry by pressing (ENTER). Use (2nd) (MODE [QUIT]) to return to the home screen.

Note: On the TI-83, replace the keystrokes (2nd) $(x^{-1}$ [MATRIX]) with (MATRIX).

Reduced Row-Echelon Form

In Example 8, Section 6.3, the reduced row-echelon form of a matrix is found. To find this reduced row-echelon form, use the following keystrokes from the home screen on the TI-83 Plus.

$$\text{(2nd) } (x^{-1} \text{ [MATRIX]}) \ (\triangleright) \ \text{(ALPHA) (APPS [B]) (2nd) } (x^{-1} \text{ [MATRIX]}) \ (1) \ (\ ) \ ) \ \text{(ENTER)}$$

The resulting matrix is shown in Figure 6.40. On the TI-83 graphing calculator use the following keystrokes to find the reduced row-echelon form.

$$\text{(MATRIX) } (\triangleright) \ \text{(ALPHA) (MATRIX [B]) (MATRIX) (1) (\ )) (ENTER)}$$

Summary: Finding reduced row-echelon form of a matrix

1. To make rref([A]) appear on the home screen, use the following keystrokes for the TI-83 Plus graphing calculator.

$$\text{(2nd) } (x^{-1} \text{ [MATRIX]}) \ (\triangleright) \ \text{(ALPHA) (APPS [B]) (2nd) } (x^{-1} \text{ [MATRIX]}) \ (1) \ (\ ) \ ) \ \text{(ENTER)}$$

2. Press (ENTER) to calculate the reduced row-echelon form.
3. Use arrow keys to access elements that do not appear on the screen.

Note: On the TI-83, replace the keystrokes (2nd) $(x^{-1}$ [MATRIX]) with (MATRIX) and (APPS [B]) with (MATRIX [B]).

Performing Arithmetic Operations on Matrices

In Example 8, Section 6.4, the matrices A and B are multiplied. Begin by entering the elements for the matrices A and B. The following keystrokes can be used to define a matrix A with dimension 3×3.

$$\text{(2nd) } (x^{-1} \text{ [MATRIX]}) \ (\triangleright) \ (\triangleright) \ (1) \ (3) \ \text{(ENTER) (3) (ENTER)}$$

Next input the 9 elements in the matrix A, row by row. Finish each entry by pressing (ENTER). See Figure 6.57. Repeat this process to define a matrix B with dimension 3×3.

Enter the 9 elements in B. See Figure 6.58. After the elements of A and B have been entered, press

(2nd) (MODE [QUIT])

to return to the home screen. To multiply the expression AB, use the following keystrokes from the home screen.

(2nd) (x^{-1}[MATRIX]) (1) (×) (2nd) (x^{-1}[MATRIX]) (2) (ENTER)

The result is shown in Figure 6.59.

Summary: Performing arithmetic operations on matrices

1. Enter the elements of each matrix beginning with the keystrokes,
   (2nd) (x^{-1}[MATRIX]) (▷) (▷) (k) (m) (ENTER) (n) (ENTER), where k is the menu number of the matrix and the dimension of the matrix is $m \times n$.
2. Return to the home screen by pressing (2nd) (MODE [QUIT]).
3. Enter the matrix expression followed by (ENTER). Use the keystrokes (2nd) (x^{-1}[MATRIX]) (k) to access the matrix with menu number k on the TI-83 Plus.

Note: On the TI-83, replace the keystrokes (2nd) (x^{-1}[MATRIX]) with (MATRIX).

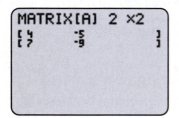

FIGURE B.59

Finding the Inverse of a Matrix

In Example 4, Section 6.5, the inverse of A, denoted A^{-1}, is displayed in Figure 6.63. To calculate A^{-1}, start by entering the elements of the matrix A, as shown in Figure B.59. To compute A^{-1}, perform the following keystrokes from the home screen.

(2nd) (x^{-1}[MATRIX]) (1) (x^{-1}) (ENTER)

The results are shown in Figure 6.63.

Summary: Finding the inverse of a square matrix

1. Enter the elements of the square matrix A.
2. Return to the home screen by pressing

   (2nd) (MODE [QUIT]).

3. Perform the following keystrokes from the home screen to display A^{-1}.

   (2nd) (x^{-1}[MATRIX]) (1) (x^{-1}) (ENTER)

Note: On the TI-83, replace the keystrokes (2nd) (x^{-1}[MATRIX]) with (MATRIX).

Solving a Linear System with a Matrix Inverse

In Example 4, Section 6.5, the solution to a system of equations is found. The matrix equation $AX = B$ has the solution $X = A^{-1}B$ provided A^{-1} exists, and is given by

$$AX = \begin{bmatrix} 4 & -5 \\ 7 & -9 \end{bmatrix} \begin{bmatrix} x \\ y \end{bmatrix} = \begin{bmatrix} 8.1 \\ -4.7 \end{bmatrix} = B.$$

To solve this equation, start by entering the elements of the matrices A and B. To compute the solution $A^{-1}B$, perform the following keystrokes from the home screen.

(2nd) (x⁻¹ [MATRIX]) (1) (x⁻¹) (×) (2nd) (x⁻¹ [MATRIX]) (2) (ENTER)

The results are shown in Figure 6.64.

Summary: Solving a linear system with a matrix inverse

1. Write the system of equations as $AX = B$.
2. Enter the elements of the matrices for A and B.
3. Return to the home screen by pressing

(2nd) (MODE [QUIT]).

4. Perform the following keystrokes.

(2nd) (x⁻¹ [MATRIX]) (1) (x⁻¹) (×) (2nd) (x⁻¹ [MATRIX]) (2) (ENTER)

Note: On the TI-83, replace the keystrokes (2nd) (x⁻¹ [MATRIX]) with (MATRIX).

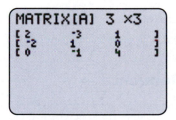

FIGURE B.60

Evaluating a Determinant

In Example 4(a), Section 6.6, a graphing calculator is used to evaluate a determinant of a matrix. Start by entering the 9 elements of the 3×3 matrix A, as shown in Figure B.60. To compute det A, perform the following keystrokes from the home screen.

(2nd) (x⁻¹ [MATRIX]) (▷) (1) (2nd) (x⁻¹ [MATRIX]) (1) ()) (ENTER)

The results are shown in Figure 6.68.

Summary: Evaluating a determinant of a matrix

1. Enter the elements of the matrix A.
2. Return to the home screen by pressing

(2nd) (MODE [QUIT]).

3. On the TI-83 Plus perform the following keystrokes.

(2nd) (x⁻¹ [MATRIX]) (▷) (1) (2nd) (x⁻¹ [MATRIX]) (1) ()) (ENTER)

Note: On the TI-83, replace the keystrokes (2nd) (x⁻¹ [MATRIX]) with (MATRIX).

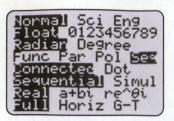

FIGURE B.61

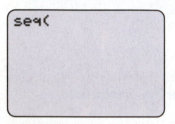

FIGURE B.62

Creating a Sequence

A graphing calculator can be used to calculate the terms of the sequence $f(n) = 2n - 5$ for $n = 1, 2, 3, 4$. See Example 2(a), Section 8.1. Start by setting the mode of the calculator to sequence "Seq" using the following keystrokes. See Figure B.61.

MODE ▽ ▽ ▽ ▷ ▷ ▷ ENTER 2nd MODE [QUIT]

Then enter the following from the home screen.

2nd STAT [LIST] ▷ 5

On the home screen "seq(" will appear, as shown in Figure B.62. This sequence utility requires four things to be entered—all separated by commas. They are the formula, the variable, the subscript of first term, and the subscript of the last term. Use the following keystrokes to obtain the first four terms (a_1, a_2, a_3, a_4) of the sequence $a_n = 2n - 5$, as shown in Figure 8.1.

Summary: Creating a sequence

1. To create a sequence, use the keystrokes

   2nd STAT [LIST] ▷ 5 .

2. Enter the formula, the variable, the subscript of the first term, and the subscript of the last term—all separate by commas. For example, if you want the first 10 terms of $a_n = n^2$, (a_1, a_2, a_3, ..., a_{10}), enter seq(n^2, n, 1, 10). Be sure to set your calculator in sequence mode.

3. Pressing ENTER causes the terms of the sequence to appear.

Entering, Tabling, and Graphing a Sequence

In Example 5, Section 8.1, a table and a graph of a sequence are created with a graphing calculator. The calculator should be set to sequence mode by entering the following keystrokes.

MODE ▽ ▽ ▽ ▷ ▷ ▷ ENTER

To enter the formula for a sequence, press Y= . See Figure B.63. Let nMin $= 1$ since the initial value of n is equal to 1. To enter the expression $a_n = 2.85a_{n-1} - .19a_{n-1}^2$, use the following keystrokes after clearing out any old formula. (Notice that the graphing calculator uses u instead of a to denote a term of the sequence.)

Since $a_1 = 1$, let $u(n$Min$) = \{1\}$. This can be performed as follows.

CLEAR 2nd (1 2nd)

See Figure B.63.

To create a table for this sequence, starting with a_1 and incrementing n by 1, perform the following keystrokes.

[2nd] [WINDOW [TBLSET]] [1] [ENTER] [1] [2nd] [GRAPH [TABLE]]

See Figure B.64 and Figure 8.7.

To graph the first 20 terms of this sequence, start by selecting [WINDOW]. Since we want the first 20 terms plotted, let nMin = 1, nMax = 20, PlotStart = 1, and PlotStep = 1. The window can be set as [0, 21, 1] by [0, 14, 1]. See Figure B.65. To graph the sequence, press [GRAPH]. The resulting graph is shown in Figure 8.9.

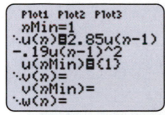

FIGURE B.63

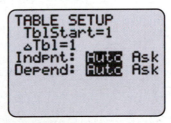

FIGURE B.64

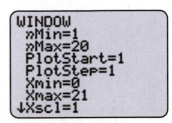

FIGURE B.65

Summary: Entering, tabling, and graphing a sequence

1. Set the mode to "Seq" by using the [MODE] menu.
2. Enter the formula for the sequence by pressing [Y =].
3. To create a table of a sequence, set the start and increment values with

    [2nd] [WINDOW [TBLSET]].

    A table is created by pressing

    [2nd] [GRAPH [TABLE]].

4. To graph a sequence, set the viewing rectangle by using [WINDOW]. Press [GRAPH] to create a graph of the sequence.

Finding the Sum of the Terms of a Series

In Example 3, Section 8.2, the sum of the series $\sum\limits_{n=1}^{50} \left(\dfrac{1}{n^4}\right)$ is found by using a graphing calculator. Use the following keystrokes from the home screen.

[2nd] [STAT [LIST]] [▷] [▷] [5] [2nd] [STAT [LIST]] [▷] [5]

The results are shown in the first three lines of Figure 8.12.

Summary: Summing a series

1. Use [2nd] [STAT [LIST]] [▷] [▷] [5] to access the sum utility.
2. Use [2nd] [STAT [LIST]] [▷] [5] to access the sequence utility.
3. To use the sequence utility, see "Creating a Sequence" in this appendix.

Calculating Factorial Notation

In Example 5, Section 8.3, factorial notation is evaluated with a graphing calculator. The factorial utility is found under the MATH NUM menus. To calculate 8! use the following keystrokes from the home screen.

$$\boxed{8} \;\boxed{\text{MATH}}\; \boxed{\triangleright}\; \boxed{\triangleright}\; \boxed{\triangleright}\; \boxed{4}\; \boxed{\text{ENTER}}$$

The results are shown in the first two lines of Figure 8.19.

Summary: Calculating factorial notation

To calculate *n* factorial, use the following keystrokes.

$$\boxed{n} \;\boxed{\text{MATH}}\; \boxed{\triangleright}\; \boxed{\triangleright}\; \boxed{\triangleright}\; \boxed{4}\; \boxed{\text{ENTER}}$$

The value of *n* should be entered as a number, not a variable.

Calculating Permutations and Combinations

In Example 6(a), Section 8.3, the permutation $P(7, 3)$ is evaluated. To perform this calculation, use the following keystrokes from the home screen.

$$\boxed{7} \;\boxed{\text{MATH}}\; \boxed{\triangleright}\; \boxed{\triangleright}\; \boxed{\triangleright}\; \boxed{2}\; \boxed{3}\; \boxed{\text{ENTER}}$$

The results are shown in the first two lines of Figure 8.21.

In Example 9(a), Section 8.3, the combination $C(7, 3)$ can be calculated by using the following keystrokes.

$$\boxed{7} \;\boxed{\text{MATH}}\; \boxed{\triangleright}\; \boxed{\triangleright}\; \boxed{\triangleright}\; \boxed{3}\; \boxed{3}\; \boxed{\text{ENTER}}$$

The results are shown in the first two lines of Figure 8.22.

Summary: Calculating permutations and combinations

1. To calculate $P(n, r)$, use $\boxed{\text{MATH}}$ and select "PRB" followed by $\boxed{2}$.
2. To calculate $C(n, r)$, use $\boxed{\text{MATH}}$ and select "PRB" followed by $\boxed{3}$.

Bibliography

Acker, A., and C. Jaschek. *Astronomical Methods and Calculations.* New York: John Wiley and Sons, 1986.

Andre-Pascal, R. *Global Energy: The Changing Outlook.* Paris, France: Organization for Economic Cooperation and Development/International Energy Agency, 1992.

Battan, L. *Weather in Your Life.* San Francisco: W. H. Freeman, 1983.

Beckmann, P. *A History of PI.* New York: Barnes and Noble, Inc., 1993.

Brearley, J., and A. Nicholas. *This Is the Bichon Frise.* Hong Kong: TFH Publication, 1973.

Brown, D., and P. Rothery. *Models in Biology: Mathematics, Statistics and Computing.* West Sussex, England: John Wiley and Sons Ltd., 1993.

Carlson, T. "Über Geschwindigkeit und Grösse der Hefevermehrung in Würze." *Biochem. A.* 57:313–334.

Cheney, W., and D. Kincaid. *Numerical Mathematics and Computing.* 3rd ed. Pacific Grove, Calif.: Brooks/Cole Publishing Company, 1994.

Clime, W. *The Economics of Global Warming.* Washington, D.C.: Institute for International Economics, 1992.

Cotton, W., and R. Pielke. *Human Impacts on Weather and Climate.* Geophysical Science Series, vol. 2. Fort Collins, Colo.: *ASTeR Press, 1992.

Easteal, S., N. McLeod, and K. Reed. *DNA Profiling: Principles, Pitfalls and Potential.* Philadelphia: Harwood Academic Publishers, 1991.

Eves, H. *An Introduction to the History of Mathematics,* 5th ed. Philadelphia: Saunders College Publishing, 1983.

Foster, R., and J. Bates: "Use of mussels to monitor point source industrial discharges." *Environ. Sci. Technol.* 12:958–962.

Freedman, B. *Environmental Ecology: The Ecological Effects of Pollution, Disturbance, and Other Stresses.* 2nd ed. San Diego: Academic Press, 1995.

Garber, N., and L. Hoel. *Traffic and Highway Engineering.* Boston, Mass.: PWS Publishing Co., 1997.

Grigg, D. *The World Food Problem.* Oxford: Blackwell Publishers, 1993.

Haber-Schaim, U., J. Cross, G. Abegg, J. Dodge, and J. Walter. *Introductory Physical Science.* Englewood Cliffs, N.J.: Prentice Hall, Inc., 1972.

Haefner, L. *Introduction to Transportation Systems.* New York: Holt, Rinehart and Winston, 1986.

Harrison, F., F. Hills, J. Paterson, and R. Saunders. "The measurement of liver blood flow in conscious calves." *Quarterly Journal of Experimental Physiology* 71:235–247.

Heinz-Otto, P., H. Jürgens, and D. Saupe. *Chaos and Fractals: New Frontiers in Science.* New York: Springer-Verlag, 1993.

Hines, A., T. Ghosh, S. Loyalka, and R. Warder, Jr. *Indoor Air Quality and Control.* Englewood Cliffs, N.J.: Prentice Hall, 1993.

Historical Topics for the Mathematics Classroom, Thirty-first Yearbook. National Council of Teachers of Mathematics, 1969.

Hoggar, S. *Mathematics for Computer Graphics.* New York: Cambridge University Press, 1993.

Hoppensteadt, F., and C. Peskin. *Mathematics in Medicine and the Life Sciences.* New York: Springer-Verlag, 1992.

Howells, G. *Acid Rain and Acid Waters.* 2nd ed. New York: Ellis Horwood, 1995.

Huffman, R. *Atmospheric Ultraviolet Remote Sensing.* San Diego: Academic Press, 1992.

Huxley, J. *Problems of Relative Growth.* London: Methuen and Co. Ltd., 1932.

Karttunen, H., P. Kroger, H. Oja, M. Poutanen, K. Donner, eds. *Fundamental Astronomy.* 2nd ed. New York: Springer-Verlag, 1994.

Kincaid, D., and W. Cheney. *Numerical Analysis.* Pacific Grove, Calif.: Brooks/Cole Publishing Company, 1991.

Kraljic, M. *The Greenhouse Effect.* New York: The H. W. Wilson Company, 1992.

Kress, S. *Bird Life—A Guide to the Behavior and Biology of Birds.* Racine, Wisc.: Western Publishing Company, 1991.

Lack, D. *The Life of a Robin.* London: Collins, 1965.

Lancaster, H. *Quantitative Methods in Biological and Medical Sciences: A Historical Essay.* New York: Springer-Verlag, 1994.

Leder, J. *Martina Navratilova.* Mankato, Minn.: Crestwood House, 1985.

Loh, W. *Dynamics and Thermodynamics of Planetary Entry.* Englewood Cliffs, N.J.: Prentice-Hall, 1963.

Mannering, F., and W. Kilareski. *Principles of Highway Engineering and Traffic Analysis.* New York: John Wiley and Sons, 1990.

Mar, J., and H. Liebowitz. *Structure Technology for Large Radio and Radar Telescope Systems.* Cambridge, Mass.: The MIT Press, 1969.

Mason, C. *Biology of Freshwater Pollution.* New York: Longman and Scientific and Technical, John Wiley and Sons, 1991.

Mehrotra, A. *Cellular Radio: Analog and Digital Systems.* Boston: Artech House, 1994.

Miller, A., and J. Thompson. *Elements of Meteorology.* 2nd ed. Columbus, Ohio: Charles E. Merrill Publishing Company, 1975.

Miller, A., and R. Anthes. *Meteorology.* 5th ed. Columbus, Ohio: Charles E. Merrill Publishing Company, 1985.

Motz, L., and J. Weaver. *The Story of Mathematics.* New York: Plenum Press, 1993.

Nilsson, A. *Greenhouse Earth.* New York: John Wiley and Sons, 1992.

Paetsch, M. *Mobile Communications in the U.S. and Europe: Regulation, Technology, and Markets.* Norwood, Mass.: Artech House, Inc., 1993.

Pennycuick, C. *Newton Rules Biology.* New York: Oxford University Press, 1992.

Pielou, E. *Population and Community Ecology: Principles and Methods.* New York: Gordon and Breach Science Publishers, 1974.

Pokorny, C., and C. Gerald. *Computer Graphics: The Principles behind the Art and Science.* Irvine, Calif.: Franklin, Beedle, and Associates, 1989.

Raggett, G. "Modeling the Eyam plague." *The Institute of Mathematics and Its Applications* 18: 221–226.

Rezvan, R. L. "Effectiveness of Local Ventilation in Removing Simulated Pollutants from Point Sources." *Proceedings of the Third International Conference on Indoor Air Quality and Climate,* 1984.

Rist, Curtis. "The Physics of Foul Shots." *Discover,* October 2000.

Rodricks, J. *Calculated Risk.* New York: Cambridge University Press, 1992.

Ronan, C. *The Natural History of the Universe.* New York: MacMillan Publishing Company, 1991.

Sharov, A., and I. Novikov. *Edwin Hubble: The Discoverer of the Big Bang Universe.* New York: Cambridge University Press, 1993.

Sinkov, A. *Elementary Cryptanalysis: A Mathematical Approach*. New York: Random House, 1968.

Smith, J. *Design and Analysis of Algorithms*. Boston: PWS Publishing Company, 1989.

Speed, William. "Downloading Your Body." *Discover,* September 2000.

Stent, G. S. *Molecular Biology of Bacterial Viruses*. San Francisco: W. H. Freeman, 1963.

Teutsch, S., and R. Churchill. *Principles and Practice of Public Health Surveillance*. New York: Oxford University Press, 1994.

Thomas, D. *Swimming Pool Operators Handbook*. National Swimming Pool Foundation of Washington, D.C., 1972.

Thomas, V. *Science and Sport*. London: Faber and Faber, 1970.

Thomson, W. *Introduction to Space Dynamics*. New York: John Wiley and Sons, 1961.

Turner, R. K., D. Pierce, and I. Bateman. *Environmental Economics, An Elementary Approach*. Baltimore: The Johns Hopkins University Press, 1993.

Walker, A. *Observation and Inference: An Introduction to the Methods of Epidemiology*. Newton Lower Falls, Mass.: Epidemiology Resources Inc., 1991.

Weidner, R., and R. Sells. *Elementary Classical Physics,* vol. 2. Boston: Allyn and Bacon, Inc., 1965.

Williams, J. *The Weather Almanac 1995*. New York: Vintage Books, 1994.

Wolff, R., and L. Yaeger. *Visualization of Natural Phenomena*. New York: Springer-Verlag, 1993.

Wuebbles, D., and J. Edmonds. *Primer on Greenhouse Gases*. Chelsea, Mich.: Lewis Publishers, 1991.

Zeilik, M., S. Gregory, and D. Smith. *Introductory Astronomy and Astrophysics*. 3rd ed. Philadelphia: Saunders College Publishers, 1992.

Answers to Selected Exercises

CHAPTER 1: INTRODUCTION TO FUNCTIONS AND GRAPHS

SECTION 1.1

1. Natural number, integer, rational number, real number
3. Integer, rational number, real number
5. Rational number, real number
7. Real number
9. Natural number, $\sqrt{9}$; integers, -3, $\sqrt{9}$; rational numbers, $-3, \frac{2}{9}, \sqrt{9}, 1.\overline{3}$; irrational numbers: $\pi, -\sqrt{2}$
11. Rational numbers   **13.** Rational numbers
15. Integers   **17.** 62.5%
19. -39.3% **21.** 1.876×10^5 **23.** 3.983×10^{-2}
25. 2.45×10^3 **27.** 5.6×10^{-1} **29.** 0.000001
31. 200,000,000   **33.** 156.7   **35.** -0.568
37. 3.4×10^{19} **39.** 1.8×10^{-1}
41. -8.075×10^3 **43.** 5.769   **45.** 0.058
47. 0.419   **49.** -0.549 **51.** 15.819
53. Tuition and fees: 137.0%, CPI: 51.5%
55. $798   **57.** About 53,794 miles per hour
59. **(a)** Approximately $1820 per person
(b) Approximately $19,715 per person
(c) Estimates will vary.
61. 2.9×10^{-4} cm
63. **(a)** 16,666,667 feet   **(b)** Yes
65. **(a)** Increased from 1940 to 1980; decreased from 1980 to 1994
(b) 4.9×10^8 gallons   **(c)** -10.9%

SECTION 1.2

1.

-30	-30	-10	5	15	25	45	55	61

(a) Max: 61; min: -30
(b) Mean: $\frac{136}{9} \approx 15.11$; median: 15; range: 91
3. $\sqrt{15} \approx 3.87$, $2^{2.3} \approx 4.92$, $\sqrt[3]{69} \approx 4.102$, $\pi^2 \approx 9.87$, $2^\pi \approx 8.82$, 4.1

$\sqrt{15}$	4.1	$\sqrt[3]{69}$	$2^{2.3}$	2^π	π^2

(a) Max: π^2; min: $\sqrt{15}$

(b) Mean: 5.95; median: 4.51; range: $\pi^2 - \sqrt{15} \approx 6.00$
5. **(a)**

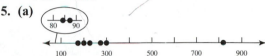

(b) Mean: 273; median: 212; range: 756
The average area of the eight largest islands in the world is 273,000 square miles. Half of the islands have areas of less than 212,000 square miles and half have more. The largest difference in area between any two islands is 756,000 square miles.
(c) Greenland.
7. **(a)**

(b) Mean: 19.0; median: 19.3; range: 21.7
The average of the maximum elevations for the seven continents is 19,000 feet. About half of these elevations are below 19,300 feet and about half are above. Their largest difference is 21,700 feet.
(c) Mount Everest
9. **(a)** $2,786,774   **(b)** $385,000
(c) One large salary can raise the average considerably, while the majority of the players have considerably lower salaries.
11. 16, 18, 26; no   **13.** 5   **15.** $\sqrt{29} \approx 5.39$
17. $\sqrt{133.37} \approx 11.55$ **19.** 8
21. $\frac{\sqrt{17}}{4} \approx 1.03$ **23.** $\sqrt{(a-b)^2} = |a-b|$
25. Yes
27. **(a)**

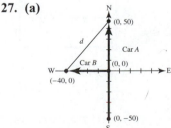

(b) $d = \sqrt{4100} \approx 64.0$ miles.
29. **(a)** $D = \{0, -3, -2, 7\}$, $R = \{5, 4, -5, -3, 0\}$
(b) x-min: -3; x-max: 7; y-min: -5; y-max; 5

(c) & (d)

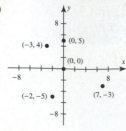

31. (a) $D = \{10, -35, 20, 75, -40, -25\}$,
$R = \{50, 45, -55, 25, 60, -25\}$
(b) x-min: -40; x-max: 75; y-min: -55; y-max: 60
(c) & (d)

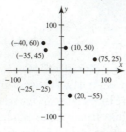

33. (a) $D = \{0.1, 0.5, -0.7, 0.8, 0.9\}$,
$R = \{-0.3, 0.4, 0.5, -0.1, 0.9\}$
(b) x-min: -0.7; x-max: 0.9; y-min: -0.3; y-max: 0.9
(c) & (d)

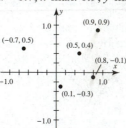

35. x-axis: 10; y-axis: 10   **37.** x-axis: 10; y-axis: 5

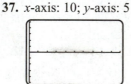

39. x-axis: 16; y-axis: 5   **41.** x-axis: 3; y-axis: 2

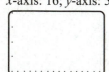

43. b.   **45.** a.
47. $[-5, 5, 1]$ by $[-5, 5, 1]$

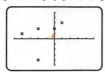

49. $[-100, 100, 10]$ by $[-100, 100, 10]$

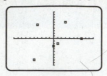

51. $[-6, 6, 2]$ by $[-12, 12, 2]$

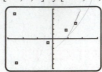

53. (a) x-min: 1979; x-max: 1997; y-min: 15;
y-max: 37
(b) $[1978, 1998, 2]$ by $[10, 40, 5]$. Answers may vary.
(c)   **(d)**

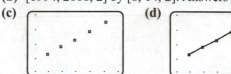

55. (a) x-min: 1996; x-max: 2004; y-min: 9.7; y-max:
12.8
(b) $[1994, 2006, 2]$ by $[8, 14, 2]$. Answers may vary.
(c)   **(d)**

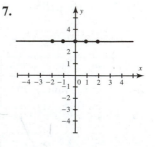

57. Answers may vary.

Year	1950	1960	1970
Doctorate Degrees	6000	10,000	30,000

Year	1980	1990	1995
Doctorate Degrees	33,000	38,000	41,000

SECTION 1.3

1. $(-2, 3)$ **3.** $f(7) = 8$
5.   **7.**

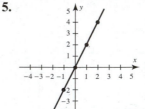

9. 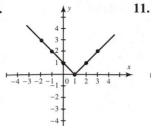 **11.**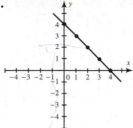

13.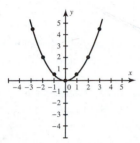

15. (a) $f(-2) = -8$, $f(5) = 125$
(b) All real numbers
17. (a) $f(-1) = $ undefined, $f(a + 1) = \sqrt{a + 1}$
(b) Nonnegative real numbers
19. (a) $f(-1) = -\dfrac{1}{2}$, $f(a + 1) = \dfrac{1}{a}$ **(b)** $x \neq 1$
21. (a) $f(6) = -7$, $f(a - 1) = -7$
(b) All real numbers
23. (a) $f(4) = \dfrac{1}{16}$, $f(-7) = \dfrac{1}{49}$ **(b)** $x \neq 0$
25. (a) $f(4) = \dfrac{1}{7}$, $f(a - 5) = \dfrac{1}{a^2 - 10a + 16}$
(b) $x \neq 3$, $x \neq -3$
27. $D = \{x \mid -3 \leq x \leq 3\}$, $R = \{y \mid 0 \leq y \leq 3\}$;
$f(0) = 3$
29. $D = \{x \mid -2 \leq x \leq 4\}$, $R = \{y \mid -2 \leq y \leq 2\}$;
$f(0) = -2$
31. $D = $ all real numbers, $R = \{y \mid y \leq 2\}$; $f(0) = 2$
33. $D = \{x \mid x \geq -1\}$, $R = \{y \mid y \leq 2\}$; $f(0) = 0$
35. (a) $f(0) = -2$, $f(2) = 2$ **(b)** $x = 1$
37. (a) $f(0) = 0$, $f(2) = 4$ **(b)** $x = 0$
39. (a) $f(0) = 0$, $f(2) = 0$ **(b)** $x = -2, 0,$ or 2
41. (a) $[-4.7, 4.7, 1]$ by $[-3.1, 3.1, 1]$ **(b)** $f(2) = 1$

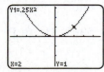

43. (a) $[-4.7, 4.7, 1]$ by $[-3.1, 3.1, 1]$ **(b)** $f(2) = 2$

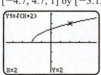

45. Verbal: Square the input x.
Graphical: Graph $Y_1 = X^2$.
$[-10, 10, 1]$ by $[-10, 10, 1]$

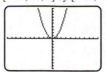

Numerical:

x	-2	-1	0	1	2
y	4	1	0	1	4

47. Verbal: Multiply the input by 2, add 1, and then take the absolute value.
Numerical:

x	-2	-1	0	1	2
y	3	1	1	3	5

Graphical: Graph $Y_1 = \text{abs}(2X + 1)$.
$[-6, 6, 1]$ by $[-4, 4, 1]$

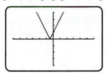

49. Verbal: Compute the absolute value of the input x.
Graphical: Graph $Y_1 = \text{abs}(X)$.
$[-6, 6, 1]$ by $[-4, 4, 1]$

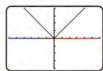

Numerical:

x	-2	-1	0	1	2
y	2	1	0	1	2

51. Verbal: Add 1 to the input and then take the square root of the result.
Numerical:

x	-2	-1	0	1	2
y	Error	0	1	$\sqrt{2}$	$\sqrt{3}$

Graphical: Graph $Y_1 = \sqrt{(X + 1)}$.
$[-6, 6, 1]$ by $[-4, 4, 1]$

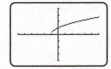

53.

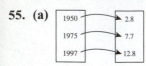

Bills (millions)	0	1	2	3	4	5	6
Counterfeit Bills	0	9	18	27	36	45	54

55. (a)

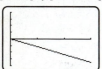

(b) $f(1975) = 7.7$; in 1975 there were 7700 radio stations on the air.
(c) $D = \{1950, 1975, 1997\}$, $R = \{2.8, 7.7, 12.8\}$
57. Yes. Domain and range include all real numbers.
59. No
61. Yes. D: $-4 \le x \le 4$; R: $0 \le y \le 4$
63. Yes. Each real number input x has exactly one cube root.
65. No. Usually, several students pass a given exam.
67. No   **69.** Yes   **71.** No   **73.** No   **75.** Yes
77. $f(x) = \dfrac{115}{528}x$, $f(15) \approx 3.3$; with a 15-second delay the lightning bolt was about 3.3 miles away.
79. Verbal: Multiply the input x by -5.8 to obtain the change in temperature.
Symbolic: $f(x) = -5.8x$.
Graphical: Graph $Y_1 = -5.8X$.
$[0, 3, 1]$ by $[-20, 20, 5]$

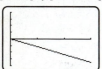

Numerical: Table $Y_1 = -5.8X$.

Section 1.4

1. $\dfrac{1}{2}$ **3.** -8 **5.** $-\dfrac{23}{30} \approx -0.7667$ **7.** Undefined

9. $-\dfrac{39}{35} \approx -1.1143$

11. Slope $= 2$; the graph rises 2 units for every unit increase in x.

13. Slope $= -\dfrac{3}{4}$; the graph falls $\dfrac{3}{4}$ unit for every unit increase in x, or equivalently, the line falls 3 units for every 4-unit increase in x.

15. Slope $= 0$; the graph neither falls nor rises for every unit increase in x.
17. (a) 1980: 30.34 years; 2000: 35.2 years
(b) Slope $= 0.243$; the median age in the U.S. is increasing by 0.243 year each year.
19. (a) $[1948, 1996, 4]$ by $[21, 25, 1]$

(b) $m = -0.0635$. The time to run the 200-meter dash is *decreasing* by 0.0635 second per year on average.
(c) -0.254 seconds
21. Linear, but not constant   **23.** Nonlinear
25. Linear and constant   **27.** Nonlinear
29. Nonlinear   **31.** Nonlinear
33. Linear, slope $= 4$ **35.** Nonlinear

37. Linear, slope $= -\dfrac{1}{2}$

39. (a) $[0, 4000, 500]$ by $[0, 280, 35]$ **(b)** Linear

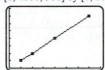

41. (a) $[1965, 2000, 5]$ by $[0, 30{,}000, 5000]$

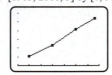

(b) Nonlinear
43. (a) No   **(b)** $f(x) = 7$
(c) $[0, 15, 3]$ by $[0, 10, 1]$

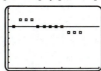

45. (a) f: linear; g: nonlinear; h: linear
(b) f: circumference; g: area; h: diameter
47. The velocity of the car is zero.

49. $f(x) = \dfrac{x}{16}$ **51.** $f(x) = 50x$ **53.** $f(x) = 500$
55. $f(x) = 6x + 1$

For Exercises 57–61, answers may vary.

57.

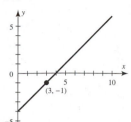

59.

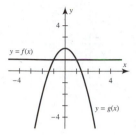

61.

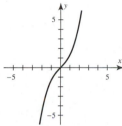

63. Average rate of change from -3 to -1 is 0. Average rate of change from 1 to 3 is 0.
65. Average rate of change from -3 to -1 is 1.2. Average rate of change from 1 to 3 is -1.2.
67. 7; the slope of the line passing through $(1, f(1))$ and $(4, f(4))$ is 7.
69. 20.75; the slope of the line passing through $(0.5, f(0.5))$ and $(4.5, f(4.5))$ is 20.75.
71. 0.62; the slope of the line passing through $(1, f(1))$ and $(3, f(3))$ is approximately 0.62.
73. (a) $[-10, 90, 10]$ by $[0, 80, 10]$

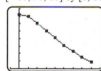

(b) $-0.24, -0.98, -0.97, -0.95, -0.93, -0.85,$ $-0.76, -0.63$; as a woman becomes older, her *remaining* life expectancy is reduced each year by the average rate of change.
(c) 80.1 years, 85.5 years; there is risk associated with living from 20 years old to 70 years old that does not apply to someone who is already 70 years old.
75. (a) 1705, 4864.2, 2899.2, $-15,713.5$
(b) From 1981 to 1985, AIDS deaths increased by 1705 per year. Other values can be interpreted similarly except a negative sign indicates a decrease.

CHAPTER 1   REVIEW EXERCISES

1. Natural number: $\sqrt{16}$; integer: $-2, 0, \sqrt{16}$; rational number: $-2, \dfrac{1}{2}, 0, 1.23, \sqrt{16}$; real number: $-2, \dfrac{1}{2}, 0, 1.23, \sqrt{7}, \sqrt{16}$

3. 1.891×10^6 **5.** 4.39×10^{-5} **7.** 15,200
9. (a) 32.07   **(b)** 2.62   **(c)** 5.21   **(d)** 49.12
11. 760 seconds   **13.** About 0.000796 inch
15.

-23	-5	8	19	24

(a) Max: 24; min: -23
(b) Mean: 4.6; median: 8; range: 47
17. (a) $D = \{-1, 4, 0, -5, 1\}, R = \{-2, 6, -5, 3, 0\}$
(b) x-max: 4, x-min: -5, y-max: 6, y-min: -5
(c) & (d)

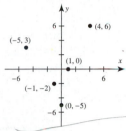

19. Not a function
$[-50, 50, 10]$ by $[-50, 50, 10]$

21. 10
23. (a) The number of Catholic schools has declined.
(b) from 1960 to 1970: $m = -0.04$; from 1970 to 1980: $m = -0.05$; from 1980 to 1990: $m = -0.02$; from 1990 to 1998: $m = -0.01$; answers may vary slightly. Because units are in thousands, $m = -0.04$ indicates that the number of Catholic schools decreased, on average, by 40 schools per year from 1960 to 1970. Other slopes may be interpreted similarly.
25. $f(x) = 16x$
$[0, 100, 10]$ by $[0, 1800, 300]$

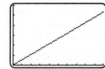

x	0	25	50	75	100
$f(x)$	0	400	800	1200	1600

27. (a) $f(-3) = 2, f(1) = -2$
(b) $x = -1, 3$
29. (a) $f(-8) = -2, f(1) = 1$
(b) All real numbers
31. (a) $f(-3) = \dfrac{1}{5}, f(a + 1) = \dfrac{1}{a^2 + 2a - 3}$
(b) $D = \{x \mid x \neq \pm 2\}$

33. No   **35.** Yes   **37.** $-\dfrac{3}{4}$ **39.** 0

41. Linear   **43.** Nonlinear

45. Constant (and linear)

47.

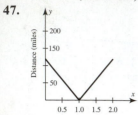

49. (a) The data decrease rapidly, indicating a very high mortality rate during the first year.

$[-1, 5, 1]$ by $[0, 110, 10]$

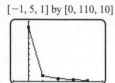

(b) Yes

(c) From 0 to 1: -90; from 1 to 2: -4; from 2 to 3: -3; from 3 to 4: -1

During the year the population decreased, on average, by 90 birds. The other average rates of change can be interpreted similarly.

51. Linear, but not constant

CHAPTER 2: LINEAR FUNCTIONS AND EQUATIONS

SECTION 2.1

1. Exactly   **3.** Approximately

5. (a) y-intercept: 25.3; in 1970 f estimates that a new home cost \$25,300.   **(b)** A slope of 5.75 indicates that new home prices have risen, on average, by \$5750 per year.

7. (a) Slope: 2; y-intercept: -1; x-intercept: 0.5

(b) $f(x) = 2x - 1$ **(c)** 0.5

9. (a) Slope: $\dfrac{3}{4}$; y-intercept: -3; x-intercept: 4

(b) $f(x) = \dfrac{3}{4}x - 3$ **(c)** 4

11. (a) Slope: 20; y-intercept: -50; x-intercept: 2.5

(b) $f(x) = 20x - 50$ **(c)** 2.5

13.

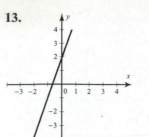

15.

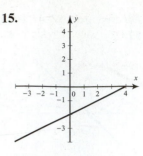

17.

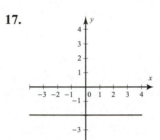

19. $f(x) = -\dfrac{3}{4}x + \dfrac{1}{3}$ **21.** $f(x) = 15x$

23. $f(x) = 0.5x + 4$

25. (a) $f(x) = 6x + 200$

(b)

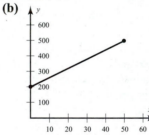

An appropriate domain is $D = \{x \mid 0 \le x \le 50\}$.

(c) y-intercept $= 200$, which indicates that the tank initially contains 200 gallons of fuel oil.

(d) No, the x-intercept of $-\dfrac{100}{3}$ corresponds to negative time, which does not apply to this problem.

27. (a) $f(x) = 16.7 - 0.326x$ **(b)** $f(7) = 14.418$, which compares favorably to 14.5.

29. $f(x) = 3x - 7$ **31.** $f(x) = -2.5x$

33. (a) No   **(b)** $f(x) = 0.08x + 2.2$; answers may vary slightly.

(c) $[-2, 28, 4]$ by $[0, 5, 1]$

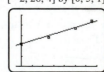

Slope $= 0.08$ indicates that the number of miles traveled has been increasing by about 0.08 trillion miles per year.   **(d)** About 4.8 trillion miles; answers may vary slightly.

35. No, because with f, if the speed doubles, the braking distance only doubles.

SECTION 2.2

1. c   **3.** b   **5.** e   **7.** $y = \frac{2}{3}x - 1$

9. $y = -\frac{3}{5}x + \frac{3}{5}$ **11.** $y = -2.4(x - 4) + 5$

13. $y = -\frac{1}{2}(x - 1) - 2$ or $y = -\frac{1}{2}(x + 9) + 3$

15. $y = 2(x - 1980) + 5$ or $y = 2(x - 1990) + 25$

17. $y = -7.8x + 5$ **19.** $y = -\frac{1}{2}x + 45$

21. $y = 4x + 9$ **23.** $y = \frac{3}{2}x - 2960$

25. $y = -3x + 5$ **27.** $y = \frac{3}{2}x - 6$

29. $y = \frac{2}{3}x - 2.1$ **31.** $y = \frac{1}{2}x + 6$

33. $y = x - 20$ **35.** $x = -5$ **37.** $y = 6$
39. $x = 4$ **41.** $x = 19$

43. x-intercept: $\frac{5}{8}$; y-intercept: -5

45. x-intercept: $\frac{11}{3}$; y-intercept: -11

47. x-intercept: $-\frac{7}{3}$; y-intercept: 7

49. x-intercept: 4; y-intercept: 2
51. x-intercept: 5; y-intercept: 7
53. **(a)** 11 miles per hour   **(b)** $y = 11x + 117$
(c) 117 miles   **(d)** 130.75 miles
55. **(a)** In 1984 the average cost of tuition and fees was \$1225. In the year 1987, the average cost of tuition and fees was \$1621.
(b) $y = 132(x - 1984) + 1225$. Tuition and fees increased, on average, \$132 per year.   **(c)** \$3337
57. **(a)** $k = 0.01$ **(b)** 1.1 mm   **(c)** No, it was a substantial thinning of the ozone layer.
59. **(a)** $f(x) = 0.29x + 4200$ **(b)** It represents the annual fixed costs.
61. $f(x) = 2(x - 1) + 3$
63. $f(x) = -3(x + 6) + 1$
65. **(b)** $f(x) = 3.4(x - 1960) + 87.8$. Answers may vary.
[1955, 1995, 5] by [70, 220, 5]

(c) Municipal solid waste is increasing by approximately 3.4 million tons each year.

(d) $f(1995) = 206.8$ million tons.
67. **(b)** $f(x) = 0.38(x - 1996) + 9.7$. Answers may vary.
[1995, 2003, 1] by [9, 13, 1]

(c) The Asian-American population is predicted to increase by about 0.38 million each year.
(d) $f(2005) = 13.12$ million
69. **(a)** No, the slope is not zero.
[0, 3, 1] by [−2, 2, 1]

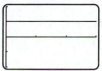

(b) The resolution of most calculator screens is not high enough to show the slight increase in the y-values.
71. **(a)** They do not appear to be perpendicular in the standard viewing rectangle. (Answers may vary for different calculators.)
(b) In [−15, 15, 1] by [−10, 10, 1] and [−3, 3, 1] by [−2, 2, 1] they appear to be perpendicular.

[−10, 10, 1] by [−10, 10, 1]   [−15, 15, 1] by [−10, 10, 1]

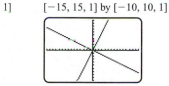

[−10, 10, 1] by [−3, 3, 1]   [−3, 3, 1] by [−2, 2, 1]

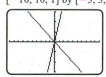

(c) The lines appear perpendicular when the distance shown along the x-axis is approximately 1.5 times farther than the distance along the y-axis.
73. $y_1 = x, y_2 = -x, y_3 = x + 2, y_4 = -x + 4$
75. $y_1 = x + 4, y_2 = x - 4, y_3 = -x + 4,$
$y_4 = -x - 4$
77. $k = 2.5, y = 20$ when $x = 8$
79. $k = 0.06, x = \$85$ when $y = \$5.10$
81. \$1048, $k = 65.5$

83. **(a)** $k = \frac{8}{15}$ **(b)** $13\frac{1}{3}$ inches

85. No. Doubling the speed more than doubles the stopping distance.

SECTION 2.3

1. Linear   **3.** Nonlinear   **5.** Linear
7. $x = 3$ **9.** $x = 1.3$ **11.** $x = 0.675$
13. $x \approx 3.621$ **15.** $x \approx 2.294$ **17.** $x \approx -2.405$
19. (a) $x = 5$ **(c)** Conditional
21. (a) No solutions   **(c)** Contradiction
23. (a) $x = \dfrac{8}{3}$ **(c)** Conditional
25. (a) All real numbers   **(c)** Identity
27. (a) $x = 3$ **(c)** Conditional
29. (a) All real numbers   **(c)** Identity
31. (a) $x = 1990$ **(c)** Conditional
33. $x = -\dfrac{1}{3}$ **35.** $x = \dfrac{4\sqrt{2} + 1.5}{0.5 - \sqrt{2}} \approx -7.828$
37. No solutions   **39.** $x = -2$ **41.** $x = 8.6$
43. $x = 3.5$ **45.** $x \approx -0.2$ **47.** $x = 2$
49. $x = 7.5$ **51.** 1993
53. (a) $V(x) = 900x$ **(b)** $V(50) = 45,000$ cubic
feet per hour   **(c)** 4.5   **(d)** $3\dfrac{1}{3}$ times
55. 1994   **57.** 41.25 feet   **59.** 189
61. 268.6 million
63. (a) Approximately 264,800 tons   **(b)** Approximately 174,700 tons   **(c)** Approximately 1.40×10^9 orders   **(d)** Approximately 11.2 orders per person
65. $f(x) = 0.75x$, $42.18
67. 3.2 hours at 55 mph and 2.8 hours at 70 mph
69. (a) About 2 hours; answers may vary.
(b) $\dfrac{15}{8} = 1.875$ hours
71. 1.6 pounds of $2.50 candy and 3.4 pounds of $4.00 candy
73. 36 inches by 54 inches   **75.** 25 feet by 50 feet
77. $1250 at 5%, $3750 at 7%
79. $-40°F$ is equivalent to $-40°C$
81. The point $(4, 3.5)$ lies on the graph of f. The graph of f is continuous, so there must be a point where it crosses the line $y = 3.5$.
83. Because $f(2) = -1 < 0$ and $f(3) = 5 > 0$, the intermediate value property states that there exists an x-value between 2 and 3 where $f(x) = 0$.
85. Because $f(0) = -1 < 0$ and $f(1) = 1 > 0$, the intermediate value property states that there exists an x-value between 0 and 1 where $f(x) = 0$.

SECTION 2.4

1. (a) A loan amount of $1000 results in annual interest of $100.   **(b)** A loan amount greater than $1000 results in annual interest of more than $100.   **(c)** A loan amount less than $1000 results in annual interest

of less than $100.   **(d)** A loan amount greater than or equal to $1000 results in annual interest of $100 or more.
3. (a) $f(x) = 100$ when $x = 1000$
(b) $f(x) > 100$ when $x > 1000$
(c) $f(x) < 100$ when $x < 1000$
(d) $f(x) \geq 100$ when $x \geq 1000$
5. (a) Car A is traveling faster since its graph has the greater slope.   **(b)** 2.5 hours, 225 miles
(c) $0 \leq x < 2.5$
7. (a) Away   **(b)** $x = 1$ or $x = 3$ **(c)** $1 \leq x \leq 3$
(d) $x > 1$
9. $x > 2.8$ **11.** $x \leq 1987.5$
13. $x > -1.82$ (approximately)   **15.** $4 \leq x < 6.4$
17. $4.6 \leq x \leq 15.2$ **19.** $1 < x < 5.5$ **21.** $[5, \infty)$
23. $[4, 19)$ **25.** $(-\infty, -37]$ **27.** $[-1, \infty)$
29. $(-3, 5]$ **31.** $(-\infty, -2)$ **33.** $[2, \infty)$
35. $(-\infty, 10.5)$ **37.** $[13, \infty)$ **39.** $[-10, \infty)$
41. $(1, \infty)$ **43.** $\left(\dfrac{7}{3}, \infty\right)$ **45.** $\left[-\dfrac{1}{2}, 2\right]$
47. $[-16, 1]$ **49.** $\left(\dfrac{5}{7}, \dfrac{17}{7}\right]$ **51.** $(-4, 1)$
53. $\left[\dfrac{9}{2}, \dfrac{21}{2}\right]$ **55.** $(-\infty, -4)$
57. $Y_1 > 0$ when $x < 4$, and $Y_1 \leq 0$ when $x \geq 4$.
59. $Y_1 \geq Y_2$ when $x \leq 50$, and $Y_1 < Y_2$ when $x > 50$.
61. $x < -\dfrac{3}{2}$ **63.** $x < 25.3$ **65.** $\left(\dfrac{13}{2}, \infty\right)$
67. $[0.717, \infty)$
69. (a) $1.8 < x \leq 6$, where 1.8 is approximate.
(b) The x-intercept represents the altitude where the temperature is $0°F$.   **(c)** $\dfrac{53}{29} < x \leq 6$
71. (a) The median price of a single-family home has increased, on average, by $3421 per year.   **(b)** From 1983 to 1988 (approximately)
73. (a) Increasing the ventilation increases the percentage of pollutants removed. A slope of 1.06 means that for each additional liter of air removed per second, the amount of pollutants decreases by 1.06%.
(b) From approximately 40.4 to 59.3 liters per second
75. $3.98\pi \leq C \leq 4.02\pi$
77. (a) $f(x) = 3x - 1.5$ **(b)** $x > 1.25$
79. (a) $f(x) = 675(x - 1997) + 47,975$ or $f(x) = 675x - 1,300,000$ **(b)** $x = 1997, 1998, \ldots, 2004$; from 1997 through 2004, this person's salary was less than or equal to $52,700.

SECTION 2.5

1. (a) Max: 55 mph; min: 30 mph   **(b)** 12 miles
(c) $f(4) = 40$, $f(12) = 30$, $f(18) = 55$
(d) $x = 4, 6, 8, 12$, and 16. The speed limit changes at each discontinuity.
3. (a) Initial: 50,000 gallons; final: 30,000 gallons
(b) $0 \le x \le 1$ or $3 \le x \le 4$ **(c)** $f(2) = 45$ thousand, $f(4) = 40$ thousand
(d) 5000 gallons per day
5. (a) $-5 \le x \le 5$
(b) $f(-2) = 2$, $f(0) = 3$, $f(3) = 6$
(c)   **(d)** f is continuous.

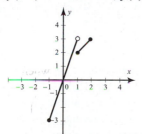

7. (a) $-1 \le x \le 2$
(b) $f(-2) = $ undefined, $f(0) = 0$, $f(3) = $ undefined
(c)   **(d)** f is not continuous.

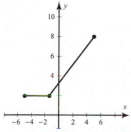

9. (a) $-3 \le x \le 3$
(b) $f(-2) = -2$, $f(0) = 1$, $f(3) = -1$
(c)   **(d)** f is not continuous.

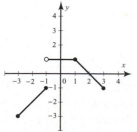

11. $f(-3) = -10$, $f(1) = 4$, $f(2) = 4$, $f(5) = 1$
13. f is not continuous.

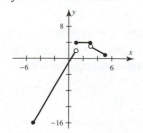

15. $g(-8) = 10$, $g(-2) = -2$, $g(2) = 2$, $g(8) = 5$
17. g is continuous.

19. (a) $[-10, 10, 1]$ by $[-10, 10, 1]$

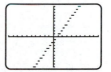

(b) $f(-3.1) = -8$, $f(1.7) = 2$
21. (a) $[-10, 10, 1]$ by $[-10, 10, 1]$

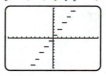

(b) $f(-3.1) = -7$, $f(1.7) = 3$
23. (a) $f(x) = 0.8[\![x/2]\!]$ for $6 \le x \le 18$
(b) $[6, 18, 1]$ by $[0, 8, 1]$

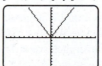

(c) $f(8.5) = \$3.20$, $f(15.2) = \$5.60$
25. (a) $[-10, 10, 1]$ by $[-10, 10, 1]$

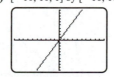

(b) $[-10, 10, 1]$ by $[-10, 10, 1]$ **(c)** $x = 0$

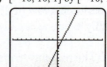

27. (a) $[-10, 10, 1]$ by $[-10, 10, 1]$

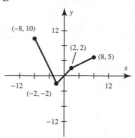

(b) $[-10, 10, 1]$ by $[-10, 10, 1]$ **(c)** $x = 1$

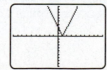

29. (a) $[-10, 10, 1]$ by $[-10, 10, 1]$

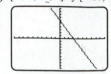

(b) $[-10, 10, 1]$ by $[-10, 10, 1]$ **(c)** $x = 3$

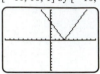

31. (a) $x = -1$ or 7 **(b)** $-1 < x < 7$
(c) $x < -1$ or $x > 7$
33. $x = -2.5$ or 7.5; $-2.5 < x < 7.5$
35. $x = \dfrac{7}{3}$ or 1; $x < 1$ or $x > \dfrac{7}{3}$
37. $x = -\dfrac{17}{21}$ or $\dfrac{31}{21}$; $x \le -\dfrac{17}{21}$ or $x \ge \dfrac{31}{21}$
39. $x = -\dfrac{1}{3}$ or $\dfrac{1}{3}$; $x < -\dfrac{1}{3}$ or $x > \dfrac{1}{3}$
41. There are no solutions for the equation or inequality.
43. $x = \pm 2$ **45.** No solutions   **47.** $x = -1$ or $\dfrac{17}{5}$

49. $x = -1$ or 1 **51.** $\left(-\dfrac{7}{3}, 3\right)$ **53.** $\left[-1, \dfrac{9}{2}\right]$

55. $\left(-\dfrac{5}{2}, \dfrac{11}{2}\right)$ **57.** $(-\infty, 1)\cup(2, \infty)$

59. $\left(-\infty, \dfrac{5}{3}\right]\cup\left[\dfrac{11}{3}, \infty\right)$ **61.** $(-\infty, -8)\cup(16, \infty)$

63.

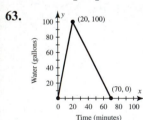

65.

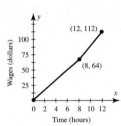

67. (a) $19 \le T \le 67$ **(b)** The monthly average temperatures in Marquette vary between a low of 19°F and a high of 67°F. The monthly averages are always within 24 degrees of 43°F.
69. (a) $28 \le T \le 72$ **(b)** The monthly average temperatures in Boston vary between a low of 28°F and a high of 72°F. The monthly averages are always within 22 degrees of 50°F.
71. (a) $49 \le T \le 74$.   **(b)** The monthly average temperatures in Buenos Aires vary between a low of 49°F (possibly in July) and a high of 74°F (possibly in January). The monthly averages are always within 12.5 degrees of 61.5°F.

73. (a) $y = x - 1$
(b)

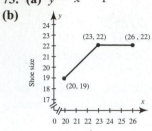

75. (a)

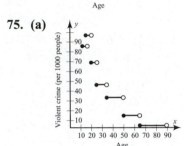

(b) The likelihood of being a victim of crime peaks from age 16 up to age 19, and then it decreases.
77. (a) $[1968, 1997, 5]$ by $[460, 490, 5]$

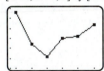

(b) It appears that scores fell dramatically from 1970 to 1980 and then recovered slightly from 1980 to 1995.
(c) In this viewing rectangle, it appears that the scores have not changed significantly. Altering the y-scale can affect the way a person interprets the data.
$[1968, 1997, 5]$ by $[200, 800, 100]$

79. (a) Cesarean births increased at a constant rate of 6% every 5-year period between 1970 and 1985, which then leveled off at 23%.
$[1965, 1995, 5]$ by $[0, 25, 5]$

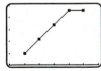

(b) $a = 1.2$, $b = 5$, $c = 23$ **(c)** $f(1978) = 14.6\%$, $f(1988) = 23\%$. In 1978, 14.6% of births were Cesarean births, and in 1988 the percentage was 23%.

SECTION 2.6

1. $530 billion   **3.** 543,949 inmates
5. 10 seconds   **7.** $(3, -0.5)$ **9.** $(-2.1, -0.35)$

11. $(\sqrt{2}, 0)$ **13.** $(0, 2b)$
15. (a) $y = 1.5x - 3.2$ **(b)** When $x = -2.7$,
$y = -7.25$ (interpolation); when $x = 6.3$, $y = 6.25$
(extrapolation)
17. (a) $y = -2.1x + 105.2$ **(b)** When $x = -2.7$,
$y = 110.87$ (extrapolation); when $x = 6.3$, $y = 91.97$
(interpolation)
19. (a) $f(x) = -108(x - 1998) + 3305$ or
$f(x) = -108x + 219,089$; answers may vary.
(b) 3521; the estimated value is too high; extrapolation
21. (a) Positive   **(b)** $y = ax + b$, where $a \approx 3.0929$
and $b \approx -2.2143$; $r \approx 0.9989$ **(c)** $y \approx 5.209$
23. (a) Negative   **(b)** $y = ax + b$, where
$a \approx -3.8857$ and $b \approx 9.3254$; $r \approx -0.9996$
(c) $y \approx -0.00028$; due to rounding, answers may
vary slightly.
25. (a) [1982, 1994, 1] by [200, 500, 100]

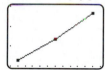

(b) High   **(c)** 347
27. (a) Not exactly linear
(b) $f(x) = 263.25(x - 1985) + 1828$
(c) $f(1987) = 2354.5$, which equals the midpoint estimate. The midpoint lies on the graph of f.
29. (a) [−100, 1800, 100] by [−1000, 28,000, 1000]

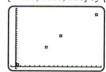

(b) $y = ax + b$, where $a \approx 14.680$ and $b \approx 277.82$
(c) 2500 light-years
31. (a) [1973, 1997, 2] by [0, 25, 1]

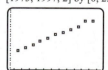

(b) $y = ax + b$, where $a \approx 0.66045$ and
$b \approx -1296.5$ **(c)** The percentage of women in state
legislatures has increased by about 0.66% per year
(d) 23.1%

Chapter 2 Review Exercises

1. (a) Slope: -2; y-intercept: 6; x-intercept: 3
(b) $f(x) = -2x + 6$ **(c)** $x = 3$
3. $f(x) = -2.5x + 5$ **5.** $x = 6.4$
7. $x = \dfrac{15}{7} \approx 2.143$ **9.** $x = \dfrac{1}{17}$

11. $x \approx -2.9$
13. (a) $x \approx 1983$. In 1983 the median U.S. family income was about \$25,000.
(b) $x = \dfrac{3847}{1321.7} + 1980 \approx 1983$
15. (a) $x = 2$ **(b)** $x > 2$ **(c)** $x < 2$
17. $(-\infty, 3]$ **19.** $(-\infty, \infty)$ **21.** $(-1, 3.5]$
23. From 2001 to 2006   **25.** $y = 7x + 30$
27. $y = -3x + 2$ **29.** $y = -\dfrac{31}{57}x - \dfrac{368}{57}$
31. $x = 6$ **33.** $y = 3$ **35.** $y = -8$
37. Initially, the car is at home. After traveling 30 mph
for 1 hour, the car is 30 miles away from home. During
the second hour the car travels 20 mph until it is 50
miles away. During the third hour the car travels toward
home at 30 mph until it is 20 miles away. During the
fourth hour the car travels away from home at 40 mph
until it is 60 miles away from home. During the last
hour, the car travels 60 miles at 60 mph until it arrives
home.
39. (a) $f(-2) = 4$, $f(-1) = 6$, $f(2) = 3$, $f(3) = 4$
(b) f is continuous

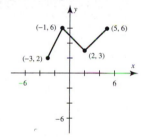

(c) $x = -2.5$ or 2
41.

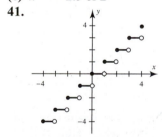

43. $x = \pm 3$; $x < -3$ or $x > 3$
45. $x = \dfrac{17}{3}$ or $x = -1$; $x < -1$ or $x > \dfrac{17}{3}$
47. $x = -0.5$ or $x = -0.75$
49. (a) [1988, 1995, 1] by [20.5, 20.9, 0.1]

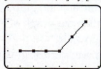

(b)
$$f(x) = \begin{cases} 20.6 & \text{if } 1989 \le x \le 1992 \\ 0.1(x - 1992) + 20.6 & \text{if } 1992 < x \le 1994 \end{cases}$$

51. 155,590 **53.** 18.75 minutes
55. 0.9 hour at 7 mph and 0.9 hour at 8 mph
57. (a) $f(x) = \dfrac{5}{3}(x - 1994) + 504$ or

$f(x) = \dfrac{5}{3}x - \dfrac{8458}{3}$

(b) 509 (answers may vary); interpolation
59. The tank initially contains 25 gallons. When $0 \leq x \leq 4$, only the 5 gal/min inlet pipe is open. When $4 < x \leq 8$, only the outlet pipe is open. When $8 < x \leq 12$, both inlet pipes are open. When $12 < x \leq 16$, all pipes are open. When $16 < x \leq 24$, the 2 gal/min inlet pipe and the outlet pipe are open. When $24 < x \leq 28$, all pipes are closed.
61. (a) 1973 **(b)** The function f will not continue to be a good model far into the future. Once the water is no longer polluted, the number of species in the river will reach its natural level.
63. (a) Positive **(b)** $y = 143.45x - 280,361.2$; $r \approx 0.978$ **(c)** $5821.55

CHAPTER 3: QUADRATIC FUNCTIONS AND EQUATIONS

SECTION 3.1

1.
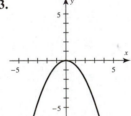

3.

5.
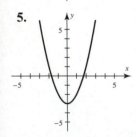

7. Quadratic; leading coefficient: 3; $f(-2) = 17$
9. Neither linear nor quadratic **11.** Linear
13. Quadratic; leading coefficient: -3; $f(-2) = -3$
15. The graph of g is narrower than the graph of f.
17. The graph of g is wider than the graph of f and opens downward rather than upward.
19. Leading coefficient: positive; vertex: $(1, 0)$; axis of symmetry: $x = 1$

21. Leading coefficient: negative; vertex: $(-3, -2)$; axis of symmetry: $x = -3$
23. Vertex: $(1, 2)$; leading coefficient: -3; $f(x) = -3x^2 + 6x - 1$
25. Vertex: $(4, 5)$; leading coefficient: -2; $f(x) = -2x^2 + 16x - 27$
27. Vertex: $\left(-5, -\dfrac{7}{4}\right)$; leading coefficient: $\dfrac{3}{4}$; $f(x) = \dfrac{3}{4}x^2 + \dfrac{15}{2}x + 17$
29. $(0, 6)$ **31.** $(3, -9)$ **33.** $(1.9, -5.61)$
35. $(-0.25, 1.875)$
37. Leading coefficient: 1; vertex: $(-2, -9)$; $f(x) = (x + 2)^2 - 9$
39. Leading coefficient: 2; vertex: $(2, -9)$; $f(x) = 2(x - 2)^2 - 9$
41. Leading coefficient: -3; vertex: $(-1.5, 8.75)$; $f(x) = -3(x + 1.5)^2 + 8.75$
43. $f(x) = (x - 2)^2 - 2$
45. $f(x) = 2(x - 1)^2 - 3$
47. $f(x) = -2(x + 1)^2 + 4$
49. (a) 32 feet **(b)** 34.25 feet
51. d **53.** a **55.** 146 feet after 2.75 seconds
57. 322.8 feet after 6.77 seconds
59. (b) $f(x) = 5.8(x - 1960)^2 + 1490$. Answers may vary.

[1955, 2000, 5] by [1000, 8000, 1000]
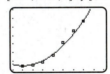

(c) $f(1995) = \$8595$
61. (b) $f(x) = 3100(x - 1982)^2 + 1586$. Answers may vary.

[1980, 1996, 2] by [−50,000, 500,000, 10,000]
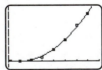

(c) $f(1998) = 795,186$ cases
63. (a) $R(2) = 72$; the company receives \$72,000 for producing 2000 CD players. **(b)** 10,000
(c) \$200,000
65. 20 feet by 20 feet
67. $f(x) = 3.125x^2 + 2.05x - 0.9$; $f(3.5) \approx 44.56$
69. $f(x) = 1.415x^2 + 1.731x + 0.995$; $f(3.5) \approx 24.39$
71. (a) $f(x) = 0.00021676x^2 - 0.86375x + 862.40$
(b) $f(1975) \approx 1.99$, which compares favorably with the actual value of 2.01.

SECTION 3.2

1. (a) $a > 0$ **(b)** $x = -6$ or 2 **(c)** Positive
3. (a) $a > 0$ **(b)** $x = -4$ **(c)** Zero
5. $x = -2$ or 3 **7.** $x = \pm\sqrt{3} \approx \pm 1.7$
9. $x = 1.5$ **11.** $x = -0.75$ or 0.2
13. $x = 0.7$ or 1.2 **15.** $x = -96$ or 72
17. (a) $3x^2 - 12 = 0$ **(b)** $b^2 - 4ac = 144 > 0$.
There are two real solutions. **(c)** $x = \pm 2$
19. (a) $x^2 - 2x + 1 = 0$ **(b)** $b^2 - 4ac = 0$.
There is one real solution. **(c)** $x = 1$
21. (a) $x^2 - 4x + 2 = 0$ **(b)** $b^2 - 4ac = 8 > 0$.
There are two real solutions. **(c)** $2 \pm \sqrt{2}$
23. (a) $x^2 - x + 1 = 0$
(b) $b^2 - 4ac = -3 < 0$. There are no real solutions.
25. (a) $2x^2 + 5x - 12 = 0$
(b) $b^2 - 4ac = 121 > 0$. There are two real solutions. **(c)** $x = -4$ or 1.5
27. (a) $\frac{1}{4}x^2 + 2x + 4 = 0$ **(b)** $b^2 - 4ac = 0$.

There is one real solution. **(c)** $x = -4$
29. (a) $\frac{1}{2}x^2 + x + \frac{13}{2} = 0$

(b) $b^2 - 4ac = -12 < 0$. There are no real solutions.
31. (a) $3x^2 + x - 1 = 0$
(b) $b^2 - 4ac = 13 > 0$. There are two real solutions.

(c) $x = -\frac{1}{6} \pm \frac{1}{6}\sqrt{13} \approx 0.4343, -0.7676$

33. $-2 \pm \sqrt{10}$ **35.** $-\frac{5}{2} \pm \frac{1}{2}\sqrt{41}$

37. $1 \pm \sqrt{\dfrac{5}{3}}$ **39.** $4 \pm \sqrt{26}$

41. (a) R quadruples **(b)** $a = 0.5$
(c) $x = \sqrt{1000} \approx 31.6$. The safe speed for a curve
with radius 500 feet is about 31 mph or slower.
43. (a) $[-5, 45, 5]$ by $[0, 1500, 100]$

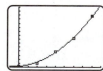

(b) and **(c)** $x \approx 35.2$ or in 1985
45. (a) $f(x) = -0.00184x^2 + 50$. Answers may vary.
(b) $x \approx \pm 165$. Only the positive solution has meaning. It represents the number of days (165) for all the
worms to die.
47. 8.5 inches by 11 inches
49. 15 inches by 25 inches
51. 20 inches by 24 inches **53.** About 18.23 inches

SECTION 3.3

1. (a) $-3 < x < 2$ **(b)** $x \leq -3$ or $x \geq 2$
3. (a) $x = -2$ **(b)** $x \neq -2$

5. (a) No solutions **(b)** All real numbers
7. (a) $x < -1$ or $x > 1$ **(b)** $-1 \leq x \leq 1$
9. $-1 < x < 4$ **11.** $x < -3$ or $x > 2$
13. $-2 \leq x \leq 2$ **15.** All real numbers

17. $x \leq -2$ or $x \geq 3$ **19.** $-\dfrac{1}{3} < x < \dfrac{1}{2}$

21. $-4 \leq x \leq 10$ **23.** $-3 < x < -1$
25. $x \leq -7$ or $x \geq 5$ **27.** $x \leq 0$ or $x \geq 1$
29. $x \leq -2$ or $x \geq 3$ **31.** $-\sqrt{5} \leq x \leq \sqrt{5}$
33. $x \leq k$ or $x \geq 22.4$, where $k \approx 3.3$
35. From 1989 to 1992
37. (a) $163x^2 - 146x + 205 > 2000$ or
$163x^2 - 146x - 1795 > 0$, where x is positive
(b) $x > k$, where $k \approx 3.8$. This corresponds to 1989
or later.
39. (a) The height does not change by the same
amount each 15-second interval. **(b)** From 43 to 95
seconds (approximately)
(c) $[-25, 200, 25]$ by $[-5, 20, 5]$

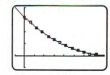

(d) From 43.8 to 93.1 seconds (approximately)

SECTION 3.4

1. $y = (x + 2)^2$ **3.** $y = \sqrt{x + 3}$
5. $y = |x + 2| - 1$ **7.** $y = (x - 2)^2 + 1$
9. $y = |x| - 3$

11.

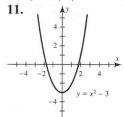

13. $y = (x+4)^2$

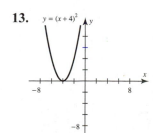

15.

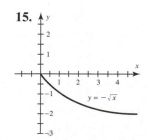

17. $y = -x^2 + 4$

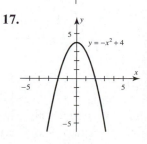

19.

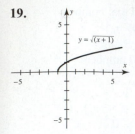

$y = \sqrt{x+1}$

21.

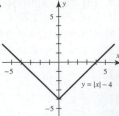

$y = |x| - 4$

23.

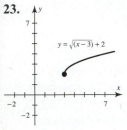

$y = \sqrt{x-3} + 2$

25.

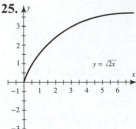

$y = \sqrt{2x}$

27.

(a)

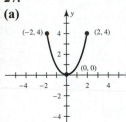

$(-2, 4)$ $(2, 4)$ $(0, 0)$

(b)

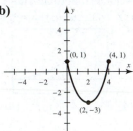

$(0, 1)$ $(4, 1)$ $(2, -3)$

(c)

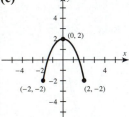

$(0, 2)$ $(-2, -2)$ $(2, -2)$

29.

(a)

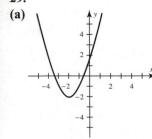

(b)

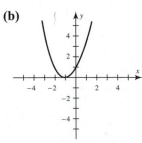

(c)

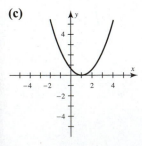

31.

(a)

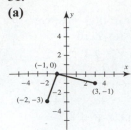

$(-1, 0)$ $(-2, -3)$ $(3, -1)$

(b)

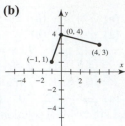

$(0, 4)$ $(4, 3)$ $(-1, 1)$

(c)

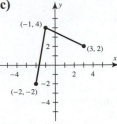

$(-1, 4)$ $(3, 2)$ $(-2, -2)$

33. $y = (x - 2)^2 - 3$
$[-10, 10, 1]$ by $[-10, 10, 1]$

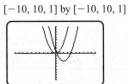

35. $y = (x + 6)^2 - 4(x + 6) + 5$
$[-10, 10, 1]$ by $[-10, 10, 1]$

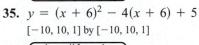

37. x-axis: $y = -(x^2 - 2x - 3)$
$[-9, 9, 1]$ by $[-6, 6, 1]$

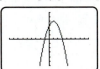

y-axis: $y = (-x)^2 - 2(-x) - 3$
$[-9, 9, 1]$ by $[-6, 6, 1]$

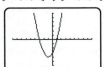

39. x-axis:

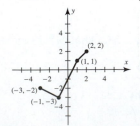

$(2, 2)$ $(1, 1)$ $(-3, -2)$ $(-1, -3)$

y-axis:

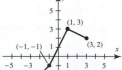

$(1, 3)$ $(3, 2)$ $(-1, -1)$ $(-2, -2)$

41.

x	1	2	3	4	5
$g(x)$	12	8	13	9	14

43.

x	-2	0	2	4	6
$g(x)$	5	2	-3	-5	-9

45.

x	0	1	2	3	4
$g(x)$	0	2	1	5	6

47. (a)

x	0	1	2	3
$g(x)$	$360	$380	$410	$430

x	4	5	6
$g(x)$	$460	$490	$520

(b)
$D_f = \{1990, 1991, 1992, 1993, 1994, 1995, 1996\}$,
$D_g = \{0, 1, 2, 3, 4, 5, 6\}$
(c) $g(x - 1990) = f(x)$ or $g(x) = f(x + 1990)$
49. (a) When $x = 2$, the y-value is about 9. There are 9 hours of daylight on February 21 at 60°N latitude.
(b) Since February 21 is 2 months after the shortest day and October 21 is 2 months before, they both have approximately the same number of daylight hours. A reasonable conjecture would be 9 hours.
(c) There are approximately the same number of daylight hours either x months before or after December 21. The graph should be symmetric about the y-axis, as shown in the figure.

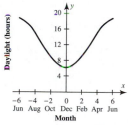

51. $y = -0.4(x - 3)^2 + 4$ (mountain)
$[-4, 4, 1]$ by $[0, 6, 1]$

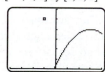

53. (a) $y = \dfrac{1}{20}x^2 - 1.6$
$[-15, 15, 1]$ by $[-10, 10, 1]$

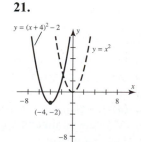

(b) $y = \dfrac{1}{20}(x - 2.1)^2 - 2.5$. The front reaches Columbus by midnight.
$[-15, 15, 1]$ by $[-10, 10, 1]$

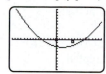

CHAPTER 3 REVIEW EXERCISES

1. Leading coefficient: negative; vertex: $(2, 4)$; axis of symmetry: $x = 2$
3. $f(x) = -2x^2 + 20x - 49$; leading coefficient: -2
5. $f(x) = (x + 3)^2 - 10$; vertex: $(-3, -10)$
7. $\left(\dfrac{1}{3}, -\dfrac{11}{3}\right)$
9. (a) $a > 0$ **(b)** $-3, 2$ **(c)** Positive
11. $x = -4$ or 5   **13.** $x < 1$ or $x > 2$
15. (a) $-3 < x < 2$ **(b)** $x \le -3$ or $x \ge 2$
17. $y = -f(x) = -(2x^2 - 3x + 1)$
$[-6, 6, 1]$ by $[-4, 4, 1]$

$y = f(-x) = 2(-x)^2 - 3(-x) + 1$
$[-6, 6, 1]$ by $[-4, 4, 1]$

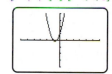

19.

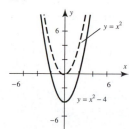

21.

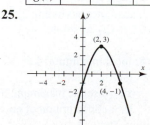

23.

x	1	2	3	4
$g(x)$	-1	5	6	9

25.

27. (a) $h(0) = 5$; the stone was held 5 feet above the ground when it was released.   **(b)** 117 feet
(c) 126 feet   **(d)** After 2 seconds and 3.5 seconds
29. (a) $f(1985) \approx 4.7$. In 1985 there were about 4.7 billion people in the world.   **(b)** 6.1 billion
(c) About 2009 or late 2008
31. (a) $C(x) = x(103 - 3x)$ **(b)** $C(6) = \$510$
(c) 10 rooms (73/3 rooms cannot be rented)
(d) 17 rooms ($884)
33. (a) The change in debt is not constant each 4-year period.   **(b)** $x = 1990$; Visa and MasterCard debt reached $212 billion in 1990.

CHAPTER 4: NONLINEAR FUNCTIONS AND EQUATIONS

SECTION 4.1

1. Yes; degree: 3; leading coefficient: 2   **3.** No
5. Yes; degree: 4; leading coefficient: -5 **7.** No
9. Yes; degree: 0; leading coefficient: 22
11. Increasing: $x \geq 2$; decreasing: $x \leq 2$
13. Increasing: $x \leq -2$ or $x \geq 1$; decreasing: $-2 \leq x \leq 1$
15. Increasing: $-8 \leq x \leq 0$ or $x \geq 8$; decreasing: $x \leq -8$ or $0 \leq x \leq 8$
17. Increasing for all x
19. Increasing: $x \leq 0$; decreasing: $x \geq 0$
21. Increasing: $x \leq -0.82$ or $x \geq 0.82$; decreasing: $-0.82 \leq x \leq 0.82$
23. Increasing: $x \geq -1$; decreasing: $x \leq -1$
25. Local maxima: approximately 800 and 900; local minima: approximately 600 and 700
27. (a) Local maximum: approximately 5.5; local minimum: approximately -5.5 **(b)** No absolute extrema
29. (a) Local maxima: approximately 17 and 27; local minima: approximately -10 and 24   **(b)** No absolute extrema
31. (a) Local minimum: approximately -3.2
(b) Absolute minimum: approximately -3.2; absolute maximum: 3
33. (a) Local minimum: -1; local maximum: 1
(b) Absolute minimum: -1; absolute maximum: 1
35. (a) Local minimum: -2; local maximum: 1, 2
(b) Absolute minimum: -2; absolute maximum: 2
37. (a) Local maximum: 5.25   **(b)** Absolute maximum: 5.25
39. (a) Local minimum: -8; local maximum: 4.5
(b) Absolute minimum: -8
41. (a) Local maximum: 8   **(b)** Absolute maximum: 8
43. (a) $f(5) \approx 8.32$; in 1955 U.S. consumption of energy was about 8.32 quadrillion Btu.

(b) The energy consumption increased, reached a maximum value, and then started to decrease.

[0, 30, 5] by [6, 16, 1]   [0, 30, 5] by [6, 16, 1]

          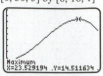

(c) Local maximum: approximately 14.5. In 1973 maximum energy use peaked at 14.5 quadrillion Btu.
45. (a) Possible absolute maximum in January and absolute minimum in July
(b) Absolute maximum of $140 in January and an absolute minimum of $15 in July

[1, 12, 1] by [0, 150, 10]   [1, 12, 1] by [0, 150, 10]

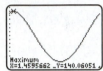   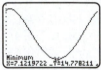

47. f is odd and the point $(8, -6)$ lies on its graph.
49. f is odd.
51. Note that $f(0)$ can be any number.

x	-3	-2	-1	0	1	2	3
$f(x)$	21	-12	-25	1	-25	-12	21

53. Answers may vary.

x	-2	-1	0	1	2
$f(x)$	5	1	2	3	4

55.

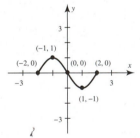

57. (a) Even   **(b)** 83°F   **(c)** They are equal.
(d) Monthly average temperatures are symmetric about July. July has the highest average and January the lowest. The pairs June-August, May-September, April-October, March-November, and February-December have approximately the same average temperatures.
59. (a) $h(-2) = h(2) = 336$. Two seconds before and 2 seconds after the time when the maximum height is reached, the projectile's height is 336 feet.
(b) $h(-5) = h(5) = 0$. Five seconds before and 5 seconds after the time when the maximum height is reached, the projectile is on the ground.

(c) h is an even function.

[−5, 5, 1] by [0, 500, 100]

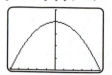

(d) $h(-x) = h(x)$ when $-5 \le x \le 5$. This means that the projectile is at the same height t seconds either before or after the time when the maximum height is attained.

61. Even   **63.** Odd   **65.** Neither   **67.** Even
69. Even
71. Answers may vary.

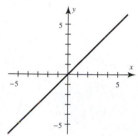

73. No. If $(2, 5)$ is on the graph of an odd function f, then so is $(-2, -5)$. Since f would pass through $(-3, -4)$ and then $(-2, -5)$, it could not always be increasing.

75.

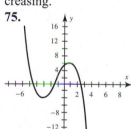

77.

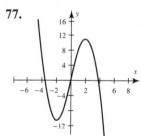

SECTION 4.2

1. (a) The turning points are approximately $(1.6, 3.6)$, $(3, 1.2)$, $(4.4, 3.6)$.   **(b)** After 1.6 minutes the runner is 360 feet from the starting line. The runner turns and jogs toward the starting line. After 3 minutes the runner is 120 feet from the starting line, turns, and jogs away from the starting line. After 4.4 minutes the runner is again 360 feet from the starting line. The runner turns and jogs back to the starting line.

3. (a) 0.5   **(b)** Positive   **(c)** 1
5. (a) $-6, -1$, and 6   **(b)** Negative   **(c)** 4
7. (a) $-3, -1, 0, 1$, and 2   **(b)** Positive   **(c)** 5
9. (a) d   **(b)** $(1, 0)$ **(c)** $x = 1$ **(d)** Local minimum: 0   **(e)** Absolute minimum: 0
11. (a) b   **(b)** $(-3, 27)$, $(1, -5)$ **(c)** $x \approx -4.9$, $x = 0$, $x \approx 1.9$ **(d)** Local minimum: -5; local maximum: 27   **(e)** Absolute minimum: none; absolute maximum: none

13. (a) a   **(b)** $(-2, 16)$, $(0, 0)$, $(2, 16)$
(c) $x \approx -2.8$, $x = 0$, $x \approx 2.8$ **(d)** Local minimum: 0; local maximum: 16   **(e)** Absolute minimum: none; absolute maximum: 16

15. (a) [−10, 10, 1] by [−10, 10, 1]

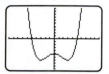

(b) $(-3, 6)$, $(3, -6)$ **(c)** Local maximum: 6; local minimum: -6

17. (a) [−10, 10, 1] by [−10, 10, 1]

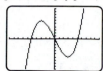

(b) There are three turning points located at $(-3, -7.025)$, $(0, -5)$, and $(3, -7.025)$.
(c) Local minimum: -7.025; local maximum: -5

19. (a) [−10, 10, 1] by [−10, 10, 1]

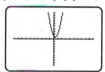

(b) $\left(\dfrac{1}{3}, \dfrac{2}{3}\right) \approx (0.333, 0.667)$ **(c)** Local minimum: $\dfrac{2}{3} \approx 0.667$

21. (a) [−10, 10, 1] by [−10, 10, 1]

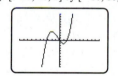

(b) $\left(-2, \dfrac{10}{3}\right) \approx (-2, 3.333)$, $\left(1, -\dfrac{7}{6}\right) \approx (1, -1.167)$

(c) Local minimum: $-\dfrac{7}{6} \approx -1.167$; local maximum: $\dfrac{10}{3} \approx 3.333$

23. (a) Degree: 3; leading coefficient: -1 **(b)** Up on left side, down on right side
(c) [−10, 10, 1] by [−10, 10, 1]

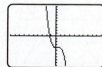

25. (a) Degree: 5; leading coefficient: 0.1
(b) Down on left side, up on right side
(c) $[-10, 10, 1]$ by $[-10, 10, 1]$

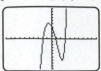

27. (a) Degree: 2; leading coefficient: $-\dfrac{1}{2}$

(b) Down on both sides
(c) $[-10, 10, 1]$ by $[-10, 10, 1]$

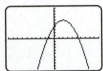

29. As the viewing rectangle increases in size, the graphs begin to look alike. Each formula contains the term $2x^4$, which determines the end behavior of the graph for large values of $|x|$.

$[-4, 4, 1]$ by $[-4, 4, 1]$ $[-10, 10, 1]$ by $[-100, 100, 10]$

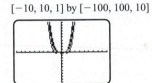

$[-100, 100, 10]$ by $[-10^6, 10^6, 10^5]$

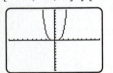

31. (a) $[-5, 5, 1]$ by $[-2, 12, 1]$ **(b)** Degree 1

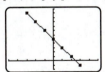

33. (a) $[-5, 5, 1]$ by $[-5, 15, 5]$ **(b)** Degree 2

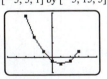

35. Answers may vary.   **37.** Answers may vary.

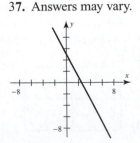

39. Answers may vary.   **41.**

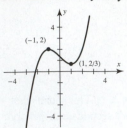
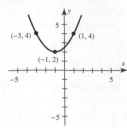

43. (a) As the speed x increases, so does the minimum sight distance y.   **(b)** Degree: 1; $D(x) = 34x + 130$. Answers may vary.

$[15, 75, 5]$ by $[500, 2800, 100]$

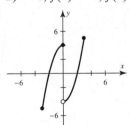

(c) $D(43) = 1592$ feet. Answers may vary.
45. (a) Approximately $(1, 13)$ and $(7, 72)$ **(b)** The monthly average low temperature of 13°F occurs in January. The monthly average high temperature of 72°F occurs in July.
47. (a) $f(x) \approx 0.02352x^3 - 0.7088x^2 + 4.444x + 27.46$ **(b)** $f(22) \approx 32.6\%$; the estimate is too high.
49. $f(-2) \approx 5, f(1) \approx 0$
51. $f(-2) = 6, f(1) = 7, f(2) = 9$
53. (a)

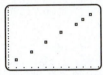

(b) f is not continuous.
55. (a)

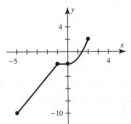

(b) f is continuous.
57. (a) 4 seconds

$[-1, 8, 1]$ by $[-10, 170, 10]$

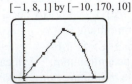

(b) From 0 to 4 seconds; from 4 to 7 seconds
(c) $m = 36, a = -16, b = 144$
(d) $x \approx 2.8$ or $x \approx 5.7$; the object is 100 feet at 2.8 seconds and at 5.7 seconds.

SECTION 4.3

1. $\dfrac{x^3}{2} - \dfrac{3}{2x}$ **3.** $x - \dfrac{2}{3x} - \dfrac{1}{3x^3}$ **5.** $-\dfrac{1}{x^3} + \dfrac{1}{4}$

7. Quotient: $x^2 + x - 2$; remainder: 0
9. Quotient: $2x^3 - 9x^2 + 4x - 23$; remainder: 40
11. Quotient: $3x^2 + 3x - 4$; remainder: 6
13. $x^3 - 8x^2 + 15x - 6$ **15.** $x^3 - 1$

17. $4x^2 + 3x - 2 + \dfrac{4}{x - 1}$

19. $x^2 - x + 1$ **21.** Yes   **23.** No

25. $f(x) = 2\left(x - \dfrac{11}{2}\right)(x - 7)$

27. $f(x) = (x + 2)(x - 1)(x - 3)$

29. $f(x) = -2(x + 5)\left(x - \dfrac{1}{2}\right)(x - 6)$

31. $f(x) = (x + 4)(x - 2)(x - 8)$
33. $f(x) = -1(x + 8)(x + 4)(x + 2)(x - 4)$

35. $f(x) = 10\,(x + 2)\left(x - \dfrac{3}{10}\right)$

37. $f(x) = -3x(x - 2)(x + 3)$

39. $f(x) = x(x + 1)(x + 3)\left(x - \dfrac{3}{2}\right)$

41. $f(x) = (x - 1)(x - 3)(x - 5)$

43. $f(x) = -4(x + 4)\left(x - \dfrac{3}{4}\right)(x - 3)$

45. $f(x) = 2x(x + 2)\left(x + \dfrac{1}{2}\right)(x - 3)$

47. -2 (odd), 4 (even); minimum degree: 5
49. -6 (even), -1 (even), 4 (odd); minimum degree: 7
51. $f(x) = (x + 1)^2(x - 6)$
53. $f(x) = (x - 2)^3(x - 6)$
55. $f(x) = (x + 2)^2(x - 4)$
57. $f(x) = -1(x + 3)^2(x - 3)^2$

59. (a) $-3, \dfrac{1}{2}, 1$

(b) $f(x) = 2(x + 3)\left(x - \dfrac{1}{2}\right)(x - 1)$

61. (a) $-2, -1, 1, \dfrac{3}{2}$

(b) $f(x) = 2(x + 2)(x + 1)(x - 1)\left(x - \dfrac{3}{2}\right)$

63. (a) $\dfrac{1}{3}, 1, 4$

(b) $f(x) = 3\left(x - \dfrac{1}{3}\right)(x - 1)(x - 4)$

65. (a) 1
(b) $f(x) = (x + \sqrt{7})(x - 1)(x - \sqrt{7})$

67. $-3, 0, 2$ **69.** $-1, 1$ **71.** $-2, 0, 2$
73. $x = -5, 0,$ or 5   **75.** $x = \pm 2$
77. $x = -3, 0,$ or 6   **79.** $x = 0$ or 1

81. $x = -\dfrac{1}{4}, 0,$ or $\dfrac{5}{3}$ **83.** $x = \pm\dfrac{2}{3}$ or ± 1

85. $x \approx -2.01, 0.12,$ or 2.99
87. $x \approx -4.05, -0.52,$ or 1.71
89. $x \approx -2.69, -1.10, 0.55,$ or 3.98
91. (a) As x increases, C decreases.
(b) $[0, 70, 10]$ by $[0, 22, 5]$

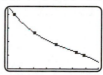

(c) $0 \le x < 32.1$ (approximately)
93. Approximately 11.34 centimeters
95. (a) $f(x) \approx -0.184(x + 6.01)(x - 2.15)(x - 11.7)$
(b) The zero of -6.01 has no significance. The zeros of $2.15 \approx 2$ and $11.7 \approx 12$ indicate that during February and December the average temperature is $0°F$.
97. June 2, June 22, and July 12

SECTION 4.4

1. $2i$ **3.** $10i$ **5.** $i\sqrt{23}$ **7.** $i\sqrt{12} = 2i\sqrt{3}$
9. $8i$ **11.** $-2 - i$ **13.** $13 - 16i$ **15.** $-1 + 6i$

17. $4 + 8i$ **19.** $5 - i$ **21.** $4 - 7i$ **23.** $\dfrac{1}{2} - \dfrac{1}{2}i$

25. $\dfrac{19}{26} + \dfrac{9}{26}i$ **27.** $-\dfrac{2}{25} + \dfrac{4}{25}i$

29. $-18.5 + 87.4i$ **31.** $8.7 - 6.7i$
33. $-117.27 + 88.11i$ **35.** $-0.921 - 0.236i$
37. $Z = 10 + 6i$ **39.** $V = 11 + 2i$

41. $I = 1 + i$ **43.** $x = \pm i\sqrt{5}$ **45.** $x = \pm i\sqrt{\dfrac{1}{2}}$

47. $x = \dfrac{3}{10} \pm \dfrac{i\sqrt{11}}{10}$ **49.** $x = 2 \pm i$

51. $x = \dfrac{3}{2} \pm \dfrac{i\sqrt{11}}{2}$ **53.** $x = -1 \pm i\sqrt{3}$

55. (a) Two real zeros   **(b)** $x = -1$ or $\dfrac{3}{2}$

57. (a) Two imaginary zeros   **(b)** $x = -\dfrac{1}{2} \pm \dfrac{i\sqrt{7}}{2}$

59. (a) Two imaginary zeros   **(b)** $x = -\dfrac{2}{5} \pm \dfrac{1}{5}i$

61. One real zero; two imaginary zeros
63. Two real zeros; two imaginary zeros
65. Three real zeros; two imaginary zeros
67. (a) $f(x) = (x - 6i)(x + 6i)$
(b) $f(x) = x^2 + 36$

69. (a) $f(x) = -1(x + 1)(x - 2i)(x + 2i)$
(b) $f(x) = -x^3 - x^2 - 4x - 4$
71. (a) $f(x) = 10(x - 1)(x + 1)(x - 3i)(x + 3i)$
(b) $f(x) = 10x^4 + 80x^2 - 90$
73. $f(x) = (x - 5i)(x + 5i)$
75. $f(x) = 3(x - 0)(x - i)(x + i)$ or $3x(x - i)(x + i)$
77. $f(x) = (x - i)(x + i)(x - 2i)(x + 2i)$
79. $f(x) = (x - 1)(x + 3)(x - 2i)(x + 2i)$
81. $x = 0, \pm i$ **83.** $x = 2, \pm i\sqrt{7}$
85. $x = 0, \pm i\sqrt{5}$ **87.** $x = 0, \dfrac{1}{2} \pm \dfrac{i\sqrt{15}}{2}$

89. $x = -2, 1, \pm i\sqrt{8}$ **91.** $x = -2, \dfrac{1}{3} \pm \dfrac{i\sqrt{8}}{3}$

SECTION 4.5

1. Yes; $D = \left\{x \,\middle|\, x \neq \dfrac{5}{4}\right\}$

3. Yes; $D =$ all real numbers   **5.** No
7. Horizontal: $y = 4$; vertical: $x = 2$; $D = \{x \mid x \neq 2\}$
9. Horizontal: $y = -4$; vertical: $x = \pm 2$; $D = \{x \mid x \neq \pm 2\}$
11. Horizontal: $y = 2$; vertical: $x = 3$
13. Horizontal: $y = 0$; vertical: $x = \pm\sqrt{5}$
15. Horizontal: none; vertical: $x = -5$ or 2
17. Horizontal: $y = 3$; vertical: $x = 1$ or -2
19. Horizontal: none; vertical: none, since $f(x) = x - 3$ for $x \neq -3$
21. b   **23.** d
25. (a) $D = \{x \mid x \neq 2\}$ **(c)** Horizontal: $y = 1$; vertical: $x = 2$
(d)

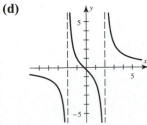

27. (a) $D = \{x \mid x \neq \pm 2\}$ **(c)** Horizontal: $y = 0$; vertical: $x = \pm 2$
(d)

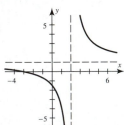

29. (a) $D = \{x \mid x \neq \pm 2\}$ **(c)** Horizontal: $y = 0$; vertical: $x = \pm 2$
(d)

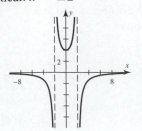

31. (a) $D = \{x \mid x \neq 2\}$ **(c)** Horizontal: none; vertical: none, since $f(x) = x + 2$ for $x \neq 2$
(d)

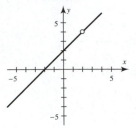

33. $y = 3$ **35.** $f(x) = \dfrac{x + 1}{x + 3}$. Answers may vary.

37. $f(x) = \dfrac{1}{x^2 - 9}$. Answers may vary.

39. Slant: $y = x - 1$; vertical: $x = -1$

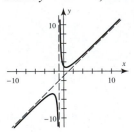

41. Slant: $y = \dfrac{1}{2}x - 3$; vertical: $x = -2$

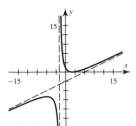

43. Slant: $y = x + 3$; vertical: $x = 1$

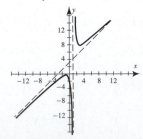

45. (a) $y = 10$

[0, 14, 1] by [0, 14, 1]

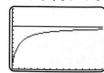

(b) When $x = 0$, there are 1 million insects.   **(c)** It starts to level off at 10 million.   **(d)** The horizontal asymptote $y = 10$ represents the limiting population after a long time.

47. $x = -3$ **49.** $x = \dfrac{1}{2}$ or 2   **51.** $x = \dfrac{1}{2}$

53. $x = -1$ **55.** $x = \dfrac{13}{8}$ **57.** $x = \pm\sqrt{2}$

59. No real solutions   **61.** $x = 0$ or ± 2
63. (a) About 12.4 cars per minute   **(b)** 3
65. Two possible solutions: width = 7 inches, length = 14 inches, height = 2 inches; width ≈ 2.266 inches, length ≈ 4.532 inches, height ≈ 19.086 inches
67. (a) $f(400) = \dfrac{2540}{400} = 6.35$ inches. A curve

designed for 60 miles per hour with a radius of 400 feet should have the outer rail elevated 6.35 inches.
(b) As the radius x of the curve increases, the elevation of the outer rail decreases.

[0, 600, 100] by [0, 50, 5]

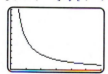

(c) The horizontal asymptote is $y = 0$. As the radius of the curve increases without bound ($x \to \infty$), the tracks become straight and no elevation or banking ($y \to 0$) is necessary.   **(d)** 200 feet
69. (a) $P(9) = \dfrac{9 - 1}{9} \approx 0.89 = 89\%$

(b) $P(1.9) = \dfrac{1.9 - 1}{1.9} \approx 0.47 = 47\%$

71. (a) $D(0.05) \approx 238$; the braking distance for a car traveling at 50 miles per hour on a 5% uphill grade is about 238 feet.   **(b)** As the uphill grade x increases, the braking distance decreases, which agrees with driving experience.
(c) $x = \dfrac{13}{165} \approx 0.079$ or 7.9%
73. (a) The vertical asymptote is $x = 0$. As the coefficient of friction x decreases to 0, the braking distances become larger without a maximum.
(b) $\dfrac{25}{102}$

75. $k = 6$ **77.** $k = 8$ **79.** $T = 160$ **81.** $y = 2$
83. Becomes half as much
85. Becomes 27 times as much   **87.** $k = 0.5, n = 2$
89. $k = 3, n = 1$ **91.** 1.18 grams

93. $\sqrt{50} \approx 7$ times farther   **95.** $\dfrac{2}{9}$ ohm

SECTION 4.6

1. (a) $x = -4, -2,$ or 2   **(b)** $(-4, -2)\cup(2, \infty)$
(c) $(-\infty, -4)\cup(-2, 2)$
3. (a) $x = -4, -2, 0,$ or 2   **(b)** $(-4, -2)\cup(0, 2)$
(c) $(-\infty, -4)\cup(-2, 0)\cup(2, \infty)$
5. (a) $x = -2, 1,$ or 2   **(b)** $(-\infty, -2)\cup(-2, 1)$
(c) $(1, 2)\cup(2, \infty)$
7. (a) $x = 0$ **(b)** $(-\infty, 0)\cup(0, \infty)$ **(c)** No solutions
9. (a) $x = 0$ or 1   **(b)** $(-\infty, 0)\cup(1, \infty)$
(c) $(0, 1)$
11. (a) $x = -2, 0,$ or 2   **(b)** $(-\infty, -2)\cup(2, \infty)$
(c) $(-2, 0)\cup(0, 2)$
13. $(-1, 0)\cup(1, \infty)$ **15.** $[-2, 0]\cup[1, \infty)$
17. $(-3, -2)\cup(2, 3)$ **19.** $(-\infty, -\sqrt{2}]\cup[\sqrt{2}, \infty)$
21. $[-2, 1]\cup[2, \infty)$ **23.** $[-3, 2]$
25. $(-\infty, 1]\cup[2, 4]$

27. $(-\infty, -1)\cup\left(0, \dfrac{4}{3}\right)\cup(2, \infty)$

29. $(-\infty, 0)$ **31.** $(-3, \infty)$ **33.** $(-2, 2)$

35. $(-\infty, 2)$ **37.** $(-\infty, -1)\cup\left(\dfrac{3}{2}, \infty\right)$ **39.** $(3, \infty)$

41. $(-\infty, 0)\cup\left(0, \dfrac{1}{2}\right]\cup[2, \infty)$

43. (a) $x \leq 36$ (approximately)   **(b)** The average line length is less than or equal to 8 cars when the average arrival rate is about 36 cars per hour or less.
45. (a) The braking distance increases.
(b) $0 < x \leq 0.3$
47. $\sqrt[3]{212.8} \leq x \leq \sqrt[3]{213.2}$ or (approximately) $5.97022 \leq x \leq 5.97396$ inches

SECTION 4.7

1. 4   **3.** $\dfrac{1}{8}$ **5.** -9 **7.** 2   **9.** 27   **11.** 16

13. $\dfrac{9}{4}$ **15.** $\dfrac{11}{6}$ **17.** $(2x)^{1/3}$ **19.** $z^{5/3}$ **21.** $\dfrac{1}{y^{3/4}}$

23. $x^{5/6}$ **25.** $y^{3/4}$ **27.** $f(1.2) = 1.2^{1.62} \approx 1.34$
29. $f(50) = 50^{3/2} - 50^{1/2} \approx 346.48$ **31.** b   **33.** 2
35. 81   **37.** $x = \pm 32$ **39.** $x = 32$ **41.** $x = 27$

43.
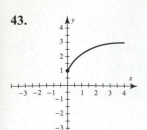

45.
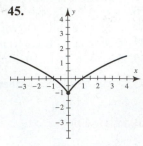

47.
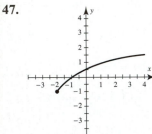

49. 7   **51.** -1

53. $x = 2$ or 3 **55.** $x = 15$ **57.** $x = -28$
59. $x = 2$ **61.** $w \approx 3.63$ pounds
63. About 58.1 years
65. (a) $a = 1960$ **(b)** $b \approx -1.2$
(c) $f(4) = 1960(4)^{-1.2} \approx 371$. If the zinc ion concentration reaches 371 milligrams per liter, a rainbow trout will survive, on average, 4 minutes.
67. (a) $f(2) \approx 1.06$ grams   **(b) & (c)** Approximately 1.1 grams
69. $a \approx 3.20$, $b \approx 0.20$, or $f(x) = 3.20x^{0.20}$

$[1, 9, 1]$ by $[0, 6, 1]$
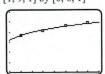

71. $a \approx 874.54$, $b \approx -0.49789$

CHAPTER 4 REVIEW EXERCISES

1. (a) $[-10, 10, 1]$ by $[-10, 10, 1]$
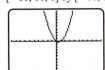

(b) Decreasing when $x \le 0.5$, increasing when $x \ge 0.5$
3. f: increasing; g: decreasing; h: neither   **5.** Odd
7. Even   **9.** Odd
11. $f(x) = \dfrac{1}{2}(x - 1)(x - 2)(x - 3)$
13. $f(x) = 2(x + 4)\left(x - \dfrac{1}{2}\right)(x - 2)$ **15.** 216
17. 8   **19.** $x^{4/3}$ **21.** $y^{1/2}$

23. $D = \{x \mid x \ge 0\}$; $f(3) \approx 15.59$
25. (a) Local minima: -4.5, -0.5; local maximum: 0
(b) Absolute minimum: -4.5; absolute maximum: none
27. (a) Local minimum: -60.08; local maxima: 15, 75.12   **(b)** Absolute minimum: none; absolute maximum: 75.12   **(c)** Increases on $(-\infty, -3.996] \cup [0.9995, 5.007]$; decreases on $[-3.996, 0.9995] \cup [5.007, \infty)$, where endpoints are approximate.
29. (a) One local maximum, two local minima, two x-intercepts   **(b)** Two local maxima, one local minimum, three x-intercepts   **(c)** One local maximum, two local minima, four x-intercepts

$[-10, 10, 1]$ by $[-100, 100, 10]$ $[-10, 10, 1]$ by $[-10, 10, 1]$
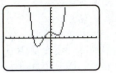   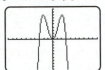

$[-10, 10, 1]$ by $[-100, 100, 10]$
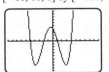

31. Answers may vary.
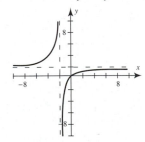

33. $x = 0, \pm\sqrt{3}$
35. $f(x) = (x + 2)^2(x - 2)^3$. The leading coefficient may vary.
37. Answers may vary.
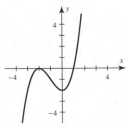

39. (a) 1   **(b)** 2   **(c)** 3
41. -1 **43.** $-10 - 11i$ **45.** $x = \pm\dfrac{3}{2}i$
47. $x = 0$ or $\pm i$
49. One real zero, two imaginary zeros

51. $f(x) = 2(x - i\sqrt{2})(x + i\sqrt{2})$

53. Zeros: $-3, \dfrac{1}{2}, 2$; $f(x) = 2(x + 3)\left(x - \dfrac{1}{2}\right)(x - 2)$

55. $x = 4$ **57.** $x = 6$ **59.** $x = \dfrac{81}{16}$ **61.** $x = 15$

63. $x = 4$ **65.** $x = \dfrac{1}{2}$ **67.** $(-3, 0) \cup (2, \infty)$

69. $(-2, -1) \cup (1, 2)$ **71.** $(-\infty, -2) \cup \left(\dfrac{1}{2}, \infty\right)$

73. (a) Dog: 148; person: 69   **(b)** 6.4 inches

75. (a) f is odd. From September to March 21, there are fewer than 12 daylight hours.

(b)

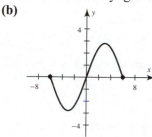

77. (a) $a \approx 1.8342$, $b \approx -0.4839$ **(b)** 1.08 steps per second

CHAPTER 5: EXPONENTIAL AND LOGARITHMIC FUNCTIONS

SECTION 5.1

1. (a) 5   **(b)** 5   **(c)** 0   **(d)** Undefined

3.

x	-2	0	2	4
$(f + g)(x)$	6	5	5	15
$(f - g)(x)$	-6	5	9	5
$(fg)(x)$	0	0	-14	50
$(f/g)(x)$	0	—	-3.5	2

5. (a) 2   **(b)** 4   **(c)** 0   **(d)** $-\dfrac{1}{3}$

7. (a) 2   **(b)** -3 **(c)** 2   **(d)** -2

9. (a) -5 **(b)** -5 **(c)** -3 **(d)** $-\dfrac{1}{3}$

11. (a) $\dfrac{11}{2}$ **(b)** 0   **(c)** $\dfrac{9}{4}$ **(d)** Undefined

13. (a) $(f + g)(x) = 2x + x^2$; all real numbers
(b) $(f - g)(x) = 2x - x^2$; all real numbers
(c) $(fg)(x) = 2x^3$; all real numbers
(d) $(f/g)(x) = \dfrac{2}{x}$; $D = \{x | x \neq 0\}$

15. (a) $(f + g)(x) = 2x$; $D = \{x | x \geq 1\}$
(b) $(f - g)(x) = -2\sqrt{x - 1}$; $D = \{x | x \geq 1\}$
(c) $(fg)(x) = x^2 - x + 1$; $D = \{x | x \geq 1\}$
(d) $(f/g)(x) = \dfrac{x - \sqrt{x - 1}}{x + \sqrt{x - 1}}$; $D = \{x | x \geq 1\}$

17. (a) $(f + g)(x) = \dfrac{4}{x + 1}$; $D = \{x | x \neq -1\}$

(b) $(f - g)(x) = -\dfrac{2}{x + 1}$; $D = \{x | x \neq -1\}$

(c) $(fg)(x) = \dfrac{3}{(x + 1)^2}$; $D = \{x | x \neq -1\}$

(d) $(f/g)(x) = \dfrac{1}{3}$; $D = \{x | x \neq -1\}$

19. (a) $(f + g)(x) = 3 - 2x$; all real numbers
(b) $(f - g)(x) = 6x - 3$; all real numbers
(c) $(fg)(x) = 6x - 8x^2$; all real numbers
(d) $(f/g)(x) = \dfrac{2x}{3 - 4x}$; $D = \left\{x | x \neq \dfrac{3}{4}\right\}$

21. (a) $(f + g)(x) = \dfrac{x + 1}{2x - 4}$; $D = \{x | x \neq 2\}$

(b) $(f - g)(x) = \dfrac{1 - x}{2x - 4}$; $D = \{x | x \neq 2\}$

(c) $(fg)(x) = \dfrac{x}{(2x - 4)^2}$; $D = \{x | x \neq 2\}$

(d) $(f/g)(x) = \dfrac{1}{x}$; $D = \{x | x \neq 0 \text{ or } x \neq 2\}$

23. (a) $(f + g)(x) = x^2 + 3x + 2$; all real numbers
(b) $(f - g)(x) = -x^2 - x + 2$; all real numbers
(c) $(fg)(x) = (x + 2)(x^2 + 2x) = x^3 + 4x^2 + 4x$; all real numbers
(d) $(f/g)(x) = \dfrac{1}{x}$; $D = \{x | x \neq -2 \text{ or } x \neq 0\}$

25. (a) $Y_1 = \sqrt{(X)}$, $Y_2 = X + 1$, and $Y_3 = Y_1 + Y_2$

[0, 9, 1] by [0, 15, 1]

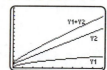

(b) To determine the graph $f + g$ add corresponding y-coordinates on the graphs of f and g.

27. (a) 5   **(b)** Undefined   **(c)** 4

29. 4; 2

31. (a) -4 **(b)** 2   **(c)** -4

33. (a) -3 **(b)** -2 **(c)** 0

35. (a) 3   **(b)** 4   **37. (a)** 18   **(b)** 23

39. (a) $(f \circ g)(x) = (x^2 + 3x - 1)^3$; all real numbers
(b) $(g \circ f)(x) = x^6 + 3x^3 - 1$; all real numbers
(c) $(f \circ f)(x) = x^9$; all real numbers

41. (a) $(f \circ g)(x) = 1 - x$; $D = \{x | x \leq 1\}$
(b) $(g \circ f)(x) = \sqrt{1 - x^2}$; $D = \{x | -1 \leq x \leq 1\}$
(c) $(f \circ f)(x) = x^4$; all real numbers

43. (a) $(f \circ g)(x) = 2 - 3x^3$; all real numbers
(b) $(g \circ f)(x) = (2 - 3x)^3$; all real numbers
(c) $(f \circ f)(x) = 9x - 4$; all real numbers

45. (a) $(f \circ g)(x) = 1 - x^2$; all real numbers
(b) $(g \circ f)(x) = 5 - (x - 4)^2$; all real numbers
(c) $(f \circ f)(x) = x - 8$; all real numbers
47. (a) $(f \circ g)(x) = \dfrac{1}{5x + 1}$; $D = \left\{x \middle| x \neq -\dfrac{1}{5}\right\}$
(b) $(g \circ f)(x) = \dfrac{5}{x + 1}$; $D = \{x | x \neq -1\}$
(c) $(f \circ f)(x) = \dfrac{x + 1}{x + 2}$; $D = \{x | x \neq -1 \text{ or } x \neq -2\}$
49. (a) $(f \circ g)(x) = \sqrt{4 - x^2} + 4$;
$D = \{x | -2 \leq x \leq 2\}$
(b) $(g \circ f)(x) = \sqrt{4 - (x + 4)^2}$;
$D = \{x | -6 \leq x \leq -2\}$ **(c)** $(f \circ f)(x) = x + 8$;
all real numbers
51. (a) $(f \circ g)(x) = \sqrt{3x - 1}$;
$D = \left\{x \middle| x \geq \dfrac{1}{3}\right\}$ **(b)** $(g \circ f)(x) = 3\sqrt{x - 1}$;
$D = \{x | x \geq 1\}$ **(c)** $(f \circ f)(x) = \sqrt{\sqrt{x - 1} - 1}$;
$D = \{x | x \geq 2\}$

Answers may vary for Exercises 53–60.
53. $f(x) = x - 2, g(x) = \sqrt{x}$
55. $f(x) = x + 2, g(x) = 5x^2 - 4$
57. $f(x) = 2x + 1, g(x) = 4x^3$
59. $f(x) = x^3 - 1, g(x) = x^2$
61. (a) $C(x) = 1.5x + 150,000$ **(b)** $R(x) = 6.5x$;
$R(8000) = \$52,000$ **(c)** $P(x) = 5x - 150,000$;
$P(40,000) = \$50,000$ **(d)** 30,000 videos
63. (a) $(g \circ f)(1) = 5.25$; a 1% decrease in the ozone
layer could result in a 5.25% increase in skin cancer.
(b) Not possible using the given tables
65. (a) $(g \circ f)(1960) = 1.98$
(b) $(g \circ f)(x) = 0.165(x - 1948)$
(c) $f, g,$ and $g \circ f$ are all linear functions.
67. $C = 12\pi t$ **69.** $S = \pi r^2 \sqrt{5}$
71. (a)

x (yr)	1991	1992	1993	1994	1995	1996
h(x) ($ million)	38	42	44	50	54	61

(b) $h(x) = f(x) + g(x)$
73. (a) 50   **(b)** $(f + g)(x)$ computes the total SO_2
emissions from coal and oil during year x.
(c)

x	1860	1900	1940	1970	2000
(f + g)(x)	2.4	12.8	26.5	50.0	78.0

75. (a)

x	1990	2000	2010	2020	2030
h(x)	32	35.5	39	42.5	46

(b) $h(x) = f(x) + g(x)$

77. $h(x) = 0.35x - 664.5$
79. (a) $h(x) = g(x) - f(x)$ **(b)** $h(1995) = 200$,
$h(2000) = 250$ **(c)** $h(x) = 10(x - 1995) + 200$ or
$h(x) = 10x - 19,750$
81. (a) 30,000 gallons   **(b)** $(g \circ f)(x)$ computes the
gallons of water in the pool after x days.
83. (a) $(f \circ g)(x) = k$; a constant function
(b) $(g \circ f)(x) = ak + b$; a constant function

Section 5.2

1. Closing a window
3. Closing a book, standing up, and walking out of
the classroom
5. Subtract 2 from x; $x + 2$ and $x - 2$
7. Divide x by 3 and then add 2; $3(x - 2)$ and $\dfrac{x}{3} + 2$
9. Subtract 1 from x and cube the result; $\sqrt[3]{x} + 1$ and
$(x - 1)^3$
11. Take the reciprocal of x; $\dfrac{1}{x}$ and $\dfrac{1}{x}$
13. One-to-one   **15.** Not one-to-one
17. Not one-to-one   **19.** Not one-to-one
21. One-to-one   **23.** One-to-one
25. Not one-to-one   **27.** Not one-to-one
29. Not one-to-one   **31.** No
33. Yes, the population has increased each fiscal year.

35. $f^{-1}(x) = x^3$ **37.** $f^{-1}(x) = -\dfrac{1}{2}x + 5$
39. $f^{-1}(x) = \dfrac{x + 1}{3}$ **41.** $f^{-1}(x) = \sqrt[3]{\dfrac{x + 5}{2}}$
43. $f^{-1}(x) = \sqrt{x + 1}$ **45.** $f^{-1}(x) = \dfrac{1}{2x}$
47. If the domain of f is restricted to $x \geq 0$, then
$f^{-1}(x) = \sqrt{4 - x}$.
49. If the domain of f is restricted to $x \geq 2$, then
$f^{-1}(x) = 2 + \sqrt{x - 4}$.
51. $f^{-1}(x) = \dfrac{x + 15}{5}$. D and R are all real numbers.
53. $f^{-1}(x) = \dfrac{1}{x} - 3$. $D = \{x | x \neq 0\}$ and
$R = \{y | y \neq -3\}$
55. $f^{-1}(x) = \sqrt[3]{\dfrac{x}{2}}$. D and R are all real numbers.
57. $f^{-1}(x) = \sqrt{x}$. D and R include all nonnegative
real
59.

x	5	7	9
f⁻¹(x)	1	2	3

For f: $D = \{1, 2, 3\}, R = \{5, 7, 9\}$;
for f^{-1}: $D = \{5, 7, 9\}, R = \{1, 2, 3\}$

61.

x	0	1	4	9	16
$f^{-1}(x)$	0	1	2	3	4

For f: $D = \{0, 1, 2, 3, 4\}$, $R = \{0, 1, 4, 9, 16\}$;
for f^{-1}: $D = \{0, 1, 4, 9, 16\}$, $R = \{0, 1, 2, 3, 4\}$

63.

x	-3	0	3	6
$f^{-1}(x)$	-8	-5	-2	1

65.

x	-8	-1	8	27
$f^{-1}(x)$	-2	-1	2	3

67. 1   **69.** 3   **71.** 5   **73.** 1

75. **(a)** $f(1) \approx \$110$ **(b)** $f^{-1}(110) \approx 1$ year
(c) $f^{-1}(160) \approx 5$ years
$f^{-1}(x)$ computes the years necessary for the account to
accumulate x dollars.

77. **(a)** 2   **(b)** 3   **(c)** 1   **(d)** 3

79.

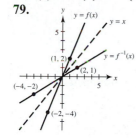

81.

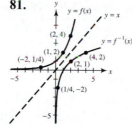

83. $Y_1 = 3X - 1$, $Y_2 = (X + 1)/3$, $Y_3 = X$
$[-4.7, 4.7, 1]$ by $[-3.1, 3.1, 1]$

85. $Y_1 = X^3/3 - 1$, $Y_2 = \sqrt[3]{(3X + 3)}$, $Y_3 = X$
$[-4.7, 4.7, 1]$ by $[-3.1, 3.1, 1]$

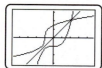

87. **(a)** $f(1995) = 1$ **(b)** 1995   **(c)** $f^{-1}(1) = 1995$
In 1995 a 30-second Super Bowl commercial cost
$1 million.

89. **(a)** Yes   **(b)** The radius r of a sphere with
volume V **(c)** $r = \sqrt[3]{\dfrac{3V}{4\pi}}$ **(d)** No. If V and r were
interchanged, then r would represent the volume and V
would represent the radius.

91. OCVJ; FUNCTIONS; if f is not one-to-one, then
f^{-1} does not exist and f^{-1} is needed to decode a message.

93. **(a)** $f^{-1}(x) = x^{2/3}$, $x \geq 0$ **(b)** The inverse calculates how many times farther from the sun a planet is
than Earth, if it takes x years to orbit the sun.

95. **(a)** $(f \circ g)(2) = 10{,}560$ represents the number of
feet in 2 miles.   **(b)** $f^{-1}(26{,}400) = 8800$ represents
the number of yards in 26,400 feet.
(c) $(g^{-1} \circ f^{-1})(21{,}120) = 4$ represents the number of
miles in 21,120 feet.

97. **(a)** $(g \circ f)(96) = 1.5$ represents the number of
quarts in 96 tablespoons.   **(b)** $g^{-1}(2) = 8$ represents
the number of cups in 2 quarts.
(c) $(f^{-1} \circ g^{-1})(1.5) = 96$ represents the number of
tablespoons in 1.5 quarts.

99. **(a)** $f(1930) = 62.5$, $f(1980) = 65.5$; increased
by 3%.   **(b)** $f^{-1}(x)$ computes the year when the cloud
cover was x percent.   **(c)** $f^{-1}(62.5) = 1930$,
$f^{-1}(65.5) = 1980$
(d) $f^{-1}(x) = \dfrac{50}{3}(x - 62.5) + 1930$

SECTION 5.3

1. $\dfrac{1}{8}$ **3.** 2   **5.** 9   **7.** 8   **9.** 2   **11.** e^{2x} **13.** 1

15. 5

17. **(a)** 15.59   **(b)** 8.82   **(c)** -5.65

19. Linear: $f(x) = -1.25x + 2$

21. Exponential; $f(x) = 8\left(\dfrac{1}{2}\right)^{x}$

23. Exponential; $f(x) = 5(2^{x})$

25. Exponential; $f(n) = 40{,}000(1.08)^{n-1}$

27. $f(x) = 2^x$ **29.** $C = 5$, $a = 1.5$

31. $C = 10$, $a = 2$ **33.** $C = 3$, $a = 3$

35. About 11 pounds per square inch

37.

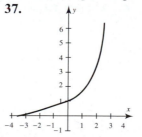

39.

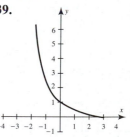

41.

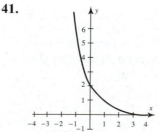

43. $C = 1$, $a = \dfrac{1}{2}$ **45.** $C = \dfrac{1}{2}$, $a = 4$

47. **(i)** b   **(ii)** d   **(iii)** a   **(iv)** c

49. (a)   **(b)**

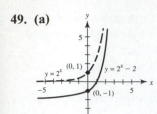

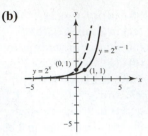

(c)   **(d)**

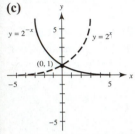

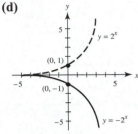

51. 22.1980   **53.** 71.2571   **55.** −0.7586
57. $841.53   **59.** $1730.97   **61.** $4451.08
63. $2072.76
65. 10%: $14,656.15; 13%: 26,553.58; a 13% rate results in considerably more interest than a 10% rate.
67. $14,326.78
69. The account with $200 will have double the money since twice the money was deposited.
71. Approximately $778.7 billion   **73.** $19,870.65
75. About 27.3%   **77.** About 18,935 years
79. (a) 31.7 mg. The half-life is less than 50 years.
(b) 30.2 years
81. (a) $C \approx 0.72$, $a \approx 1.041$; answers may vary slightly   **(b)** 1.21 ppb; answers may vary slightly
83. (a) 2.5 parts per million   **(b)** About 1.4 days
85.

x (inches)	6	7.5	9	10.5	12
y (lb × 1000)	80	116	168	242	350

87. (a) H is increasing.
[0, 100, 10] by [0, 5, 1]

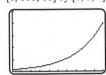

(b) $H(30) \approx 0.42$. About 0.42 horsepower is required for each ton pulled at 30 miles per hour.   **(c)** About 2100 horsepower   **(d)** 2
89. (a) [1968, 1998, 2] by [−1, 6, 1]

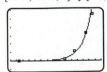

(b) 39%

91. (a) $P(2) \approx 0.20$ and $P(20) \approx 0.90$. There is a 20% chance that at least one tree is located within a circle having radius 2 feet and a 90% chance for a circle having radius 20 feet.
(b) The larger the circle, the more likely it is to contain a tree.
[0, 25, 5] by [0, 1, 0.1]

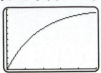

(c) $x \approx 6.1$. A circle of radius 6.1 feet has a 50–50 chance of containing at least one tree.

SECTION 5.4

1.

x	10^0	10^4	10^8	$10^{1.2}$
$\log x$	0	4	8	1.2

3. (a) 1   **(b)** 4   **(c)** −20   **(d)** −2
5. (a) 2   **(b)** $\dfrac{1}{2}$ **(c)** 3   **(d)** Undefined
7. (a) $n = 1, \log 79 \approx 1.898$
(b) $n = 2, \log 500 \approx 2.699$
(c) $n = 0, \log 5 \approx 0.6990$
(d) $n = -1, \log 0.5 \approx -0.3010$
9. (a) 3   **(b)** 6   **(c)** −2   **(d)** −4
11. (a) $\dfrac{3}{2}$ **(b)** $\dfrac{1}{3}$ **(c)** $\dfrac{3}{10}$ **(d)** −1
13. 6   **15.** $\dfrac{1}{2}$ **17.** 0   **19.** −1   **21.** 0   **23.** −4
25. k
27.

x	$\dfrac{1}{16}$	1	8	32
$f(x)$	−4	0	3	5

29.

x	6	7	21
$f(x)$	0	2	8

31. $x = \log_2 72$ **33.** $x = \log_5 0.25$ **35.** $x = \ln 25$
37. $x = -\ln 3$ **39.** $x = 2$ **41.** $x = \dfrac{2}{3} \approx 0.6667$
43. $x = \ln 23$ or $x \approx 3.1355$
45. $x = \log_2 14$ or $x \approx 3.8074$
47. $x = 10^{2.3} \approx 199.5$ **49.** $x = 2^{1.2} \approx 2.297$
51. $x = e^{-2} \approx 0.1353$ **53.** $x = 1000$ **55.** $x = 20$
57. $x = e^{3/4} \approx 2.1170$ **59.** $x = e^{7/5} \approx 4.0552$
61. $x = 16$ **63.** $a = 5, b = 2$

65. $f^{-1}(x) = \ln x$

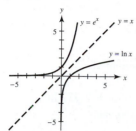

67. d   **69.** c
71. $D = \{x \mid x > -1\}$ **73.** $D = \{x \mid x < 0\}$
$[-6, 6, 1]$ by $[-4, 4, 1]$ $[-6, 6, 1]$ by $[-4, 4, 1]$

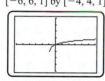

75. $a = 7$, $b = 4$; about 178 km^2
77. (a) $C = 3$, $a = 2$ **(b)** After about 2.4 days
79. (a) Since L is increasing, heavier planes generally require longer runways.
$[0, 50, 10]$ by $[0, 6, 1]$

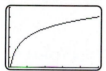

(b) No. It increases by 3000 feet.   **(c)** If the weight increases tenfold, the runway length increases by 3000 feet.
81. (a) $10^{-4.92} \approx 0.000012$ **(b)** $10^{-3.9} \approx 0.000126$
83. (a) Yugoslavia: $x = 1,000,000$;
Indonesia: $x = 100,000,000$ **(b)** 100
85. (a) $f(0) = 27$, $f(100) \approx 29.2$ inches. At the eye, the barometric air pressure is 27 inches; 100 miles away it is 29.2 inches.   **(b)** The air pressure rises rapidly at first and then starts to level off.
$[0, 250, 50]$ by $[25, 30, 1]$

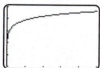

(c) About 7 miles
87. (a) About 49.4°C   **(b)** About 0.69 hour or 41.4 minutes
89. (a) About 0.189 or an 18.9% chance
(b) $x = -3 \ln (0.3) \approx 3.6$ minutes

SECTION 5.5

1. $\log 4 + \log 7 = \log 28 \approx 1.447$; Property 2
3. $\ln 72 - \ln 8 = \ln 9 \approx 2.197$; Property 3

5. $10 \log 2 = \log 1024 \approx 3.010$; Property 4
7. $\log 6 - \log z$ **9.** $2 \log x - \log 3$
11. $\ln 2 + 7 \ln x - \ln 3 - \ln k$
13. $2 + 2 \log_2 k + 3 \log_2 x$
15. $3 \log_5 x - 4 \log_5 (x - 4)$
17. $\dfrac{1}{2} \log_2 x - 2 \log_2 z$
19. $\dfrac{1}{3} \ln (2x + 6) - \dfrac{5}{3} \ln (x + 1)$ **21.** $\log 6$
23. $-\dfrac{3}{2} \ln 5$ **25.** $\dfrac{3}{2} \log x$ **27.** $\ln \dfrac{2}{e}$ **29.** $\ln \dfrac{x^2 \sqrt{z}}{y^4}$
31. $\log (4x^{5/2})$ **33.** $\log \dfrac{(x^2 - 1)^2 (x - 2)^4}{\sqrt{y}}$
35. (a) Yes

X	Y₁	Y₂
1	.77815	.77815
2	1.3802	1.3802
3	1.7324	1.7324
4	1.9823	1.9823
5	2.1761	2.1761
6	2.3345	2.3345
7	2.4683	2.4683

X=1

(b) By Property 2: $\log 3x + \log 2x = \log (3x \cdot 2x) = \log 6x^2$
37. (a) No

X	Y₁	Y₂
1	.40547	0
2	.40547	.69315
3	.40547	1.0986
4	.40547	1.3863
5	.40547	1.6094
6	.40547	1.7918
7	.40547	1.9459

X=1

39. (a) Yes

X	Y₁	Y₂
1	0	0
2	1.2041	1.2041
3	1.9085	1.9085
4	2.4082	2.4082
5	2.7959	2.7959
6	3.1126	3.1126
7	3.3804	3.3804

X=1

(b) By Property 4: $\log (x^4) = 4 \log x$
41. (a) Yes

X	Y₁	Y₂
1	0	0
2	1.3863	1.3863
3	2.1972	2.1972
4	2.7726	2.7726
5	3.2189	3.2189
6	3.5835	3.5835
7	3.8918	3.8918

X=1

(b) By Property 4: $\ln x^4 - \ln x^2 = 4 \ln x - 2 \ln x = 2 \ln x$
43. $\dfrac{\log 25}{\log 2} \approx 4.644$ **45.** $\dfrac{\log 130}{\log 5} \approx 3.024$
47. $\dfrac{\log 5}{\log 2} + \dfrac{\log 7}{\log 2} \approx 5.129$ **49.** $\sqrt{\dfrac{\log 46}{\log 4}} \approx 1.662$
51. $\dfrac{\log 12 / \log 2}{\log 3 / \log 2} = \dfrac{\log 12}{\log 3} \approx 2.262$
53. $x \approx 4.714$ **55.** $x \approx \pm 2.035$
57. $L(x) = \dfrac{3 \ln x}{\ln 10}$; $L(50) \approx 5.097$; yes, the answers agree.

59. $f(x) = 160 + \dfrac{10 \ln x}{\ln 10}$; $f(5 \times 10^{-8}) \approx 86.99$; yes, the answers agree.

61. $I = I_0 e^{-kx}$

SECTION 5.6

1. **(a)** $x \approx 2$ **(b)** $x = \ln 7.5 \approx 2.015$

3. **(a)** $x \approx 2$ **(b)** $x = 5 \log 2.5 \approx 1.990$

5. $x = \ln 1.25 \approx 0.2231$

7. $x = \log 20 \approx 1.301$

9. $x = -\dfrac{5}{6} \ln 0.4 \approx 0.7636$

11. $x = \dfrac{\ln 0.5}{\ln 0.9} \approx 6.579$

13. $x = 1 + \dfrac{\log 4}{\log 1.1} \approx 15.55$ **15.** $x = 1$ or 2

17. No solutions

19. $x = \dfrac{\log 4}{\log 4 - 2 \log 3} \approx -1.710$

21. $x = \dfrac{\log (64/3)}{\log 1.4} \approx 9.095$

23. $x = 1980 + \dfrac{\log (8/5)}{\log 1.015} \approx 2012$

25. $x = 10^{2/3} \approx 4.642$ **27.** $x = \dfrac{1}{2} e^5 \approx 74.207$

29. $x = \pm\sqrt{50} \approx \pm 7.071$ **31.** $x = 10^{-11}$

33. $x = e \approx 2.718$ **35.** $x = 2^{2.1} \approx 4.287$

37. $x = \sqrt{50} \approx 7.071$ **39.** $x = 1 + e \approx 3.718$

41. $x = 1 + \sqrt{2} \approx 2.414$ **43.** $x = 2$

45. $x \approx 0.31$ **47.** $x \approx 1.71$ **49.** $x \approx 2.10$

51. $x \approx 3$ or 1990 **53.** In about 1994

55. **(a)** $C = 1$, $a \approx 1.01355$ **(b)** $f(2010) \approx 1.14$ billion   **(c)** In about 2030

57. **(a)** $x = \dfrac{\ln 1.5}{0.09} \approx 4.505$ **(b)** 500 invested at 9% compounded continuously results in 750 after 4.5 years.

59. In 8.25 years, or 8 years and 3 months

61. About 8633 years

63. **(a)** 0.993, or a 99.3% chance   **(b)** $x \approx 0.51$, or about half a minute

65. $\dfrac{\ln (12/3.4)}{\ln 1.07} + 1995 \approx 2014$

67. **(a)** $f(30) \approx 34$; after 30 minutes the soda can had cooled to 34°F.   **(b)** About 9.3 minutes

69. $100 - 10^{1.16} \approx 85.5\%$ **71.** About 2.8 acres

73. $P = 1 - e^{-0.3214} \approx 0.275$. If a 60 tax is placed on each ton of carbon burned, carbon dioxide emissions could decrease by 27.5%.

75. $n = \dfrac{\log (A_n/A_0)}{m \log (1 + r/m)}$

SECTION 5.7

1. Logarithmic   **3.** Exponential

5. Exponential: $f(x) = 1.2(1.7)^x$

7. Logarithmic: $f(x) = 1.088 + 2.937 \ln x$

9. **(a)** $[25, 75, 5]$ by $[-100, 2100, 200]$

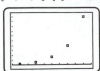

(b) $f(x) = 1.568(1.109)^x$ **(c)** About 6164 deaths per 100,000

11. **(a)** $a \approx 1.4734$, $b \approx 0.99986$, or $f(x) = 1.4734(0.99986)^x$ **(b)** Approximately 0.55 kg/m^3

13. **(a)** The data is not linear.

$[-2, 32, 5]$ by $[0, 80, 10]$

(b) $a \approx 12.42$, $b \approx 1.066$, or $f(x) = 12.42(1.066)^x$
(c) $f(39) \approx 150$. Chemical fertilizer use increased but at a slower rate than predicted by f.

15. **(a)** $f(x) = \dfrac{4.9955}{1 + 49.7081 e^{-0.6998x}}$

(b) About 5 thousand per acre

17. **(a)** $f(25) \approx 0.065$ and $f(65) \approx 0.82$. For people age 25, 6.5% have some CHD, whereas for people age 65, 82% have some CHD.   **(b)** 48 years (approximately)

CHAPTER 5 REVIEW EXERCISES

1. **(a)** 8   **(b)** 0   **(c)** -6 **(d)** Undefined

3. **(a)** 7   **(b)** 1   **(c)** 0   **(d)** $-\dfrac{9}{2}$

5. **(a)** 2   **(b)** 1

7. **(a)** $\sqrt{6}$ **(b)** 12

9. Subtract 6 from x and then multiply the result by 10; $\dfrac{x}{10} + 6$ and $10(x - 6)$.

11. f is one-to-one.   **13.** f is not one-to-one.

15. **(a)** $f(1) \approx \$1050$ **(b)** $f^{-1}(1200) \approx 4$ years; f^{-1} computes the number of years it takes to accumulate x dollars.

17. $f^{-1}(x) = \dfrac{x + 5}{3}$ **19.** $f^{-1}(x) = \dfrac{1}{x} - 7$

21. $1562.71   **23.** 11.023   **25.** $\ln 19 \approx 2.9444$

27. 3   **29.** -21 **31.** 2   **33.** 1

35. $\dfrac{\log 18}{\log 3} \approx 2.631$ **37.** $x = \log 125 \approx 2.097$

39. $x = 10 \ln 5.2 \approx 16.49$

41. $x = -\dfrac{1}{\log 5} \approx -1.431$

43. $x = 10 + \dfrac{\log (29/3)}{\log (0.78)} \approx 0.869$

45. $f(x) = 1.5(2)^x$ **47.** $x = 10^{1.5} \approx 31.62$

49. $x = e^{3.4} \approx 29.96$ **51.** $x = 126$ **53.** $\log 30x$

55. $\ln 4 - 2 \ln x$ **57.** $x = 10^{1/4} \approx 1.778$

59. $x = \dfrac{100,000}{3} \approx 33,333$ **61.** $x = 1$

63. The x-intercept is b. If $(0, b)$ is on the graph of f, then $(b, 0)$ is on the graph of f^{-1}.

65. $x \approx 13$ or 2008

67. (a) $(g \circ f)(32) = 1$ represents the number of quarts in 32 fluid ounces.   (b) $f^{-1}(1) = 16$ represents the number of fluid ounces in 1 pint.
(c) $(f^{-1} \circ g^{-1})(1) = 32$ represents the number of fluid ounces in 1 quart.

69. (a) $W(1) \approx 19.5$ (b) $x \approx 2.74$. The fish weighs 50 milligrams at about 3 weeks.

(c) $x = -\dfrac{1}{0.24} \ln\left(\dfrac{1 - (50/175.6)^{1/3}}{0.66}\right) \approx 2.74$

71. After about 3.8 minutes

73. (a)

x	0	15	30	45	60	75	90	125
$g(x)$	7	21	57	111	136	158	164	178

(b) $g(x) = 261 - f(x)$ (c) y_1 models $g(x)$ better.
$[-10, 140, 10]$ by $[-20, 200, 10]$

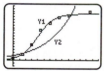

(d) $f(x) = 261 - \dfrac{171}{1 + 18.6e^{-0.0747x}}$

75. $a \approx 3.50, b \approx 0.74$, or $f(x) = 3.50(0.74)^x$
$[0, 5, 1]$ by $[0, 3, 1]$

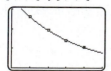

CHAPTER 6: SYSTEMS OF EQUATIONS AND INEQUALITIES

SECTION 6.1

1. $f(5, 8) = 20$. The area of a triangle with base 5 and height 8 is 20.

3. $S(86, 185) \approx 2.1$ m^2

5. $S(60, 157.48) \approx 1.6$ m^2

7. $f(x, y) = y + 2x$ **9.** $f(x, y) = \dfrac{xy}{1 + x}$

11. $f(1960, \text{Male}) = 66.6$. A male born in 1960 had a life expectancy at birth of 66.6 years.

13. (a) $b = \dfrac{2A}{h}$ (b) $h = \dfrac{2A}{b}$

15. (a) $x = \pm\sqrt{\dfrac{z(y + 1)}{2}}$ (b) $y = \dfrac{2x^2}{z} - 1$

17. $(2, 1)$ **19.** $(4, -3)$ **21.** $(2, 2)$ **23.** $(1, -1)$

25. $(1, 2)$ **27.** $(-2, 1)$ **29.** $(-6, 4)$ **31.** $(0.5, 2)$

33. $(6, 8)$ **35.** $(4, 3)$ **37.** $(2, 1)$

39. $(-\sqrt{8}, -\sqrt{8}), (\sqrt{8}, \sqrt{8})$ **41.** $(-7, -9), (9, 7)$

43. $(-2, 4), (0, 0)$ **45.** $(2, 4), (4, 2)$

47. $(2, 4), (-2, -4)$ **49.** $(2, \sqrt{2})$

51. $(-2, 0), (2, 0)$

53. $(-2, -2), (0, 0), (2, 2)$

55. $l = 7, w = 5$

57. $(-1.588, 0.240), (0.164, 1.487), (1.924, -0.351)$

59. $(1.220, 0.429)$ **61.** $(0.714, -0.169)$

63. (a) Let x represent the number of bank robberies in 1994 and y the number in 1995.
$x + y = 13,787$
$x - y = 271$
(b) & (c) $(7029, 6758)$

65. 1.2 million foreign tourists, 5.0 million American tourists   **67.** $r \approx 1.538, h \approx 6.724$

69. $x \approx 177.1, y \approx 174.9$; if an athlete's maximum heart rate is 180 beats per minute, then it will be about 177 beats per minute after 5 seconds and 175 beats per minute after 10 seconds.

71. $W_1 = \dfrac{300}{1 + \sqrt{3}} \approx 109.8$ lb,

$W_2 = \dfrac{300\sqrt{3}}{\sqrt{6} + \sqrt{2}} \approx 134.5$ lb

73. 0.51

75. Approximately 10,823 watts

77. Approximately 2.54 cords

79. $S(w, h) = 0.0101(w^{0.425})(h^{0.725})$

SECTION 6.2

1. The system is consistent with solution $(2, 2)$.

3. The system is inconsistent.

5. (14, 6); consistent and independent
7. (1, 3); consistent and independent
9. $\{(x, y) \mid x + y = 500\}$; consistent and dependent
11. $(-2, 5)$; consistent and independent
13. Inconsistent
15. $\left(-\dfrac{11}{7}, \dfrac{12}{7}\right)$; consistent and independent
17. $(2, -4)$ **19.** $\left(\dfrac{1}{3}, -\dfrac{1}{3}\right)$
21. Infinitely many solutions of the form $\{(x, y) \mid 7x - 3y = -17\}$
23. No solutions   **25.** $\left(-2, \dfrac{5}{2}\right)$ **27.** (10, 20)
29. 10.5 inches wide, 8.5 inches high
31. 23 child tickets, 52 adult tickets
33. c; one solution is (2, 3). Answers may vary.
35. d; one solution is $(-1, -1)$. Answers may vary.
37.
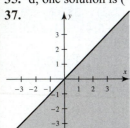
39.
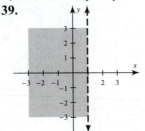
41.
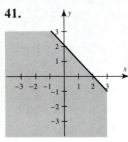
43.
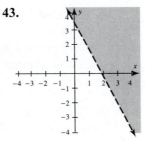
45. One solution is (0, 2). Answers may vary.
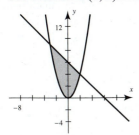
47. One solution is (0, 0). Answers may vary.
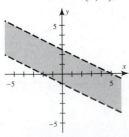

49. One solution is $(-1, 1)$. Answers may vary.
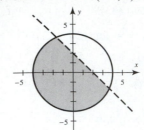
51.
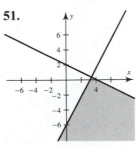
53.
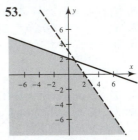
55.
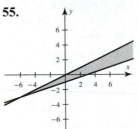
57.
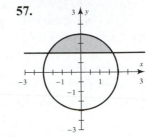
59. **(a)** $x + y = 3000, 0.08x + 0.10y = 264$
(b) $1800 at 8% and $1200 at 10%
61. There are no solutions. If loans totaling $3000 are at 10%, then the interest must be $300.
63. **(a)** Let x represent the number of television sets sold with stereo sound and y represent the number sold without stereo sound.
$$x + y = 32$$
$$\frac{19}{10}x - y = 0$$
(b) Approximately (11.03, 20.97); in 1999 about 11 million television sets were sold with stereo sound and about 21 million were sold without stereo sound.
65. 63.4 pounds of beef, 14.4 pounds of turkey
67. Air speed: 520 mph; wind speed: 40 mph
69. $x = 300, y = 350$
71. This individual weighs less than recommended for his or her height.
73. $25h - 7w \le 800$
       $5h - w \ge 170$
75. Maximum: 65; minimum: 8
77. Maximum: 66; minimum: 3
79. Maximum: 100; minimum: 0
81. $x + y \le 4, x \ge 0, y \ge 0$
83. Minimum: 6
85. Maximum: $z = 56$; minimum: $z = 24$

87. 25 radios, 30 CD players
89. 200,000 gallons of gasoline, 400,000 gallons of fuel oil   **91.** Part X: 9; part Y: 4

SECTION 6.3

1. (a) 3×1 **(b)** 2×3 **(c)** 2×2
3. Dimension: 2×3 **5.** Dimension: 3×4

$$\begin{bmatrix} 5 & -2 & | & 3 \\ -1 & 3 & | & -1 \end{bmatrix}$$
$$\begin{bmatrix} -3 & 2 & 1 & | & -4 \\ 5 & 0 & -1 & | & 9 \\ 1 & -3 & -6 & | & -9 \end{bmatrix}$$

7. $\begin{aligned} 3x + 2y &= 4 \\ y &= 5 \end{aligned}$ **9.** $\begin{aligned} 3x + y + 4z &= 0 \\ 5y + 8z &= -1 \\ -7z &= 1 \end{aligned}$

11. (a) Yes   **(b)** No   **(c)** Yes
13. $(5, -1)$ **15.** $(2, 0)$ **17.** $(2, 3, 1)$
19. $\{(3 - 3z, 1 + 2z, z) \mid z$ is a real number$\}$

21. $\begin{bmatrix} 1 & -2 & 3 & | & 5 \\ -3 & 5 & 3 & | & 2 \\ 1 & 2 & 1 & | & -2 \end{bmatrix}$ **23.** $\begin{bmatrix} 1 & -1 & 1 & | & 2 \\ 0 & 1 & -1 & | & 2 \\ 0 & 8 & -1 & | & 3 \end{bmatrix}$

25. $(-17, 10)$ **27.** $(0, 2, -1)$ **29.** $(3, 2, 1)$
31. No solutions   **33.** $(1, 1, 1)$ **35.** $\left(\dfrac{1}{2}, -\dfrac{1}{2}, 4\right)$
37. $(-2, 5, -1)$ **39.** No solutions   **41.** $(3, 2)$
43. $(-2, 1, 3)$ **45.** $(-2, 5, 7)$
47. $(-9.226, -9.167, 2.440)$
49. $(5.211, 3.739, -4.655)$
51. $(7.993, 1.609, -0.401)$
53. (a) $F = 0.5714N + 0.4571R - 2014$ **(b)** \$5700
55. Pump 1: 12 hours; pumps 2 and 3: 24 hours
57. (a)
$$\begin{aligned} x + y + z &= 5000 \\ x + y - z &= 0 \\ 0.08x + 0.11y + 0.14z &= 595 \end{aligned}$$
(b) \$1000 at 8%; \$1500 at 11%; 2500 at 14%
59. (a) At intersection A, incoming traffic is equal to $x + 5$. The outgoing traffic is given by $y + 7$. Therefore, $x + 5 = y + 7$. The other equations can be justified in a similar way.
(b) The three equations can be written as
$$\begin{aligned} x - y &= 2 \\ x - z &= 3 \\ y - z &= 1 \end{aligned}$$
The solution can be written as $\{(z + 3, z + 1, z) \mid z \geq 0\}$.
(c) There are infinitely many solutions since some cars could be driving around the block continually.
61. (a)
$$\begin{aligned} 1990^2 a + 1990b + c &= 11 \\ 2010^2 a + 2010b + c &= 10 \\ 2030^2 a + 2030b + c &= 6 \end{aligned}$$
(b) $f(x) = -0.00375x^2 + 14.95x - 14{,}889.125$

(c) $[1985, 2035, 5]$ by $[5, 12, 1]$

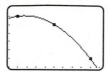

(d) Answers may vary. For example, in 2015 the predicted ratio is $f(2015) \approx 9.3$.
63. (a)
$$\begin{aligned} 3^2 a + 3b + c &= 2.2 \\ 18^2 a + 18b + c &= 10.4 \\ 26^2 a + 26b + c &= 12.8 \end{aligned}$$
(b) $f(x) = -0.010725x^2 + 0.77188x - 0.019130$
(c) $[0, 30, 5]$ by $[0, 14, 1]$

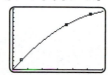

(d) Answers may vary. For example, in 1990 this percentage was $f(20) \approx 11.1$.

SECTION 6.4

1. (a) $a_{12} = 2$, $a_{21} = 4$, a_{32} is undefined.
(b) $a_{11}a_{22} + 3a_{23} = 23$
3. (a) $a_{12} = -1$, $a_{21} = 3$, $a_{32} = 0$
(b) $a_{11}a_{22} + 3a_{23} = 13$

5. (a) $A + B = \begin{bmatrix} 1 & -6 \\ 1 & 4 \end{bmatrix}$ **(b)** $3A = \begin{bmatrix} 6 & -18 \\ 9 & 3 \end{bmatrix}$

(c) $2A - 3B = \begin{bmatrix} 7 & -12 \\ 12 & -7 \end{bmatrix}$
7. (a) $A + B$ is undefined.

(b) $3A = \begin{bmatrix} 3 & -3 & 0 \\ 3 & 15 & 27 \\ -12 & 24 & -15 \end{bmatrix}$

(c) $2A - 3B$ is undefined.

9. (a) $A + B = \begin{bmatrix} 0 & -2 \\ -2 & 2 \\ 9 & -8 \end{bmatrix}$

(b) $3A = \begin{bmatrix} -6 & -3 \\ -15 & 3 \\ 6 & -9 \end{bmatrix}$

(c) $2A - 3B = \begin{bmatrix} -10 & 1 \\ -19 & -1 \\ -17 & 9 \end{bmatrix}$

11. $\begin{bmatrix} 0 & 2 \\ 13 & -5 \\ 0 & 1 \end{bmatrix}$ **13.** $\begin{bmatrix} 2 & 6 \\ 11 & -9 \end{bmatrix}$ **15.** $\begin{bmatrix} 7 & 4 & 7 \\ 4 & 7 & 7 \\ 4 & 7 & 7 \end{bmatrix}$

17. $A = \begin{bmatrix} 1 & 2 & 1 \\ 1 & 2 & 1 \\ 1 & 2 & 1 \end{bmatrix}$

19. $B = \begin{bmatrix} 1 & 1 & 1 \\ 1 & 1 & 1 \\ 1 & 1 & 1 \end{bmatrix}, A - B = \begin{bmatrix} 0 & 1 & 0 \\ 0 & 1 & 0 \\ 0 & 1 & 0 \end{bmatrix}$

21. $B = \begin{bmatrix} 3 & 3 & 3 \\ 3 & 3 & 3 \\ 3 & 3 & 3 \end{bmatrix} \quad B - A = \begin{bmatrix} 3 & 0 & 3 \\ 3 & 0 & 3 \\ 3 & 0 & 3 \end{bmatrix}$

23. $A = \begin{bmatrix} 3 & 3 & 3 & 3 \\ 3 & 0 & 0 & 0 \\ 3 & 3 & 3 & 0 \\ 3 & 0 & 0 & 0 \\ 3 & 0 & 0 & 0 \end{bmatrix}$

25. (a) One possibility is $A = \begin{bmatrix} 3 & 3 & 3 & 3 \\ 0 & 0 & 3 & 0 \\ 0 & 3 & 0 & 0 \\ 3 & 3 & 3 & 3 \end{bmatrix}$.

(b) $B = \begin{bmatrix} 3 & 3 & 3 & 3 \\ 3 & 3 & 3 & 3 \\ 3 & 3 & 3 & 3 \\ 3 & 3 & 3 & 3 \end{bmatrix}$

27. (a) One possibility is $A = \begin{bmatrix} 3 & 0 & 0 & 0 \\ 3 & 0 & 0 & 0 \\ 3 & 0 & 0 & 0 \\ 3 & 3 & 3 & 3 \end{bmatrix}$.

(b) $B = \begin{bmatrix} 3 & 3 & 3 & 3 \\ 3 & 3 & 3 & 3 \\ 3 & 3 & 3 & 3 \\ 3 & 3 & 3 & 3 \end{bmatrix}$

29. (a) $A = \begin{bmatrix} 12 & 4 \\ 8 & 7 \end{bmatrix}, B = \begin{bmatrix} 55 \\ 70 \end{bmatrix}$

(b) $AB = \begin{bmatrix} 940 \\ 930 \end{bmatrix}$. Tuition for Student 1 is $940, and for Student 2 tuition is $930.

31. (a) $A = \begin{bmatrix} 10 & 5 \\ 9 & 8 \\ 11 & 3 \end{bmatrix}, B = \begin{bmatrix} 60 \\ 70 \end{bmatrix}$

(b) $AB = \begin{bmatrix} 950 \\ 1100 \\ 870 \end{bmatrix}$. The product AB calculates tuition for each student.

33. $AB = \begin{bmatrix} -3 & 1 \\ -4 & 6 \end{bmatrix}, BA = \begin{bmatrix} 4 & 2 \\ 5 & -1 \end{bmatrix}$

35. AB and BA are undefined.

37. AB and BA are undefined.

39. $AB = \begin{bmatrix} -1 & 15 & -2 \\ 0 & 1 & 3 \\ -1 & 14 & -5 \end{bmatrix}, BA = \begin{bmatrix} 0 & 6 & 0 \\ 6 & -5 & 9 \\ 0 & 1 & 0 \end{bmatrix}$

41. $AB = \begin{bmatrix} -1 \\ 6 \end{bmatrix}$. BA is undefined.

43. $AB = \begin{bmatrix} -7 & 8 & 7 \\ 18 & -32 & -8 \end{bmatrix}$, BA is undefined.

45. $AB = \begin{bmatrix} -5 \\ 10 \\ 17 \end{bmatrix}$, BA is undefined.

47. $AB = \begin{bmatrix} 350 \\ 230 \end{bmatrix}$; the total cost of order 1 is $350, and the total cost of order 2 is $230.

49. They both equal $\begin{bmatrix} 36 & 36 & 8 \\ -15 & -38 & -4 \\ -11 & 13 & 10 \end{bmatrix}$. The distributive property appears to hold for matrices.

51. They both equal $\begin{bmatrix} 50 & 3 & 12 \\ -6 & 55 & 8 \\ 27 & -3 & 29 \end{bmatrix}$. Matrices appear to conform to rules of algebra except that $AB \neq BA$.

SECTION 6.5

1. B is the inverse of A.   **3.** B is the inverse of A.
5. $k = 1$ **7.** $k = 2.5$ **9.** A **11.** A
13. (a) $(2, 4)$ **(b)** It will translate $(2, 4)$ to the left 2 units and downward 3 units, back to $(0, 1)$;

$A^{-1} = \begin{bmatrix} 1 & 0 & -2 \\ 0 & 1 & -3 \\ 0 & 0 & 1 \end{bmatrix}$ **(c)** I_3

15. $A = \begin{bmatrix} 1 & 0 & -3 \\ 0 & 1 & -5 \\ 0 & 0 & 1 \end{bmatrix}$ and $A^{-1} = \begin{bmatrix} 1 & 0 & 3 \\ 0 & 1 & 5 \\ 0 & 0 & 1 \end{bmatrix}$.

A^{-1} will translate a point 3 units to the right and 5 units upward.

17. (a) $BX = \begin{bmatrix} -2 \\ 0 \\ 1 \end{bmatrix} = Y$

(b) $B^{-1}Y = \begin{bmatrix} -\sqrt{2} \\ -\sqrt{2} \\ 1 \end{bmatrix} = X$

B^{-1} rotates the point represented by Y counterclockwise 45° about the origin.

19. (a) $ABX = \begin{bmatrix} 2 \\ 2 \\ 1 \end{bmatrix} = Y.$

(b) The net result of A and B is to translate a point 1 unit right and 1 unit upward.

$$AB = \begin{bmatrix} 1 & 0 & 1 \\ 0 & 1 & 1 \\ 0 & 0 & 1 \end{bmatrix}$$

(c) Yes **(d)** Since AB translates a point 1 unit right and 1 unit upward, the inverse of AB would translate a point 1 unit left and 1 unit downward. Therefore

$$(AB)^{-1} = \begin{bmatrix} 1 & 0 & -1 \\ 0 & 1 & -1 \\ 0 & 0 & 1 \end{bmatrix}$$

21. $A^{-1} = \begin{bmatrix} -10 & 30 \\ -4 & 10 \end{bmatrix}$

23. $A^{-1} = \begin{bmatrix} 0.2 & 0 & 0.4 \\ 0.4 & 0 & -0.2 \\ 1.4 & -1 & -1.2 \end{bmatrix}$

25. $A^{-1} = \begin{bmatrix} 0.5 & 0.2 & 2.1 \\ 0 & 0.2 & 1.6 \\ 0 & 0 & -1 \end{bmatrix}$

27. $A^{-1} = \begin{bmatrix} 0.5 & 0.25 & 0.25 \\ 0.25 & 0.5 & 0.25 \\ 0.25 & 0.25 & 0.5 \end{bmatrix}$

29. $A^{-1} = \begin{bmatrix} 3 & -2 \\ -1 & 1 \end{bmatrix}$ **31.** $A^{-1} = \begin{bmatrix} 5 & 2 \\ 3 & 1 \end{bmatrix}$

33. $A^{-1} = \begin{bmatrix} 0 & 1 & 0 \\ 0 & 0 & 1 \\ 1 & 0 & 0 \end{bmatrix}$

35. $A^{-1} = \begin{bmatrix} -2 & 1 & -1 \\ -5 & 2 & -1 \\ 3 & -1 & 1 \end{bmatrix}$

37. (a) $AX = \begin{bmatrix} 1.5 & 3.7 \\ -0.4 & -2.1 \end{bmatrix}\begin{bmatrix} x \\ y \end{bmatrix} = \begin{bmatrix} 0.32 \\ 0.36 \end{bmatrix} = B$

(b) $X = \begin{bmatrix} 1.2 \\ -0.4 \end{bmatrix}$

39.
(a) $AX = \begin{bmatrix} 0.08 & -0.7 \\ 1.1 & -0.05 \end{bmatrix}\begin{bmatrix} x \\ y \end{bmatrix} = \begin{bmatrix} -0.504 \\ 0.73 \end{bmatrix} = B$

(b) $X = \begin{bmatrix} 0.7 \\ 0.8 \end{bmatrix}$

41.
(a) $AX = \begin{bmatrix} 3.1 & 1.9 & -1 \\ 6.3 & 0 & -9.9 \\ -1 & 1.5 & 7 \end{bmatrix}\begin{bmatrix} x \\ y \\ z \end{bmatrix} = \begin{bmatrix} 1.99 \\ -3.78 \\ 5.3 \end{bmatrix} = B$

(b) $X = \begin{bmatrix} 0.5 \\ 0.6 \\ 0.7 \end{bmatrix}$

43.
(a) $AX = \begin{bmatrix} 3 & -1 & 1 \\ 5.8 & -2.1 & 0 \\ -1 & 0 & 2.9 \end{bmatrix}\begin{bmatrix} x \\ y \\ z \end{bmatrix} = \begin{bmatrix} 4.9 \\ -3.8 \\ 3.8 \end{bmatrix} = B$

(b) $X \approx \begin{bmatrix} 9.26 \\ 27.39 \\ 4.50 \end{bmatrix}$

45. $(5, 13, -12)$
47. Type A: \$10.99; type B: \$12.99; type C: \$14.99
49. (a) Intersection A: incoming traffic is $x_1 + 5$ and outgoing traffic is $4 + 6$, so $x_1 + 5 = 4 + 6$. Intersection B: incoming traffic is $x_2 + 6$ and outgoing traffic is $x_1 + 3$, so $x_2 + 6 = x_1 + 3$. Intersection C: incoming traffic is $x_3 + 4$ and outgoing traffic is $x_2 + 7$, so $x_3 + 4 = x_2 + 7$. Intersection D: incoming traffic is $6 + 5$ and outgoing traffic is $x_3 + x_4$, so $6 + 5 = x_3 + x_4$.

(b) $AX = \begin{bmatrix} 1 & 0 & 0 & 0 \\ -1 & 1 & 0 & 0 \\ 0 & -1 & 1 & 0 \\ 0 & 0 & 1 & 1 \end{bmatrix}\begin{bmatrix} x_1 \\ x_2 \\ x_3 \\ x_4 \end{bmatrix} = \begin{bmatrix} 5 \\ -3 \\ 3 \\ 11 \end{bmatrix} = B$

The solution is $x_1 = 5$, $x_2 = 2$, $x_3 = 5$, and $x_4 = 6$.
(c) The traffic traveling west from intersection B to intersection A has a rate of $x_1 = 5$ cars per minute. The other values for x_2, x_3, and x_4 can be interpreted in a similar manner.

51. (a) $\begin{bmatrix} 1 & 1500 & 8 \\ 1 & 2000 & 5 \\ 1 & 2200 & 10 \end{bmatrix}\begin{bmatrix} a \\ b \\ c \end{bmatrix} = \begin{bmatrix} 122 \\ 130 \\ 158 \end{bmatrix}$

(b) \$130,000

SECTION 6.6

1. $\det A = 1 \neq 0$. A is invertible.
3. $\det A = 0$. A is not invertible.
5. $M_{12} = 10$, $A_{12} = -10$
7. $M_{22} = -15$, $A_{22} = -15$
9. $\det A = 3 \neq 0$. A^{-1} exists.
11. $\det A = 0$. A^{-1} does not exist.
13. 30   **15.** 0   **17.** -32 **19.** 0   **21.** 643.4
23. -4.484 **25.** 7   **27.** 6.5   **29.** $(5, -3)$
31. $(0.45, 0.67)$ **33.** The points are collinear.
35. The points are not collinear.

CHAPTER 6 REVIEW EXERCISES

1. $f(3, 6) = 9$

3. (a) $\dfrac{249}{99} \approx 2.5$ **(b)** They are approximately equal.

5. $(1, -2)$ **7.** $l = 11, w = 7$

9. $(2, 3)$, consistent

11. \$1,200 at 7%, \$800 at 9%

13. One solution is $(2, 2)$. Answers may vary.

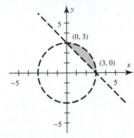

15. $(-9, 3)$ **17.** $(-1, 3)$

19. (a) $A + 2B = \begin{bmatrix} 7 & 1 \\ -8 & 1 \end{bmatrix}$

(b) $A - B = \begin{bmatrix} -2 & -5 \\ 7 & -2 \end{bmatrix}$

21. $AB = \begin{bmatrix} -2 & -4 \\ 17 & 31 \end{bmatrix}, BA = \begin{bmatrix} 8 & -6 \\ -27 & 21 \end{bmatrix}$

23. $AB = \begin{bmatrix} 3 & 7 \\ -2 & 8 \end{bmatrix}, BA = \begin{bmatrix} 2 & -1 & 3 \\ 2 & 9 & -3 \\ 6 & 12 & 0 \end{bmatrix}$

25. $\det A = -1951 \neq 0$. A is invertible.

27. B is the inverse of A.

29. (a) $AX = \begin{bmatrix} 11 & 31 \\ 37 & -19 \end{bmatrix}\begin{bmatrix} x \\ y \end{bmatrix} = \begin{bmatrix} -27.6 \\ 240 \end{bmatrix} = B$

(b) $X = \begin{bmatrix} 5.1 \\ -2.7 \end{bmatrix}$

31. (a) $x \approx 1.838, y \approx 0.168$; consistent and independent

(b) All points (x, y) satisfying $y = \dfrac{4.2x - 6.2}{3.4}$, where x is a real number; consistent and dependent

(c) No solutions; inconsistent

33. $A^{-1} = \begin{bmatrix} -1 & -2 \\ -1 & -1 \end{bmatrix}$ **35.** $\det A = 25$

37. $A = \begin{bmatrix} 3 & 3 & 3 \\ 0 & 3 & 0 \\ 0 & 3 & 0 \end{bmatrix}, B = \begin{bmatrix} 0 & 0 & 0 \\ 1 & 0 & 1 \\ 1 & 0 & 1 \end{bmatrix}$

39. $a \approx -0.010764, b \approx 1.2903; c \approx 24.133$
[20, 100, 20] by [45, 65, 5]

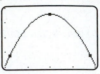

41. There were 31 fatalities through February of 1997. The annual average is 26 fatalities.   **43.** 4500

CHAPTER 7: CONIC SECTIONS

SECTION 7.1

1.

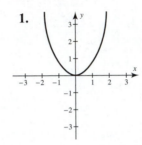

3.

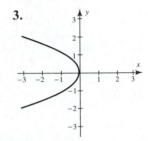

5.

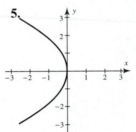

7. e   **9.** a   **11.** d

13. Vertex: $V(0, 0)$; focus: $F(0, 4)$; directrix: $y = -4$

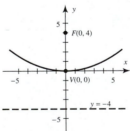

15. Vertex: $V(0, 0)$; focus: $F(2, 0)$; directrix: $x = -2$

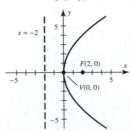

17. Vertex: $V(0, 0)$; focus: $F(-1, 0)$; directrix: $x = 1$

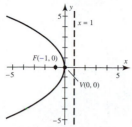

19. Vertex: $V(0, 0)$; focus: $F(0, -2)$; directrix: $y = 2$

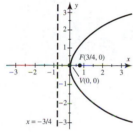

21. Vertex: $V(0, 0)$; focus: $F\left(\dfrac{3}{4}, 0\right)$; directrix: $x = -\dfrac{3}{4}$

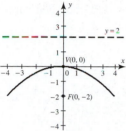

23. $x^2 = 4y$ **25.** $y^2 = -12x$

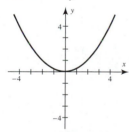

27. $x^2 = 3y$ **29.** $y^2 = -8x$

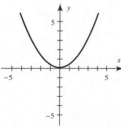

31. $y^2 = 4x$ **33.** $y^2 = -x$

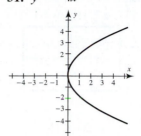
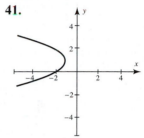

35. $x^2 = -12y$ **37.** $y^2 = -4x$

39.   **41.**

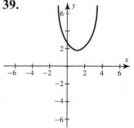
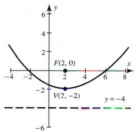

43. c   **45.** a

47. Vertex: $V(2, -2)$; focus: $F(2, 0)$; directrix; $y = -4$

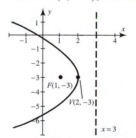

49. Vertex: $V(2, -3)$; focus: $F(1, -3)$; directrix: $x = 3$

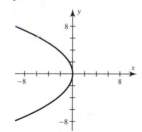

51. Vertex: $V(-2, 0)$; focus: $F(-2, -1)$; directrix: $y = 1$

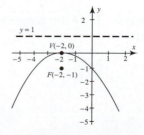

53. $x^2 = 4(y - 1)$ **55.** $y^2 = 4(x + 1)$
57. $(x + 1)^2 = -8(y - 5)$

59. $y = \pm\sqrt{2x}$
$[-6, 6, 1]$ by $[-4, 4, 1]$

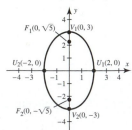

61. $y = -0.75 \pm \sqrt{-3x}$
$[-6, 6, 1]$ by $[-4, 4, 1]$

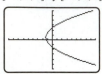

63. $y = 0.5 \pm \sqrt{3.1(x + 1.3)}$
$[-9, 9, 1]$ by $[-6, 6, 1]$

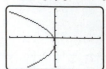

Note: If a break in the graph appears near the vertex, it should not be there. It is a result of the low resolution of the graphing calculator screen.

65. $y = -1 \pm \sqrt{\dfrac{x}{2.3}}$

$[-6, 6, 1]$ by $[-4, 4, 1]$

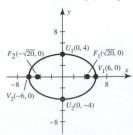

67. $p = 3$ feet

69. **(a)** $y = \dfrac{32}{11{,}025}x^2$ **(b)** About 86.1 feet

71. $\dfrac{25}{64}$ inches

SECTION 7.2

1. Foci: $F(0, \pm\sqrt{5})$; vertices: $V(0, \pm3)$; endpoints of the minor axis $U(\pm2, 0)$

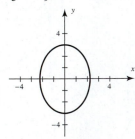

3. Foci: $F(\pm\sqrt{20}, 0)$; vertices: $V(\pm6, 0)$; endpoints of the minor axis $U(0, \pm4)$

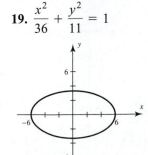

5. Foci: $F(0, \pm2)$; vertices: $V(0, \pm3)$; endpoints of the minor axis $U(\pm\sqrt{5}, 0)$

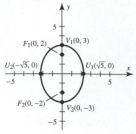

7. Foci; $F(0, \pm4)$; vertices: $V(0, \pm5)$; endpoints of the minor axis: $U(\pm3, 0)$

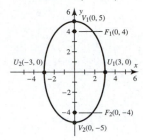

9. b   **11.** c

13. $\dfrac{x^2}{25} + \dfrac{y^2}{9} = 1$ **15.** $\dfrac{x^2}{5} + \dfrac{y^2}{9} = 1$

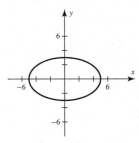

17. $\dfrac{x^2}{12} + \dfrac{y^2}{16} = 1$ **19.** $\dfrac{x^2}{36} + \dfrac{y^2}{11} = 1$

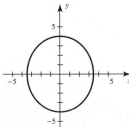

21. $\dfrac{x^2}{16} + \dfrac{y^2}{9} = 1$

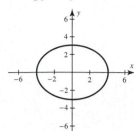

23. $\dfrac{x^2}{9} + \dfrac{y^2}{5} = 1$

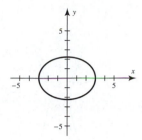

25.

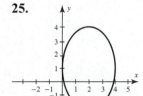

27.

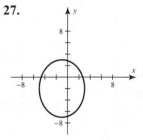

29.

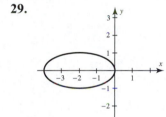

31. d   **33.** c

35. Foci: $(1, 1 \pm 4)$; vertices: $(1, 1 \pm 5)$

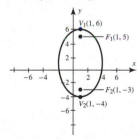

37. Foci: $(-4 \pm \sqrt{7}, 2)$; vertices: $(-4 \pm 4, 2)$

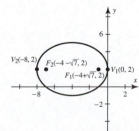

39. $\dfrac{(x-2)^2}{5} + \dfrac{(y-1)^2}{9} = 1$

41. $\dfrac{x^2}{9} + \dfrac{(y-2)^2}{5} = 1$

43. $\dfrac{(x-2)^2}{16} + \dfrac{(y-4)^2}{4} = 1$

45. $x^2 + y^2 = 16$

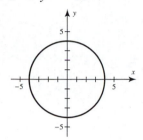

47. $(x-3)^2 + (y+4)^2 = 1$

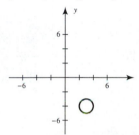

49. Center: $(2, 1)$; radius: $r = 4$

51. Center: $(0, -5)$; radius: $r = 5$

53. $y = \pm\sqrt{10\left(1 - \dfrac{x^2}{15}\right)}$

$[-6, 6, 1]$ by $[-4, 4, 1]$

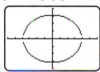

55. $y = \pm\sqrt{\dfrac{25 - 4.1x^2}{6.3}}$

$[-4.7, 4.7, 1]$ by $[-3.1, 3.1, 1]$

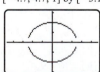

57. $(0, 3)$, $\left(\dfrac{24}{13}, \dfrac{15}{13}\right)$

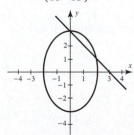

59. Four solutions: $\left(\pm\sqrt{\dfrac{20}{3}},\ \pm\sqrt{\dfrac{7}{3}}\right)$

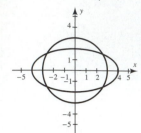

61. $(3, 0), (-3, 0)$ **63.**

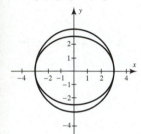   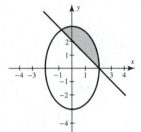

65.   **67.**

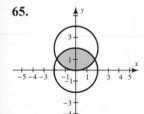   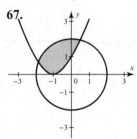

69. $\dfrac{x^2}{0.387^2} + \dfrac{y^2}{0.379^2} = 1$; sun: $(0.0797, 0)$

$[-0.6, 0.6, 0.1]$ by $[-0.4, 0.4, 0.1]$

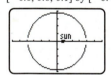

71. 6.245 inches

73. **(a)** $\dfrac{x^2}{17.95^2} + \dfrac{y^2}{4.44^2} = 1$ **(b)** $(17.39, 0)$

(c) Maximum: 35.34 units or about 3.3 billion miles; Minimum: 0.56 unit or about 52 million miles

75. About 21.65 feet

77. Maximum: 668 miles; minimum: 340 miles

SECTION 7.3

1. Asymptotes: $y = \pm\dfrac{7}{3}x$; $F(\pm\sqrt{58}, 0)$

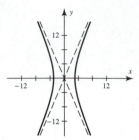

3. Asymptotes: $y = \pm\dfrac{3}{2}x$; $F(0, \pm\sqrt{52})$

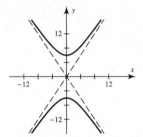

5. Asymptotes: $y = \pm x$; $F(\pm\sqrt{18}, 0)$

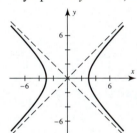

7. Asymptotes: $y = \pm\dfrac{4}{3}x$; $F(0, \pm 5)$

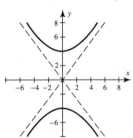

9. d   **11.** a

13. $\dfrac{x^2}{16} - \dfrac{y^2}{9} = 1$

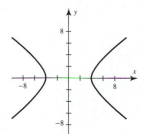

15. $\dfrac{y^2}{36} - \dfrac{x^2}{64} = 1$

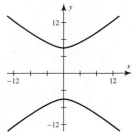

23. $\dfrac{x^2}{16} - \dfrac{y^2}{9} = 1$; asymptotes: $y = \pm \dfrac{3}{4}x$

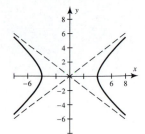

17. $\dfrac{y^2}{144} - \dfrac{x^2}{25} = 1$; asymptotes: $y = \pm \dfrac{12}{5}x$

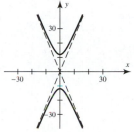

19. $\dfrac{y^2}{4} - \dfrac{x^2}{21} = 1$; asymptotes: $y = \pm \dfrac{2}{\sqrt{21}}x$

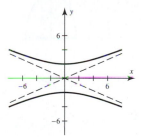

21. $\dfrac{x^2}{9} - \dfrac{y^2}{4} = 1$; asymptotes: $y = \pm \dfrac{2}{3}x$

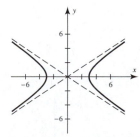

25.

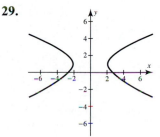

27.

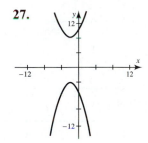

29.

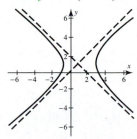

31. b   **33.** c   **35.** $\dfrac{(y+4)^2}{16} - \dfrac{(x-4)^2}{4} = 1$

37. Vertices: $(1 \pm 2, 1)$; foci: $(1 \pm \sqrt{8}, 1)$; asymptotes: $y = \pm(x - 1) + 1$

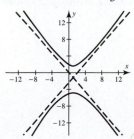

39. Vertices: $(1, -1 \pm 4)$; foci: $(1, -1 \pm 5)$; asymptotes: $y = \pm \dfrac{4}{3}(x - 1) - 1$

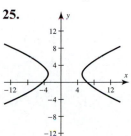

41. $(x - 2)^2 - \dfrac{(y + 2)^2}{3} = 1$

43. $y^2 - \dfrac{(x + 1)^2}{8} = 1$

45. $y = 1 \pm \sqrt{11\left(1 + \dfrac{x^2}{5.9}\right)}$

$[-15, 15, 5]$ by $[-10, 10, 5]$

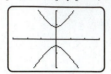

47. $y = \pm\sqrt{\dfrac{4x^2 + 15}{3}}$

$[-9, 9, 1]$ by $[-6, 6, 1]$

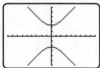

49. Four solutions: $\left(\pm\sqrt{\dfrac{13}{2}}, \pm\sqrt{\dfrac{5}{2}}\right)$

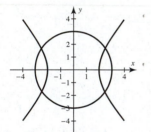

51. $(2, 0), (-5.2, 7.2)$

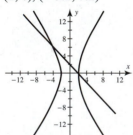

53. **(a)** Elliptic   **(b)** Its speed should be 4326 meters per second or greater.

(c) If D is larger, then $\dfrac{k}{\sqrt{D}}$ is smaller, so smaller values for V satisfy $V > \dfrac{k}{\sqrt{D}}$.

CHAPTER 7 REVIEW EXERCISES

1.

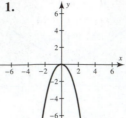

3.

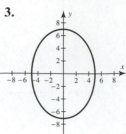

5.

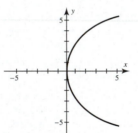

7. d   **9.** a   **11.** e

13. $y^2 = 8x$ **15.** $\dfrac{x^2}{16} + \dfrac{y^2}{49} = 1$

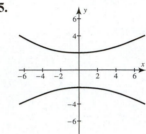   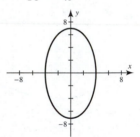

17. $(x - 5)^2 = -8(y - 2)$

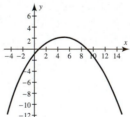

19. $F(0, -1)$ **21.** $F(\pm\sqrt{21}, 0)$

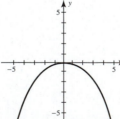   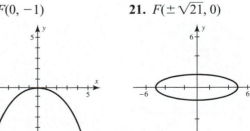

23. $F(\pm 5, 0)$

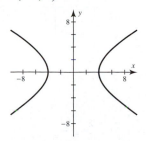

25. Both foci are located at $(0, 0)$.

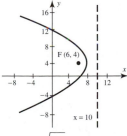

27. Center: $(-1, 2)$ **29.** Center $(1, -1)$

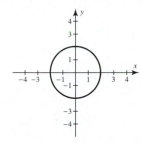

31. Focus: $F(6, 4)$; directrix: $x = 10$

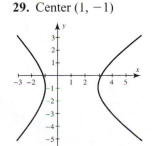

33. $y = \pm\sqrt{\dfrac{3}{4}x}$

$[-6, 6, 1]$ by $[-4, 4, 1]$

35. $y = \pm\sqrt{\dfrac{9x^2 - 17}{2}}$

$[-6, 6, 1]$ by $[-4, 4, 1]$

Note: The breaks in the graph near the vertices should not be there. It is a result of the low resolution of the graphing calculator screen.

37. $x^2 + y^2 = 49$.

39. Four solutions: $\left(\pm\sqrt{\dfrac{8}{3}}, \pm\sqrt{\dfrac{4}{3}}\right)$

41.

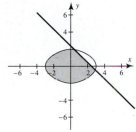

43. **(a)** Minimum: 4.92 million miles; maximum: 995.08 million miles   **(b)** $2\pi\sqrt{\dfrac{500^2 + 70^2}{2}} \approx 2243$ million miles or 2.243 billion miles

45. About 29.05 feet

CHAPTER 8: FURTHER TOPICS IN ALGEBRA

SECTION 8.1

1. $a_1 = 3, a_2 = 5, a_3 = 7, a_4 = 9$

3. $a_1 = 4, a_2 = -8, a_3 = 16, a_4 = -32$

5. $a_1 = \dfrac{1}{2}, a_2 = \dfrac{2}{5}, a_3 = \dfrac{3}{10}, a_4 = \dfrac{4}{17}$

7. $a_1 = -\dfrac{1}{2}, a_2 = \dfrac{1}{4}, a_3 = -\dfrac{1}{8}, a_4 = \dfrac{1}{16}$

9. $a_1 = \dfrac{2}{3}, a_2 = -\dfrac{4}{5}, a_3 = \dfrac{8}{9}, a_4 = -\dfrac{16}{17}$

11. $a_1 = 3, a_2 = 8, a_3 = 17, a_4 = 32$

13. $2, 4, 3, 5, 3, 6, 4$

15. **(a)** $a_1 = 1, a_2 = 2, a_3 = 4, a_4 = 8$

(b) $[0, 5, 1]$ by $[0, 9, 1]$

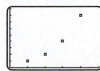

17. (a) $a_1 = 2$, $a_2 = 5$, $a_3 = 3$, $a_4 = -2$
(b) $[0, 5, 1]$ by $[-3, 6, 1]$

19. (a) $a_1 = 2$, $a_2 = 4$, $a_3 = 16$, $a_4 = 256$
(b) $[0, 5, 1]$ by $[0, 300, 50]$

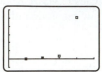

21. (a) $a_1 = 1$, $a_2 = 3$, $a_3 = 6$, $a_4 = 10$
(b) $[0, 5, 1]$ by $[0, 10, 1]$

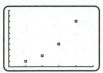

23. (a) $a_1 = 2$, $a_2 = 3$, $a_3 = 6$, $a_4 = 18$
(b) $[0, 5, 1]$ by $[0, 20, 2]$

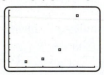

25. The insect population density increases rapidly and then levels off near 5000 per acre.
27. (a) $a_n = 0.8a_{n-1}$, $a_1 = 500$ **(b)** $a_1 = 500$, $a_2 = 400$, $a_3 = 320$, $a_4 = 256$, $a_5 = 204.8$, and $a_6 = 163.84$. The population density decreases by 20% each year.   **(c)** $a_n = 500(0.8)^{n-1}$
29. (a) $a_1 = 300$
$a_n = 2a_{n-1}$, $n > 1$
$a_1 = 300$, $a_2 = 600$, $a_3 = 1200$, $a_4 = 2400$, $a_5 = 4800$
(b) $a_{16} = 9,830,400$ bacteria per milliliter
(c) Geometric, since each term is found by multiplying the previous term by 2
31. (a)

n	1	2	3	4	5	6	7	8
a_n	1	3	5	7	9	11	13	15

(b) $[0, 10, 1]$ by $[0, 16, 1]$

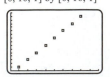

(c) $a_n = 2n - 1$

33. (a)

n	1	2	3	4	5	6	7	8
a_n	7.5	6	4.5	3	1.5	0	-1.5	-3

(b) $[0, 12, 1]$ by $[-4, 8, 1]$

(c) $a_n = -1.5n + 9$
35. (a)

n	1	2	3	4	5	6	7	8
a_n	8	4	2	1	$\frac{1}{2}$	$\frac{1}{4}$	$\frac{1}{8}$	$\frac{1}{16}$

(b) $[0, 10, 1]$ by $[-1, 9, 1]$

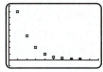

(c) $a_n = 8\left(\frac{1}{2}\right)^{n-1}$

37. (a)

n	1	2	3	4	5	6	7	8
a_n	$\frac{3}{4}$	$\frac{3}{2}$	3	6	12	24	48	96

(b) $[0, 10, 1]$ by $[-10, 110, 10]$

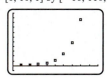

(c) $a_n = \frac{3}{4}(2)^{n-1}$
39. $a_n = -2n + 7$ **41.** $a_n = 3n - 8$
43. $a_n = 2n + 1$ **45.** $a_n = 2\left(\frac{1}{2}\right)^{n-1}$
47. $a_n = \frac{1}{2}\left(-\frac{1}{4}\right)^{n-1}$ **49.** $a_n = 8\left(\frac{1}{2}\right)^{n-1}$
51. Arithmetic   **53.** Geometric   **55.** Neither
57. Arithmetic, $d < 0$, $d = -1$
59. Geometric, $r < 0$, $|r| < 1$
61. (a) 1, 1, 2, 3, 5, 8, 13, 21, 34, 55, 89, 144

(b) $\dfrac{a_2}{a_1} = 1, \dfrac{a_3}{a_2} = 2, \dfrac{a_4}{a_3} = 1.5, \dfrac{a_5}{a_4} = \dfrac{5}{3} \approx 1.6667,$

$\dfrac{a_6}{a_5} = \dfrac{8}{5} = 1.6, \dfrac{a_7}{a_6} = \dfrac{13}{8} = 1.625, \dfrac{a_8}{a_7} = \dfrac{21}{13} \approx 1.6154,$

$\dfrac{a_9}{a_8} = \dfrac{34}{21} \approx 1.6190, \dfrac{a_{10}}{a_9} = \dfrac{55}{34} \approx 1.6176,$

$\dfrac{a_{11}}{a_{10}} = \dfrac{89}{55} = 1.6182,$ and $\dfrac{a_{12}}{a_{11}} = \dfrac{144}{89} \approx 1.6180.$

These ratios appear to approach a number near 1.618.
(c) $n = 2$:
$a_1 \cdot a_3 - a_2^2 = (1)(2) - (1)^2 = 1 = (-1)^2$
$n = 3$:
$a_2 \cdot a_4 - a_3^2 = (1)(3) - (2)^2 = -1 = (-1)^3$
$n = 4$:
$a_3 \cdot a_5 - a_4^2 = (2)(5) - (3)^2 = 1 = (-1)^4$
63. (a) $a_n = 2000n + 28{,}000$ or
$a_n = 30{,}000 + 2000(n - 1)$; arithmetic
(b) $b_n = 30{,}000(1.05)^{n-1}$; geometric
(c) Since $a_{10} = \$48{,}000 > b_{10} \approx \$46{,}540$, the first salary is larger after 10 years. Since
$a_{20} = \$68{,}000 < b_{20} \approx \$75{,}809$, the second salary is larger after 20 years.
(d) With time the geometric sequence with $r > 1$ overtakes the arithmetic sequence.
[0, 30, 10] by [0, 150,000, 50,000]

65. $a_6 \approx 1.414213562, \sqrt{2} \approx 1.414213562$

SECTION 8.2

1. (a) $8518 + 9921 + 10{,}706 + 14{,}035 + 14{,}307 + 12{,}249 = 69{,}736$ **(b)** Finite

3. $S_8 = 8\left(\dfrac{3 + 17}{2}\right) = 80$

5. $S_{20} = 50\left(\dfrac{1 + 50}{2}\right) = 1275$

7. $S_{37} = 37\left(\dfrac{-7 + 101}{2}\right) = 1739$ **9.** 4100

11. $S_8 = 255$ **13.** $S_7 = 546.5$ **15.** 3,145,725

17. $S = \dfrac{3}{2}$ **19.** $S = \dfrac{18}{5}$ **21.** 2 **23.** \$91,523.93

25. \$62,278.01

27. $\dfrac{2}{3} = 0.6 + 0.06 + 0.006 + 0.0006 +$
$\qquad 0.00006 + \cdots$

29. $\dfrac{9}{11} = 0.81 + 0.0081 + 0.000081 +$
$\qquad 0.00000081 + \cdots$

31. $\dfrac{8}{9}$ **33.** $\dfrac{5}{11}$

35. $\displaystyle\sum_{k=1}^{4} (k + 1) = 2 + 3 + 4 + 5$

37. $\displaystyle\sum_{k=1}^{8} 4 = 4 + 4 + 4 + 4 + 4 + 4 + 4 + 4$

39. $\displaystyle\sum_{k=1}^{7} k^3 = 1 + 8 + 27 + 64 + 125 + 216 + 343$

41. $\displaystyle\sum_{k=4}^{5} (k^2 - k) = 12 + 20$ **43.** $\displaystyle\sum_{k=1}^{6} k^4$

45. $\displaystyle\sum_{k=1}^{\infty} \left(\dfrac{1}{k^2}\right)$ **47.** 540 **49.** 600 **51.** 1395

53. 5525 **55.** 1360 **57.** 290

59. $\displaystyle\sum_{k=1}^{n} k = 1 + 2 + 3 + 4 + \cdots + n$ is an arithmetic series with $a_1 = 1$ and $a_n = n$. Its sum equals

$$S_n = n\left(\dfrac{a_1 + a_n}{2}\right) = n\left(\dfrac{1 + n}{2}\right) = \dfrac{n(n + 1)}{2}.$$

61. $S_9 = 9\left(\dfrac{7 + 15}{2}\right) = 99$ logs

63. (a) 2500, 2500, 2500, 2500
4000, 3000, 2000, 1000
(b) $2500 + 2500 + 2500 + 2500$
$4000 + 3000 + 2000 + 1000$

65. $1 + 1 + \dfrac{1}{2} + \dfrac{1}{6} + \dfrac{1}{24} + \dfrac{1}{120} + \dfrac{1}{720} +$
$\dfrac{1}{5040} \approx 2.718254; e \approx 2.718282$

67. $S_2 = \dfrac{4}{3} \approx 1.3333, S_4 = \dfrac{40}{27} \approx 1.4815,$
$S_8 \approx 1.49977, S_{16} \approx 1.49999997; S = 1.5$
As n increases, the partial sums approach S.
69. $S_1 = 4, S_2 = 3.6, S_3 = 3.64, S_4 = 3.636,$
$S_5 = 3.6364, S_6 = 3.63636; S = \dfrac{40}{11} = 3.\overline{63},$
As n increases, the partial sums approach S.

SECTION 8.3

1. $2^{10} = 1024$ **3.** $2^5 4^{10} = 33{,}554{,}432$
5. $10^3 \cdot 26^3 = 17{,}576{,}000$
7. $26^3 \cdot 36^3 = 820{,}025{,}856$ **9.** $3^5 = 243$ **11.** 2
13. 24 **15.** 1,000,000 **17.** $2^{12} = 4096$
19. No, there are 35,152 call letters possible. **21.** 24
23. 200 **25.** $P(5, 3) = 60$ **27.** $P(8, 1) = 8$

29. $P(7, 3) = 210$ **31.** $P(4, 4) = 4! = 24$
33. $P(5, 5) = 5! = 120$ **35.** $7 \cdot 6 \cdot 5 = 210$
37. 720 **39.** $C(3, 1) = 3$ **41.** $C(6, 3) = 20$
43. $C(5, 0) = 1$ **45.** $C(39, 5) = 575,757$
47. $C(5, 2) \cdot C(3, 2) = 30$
49. $\binom{5}{3} \cdot \binom{5}{4} = 50$ **51.** $\binom{10}{3} \cdot \binom{12}{2} = 7920$

SECTION 8.4

1. 5 **3.** 1 **5.** 6 **7.** 1 **9.** $C(5, 2) = 10$
11. $C(8, 4) = 70$ **13.** $x^2 + 2xy + y^2$
15. $m^3 + 6m^2 + 12m + 8$
17. $8x^3 - 36x^2 + 54x - 27$
19. $p^6 - 6p^5q + 15p^4q^2 - 20p^3q^3 + 15p^2q^4 - 6pq^5 + q^6$
21. $8m^3 + 36m^2n + 54mn^2 + 27n^3$
23. $x^2 + 2xy + y^2$
25. $81x^4 + 108x^3 + 54x^2 + 12x + 1$
27. $32 - 80x + 80x^2 - 40x^3 + 10x^4 - x^5$
29. $x^8 + 8x^6 + 24x^4 + 32x^2 + 16$
31. $256x^4 - 768x^3y + 864x^2y^2 - 432xy^3 + 81y^4$
33. $84a^6b^3$ **35.** $70x^4y^4$ **37.** $40x^2y^3$

SECTION 8.5

1. Yes **3.** No **5.** Yes **7.** No **9.** $\frac{1}{2}$ **11.** $\frac{1}{6}$
13. $\frac{1}{2}$ **15.** $\frac{4}{52} = \frac{1}{13}$ **17.** $\frac{1}{10,000}$
19. **(a)** 0.57 or 57% **(b)** 0.33 or 33%
21. $\frac{1}{4}$ **23.** $\frac{1}{27}$ **25.** $\frac{6}{36} = \frac{1}{6}$ **27.** $\frac{625}{1296} \approx 0.482$
29. $\frac{1}{270,725}$
31. $\dfrac{\binom{13}{3} \cdot \binom{13}{2}}{\binom{52}{5}} \approx 0.0086$ or a 0.86% chance
33. $\frac{4}{20} = 0.2$
35. **(a)**

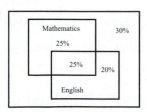

(b) 0.7 or 70% **(c)** Let M denote the event of needing help with math and E the event of needing help with English. Then $P(M \cup E) = P(M) + P(E) - P(M \cap E) = 0.5 + 0.45 - 0.25 = 0.7$

37. **(a)** $\dfrac{5427}{100,405} \approx 0.054$ **(b)** $\dfrac{89,189}{100,405} \approx 0.888$
39. $\dfrac{95.3}{100,000} = 0.000953.$
41. The probability is $\left(\frac{1}{2}\right)^n$. As n increases the probability decreases. This agrees with intuition. The probability of tossing a long string of consecutive heads is small. The longer the string, the lower the chance that it will happen.
43. **(a)** $\dfrac{3}{16}$ **(b)** $\dfrac{9}{16}$ **(c)** $\dfrac{9}{64}$ **(d)** $\dfrac{3}{64}$
45. **(a)** 0.09 **(b)** 0.12 **47.** $\dfrac{1}{1000}$
49. **(a)** $\dfrac{22}{50} = 0.44$ **(b)** $\dfrac{28}{50} = 0.56$
(c) $\dfrac{28}{50} = 0.56$
51. **(a)** About 0.12 or 12% **(b)** There is a smaller lifetime risk of bladder cancer when the percentage is under 4%. Higher percentages result in a dramatic increase in risk.
53. $\dfrac{3}{51}$ **55.** $\dfrac{1}{3}$ **57.** $\dfrac{19}{27}$ **59.** 40% **61.** $\dfrac{1}{3}$
63. **(a)** $\dfrac{18}{235}$ **(b)** $\dfrac{7}{18}$ **(c)** $\dfrac{7}{235}$
65. **(a)** $\dfrac{8}{15}$ **(b)** $\dfrac{7}{15}$ **(c)** $\dfrac{2}{5}$ **(d)** $\dfrac{1}{3}$ **(e)** $\dfrac{1}{15}$

CHAPTER 8: REVIEW EXERCISES

1. $-1, -4, -7, -10$ **3.** $0, 1, 3, 7$
5. $5, 3, 1, 2, 4, 6$
7. **(a)**

n	1	2	3	4	5	6	7	8
a_n	2	4	6	8	10	12	14	16

(b) $[0, 17, 1]$ by $[0, 17, 1]$

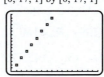

(c) $a_n = 2n$

9. (a)

n	1	2	3	4	5	6	7	8
a_n	81	27	9	3	1	$\frac{1}{3}$	$\frac{1}{9}$	$\frac{1}{27}$

(b) $[0, 10, 1]$ by $[-10, 90, 10]$

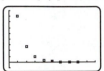

(c) $a_n = 81\left(\frac{1}{3}\right)^{n-1}$

11. $a_n = 4n - 15$

13. (a) 4, 3.6, 3.24, 2.916, 2.6244; geometric

(b) $[0, 6, 1]$ by $[0, 6, 1]$

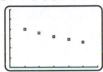

(c) $a_n = 4(0.9)^{n-1}$

15. $S_9 = 9\left(\dfrac{-2 + 22}{2}\right) = 90$

17. $S_8 = 1\left(\dfrac{1 - 3^8}{1 - 3}\right) = 3280$

19. $S = 4$ **21.** $S = \dfrac{2}{9}$

23. $6 + 11 + 16 + 21 + 26$ **25.** $\displaystyle\sum_{k=1}^{6} k^3$

27. $4^{20} \approx 1.1 \times 10^{12}$ **29.** $50^4 = 6{,}250{,}000$

31. $P(6, 3) = 120$ **33.** $P(5, 5) = 120$

35. $C(6, 3) = 20$ **37.** $C(10, 6) = 210$

39. $\dfrac{1}{2}$ **41.** $\dfrac{91}{120} \approx 0.758$

43. (a) $\dfrac{27}{60} = 0.45$ **(b)** $\dfrac{33}{60} = 0.55$

(c) $\dfrac{13}{60} \approx 0.217$

45. Initially, the population density grows slowly, then increases rapidly, and finally levels off near 4,000,000 per acre.

$[0, 16, 1]$ by $[0, 5000, 1000]$

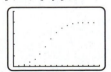

CHAPTER R

Section R.1

1. Area: 105 feet^2; perimeter: 44 feet

3. Area: 3500 meters^2; perimeter: 270 meters

5. Area: $3xy$; perimeter: $6x + 2y$

7. Area: $2W^2$; perimeter: $6W$

9. Area: $W(W + 5)$; perimeter: $4W + 10$

11. 20 centimeters^2 **13.** 20 inches^2

15. 3686.5 meters^2 **17.** $6x^2$ **19.** $\dfrac{5}{2}z^2$

21. Circumference: $8\pi \approx 25.1$ meters; area: $16\pi \approx 50.3$ meters^2

23. Circumference: $38\pi \approx 119.4$ inches; area: $361\pi \approx 1134.1$ inches^2

25. Circumference: $4\pi x$; area: $4\pi x^2$

27. $c = 61$ feet; perimeter: 132 feet

29. $c = \sqrt{170} \approx 13.0$ yards; perimeter: $18 + \sqrt{170} \approx 31.0$ yards

31. $b = \sqrt{189} \approx 13.7$ meters; perimeter: $21 + \sqrt{189} \approx 34.7$ meters

33. $a = 1.6$ miles; perimeter: 4.8 miles

35. Area: 9 feet^2

37. Area: $\dfrac{11}{2}\sqrt{104} \approx 56.1$ inches^2

39. Volume: 24 feet^3; surface area: 52 feet^2

41. Volume: 216 inches^3; surface area: 240 inches^2

43. Volume: $6x^3$; surface area: $22x^2$

45. Volume: $6xyz$; surface area: $4xy + 6xz + 12yz$

47. W^3

49. Volume: $36\pi \approx 113.1$ feet^3; surface area: $36\pi \approx 113.1$ feet^2

51. Volume: $\dfrac{500\pi}{3} \approx 523.6$ centimeters^3; surface area: $100\pi \approx 314.2$ centimeters^2

53. Volume: $\dfrac{500\pi}{3} \approx 523.6$ millimeter^3; surface area: $100\pi \approx 314.2$ millimeters^2

55. Volume: $\dfrac{1}{2}\pi \approx 1.6$ feet^3; side surface area: $2\pi \approx 6.3$ feet^2; total surface area: $\dfrac{5}{2}\pi \approx 7.9$ feet^2

57. Volume: $450\pi \approx 1413.7$ inches^3; side surface area: $180\pi \approx 565.5$ inches^2; total surface area: $230\pi \approx 722.6$ inches^2

59. Volume: $3456\pi \approx 10{,}857.3$ millimeters^3; side surface area: $576\pi \approx 1809.6$ millimeters^2; total surface area: $864\pi \approx 2714.3$ millimeters^2 **61.** $2\pi r^3$

63. Volume: 157.1 centimeters^3; side surface area: 122.7 centimeters^2

65. Volume: 12.6 feet^3; side surface area: 22.7 feet^2

67. Volume: 43.4 feet^3; side surface area: 57.2 feet^2

69. $x = \dfrac{20}{3} \approx 6.7$ **71.** $x = \dfrac{21}{2} = 10.5$

73. $x = \dfrac{15}{2} = 7.5$

SECTION R.2

1. Center: $(0, 0)$; radius: 5
3. Center: $(0, 0)$; radius: $\sqrt{7}$
5. Center: $(2, -3)$; radius: 3
7. Center: $(0, -1)$; radius: 10
9. $(x - 1)^2 + (y + 2)^2 = 1$
11. $(x + 2)^2 + (y - 1)^2 = 4$
13. $(x - 3)^2 + (y + 5)^2 = 64$
15. $(x - 3)^2 + y^2 = 49$ **17.** $x^2 + y^2 = 10$
19. $(x + 2)^2 + (y + 3)^2 = 25$
21. Center: $(0, 0)$; radius: 3

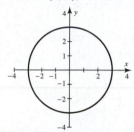

23. Center: $(-1, 1)$; radius: 2

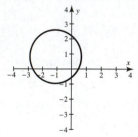

25. Center: $(0, 1)$; radius: 1

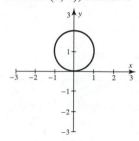

27. Center: $(-9, -5)$; radius: 3

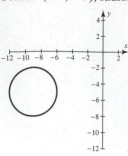

SECTION R.3

1. Base: 8; exponent: 3   **3.** 7^3

5. No, $2^3 = 8$ and $3^2 = 9$.   **7.** $\dfrac{1}{7^n}$ **9.** 5^{m-n}

11. 2^{mk} **13.** 5000   **15.** 2^3 **17.** 4^4 **19.** 3^0

21. 125   **23.** -16 **25.** 1   **27.** $\dfrac{8}{27}$ **29.** $\dfrac{1}{16}$

31. $6^{-1} = \dfrac{1}{6}$ **33.** $\dfrac{1}{5^3} = \dfrac{1}{125}$ **35.** $6x^3$

37. $10^8 = 100{,}000{,}000$ **39.** $\dfrac{5}{2}$ **41.** $8a^{-1}b^{-3} = \dfrac{8}{ab^3}$

43. $5^2 = 25$ **45.** $10^6 = 1{,}000{,}000$ **47.** $\dfrac{1}{a^6}$

49. $4x^2$ **51.** $\dfrac{2b}{3a^2}$ **53.** $\dfrac{3y^6}{x^7}$ **55.** $\dfrac{1}{5^3} = \dfrac{1}{125}$

57. $\dfrac{1}{3^2 y^8} = \dfrac{1}{9y^8}$ **59.** $4^3 y^6 = 64y^6$ **61.** $\dfrac{4^3}{x^3} = \dfrac{64}{x^3}$

63. $\dfrac{z^{20}}{2^5 x^5} = \dfrac{z^{20}}{32x^5}$

SECTION R.4

1. $3x^2$; answers may vary.
3. No, the powers must match for each variable.
5. No, the opposite is $-x^2 - 1$.   **7.** $a^2 - b^2$
9. $8x^3$ **11.** $-3y^7$ **13.** $6x^2 + 8x$ **15.** $3x^2 + 3x$
17. $5x^2 + 13x - 2$
19. Degree: 2; leading coefficient: 5

21. Degree: 3; leading coefficient: $-\dfrac{2}{5}$

23. Degree: 5; leading coefficient: 1   **25.** $3x + 12$
27. $3x - 2$ **29.** $-0.5x + 1$

31. $-7x^4 - 2x^2 - \dfrac{9}{2}$ **33.** $-7x^3$

35. $-19z^5 + 5z^2 - 3z$ **37.** $-z^4 + z^2 + 9$
39. $3x - 7$ **41.** $6x^2 - 5x + 5$
43. $3x^4 + 4x^2 - 4$ **45.** $-3x^4 - 3x - 8$

47. $5x^2 - 25x$ **49.** $-15x - 5$ **51.** $5y + 10$
53. $-10x - 18$ **55.** $6y^2 - 18y$
57. $y^2 - 2y - 35$ **59.** $-49x^2 + 49x - 12$
61. $-2x^2 + 7x - 6$ **63.** $x^2 - \dfrac{1}{4}x - \dfrac{1}{8}$
65. $2x^4 + x^2 - 1$ **67.** $x^2 - xy - 2y^2$
69. $6x^3 - 3x^2 - 3x$ **71.** $-2x^5 + x^3 - 10x$
73. $6x^4 - 12x^3 + 3x^2$ **75.** $x^3 + 3x^2 - x - 3$
77. $x^2 - 49$ **79.** $9x^2 - 16$ **81.** $4x^2 - 9y^2$
83. $x^2 + 8x + 16$ **85.** $4x^2 + 4x + 1$
87. $x^2 - 2x + 1$ **89.** $9x^2 - 12x + 4$
91. $3x^3 - 3x$

Section R.5

1. $5(2x - 3)$ **3.** $x(2x^2 - 5)$
5. $4x(2x^2 - x + 4)$ **7.** $5x^2(x^2 - 3x + 3)$
9. $5x(3x^2 + 2x - 5)$ **11.** $(x + 3)(x^2 + 2)$
13. $(3x - 2)(2x^2 + 3)$ **15.** $(z - 5)(z^2 + 1)$
17. $y(y + 2)(y^2 - 5)$ **19.** $(x + 2)(x + 5)$
21. $(x + 2)(x + 6)$ **23.** $(z - 6)(z + 7)$
25. $(z + 3)(z + 8)$ **27.** $(4x + 3)(6x - 1)$
29. $(2x + 1)(3x - 2)$ **31.** $(1 - x)(1 + 2x)$
33. $(5 - 2x)(4 + 3x)$ **35.** $x(x - 1)(5x + 6)$
37. $3x(x + 3)(2x + 1)$ **39.** $(x - 5)(x + 5)$
41. $(2x - 5)(2x + 5)$
43. $4(3x - 5)(3x + 5)$
45. $z^2(8 - 5z)(8 + 5z)$
47. $(2x - y)(2x + y)(4x^2 + y^2)$ **49.** $(x + 1)^2$
51. $(2x + 5)^2$ **53.** $(x - 6)^2$ **55.** $z(3z - 1)^2$
57. $(x - 1)(x^2 + x + 1)$
59. $(y + z)(y^2 - yz + z^2)$
61. $(2x - 3)(4x^2 + 6x + 9)$

63. $x(x + 5)(x^2 - 5x + 25)$
65. $(4x - 5)(4x + 5)$
67. $(x - 4)(x^2 + 4x + 16)$ **69.** $(x + 8)^2$
71. $(x - 8)(5x + 2)$ **73.** $x(x + 2)(x^2 - 2x + 4)$
75. $8(2x + y)(4x^2 - 2xy + y^2)$

Section R.6

1. $2x$ **3.** $x + 5$ **5.** $x + 4$ **7.** $\dfrac{1}{2x - 1}$ **9.** $\dfrac{3}{2x}$

11. 1 **13.** $\dfrac{x}{2x - 5}$ **15.** $x(x + 3)$

17. $\dfrac{(x - 1)(x + 5)}{2(2x - 3)}$ **19.** $\dfrac{4}{b^3}$ **21.** $\dfrac{3a(3a + 1)}{a + 1}$

23. $\dfrac{x^2}{(x - 5)(x^2 - 1)}$ **25.** $7(x + 1)$

27. $(x + 4)(x - 4)$ **29.** $2(2x + 1)(x - 2)$

31. $\dfrac{7}{x + 1}$ **33.** $-\dfrac{1}{x + 1}$

35. $\dfrac{x^2 - x - 1}{x(x + 4)}$ **37.** $\dfrac{-4x^2 + x + 2}{x^2}$

39. $\dfrac{x^2 + 5x - 34}{(x - 5)(x - 3)}$ **41.** $\dfrac{-2(x^2 - 4x + 2)}{(x - 5)(x - 3)}$

43. $\dfrac{x(5x + 16)}{(x - 3)(x + 3)}$ **45.** $-\dfrac{1}{2}$ **47.** $\dfrac{6x^2 - 1}{(x - 5)(3x - 1)}$

49. $x + 3$ **51.** $5x^2 + 2x - 1$ **53.** $3x + 3$
55. $3x^2 + 2x - 6$ **57.** $3x^3 - 15x^2 - 5x - 25$

59. $\dfrac{x + 1}{x - 1}$ **61.** $\dfrac{x}{3x - 20}$ **63.** $\dfrac{2}{3 - x}$

65. $\dfrac{3(x + 1)}{(x + 3)(2x - 7)}$ **67.** $\dfrac{4x(x + 5)}{(x - 5)(2x + 5)}$

Index of Applications

Index